AF615974

CONCISE ENCYCLOPEDIA OF MINERAL RESOURCES

ADVANCES IN MATERIALS SCIENCE AND ENGINEERING

This is a new series of Pergamon Scientific reference works, each volume providing comprehensive, self-contained and up-to-date coverage of a selected area in the field of materials science and engineering. The series is being developed primarily from the highly acclaimed *Encyclopedia of Materials Science and Engineering*, published in 1986. Other titles in the series are listed below.

BEVER (ed.)
Concise Encyclopedia of Materials Economics, Policy & Management

BROOK (ed.)
Concise Encyclopedia of Advanced Ceramic Materials

CORISH (ed.)
Concise Encyclopedia of Polymer Processing & Applications

EVETTS (ed.)
Concise Encyclopedia of Magnetic & Superconducting Materials

KELLY (ed.)
Concise Encyclopedia of Composite Materials

KIMERLING (ed.)
Concise Encyclopedia of Electronic & Optoelectronic Materials

MOAVENZADEH (ed.)
Concise Encyclopedia of Building & Construction Materials

SCHNIEWIND (ed.)
Concise Encyclopedia of Wood & Wood-Based Materials

WILLIAMS (ed.)
Concise Encyclopedia of Medical & Dental Materials

CONCISE ENCYCLOPEDIA OF
MINERAL RESOURCES

Editors

DONALD D CARR

Indiana Geological Survey
Bloomington, IN, USA

&

NORMAN HERZ

University of Georgia
Athens, GA, USA

Executive Editor

ROBERT W CAHN

University of Cambridge, UK

Senior Advisory Editor

MICHAEL B BEVER

MIT, Cambridge, MA, USA

PERGAMON PRESS

OXFORD • NEW YORK • BEIJING • FRANKFURT
SÃO PAULO • SYDNEY • TOKYO • TORONTO

U.K.	Pergamon Press plc, Headington Hill Hall, Oxford OX3 0BW, England
U.S.A.	Pergamon Press, Inc., Maxwell House, Fairview Park, Elmsford, New York 10523, U.S.A.
PEOPLE'S REPUBLIC OF CHINA	Pergamon Press, Room 4037, Qianmen Hotel, Beijing, People's Republic of China
FEDERAL REPUBLIC OF GERMANY	Pergamon Press GmbH, Hammerweg 6, D-6242 Kronberg, Federal Republic of Germany
BRAZIL	Pergamon Editora Ltda, Rua Eça de Queiros, 346, CEP 04011, Paraiso, São Paulo, Brazil
AUSTRALIA	Pergamon Press Australia Pty Ltd., P.O. Box 544, Potts Point, N.S.W. 2011, Australia
JAPAN	Pergamon Press, 5th Floor, Matsuoka Central Building, 1-7-1 Nishishinjuku, Shinjuku-ku, Tokyo 160, Japan
CANADA	Pergamon Press Canada Ltd., Suite No. 241, 253 College Street, Toronto, Ontario, Canada M5T 1R5

First edition 1989

Library of Congress Cataloging in Publication Data
Concise encyclopedia of mineral resources / editors, Donald D. Carr & Norman Herz. — 1st ed.
p. cm. — (Advances in materials science and engineering: 8)
Includes index.
1. Materials—Dictionaries. 2. Nonmetallic minerals—Dictionaries. I. Carr, Donald D. II. Herz, Norman, 1923–
III. Encyclopedia of materials science and engineering. IV. Series.
TA402.C66 1989
620.1'1—19 88–38478

British Library Cataloguing in Publication Data
Concise encyclopedia of mineral resources
1. Mineral resources
I. Carr, Donald D. II. Herz, Norman
III. Cahn, Robert W.
333.8'5

ISBN 0–08–034734–7

Distributed in North and South America by The MIT Press, Cambridge, Massachusetts, USA

Printed in Great Britain by BPCC Wheatons Ltd, Exeter

CONTENTS

HONORARY EDITORIAL ADVISORY BOARD

FOREWORD

In the short time since its publication, the *Encyclopedia of Materials Science and Engineering* has been accepted throughout the world as the standard reference about all aspects of materials. This is a well-deserved tribute to the scholarship and dedication of the Editor-in-Chief, Professor Michael Bever, the Subject Editors and the numerous contributors.

During its preparation, it soon became clear that change in some areas is so rapid that publication would have to be a continuing activity if the Encyclopedia were to retain its position as an authoritative and up-to-date systematic compilation of our knowledge and understanding of materials in all their diversity and complexity. Thus, the need for some form of supplementary publication was recognized at the outset. The Publisher has met this challenge most handsomely: both a continuing series of Supplementary Volumes to the main work and a number of smaller encyclopedias, each covering a selected area of materials science and engineering, will be published in the next few years.

Professor Robert Cahn, the Executive Editor, was previously the editor of an important subject area of the main work and many other people associated with the Encyclopedia will contribute to its Supplementary Volumes and derived Concise Encyclopedias. Thus, continuity of style and respect for the high standards set by the *Encyclopedia of Materials Science and Engineering* are assured. They have been joined by some new editors and contributors with knowledge and experience of important subject areas of particular interest at the present time. Thus, the Advisory Board is confident that the new publications will significantly add to the understanding of emerging topics wherever they may appear in the vast tapestry of knowledge about materials.

The appearance of Supplementary Volumes and the new series *Advances in Materials Science and Engineering* is an event which will be welcomed by scientists and engineers throughout the world. We are sure that it will add still more luster to a most important enterprise.

Walter S Owen
Chairman
Honorary Editorial Advisory Board

EXECUTIVE EDITOR'S PREFACE

As the publication of the *Encyclopedia of Materials Science and Engineering* approached, Robert Maxwell resolved to build upon the immense volume of work which had gone into its creation by embarking on a follow-up project. This project had two components. The first was the creation of a series of Supplementary Volumes to the Encyclopedia itself. The second component of the new project was the creation of a series of Concise Encyclopedias on individual subject areas included in the Main Encyclopedia to be called *Advances in Materials Science and Engineering*.

These Concise Encyclopedias are intended, as their name implies, to be compact and relatively inexpensive volumes (typically 300-600 pages in length) based on the relevant articles in the Encyclopedia (revised where need be) together with some newly commissioned articles, including appropriate ones from the Supplementary Volumes. Some Concise Encyclopedias will offer combined treatments of two subject fields which were the responsibility of separate Subject Editors during the preparation of the parent Encyclopedia (e.g., dental and medical materials).

At the time of writing, ten Concise Encyclopedias have been contracted and others are being planned. These and their editors are listed below.

Concise Encyclopedia of Advanced Ceramic Materials	Prof. Richard J Brook
Concise Encyclopedia of Building & Construction Materials	Prof. Fred Moavenzadeh
Concise Encyclopedia of Composite Materials	Prof. Anthony Kelly CBE, FRS
Concise Encyclopedia of Electronic & Optoelectronic Materials	Dr Lionel C Kimerling
Concise Encyclopedia of Magnetic & Superconducting Materials	Dr Jan E Evetts
Concise Encyclopedia of Materials Economics, Policy & Management	Prof. Michael B Bever
Concise Encyclopedia of Medical & Dental Materials	Prof. David F Williams
Concise Encyclopedia of Mineral Resources	Dr Donald D Carr & Prof. Norman Herz
Concise Encyclopedia of Polymer Processing & Applications	Mr P J Corish
Concise Encyclopedia of Wood & Wood-Based Materials	Prof. Arno P Schniewind

All new or substantially revised articles in the Concise Encyclopedias will be published in one or other of the Supplementary Volumes, which are designed to be used in conjunction with the Main Encyclopedia. The Concise Encyclopedias, however, are "free-standing" and are designed to be used without necessary reference to the parent Encyclopedia.

The Executive Editor is personally responsible for the selection of topics and authors of articles for the Supplementary Volumes. In this task, he has the benefit of the advice of the Senior Advisory Editor and of other members of the Honorary Editorial Advisory Board, who also exercise general supervision of the entire project. The Executive Editor is responsible for appointing the Editors of the various Concise Encyclopedias and for supervising the progress of these volumes.

Robert W Cahn
Executive Editor

EDITORS' PREFACE

The *Concise Encyclopedia of Mineral Resources* is mainly a compilation of articles from the subject areas of industrial minerals and mineral resources that appeared in the *Encyclopedia of Materials Science and Engineering*, published in 1986, but selected articles from closely aligned subject areas and newly commissioned articles are also included. Many authors of articles in the *Encyclopedia* of *Materials Science and Engineering*, particularly those in fields of rapid development, updated articles for this volume. Although the articles selected for the latter were originally written for the materials scientist, it is believed that the *Concise Encyclopedia of Mineral Resources* will be useful to geologists, engineers, teachers, managers and scientists in all fields who do not have access to the *Encyclopedia of Materials Science and Engineering* or who would benefit from a single encyclopedia volume devoted to economic geology.

Materials science and engineering is concerned with the fundamental nature of materials and their application. Mineral resources fill a basic need in that they are raw materials from which all other products are made. The articles selected from the *Encyclopedia of Materials Science and Engineering* are among 1580 articles in the 8-volume set consisting of 45 specialized subject areas.

Although there are a number of modern texts that deal fairly with the metals, industrial minerals and fuels, we believe the *Concise Encyclopedia of Mineral Resources* to be the first encyclopedia to give comprehensive coverage of the combined topics of economic geology in a single volume.

The authors of the *Concise Encyclopedia of Mineral Resources* are recognized authorities in their fields, coming from academia, industry, and government agencies. Although encyclopedias are flexible in terms of the articles selected and the detail presented, the editors sought to give comprehensive coverage to the part of economic geology that includes industrial minerals, metals and fuels. The authors were given guidelines on length of articles and content, and they were encouraged to place emphasis on the areas considered to be most important. Editing was not so extensive as to stylize the work. An important part of each article is the list of references. The references are selective rather than comprehensive, but they provide the reader with avenues for further study.

Donald D Carr
Norman Herz
Editors

GUIDE TO USE OF THE ENCYCLOPEDIA

This Encyclopedia is a comprehensive reference work covering all aspects of mineral resources. Information is presented in a series of alphabetically arranged articles which deal concisely with individual topics in a self-contained manner. This guide outlines the main features and organization of the Encyclopedia, and is intended to help the reader to locate the maximum amount of information on a given topic.

Accessibility of material is of vital importance in a reference work of this kind and article titles have therefore been selected, not only on the basis of article content, but also with the most probable needs of the reader in mind. On this basis, articles in the area of industrial minerals have, in general, been titled according to their use; for instance, *Fillers and Coatings: Clay Minerals*. When the term "Mineral Resources" appears in a title, this refers to metallic mineral resources. An alphabetical list of all the articles contained in this Encyclopedia is provided.

Articles are linked by an extensive cross-referencing system. Cross-references to other articles in the Encyclopedia are of two types: in-text and end-of-text. Those in the body of the text are designed to refer the reader to articles that present in greater detail material on the specific topic under discussion. They generally take one of the following forms:

...as described in the article *Open-Cut Mining*.

...a distinct deposit type (see *Ocean Resources and Mining*).

The cross-references listed at the end of an article serve to identify broad background reading and to direct the reader to articles that cover different aspects of the same topic.

The nature of an encyclopedia demands a higher degree of uniformity in terminology and notation than many other scientific works. The widespread use of the International System of Units has determined that such units be used in this Encyclopedia. It has been recognized, however, that in some fields Imperial units are more generally used. Where this is the case, Imperial units are given with their SI equivalent quantity and unit following in parentheses. Where possible the symbols defined in *Quantities, Symbols and Units*, published by the Royal Society of London, have been used.

All articles in the Encyclopedia include a bibliography giving sources of further information. Each bibliography consists of general items for further reading and/or references which cover specific aspects of the text. Where appropriate, authors are cited in the text using a name/date system as follows:

...as was recently reported (Smith 1988).

Jones (1984) describes...

The contributor's name and the organization to which they are affiliated appear at the end of an article. All contributors can be found in the alphabetical List of Contributors, along with their full postal address and the titles of the articles of which they are authors or co-authors.

The Introduction to this Concise Encyclopedia provides an overview of the general areas of mineral resources covered and lists all articles within these areas.

The most important information source for locating a particular topic in the Encyclopedia is the multilevel Subject Index, which has been made as complete and fully self-consistent as possible.

ALPHABETICAL LIST OF ARTICLES

AN INTRODUCTION TO MINERAL RESOURCES

by Donald D Carr and Norman Herz

The term "mineral resources" is used here in the broad sense to mean mineral materials from the earth that are necessary or useful to man. These materials include the metals, industrial minerals (nonmetallics) and fuels, which are covered in the Encyclopedia, and water, which is not.

The articles in the Encyclopedia fall under five general headings: (a) principles, definitions and genesis of mineral resources and deposits; (b) exploration and development; (c) metallic mineral resources by commodity; (d) industrial minerals; and (e) fuels.

1. Principles, Definitions and Genesis

The articles in the area of principles, definitions and genesis of mineral resources and deposits are listed in Table 1. Basic definitions and concepts of mineral resources are given in the article *Mineral Resources: Definitions, Uses, Classification and Future Availability*. Mineral resources include deposits of energy resources, and metallic and nonmetallic minerals and rocks. Ferrous, base and precious metals are considered metallic resources; phosphates, saline minerals, gravel, gemstones and the like are considered nonmetallic. Deposits that are recoverable economically (i.e., at a profit) are termed ores. Traditional mineral deposits are nonrenewable, but as high-grade ores are mined out, lower grade ores become economic, albeit at higher costs. Mineral products can be used either directly as mined, as in the case of sand, gems and abrasives, or must be processed to obtain metal or chemical products. The growth in world population and affluence has been due in large part to the increased use of mineral products, unfortunately concomitant with serious problems of pollution and long-term stability of the global ecological system.

Table 1
Classification of general articles

Principles, definitions and genesis of mineral resources and deposits
Critical and Strategic Materials
Economic Theory of Mineral Resources and Markets
Geochemical Distribution of the Elements
Igneous Ore Deposits: Formation Processes
Metallogenic Provinces
Metamorphic Ore Deposits: Formation Processes
Mineral Resources: Definition, Uses, Classification and Future Availability
Ocean Resources and Mining
Ore Minerals
Ores of the Future
Plate Tectonic Settings for Metallic Ore Resources
Prices of Metals: History
Recycling of Metals: Technology
Resource Appraisal
Sedimentary Ore Deposits: Formation Processes
Stockpiling
Stockpiling in the USA
Exploration and development
Geochemical Exploration for Mineral Deposits
Geological Exploration for Mineral Deposits
Geophysical Exploration for Mineral Deposits
Mineral Deposits: Examination and Evaluation
Mining and Processing: Environmental Problems
Ocean Resources and Mining
Open-Cut Mining
Remote Sensing in the Search for Mineral Deposits
Statistical and Computer Models in Mining and Exploration
Underground Mining

Mineral resources can be classified on the basis of economic and technological feasibility of recovery. The term *reserves* refers only to known deposits that can be produced economically at the time of the estimate, whereas *resources* denotes the sum of all categories: reserves as well as deposits that are subeconomic or as yet undiscovered. The future availability of mineral resources to meet world needs has been a concern since at least the beginning of this century. Many factors contribute to make precise estimates of the future availability of individual commodities difficult, such as projections of the increase in rate of consumption, discovery of new economic deposits, improvement of mineral technology including mining, beneficiation and refining, and substitution of disappearing or of more expensive minerals by others that are more plentiful and cheaper (e.g., copper by aluminum, and iron for more exotic metals). The USA has changed from a position of self-reliance in most minerals to becoming a net importer, a position which is exacerbated with each passing year.

Resource appraisal (see *Resource Appraisal*) is concerned with evaluation of the form, concentration, location and extractability of minerals in the ground, both discovered and undiscovered. Resource base refers to the total amount of mineral material presumed to exist in the earth's crust; those parts that are extractable under current and foreseeable technology and economic conditions are the reserves. The grade or weight percent distribution of metallic elements in ores and rocks forms a low-to-high spectrum; in the case of the more abundant elements, such as iron, aluminum and titanium, the abundance pattern of ore grades is a symmetrical, bell-shaped curve. In the case of the scarcer metals, such as copper, lead and mercury, the curve has two distinct peaks with one at higher grade but lower total abundance, where the element is presumably contained in the relatively uncommon ore metals. The other peak, at lower grade but higher total abundance, suggests the element is contained in the rock-forming silicate minerals. In the first case, the element is relatively easy to extract; in the second, it cannot be economically extracted for at least the foreseeable future. Typically, more than half of the known recoverable amounts of a metal are concentrated in the largest 10% of deposits of that metal. Lasky's law as applied to porphyry deposits states this in another way: whereas the cumulative tonnage of mineralized rock increases geometrically, the associated average grade decreases arithmetically. A clear line must always be drawn between the existence of a resource and the current availability of that resource.

The metallic ore minerals are defined in the article *Ore Minerals*, and their types, distribution, identification and characterization are shown in tables listing chemical formulae, metal(s) extracted and principal physical properties. Generally, the precious metals occur in the native state, the base metals as sulfides, ferrous and ferroalloys and aluminum as oxides, other light metals as silicates, and rare earths as phosphates and carbonates.

Terrestrial abundances and distribution of the elements (see *Geochemical Distribution of the Elements*) are necessary information in understanding the factors that influence the formation of mineral deposits. This information includes the occurrence, distribution and abundances of naturally occurring chemical elements in rocks and in their principal minerals. The rocks composing the crust and mantle consist largely of silicate minerals with smaller amounts of carbonates, oxides, phosphates and sulfides. These minerals form solid solution series which provide the sites for major element cation components of the correct size and charge. Trace elements are also picked up in minerals as non-structure-determining elements, and also according to the principles of solution chemistry and ionic size and charge. Because of economic and technological factors, natural elemental abundances cannot be related simply and directly to commercial exploitable ore grades.

Metallogenic provinces (see *Metallogenic Provinces*) are those parts of the earth's crust that have formed high concentrations of a metal or metals at certain geological times. These provinces display an unusually large number of ore deposits, indicating that conditions were appropriate for significant concentrations of certain metals. Recognition of metallogenic provinces and epochs adds order to what otherwise appears to be an irregular distribution of metals in the crust and thus relates metallogenesis to processes of crustal evolution. These provinces also correspond to clearly defined tectonic regions which have undergone similar geological processes in their evolution. Some typical metallogenic provinces in the USA include mid-continent Pb–Zn, southeast Cu–Fe, Colorado Plateau U–V, the western Rocky Mountains Cu-Mo and California gold.

Plate tectonic settings for metallic ore deposits are now well established (See *Plate Tectonic Settings for Metallic Ore Resources*). There are definite correlations between plate tectonic environments and mineral deposits; this information is useful in explaining the genesis of the deposits as well as in assisting in exploration for new deposits. The crust of the earth is the principal reservoir for metals but the upper mantle dominates crustal dynamics. The plates that make up the upper 100 km or so of the earth are the lithosphere which consists of the crust and upper mantle. This is separated from the underlying lower mantle by the esthenosphere, a layer of plastic material. Most mineral deposits form in relation to constructive or destructive plate tectonic boundaries. A third type of boundary, conservative, has no associated ore deposits. Many deposits also form in an intraplate environment. The characteristics of these regions are as follows.

(a) Constructive plate boundaries—these occur at oceanic spreading ridges where new lithosphere is created by upwelling of hot mantle material as two plates move apart. Massive sulfide deposits of Cu–Fe–Zn and magmatic deposits of podiform chromite or of Cu–Ni–Pt sulfides are typical.

(b) Destructive plate boundaries—these are found at oceanic trenches where one lithospheric slab is drawn beneath another by subduction. Magmas are generated at depth along the subducted plate or by upwelling diapirs (piercement folds). Typical deposits formed include massive sulfides of Cu–Zn or Pb–Zn–Cu, disseminated porphyry, or porphyry U–W–Sn–Mo.

(c) Conservative plate boundaries—these are strike slip or transform faults where plates move laterally past one another.

(d) Intraplate environments—these are associated with the formation of many types of deposits. Some may have been linked with early Precambrian plate tectonic regimes, including the gold deposits of Archean greenstone belts.

Mineral deposits are developed by various processes of formation in igneous, sedimentary and metamorphic rocks (see *Igneous Ore Deposits: Formation Processes*; *Sedimentary Ore Deposits: Formation Processes; Metamorphic Ore Deposits: Formation Processes*). Igneous rocks form from molten magma and are either emplaced below the surface of the earth (intrusive) or on the surface (extrusive). Sedimentary rocks form by deposition of chemical or clastic material in bodies of water or as thin layers on the surface of the earth. Metamorphic rocks are formed largely by solid-state reactions from preexisting rocks as a result of deformation and rise of temperature.

Mineral deposits associated with igneous rocks may form:

(a) by simple crystallization from mafic or ultramafic magmas (e.g., Ni, Ti–Fe and chromite deposits);

(b) by hydrothermal processes from granitic and other felsic magmas: ore mineral components are concentrated in late stage volatiles and form deposits of Cu, Pb and Ag;

(c) by hydrothermal alteration of existing minerals by released H^+ ions: this commonly results in alteration patterns associated with porphyry Cu deposits;

(d) replacement and vein deposits including Pb–Zn;

(e) extrusive igneous rocks such as the Nevada Au–Ag deposits; and

(f) exhalative submarine volcanic deposits by shallow submarine volcanism. These include massive stratabound sulfide deposits.

Deposits form in sedimentary rocks by the weathering and transport of material from older rocks including banded iron formations and manganese and phosphorus deposits. Bacteria are responsible for the formation of some deposits including sulfur by the reduction of sulfates in solution. Evaporation of brines may form economic concentrations of halides and magnesium. Other sedimentary processes may also be responsible for important copper deposits, as in Zambia, and for the uranium deposits of the Colorado Plateau.

Deposits formed by metamorphic processes include those of the aluminum silicates, sillimanite, kyanite and andalusite; asbestos, talc and pyrophyllite; graphite; and refractory and grinding materials, such as garnet and emery.

Unconventional future ores (see *Ores of the Future*) must be considered to replace current traditional nonrenewable ores. Some are already being developed although most of their occurrences are still subeconomic. As economic conditions change, the principal constraints on replacement ores may be only human ingenuity to develop the technological skills needed to discover, recover and process unconventional materials. Sources for unconventional materials include (a) conservation opportunities—by recycling or reprocessing used materials and recovering coproducts or by-products; (b) technology dependence—technological breakthroughs which expand the mineral resource base by using lower quality or less conventional and synthetic materials; (c) substitutes for conventional types of materials; and (d) unconceived materials, which depend on geological and chemical research, new exploration methods, and future recovery and product-development technology.

2. *Exploration and Development*

The articles discussing exploration and development of mineral deposits are listed in Table 1. Modern exploration for mineral deposits is carried out by geophysical techniques, geochemical methods, geological field methods and remote sensing (see *Geophysical Exploration for Mineral Deposits*; *Geochemical Exploration for Mineral Deposits*; *Geological Exploration for Mineral Deposits*; *Remote Sensing in the Search for Mineral Deposits*).

Techniques used in geophysical exploration include electrical, magnetic, gravimetric, seismic, radiometric, thermal, spectral and chemical methods and are carried out on the ground, in the air, or at sea. The principal methods used in exploration today are:

(a) electrical techniques, based on either naturally occurring influence fields (such as self-potential, telluric-current and magnetotelluric), or on fields introduced into the earth artificially, as potential drop (including resistivity) electromagnetic and induced polarization;

(b) magnetic techniques, depending on the earth's magnetic field and local variations in the natural remanent magnetic properties of the earth's crust which are generally due to buried magnetic bodies;

(c) gravimetric techniques, which measure differences in density—positive and negative values can be correlated with denser and less dense rock bodies;

(d) seismic techniques, which measure the reflectance and refraction of elastic waves within the earth by subsurface rock bodies; and

(e) Radiometric techniques, which measure gamma emissions by near surface rocks—scintillation counters can differentiate U, Th and K emissions.

Geochemical exploration is based on analysis of natural materials to detect anomalous concentrations of chemical elements derived from hidden mineral deposits (see *Geochemical Exploration for Mineral Deposits*). Typical materials sampled include rocks, soils, stream sediments, vegetation and water. Anomalous concentrations of an element are amounts that are significantly higher than the background value for that element, and may be due to primary processes occurring during ore emplacement, or secondary processes, such as weathering or transport of material.

Geological exploration is the fundamental method in the search for ore deposits (see *Geological Exploration for Mineral Deposits*). Conceptual models are developed for ore deposits based upon reconnaissance of a region or a single prospect, using genetic characteristics of well-studied ore bodies in geologically similar terrain to develop an exploration model. The exploration program may then use all available techniques, including geochemical and geophysical methods, once the exploration targets (i.e., the hypothetical orebodies) have been evidenced by specific anomalies.

Remote sensing for mineral deposits refers to the detection and analysis of phenomena through the use of aerial and satellite sensor systems (see *Remote Sensing in the Search for Mineral Deposits*). The sensors include photographic cameras, electronic side-looking radar, electroscanning radiometers and line or area array cameras. The data are analyzed by visual and computer-assisted analysis and yield information for geological reconnaissance, maps of topographic features and structures associated with mineral deposits, maps of lineaments indicating regional stress fields, reflectance patterns unique to altered rocks associated with mineral deposits and basic control for detailed geological mapping.

Mining is carried out underground, by open-cut methods, or in the ocean. Underground mining generally follows similar developmental methods and mining sequences (see *Underground Mining*). The principal production methods vary for competent rocks (i.e., those that will support openings), or weaker or incompetent rocks that will cave naturally. Other underground systems involve in situ leaching and borehole mining by hydrometallurgical extraction within the orebody itself.

Open-cut mining is concerned with methods of open-pit (open-cast) mining, strip mining, quarrying and placer mining. It accounts for approximately 60% of the world's metallic mineral production (see *Open-Cut Mining*). Open cuts are used for mineral deposits that occur close to the surface, generally related to large and continuous deposits where highly mechanized and low-cost operations can be used on a large scale. Such techniques allow the recovery of large deposits of low-grade ore. Reclamation is a serious concern for open-cut mines with laws now requiring lands to be restored to nearly their condition before mining.

Ocean resources and mining covers deposits associated with the continental margins and coastal zones as well as those of the deep sea (see *Ocean Resources and Mining*). The continental margin forms about 50% of the total land area and 20% of the sea. Oil and gas are the most important resources extracted in this zone but dredging of placer deposits for diamonds, tin, zircon and titanium is also significant. The principal mineral resource interest in the deep sea centers around manganese nodules which contain 5–15% Fe, 10–30% Mn and more important commercial amounts (1–3%) of Cu, Ni and Co. Metalliferous sediments with high amounts of Fe, Zn, Cu, Ag, Ni, Co and V are found in the Red Sea.

Examination and evaluation of mineral deposits (see *Mineral Deposits: Examination and Evaluation*) is carried out by detailed sampling, calculation of ore reserves, and technical and economic consideration. Evaluation usually involves a brief preliminary study, a more detailed follow-up evaluation, and finally a feasibility study in which the decision is made to develop a mine.

Environmental problems associated with mining and processing that are common in both coal and metal mining include air and water pollution, land subsidence and land reclamation, concurrent with or shortly after mining activities (see *Mining and Processing: Environmental Problems*). Many of these same problems, plus a need for large amounts of water, are also anticipated in oil shale mining. Problems that are specific to metal mining and processing are the disposal of tailings, and smelter stack emissions which may include significant amounts of particulate matter and gases. The gases contain SO_2, CO and CO_2 as well as metals, asbestos and fluorides, all of which may be damaging to human health and environment.

Table 2
Classification of articles on metallic mineral resources

Major base, light and alloying metals
Aluminum Resources
Chromite Resources
Cobalt Resources
Copper Resources
Iron Resources
Lead Resources
Magnesium Resources
Manganese Resources
Molybdenum Resources
Nickel Resources
Tin Resources
Titanium Resources
Tungsten Resources
Vanadium Resources
Zinc Resources
Other metals
Antimony Resources
Arsenic Resources
Beryllium Resources
Bismuth Resources
Cadmium Resources
Gallium, Germanium and Indium Resources
Gold Resources
Lithium, Cesium and Rubidium Resources
Mercury Resources
Niobium and Tantalum Resources
Platinum Group Metals Resources
Rare-Earth Metals Resources
Silver Resources
Thorium Resources
Uranium Resources
Zirconium and Hafnium Resources

3. *Metallic Mineral Resources*

The Encyclopedia also includes a large number of articles concerned with individual metal resources (see Table 2). The length of each article and the amount of detail varies, depending on the importance of the commodity in the economic society of today. Generally, most articles include discussions of (a) crustal distribution, geochemistry and ore tenors, (b) uses, (c) present production patterns and reserves, (d) global resources and future production patterns and (e) principal types of ore deposits with examples. The metals covered include:

(a) iron and ferrous alloy resources—iron, manganese, nickel, chromium, molybdenum, tungsten, vanadium and cobalt;

(b) nonferrous metals resources—copper, lead, zinc, cadmium, tin and aluminum;

(c) precious metal resources—gold, silver and platinum group elements;

(d) other metals—lithium and other rare alkali metals, gallium, germanium and rubidium, niobium and tantalum, rare earths, mercury, beryllium, zirconium, titanium, bismuth, antimony, magnesium, uranium, and thorium.

4. Industrial Minerals

Industrial minerals, including stone and rocks, may be defined as those naturally occurring materials used to build structures or supply products that are useful to an industrialized society. Since industrial minerals exclude the ores of metals, they have been called the "nonmetallics." This term, however, is not now widely used. Neither is it properly inclusive, as fuels and water are also excluded from industrial minerals.

Industrial minerals are extensively used, although the average individual is unaware of their importance. Roads and buildings are conspicuous signboards of industrial minerals, but minerals used in processing or making products are not so apparent. For example, automobiles have metal parts molded in foundry sand, bodies polished by abrasives and covered with paint containing mineral fillers and pigments, and engines fuelled by petroleum products which are produced and refined with the help of clay.

4.1 Classification

The industrial minerals covered in this Encyclopedia include more than 50 rocks and minerals from igneous, metamorphic and sedimentary sources. They have been classified as to principal use (see Table 3).

4.2 Characteristics

Most Classifications draw sharp boundaries that are more apparent than real. The classification for industrial minerals is no exception. Some broad generalizations can be made about the characteristics of industrial minerals (Table 4), but a classification based on use has inconsistencies because of the great number and diversity of the minerals.

Although industrial minerals can be thought of as a group of "nonmetallics," some ores of metals are included in the classification. For example, iron oxide, an ore of iron, is listed as a pigment, a usage that qualifies it as an industrial mineral, and so it is placed in the classification. Some minerals may have more than one use and are included under multiple entries. Discussions of clay minerals, because of their many applications, are found under binders, construction materials, fillers and coatings, filters, sorbents and ion exchangers; and well-drilling materials. Clay is also an important abrasive and a principal component in most fill and soil.

Industrial minerals as used here exclude those that undergo major chemical conversion, a definition in itself that does not bear close analysis. Zeolites, for example, undergo ionic substitution, a chemical process, but because these minerals also have physical properties that make them useful as filters and sorbents, they are included as industrial minerals. Most potash and phosphate minerals used in commerce are chemically converted into fertilizer products, but some are not, and so they qualify for inclusion as additives to soil. In this classification, the quantity of material used is not so important as how the material is used. Gems and art objects are valuable for their intrinsic

Table 3
Classification of industrial minerals according to use

Abrasives
Abrasives
Abrasives: Nonsiliceous Minerals
Abrasives: Precious and Semiprecious Minerals
Abrasives: Siliceous Minerals

Binders
Binders: Clay Minerals
Binders: Industrial Minerals

Construction materials
Asbestos: Alternatives
Asbestos Fibers
Asbestos Hazards
Calcium Aluminate Cements
Construction Materials: Crushed Stone
Construction Materials: Dimension Stone
Construction Materials: Fill and Soil
Construction Materials: Granules
Construction Materials: Industrial Minerals
Construction Materials: Lightweight Aggregate
Construction Materials: Sand and Gravel
Gypsum and Anhydrite
Insulation Raw Materials
Lime
Portland Cement Raw Materials
Portland Cements, Blended Cements and Mortars
Recycling of Demolition Wastes
Slag
Structural Clay Products
Traditional Ceramics: An Overview
Whiteware: Component Minerals

Electronic and optical materials
Electronic and Optical Minerals

Fillers and coatings
Fillers and Coatings: Carbonate Minerals
Fillers and Coatings: Clay Minerals
Fillers and Coatings: Industrial Minerals
Fillers and Coatings: Siliceous Minerals
Fillers and Coatings: Sulfate Minerals

Filters, sorbents and ion exchangers
Filters, Sorbents and Ion Exchangers
Filters, Sorbents and Ion Exchangers: Clay Minerals
Filters, Sorbents and Ion Exchangers: Siliceous Minerals
Filters, Sorbents and Ion Exchangers: Zeolite Minerals

Foundry sands
Foundry Sands: Classification and Sources
Foundry Sands: Silica Sand and Nonsilica Minerals

Glass
Glass: An Overview

Pigments
Pigments: Industrial Minerals
Pigments: Titania, Iron Oxides and Other Minerals

Soil additives
Soil Additives: Industrial Minerals
Soil Additives: Phosphate, Potash and Sulfur

Well-drilling materials
Well-drilling: Clay and Nonclay Minerals
Well-drilling: Industrial Minerals

Table 4
Characteristics of natural industrial minerals (excluding those undergoing chemical conversion)

Mineral and rock types
Igneous, metamorphic, and sedimentary rocks and minerals of many different types. Those of low unit value are generally rocks of widespread occurrence and associated with simple geological conditions; those of high unit value are generally rocks and minerals of restricted occurrence and associated with complex geological conditions

Exploration
Those of low unit value generally cannot bear the cost of intensive geological evaluation; exploration is simple or nonexistent, relying on use of regional studies or production trial and error. Those of high unit value, which require large expenditures for mine and plant development, can bear the cost of regional studies and site evaluations

Mining
Either surface or underground mining may be used, but underground mining is generally more expensive and most often used for deposits that are not near the surface, for special structural conditions, for selective removal of particular beds, and for accommodation of weather or land-use considerations. In some places the extra cost of underground mining is compensated by the usable space that is created

Processing
Those of low unit value require simple processing, but those of high unit value can bear the cost of complex processing

Transportation
Those of low unit value are generally used near the point of production, require short haulage and use trucks extensively. Those of high unit value generally can absorb higher transportation costs, make use of rail and water haulage and be traded in import and export markets

Marketing
Those of low unit value are generally marketed near the deposit by the producing company; price is generally more important than specification to the consumer. Those of high unit value may be marketed far from the deposit by the producer, sales agents, or other companies; specification is generally more important than price to the consumer. For both those of high and low unit value, one mineral or synthetic product may be substituted for the same use

properties, but because they are not used in the sense of structures or products, they are not included. Industrial-grade diamonds and semiprecious minerals, however, are useful to industry because of their hardness and are included under abrasives.

Although chemical properties may affect physical properties of rocks and minerals, they are a secondary consideration for most uses of industrial minerals. Some industrial minerals are used just as they are taken from the ground; processing, compared with that of metals, is simple, and, if needed at all, usually consists of crushing, sizing, or washing. The physical properties inherent in the rock are the same properties that make it useful to industry, and processing seldom appreciably enhances these properties. Processing may be used to beneficiate the product. At times processing may be a sophisticated operation to remove unwanted material or produce particle sizes within close limits, but it does not alter or refine the material in the way that metals are altered or refined. Clay for filler and coating applications, for example, requires special processing, which may include dispersion, sizing, leaching, magnetic separation, drying and pulverizing. But even after processing, the clay minerals are still similar in character to what they were before processing.

The wide range in values of industrial minerals indicates something of their diversity in occurrence and in properties. The cost of some construction materials is little more than the cost of the labor and materials to produce them; that is, the minerals have a value no greater than the surface value of the land. For example, in some parts of the USA, land can be purchased and stone can be quarried, crushed, sized and loaded for shipment for no more than $4 per ton. Other industrial-mineral deposits (for example, industrial diamonds) may be much more valuable than the surface value of the land and on a weight basis sell for a price a million or more times that of crushed stone.

The industrial minerals of low unit value, such as the construction materials, are generally produced near the place where they are used, which is in or near populated areas. Reserves of these materials are widespread and abundant, and because they are heavy and bulky they cannot be shipped far before transportation costs exceed production costs. Operations that produce these types of materials generally have production capacities that far exceed product demand. Materials of high unit value generally have properties that are not so widespread; they can be shipped farther and can bear the cost of higher transportation in a competitive market. In general, as the demand grows, so does the distance a product can be shipped, but this increase is not without limits. More than one mineral may have the same use, and the one that is used will be the one that is more competitively priced.

Surface mining is more generally used than underground mining for industrial minerals because it is usually cheaper, but both methods have advantages and disadvantages. For the industrial minerals of low unit value, cost of production is an important factor in a competitive market, and so, if possible, surface mining will be the method of choice. But in some places where surface deposits are not available or where such factors as unfavorable geological structures, thin mining seams, all-weather mining, environmental problems, governmental incentives, and postmining use are considerations, underground mining may be preferred. For industrial minerals of high unit value, underground mining may be the more efficient method of extraction because their mineral occurrence is generally geologically more complex.

5. *Fuels*

The articles on fuels include the fossil fuels (coal and petroleum) and nuclear energy (uranium) resources (see Table 5).

Table 5
Classification of articles on fuels

Coal
Coal: Geology
Coal: Mining
Coal: World Resources
Petroleum
Hydrocarbons: Origin, Migration and Accumulation
Petroleum: Oil and Gas Fields
Petroleum: World Resources
Nuclear
Uranium Resources

Our civilization has been enabled to reach its present high level because of the abundance of relatively cheap fossil fuels, and if it is to be maintained and expanded, new sources of fossil fuels and nuclear energy will be required. It is essential that the search for coal, oil, gas and uranium be continued until alternative energy sources can be developed that are abundant and environmentally acceptable. In many countries high-grade mineral resources have been already expended, leaving lower grades of ores for sources of metals and industrial minerals. Extraction of the metals and minerals is still possible from these lower-grade deposits, but only at the expense of additional energy. Energy is an important key to society's future well-being.

Bibliography

Bates R L 1969 *Geology of the Industrial Rocks and Minerals.* Dover, New York

Berthoumieux G 1978 The industrial minerals of France. *Proc. 3rd Industrial Minerals Int. Congr.* Metal Bulletin, Worcester Park, Surrey, pp. 15–24

Blunden J 1975 *The Mineral Resources of Britain: A Study in Exploitation and Planning.* Hutchinson, London

Borzunov V M 1969 *Nonmetallic Deposits, Their Exploration and Economic Appraisal.* Nedra, Moscow

Bowie S H U, Kvalheim A, Haslam H W (eds.) 1978 *Mineral Deposits of Europe*, Vol. 1: *Northwest Europe.* The Institution of Mining and Metallurgy and the Mineralogical Society, London

Brazilian Ministry of Mines and Energy (1973 onwards) *Analytical Profile Series.* Brazilian National Department of Mineral Production, Brasilia. (Each bulletin deals with a specific industrial mineral commodity)

Brobst D A, Pratt W P (eds.) 1973 *United States Mineral Resources*, US Geological Survey Professional Paper 820. US Government Printing Office, Washington, DC

Council of Commonwealth Mining and Metallurgical Institutions 1957 Geology of Canadian industrial mineral deposits. *Proc. Commonwealth Mining and Metallurgical Congr.* Council of Commonwealth Mining and Metallurgical Institutions, London

Dunn J R 1973 A matrix classification for industrial minerals and rocks. *Proc. 8th Forum on Geology of Industrial Minerals*, Iowa Geological Survey Public Information Circular No. 5. IGS, Iowa City, Iowa, pp. 185–89

Guilbert J M, Park C F Jr 1986 *The Geology of Ore Deposits*, Freeman, New York

Harben P W, Bates R L 1984 *Geology of the Nonmetallics.* Metal Bulletin, Worcester Park, Surrey

Jensen M L, Bateman A M 1979 *Economic Mineral Deposits*, 3rd edn. Wiley, New York

Knight C L (ed.) 1976 *Economic Geology of Australia and Papua New Guinea—4: Industrial Minerals and Rocks*, Australasian Institute of Mining and Metallurgy Monograph Series No. 8. AIMM, Parkville, Victoria

Kuzvart M 1984 *Industrial Minerals and Rocks.* Elsevier, Amsterdam

Lefond S J (ed.) 1983 *Industrial Minerals and Rocks*, 5th edn. American Institute of Mining, Metallurgical and Petroleum Engineers, New York

Lüttig G et al. 1977 The industrial minerals of the Federal Republic of Germany, and the role of the industry in the public scene. *Proc. 2nd Industrial Minerals Int. Congr.* Metal Bulletin, Worcester Park, Surrey; pp. 3–16

Mineral Resources Consultative Committee (1971 onwards) *Mineral Dossier Series.* HMSO, London. (Each volume deals with a specific industrial mineral commodity)

Mitchell A H G, Garson M S 1981 *Mineral Deposits and Global Tectonic Settings.* Academic Press, London

North F K 1985 *Petroleum Geology.* Allen and Unwin, Boston

Peters W C 1978 *Exploration and Mining Geology.* Wiley, New York

Selley R C 1985 *Elements of Petroleum Geology.* Freeman, New York

Skinner B J (ed.) 1981 *Economic Geology Seventy-Fifth Anniversary Volume 1905–1980.* Economic Geology, Lancaster, Pennsylvania

Smith V K (ed.) 1979 *Scarcity and Growth Reconsidered.* Johns Hopkins University Press, Baltimore, Maryland

US Bureau of Mines 1985 *Mineral Facts and Problems*, US Bureau of Mines Bulletin 675. USBM, Washington, DC

US Bureau of Mines 1988 *Mineral Commodity Summaries 1988.* USBM, Washington, DC

Ward C R 1984 *Coal Geology and Coal Technology.* Blackwell, Melbourne

Woodland A W, Highley D E 1975 Britain's industrial minerals: Their geological framework, their production today, and some aspects of their development. *Proc. 1st Industrial Minerals Int. Congr.* Metal Bulletin, Worcester Park, Surrey, pp. 2–11

Abrasives

Abrasives are materials that are used to smooth, roughen, polish, or clean surfaces, or simply to remove material to alter surface shape or dimensions. Abrasives generally accomplish their purpose through a rubbing action but may also work by impact (e.g., sand blasting). Certain chemical cleaners, such as detergents and organic solvents, are not abrasives, although many polishes on the market combine abrasives with chemical cleaners. Only naturally occurring minerals and rocks are discussed here though it should be understood that synthetic abrasives account for the major part of the abrasives market.

1. Properties

The most important physical properties of abrasives are hardness, toughness/brittleness, grain shape and size, character of fracture, and purity or uniformity. No single property is of paramount importance for every use. For some uses, such as diamonds for drill bits, extreme hardness and toughness are important; in other uses, hardness may be best matched with brittleness so that fresh cutting surfaces are created by the breaking of worn grains. Inferior hardness may be desired in dentifrices and household cleansers. Grain shape is important because an angular grain will cut better than a rounded one, other things being equal. Grain size, of course, controls the depth of cut that a grain makes; in general, coarser grains are used for rapid reduction of a surface, and finer grains for a smooth finish. Character of fracture affects cutting power of a grain; a grain that breaks with a rough fracture will have superior cutting power to one that cleaves with a smooth surface, other factors being comparable. Purity and uniformity relate primarily to the performance that the user expects from an abrasive and become increasingly important as the conditions of use become more exacting.

Hardness is the basic property by which all abrasive materials may be compared, but it has also proved difficult to standardize. The Mohs scale of hardness, long used by mineralogists as a tool for mineral identification, provides a relative measure of scratch hardness. It is based on the relative hardness of a series of well-known minerals arranged on a scale of 1 to 10, from talc, the softest (1), to diamond, the hardest natural substance known (10). While the Mohs scale has proved an effective tool for mineral identification on both a macro- and microscopic scale, it has not lent itself to precise measurement and standardization for industrial use. Not only are the methods of testing crude, but the intervals between steps in the scale are of unequal magnitude: for example, the difference in hardness between 9 (corundum) and 10 (diamond) is far greater than for any other single step. Probably the most widely accepted numerical measure of hardness for abrasives now in use is the Knoop scale (Knoop et al. 1939). Knoop hardness numbers are derived by a method of diamond indentation and are reproducible for a wide variety of materials up to the hardness of diamond. The Knoop scale is of particular value for measuring the hardness of synthetic abrasives, such as silicon carbide and boron carbide, that range in hardness between corundum and diamond. Table 1 shows comparative hardness values for some common natural and synthetic abrasives.

2. Types of Abrasive Products

The forms in which abrasives are used may be conveniently grouped as (a) loose abrasive grains; (b) bonded abrasives; (c) coated abrasives; (d) fine grains and powders for buffing and polishing compounds, soaps, and cleaners; (e) natural stones for such purposes as grinding, sharpening, rubbing and polishing; and (f) natural stones for use in grinding mills.

Loose abrasive grains are used in pressure blasting for cleaning stone and concrete surfaces, cleaning metal castings, and etching glass, among other purposes. Quartz sand enjoys the largest use by volume for this purpose, but corundum, garnet, flint and chert may also be used. Quartz sand is also the common loose abrasive grain used for grinding and rough polishing of a variety of materials, such as building and ornamental stone, flooring, glass, metals, wood, and gem stones. Most of the harder natural materials, such as quartz, garnet and emery, may be used for these purposes.

Table 1
Hardness numbers of some common abrasives

Abrasive	Mohs scale	Knoop scale (average)
Diamond	10	8200
Boron carbide	9.7–9.8	2800
Silicon carbide	9.5–9.6	2500
Corundum (aluminum oxide)	9	
Aluminum oxide, manufactured		1950
Garnet	7.5–8	1360
Quartz	7	820
Feldspar	6	
Pumice	5.5–6	
Opal (diatomite)	5	

Bonded abrasives are carefully sized grains molded in a cementing medium into wheels, disks, and other shapes. Various types of bonds, depending on severity or delicacy of end use, are employed. Vitrified bond, fired in a ceramic kiln, is generally the most durable and has the best dimensional stability; other bonds commonly used are hard synthetic resin, rubber, shellac and silicate. Wheels and disks are the most common shapes for grinding, polishing and cutoff, but other shapes include blocks and sticks for sharpening and polishing, and specially curved shapes for polishing interior and exterior surfaces. Originally, the natural products, corundum and emery, were widely used in grinding wheels, but electric-furnace products, such as aluminum oxide and silicon carbide, have largely supplanted them. However, diamond, because of its great hardness and durability, is experiencing increasing demand in many industrial applications, especially high-speed and high-precision applications.

Coated abrasives are simply abrasive grains cemented to a paper or cloth backing (e.g., sandpaper, garnet paper and emery cloth). Today most sandpaper and flint paper are coated with crushed quartz, although true flint is still used in some parts of Europe. Garnet paper is still coated with garnet. Most emery cloth is now coated with the manufactured abrasives, aluminum oxide and silicon carbide. Garnet and flint papers are used mainly on wood and other soft materials, whereas emery cloth is used chiefly on metal surfaces.

Fine grains and powders for polishing and cleaning purposes embrace such a wide variety of materials and end uses that any discussion is best left to the articles on specific abrasive materials.

Most natural stones used for abrasive purposes and in grinding mills have been replaced by synthetic and manufactured products. However, there is still a small but significant demand for certain types of stones for specialized uses.

See also: Abrasives: Nonsiliceous Minerals; Abrasives: Precious and Semiprecious Minerals; Abrasives: Siliceous Minerals

Bibliography

Harben P 1978 Abrasives. *Ind. Miner. (London)* 134: 49–73

Hight R P 1983 Abrasives. In: Lefond S J (ed.) 1983 *Industrial Minerals and Rocks*, 5th edn. American Institute of Mining, Metallurgical, and Petroleum Engineers, New York, pp. 11–32

Knoop F, Peters C G, Emerson W B 1939 A sensitive pyramidal-diamond tool for indentation measurements. *J. Res. Natl. Bur. Stand. Sect. C* 23: 39–61

Ladoo R B, Myers W M 1951 *Nonmetallic Minerals*. McGraw-Hill, New York

US Bureau of Mines 1986 Abrasive materials. Preprint from *US Bureau of Mines Minerals Yearbook 1985*. US Bureau of Mines, Washington, DC

J. C. Bradbury

[Illinois Department of Energy and Natural Resources, Champaign, Illinois, USA]

Abrasives: Nonsiliceous Minerals

Nonsiliceous abrasives include feldspar, nepheline syenite, olivine, bauxite, clay, emery, calcite, dolomite, iron oxides, staurolite and talc. These are all natural abrasives except for clay and bauxite, which are calcined to produce a harder mineral phase. Staurolite, calcined clay and calcined bauxite are relatively hard, but the other minerals are of intermediate or inferior hardness. However, as long as a material can scratch another it may in the broadest sense be termed an abrasive. No single abrasive is all-purpose, and the choice depends on which abrasive will do the job most efficiently and economically. Sometimes soft materials are used to remove scale or similar coatings without affecting the actual product.

Many of the nonsiliceous abrasives are used in making soaps, cleansers and polishes. Often low price is the primary factor in the selection of relatively soft abrasives for use in soaps and cleansers. Any household cleanser should not contain abrasive material with a Mohs hardness greater than 5–5.5, because glass and enamel are scratched by minerals with a hardness of about 6. Soaps and cleansers should not contain calcite, gypsum or dolomite because calcium reacts with most soaps to form insoluble compounds which are very difficult to remove from glass and enamel surfaces. The grain size of abrasives ranges from about 100 μm or even coarser for heavy-duty scouring soaps to extremely fine particles less than 2 μm for use in certain polishing compounds. The material must be sized accurately and closely to avoid large grains scratching the surface, or small grains not contributing to the abrasive action.

1. Feldspar and Nepheline Syenite

Feldspars are a group of aluminum silicate minerals that contain either potassium, sodium or calcium, or a mixture of these elements. The relatively common feldspar minerals are orthoclase and microcline (potassium), albite (sodium), and anorthite (calcium). Feldspar minerals are a major component of igneous and metamorphic rocks. The hardness of the feldspar minerals on the Mohs scale is about 6. They break with a semiconchoidal fracture, which makes them useful as mild abrasives. Feldspars of commercial significance are found mainly in pegmatites as large crystals usually free of iron-bearing impurities. Pegmatites are widely distributed coarse-grained igneous rocks, which frequently occur in association with granites. In pegmatites consisting of large mineral crystals, the feldspar can be easily separated by hand, but when the feldspar is finer it must be ground to less than 20 mesh (0.84 mm) and beneficiated by using flotation and/or magnetic separation.

Two terms relating to both mineralogy and economics are in common usage in the feldspar industry. These are aplite and alaskite. Aplite is a feldspathic rock, mined and beneficiated in Virginia, in which both titanium and feldspar minerals are present. Alaskite is a relatively coarse-grained granite-like rock found in the vicinity of Spruce Pine, North Carolina, which contains both potash feldspar and sodium and calcium feldspar (plagioclase), plus minor amounts of quartz and muscovite.

A major use of feldspar is in scouring powders where it has largely replaced silica sand in most parts of the world. Feldspars are not as hard as quartz, and thus are safer to use on glass and enamel surfaces. Feldspar is mined in many countries in the world with the USA being the largest producer. In the USA, feldspar is produced in North Carolina, Virginia, Connecticut, Georgia, South Dakota, Wyoming, Arizona and California. Other countries producing feldspar are Canada, Mexico, Argentina, Colombia, Finland, France, the Federal Republic of Germany, Italy, Norway, Poland, Spain, Sweden, the USSR, the UK, Yugoslavia, South Africa, India, Japan, Korea and the Philippines. The total amount used by the abrasive industry is very small, constituting less than 1% of the total production of feldspar.

Nepheline syenite is a sodium–potassium aluminum silicate rock consisting of the minerals albite, microcline, nepheline, and small amounts of iron-bearing and other accessory minerals. Nepheline syenites are widely distributed around the world, but are mined only in Canada, Norway and the USSR. Nepheline syenite is a coarse- to medium-grained igneous rock and has a hardness on the Mohs scale of 5.5–6. Nepheline syenite competes with feldspar in most applications. The rock is mined, crushed, screened, ground, and the iron- and titanium-bearing minerals are removed by high-intensity magnetic separators.

2. *Olivine*

Olivine is a group term used to indicate an isomorphous series of magnesium–iron silicates of which forsterite (Mg_2SiO_4) and fayalite (Fe_2SiO_4) are end-members. In commercial practice, the use of the term olivine applies to deposits containing a mixture of forsterite and fayalite in solid solution with the fayalite content usually less than 15%. Olivine occurs in dunites which are of igneous origin; large deposits of olivine-bearing dunites are known in North Carolina and Washington in the USA, and in Norway, Sweden, the USSR, Austria, Japan, New Zealand, Rhodesia, South Africa and New Caledonia. Olivine has a hardness on the Mohs scale of 6–7. The grains have a high mechanical strength and are therefore quite durable, and the grain surfaces are more angular than silica sands. Olivine is used extensively in Scandinavia as a blasting medium, where it is estimated as much as 40–50 kt are being used annually for this purpose. The production of olivine is essentially a dry crushing, grinding and sizing operation. In some deposits where there is a relatively high impurity content, beneficiation by flotation or gravity separation may be necessary.

3. *Bauxite*

Aluminum oxide abrasives are made from bauxite, which is the major ore of aluminum. Bauxite is a rock primarily comprised of gibbsite ($Al_2O_3.3H_2O$) and boehmite ($Al_2O_3.H_2O$), although a few bauxites contain the mineral diaspore ($Al_2O_3.H_2O$). Boehmite and diaspore are both alumina monohydrates, but diaspore has a higher specific gravity and a greater hardness. Silica, iron oxides and titania are major impurities always present in bauxite deposits. Bauxite is a product of tropical weathering (laterization) of rocks containing aluminum. The more soluble minerals are leached and the less soluble compounds retained, resulting in an accumulation of aluminum and iron oxides. Bauxite is known to occur on all continents except Antarctica. The major producing countries are Australia, Jamaica, Surinam, Guyana, the USSR, France, Guinea, Greece, Hungary and Yugoslavia. Substantial reserves are known to be present in Brazil, Indonesia and China.

Bauxite is mined by open-pit or strip-mining methods using large excavators. The bauxite is transported to the processing plant where it is crushed, washed and screened to remove as much of the fine clay minerals and quartz as possible. To produce abrasive grades, the bauxite is calcined to $\geqslant 1300\,°C$; this results in the formation of hard aluminum oxide grains which are essentially corundum and have a hardness of 8.5–9 on the Mohs scale. These hard, tough grains can then be sized into coarse, medium and fine products depending on the abrasive need. The abrasive product can be used as a loose-grain abrasive, coated abrasive, or molded into grinding wheels and bar compounds. Calcined bauxites for abrasive use are produced principally in Guyana, Surinam, Australia, China and the USA. Those bauxites with the highest alumina content, and the lowest iron and silica value produce the highest quality abrasive products.

The first aluminum oxide abrasive was produced commercially in 1904 from fused alumina. Aluminum oxide abrasives are made from bauxites by three different methods: (a) bauxite may be fused directly in an electric furnace, (b) it may be processed to form alumina using the Bayer process and then fused, and (c) it may be sintered. This latter process involves calcining the processed bauxite. Abrasive-grade bauxite constitutes only about 2–3% of the total market for bauxite.

4. *Clay*

Calcined kaolin is the major clay material used as an abrasive. Kaolin is a rock made up primarily of the mineral kaolinite. Kaolinite is a hydrated aluminum silicate ($Al_2O_3.2SiO_2.2H_2O$) which is naturally very fine in particle size, with most of the particles being less than 2 μm. Kaolinite has a hardness of about 1.5 on the Mohs scale, and is thus much too soft to be used as an abrasive. However, when kaolinite is heated to 1000 °C, the mineral is transformed into mullite (Al_2SiO_5) and cristobalite (SiO_2) which have a hardness in the range of 6–7 on the Mohs scale. Mullite is a dense lath-like mineral that is tough and hard. Being extremely fine in particle size, the calcined kaolin makes an excellent polishing agent because any scratches are too small to be detected by the eye. Calcined kaolin is used in automotive polishes, several types of metal polishes (primarily silver) and toothpaste. Kaolin is white or near white in color and so can be readily used with various colors. It is also used in bar compounds when a very fine abrasive is required primarily to clean the surface prior to polishing.

Kaolins are mined and processed in many areas of the world but the two major areas are Georgia in the USA and Cornwall in the UK. The major abrasive production is from Georgia, although in comparison to the total amount produced the quantity is very small.

Another clay that was once used as an abrasive was meerschaum, which was primarily used to make pipe bowls. Meerschaum is relatively hard and dense. The clay mineral which comprises meerschaum is sepiolite, a hydrated magnesium silicate. Meerschaum is mined in Turkey and Spain. It was used in the past as the standard polish for naval and military tunic buttons.

5. *Emery*

Emery is an impure form of corundum, a natural aluminum oxide with a hardness of 9 on the Mohs scale. The impurities are iron, titania, silica and magnesia. Depending on the amount of impurities the hardness ranges from about 7.5–8.5. The major world producer of emery is Turkey with an annual output of about 80 kt. The second most important producer is Greece, where emery is mined on the island of Naxos. Emery has been mined on Naxos for about 4000 years and at present the production is estimated at about 7 kt per year. Most of the raw emery from Turkey and Greece is shipped to the Netherlands where it is processed. Processing involves washing, crushing and heat treatment to produce some 35 different grain sizes. The coarse products are used for wear-resistant top layers on roads, concrete tiles, epoxies and other resins, and to produce millstones for rice milling. The medium and finer grades are used to make emery paper and cloth, and the very fine flour goes into the production of polishing pastes.

6. *Calcite and Dolomite*

Fine-grained powders are used for polishing and lapping a variety of materials including plastics, wood surfaces and semiprecious stones. Very fine particle size calcite ($CaCO_3$) and dolomite [$CaMg(CO_3)_2$] are sometimes used for this purpose. Calcite has a hardness of 3 on the Mohs scale and dolomite a hardness of 3.5–4.0. The fine-grained carbonates known as chalks or whiting from the UK and France are those most commonly used for polishing. Only very small quantities are used by the abrasives industry.

7. *Iron Oxides*

The chief iron oxide minerals are *hematite* (Fe_2O_3), magnetite (Fe_3O_4), goethite ($Fe_2O_3.H_2O$) and limonite ($2Fe_2O_3.3H_2O$). Very little natural iron oxide is used as an abrasive or polishing agent, but chemically precipitated oxides (rouges) are employed because all have a very fine grain size. Some small quantities of very finely ground natural hematite and magnetite are used as polishing agents, and it is expected that this use will increase in the future because of the rising relative costs of producing the precipitated iron oxides. The major use of rouge is in polishing plate glass; other uses include polishes for precious metals and stone to give a high luster.

8. *Apatite*

Apatite is a calcium phosphate [$Ca_3(PO_4)_2$] with a hardness of 5 on the Mohs scale. Some coarse-grained apatites found in igneous rocks in Russia and South Africa are used as mild abrasives such as scouring powders, but only in small quantities and on a local level.

9. *Perlite*

Perlite is glassy volcanic rock which on rapid controlled heating will expand or pop into a very fine lightweight product with a low bulk density (32–320 $kg\,m^{-3}$). Most perlites have a silica content greater than 70% and a hardness ranging from 5.5 to 7 on the Mohs scale. A miscellaneous use of the expanded perlite is as a polishing or scouring powder. The USA is the largest producer of perlite with 90% of the production coming from New Mexico. Other countries producing perlite are Greece, Hungary, Italy, Japan, Mexico, the Philippines and the USSR.

10. *Staurolite*

Staurolite ($Al_2O_3.SiO_2$) is an anhydrous aluminum silicate with a hardness on the Mohs scale of 7.0–7.5. Staurolite is found in metamorphic rocks and is a resistant mineral that persists in beach and river sands derived from metamorphic rocks. Staurolite is mined from various river and beach placer deposits in the southeastern USA, where it is recovered as a heavy mineral concentrate. It is separated by gravity and/or flotation, and is sold as an abrasive for pressure blasting.

11. Talc

Talc [$Mg_6Si_8O_{20}(OH_4)$] is a hydrated magnesium silicate, which is soft (Mohs hardness of 1) and white. It is found in metamorphic rocks associated with dolomites. Talc is used primarily as a filler in paint and paper, in ceramics, and as a constituent in cosmetics. Despite its low hardness, it is used as a polish for cereals such as rice, corn and barley, in shoe polish, and for polishing synthetic fibers such as nylon.

12. Summary

As indicated in the discussions of the various nonsiliceous abrasive minerals one of the most important properties is grain-size distribution. The material has to be sized accurately and closely to avoid large grains scratching the surface or small grains not contributing to the abrasive action. Another important characteristic is toughness which is the ability to resist fracture, and when a mineral does fracture, the type of fracture has a bearing on its utilization as an abrasive. The cost of the various abrasives must be weighed against performance and convenience. Synthetic abrasive materials over the years have seriously encroached on the use of natural abrasives but there is still a market for the quality produced natural abrasive mineral.

See also: Abrasives; Abrasives: Precious and Semiprecious Minerals; Abrasives: Siliceous Minerals

Bibliography

Harben P 1978 Abrasives. *Ind. Miner. (London)* 134: 49–73

Hight R P 1983 Abrasives. In: Lefond S J (ed.) 1983 *Industrial Minerals and Rocks: (Nonmetallics Other than Fuels)*, 5th edn., Vol. 1. American Institute of Mining, Metallurgical, and Petroleum Engineers, New York, pp. 11–32

H. H. Murray
[Indiana University, Bloomington, Indiana, USA]

Abrasives: Precious and Semiprecious Minerals

Prior to the development of synthetic abrasives with hardness greater than 8, only natural minerals, such as quartz sand and feldspars, were used as abrasives. However, these minerals were restricted as they were not able to cut the harder materials such as garnets, sapphires, rubies and other gem stones. Because a hard stone can be cut by a harder stone (see *Abrasives*), nongem-quality stones became used for abrasive materials to cut and polish gem-quality stones. Today, the precious and semiprecious abrasives used industrially are corundum, emery, diamonds and garnets.

1. Corundum and Emery

Corundum is a naturally occurring aluminum oxide with the formula Al_2O_3 (Al: 52.91 wt%; O: 47.01 wt%). Emery is a corundum which contains varying amounts of hematite or magnetite, quartz and spinel.

Corundum varies considerably in color from blue, green-gray, brown, black to colorless. The transparent blue sapphire and red ruby are the valuable gem-stone varieties of corundum. Corundum crystallizes in the rhombohedral system with steep pyramidal or rough and rounded barrel-shaped habits. Some of the crystals are of considerable size. Corundum breaks with an uneven to conchoidal fracture. It does not have cleavage but often has perfect parting, frequently having a pearly luster. Corundum has a hardness of about 9, which is next to diamond. However, the hardness of different types may vary considerably. Emery has a hardness in the range 7.5–8.5.

Corundum deposits occur in as many as 35 different countries in over 20 different rock types, but many of the reported deposits are small and not of economic importance. Corundum is generally believed to be one of the first constituents to crystallize out from a molten magma. In certain supersaturated magmas, the excess alumina separates out as corundum if the iron, magnesia and silica contents are low. If they are high, the alumina separates out as spinel, cordierite, and other minerals. Corundum does not occur in association with free silica.

In the USA, corundum is commonly associated with peridotite; in Canada, with nepheline syenites; and in India and South Africa, with feldspathic rocks. However, the main source of corundum today is from detrital or alluvial deposits. India contains some of the most extensive deposits in the world. Massive corundum (some large blocks weigh several tonnes), disseminated crystals in rock, and grains in sands and clays have all been exploited. In the USSR, coarse-grained corundum occurs as lenses in the Ilmen Mountains. In South Africa, corundum occurs in lenses or veins in basic rocks and in detrital or eluvial deposits, usually very near the primary corundum-bearing rocks.

The important corundum producers today are Zimbabwe, the USSR, India, Uruguay and the Republic of South Africa. Total world production in 1984 was an estimated 18 500 t. None has been mined in the USA since the early 1900s.

Large-scale mining of corundum is not very practical due to its occurrence being in small, randomly distributed deposits, and to the lack of major markets. Most mining and processing are done by primitive hand methods by small producers. The hand-cobbed or concentrated material is usually sold to a central agency that prepares the corundum for transport.

To some degree, the same problems exist with emery. The bulk of the world's emery production comes from Greece and Turkey. Greece produced an estimated 9000 t of emery in 1980, while Turkey produced 39 900 t in 1980 from twenty small mines. Two small producers in the USA, near Peekskill in Westchester County, New York, were active in 1980.

After the higher quality is selected for use in abrasives, it is washed, screened, crushed and heat treated to "enhance the purity and make screening more accurate." Some thirty-five different grain sizes are produced. The coarser material is used in nonskid wear-resistant surfaces and to produce millstone for rice polishing. The medium and finer grades are used in the production of emery paper and cloth, while the flour goes into polishing pastes and compounds.

Corundum is used primarily as an abrasive material for use in the grinding of lenses and in a few special types of grinding wheels. Artificial materials such as fused alumina and industrial diamonds have replaced corundum in many of its applications.

2. *Diamonds*

Industrial diamonds are one of the most important of materials used in modern industry. In addition to use as an abrasive, they are used for a multitude of purposes, including: bits; dies for wire drawing, turning and boring equipment; hardness testers; saws; and accurate shaping, truing and dressing of abrasive wheels.

Diamond, which is composed of carbon crystallized in the cubic system, is the hardest known natural mineral. Its unique properties have led to two unrelated and diverse end uses. Gem diamonds are jewels of great beauty and high value per carat, while industrial diamonds used for abrasives are normally nondescript and have only a nominal value per carat. Industrial diamonds are those that because of color, size, shape or structural defects do not meet the standards for gem diamonds; they may come from either natural or manufactured sources.

2.1 Natural Diamonds

Natural diamonds are normally found in a special type of igneous rock known as kimberlite. Kimberlite is a highly serpentinized peridotite, usually a breccia because of inclusions of the surrounding rocks it has penetrated. Evidently, the molten kimberlite intruded and rose along fracture intersections or other planes of weakness in the country rock. Once the molten kimberlite started to rise, no explosive reactions took place and the kimberlite continued to rise without any unusual violence. The intrusion may have occurred in stages and over a long period of time. This mass is called a "pipe." The size and shape of the pipe are determined by the position of the planes of structural weakness in the country rock through which the molten kimberlite passed. The pipe may be columnar, tabular or irregular in shape. Mining at depth has shown the pipes to decrease in area and assume a dike-like habit.

Alteration of the kimberlite produces products known as yellow ground, blue ground and hardebank. The yellow ground, often 10–43 m thick, is usually richer in diamonds than unweathered kimberlite in the same pipe. Blue ground is a less altered kimberlite that underlies the yellow ground to as much as 1.1 km. Diamonds are usually bedded in the blue ground without much adhesion. Hardebank is apparently a phase or type of kimberlite that resists disintegration when exposed to the atmosphere. It is occasionally found in the blue ground.

Most kimberlite deposits become poorer in diamonds at depth. Thus, it is assumed that conditions at certain depths were more conducive to the formation of diamonds than at others. It is possible the diamonds were formed under the same conditions of temperature and pressure as the primary minerals in the kimberlite. These minerals include ilmenite, diopside, magnetite, pyrope garnet, zircon, rutile, corundum, monazite, epidote, topaz, spinels, staurolite, tourmaline and gorceixite. Diamonds often contain inclusions of some of these minerals, and many of these minerals also contain inclusions of diamonds. These minerals are also known as satellites of diamonds. Their survival in an alluvial deposit is indicative of transport from a primary diamond pipe. However, there is no correlation between the amount or presence of these minerals and the diamond content of the area.

While there are over 1000 known occurrences of kimberlite reported throughout the world, diamonds have only been found in a few. The name diamond pipe is given to kimberlite deposit or pipe that is large enough and contains a sufficient number of diamonds to be economically mined.

Diamonds are also recovered from alluvial deposits. These can be eluvial placers; recent, elevated or fossil fluvial placers; recent or elevated marine beach placers; and glacial deposits. It is estimated that approximately 90% of diamonds from primary sources come from alluvial deposits.

Prospecting for and evaluation of a diamond property is extremely difficult. To detect a valuable mineral in which the ratio is 1 diamond to 15–30 million parts of waste can be most tedious. After an area is located in which diamonds are believed to be present, prospecting pits are sunk to bedrock and the various horizons examined for the satellite minerals. If all the signs are positive, a large area may be excavated to bedrock and the material examined. This often requires the setting up and operation of a small size, or pilot, diamond processing plant.

Mining methods for diamonds vary from crude hand mining and panning methods to large sophisticated operations using continuous bucket-wheel excavators, large shovels, draglines, scraper–loaders and large trucks.

The processing of a diamond-bearing ore is as varied as the mining, depending upon the type of ore, size of operation, location and other factors. Crushing of the ore is usually done with gyratory and roll crushers to avoid damaging the diamonds. Attrition and ball mills are used to break up the gangue material and to liberate the diamonds from the kimberlite or conglomerate. Log washers are used to process clayey ore. Concentration of the diamonds is done by using diamond washing pans, jigs, heavy-media separators and hydrocyclones.

The diamonds are further concentrated by means of grease tables and belts, electromagnetic or electrostatic separators, optical and x-ray sorters, and other devices. The method chosen depends upon the type, shape and size of the diamonds present. However, in spite of all the sophisticated equipment and processing facilities, the final recovery process is done by hand sorting.

Wastes from the use of diamond saws, grinders, diamond drill bits and other operations using diamonds are processed to recover the diamonds. Reclamation methods include distillation to remove coolants and lubricants, accompanied by such normal processing as electrostatic, magnetic or flotation techniques. Diamond waste from wet grinding is called sludge; waste from dry grinding is called swarf. Drill bits are usually placed in an acid to dissolve the enclosing metal and the released diamonds are further processed.

South Africa, Botswana, Ghana, the USSR, Zaire and Namibia are the major diamond producing countries of the world. Lesser amounts come from Angola, Brazil, the Central African Empire, Guinea, Guyana, India, Indonesia, Ivory Coast, Lesotho, Sierra Leone and Venezuela.

2.2 Manufactured Diamonds

Manufactured diamonds are made by two different processes. One involves applying ultrahigh temperatures and pressures to a carbon–metal catalyst mixture by the use of powerful hydraulic presses. This method is the one most widely used. The second method submits a carbon–metal catalyst mixture to an explosive shock.

So far, manufactured diamonds have only been made in sizes of 25–30 mesh (US). Thus, they are usually limited to grinding and finishing applications in metalworking, glass and concrete. In 1984, some 76 million carats of man-made diamonds were produced in the USA.

2.3 Diamond Uses

Industrial diamonds come in many different sizes and shapes, and are classified according to use:

(a) tool and die stones;

(b) drill material; and

(c) grits and powders.

They are also classified according to the type of material:

(a) Industrial stones—large stones not suitable for gem use due to shape, imperfections or poor color.

(b) Bort (boart, boort, bortz, bour)—small stones with flaws or inclusions, or fine crystalline aggregate that renders them unsuitable for gem use. These can be subdivided into: (i) drilling bort—stones whose soundness is such they can be used in diamond drill bits; (ii) crushing bort—crystal and crystal aggregate of lower grade that can be crushed into grit, powder and dust; (iii) carbonado (carbon, black diamond)—gray to black crystalline material composed of diamond, graphite and perhaps some amorphous carbon; (iv) ballas (bort-ballas, short-bort)—dense, globular, aggregates of many very small diamonds arranged radially and more-or-less concentrically around a central point.

The metric carat is the unit of measure for diamonds. One kilogram contains 5000 carats, while 1 lb contains 2268 carats. There are 100 points in a carat.

The sizes of diamond grits and powder are identified by US standard screen sizes. There is a specific diamond-grit type, shape and size for each specific application. Diamond grits varying in size from very coarse material (8/10 (+8 to −10 mesh)) to fine material (325/400 (−325 to +400 mesh)). Very fine diamond powder is referred to as graded micron powder, and is used in lapping compounds and in fine-finishing tools. Diamond micron powders are graded by a combination of sedimentation and centrifugation. Sand, powder and dust describe the decreasing order of fine diamond sizes, and grade numbers are used to indicate sizes of diamond micron powders (Table 1).

3. Garnets

Garnet is the general name for a series of complex minerals with the same general formula—iron aluminum silicate—and similar physical properties and

Table 1
Standard grade numbers for diamond powders[a]

Grade numbers	Range	Approximate mesh equivalent
$\frac{1}{2}$	0–1	50000
1	0–2	14000
3	2–4	8000
6	4–8	3000
9	8–12	1800
15	12–22	1200
30	22–36	800
45	36–54	500/600
60	54–80	400/500

[a] Source: US Bureau of Standards, C/S 261/63

crystal form. Almandite, $Fe_3Al_2(SiO_4)_3$, the most important of the industrial garnets, has a hardness in the range 7.5–8.5 and a color of various shades of red; it crystallizes in the cubic system. Pyrope garnet, $Mg_3Al_2(SiO_4)_3$, is also important commercially.

The garnet mined in New York is basically pyrope–almandite. It exhibits incipient lamellar parting planes which break into sharp, chisel-edged plates. Even when crushed to a very fine size, it retains its natural sharp, irregular grain shape. This feature is important in coated abrasives as well as other applications.

Garnets occur in a great variety of different rock types, such as in gneisses and schists, contact metamorphic deposits, pegmatites, serpentines, and as a gangue mineral in veins formed at high temperatures. Most garnets are highly resistant to chemical and mechanical erosion. Thus, they also tend to be concentrated in alluvial deposits of present day or preexisting beaches and streams.

In the Adirondack mine of Barton Mines, the ore is mined in 12 m benches and trucked to the processing mill. The ore is dumped into a crushing system designed to liberate the garnets from the enclosing matrix. Further processing consists of such concentration methods as heavy media and magnetic separation, flotation, screening, tabling, and air and water separation.

In Idaho the garnet-bearing gravel is mined by dragline, and concentrated by jigs and tables. It is then dried, crushed and screened to produce the sizes required by industry.

Today, nearly all garnets are heat treated as a processing function, and not to improve inherent abrasive quality. Heat treating helps to remove impurities picked up during processing. These particles tend to destroy the capillary action needed to give adhesion for bonding.

The USA is the world's largest producer and consumer of garnets. In 1982 a total of about 24 000 t was produced by one company in New York, one in Maine, and two in Idaho. Garnet is also produced as a by-product from a wollastonite mine in New York. Australia has a garnet production of about 5000 t; India, about 1900 t; Sri Lanka, about 18 t; and the USSR about 4500 t.

The following end uses accounted for the consumption of garnets (1982): transportation (aircraft 10%, other 26%), water filtration 28%, wood furniture finishing 13%, electric components 11%, ceramics and glass 4%, and others 8%.

Garnet, as an abrasive, competes in all its applications with natural and manufactured abrasives such as diamonds, boron nitride, fused aluminum oxides and silicon carbide.

See also: Abrasives; Abrasives: Nonsiliceous Minerals; Abrasives: Siliceous Minerals

Bibliography

Baskin G D 1980 Diamond—Industrial. *Mineral Facts and Problems*, US Bureau of Mines Bulletin 671. US Bureau of Mines, Washington, DC, pp. 257–70

Clarke R G 1976 Corundum and emery. *Mineral Facts and Problems*. US Bureau of Mines, Washington, DC, pp. 311–23

Clarke R G 1976 Diamond—Industrial. *Mineral Facts and Problems*. US Bureau of Mines, Washington, DC, pp. 323–38

Clarke R G 1976 Garnet. *Mineral Facts and Problems*. US Bureau of Mines, Washington, DC, pp. 407–18

Hight R P 1983 Abrasives. In: Lefond S J (ed.) *Industrial Minerals and Rocks*, 5th edn., Vol. 1. American Institute of Mining, Metallurgical and Petroleum Engineers, New York, pp. 11–32

Reckling K, Hoy R B, Lefond S J 1983 Diamonds. In: Lefond S J (ed.) 1983 *Industrial Minerals and Rocks*, 5th edn., Vol. 1. American Institute of Mining, Metallurgical and Petroleum Engineers, New York, pp. 653–76

Smoak J F 1982 *Abrasive Materials*. US Bureau of Mines, Washington, DC

Taylor H A 1980 Corundum and emery. *Mineral Facts and Problems*. US Bureau of Mines Bulletin 671. US Bureau of Mines, Washington, DC, pp. 245–55

Taylor H A 1981 Garnet. *Mineral Facts and Problems*, US Bureau of Mines Bulletin 671. US Bureau of Mines, Washington, DC, pp. 329–38

Thaden R E 1973 Abrasives. In: Brobst D A, Pratt W P (eds.) 1973 *United States Mineral Resources*, Professional Paper 820. US Geological Survey, Washington, DC, pp. 27–33

S. J. Lefond
[Industrial Minerals Inc., Evergreen, Colorado, USA]

Abrasives: Siliceous Minerals

Siliceous abrasives are chiefly those materials composed of silica (SiO_2), but also, for convenience, include glassy materials of volcanic origin (pumice) and some impure clay rocks, such as rottenstone. The most common mineral composed of silica is quartz, which is the essential mineral of most siliceous rocks. The less common minerals, cristobalite and opal, may be found in volcanic rocks and in certain sedimentary rocks, such as diatomite. Silica is of intermediate hardness (5–7 on the Mohs scale, depending on its mineral variety). Its modes of occurrence are so varied (from massive quartzite to powdery diatomite and tripoli) that its uses range from rough grinding to relatively mild applications such as buffing compounds and dentifrices. Since the specific uses of siliceous abrasives depend largely on the natural state of the material, it is best to discuss the different sources of silica by the physical nature of the deposit rather than by its mineralogy, particularly as most silica deposits are composed of the same mineral (i.e., quartz). The designation of quartz, in the following discussion, as a deposit type (and mine product) is only a convenience and does not imply that other silica deposits are not also composed of the mineral quartz.

1. Quartz

Crushed quartz is the main source of abrasive grain for sandpaper and US-made "flint" paper. Originally, sandpaper was coated with sand grains, but the superior cutting power of broken fragments, compared with more-or-less rounded sand, has made crushed quartz the favored material. Massive, crystalline quartz, white or very light-colored, is the preferred raw material. Common sources of massive quartz are pegmatites (extremely coarse-grained rock in veins, lenses or pods, commonly associated with granite and consisting of large crystals of feldspar and quartz), quartz veins, and some quartzites (metamorphic rocks commonly formed by the recrystallization of sandstones).

2. Silica Sand

Silica sand is a high-purity quartz sand, mined chiefly for glass making and as molding sand. Its use as an abrasive accounts for about 10% of its annual production and makes it the single most important natural abrasive (in terms of tonnage). Silica sand finds extensive use in pressure blasting, for the initial grinding of plate glass, and as a cutting medium in the sawing of stone. Ground sand (silica flour) is used in buffing and polishing compounds.

The major sources of glass sand are also major suppliers of silica-sand abrasives. The two most important glass-sand sources in the USA are the Oriskany sandstone (Devonian age) in Virginia, West Virginia and Pennsylvania, and the St. Peter sandstone (Ordovician age) in northern Illinois and southeastern Missouri and its age equivalent in Oklahoma and Arkansas. No sandstones of comparable purity are found in the western USA, but the long distance (and high transportation costs) from eastern and midwestern sources allow western producers to go to the added expense of processing their raw sands to meet glass grade specifications. Many abrasive uses do not require great chemical purity, and local markets for abrasives may be supplied by local sands that meet particle-size requirements.

Sand for pressure blasting must be clean and well graded; both angular and rounded grains may be used. The coarsest grain size in general use (No. 4) is between 4 and 8 mesh (4.76–2.38 mm) and is used for heavy cast iron and steel work. The finest size (No. 1), between 20 and 35 mesh (0.84–0.50 mm), is used for light work. Important sources for blast sand are the St. Peter sandstone and the Cohansey sand (a weakly consolidated coastal plain sand in the Cape May area of southern New Jersey). The Cohansey sand is composed of subangular grains, is more erratic in composition than the St. Peter, and provides a larger percentage of the coarser grain sizes. A decomposed rock containing friable quartz and china clay is processed at several localities in Ontario (Canada) and yields an angular grain, as does a friable quartzite at East Templeton, Quebec, a few miles northeast of Ottawa in Ontario. Also, many local markets may use nearby sources of common alluvial or glacial sand if the requisite particle sizes are available.

The initial grinding of crude, rolled plate glass to achieve a flat surface prior to polishing is accomplished with silica sand. Many glass companies use the same silica sand as in the glass batch and perform their own particle sizing to achieve the required size grades for grinding. Glass-sand producers also market a glass-grinding product, commonly screened to pass 20 mesh (0.84 mm) with a lower size limit of about 150 mesh (0.10 mm) to eliminate unwanted fines. No tonnage figures are available for consumption of glass-grinding sand. However, demand for this product is on the decline owing to the installation of the float glass process by a growing number of glass manufacturers. By carefully cooling the glass melt on a batch of molten tin, a perfectly flat, finished sheet of the desired thickness can be produced without the need for grinding and polishing.

Stone-sawing sand is the cutting agent employed in the quarrying and processing of dimension (building) stone. The abrasive action is accomplished by feeding the sand under a moving steel cable (wire saw) or under a set of parallel steel blades (gang saw). The wire saw is generally used for large-scale work in the quarry, and the gang saw is used for cutting the quarry blocks into slabs in the mill. The main requirements for stone-sawing sand are a high percentage of quartz and low percentage of soft rock and mineral fragments. Silica sand meets these requirements easily, but common river and glacial sands of the required particle size are often used because of local availability and lower cost. (Particle-size specifications generally approximate those of a No. 1 blast sand.) Quartz sand is commonly used in limestone and marble quarries, but for harder rocks, such as granite, either garnet or the manufactured abrasive silicon carbide are generally used.

Silica sand may also be finely ground in tube mills to a product commonly called silica flour. The pulverized sand is used in hand soaps, scouring compounds, and buffing and polishing compounds. As iron contamination cannot be tolerated, mill liners and grinding balls must be nonmetallic; high-density aluminum oxide is a common material for both liners and balls, but other ceramic products may be used. Mill discharge is normally air-separated and marketed as size grades ranging from about 90% passing 200 mesh (0.074 mm) to the finest at 99.9% passing 325 mesh (0.044 mm). The finest grade normally goes into metal polishes. Sources of most silica flour are the high-grade glass sands of the eastern and midwestern USA. Most of the beneficiated glass sands of the western states will not yield a pulverized product of suitable whiteness and cannot compete in this market with the eastern products (Murphy 1983).

Tepordei (1986) reported the following tonnages and values of silica sand sold or used for abrasives in 1985: blasting sand, 1 431 000 tons valued at $22 762 000; sand for sawing and sanding, 17 000 tons valued at $355 000; and scouring cleansers (ground sand), 233 000 tons valued at $5 358 000.

3. *Tripoli*

Tripoli is a microcrystalline, more or less friable form of quartz that occurs in bedded deposits. It is believed to have formed through the leaching, over a very long period of time, of beds of siliceous limestone or calcareous (calcite-containing) chert. Calcite ($CaCO_3$), the essential mineral of limestone, is relatively soluble under conditions of weathering, and its removal from siliceous limestone or from chert that contained an intimately mixed calcite component would leave a residual accumulation of powdery silica or a weak, minutely porous mass of extremely fine-grained quartz.

Virtually all commercial tripoli production in the USA comes from three areas: southern Illinois (Alexander County), the Missouri–Oklahoma district (Newton County and Ottawa County, respectively), and west-central Arkansas near Hot Springs. Other deposits, some with intermittent production, are known in northwestern Arkansas, western Tennessee and adjacent parts of Alabama and Mississippi, northeastern Alabama, and northwestern Georgia.

Slight differences in the character of the tripoli from the various localities has prompted the industry to designate two different types of product. Commercially, Missouri–Oklahoma material is known as tripoli, and southern Illinois material is called amorphous silica. The chief differences between the two types of deposits are degree of coherence and color. The Missouri–Oklahoma material is firmer, generally quarried in blocks, and is characteristically iron-stained. The Illinois material is mostly incoherent and powdery and is white in color (iron-stained material is by-passed in mining). All tripoli producers, regardless of type of deposit, process the mine product through grinding mills and sell the finished product by particle-size designations. Thus, Missouri tripoli is marketed as "rose" or "cream" in three size grades: once-ground, twice-ground, and air-float. Illinois amorphous silica is sold in nine size grades from 90 to 95% through 200 mesh (0.074 mm) to the finest at 99.9% less than 10 μm. However, Illinois silica may also be marketed as tripoli, in which case it is listed as "white, air-floated through 200 mesh." West-central Arkansas material is generally listed under amorphous silica.

About two-fifths of the tripoli produced is for abrasive use, the remainder being consumed as fillers in paints, plastics, rubber, and other industrial products. Abrasive uses are much the same as for ground silica sand, that is, hand soaps, dentifrices, scouring compounds, and buffing and polishing compounds. However, tripoli particles are aggregates of quartz crystallites, in contrast to ground sand, which consists of single crystal fragments. Thus, tripoli particles will break down during use and, consequently, are a milder abrasive than silica flour of the same particle size; tripoli also constantly renews its sharp cutting edges as the particles crumble during use. The industrial user of abrasives or the manufacturer of cleaner–polisher formulations is aware of these differences between silica flour and tripoli and purchases the raw material that will best suit his purpose. Certain advantages are also claimed as unique to Missouri tripoli or to Illinois amorphous silica and are probably based on user preference for one or the other material.

Rottenstone is similar to tripoli in that its chief component is very fine particulate quartz. However, rottenstone also contains a substantial percentage of clay and appears to be the weathering product of shale. As for tripoli, rottenstone is used for fillers and abrasives. In this latter function, it is an ingredient of polishing compounds, such as automobile polish. There is only one producer of rottenstone in the USA, namely the Keystone Filler and Manufacturing Company in Pennsylvania.

The US Bureau of Mines (1986) reported that 40 022 tons of tripoli, valued at $3 670 000, was sold for abrasive uses in 1985, accounting for 37% of USA tripoli consumption in that year.

4. *Diatomite*

Diatomite is a siliceous sedimentary rock formed by the accumulation of the siliceous skeletons of microscopic marine plants called diatoms. The silica mineral composing the skeletons is generally considered to be opal, an amorphous form of silica, but x-ray studies indicate a poorly crystallized form of cristobalite. The ultimate hardness of diatomaceous silica is about 5 on the Mohs scale, somewhat softer than that of quartz (7). However, individual skeletons, most of which are in the 50–150 μm size range, are composed of fragile networks that disintegrate easily during use to extremely fine particles, so that the abrasive character of diatomite is in effect quite mild.

Diatomite was formed in broad, shallow basins that had an abundant supply of silica, such as volcanic ash, for skeletal growth. Contaminants in the deposits are commonly sand, clay, carbonate minerals and volcanic ash. Large commercial deposits are generally restricted to relatively young volcanic terranes (less than about 60 million years old) and are found at several localities throughout the world. The most productive deposits, because of both size and degree of purity, are in the Lompoc area near Santa Barbara in California, as well as in some parts of Nevada.

Processing of diatomite must preserve the delicate structure of the diatom skeletons (of importance for filtration uses) and generally consists of a combination milling–drying scheme. Following primary crushing, the diatomite is suspended in a stream of hot gases and passed through a series of separators, which facilitate disaggregation of the diatomite and separation of impurities. Use as an abrasive consumes only a small part of total diatomite production, but the mild abrasive action of the milled product makes it particularly well suited for use in silver polishes. Calcination of the milled product increases the strength of the individual particles and leads to greater abrasiveness; calcined diatomite is used in automobile polishes and other buffing and polishing compounds.

No information is available on actual amounts of diatomite sold for abrasive purposes, but Meisinger (1983) reported that in 1982 11% (about 67 000 tons) went into several miscellaneous uses, one of which was abrasives. Thus, one can estimate that a few to several thousand tons of diatomite are probably used as abrasives. Diatomite for abrasive uses sold at an average of $174.09 per ton in 1979.

5. *Flint and Chert*

Flint and chert are hard, dense, massive forms of microcrystalline quartz, formed in a sedimentary environment as continuous beds or as individual nodules, normally in limestones. The distinction between the two is not precise, but flint is generally recognized as having a porcelaneous luster, a smooth fracture, and a dark gray or black color. Chert, on the other hand, has a dull luster, is less likely to have a smooth fracture, and is light in color. Flint is also said to be composed of chalcedony, a fibrous form of quartz, while chert is granular quartz.

European flints, particularly in the countries bordering the English channel, are crushed for use in the manufacture of flint paper. Flints for use in flintlock rifles are produced on a small scale in England. The use of flint for mill linings and grinding media is described in Sect. 8.

In the USA, flint is not a term commonly used by geologists, and dense, massive, microcrystalline quartz is generally referred to as chert, including the more or less porcelaneous, dark-colored varieties. The only chert that is known to be used commercially to any extent in the USA is Arkansas novaculite, a thick, extensive deposit of dense microcrystalline quartz that is believed to be the product of metamorphism of ordinary, sedimentary chert. The chief use of novaculite is for hand-held sharpening stones (see Sect. 7).

6. *Pumice*

Pumice is a light-colored, highly vesicular volcanic glass produced by explosive volcanic eruptions. It may occur in massive blocks or in a fragmented state. Finely divided pumice, which is carried farther from the vent as volcanic ash, is called pumicite. A useful particle-size limit for classification purposes is 4 mm, below which the material is designated pumicite (Peterson and Mason 1983).

Pumice occurs in areas of geologically young active volcanoes. The major USA production comes from the western States, with minor production of pumicite from volcanic ash beds in the Great Plains. Greece and Italy (especially Sicily) are the major European producers.

Pumice is a relatively minor abrasive. Most of it is processed by fine grinding and air classification for use in cleaning and scouring compounds. In the natural state, small blocks of pumice may be used for removal of callouses, dressing metal-polishing belts, and other minor specialized uses. In 1985, 28 000 tons of pumice (5.5% of total pumice produced in the USA), valued at $497 000, were used chiefly in cleaning and scouring compounds (Meisinger 1986). European pumice producers report similarly small percentages used for abrasives (Robbins 1984).

7. *Natural Stones*

Bonded synthetics have largely replaced natural stones as grinding wheels and sharpening stones, but there is still some demand for natural hand-held stones, such as oilstones, water stones, whetstones, razor hones, scythe stones, holystones and rubbing stones. Arkansas novaculite, a hard, dense rock composed of microcrystalline quartz, is a popular material for whetstones and oilstones and is the preferred medium for sharpening fine-edged tools for surgeons, carvers and engravers. Whetstones are produced from an Indiana sandstone, and whetstones and scythe stones from New Hampshire are produced from a quartz-mica schist (a laminated metamorphic rock). Holystones, used for scrubbing wooden decks of ships, are a soft sandstone. For a discussion of other sources of natural abrasive stones, both in the United States and abroad, the interested reader is referred to Ladoo and Myers (1951 pp. 5–12).

8. *Stones for Milling*

Although milling (size-reduction or comminution, in modern technical terminology) is accomplished more by crushing or breaking than by true abrasion, the materials used in grinding mills must be tough and abrasion resistant, the same properties that are required in abrasives. It is appropriate, therefore, to include with abrasive materials those natural stones that are used for milling purposes.

Most industrial grinding mills are made of iron or steel, and hardened steel balls (or rods) are the preferred grinding medium because of their heavy weight and superior impact on the material being ground.

However, many industrial materials will not tolerate any contamination by iron, whether for chemical reasons or simply a matter of color, and the finer the material that must be ground, the more serious is contamination from the grinding medium. Therefore, a great many industrial minerals for paints, enamels, chemicals, ceramic bodies and glazes, and similar end uses require nonmetallic grinding mills.

The most favored natural material for mill linings is Belgian silex, a hard, tough quartzite. Also used to some extent in the USA are domestic quartzites from Minnesota, Wisconsin and Tennessee, and granites from North Carolina. The preferred grinding media are Danish flint pebbles, which actually originate from the shores of Greenland. Similar flint pebbles are produced along the seashores of France, Belgium, Norway and the UK. Domestic materials that have been used in the USA are natural pebbles of flint, quartz and quartzite, and artificially rounded blocks of quartzite, granite and other rocks.

However, manufactured materials are now making inroads on the natural stones. Mill liners and pebbles are made from dense porcelain and alumina; a dense, heavy pebble, with a specific gravity of 3.7, is made from zircon (Hight 1983).

See also: Abrasives; Abrasives: Nonsiliceous Minerals; Abrasives: Precious and Semiprecious Minerals

Bibliography

Bradbury J C, Ehrlinger H P 1983 Tripoli. In: Lefond S J (ed.) 1983, pp. 1363–74

Harben P 1978 Abrasives: Taking the rough with the smooth. *Ind. Miner. (London)* 134: 49–73

Hight R P 1983 Abrasives. In: Lefond S J (ed.) 1983, pp. 11–32

Kadey F L 1983 Diatomite. In: Lefond S J (ed.) 1983, pp. 605–36

Ladoo R B, Myers W M 1951 *Nonmetallic Minerals*, 2nd edn. McGraw-Hill, New York

Lefond S J (ed.) 1983 *Industrial Minerals and Rocks*, 5th edn. American Institute of Mining, Metallurgical, and Petroleum Engineers, New York

Meisinger A C 1983 Diatomite. *Minerals Yearbook 1982*. US Bureau of Mines, Washington, DC, pp. 309–11

Meisinger A C 1986 Pumice and pumicite. Preprint from *Minerals Yearbook 1985*. US Bureau of Mines, Washington, DC

Murphy T D 1983 Silica and silicon. In: Lefond S J (ed.) 1983, pp. 1167–85

Peterson N V, Mason R S 1983 Pumice, pumicite, and volcanic cinders. In: Lefond S J (ed.) 1983, pp. 1079–84

Robbins J 1984 Pumice—The ins and outs of a reinforced market. *Ind. Miner. (London)* 200: 31–51

Tepordei V V 1986 Sand and gravel. Preprint from *Minerals Yearbook 1985*. US Bureau of Mines, Washington, DC

US Bureau of Mines 1986 Abrasive materials. Preprint from *Minerals Yearbook 1985*. US Bureau of Mines, Washington, DC

J. C. Bradbury

[Illinois Department of Energy and Natural Resources, Champaign, Illinois, USA]

Aluminum Resources

Aluminum, surpassed only by iron in terms of the quantity of metal used by man, is recovered exclusively from bauxite containing 35–55% Al_2O_3 (18.9–35.1% metal), except in the USSR. Aluminum is made by extracting alumina (Al_2O_3) from bauxite by the Bayer process, an alkaline leach under moderate temperature and pressure, and reducing the alumina to metal by the Hall–Héroult electrolytic process. World bauxite reserves are adequate to meet demands well into the twenty-first century. However, geographic, economic and political considerations complicate the supply picture. The large, high-grade bauxite reserves are in Australia, and in developing nations in Asia, Africa, South America and the Caribbean region, whereas all but a small percentage of the aluminum is made and consumed in North America, Europe and Japan.

The major cost items in the production of aluminum include bauxite acquisition and mining, transportation and electrical power. The largest bauxite cost in recent years has been the increasing revenue paid to bauxite-exporting countries. The increases have been brought about by the International Bauxite Association (IBA), an organization of eleven major bauxite-producing countries. Not only have these revenues increased, but trends have developed that will influence supplies and costs of bauxite in the future. In addition to the problem of inflation, IBA nations and nonmember developing countries having large reserves are moving toward nationalization by direct ownership of, or major participation in, bauxite, alumina and metals production; these moves will affect cost to the consumers. High transportation costs result from increasing prices of fuel and from the global distribution of the bauxite reserves, alumina plants, smelters and markets. The 43.2–57.6 MJ of electricity required to smelt 2 kg of alumina to 1 kg of metal is the largest single cost item in producing aluminum. Increasing energy prices and uncertainties about future supplies complicate forecasts of aluminum availability.

1. Aluminum Ore (Bauxite)

Aluminum is the third most abundant element in the earth's crust, of which it constitutes about 8% by weight. Bauxite deposits that are minable at a profit represent natural concentrations approximately two to four times the average crustal abundance of aluminum. Aluminum is less mobile than many other elements in rocks, and was concentrated geologically in the form of bauxite as intense tropical weathering leached out silica and elements such as the alkali and alkaline-earth groups.

Bauxite deposits are commonly classified into the laterite and karst types. The very large tabular deposits in subsoils are the laterite type; the karst-type deposits are concentrated in solution depressions on carbonate rock. Laterite deposits generally are on erosional remnants of peneplains that are rarely older than Tertiary in age. Typical karst bauxite deposits, like those in several European countries, are Mesozoic or older in age, and many of them are overlain by younger carbonate beds. Much younger karst bauxite deposits occur in surficial residual accumulations in the Caribbean and southwest Pacific regions.

The principal bauxite mineral in laterite bauxite is gibbsite, but minor quantities of boehmite are commonly present. The major impurities include the iron-bearing minerals goethite and hematite, titanium-bearing anatase, quartz and kaolinite. Boehmite is the principal bauxite mineral in the Mesozoic karst bauxite, but the younger karst deposits are similar to the laterite type in mineral composition.

2. *Uses*

Aluminum metal is used in many different products, and, on an elemental basis, aluminum has many nonmetallic uses. The uses of aluminum in the USA include the following approximate shares of the production: (a) 22% in construction of buildings, mobile homes, heating and air-conditioning equipment, bridge and guard rails, and other structural materials; (b) 22% by the transportation industry in parts for such products as cars, buses, trucks, aircraft and boats; (c) 10% by the electrical and communications industries for transmission lines and towers, tubing, cable, fixtures, lamps and other components; (d) 29% in cans, containers and packaging; (e) 8% for durable goods and appliances; (f) 9% for agricultural and other specialized uses in areas such as machinery and irrigation pipe; (g) 9% for other miscellaneous applications; and (h) on an elemental basis, 11% of the total aluminum is consumed in the form of alumina and bauxite in such nonmetallic products as refractories, abrasives and chemicals. Alumina also has a long list of special applications, being used in the manufacture of fire retardants, adsorbents, ceramics, glass and other products.

3. *Resource and Supply Adequacy*

World aluminum resources are adequate to supply the significant long-range increases in demands that are forecast for aluminum. In 1987, about 87 Mt of bauxite were mined, and world aluminum production was 16 Mt. World demands for aluminum are expected to have an average increase of nearly 5% per year, and may be about 60 Mt annually by the year 2000. World bauxite reserves are estimated to be 22–23 $\times 10^3$ Mt, the largest reserves being in the Republic of Guinea, estimated as 5.6 $\times 10^3$ Mt. In order of rank of bauxite reserves (in 10^3 Mt), Guinea is followed by Australia (4.4), Brazil (2.8), Jamaica (2), and India and Cameroon (1 each). Indonesia and Guyana each have an estimated 700 Mt of reserves, Greece has 600 Mt, Surinam has 575 Mt, Ghana has 570 Mt, Yugoslavia has 350 Mt, and Venezuela has 320 Mt (US Bureau of Mines 1988). Countries having reserves in the 100–300 Mt range include Costa Rica, Hungary, the USSR, Sierra Leone and the People's Republic of China. In addition to reserves, several countries have very large subeconomic and hypothetical bauxite resources, as well as virtually inexhaustible quantities of potential nonbauxite sources of aluminum. The nonbauxite resources include nepheline syenite and alunite (which are used as ores of aluminum in the USSR) and anorthosite, leucite, high-alumina clay, dawsonite, aluminum phosphate rock, coal-mine waste, and fly ash.

The resources of refractory, abrasive and chemical grades of bauxite are far more limited than those of bauxite used for aluminum, because each of the nonmetal grades must fulfill rigid specifications for the specialized use. Accordingly, the bauxite in each grade must be of exceptional quality, and very little is known about the resources of such deposits. The demand of the industrialized countries for high-grade refractory bauxite is satisfied mainly by Guyana, Surinam and the People's Republic of China. Abrasive and chemical grades of bauxite are exported by Guyana, Surinam, Australia and the Republic of Guinea.

Bibliography

Crockett R N 1978 *Bauxite, Alumina and Aluminum*, Mineral Dossier No. 20. Great Britain Institute of Geological Sciences, London

Kurtz H F, Baumgardner L H 1981 Aluminum. *Mineral Facts and Problems*, US Bureau of Mines Bulletin 671. USBM, Washington, DC

Patterson S H 1967 *Bauxite Reserves and Potential Aluminum Resources of the World*, US Geological Survey Bulletin 1228, USGS, Reston, Virginia

Patterson S H, Kurtz H F, Olson J C, Neeley C 1986 *World Bauxite Resources*, US Geological Survey Professional Paper 1076-B. US Government Printing Office, Washington, DC

US Bureau of Mines 1988 *Mineral Commodity Summaries 1988*. USBM, Washington, DC

S. H. Patterson
[US Geological Survey, Reston, Virginia, USA]

Antimony Resources

Antimony, a brittle silver-white metal, was used as early as 4000 BC in containers, mirrors and bells. It is now used in storage batteries, semiconductors, chemical pumps, tank linings, plumbing fixtures and pipes, foil, and bullets. It also is a constituent of antifriction bearings, type metal, solder, and decorative castings of Britannia metal and some pewter. The sharp definition of type metal is caused by the expansion of antimony when it cools. Antimony oxides are used in paints and plastics as a pigment and fire-retarding agent and in white ceramic enamels as an opacifying agent. Antimony sulfides are used in the manufacture of ammunition primers, smoke generators, visual range-finding shells, tracer bullets, fireworks and as a vulcanizing agent in the rubber industry. Organic antimony compounds are used in the treatment of certain parasitic diseases.

Most of the world production of antimony comes from South Africa, Bolivia, the People's Republic of China, Mexico, Canada, the USSR, Turkey and Thailand. Major world resources of antimony are in the People's Republic of China, Bolivia, the USSR, South Africa and Mexico.

Stibnite (Sb_2S_3) is the main ore mineral of antimony. Other important antimony minerals are: valentinite (Sb_2O_3); senarmontite (Sb_2O_3); stibiconite ($Sb_2O_4.H_2O$); bindheimite ($Pb_2Sb_2O_7.nH_2O$); kermesite ($2Sb_2S_3.Sb_2O_3$); tetrahedrite ($Cu_8Sb_2S_7$); and jamesonite ($Pb_2Sb_2S_5$).

Most antimony deposits are mineralogically simple. They consist principally of stibnite or, rarely, native antimony in a siliceous gangue, commonly with some pyrite, and, in places, a little gold and small amounts of sulfides of other metals, principally silver and mercury.

More complex deposits consist of stibnite associated with pyrite, arsenopyrite, cinnabar or scheelite; still others have antimony sulfosalts with varying amounts of copper, lead and silver, as well as the common sulfides of these metals and of zinc. Most antimony produced in the USA, Australia and Canada is a by-product from complex deposits.

The major antimony deposits are generally regarded as having formed from hydrothermal solutions at relatively low temperature and shallow depth. They occur as filled fissures and joints, and irregular replacement bodies.

The Consolidated Murchison mine in the Republic of South Africa is an example of a large mine operated by shrinkage stoping. Lump ore is hand cobbed, but the fines are milled to concentrate the gold and antimony. At other mines, the ores may be roasted or liquated or both to produce a marketable product. The Sunshine Mining Company, Coeur d'Alene district, Idaho, recovers antimony by electrolysis of antimony-enriched solution derived from leaching of tetrahedrite concentrates.

Bibliography

Llewellyn T O 1988 Antimony. *US Bureau of Mines Mineral Commodity Summaries 1988*. USBM, Washington, DC, pp. 12–13

Miller M H 1973 Antimony. In: Brobst D A, Pratt W P (eds.) 1973 *United States Mineral Resources*, US Geological Survey Professional Paper 820. US Government Printing Office, Washington, DC, pp. 45–50

Plunkett P A 1981 Antimony. *US Bureau of Mines and US Geological Survey Mineral Commodity Summaries 1981*. US Government Printing Office, Washington, DC, pp. 8–9

Rathjen J A 1979 *Antimony*, US Bureau of Mines Mineral Commodity Profile. US Government Printing Office, Washington, DC

US Bureau of Mines 1988 *Mineral Commodity Summaries 1988*. USBM, Washington, DC, pp. 12–13

M. H. Miller
[US Geological Survey, Denver, Colorado, USA]

Arsenic Resources

Arsenic is a silver-gray metalloid, moderately hard but brittle, having a specific gravity of ~5.73. It is used principally in agriculture and to a much lesser degree in industry. In the future, arsenic may be utilized in the production of gallium arsenide high-speed integrated circuits.

1. Geochemistry

The chemical and physical properties of arsenic are similar to those of antimony and bismuth. Arsenic is relatively rare, occurring sparingly in the early and main stages of magmatic crystallization and relatively abundantly in the late magmatic stage. It combines with sulfur, selenium, tellurium, copper, iron, nickel, cobalt and some platinum-group elements.

2. Principal Types of Ore Deposits and Examples

Arsenic-bearing deposits are of several types and are classed as follows in approximately decreasing order of importance, with examples of each.

(a) Enargite-bearing copper–zinc–lead deposits: Butte, Montana, USA; Chuquicamata, Chile; and Cerro de Pasco, Peru.

(b) Arsenical pyrite copper deposits: Boliden, Sweden; Rammelsberg, Federal Republic of Germany; and Jerome, Arizona, USA.

(c) Native silver and nickel–cobalt arsenide deposits: Cobalt, Ontario, Canada; Harz Mountains, Federal Republic of Germany; and Kongsberg district, Norway.

(d) Arsenical gold and arsenic sulfide gold deposits: Jardine mine, Montana, USA; and Monte Cristo mine, Washington, USA.

(e) Arsenical tin deposits: Aberfoyle mine, Australia; and Stavoren Fides tin mines, Republic of South Africa.

3. Present Production Patterns and Reserves

Arsenic is obtained as a by-product from the smelting of ores of other metals, mostly copper. It is collected in flue dusts and is further treated to produce a purified

Table 1
World arsenic production and reserves

	Production in 1983 (t)	Reserves (t)
Belgium	3	
Canada	3	
Chile	6	
France	10	
Mexico	6	
Namibia	2	
Peru	1	
Philippines	5	
Sweden	10	
USA (1985)	2.2	50
USSR	8	
Other market economies	1	
World total	55	1000

product, commonly arsenic trioxide (As_2O_3). Arsenic is produced in about twenty countries (Table 1). Sweden and France are the world's leading producers and currently supply more than a third of the world total.

Arsenic reserves (the quantities in known sources that are currently economically recoverable) are difficult to estimate accurately because most arsenic is associated with complex base-metal ores for which mine assay data are lacking. To obtain approximate figures for arsenic reserves (Table 1), an arbitrary function of prime product ore reserves is assumed.

See also: Ore Minerals

Bibliography

Gualtieri J L 1973 Arsenic. In: Brobst D A, Pratt W P (eds.) 1973 *United States Mineral Resources*, US Geological Survey Professional Paper 820. US Government Printing Office, Washington, DC, pp. 51–61

Johnsen G 1984 Gallium arsenide chips emerge from the lab. *High Technol.* 4(7): 44–51

Loebenstein J R 1980 Arsenic. *Mineral Facts and Problems* US Bureau of Mines Bulletin 671. USBM, Washington, DC, pp. 47–54

US Bureau of Mines 1988 *Mineral Commodity Summaries 1988*. USBM, Washington, DC

J. L. Gualtieri
[US Geological Survey, Denver, Colorado, USA]

Asbestos: Alternatives

In 1979 the report of the UK Advisory Committee on Asbestos (UK Health and Safety Executive 1979) recommended, among many other important matters, that manufacturers of asbestos products should be obliged to consider the substitution of asbestos by other materials. Much criticism has been levelled at asbestos producers and users, although the industry has now learned how to handle asbestos safely. In the light of these pressures it is not surprising that the subject of asbestos substitution has emerged as a minibranch of materials science and technology, involving intensive development work, novel uses of raw materials, and for workers in the field, searches into less familiar areas of knowledge and information.

Three criteria govern the application of alternative materials in what have been traditional asbestos-based products: cost, availability and performance. The price of asbestos in most of its applications lies between US$0.40 and 0.80 per kilogram (1987), and its availability exceeds 4 Mt per annum. In comparative terms it is one of the least expensive and most abundant of all the raw materials considered here (see Tables 1 and 2). Similarly it ranks high in performance (see Table 3).

A fourth criterion, concerning health hazards, accepts a broadly held concept that fibers of any type may have biological effects if they are durable and respirable. This category embraces, among the alternatives to asbestos, man-made mineral fibers (MMMF), some of the synthetic refractory fibers, possibly aramid pulp fiber, and wollastonite, sepiolite and attapulgite. Vermiculite, talc and micas may all contain tremolite asbestos as an impurity, and diatomite and perlite may contain free silica dust. Hazards may also arise from harmful vapors released by the combustion of any of the synthetic organic fibers proposed as alternatives.

It is convenient to consider the alternatives in three broad groups, each of which refers to a specific property of the input material; that is, reinforcing strength, heat resistance, or compatibility with synthetic resin systems. Reinforcing strength relates mainly to construction materials, heat resistance to textiles and insulation products, and compatibility with synthetic resin systems relates to reinforced plastics, friction products and packings and jointings.

1. Construction Materials—Fiber-Reinforced Cements and Building Boards

Here the matrix comprises either ordinary portland cement or autoclaved calcium silicate or gypsum, and the reinforcement can be glass or mineral-wool fibers, or synthetic or natural organic fibers. Minerals such as vermiculite, mica and wollastonite enter the mix as insulating materials in insulation boards. Asbestos is not completely excluded and may still be used in conjunction with any of these materials.

Asbestos-cement products rely upon the natural network of asbestos fibers to support cement particles in the manufacturing process. This is possible because of the variable dimensions of fibers used to provide the

Table 1
Alternative materials: costs[a]

<US$2 kg^{-1}	US$2–10 kg^{-1}	US$10–20 kg^{-1}	>US$10 kg^{-1}
mineral wool glass wool cellulose pulp vegetable fiber perlite diatomite mica vermiculite talc wollastonite attapulgite sepiolite	continuous filament glass AR glass polypropylene PVA aluminosilicates steel fibers	continuous filament glass PTFE	aramid fiber pitch carbon fiber PAN carbon fiber alumina fiber silica fiber
(asbestos grades 4, 5, 6, 7)	(asbestos grade 3)		

[a] 1987 prices

Table 2
Alternative materials: annual availabilities

Availability	Minerals	Man-made mineral fibers	Synthetic organic fibers	Natural organic fibers
>1 Mt	perlite diatomite talc (asbestos)	glass wool mineral wool	polypropylene fiber	cellulose pulp
0.5–1 Mt	attapulgite vermiculite	glass filament, yarn and staple		vegetable fiber
50–500 kt	mica wollastonite			
<50 kt	sepiolite	refractory fibers (also metallic fibers)	PVA fiber aramid fiber carbon fiber PTFE fiber	

asbestos furnish. Only cellulose fibers come close to providing a similar function. Synthetic fibers such as those of poly(vinyl alcohol) (PVA) and polyacrylonitrile (PAN) are offered as replacements in finite dimensions, and as such cannot support particulate materials. They cannot be used in conventional asbestos-cement manufacturing processes without the additional support of cellulose pulp or even asbestos fibers. However, all the synthetic fibers and cellulose pulp are less dense than asbestos or glass fibers, and therefore can be used at lower weight loadings in the matrix to give equivalent volume fractions of fiber for reinforcement.

Polypropylene (PP) fibers are hydrophobic and require a surface treatment before application as reinforcement in a cement matrix. In staple form they have not been successful in fiber-reinforced cements (FRC) but in the form of prestretched multilayer networks they make good reinforcement, for example as fibrillated polypropylene, marketed under the name of Retiflex. Retiflex-reinforced cement (Netcem) is as strong as asbestos cement, although of higher density, and also has considerable impact resistance, being pseudoductile under flexural stress.

Alkali-resistant glass fibers (AR glass), developed in the 1960s for cement reinforcement, have not achieved any real measure of success. Long-term tests on AR glass-reinforced cement (GRC) have shown a decline in reinforcing effect in less than ten years, and many other tests indicate that glass fiber is not sufficiently alkali resistant. Further difficulties arise when attempting to mix glass fiber into a cement slurry, as the fibers tend to clump together, and consequently alternative methods of making glass-fiber-reinforced cement by spray-lay-up methods have been developed. However, these developments have involved replacing

Table 3
Alternative materials: review of properties

Heat resistance	Reinforcing strength	Chemical resistance
Good (above 400 °C)	*Good*	*Good*
mineral wool	aramid fiber	PBI
refractory fibers	carbon fiber	PAN
all minerals	PVA	PTFE
steel fiber	glass fiber	PVA
(asbestos)	steel fiber	PP
	(asbestos)	carbon fiber
		refractory fiber, except in alkalis
		most minerals
		(asbestos, except in acids)
Moderate (200–400 °C)	*Moderate*	*Moderate*
aramid fiber	cellulosic fibers	aramid fiber
PVA	PP	steel fiber
PTFE	PAN	
PBI	refractory fiber	
PAN	mineral wools	
carbon fiber		
glass fiber		
Poor	*Poor*	*Poor*
cellulosic fibers	all minerals	cellulosic fibers
PP	PTFE	

ordinary portland cement (OPC) by less-aggressive matrices such as high-alumina cement, supersulfated cement and gypsum. Blended cements incorporating OPC, blast furnace slag cement, pulverized fuel ash, silica fume and several types of organic modifiers have all helped to improve the long-term performance of GRC, such that after its early difficulties it could have a realistic future. GRC products are currently being made in the UK, the FRG, Japan and Australia. The properties of GRC resemble those of Netcem in that it has high impact strength, good flexural strength and pseudoductility, although its density is high compared with that of asbestos cement.

High-quality cellulose fiber has more in its favor as a replacement for asbestos than other types of potential reinforcement in cement. Cellulose is inexpensive and abundant. Its lack of reinforcement can be overcome by incorporating it at higher loadings than may be used for asbestos in FRC, or additional strength can be gained by incorporating a small proportion of synthetic fiber such as poly(vinyl alcohol). The manufacture of Duracem adopts the latter approach and uses the conventional asbestos-cement machinery and system. The product is somewhat more dense than normal asbestos-cement sheeting.

Other methods of making a cellulose-based sheeting involve high-pressure autoclave curing systems. The green sheet is made on a conventional Hatzchek machine, but its premix comprises cellulose, cement and silica, or cellulose, lime and silica. The green sheet requires a compression stage and is then passed into an autoclave for a number of hours, during which the components of the matrix react together in high-pressure steam to form a hydrated calcium silicate. The cellulose is not degraded at the autoclave temperature of 170–185 °C, and appears to be protected by the alkaline conditions. The cellulose–lime–silica product, Unicem, has lower density than asbestos cement, which is to its advantage. Cellulose-cement sheets do not have the temperature resistance of asbestos cement and the cellulose will burn, although without spread of flame.

All the organic fibers included here possess some negative properties which raise doubts as to their suitability in FRC. They have poor temperature resistance and may burn, degrade or decompose in conditions of moderate heat (150–250 °C). Furthermore, little is known about the viability of products based on replacement fibers. Manufacturers of cellulose FRC claim a design life of 30 years for these products, using accelerated and simulated weathering tests. The life of products using synthetic organic fibers is doubtful since these polymers may degrade under natural circumstances over an extended period of time.

Lightweight insulating boards for interior use have been made for many years in the UK, Western Europe, Australia and Japan using combinations of asbestos, cellulose and cement, or specifically, amosite asbestos and calcium silicate from an autoclave reaction. In the development of alternative products during recent years, manufacturers have replaced asbestos by other heat-resistant minerals, such as vermiculite, mica and wollastonite. These minerals are combined with cellulose in an autoclaved calcium silicate process, the cellulose acting as carrier to the fine particulate materials at the board forming stage, and as reinforcement in the final product. Examples of such products are Supalux and Monolux. It is not possible to combine mineral wool or glass fibers into this type of process, because in the presence of alkalis, and in the autoclaving reaction, these materials are severely degraded. The problem can be overcome by first synthesizing the calcium silicate in a stirring autoclave and then dispersing mineral-wool fibers in the slurry. The slurry is vacuum-drained to form a board or slab. The process is costly and low in output.

Patent literature reveals developments of purely inorganic boards using vermiculite, perlite or mineral wools in a matrix of gypsum. Many such boards have been marketed for some years, particularly in Japan. Reinforced gypsum products are incombustible, but they degrade at about 130 °C as the gypsum dehydrates.

Vermiculite and mineral wools are now widely used in spray insulating products. In principle these are the same as the now-defunct asbestos-cement sprays, with vermiculite or mineral wool wholly replacing the asbestos. Vermiculite spray products have both thermal and acoustic applications, and are used in protection of steelwork, both inland and on offshore oil rigs, and in mines. High-temperature block insulations based on exfoliated vermiculite bonded with potassium silicate completely replace, technically and environmentally, the amosite asbestos insulation which was bonded with sodium silicate.

Other innovations incorporate organic foams in sandwich construction with asbestos-cement sheets, glass-reinforced gypsum boards or reinforced plastic laminates, to form an insulation board. In their primary use as internal panelling these types of boards have a high fire risk, and in the event of fire they decompose to produce highly toxic fumes.

An important group of building materials comprises bitumen or asphalt-pitch-based roofing felts incorporating asbestos, and vinyl asbestos floor tiles. Alternative roofing products include those based on impregnated cellulose, paper and shoddy felts, which are not rot-proof, and those using glass-fiber felts. The latter do not support as much pitch as does asbestos. Mastics and sealants, which are also bitumen based, may use glass, mica, wollastonite or talc as alternatives to asbestos, but these materials do not have the same holding or bulking properties as asbestos. Asbestos has many advantages in this product area, since it acts as a binding agent and provides cohesion, viscosity, elasticity, plasticity and durability. Alternative fillers fall short in all of these properties. The mineral sepiolite, which has good extending, thixotropic and binding properties, is becoming more commonly used in asphalt and bitumen-based compositions, including car body undercoatings.

In vinyl asbestos floor tiles, asbestos has the advantage of providing wear and nonslip properties. Substitution of the asbestos by glass or mineral-wool fibers makes for a brittle product, and combinations of glass fibers with cellulose or synthetic fibers do not provide the ideal nonslip properties.

2. *Heat-Resistant Products: Textiles and Insulations*

Heat-resistant fibers, which can be made into textiles, extend over a range of service temperatures from 200 °C to above 1200 °C. Asbestos textiles occupy a position near the middle of this range, with a service temperature of about 600 °C. Below asbestos stand the synthetic organic fibers and glass fiber, and above it stand the inorganic refractory fibers.

Several synthetic organic fibers have been proposed as replacements for asbestos in heat-resistant textile applications, and no doubt these serve a good though limited purpose. A heat-resistant textile must withstand extreme conditions such as welding sparks in many industrial situations, molten metal splashes in steel works, and bare flame hazards in fire fighting. In such circumstances they are worn as personal protective garments. The service temperature of the organic fibers is little above 200 °C, and at temperatures greater than this they will lose strength, or char. Thus they do not constitute wholly acceptable alternatives.

The most widely promoted of the organic fibers are the aramids (Kevlar, Twaron and Nomex) and the carbon-based fibers. Kevlar fibers begin to lose strength at 180 °C, but their heat resistance is said to be good to 350 °C. At 420 °C they decompose to a char. Nomex fiber can be used up to 230 °C but deteriorates above this temperature and then degrades rapidly at 375 °C.

Kevlar and Twaron textiles are difficult to tailor, and have limited applications in industrial gloves, protective armlets and leggings, where resistance to abrasion and puncturing is important. Nomex fabrics are used in protective jackets, and aluminized Nomex textiles can provide protection at intermittent temperatures up to 400 °C.

Carbon-based fibers and textiles are made by the partial carbonization of acrylic fibers or viscose rayon textiles. These products have moderate strength, about one tenth that of Kevlar and asbestos textiles, and they are generally serviceable to about 200 °C. At higher temperatures they lose strength and convert to higher carbon contents, with consequent loss of volume. Carbon will burn in air, and its threshold temperature is about 500 °C. Carbon textiles are used in garments for military personnel and as fire barriers in upholstery applications. Fabrics based on activated carbon fiber have specific applications in protective clothing and screens where gaseous, liquid chemical and bacterial hazards arise.

After the alarms raised by fire in aircraft, new US regulations came into force in 1987 governing the assembly of aircraft seats. A fire-blocking layer between the seat base and its upholstery is required, and among the materials which conform to the specification for this purpose are fire-retardant woollen fabrics and carbon fabrics. Woollen fabrics have been traditionally used by steelworkers as protection against heat and molten metal splash.

Other synthetic organic textile fibers include polytetrafluoroethylene (PTFE), polybenzimidazole (PBI) and the polyimides. These materials have relatively low strength and a limited temperature resistance at 200–300 °C. PBI and polyimide textiles are relative newcomers, but PTFE fiber has been available for many years. PTFE has the distinction of being the most fire-resistant organic fiber in oxygen-rich and pressurized atmospheres, and for this reason it is made into coverall garments for in-flight use by astronauts.

The refractory fibers, such as the aluminosilicates and their derivatives, have a temperature limitation in continuous use which is more than twice as high as that for asbestos fibers, and take over in high-technology applications where asbestos is no longer useful. Materials such as Fiberfrax, Nextel and Refrasil are produced as blown or staple fibers, and can be made into yarns or formed as blankets and felts. Yarns can be woven into cloths, ropes and braided tapes, as in conventional textile manufacture. Aluminosilicate fibers and their products can be used freely up to temperatures of 1260 °C. Much of their use lies in such applications as heat shields and insulation seals at very high operating temperatures. A major drawback of refractory fibers lies in their poor abrasion resistance, the products of their disintegration being themselves abrasive.

Glass textiles and asbestos textiles lie in the middle of the temperature range of heat-resistant fabrics. Glass textiles can be put to continuous use at 250–400 °C, somewhat in excess of the limit for synthetic organic materials. They are not considered suitable for containment of molten metal splashes and welding sparks, but specially finished heavyweight Fortaglas fabrics have acceptable performance up to 1200 °C. Glass-fiber fabrics are not suitable for machining into fully protective garments. Asbestos cloths will sustain hot face temperatures up to 600 °C without appreciable deterioration. Aluminized asbestos cloth is the only fabric among the heat-resistant textiles which can be safely used as fully protective firefighting garments where contact with bare flame is likely. In situations involving molten metal splashes and welding sparks, asbestos cloth is safe for use at temperatures well above 1500 °C.

Asbestos cloths represent the apex of performance as heat-resistant textiles in personal protection. Glass fabrics lie close to asbestos in terms of performance (though not in all applications) and particularly in terms of cost. All synthetic organic textiles are several, if not many, times more expensive than glass or asbestos textiles, and they possess severe limitations in temperature resistance. Refractory fibers and their products are again more costly than asbestos textiles, but their main advantage lies in applications which are well above the temperature limitations of the latter.

Man-made mineral fibers (MMMF) include the glass wools, rock wools and slag wools, which, with the application of a little resin or mineral oil, can be made in preforms like the earlier asbestos insulations. MMMF insulation products have a lower density and a lower thermal conductivity than their asbestos counterparts and, moreover, they have found wide use in both industrial and domestic applications. These indeed have the distinction of being alternative products of long standing, which technically and competitively have overtaken asbestos in a large and important market.

3. Synthetic Resin Systems: Friction Products, Reinforced Plastics, Packings and Jointing

Two common factors apply to these product groups. First, the many end products which fall into these categories are constituted basically from reinforced polymers, mostly thermosetting resins, and to a lesser extent, thermoplastic resins, and synthetic and natural rubbers, Second, despite the challenges from alernative materials, they are largely dominated by asbestos as the crucial reinforcing fiber.

Friction products, that is, disk pads, brake linings and clutch facings, contain 25–60% of chrysotile asbestos, depending on the type of product, in a matrix of phenolic or cresylic resins. In the larger brake linings the fiber can be in the form of an impregnated woven coarse cloth. The products also contain a number of metallic or inorganic modifiers to control frictional properties, wear, fade and noise.

In the development of alternative friction products, manufacturers are faced with the choice of replacing the asbestos content with other materials, or making revolutionary products based on new concepts. To some extent, this second choice has been achieved in semimetallic disk pads for heavy commercial vehicles, and in cermets and carbon–carbon composites for disk pads in aircraft and racing cars. Disk pads for passenger vehicles in which the asbestos is replaced by steel fiber, glass fiber or aramid fiber are either in the course of development or have become available. Various other alternatives as reinforcements or modifiers include cellulose, mica, attapulgite clay, wollastonite, diatomaceous earth and nonfibrous serpentine. The alternative end products are claimed to have superior wear rates but they also cause rotor wear.

While the reformulation and manufacture of disk pads is possible, developments in alternative brake linings have not been promising. Combinations of materials other than asbestos are less suited to the manufacture of linings, owing to the difficulties in making the essential arcuate preforms with adequate cohesiveness. Heavy-duty linings for commercial vehicles, industrial machinery and mine hoists appear to be reliant on asbestos as the main reinforcing and frictional element for some years to come.

Reinforced plastic products, in the accepted description, include small high-strength engineering products, panels made by both sheet molding and hand-lay-up methods, extruded components, and high-technology items for aircraft and military use. The matrix in these materials may comprise a phenolic or epoxy resin, or extrudable thermoplastics such as PVC, polypropylene, nylon, polyester and PTFE. Reinforcements include glass, asbestos and synthetic organic fibers, but reinforced plastic products may also include a wide range of particulate mineral fillers which contribute only marginally to reinforcement and extend the matrix so as to reduce its initial cost.

Glass fibers are brittle and therefore present difficulties in premixing operations, particularly those associated with extruded or injection-molded thermoplastic products. Asbestos is technically more acceptable in many of these products, but glass fiber is taking an increasing share of the market. In the form of webs, mats or felts, glass fiber has an almost exclusive hold in reinforced plastic panels and hand-lay-up components such as boat hulls. As glass fiber is easily available in filament or yarn form it is widely used in the manufacture of filament wound epoxy resin pipes for applications in aggressive chemical environments. Glass-reinforced plastic pipes (GRPP) have been manufactured for some years and fulfill many purposes for which asbestos-cement pressure pipes are unsuitable. The mechanical properties of GRPP have been improved by more complex methods of filament laying, and similarly their chemical resistance has been improved by building the pipes around a core of PVC.

The manufacture of packings and jointings still relies heavily on the use of chrysotile asbestos, largely because of its absorptive capacity for resin impregnants, elastomers and mineral oils, and because of the resilience which it imparts to these products. Asbestos-based gland packings for pumps and valves are not suitable where contact with highly corrosive chemicals and liquors is likely, and under such conditions packings made from PTFE, carbon or graphite remain the sole options. Jointing and gasketing products are manufactured from compressive asbestos sheeting impregnated with natural or synthetic rubbers. Similar materials made from felts or mats based on aramid pulp, glass or ceramic fibers are available as alternatives, but they do not possess the resilience of asbestos fibers and they have certain limitations in chemical resistance. The substitution of asbestos by aramid fibers in gasket materials makes for a good product in the first instance; but on both economic and volumetric grounds the utilization of aramid-fiber reinforcement has to be limited. This then requires the inclusion of other nonreinforcing components, mainly minerals, and the end result according to manufacturers is a product which embrittles in the extremes of temperature and pressure, where otherwise asbestos-based products maintain their integrity.

4. Acknowledgement

Parts of the text, together with the tables, are reproduced, with permission, from Hodgson (1987).

See also: Asbestos Fibers; Asbestos Hazards

Bibliography

Griffiths J 1986 Asbestos—Rising in the east and setting in the west. *Ind. Miner. (London)* No. 223, April 1986: 21–37

Hodgson A A 1986 *Scientific Advances in Asbestos, 1967 to 1985.* Anjalena Publications, Crowthorne, UK

Hodgson A A 1987 *Alternatives to Asbestos and Asbestos Products,* 2nd edn. Anjalena Publications, Crowthorne, UK

Michaels L, Chissick S S (eds.) 1979 *Asbestos,* Vol. 1: *Properties, Applications and Hazards.* Wiley, Chichester

UK Health and Safety Executive 1979 *Final Report of the Advisory Committee on Asbestos.* Her Majesty's Stationery Office, London

UK Health and Safety Executive 1986 *Alternatives to Asbestos Products: A Review.* Her Majesty's Stationery Office, London

World Health Organization 1986 *Environmental Health Criteria, 53—Asbestos and Other Natural Mineral Fibres.* World Health Organization, Geneva

A. A. Hodgson
[Crowthorne, UK]

Asbestos Fibers

Asbestos is a generic name applied to certain specific minerals which occur in nature in fibrous form. The common features which make these minerals unique are their flexibility, high tensile strength and value as reinforcing fibers in composite materials.

1. Types of Asbestos

Chrysotile is the most abundant type of asbestos and accounts for 95% of total world availability of asbestos fiber. It is one of the serpentine group of minerals, and the only one of such to occur in fibrous form.

The other types of asbestos belong to the amphibole group of minerals which occur widely in many igneous and metamorphic rocks. Asbestiform amphiboles, however, have a restricted occurrence. The principal commercial varieties, crocidolite and amosite, are mined only in the Precambrian ironstone formations in South Africa.

Three other asbestiform amphiboles, anthophyllite, tremolite and actinolite, have little commercial importance nowadays.

Recent detailed availability of asbestos is given in Table 1 (the total production may be slightly in excess of reality). USBM condensed data released in January 1987 indicates an annual world production of 4.1 Mt for 1985 and 1986 (estimated). Various other sources suggest that output from the USSR will continue to exceed the deficits suffered by Canada, USA and Southern Africa. Forecasts for an increase in asbestos consumption over the next decade are in the range 3–5% per annum.

Some years ago, it was estimated that there were 3000 applications for asbestos fibers, but environmental restrictions in many countries have severely reduced the range of use mainly to those products where the fiber is said to be "locked in." About 75% of all asbestos produced goes into asbestos-cement products and the remainder is spread over such applications as friction materials, papers and felts, textiles, vinyl floor tiles and asphaltic coatings. Despite the loss of many applications, not least among these a wide variety of insulation materials, the overall consumption of asbestos has diminished only by a limited extent, principally because of the increase in recent years of the world capacity for asbestos-cement manufacture.

2. Properties of Asbestos

The structure, composition, chemistry and form of chrysotile on the one hand and amphibole fibers on the other differ considerably. Only in terms of physical properties are there close similarities between these important varieties of asbestos fiber.

Chrysotile is a hydrated magnesium silicate. It has a layered lattice structure, consisting of a sheet of Mg–OH units superimposed on a sheet of $Si_2O_5^-$ units. These layers are curved in one direction, and in chrysotile they form scrolls or concentric tubes of up to 20 turns. A chrysotile fiber consists of a number of such fibrils in parallel orientation, together with interstitial material of similar composition. The fibers possess characteristic flexibility.

The basic structure of the fibrous amphiboles consists of two parallel chains of Si_4O_{11} units linked by a ribbon of cations of Fe^{3+}, Fe^{2+}, Mg, Ca and Na, the distribution of which determines the type of amphibole. These fibers may be likened to a uniform stack of laths of the basic structure which are cross-linked to each other.

Thus, in contrast to chrysotile asbestos, amphibole fibers are considerably less flexible and are generally referred to as harsh fibers.

In the period 1970 to 1982, investigations on the microstructure of asbestos fibers at levels of atomic resolution enabled knowledge of chrysotile structures to be expanded considerably, and revealed in the amphibole fibers the nature of fibrosity. The collective term, biopyribole, has been reintroduced to explain structural relations and alteration mechanisms between sheet silicates, pyroxenes and amphiboles, notably between talc, chrysotile or vermiculite and tremolite or anthophyllite (Hodgson 1986).

The idealized chemical formulae of the three main types of asbestos are given in Table 2. Chrysotile has a predominantly basic composition, whereas the composition of the amphiboles is balanced in favor of the acidic (Si_4O_{11}) component. This has an important bearing on the chemical properties of asbestos. Whereas the amphibole varieties show considerable resistance to decomposition by acids, chrysotile is easily decomposed by loss of its MgO and OH components (Table 3). Indeed, part of the Mg^{2+} content of chrysotile can be leached out by water alone.

Degradation by heating is also distinctive for the different types of asbestos. Chrysotile loses water gradually from 100 °C upwards until at about 650 °C

Table 1
Asbestos: world production 1984

Country	Production (t)
Australia	10000
Brazil	160000
Canada (shipments)	922000
China	160000
Cyprus	16000
Greece	110000
India	25000
Indonesia	25000
Italy	140000
Korea	15000
South Africa	170000
Swaziland	30000
USSR	2300000
USA	57422
Yugoslavia	10400
Zimbabwe	165000
Others, including Argentina, Bulgaria, Columbia, Egypt, Japan, Mozambique, Taiwan, Turkey	22475
Total (approx.)	4338000

Source: R A Clifton, *Asbestos*, US Bureau of Mines Minerals Yearbook 1984

Table 2
Idealized chemical formulae of the main types of asbestos

Name	Formula
Chrysotile	$Mg_3Si_2O_5(OH)_4$
Crocidolite	$Na_2Fe_2^{3+}(Fe^{2+}Mg)_3(Si_4O_{11})_2(OH)_4$
Amosite	$(Fe^{2+}Mg)_7(Si_4O_{11})_2(OH)_4$

Table 3
Some physical and chemical properties of asbestos

	Chrysotile	Crocidolite	Amosite
Tensile strength[a] ($MN\,m^{-2}$)	3100	3500	2500
Elastic modulus[a] ($GN\,m^{-2}$)	160	190	160
Specific gravity	2.55	3.43	3.37
Magnetic susceptibility (mean χ at 10 kOe)	5.3×10^{-6}	78.7×10^{-6}	60.9×10^{-6}
% loss in tensile strength[a]			
at 300 °C	0	13	37
400 °C	(2.7)[b]	63	61
500 °C	(13.5)[b]	78	80
600 °C	84	83	96
% weight loss by acid decomposition[c] in period			
0.5 h	60	6	8
2 h		7	15
4 h		7.5	22
8 h		8.5	30

[a] Measurements made on micro fibers at 4 mm length between jaws [b] Gain in tensile strength [c] Under conditions of boiling refluxed 4 M hydrochloric acid

all 13% of its water content is lost. At about 800 °C, the residual material recrystallizes to forsterite and free silica. All the amphiboles lose up to 2% of water but at widely varying temperatures. The ferrous iron content of crocidolite oxidizes at 400 °C, that of amosite at about 650 °C, and these temperatures are coincident with dehydroxylation. Ultimate decomposition occurs at 900 °C for crocidolite and 600 °C for amosite.

Table 3 gives the main physical properties of asbestos insofar as they are relevant to composite materials. Fiber strength and modulus are high and are substantially greater than most comparable man-made fibers. The information on specific gravity and magnetic susceptibility is important for specific applications in fiber-reinforced resin systems. Table 3 also shows the percentage loss in tensile strength of asbestos fiber as it is heated up to 600 °C. Chrysotile resists changes in tensile strength as temperature increases and even gains at 500 °C, because of slight dimensional shrinkage. Thereafter, it shows catastrophic loss in strength. Crocidolite holds up well to 300 °C but both crocidolite and amosite lose a substantial proportion of strength at and above 400 °C.

3. *Reinforcing Properties of Asbestos*

It is both a theoretical and observed fact that the best dimensional form of asbestos fibers in reinforcement terms corresponds to a critical length between 1 and 2 mm and an aspect ratio between 150 and 250. Asbestos fiber of a quality within these narrow limits could give reinforcing strengths in asbestos cement, for example, some four to five times greater than actually observed for this product.

In commercially produced asbestos there is, however, a very wide spectrum of fiber lengths and diameters and it is generally the preponderance of length within certain broad limits which determines the grade of asbestos. Observations of fracture failure in asbestos-reinforced composite materials indicate that long, thick fibers, and short fibers below the critical length, pull out of the matrix whereas all thin fibers, as long as they exceed critical length, are most likely to break at the plane of fracture. Since the tensile properties of the fiber contribute more than do pull-out properties to the reinforcement strength of a composite material, the aim in preparing asbestos fiber for use in many applications is to improve the fineness of the fiber without appreciable loss of length.

A variety of both wet- and dry-milling methods are used in the preparation of asbestos fiber and their effectiveness may well vary with the type of fiber being processed. Chrysotiles in general withstand vigorous milling, but amosite requires careful preparation to avoid length reduction. Crocidolite asbestos responds in a way intermediate between chrysotile and amosite.

The most important tests for asbestos fiber concern length, diameter (as expressed by apparent surface area) and reinforcement properties. Some abridged but typical properties of asbestos fiber used in asbestos cement, and based upon the above tests, are given in Table 4.

4. *Applications of Asbestos*

The manufacture of asbestos-cement products accounts for an annual turnover of some 3 million tonnes of asbestos. Such products contain 10–17% of asbestos, mainly chrysotile. Chrysotile gives plasticity to a wet asbestos-cement system, essential to product molding. Crocidolite and amosite provide ease of drainage in the manufacturing process, and crocido-

Table 4
Typical properties of asbestos fiber used in asbestos-cement manufacture

	Chrysotile (medium length)	Chrysotile (short)	Crocidolite (medium length)	Amosite (medium length)
Length distribution fraction (%)				
>5 mm	10	1	40	15
5–1 mm	20	4	20	20
1–0.1 mm	30	35	5	25
<0.1 mm	40	60	35	40
Apparent surface area ($m^2 kg^{-1}$)	1800	1800	1100	900
MR[a] ($MN m^{-2}$)	32	20	36	25

[a] Modulus of rupture of asbestos-cement test pieces with 10% fiber content and density 1.6 $g cm^{-3}$

lite, in conjunction with medium-length chrysotiles, ensures the required circumferential strength in pressure pipes.

Asbestos is compatible with calcium silicate matrices formed by reaction between lime and silica in high-pressure steam autoclaves. The asbestos is unaffected by the conditions of the reaction. Amosite asbestos is used in such a matrix in the manufacture of fireproof insulation boards, although this industry has diminished in favor of products based on mineral wools and cellulose.

The use of asbestos in components based on reinforced plastics is small, because of the preference for glass-reinforced plastic products. Although the strongly polar nature of chrysotile interferes to some extent with its bonding with resins, this type of asbestos is a crucial requirement in friction materials, which are basically asbestos-reinforced phenolic resins, with modifiers to control friction and wear characteristics. Both types of amphibole asbestos have been used to advantage in reinforced thermosetting and thermoplastic resins, but in composites embodying thermoplastics it is necessary to use complex stabilizing systems to prevent higher-temperature depolymerization. The ferrous iron content of the asbestos is the active agent in depolymerization.

Vinyl floor tiles account for the use of a limited quantity of the less expensive short-fiber chrysotiles. The fiber is an extender in the system and also provides a measure of resilience to the product. Chrysotile asbestos is also formed into papers and felts which may be used in flooring materials and in the manufacture of electrical components based on synthetic resins.

Asbestos textiles based on chrysotile continue to be in demand, particularly for fire-protection purposes and in making heavy-duty brake linings. Modern wet yarn-drawing processes have partly replaced conventional carding and spinning methods in this industry. Some asbestos cloth, as well as ordinary chrysotile fiber, is widely used in the manufacture of heat-resistant glands, packings and seals.

5. Health and Environmental Hazards Associated with Asbestos

In the course of processing asbestos at any stage from mining and milling through to industrial processes, exceedingly fine fibers are generated to form airborne dust clouds of concentrations up to 1000 fibers per cm^3 of air. Inhalation of asbestos dust is a hazard to health and measures of environmental control are necessary to ensure minimum risk to those who work with asbestos.

Safety in the asbestos industry nowadays is governed by control measures, coupled with current dust-control technology, and by an international policy of information and awareness concerning the hazards. Stringent regulations in many countries are designed to protect workers in the industry and to ensure that dust levels in the workplace are maintained at 0.25 fiber per ml of air. Many dusty processes have been eliminated, such as spray insulation and asbestos-based thermal insulation and fireproof boards. The uses for asbestos in the 1980s have been those in which the fiber is locked in and in which the manufacturing process can be kept in tight control.

At the same time, there has been considerable activity in recent years towards the development and application on alternatives to asbestos in many product lines. While there has been success in some areas, there are undeniable constraints to the use of alternative raw materials in terms of costs and availabilities, technical acceptance, and even possible new health hazards.

See also: Filters, Sorbents and Ion Exchangers: Zeolite Minerals

Bibliography

American Society for Testing and Materials *Annual Books of ASTM Standards*. American Society for Testing and Materials, Philadelphia, Pennsylvania

Asbestos International Association Health and Safety Publications; *Proc. Biennial Confs.* Asbestos International Association, London

Deer W A, Howie R A, Zussman J 1962 *Rock Forming Minerals*, Vols. 2, 3. Longman, London

Hodgson A A 1986 *Scientific Advances in Asbestos 1967 to 1985*. Anjalena, Crowthorne, UK

Hodgson A A 1987 *Alternatives to Asbestos and Asbestos Products*, 2nd edn. Anjalena, Crowthorne, UK

International Standards Organisation ISO/TC 77: Paper on test methods for asbestos fibers. International Standards Organisation, Geneva

Michaels L, Chissick S S (eds.) 1979 *Asbestos: Properties, Applications and Hazards.* Wiley, Chicester

Roskill Information Services 1986 *The Economics of Asbestos*, 5th edn. Roskill Information Services, London

World Health Organisation 1986 *Environmental Health Criteria, 53: Asbestos and Other Mineral Fibres.* World Health Organisation, Geneva

A. A. Hodgson
[Crowthorne, UK]

Asbestos Hazards

Asbestos is a useful industrial material that has many important and essential applications—its widespread use has however increased its potential for harm. Exposure to airborne asbestos dust is recognized as a hazard to health: pulmonary fibrosis (asbestosis), malignancies of the lung and possibly of the gastrointestinal tract, and mesothelioma are effects which can result from asbestos-dust inhalation. The risk of developing asbestos-related diseases depends on factors such as the concentration of fiber in the inhaled air, the duration of exposure, and the type and dimension of the fiber.

1. Mineralogy

Asbestos is the generic term given to a group of six fibrous silicates in two classes; each one has its own characteristic morphology and chemistry with impurities depending on the mineral source (Table 1).

An important feature of asbestos is its fibrous character; in this context a fiber is considered to have a minimum aspect (length to width) ratio of 3:1. During processing, friable asbestos fiber may undergo further breakage, with the production of increasingly finer diameter fibers, referred to as fibrils. Milled, powder-like chrysolite asbestos has fibrils with an average aspect ratio as high as 50.

Table 1
Classification of asbestos

Classification	Nominal composition
Serpentine asbestos	
chrysotile (white asbestos)	$3MgO, 2SiO, 2H_2O$
Amphibole asbestos	
actinolite	$2CaO, 4MgO, FeO, 8SiO_2, H_2O$
amosite (brown asbestos)	$5.5FeO, 1.5MgO, 8SiO_2, H_2O$
anthophyllite	$7MgO, 8SiO_2, H_2O$
crocidolite (blue asbestos)	$Na_2O, Fe_2O_3, 3FeO, 8SiO_2, H_2O$
tremolite	$2CaO, 5MgO, 8SiO_2, H_2O$

2. Uses

More than a thousand uses of asbestos have been described. Asbestos cement has been manufactured in the form of roofing sheets, wallboard and pipes, including high-pressure pipes. Asbestos is also used as filler for plastics, roof shingles and floor tiling; as heat insulation and fireproofing both in ships and buildings; and for improved fire- and heat-resistant products, e.g., fireproof asbestos cloth, yarns and fabrics, brake and clutch linings, high-pressure chemical gaskets, and paints.

Asbestos possesses both high thermal stability and chemical resistance, and these highly sought-after properties strongly influence its selection for a variety of products. Over 90% of the asbestos used is chrysolite.

3. Health Hazards

The health hazards related to asbestos exposure arise from inhalation and possibly ingestion. On inhalation, most of the larger asbestos fibers ($>10\ \mu m$ in diameter) are trapped in the nasal passages by hairs and/or mucus, and may be eliminated or swallowed. Smaller fibers ($<5\ \mu m$ in diameter) may enter the bronchi, although the majority are cleared by the mucociliary process. The most significant fibers from a human health standpoint are the smallest ($<3\ \mu m$ in diameter), which may enter the terminal bronchioles and alveoli and eventually induce a variety of pathological changes. Once imbedded in the tissue, they can remain in the body for prolonged periods. Fibers of very fine diameter, even though $>10\ \mu m$ in length, can penetrate to the alveoli. The number of airborne fibers per cm^{-3} which are $>5\ \mu m$ in length is measured to indicate the asbestos dust respiratory hazard.

3.1 Asbestosis

Asbestosis is a form of pneumoconiosis resulting from the excessive inhalation of asbestos dust. It is a progressive disabling disease that is not reversible. All types of asbestos used commercially have produced asbestosis. The clinical symptoms are due to fibrosis of the lung and concomitant pathological changes, mainly emphysema and bronchiectasis. Progressive dyspnea can lead to disablement. Chronic coughing and loss of weight may also occur. Death is usually due to secondary respiratory infection or cardiac involvement. The development of asbestosis is slow and usually involves excessive exposures of five years or more.

3.2 Lung Cancer (Bronchogenic Carcinoma)

The clinical characteristics of both occupational and nonoccupational lung cancers are similar. The location, histologic type, and rate of growth of the tumor determines whether symptoms and signs develop early or late in the course of the disease. The nonspecificity of the symptoms and signs may delay a definitive diagnosis. Usually, the earliest symptom is an irritative cough, with increasing sputum production. Hemoptysis (expectoration of blood) and chest pain may be present. The radiological changes in patients with lung cancer may be attributable to the tumor itself or to abnormalities in the distal lung tissue secondary to the tumor's induced destruction of the airways. With asbestos workers, peripheral tumors are more frequent than central tumors. The onset of the disease is insidious and can reflect a prolonged period of exposure. Most lung cancer associated with asbestos is also associated with cigarette smoking (see Sect. 4).

3.3 Mesothelioma

Malignant mesotheliomas are rare tumors which spread over membranes such as the pleura, pericardium and peritoneum. A patient with pleural mesothelioma may exhibit chest pain, shortness of breath, coughing, weight loss, anorexia, fever, vomiting, lassitude, pleural effusion, thickening of the pleura and ascites. Patients with peritoneal mesothelioma have exhibited abdominal masses, peripheral edema and ascites. No successful treatment for mesothelioma has been developed. Once diagnosed, the prognosis is immutable and death follows shortly after.

3.4 Other Asbestos-Related Diseases

Pleural plaques and cancer of the larynx and gastrointestinal tract may be related to exposure to asbestos. Pleural plaques are localized fibrous thickenings of the pleura which may or may not be calcified. They are not neoplastic, but avascular structures composed of collagen and fibrocytes.

Cancer of the larynx and of the gastrointestinal tract (esophagus, stomach, colon and rectum) appear to have a higher statistical incidence in asbestos workers than that reported for the general population. However, further study is needed to confirm these associations and to establish the influence of smoking habits.

4. Smoking

Tobacco smoke has a marked synergistic effect on asbestos-induced lung cancer. It has been estimated that the risk of bronchogenic carcinoma in asbestos workers who smoke is increased ten to thirty fold over that of nonsmoking asbestos workers. There is no evidence that smoking increases the risk of mesothelioma.

5. Exposure Levels

There is overwhelming evidence, both from animal studies and epidemiology, that asbestosis, lung cancer and mesothelioma are dose-related. The higher the asbestos exposure, the greater the risk of developing one or more of these diseases. There can be a long latent period between the exposure to asbestos and the onset of the disease. In general, the higher the exposure to asbestos dust, the shorter the induction period. There is also some indication that short exposures to high levels of asbestos dust present a greater hazard than lower exposures over a longer period.

Considerable evidence indicates linear dose-response relationships between the exposure level (dose) and the incidence of asbestosis, lung cancer and probably mesothelioma. However, data at low exposure levels are limited and the relationship here is uncertain. At this time, it is not possible to state if a threshold dose for cancer exists, but lowering the exposure level lowers the risk of developing malignancies.

Epidemiologic investigations and animal studies indicate that the toxic potential of the different types of asbestos with regard to the induction of lung cancer and mesothelioma decreases in the order: crocidolite > amosite > chrysotile. The difference in toxic potential is probably related to factors such as chemical difference, the particle shape and size, and aerodynamic flow characteristics which favor deposition in the lung.

6. Environmental Exposures

A variety of environmental exposures to asbestos dust have been reported. Although some of these exposures may still occur, changes in work practices and improved dust control measures have probably either eliminated or greatly reduced their significance. The carrying home of workclothes saturated with asbestos fibers (e.g., heavy sweaters) has inadvertently produced asbestos-related diseases in the families of workers. Modern industrial hygiene practices call for a change of workclothes at the plant and their laundering by the factory. Special procedures should be followed to prevent air pollution at mines and processing plants from presenting risks to adjacent populations.

Careless construction techniques, such as unprotected sawing of asbestos board or sanding of asbestos joint cements, have exposed both the workforce and bystanders to asbestos dust. Electric saws, with a suction component and a nonasbestos joint cement, can readily eliminate such unnecessary and potentially dangerous exposures.

A variety of consumer products such as play dough, modelling clays, simulated fireplace ash, and linings for fur coats, oven mitts and ironing boards may contain asbestos fibers capable of being released into the air. In many jurisdictions the sale of such products has been discouraged or banned.

Many public and private buildings have been treated with sprayed-on asbestos or have had friable asbestos acoustic tile installed. There may be untoward exposures to airborne asbestos fibers from such installations, and techniques have been developed to resolve these problems.

7. *Control Measures*

All unnecessary exposures to airborne asbestos fibers should be avoided. Factory clean-ups should be by damp mopping or vacuum cleaning to reduce dust. Equipment such as power saws and/or grinders should be fitted with suction hoses to trap asbestos fibers and prevent their release into the air.

Waste asbestos should be disposed of with care. The waste product should be placed in heavy plastic bags and the bags placed in a sealed metal container. The indiscriminate use of mine tailings and waste asbestos should be discouraged.

Permitted exposure levels to asbestos dust in the workplace have continually been lowered. In many jurisdictions, the time weighted 40-hour exposure level varies from 1 to 5 fibers per cm^{-3} with emphasis on the lower levels. In some instances amosite and crocidolite have lower permissible exposure levels than chrysotile. The review article by Meek et al. (1984) contains a comprehensive multinational and timely listing of permitted exposure levels. If these exposure levels are exceeded, the wearing of appropriate respiratory protection is advocated and in many jurisdictions is mandatory.

See also: Asbestos Fibers

Bibliography

Acheson E O, Gardiner M J 1983 *The Control Limit for Asbestos.* Health and Safety Commission, London

Casey K R, Rom W W, Moatamed F 1981 Asbestos related disease. *Clin. Chest. Med.* 2: 179–202

Committee on Non-Occupational Health Risks of Asbestos Fibers 1982 *Non-Occupational Health Risks of Asbestos Fibers.* National Academy of Sciences, Washington, DC

Consumer Product Safety Commission 1983 *Chronic Health Advisory Panel on Asbestos.* Consumer Product Safety Commission, Washington, DC

Environmental Health Directorate 1984 *Report of the Committee of Experts Advising the Department of National Health and Welfare (Canada) concerning the Scientific Basis for Occupational Standards of Asbestos.* Environmental Health Directorate, Ottawa

International Agency for Research on Cancer 1977 *Asbestos,* IARC Monograph on the Evaluation of Carcinogenic Risks of Chemicals to Man No. 14. International Agency for Research on Cancer, Geneva

Lemen R, Dement J M (eds.) 1979 *Dust and Diseases.* Pathotox, Park Forest South, Illinois

Meek M E, Shannon H S, Toft P 1984 *Case Study—Asbestos, in Toxicological Risk Assessment.* CRC Press, Cleveland, Ohio

Michaelis L, Chissick S S (eds.) 1979 *Asbestos,* Vol. 1, *Properties, Applications and Hazards.* Wiley, New York

National Cancer Institute 1978 *Asbestos: An Informational Resource.* National Cancer Institute, Bethesda, Maryland

National Institute for Environmental Health and Safety/ Environmental Protection Agency 1974 Proceedings of the joint NIEHS/EPA conference on biological effects of ingested asbestos. *Environ. Health Perspec.* 9: 113–462

National Institute for Occupational Safety and Health 1977 *Occupational Exposure to Asbestos. NIOSH Criteria for a Recommended Standard.* NIOSH 77–169. US Department of Health and Human Services, Cincinatti, Ohio

National Research Council of Canada 1979 *Effect of Asbestos in the Canadian Environment,* NRC 16452. National Research Council, Ottawa

Preger L (ed.) 1978 *Asbestos-Related Disease.* Grune and Stratton, New York

Royal Commission on Matters of Health and Safety 1984 *The Use of Asbestos in Ontario.* Ontario Ministry of Government Services, Toronto

Selikoff I J, Lee D H (eds.) 1978 *Asbestos and Disease.* Academic Press, New York

UK Advisory Committee on Asbestos 1979 *Final Report of the UK Advisory Committee on Asbestos.* HMSO, London

US Federal Register 1979 *Commercial and Industrial Uses of Asbestos Fibers and Consumer Products Containing Asbestos,* CPSC–EPA Pt. 6. US Government Printing Office, Washington, DC

Walton W A 1982 The nature, hazards and assessment of occupational exposure to asbestos: A review. *Ann. Occup. Hyg.* 25:117

G. S. Wiberg
[Canadian Bureau of Chemical Hazards, Ottawa, Ontario, Canada]

B

Beryllium Resources

Beryllium, with an atomic number of 4 and a specific gravity of 1.85, is one of the lightest metals stable in air and is used commercially as a metal, alloy and oxide. It is recovered mainly as beryl or bertrandite from granitic pegmatites and hydrothermal–pneumatolytic skarn and greisens that also contain, for example, tin, fluorite and tungsten. Increased US demand for beryllium can be met by nonpegmatite deposits of beryl and bertrandite at least to the end of the century (Griffits 1973).

1. Uses

Beryllium has a wide range of uses as a metal, alloy and oxide. As a metal, beryllium has great stiffness and mechanical dissipation properties, as well as an ability to resist corrosion. Beryllium in metal form accounts for 40% of the total beryllium production. Beryllium–copper alloys represent 53% of the beryllium consumption. Beryllium oxide (beryllia) uses are based upon its properties of heat resistance, thermal conductivity, hardness and ability to moderate or reflect neutrons, and accounts for the remaining 7% of beryllium production.

2. Geochemistry and Occurrence

The average crustal abundance of beryllium is 2–3.5 ppm. Owing to its ability to replace silicon (similar ionic radius) it tends to be widely dispersed rather than concentrated in discrete beryllium minerals. Beryllium is an essential constituent in approximately 40 minerals and an occasional constituent in 50 others including plagioclase, micas and clays. The principal beryllium minerals are beryl, bertrandite, phenakite, barylite, chyrosoberyl and heloite.

2.1 Pegmatites

Beryl occurs in the late stage, coarsely crystalline pegmatite phase of some granitic intrusions. The pegmatite dikes are peripheral to a main intrusion and form by crystallization of residual fluids. They are composed of quartz, sodic plagioclase and microcline with or without spodumene, muscovite or lepidolite. Beryl is usually associated with the presence of spodumene. Such pegmatites are common in ancient continental shield areas of Canada, Australia and Africa; in the USA they are found in the southern Appalachians, New England and the Black Hills of South Dakota. Other important beryl localities are Brazil, the People's Republic of China, India and Argentina. In zoned pegmatites, with quartz-rich cores, beryl crystals occur along the margins and may be up to 1 m or more in width and 6 m in length. In unzoned pegmatites, beryl is generally present in small amounts and is fine grained (which has prevented its recovery). Most often, beryl is mined as a coproduct in pegmatites since it is rarely abundant enough to be mined separately.

2.2 Nonpegmatite Deposits

Bertrandite became the major source of US beryllium production in the late 1960s with the discovery of deposits at Spor Mountain, Utah. These are the largest known US beryllium deposits. The bertrandite is formed by epithermal alteration of calcareous tuffs and is commonly associated with montmorillonite, quartz, fluorite and a dularia as replacement products in calcareous hosts. Beryl and bertrandite are also found associated with fluorite in mica quartz greisen near a granitic intrusion in Park County, Colorado, Phenakite and bertrandite occur with scheelite, fluorite and pyrite in White Pine County, Nevada. Other beryllium deposits are in Sea Lake, Labrador, Canada as barylite; and in Coahuila, Mexico as bertrandite (Petkof 1980).

3. Resources and Industry Outlook

Beryllium consumption in the USA was 315 t in 1987. This included 140 t imported, largely from Brazil and China, and most of the domestic supply from bertrandite at Spor Mountain, Utah. By the year 2000 bertrandite production is expected to increase, while imported and domestic beryl will decrease. The US consumption of beryllium between 1973 and 2000 is predicted to be 27 000 t. Domestic reserves are 28 000 t and resources are 80 000 t. The rest-of-the-world demand through the forecast period is estimated at 11 100 t. Ample world supplies are present at the predicted average 3.0% growth rate in beryllium use.

Bibliography

Beus A A 1962 *Beryllium: Evaluation of Deposits during Prospecting and Exploratory Work.* Freeman, San Francisco, California

Beus A A 1966 *Geochemistry of Beryllium and Genetic Types of Beryllium Deposits.* Freeman, San Francisco, California

Griffits W R 1973 Beryllium. In: Brobst D A, Pratt W P (eds.) 1973 *United States Mineral Resources,* US Geological Survey Professional Paper 820. US Government Printing Office, Washington, DC, pp. 85–93

Petkof B 1980 Beryllium. *Mineral Facts and Problems,* US Bureau of Mines Bulletin 671. USBM, Washington, DC, pp. 85–94

US Bureau of Mines *1988. Mineral Commodity Summaries* 1988. USBM, Washington, DC, pp. 22–23

N. Herz
[University of Georgia, Athens, Georgia, USA]

Binders: Clay Minerals

The clay minerals most widely used as binders are sodium and calcium montmorillonite, and kaolinite. Binders based on these minerals are employed in the foundry, pelletizing and ceramics industries because of their plasticity when mixed with water and their fine particle size. Kaolinite ($Al_2O_3.2SiO_2.2H_2O$) is the major clay mineral constituent of the rock kaolin, and the montmorillonites are the chief constituents of bentonite.

1. Kaolin

A more complete description of kaolin and its general properties and occurrences is given elsewhere (see *Fillers and Coatings: Clay Minerals*). In addition to the pigment grade of kaolin used as a binder in ceramics and other applications, binders based on other kaolinitic materials (ball clays, fire clays and plastic underclays) are also available. Ball clay is a secondary clay commonly characterized by the presence of organic matter, high plasticity, high dry strength, long vitrification range and a light color when fired. Kaolinite is the principal mineral constituent of ball clay. Ball clays are mined in the USA, the UK, France, Germany, India and South Africa. Fire clay is a kaolinitic clay that is not white-burning and has a pyrometric cone equivalent above 19, that is, it has a fusion point above 1520 °C. The term fire clay excludes most kaolin and ball clays because they become white on burning. Plastic underclay is a nonrefractory kaolinitic clay found immediately below coal beds of carboniferous age.

In ceramic applications, kaolins, ball clays, fire clays and plastic underclays are used as binders. Determination of which is the best binder for a particular application involves matching the properties of each with the specifications required in the final product. The specifications normally considered are modulus of rupture, plasticity, pyrometric cone equivalent, casting rate, fired color and shrinkage. Other tests may be required for certain products.

The modulus of rupture (MOR) is the fracture strength of a material under a bending load. The MOR measurement is made on a long bar of either rectangular or circular cross section, supported near its ends; a load is applied to the central part of the supported span. Plasticity, the ease with which a mass can be deformed and retain the deformed shape without any elastic rebound, is measured on a plastigraph or with a liquid-limit tester. The casting rate is important; fine-grained bodies cast more slowly than coarse ones. Viscosity of the casting slip must be carefully controlled, as if the slip is too viscous it will not properly fill the mold or drain cleanly. The fired color is important in many instances; for example, in whiteware bodies the fired color must be white or cream. This is a visual test and the test specimen, fired to a predetermined temperature, is compared with a standard. Shrinkage is also important, as ceramic articles undergo shrinkage at several different points in the manufacturing sequence. During drying, articles will shrink by varying amounts, depending on their composition and the amount of water present. Further shrinkage occurs during firing; therefore it is important to know both the drying and firing shrinkage.

The particle size of kaolin products is generally controlled, as a coarse kaolin has very different ceramic properties to a fine kaolin. In general, the finer the particle size the higher the MOR, the plasticity and the shrinkage, and the lower the casting rate. A particular grade of kaolin can be matched with the properties needed in the final product. In castings a coarse kaolin is preferred, and in whiteware a plastic kaolin of fine particle size that fires white is preferred. The Georgia kaolins have pyrometric cone equivalent values ranging from 33 to 35 (i.e., fusion points of 1745–1785 °C) indicating a highly refractory clay.

Ball clays are used where high plasticity, high dry strength and a long vitrification range are required along with a white or light cream color. Fire clays are used in refractory applications when fired color is not important. Plastic underclays are used when high plasticity is required and the fired color does not have to be white.

2. Bentonite

The most important binder mineral for use in the foundry and pelletizing industries is bentonite, the chief mineral of which is sodium or calcium montmorillonite. The montmorillonites are extremely plastic over a wide moisture range and can meet specifications for moisture content, gelling index, binding properties and liquid limit. The latter is a measure of the ability of montmorillonites to hold water without flowing.

Sodium montmorillonite clays have the ability to hold moisture and are not easily dehydrated when subjected to heat. They give a very durable mold or pellet, and much less clay can be used to bind the foundry sand or the iron ore in the pellet. Sodium montmorillonites also have a high resistance to expansion defects, good refractoriness and a low moisture requirement. The hot and dry strengths and thermal stability are preferred for the production of steel and heavy iron castings in a clay-bonded foundry

Table 1
Comparison of foundry binding properties of typical sodium and calcium montmorillonites

Property	Calcium montmorillonite	Sodium montmorillonite
Green compression strength	high	slightly lower
Bond development	rapid	moderate
Hot compression strength	moderate	high
Dry compression strength	low	high
Durability	low	high
Resistance to expansion defects	low	high
Thermal stability at casting temperatures	poor	very good
Sand flowability	very good	moderate
Shakeout characteristics	easy	poor

sand. Therefore the steel casting industry uses sodium montmorillonites (Wyoming bentonite).

Calcium montmorillonites are called nonswelling bentonites or sometimes sub-bentonites. They give very rapid bond development, a high green compression strength, moderate hot strength, low dry compression strength, low durability and very good sand flowability. Because the majority of iron foundries want a low dry strength and low durability to achieve good shakeout characteristics, calcium montmorillonites are in common use. The best nonswelling bentonites are mined in Mississippi and Texas in the USA, and in the Federal Republic of Germany, Italy and Greece.

Blends of sodium and calcium montmorillonites are used in many foundries to achieve desired binding properties. Table 1 shows a summary of the properties of sodium and calcium montmorillonites with respect to foundry use. Some producers of calcium montmorillonites activate the clay by changing the calcium cation to sodium through treatment with soda ash or occasionally with a sodium polymer such as sodium polyacrylate. Some calcium bentonites respond very well to this treatment and give considerably improved properties. Although a soda ash-treated calcium montmorillonite is often called sodium montmorillonite after treatment, the properties at high casting temperatures and the durability are not as good as those of natural sodium montmorillonite.

Specifications for montmorillonites used as binders in the foundry industry are outlined by the Steel Founders Society of America and the American Foundrymen's Society. The tests include moisture content, pH, calcium oxide content, liquid limit, thermal stability at casting temperatures, and green, dry and hot strengths. Specifications for montmorillonites used for pelletizing taconite-type iron ores have not been standardized, and several tests are used as green pellets must be capable of withstanding handling, compaction and drying, and the dry pellets must be even stronger. Mixtures of bentonite, iron ore and water are tested for wet drop strength, wet compression strength, plastic deformation and dry compression strength. The final choice of a suitable bentonite for pelletizing appears to be largely logistical with locational advantage and transportation costs playing a major role. In the case of foundries, although the locational advantage and transportation costs are important, the performance characteristics play the major role in the selection of the binder.

See also: Binders: Industrial Minerals

Bibliography

American Foundrymen's Society 1963 *Foundry Sand Handbook*. 7th edn. American Foundrymen's Society, Chicago, Illinois

Norton F H 1968 *Refractories*, 4th edn. McGraw-Hill, New York

Patterson S H, Murray H H 1983 Clays. In: Lefond S J (ed.) 1983 *Industrial Minerals and Rocks*. 5th edn. American Institute of Mining, Metallurgical and Petroleum Engineers, New York, pp. 585–651

Steel Founders Society of America 1965 *Tentative Specifications for Western Bentonite*, SFSA Designation 13T-65. Steel Founders Society of America, Cleveland, Ohio

H. H. Murray
[Indiana University, Bloomington, Indiana, USA]

Binders: Industrial Minerals

Fine-particle minerals that have the property of plasticity when mixed with water are used to bind or hold other ingredients together in a more-or-less homogeneous mass or shape. This mass or shape is then dried and/or further processed to make the desired end product. The most common binders are clays such as kaolins, ball clays, fireclays, plastic underclays and bentonites (see *Binders: Clay Minerals*). The most common applications are in foundries, where the sands are bound with bentonite; in pelletizing iron ores, where bentonite is the binder; and in ceramics, where kaolinitic clays are used to bind together the feldspar, silica or other ingredients that go to make up the body of the ceramic product.

1. Foundry

The metal-casting industry uses a large quantity of binders annually as the total output of metal castings is of the order of 80 million tonnes per year. In recent years, the foundry industry has developed rapidly in a number of countries where previously it was very small—e.g., in Brazil, Spain, India, South Africa and Yugoslavia. As nations develop their manufacturing industries they erect foundries to meet the demand for castings for automobiles, domestic appliances, machine tools and general industrial requirements.

The first primitive metal-casting operations are estimated to have begun some 5000 years ago, and the metal being cast was undoubtedly copper. Today, with the advent of high-temperature furnaces, nearly all metals can be melted and poured into molds in order to make the required sizes and shapes. The most common mold is made up of a high-silica sand and bentonite mixture, the sand making up about 95% and bentonite 5% of the mix. This mixture is mulled with sufficient water to develop plasticity and to enable a mold to be formed without collapsing, to withstand filling with molten metal, and to retain its shape until the mold has solidified. Until the 1950s, naturally bonded sands were used to make the molds. The definition of a naturally bonded sand was a silica sand which contained sufficient clay as mined to be made into satisfactory molds after mulling and tempering with water. Today, a synthetic sand is used which is the proper mixture of clean high-silica sand, bentonite clay, water and whatever other additives that are necessary to give the mold its desired properties. The advantages of synthetic sand are higher refractoriness, greater durability, lower moisture content, closer control of properties, readily altered properties, wide selection of grain size, higher permeability, and, in the long run, greater economy.

For special molds for certain metals, other sand and binder compositions may be used. The sand may be composed of zircon, olivine, chromite or sillimanite, and the binding clay may be sodium (high-swelling) bentonite, calcium (low-swelling) bentonite, fire clay, kaolin or ball clay. Zircon, olivine and chromite are refractory, and are used where casting temperatures and conditions are more severe. The bentonites are extremely plastic over a wide moisture range, this property being more pronounced in the sodium variety. Fire clay, kaolin and ball clay are used only in special instances where higher refractoriness, smoother surfaces, or lower permeability is required in the mold.

2. Pelletizing

In the early 1950s, the pelletizing of taconite-type iron ore became a necessity. Taconite is a siliceous, marginal iron ore, and in order to remove a substantial portion of the silica the taconite was finely ground before the flotation, gravity-separation or magnetic-separation processes. The concentrated iron-ore dust was impossible to handle in shipping and melting, so binding the dust with sodium bentonite and forming it into pellets became a very large use for the bentonite. Both magnetite and hematite ores, as well as blended ores, may be pelletized using sodium bentonite, which has many advantages for both producers and consumers. Among these are ease of handling, high strength, good reducibility and homogeneous chemical composition. From the producers' standpoint, leaner and more friable ores can be mined and marketed, which increases the viability of marginal ores and extends the life of individual ore bodies. The handling benefits of pellets, combined with the increased value of iron ore in this form (35–45% higher than the price of lump ore with the same iron content), have made the pelletizing process particularly attractive to producers all over the world. The pelletizing capacity in the world is over 250 million tonnes per year, and over 1.5 million tonnes of bentonite are being used as the binder.

3. Ceramics

A ceramic product or processed material is a solid composed of materials which have been subjected to heat above 454 °C (875 °F). Any solid nonmetallic inorganic product processed by subjection to temperature above a red heat may be called a ceramic product. Widely used construction materials and units, domestic utensils and units, industrial parts to aid in manufacturing, technical components and artistic creations are very often of a ceramic nature because of durability, inertness, hardness, ease of fabrication, cost and availability of the ingredients.

A ceramic composition or body contains several ingredients, each of which performs a specific function. In a ceramic plate, for example, the major constituents are kaolin, feldspar or nepheline syenite, and silica. These ingredients function as follows: the silica is the skeleton former or the filler, the feldspar or nepheline syenite is the flux, and the kaolin is the binding agent. This is a simplified way of looking at a ceramic piece. In wall tile, talc may be the skeleton or filler, wollastonite or calcite the flux, and ball clay or kaolin the binder. The skeleton or filler, the flux and the binder may be different, depending on the qualities needed either in the handling, processing, firing, or the specifications of the product.

The binders used in ceramics are several, including kaolins, ball clays, fire clays, plastic underclays and smectites. Kaolins have a lower shrinkage, better fired color and higher refractoriness than the other clays. Fire clays are used in refractory applications where fired color is not important. Plastic underclays are used when high green strength and good plasticity are required, but where fired color is not important. Ball clays are very plastic, have high green strength, and fire a light-cream to white color. Bentonites have very high plasticity, have very high shrinkage and are used only in very special applications and in small quantities.

4. Others

Binders are used in other applications than those described above (i.e., foundries, pelletizing and ceramics). Some of these are to make pellets or noodles for animal feeds, to bind catalyst components or supports, to bind graphite in making pencil leads, and in binding medicines and drugs into pills. In animal feeds, bentonite is used as an inert binder to produce extruded pellets of various sizes. White- or cream-colored bentonite or kaolin is used to bind silica, alumina and other support materials used for metallic catalysts. A blend of kaolin and bentonite is mixed with graphite to make pencil leads. The percentage of binder is directly proportional to the hardness of the pencil lead. In making pills, the pharmacist commonly uses kaolin as the binder.

See also: Binders: Clay Minerals

Bibliography

Patterson S H, Murray H H 1975 Clays. In: Lefond S J (ed.) 1975 *Industrial Minerals and Rocks*, 4th edn. American Institute of Mining, Metallurgical and Petroleum Engineers, New York, pp. 519–85

H. H. Murray
[Indiana University, Bloomington, Indiana, USA]

Bismuth Resources

Bismuth, a silvery-white metallic element, was probably unrecognized before the fifteenth century. Its usefulness in treating intestinal disorders was noted about 1600, and it is still used for this purpose. However, the principal present-day uses for bismuth are in medicines, low-melting alloys, and in metallurgical additives for aluminum, carbon steel and malleable iron. Bismuth imparts fusibility and machinability to alloys.

The major bismuth-producing countries are Australia, Mexico, Japan, Peru and Bolivia. Canada and the USA have moderate reserves of bismuth, and the People's Republic of China, North Korea and South Korea have large reserves. Most bismuth is recovered as a by-product from copper and lead ores. Only Bolivian ores are mined for bismuth alone.

The principal ore minerals of bismuth are bismite (Bi_2O_3) and bismuthinite (Bi_2S_3); they occur in small amounts in ores throughout the world, especially with lead and copper, but also with molybdenum, cobalt, gold, silver, tin, tungsten and zinc. In Peru, bismuth and copper occur in replacement deposits associated with silicic intrusive rocks. Bolivian ores occur with tin and silver in fissure-vein deposits. In Australia and Japan, bismuth is associated with tin, tungsten, molybdenum and gold in replacement deposits. In the western USA and Mexico, bismuth occurs with lead in replacement deposits in carbonate rocks.

Bibliography

Bascle R J, Carlin J F Jr 1980 Bismuth. In: Bascle R J, Carlin J F Jr (eds.) 1980 *Mineral Facts and Problems.* US Bureau of Mines, Washington, DC

Carlin J F Jr 1979 *Bismuth* US Bureau of Mines, Mineral Commodity Profile. USBM, Washington, DC

Hasler J W, Miller M H, Chapman R M 1973 Bismuth. In: Brobst D A, Pratt W P (eds.) 1973 *United States Mineral Resources.* US Geological Survey Professional Paper 820. US Government Printing Office, Washington, DC

US Bureau of Mines 1988 *Mineral Commodity Summaries 1988.* USBM, Washington, DC

M. H. Miller
[US Geological Survey, Denver, Colorado, USA]

C

Cadmium Resources

Cadmium is a silver-white, soft, malleable metal when freshly cast, but on exposure to air develops a thin, highly corrosion-resistant gray oxide. Most cadmium is consumed as coatings and platings on parts used in the manufacture of transportation and defense equipment, architectural and precision hardware, electronics, home appliances, and industrial machinery. Potential new uses include the manufacture of photovoltaic cells. In 1987, 30% of cadmium produced was consumed in plating, 30% in batteries, 20% in pigments and 15% in stabilizers. Alloy, chemical, nuclear engineering and electrical applications accounted for the remaining 5% (US Bureau of Mines 1988).

1. Geochemistry, Crustal Distribution and Ore Tenors

Cadmium is strongly chalcophilic. Estimates of its crustal abundance range from 0.13 to 0.20 ppm, with igneous rocks containing from about 0.02 ppm in ultramafic rocks to 0.20 ppm in granite and rhyolite. Carbonate rocks, sandstone and conglomerate contain about 0.07 ppm cadmium, whereas deep ocean sediments (0.5 ppm), shale (1.3 ppm), oceanic manganese oxides (8.0 ppm) and phosphorites (25 ppm) are substantially enriched.

Primary cadmium is recovered almost entirely as a by-product from smelting zinc ores, in which cadmium occurs as impurities in sphalerite (Zn, Fe) S, or as surface coatings of greenockite (CdS) on sphalerite. Cadmium contents of zinc sulfides typically range from 0.02 to 1.4%, with a median value of about 0.3% (Fleischer et al. 1974) and a maximum of 5%.

2. Production

World cadmium production has been generally characterized by relatively steady and rapid growth. However, since 1969 production has been somewhat erratic, reflecting in part the repressive effects of periodic, rapid escalation of energy costs and accompanying inflation on cadmium consuming industries, as well as considerable concern about the toxicity of cadmium in the environment. Production of refined cadmium in 1987 was about 18 400 t. The principal producers were Japan (2400 t), USA (1500 t), Canada (1500 t), Belgium (1400 t), Australia (1000 t), Mexico (700 t), and the USSR and other centrally planned economies (4300 t) (US Bureau of Mines 1988). Sources of cadmium ores for refining are distributed differently. Mines in only six countries produced 61% of world cadmium in 1978: Canada 3600 t, the USSR 1900 t, Australia 1800 t, Mexico 1800 t, the USA 1550 t, and Japan 1050 t.

3. Reserves and Resources

Estimates of reserves and resources of cadmium are highly conjectural because the cadmium content of ore bodies containing recoverable zinc are seldom reported, or are even unknown. Typically, generalizations about Cd:Zn ratios from production data are applied to known zinc resources in order to estimate the cadmium by-product. The world average Cd:Zn ratio for metal produced during 1970–82 was consistently about 0.3%. The seven largest producers of refined cadmium during 1978–82, their cumulative production, and their Cd:Zn ratios were: the USSR (14 500 t) 0.27%, Japan (11 299 t) 0.31%, the USA (8937 t) 0.44%, Canada (6430 t) 0.23%, Belgium (6305 t) 0.53%, the Federal Republic of Germany (5863 t) 0.34% and Australia (4611 t) 0.31%. Lucas (1979) estimated 1.26 Mt of recoverable reserves and resources of cadmium including: Canada 240 kt; the USA 220 kt; Australia 195 kt; the USSR 100 kt; Brazil 50 kt; Peru 45 kt; Ireland 40 kt; Japan and Mexico 25 kt each; and Spain, Yugoslavia and the People's Republic of China 20 kt each. Using the average Cd:Zn ratios in many sphalerites of near 0.5%, the estimate of economic and promising subeconomic resources of zinc by Briskey and Morris (1979) predicts 1.3 Mt of cadmium. Wedow (1973) estimated that total recoverable and subeconomic, identified and undiscovered zinc resources (5085 Mt) would provide approximately 25.4 Mt of cadmium. The high Cd:Zn ratio of 0.5% in geologic materials, compared with the average recovery of 0.3%, implies that substantial amounts of cadmium are lost in extractive metallurgy operations, or are not being reported. If this lower ratio remains unchanged, the effective world estimates are reduced to 0.8 and 15.3 Mt, respectively. Scrap cadmium, mostly from batteries, provides less than 5% of the total supply.

4. Consumption and Adequacy of Supplies

Seven countries accounted for more than 78% of refined cadmium consumed in 1982, including: the USA 4127 t, the USSR 2250 t, Belgium 1652 t, the Federal Republic of Germany 1502 t, the UK 1212 t, Japan 1143 t and France 1017 t. Cadmium consumption is expected to grow 2.4% annually, from about 16.5 kt in 1982, to about 32.5 kt by the year 2000

(Lucas 1979). Identified economic and promising subeconomic resources are more than adequate to meet anticipated demand.

Bibliography

Briskey J A, Morris H T 1979 Lead and zinc resources of the United States and of the world. *US Geological Survey Research 1979*, US Geological Survey Professional Paper 1150. US Government Printing Office, Washington, DC, pp. 2–3

Fleischer M, Sarofim A F, Fassett D W, Hammond P, Shacklette H T, Nisbet C T, Epstein S 1974 Environmental impact of cadmium: A review by the panel on hazardous trace substances. *Environ. Health Perspect.* 7: 253–323.

Lucas J M 1979 *Cadmium*, US Bureau of Mines Mineral Commodity Profile. USBM, Washington, DC

Nriagu J O 1980 Production, uses, and properties of cadmium. In: Nriagu J O (ed.) 1980 *Cadmium in the Environment*. Wiley, New York, pp. 35–70

Plunkert P A 1983 Cadmium. *Mineral Commodity Summaries 1983*. US Bureau of Mines, Washington, DC, pp. 26–7

Roskill Information Services 1980 *The Economics of Cadmium*. Roskill Information Services, London

US Bureau of Mines 1988 *Mineral Commodity Summaries 1988*. USBM, Washington, DC

Wedow H Jr 1973 Cadmium. In: Brobst D A, Pratt W P (eds.) 1973 *United States Mineral Resources*, US Geological Survey Professional Paper 820. US Government Printing Office, Washington, DC, pp. 105–9

World Metal Statistics (updated monthly). World Bureau of Metal Statistics, London

J. A. Briskey
[US Geological Survey, Menlo Park, California, USA]

H. Wedow Jr.†

Calcium Aluminate Cements

Hydraulic cements consisting predominantly of monocalcium aluminate, $CaAl_2O_4$ (or CA in cement industry nomenclature—see *Portland Cements, Blended Cements and Mortars*), are referred to variously as calcium aluminate cements, aluminous cements or high-alumina cements. A major use of CA cements is as the binder in refractory concretes. Structural uses in low-temperature loadbearing applications are at present discouraged because of strength reductions due to the "conversion" reaction.

1. Composition and Manufacture

The composition of CA cements varies widely from low- to high-purity types (Table 1). The C–A–S phase diagram indicates that $C_{12}A_7$, C_2AS, C_2S and CA_2 are the only phases which can coexist with CA at equilibrium. Most of the iron is usually in a ferrite solid solution of composition between C_4AF and C_6AF_2. Ferrous iron compounds are present in cements produced under reducing conditions.

The raw materials for manufacturing the lower-purity CA cements usually consist of relatively pure limestone and low-silica bauxite. High-purity (white) CA cements are formed from the purest raw materials—limestone and Bayer alumina. After grinding and proportioning of the raw materials, the mixture is sintered or fused between 1350 and 1500 °C under either oxidizing or reducing conditions. In reductive fusion, molten iron can be separated from the molten clinker, and the CA clinker produced tends to have a low iron content. The final clinker compounds (Table 2) in high-purity cements are formed during the non-equilibrium heating cycle of a sinter process, while in lower-purity cements they are usually formed under equilibrium conditions during the cooling of a melt.

The cooled clinker is made into cement by grinding it to a specific surface area in the range of 250–1800 $m^2\,kg^{-1}$. The range of physical properties of CA cements is shown in Table 3. CA cement concrete can develop a 24 h compressive strength comparable with a 28 day strength in Portland cement concrete.

2. Reactions with Water

The phases in CA cements that react most readily with water are CA, $C_{12}A_7$ and CA_2. Although C_3AH_6 is the stable hydrate, the main hydration product of the anhydrous calcium aluminates at temperatures $<21\,°C$ is the metastable CAH_{10}:

$$CA + 10H \rightarrow CAH_{10} \qquad (1)$$

Some AH gel is also present. Above 21 °C, the main products are C_2AH_8, C_3AH_6 and AH_3. If β-C_2S or C_2AS is present in CA cements, their hydration is believed to be slow and to produce either a calcium silicate hydrate similar to that in hydrated Portland cements or gehlenite hydrate. Also, hydration products of the ferrite phases are similar to phases produced by ferrites in Portland cements.

3. Uses

Uses for CA cements (Table 4) exploit the refractory nature of these cements and their hydration products, rapid strength development, and high chemical resistance. White CA cements are used for the most demanding refractory applications. Aggregates used with CA cements in refractory concretes must also have refractory properties. They include blast furnace slags, expanded clay or shale, vermiculite, crushed firebrick, calcined flint clay or bauxite and tabular alumina. Combinations of proportioned cements and aggregates are referred to as castables. They are commercially available in dry packaged form and classified by their modulus of rupture, bulk density and volume stability at high temperature.

Table 1
Typical oxide compositions of calcium aluminate cements (%)

Cement type	SiO_2	Al_2O_3	CaO	Total Fe as Fe_2O_3	MgO	SO_3	Alkalis
Low purity	4.5–9.0	39–50	35–39	7–16	0.2–1.0	0.1–2.5	0.1–0.3
Intermediate purity	3.5–6.0	55–66	31–36	1–3	0.2–0.7	0.1–0.6	0.1–0.3
High purity	0.0–0.2	70–80	17–27	0.0–0.4	0.0–0.4	0.0–0.5	0.1–0.6

Table 2
Phases identified in calcium aluminate cements[a]

CA	β-C_2S	α-A
CA_2	C_4AF	$C_{12}A_7$[b]
C_2AS	$C_4A_3\bar{S}$	CT[b]

[a] For details of the nomenclature of these phases, see *Portland Cements, Blended Cements and Mortars* [b] Minor amounts only

Table 3
Physical properties of calcium aluminate cements

Property	Range
Color	Dark gray to white, depending on the quantity of iron and its oxidation state
Density ($kg\,m^{-3}$)	3000–3250
Bulk density ($kg\,m^{-3}$)	1500
Fineness ($m^2\,kg^{-1}$)	250–1800
Set time, initial, Vicat (ASTM C191)[a] (h)	0.5–9

[a] American Society for Testing and Materials (1982)

Table 4
Some uses of calcium aluminate cement mortars and concretes

Refractory
- Insulating walls and furnace linings
- Heat-resistant floors and foundations
- Linings for pipes and chemical reactors
- Rocket exhaust deflector pads

Nonrefractory
- Quick-setting and high 24 h strength concretes
- Grouting
- Insulators for power lines
- Cementing of bore holes
- Decorative panels and terrazzo flooring
- Jointing and repointing in railroad tunnels and sewers
- Chemically resistant floors and tanks
- Chimney (stack) linings (where condensation occurs)

4. The Conversion Reaction

"Conversion" is the name given to the change from the metastable hexagonal CAH_{10} in CA cement concretes to stable cubic C_3AH_6 and AH_3 (gibbsite). The reaction occurs slowly at low ambient temperatures and more rapidly at higher ambient temperatures under conditions of high relative humidity:

$$3CAH_{10} \rightarrow C_3AH_6 + 2AH_3 + 18H \qquad (2)$$

This reaction increases the porosity of the solid concrete and can cause a loss in strength. These occurrences can be minimized by using a low water–cement ratio and curing at $>30\,°C$ in the first 24 h after placement.

Conversion occurs in hydrated cements used for refractory applications. These reactions do not cause detrimental effects in concrete used under high-temperature conditions, because the low-temperature strength is still adequate and ceramic bonds form at high temperatures.

5. Chemical Resistance

CA cement mortars and concretes are usually more resistant than Portland cement to dilute acids (pH > 3.5) and sulfate solutions, but not to alkali hydroxides. Possible reasons for acid resistance are that the hydration of CA cements does not produce much calcium hydroxide and that hydrated alumina acts as a barrier to acid reaction. To explain sulfate resistance, it has been suggested that any calcium sulfate aluminate hydrates which form do so through solution rather than by a topochemical reaction, such as is believed to lead to expansion in sulfate attack of Portland cement concretes. The aggregates selected for use with CA cement in these applications must have a comparable chemical resistance.

See also: Portland Cements, Blended Cements and Mortars

Bibliography

American Concrete Institute 1978 *Refractory Concrete*, ACI Publication SP-57. American Concrete Institute, Detroit

American Concrete Institute 1979 *Refractory Concrete*, ACI State-of-the-Art Report 547R-79. American Concrete Institute, Detroit

American Society for Testing and Materials 1984 *Annual Book of ASTM Standards*, Sect. 15, Vol. 15.01, *Refractories; Carbon and Graphite Products; Activated Carbon.* American Society for Testing and Materials, Philadelphia, Pennsylvania

Lea F M 1971 *The Chemistry of Cement and Concrete.* Chemical Publishing, New York

Neville A M, Wainwright P J 1975 *High Alumina Cement Concrete.* Construction Press, Hornby
Robson T D 1962 *High Alumina Cements and Concretes.* Wiley, New York
Robson T D 1964 Aluminous cements and refractory concretes. In: Taylor H F W (ed.) *The Chemistry of Cements,* Vol. 2. Academic Press, New York, pp. 3–35

G. J. Frohnsdorff and J. E. Kopanda
[National Bureau of Standards, Washington, DC, USA]

Chromite Resources

Chromite is the only commercial chromium ore. It is used (a) as a chromium source in alloying (58–76%); (b) as a chemical for dyes, pigments, tanning and plating (11–23%); and (c) directly in refractories (13–18%). These ranges reflect variations in usage from country to country (De Young et al. 1984). Historically, high-Cr ores containing >45% Cr_2O_3, with Cr:Fe$\geqslant$3:1, were used in metallurgy; high-Fe ores containing 42–46% Cr_2O_3, with Cr:Fe<2:1, were used by the chemical industry; and high-Al ores, with Al_2O_3>20% and $Al_2O_3+Cr_2O_3$>60%, were used in refractories. Although high-Cr ore is still preferred for some alloys, high-Fe ore, which is the most abundant, can be used for all purposes.

The predominant sources of chromite are primary deposits, of which two major types are distinguished: stratiform and podiform. The stratiform deposits form extensive layers in stratified igneous complexes, and they are remarkably uniform in thickness and grade. The Steelpoort Main Seam in South Africa, for example, averages about 1 m in thickness over 65 km of outcrop length, and the G-chromite zone of the Stillwater Complex in Montana has been traced for more than 40 km. Furthermore, most stratiform complexes contain more than one minable chromite layer.

Podiform deposits are irregular in form and vary widely in size and grade. Most orebodies yield less than 10 kt each; approximately twelve orebodies of more than 1 Mt are known outside the USSR and Albania. Most podiform deposits are randomly scattered in the host rock, either individually or in loose groups.

1. History of Production

Total world production of chromite ore is estimated to be about 200 Mt, and since 1945 annual production has risen from about 1.1 to 10 Mt. Before 1970, 60–65% of world production came from podiform deposits, but since 1975 the percentage has dropped to approximately 50% (see Table 1). The percentage of high-Cr ore has fluctuated only between 63 and 66% of total production, but since 1965 about half of this ore has been provided by the USSR and Albania. Production of high-Fe ore has risen from 20–25% of the world total in 1945 to nearly 33% at present, while high-Al ore production has dropped from 15 to 6%. The shift in production to high-Fe ore from stratiform deposits is largely owing to the exhaustion of the large high-Cr podiform orebodies that could be mined in open pits or followed underground from outcrops.

The USA has produced about 1.9 Mt of chromite, half of which was high-Fe ore from the Stillwater Complex in Montana. Almost all the rest was mined from podiform deposits in California and Oregon, Pennsylvania and Maryland, and Alaska. Since 1900, domestic production has exceeded 3.5 kt a year only during wartime. Before 1940 only podiform deposits were mined; since 1940, however, high-Fe ore from the Stillwater Complex has predominated during US Government purchase programs. In 1957–58 production of podiform ore, at subsidized prices about triple the commercial rates, reached only about 50 kt per year.

2. Stratiform and Podiform Deposits

The extent, regularity and uniformity of stratiform deposits present great geological and economic contrasts to the random distribution, irregular forms and structural complexities of podiform deposits. In the Bushveld and Great Dyke stratiform complexes in South Africa and Zimbabwe, respectively, inexpensive surface mapping and several drill holes have proved that existing reserves far exceed any projected demand and that these can be mined by methods similar to those used in coal mining. In stratiform complexes that have been folded and faulted, the original order of chromite and other rock layers can be used to guide exploration for concealed ore zones (Jackson 1968).

Traditionally, podiform chromite deposits have been found by looking for outcrops; most known concealed deposits have been found by underground exploration for extensions of known orebodies that were originally exposed. Under exceptionally favorable conditions, geophysical surveys guided by detailed geological mapping coordinated with drilling have located large concealed orebodies (Flint et al. 1948, Hammer et al. 1945). So far, however, no economical system has been devised to find individual concealed orebodies in the average district.

In the absence of specific geological information in a given district, potential resources in podiform orebodies have been calculated as equal to production plus known reserves projected to twice the depth of the deepest mine in the area. No more than a quarter to a third of this hypothetical quantity of ore would probably be recoverable, because only a few relatively large and high-grade deposits could be mined economically at depth.

Table 1
Chromite reserves and resources, and production for 1975–87, in million metric tons of shipping-grade ore or concentrates

Country and type of ore[a]	Reserves[b]	Resources[b]	Production[b]
South Africa (S; Fe)	913	6300	3.7
USSR (P, S; Cr, Fe)	142	142	3.3
Finland (S; Fe)	19	32	0.5
Zimbabwe (S; Cr)	19	830	0.6
India (S; Cr)	15	66	0.7
Philippines (P; Al, Cr)	15	32	0.3
Brazil (S, P; Cr, Al)	9	10	0.3
Albania (P; Cr)	7	22	1.0
Turkey (P; Cr)	5	80	0.6
Other market economy countries	17	25	0.3
Other centrally planned economies	4	4	0.04
World total	1165	7500	11.4
Stratiform total	3335	30125	23.1
Podiform total	25–30	135–210	24.1

[a] S, stratiform; P, podiform; Cr, high-chromium; Fe, high-iron; Al, high-aluminum [b] US Bureau of Mines 1988

3. *World Reserves and Resources*

World chromite resources are inexhaustible, with most being found in the eastern hemisphere (Thayer 1973, Thayer and Lipin 1978, De Young et al. 1984). Immense reserves in the Bushveld and Great Dyke stratiform complexes assure long-term domination of world chromium supply by South Africa and Zimbabwe, respectively, because the discovery of comparable deposits elsewhere is extremely unlikely. Large Russian and Albanian resources appear to be mostly in podiform deposits but reliable information on them is not available. The large figures shown for Finland and India in Table 1 are based on extensive exploration of highly deformed stratiform deposits. In countries other than the top five shown in the table and in Greenland, discoveries by exploration may be important and may affect world supply for 10–20 years, but probably will not change total world resources significantly.

The chromite resources of the western hemisphere are very limited; in South America, chromite is being mined only in Brazil, mostly from complexly deformed high-Cr stratiform deposits in Bahia. The other main western resources are in three deformed stratiform complexes in the USA, Canada and Greenland. These deposits will yield only subcommercial high-Fe ore that contains between about 40% Cr_2O_3 (Montana) and 30–35% Cr_2O_3 (Greenland). The Stillwater deposits in Montana were mined during World War II and the Korean War, and are the best known (Jackson 1968). The Greenland (Fiskenaesset) deposits are the most extensive but would be the most expensive to mine. The identified podiform chromite resources of the western hemisphere probably total less than 3 Mt of ore equivalent, and are mostly in Cuba. Production of as much as 500 kt of ore from podiform deposits in the USA will probably be uneconomic even at several times world prices. Placer deposits along the Oregon coast are estimated to contain about 200 kt of recoverable high-Fe chromite.

4. *Demand and Future Supplies*

World chromium demand up to the year 2000 is projected to be 5.5–8.2 Mt per year, about 90% being from primary sources and 10% from secondary sources (scrap) (Morning et al. 1981). The estimated cumulative primary world and US demands to the year 2000 are about 105 and 16 Mt, respectively. In terms of ore averaging 45% Cr_2O_3, with 85% recovery of Cr in smelting, the cumulative primary needs of the world and the USA represent about 400 and 60 Mt, respectively. The cumulative world demand for ore is more than twice total production to 1980, and also more than twice the combined reserves and probable available chromite resources in countries other than South Africa, Zimbabwe and the USSR. If the estimates of world reserves and resources of chromite, and of world demand, are approximately correct, and if South Africa, Zimbabwe and the USSR continue to make up half of world production, the chromite resources of most other countries in Table 1 will be mined out in 20 years.

The chromium requirements of the USA could be reduced immediately by about 30% using acceptable substitutes and by another 4–5% within 5 years by developing promising new substitutes; both would entail cost penalties. In 10 years, a comprehensive research program might cut consumption to 35% of current usage (about 500 kt) at substantial economic cost (National Materials Advisory Board 1978). The production of only 175 kt of chromium, equivalent to 750–800 kt of low-grade ore from North American deposits, would still require a substantial lead time.

Bibliography

De Young J H Jr, Lee H H, Lipin B R 1984 *International Strategic Minerals Inventory Summary Report—Chromium*, US Geological Survey Circular 930-B. US Government Printing Office, Washington, DC

Flint D E, de Albear J F, Guild P W 1948 *Geology and Chromite Deposits of the Camagüey District, Camagüey Province, Cuba*, US Geological Survey Bulletin 954-B. US Government Printing Office, Washington, DC, pp. 39–63

Hammer S, Nettleton L L, Hastings W K 1945 Gravimeter prospecting for chromite in Cuba. *Geophysics* 10: 34–49

Jackson E D 1968 The chromite deposits of the Stillwater Complex, Montana. In: Ridge J D (ed.) 1968 *Ore Deposits of the United States, 1933–1967*, Vol. 2. American Institute of Mining, Metallurgical and Petroleum Engineers, New York, pp. 1495–510

Morning J L, Matthews N A, Peterson E C 1981 Chromium. *Mineral Facts and Problems*, US Bureau of Mines Bulletin 671. USBM, Washington, DC

National Materials Advisory Board 1978 *Contingency Plans for Chromium Utilization*, National Research Council Publication NMAB-335. US Government Printing Office, Washington, DC

Thayer T P 1973 Chromium, In: Brobst D A, Pratt W P (eds.) 1973 *United States Mineral Resources*, US Geological Survey Professional Paper 820. US Government Printing Office, Washington, DC, pp. 111–21

Thayer T P, Lipin B R 1978 A geological analysis of world chromite production to the year 2000 A.D. *Proc. Economic Council AIME*. American Institute of Mining, Metallurgical and Petroleum Engineers, New York, pp. 143–51

US Bureau of Mines 1977–1979 Chromium. *Minerals Yearbook*, annual editions, 1977–1979. USBM, Washington, DC, pp. 36–37

US Bureau of Mines 1988 *Mineral Commodity Summaries 1988*. USBM, Washington, DC

Von Gruenewaldt G 1977 The mineral resources of the Bushveld Complex. *Minerals Sci. Eng.* 9: 83–95

T. P. Thayer
[St. Petersburg Beach, Florida, USA]

Coal: Geology

Coal geology is a specialized discipline which applies the science of geology and related topics of chemistry, biology and physics to seek more intelligent use of coal, a unique and perhaps the most useful rock in all of nature. Coal is a sedimentary rock composed of lithified plant remains and variable amounts of dispersed mineral matter, water and gases. The parent sediment for coal is peat, deposited in a swamp environment. Plant debris begin their conversion to peat when entombed in water. Over geological time peat is altered and metamorphosed into coal. The plant debris become lithified into more dense constituents which coal scientists classify into various types of macerals depending on their microscopic properties and morphology. Macerals are lenticular, oval or irregularly particulate in shape, and they are commonly intermixed with mineral matter. Mineral matter includes all inorganic constituents: discrete grains of various minerals and isolated inorganic atoms that occur structurally combined within macerals. Water and gases fill submicroscopic pores in macerals and also between macerals and minerals. As such, coal is a distinctly heterogeneous material.

Coal and similar deposits rich in organic matter are found in many regions of the world and in strata of nearly all geological ages, from Precambrian to Recent. Economically, the most important deposits are of Pennsylvanian age in North America; Carboniferous age in Europe and Asia; Permian age in Australia, India and South Africa; Cretaceous age in North America; Jurassic age in Australia; and of Tertiary age in North America, Europe and Southeast Asia.

1. Sedimentary Associations and Structural Features of Coal Strata

The geological assessment of a coal deposit begins in the field with the recording of the thickness and physical nature of the coal seam and the underlying and overlying strata. A wide diversity of sedimentary rock associations and structural features of the coal are significant in terms of understanding how coal formed and how to predict its properties. Lastly, representative samples need to be collected for laboratory analyses in order to relate field observations with vital petrologic and geochemical assessments.

1.1 Sedimentary Associations

Coal seams are nearly always interstratified with shale or sandstone, and like many of these associated strata, some seams can be traced laterally for more than 450 km. Coal seams may contain thin beds of shale or related mineral matter which, when thinner than about 5 cm, are described as mineral partings; when thicker, the term mineral band (or ply) is applied. Mineral partings and bands composed of flint clay are recognized in some coal beds by their fineness of grain, pelletoidal texture and/or conchoidal fracture. Mineralogically, flint-clay bands are made up almost entirely of well-crystallized kaolinite. The term tonstein is applied to the type of mineral band. In some coal beds they probably formed by diagenetic alteration of volcanic ash; in others kaolinite may have been the dominant clastic component, or a product of recrystallization of other types of clay minerals in a special case of epigenetic alteration. More normal depositional types of mineral partings or bands of carbonaceous shale, siltstone or sandstone also occur within seams. They have the common detrital/clastic texture of muddy silt or sand that was deposited during a flood stage of a nearby river or a transgressive coastal sea. Both types of partings/bands are useful for lateral correlation of the seam into adjacent areas.

Commonly, a seam splits locally into two or more beds as it is traced laterally. The upper beds are described as rider seams. In some coal basins splits are associated with sandstone or siltstone channels that have a sinuous map plan, along which an ancient river traversed the swamp. Such channel-fill sediments are observed intertonguing with coal. These were deposited in rivers that were contemporaneous with the peat. During major flood stages, crevasses may cut through the river's natural levee and allow much sand, silt or mud to be deposited on the adjacent peat surface. These crevasse-splay deposits, if thick enough, appear to destroy the swamp habitat. Such deposits have been discovered to overlie coal in some coal fields. In the Illinois Basin, where crevasse-splay type shales overlie coal beds and the shales are thicker than about 6 m, the coal usually contains relatively low amounts of sulfur. Similarly, where fluvial or other types of nonmarine strata overlie coal, such as at many places in the Appalachian Basin, the coal is similarly low in sulfur. It is thought that these types of nonmarine strata provide a shield that effectively reduces the amount of sulfate-bearing seawater that filters down into the peat when the area is later covered by a transgressing sea.

Other channel sands that displace coal have the same type of sinuous pattern, but their bottom surface is entirely erosional. These were formed by erosion of the peat long after the swamp was buried and the sulfur content of the peat is unaffected by this process.

1.2 Structural Features

Geological analysis and detailed description of structural discontinuities occurring in the seam help predict their occurrences elsewhere. Roll is a term applied to a number of structures: to irregular hills and valleys at the top surface of the seam; to long and narrow depressions of the roof of the seam or to similarly shaped rises of the floor. The origin of rolls is highly variable from one locality and type to another. Many may originate from differential compaction. Peat compacts much more than sand, silt or mud, and the differential compaction can significantly alter the cross-sectional form of coal/noncoal contacts along rolls, channel deposits and other primary depositional features. So-called soft-sediment faults, observed in some mines, are likely to have been caused by differential compaction. In addition, clastic dikes occur in some seams formed by the washing of sand or mud into an open joint in subaqueous peat. The slumping off of a ridge of peat or differential compaction of underlying sediments or earthquakes may be the cause for original jointing in the coal-forming peat. In addition, tectonic faults and igneous dikes may form major structural discontinuities. In many of these cases, rock material from above or below the coal is observed to have been dragged or injected into the seam.

Joints, termed cleat, may form notable discontinuities in coal beds. Cleat comprises a pair of parallel or nearly parallel joint surfaces that are both oriented normal to the original bedding of the peat. Normal cleat ranges in height up to a meter or more, and in width from 0.1 mm to about 5 mm or so. The abundance of cleat varies within a region, as well as within an area the size of a small mine.

1.3 Underclay and Seat Rock

The rock unit immediately underlying coal beds is typically a massive claystone about 1 m thick described as underclay. The thickness of underclay may vary considerably, but its position under the coal, its lack of distinct bedding and the presence of fossilized roots are criteria that strongly suggest that this material formed the soil that supported the initial vegetation of the swamp. The abundance in underclay of kaolinite, a clay mineral stable in acidic conditions, also supports a "soil" genesis. However, many coal beds are underlain by other rock types, which are described simply as seat rock. These appear to be unrelated to soil and explanations to account for their origin are generally unsatisfactory.

2. *Field Description and Sampling of Coal*

Detailed descriptions of coal beds as observed in drill cores, exploration pits or mine faces provides much data required for geological evaluation. The presence of conspicuous shale and/or other types of mineral matter distinguishes impure coal from coal; and the description of the relative abundance and vertical and lateral distributions of mineral matter provides the first appraisal of a new site.

In the field, coals are classified into two types, banded and nonbanded. Banded coals are characteristically layered with bands typically 1–10 mm thick, dull to very bright in luster, brittle and with a very irregular fracture. Nonbanded coal (or sapropelic or canneloid coal) is massive, has a greasy luster and conchoidal fracture. Close visual examination of banded coals reveals them to consist of one or more of four different lithotypes: vitrain, clarain, durain and fusain. Vitrain is the brightest type, with a vitreous (glassy) luster and conchoidal fracture. Clarain is bright to semibright and finely laminated, typically consisting of 1–3 mm of fine granular material (termed attrital vitrain), interlayered with thin bands of dull fusainic components. Durain is dull, made tough and blocky by the abundance of attrital vitrain and intermixed fusainic components. Pure fusain is a soft, fibrous, powdery material with a silky luster like charcoal. When describing a coal bed, bands thinner than 1 cm are not usually described separately. Such bands are grouped with the upper and lower ones, depending on their overall lithotype composition. Banding is most prominent in coals of bituminous rank that are rich in vitrain and clarain. Apparently, lignites have not been metamorphosed enough to enhance the banding, and in anthracites much of the banding has been obliterated. Nonbanded coal contains an abundance of fine granular and waxy components. Most coal beds are composed entirely of either the banded or the nonbanded type; however, exceptions do occur. Some banded coals contain relatively thin layers that are themselves nonbanded. The two varieties of nonbanded coal, cannel and boghead coal, can only be distinguished with the use of a petrographic microscope (see Sect. 3.1). The term sapropelic coal is sometimes applied to nonbanded coal because of the common presence of sapropel, an ooze of algal matter. Further discussions of the field description of coal are given in Dutcher (1978) and Ward (1984).

The main task of sampling coal is to obtain a representative sample at the collection site. In the USA, the standard practice for sampling a seam in a mine is given in ASTM standard no. D4596 (American Society for Testing and Materials 1986). Briefly, the collection site is selected for good representation of the seam; the face of the seam is cleaned of oxidized coal and made normal to the bedding surface; a channel is dug with a tool from the top to the bottom of the seam and all pieces of coal removed are collected on a cloth laid at the base of the seam. When the sample is collected to determine the rank of the coal all mineral partings thicker than 1 cm and nodules larger than 1.25 by 5 cm occurring in the channel are excluded from the sample so the analytical results more accurately represent the organic matter in the coal. Collection of samples from a pile or other industrial processes are described in ASTM standard no. D2234 (1986). The British Standards Institution (1981) and the Standards Association of Australia (1982) give similar sampling procedures that are applicable in many countries.

3. *Constituents of Coal*

Major vascular plants capable of forming peat swamps did not evolve until the Devonian period, and extensive forested swamps did not develop until the Carboniferous period in Europe and Asia and the Pennsylvanian period in North America. A wide variety of plants inhabited ancient peat swamps that formed coal beds. Pennsylvanian flora are predominately lycopods, ferns, pteridosperms (seed ferns), cordaites and sphenopsids (Phillips et al., 1985). Mesozoic and Cenozoic peat swamps were predominately inhabited by conifers, angiosperms (flowering plants), ferns and horsetails (Stewart 1983).

3.1 Maceral Composition

Macerals are the dominant physical components that comprise coal. Unlike minerals, each maceral type varies in its chemical and structural properties. The chemical composition of most maceral types varies with the rank of the coal, and macerals have little or no crystallinity. The reflectivity of macerals is the main diagnostic property. On this and other bases, macerals are classified into three main groups: vitrinite, liptinite and inertinite. The relative reflectance, fluorescence and morphology are used to distinguish the various macerals within each group. Two main classifications have been proposed (Table 1).

When examined under a reflected-light microscope using an oil-immersion objective lens, the vitrinite group macerals appear medium gray in contrast to the other group macerals. Telinite and telocollinite (and pseudovitrinite, see below) appear somewhat brighter and more uniform than the other macerals in this group. The liptinite group macerals are all distinctively lower in reflectance than vitrinite in the same coal, and the individual macerals in this group are distinguished by their morphology depending on the type of plant material they were derived from: spore, leaf cuticle, alga, resin, and so on. Liptinite macerals, especially those in low-rank coals, are noted for their fluorescence characteristics when examined under blue or ultraviolet light. Inertinite group macerals all have a distinctively higher reflectance than do the vitrinite macerals. Fusinite has the highest reflectance and displays a cellular structure; semifusinite is somewhat lower in reflectance and has significantly reduced cellular forms; micrinite is bright and occurs in very fine grains enclosed in vitrinite or liptinite macerals; macrinite occurs as large and rounded grains; sclerotinite takes the form of rounded and very porous sclerotia (fungus); inertodetrinite occurs as bright and angular, clastic fragments of other inertinite macerals, surrounded by vitrinite, rarely by other macerals. Liptinite and inertinite macerals are not generally recognizable in anthracites. Stach et al. (1982) and the American Society for Testing and Materials (1986) give additional characteristics of the different macerals.

In practice, many coal petrographers in North America use a modified version of one or both of the classifications given in Table 1, depending on the source and rank of the coal they are analyzing and the purpose of the analysis. Many who evaluate coal for coking applications distinguish two types of vitrinite: normal vitrinite and pseudovitrinite; while others use some or all of the submaceral terms of the ICCP classification (Table 1). A genetic system of maceral classification that is a significant departure from the two shown in the table is used by Russian and many other east European petrographers (Ward 1984). In addition, some petrographers use a special classification for the vitrinite-like macerals in brown coals and equivalent ranks of lignite and subbituminous coals (Stach et al. 1982, Ward 1984).

Table 1
Classification of macerals by ASTM and ICCP (International Committee for Coal Petrology). The group macerals are the same for both classifications (Stach et al. 1982)

ASTM (1986)		ICCP	
Maceral group	Maceral	Maceral	Submaceral
Vitrinite	Vitrinite	Telinite	
		Collinite	Telocollinite
			Gelocollinite
			Desmocollinite
			Corpocollinite
		Vitrodetrinite	
Liptinite	Sporinite	Sporinite	
	Cutinite	Cutinite	
	Resinite	Resinite	
	Alginite	Alginite	
		Suberinite	
		Fluorinite	
		Bituminite	
		Exudatinite	
		Liptodetrinite	
Inertinite	Micrinite	Micrinite	
	Macrinite	Macrinite	
	Fusinite	Fusinite	
	Semifusinite	Semifusinite	
	Sclerotinite	Sclerotinite	
	Inertodetrinite	Inertodetrinite	

Microscopically, banded coals are observed to consist of abundant vitrinite, which is derived from woody plant debris; while nonbanded coal beds and bands are composed of abundant liptinite macerals. Those with abundant sporinite (derived from plant spores) are described as cannel coals, those with abundant alginite (derived from algal materials) are described as boghead coals.

3.2 Mineral Matter in Coal

Impurities in coal include a wide variety of minerals (Stach et al. 1982, Harvey and Ruch 1986). The abundant ones are pyrite, quartz, various types of clay minerals (mostly kaolinite, illite and smectite) and calcite. Several sulfides other than pyrite are often reported, notably marcasite and sphalerite. Pyrite, if exposed to oxidizing conditions, alters to one of several types of sulfates such as gypsum or szomolnokite. Dolomite or siderite is moderately abundant among the carbonate minerals in many coals. Cleat surfaces are commonly mineralized with pyrite, calcite and/or kaolinite.

Another type of mineralization that occurs in some coal fields is referred to as a coal ball. These are elliptical nodules of mineralized peat or lignite, commonly composed of calcite, pyrite or siderite, or mixtures of these. Most coal balls are about 2–10 cm across but in some areas abnormally elongated coal balls replace the whole coal seam for a distance of several meters. Coal balls are unique in that the mineralization has preserved much of the original plant morphology, thus enabling paleobotanists to determine the abundances of each kind of plant that inhabited the peat swamp (Phillips et al. 1985).

3.3 Chemical Composition and Coal Classification

Nearly all naturally occurring elements in the periodic table have been found in coal when modern instrumental methods of analysis are applied. However, most geological assessments are based on standardized tests of proximate analyses (moisture, volatile matter, ash and fixed carbon contents) and ultimate analyses (carbon, hydrogen, nitrogen, sulfur, oxygen and ash) (ASTM 1986). In addition to these tests, analyses of the heating value of the coal and the forms of sulfur are carried out: sulfate, pyritic and organic sulfur are almost always required for both geochemical and applied engineering purposes. Evaluations of coking and other special-use coals require additional tests: free swelling index (or crucible swelling number), Gray–King carbonization, Gieseler plasticity, Fischer carbonization, Audibert–Arnu dilatometer and the Roga index test (Ward 1984).

One of the many geological uses of the chemical data is to classify the coal according to rank. Ranking parameters include heating value, agglomerating

Table 2
ASTM classification of coals by rank and other related properties. Rank properties for ASTM are: heating value (for coals containing more than 31% volatile matter (VM) and having an agglomerating nature) and percentage of volatile matter (for coals with ⩽31% VM)

	ASTM group	Heating value: Btu/1b(000) moist, mmf	Heating value: MJ/kg^{-1} moist, mmf	Agglomerating	% VM dmmf	% Maximum reflectance vitrinite	% Moisture mmf	% C dmmf	% O dmmf	% H dmmf
Lignite	lignite B	<6.3	<14.6	no	40–65	< 0.4	45–60[a]	55–73	23–35	5–7
	lignite A	6.3–8.3	14.6–19.3	no	40–65	< 0.4	31–50[a]	55–73	23–35	5–7
Subbituminous	subbituminous C	8.3–9.5	19.3–22.1	no	35–55	<0.5	25–38[a]	60–80[b]	15–28[b]	4.5–6.0
	subbituminous B	9.5–10.5	22.1–24.4	no	35–55	<0.5	20–30[a]	60–80[b]	15–28[b]	4.5–6.0
	subbituminous A	10.5–11.5	24.4–26.7	no	35–55	< 0.5	18–25[a]	60–80[b]	15–28[b]	4.5–6.0
Bituminous	high volatile C	10.5–13.0	26.7–30.2	yes	35–55	0.4–0.7	10–25[a]	76–83[b]	8–18[b]	4.5–6.0
	high volatile B	13.0–14.0	30.2–32.5	yes	35–50	0.6–0.9[b]	5–12[a]	80–84[b]	7–12[b]	4.5–6.0
	high volatile A	>14.0	>32.5	yes	31–45	0.8–1.2[a]	1–7[a]	78–88[b]	6–10[b]	4.5–6.0
	medium volatile	>14.0	>32.5	yes	22–31	1.1–1.5[a]	<1.5	84–91	4–9	4.5–6.0
	low volatile	>14.0	>32.5	yes	14–22	1.4–2.0[a]	<1.5	87–92	3–5	4.5–6.0
Anthracite	semianthracite	>14.0	>32.5	no	8–14	2.0–3.0[b]	<1.5	89–93[b]	3–5[b]	3–5[a]
	anthracite	>14.0	>32.5	no	2–8	>3	0.5–2	90–97[b]	2–4[b]	2–4[a]
	meta-anthracite	>14.0	>32.5	no	<2	>3	1–3	>94[b]	1–2[b]	1–2[a]

[a] Well-suited for rank discrimination in the range indicated [b] Moderately well-suited for rank discrimination Key: mmf, mineral matter free; dmmf, dry mineral matter free

character, amount of volatile matter, and reflectance, moisture, carbon, oxygen and hydrogen content (Table 2). Other classification schemes are used in many countries where there is an active coal industry (Ward 1984).

See also: Coal: Mining; Coal: World Resources

Bibliography

American Society for Testing and Materials 1986 *Annual Book of ASTM Standards*, Vol. 5.05, Sect. 5. ASTM, Philadelphia, Pennsylvania

British Standards Institution 1981 *British Standards Institution Yearbook*. BSI, London

Damberger H D, Harvey R D, Ruch R R 1984 Coal characterization. In: Cooper B R, Ellingson W A (eds.) 1984 *The Science and Technology of Coal and Coal Utilization*. Plenum, New York, pp. 7–45

Dutcher R (ed.) 1978 *Field Description of Coal*, STP 661. American Society for Testing and Materials, Philadelphia, Pennsylvania

Harvey R D, Ruch R R 1986 Mineral matter in Illinois and other US coals. In: *Mineral Matter and Ash in Coal*, ACS Symposium Series 301. American Chemical Society, Washington, DC, pp. 10–40

Phillips T L, Peppers R A, DiMichele W A 1985 Stratigraphic and interregional changes in Pennsylvanian coal-swamp vegetation: Environmental inferences. *Int. J. Coal Geol.* 5: 43–109

Stach E, Mackowsky M-Th, Teichmüller M, Taylor G H, Chandra D, Teichmüller R 1982 *Stach's Textbook of Coal Petrology*, 3rd edn. Gebruder Borntraeger, Stuttgart

Standards Association of Australia 1982 *Standards Association of Australia Yearbook*. SAA, Sydney

Stewart W N 1983 *Paleobotany and the Evolution of Plants*. Cambridge University Press, Cambridge

Ward C R (ed.) 1984 *Coal Geology and Coal Technology*. Blackwell Scientific, Melbourne

R. D. Harvey
[Illinois State Geological Survey, Champaign, Illinois, USA]

Coal: Mining

Underground coal mines are among the most rugged working environments on earth: operations are often conducted in artificially illuminated and ventilated confined spaces where workers and machinery are exposed to hazardous and corrosive environmental agents such as explosive gas and dust, suffocating gases, noxious fumes from blasting and chemicals, and saline and acidic water. Work in the often cramped spaces is carried out amidst powerful moving machinery and high-voltage electrical power systems beneath rock strata that may be unstable and subject to collapse without warning.

Surface coal mines are much less labor-intensive than underground mines, but large-scale surface mining may require the use of large, powerful and expensive machinery to remove the valueless rock that overlies the coalbed. Productivity and safety are generally greater in surface mines than in underground mines, but surface mining is always limited by the thickness of the valueless overburden, so that in most regions reserves of surface-minable coal are much less than reserves of underground-minable coal. Because surface mining completely disrupts the surface, techniques of restoring the land to a useful condition have been developed.

1. Geologic Conditions Affecting Mining

Coal occurs as a layered sedimentary rock that is typically overlain and underlain by other stratified sedimentary rocks such as sandstones, shales, limestones and claybeds. Mining productivity and safety are strongly affected by the character of the overlying strata (roof) and underlying strata (floor). Roof strata that easily sag or fracture require more costly methods of support. Floor strata that are soft may squeeze into mine openings and thereby obstruct mining operations as well as creating instability within the roof strata. In some places the sediments that form the roof strata may have intruded down into and through the coalbed (roof rolls), removing the coal and creating obstructions to mining.

Coalbeds may contain layered deposits of other rock (splits or partings) that may increase the overall hardness of the coalbed and thereby increase the power required to cut away the coal. Such layered impurities may also degrade the quality of the mined product. The quality of the product may also be adversely affected by the presence of impurities in vertical veins (clay veins or clastic dikes). Mining may also be affected by the presence of hard concretions within the coalbed or in the roof strata immediately overlying the coal.

In some places coalbeds are partly or entirely removed by erosional channels that were active after deposition of the coal and that cut down into the coalbed (washouts). Such features, whose occurrence, distribution and dimensions may be difficult or impossible to predict, may require costly alterations in mine plans. Intrusion into the coalbed of extremely hard igneous rocks may also require costly alterations in mine layouts.

Coalbeds typically contain a system of fractures (joints or cleats). The orientation of mine openings may be related to this system because it may affect the ease of cutting in different directions.

Most coalbeds are relatively flat-lying, but coalbeds may be complexly folded and faulted locally or throughout certain regions. Such structural warping may affect mining productivity by requiring alterations in the normal mining patterns and through its affect on haulage and drainage.

The removal of groundwater that flows naturally from adjacent strata into mine openings may represent a major problem to mine operators. Where mine drainage is acidic, this problem is compounded by the highly corrosive effects of the water on pumps and water lines and the problem of treating or disposing of the water on the surface. Water in abandoned workings also represents a significant hazard: many lives have been lost when active mines were quickly inundated after accidentally breaking into abandoned water-filled workings. In most areas laws were passed in the late nineteenth century, requiring regular filing of mine maps with government agencies, but inundations may still occur when such records are not consulted or where maps are inaccurate or were never filed.

Because they are derived from vegetable matter, coalbeds generate (or have generated) explosive methane gas, which is colorless, odorless and tasteless. In some areas, large quantities of such gas are emitted into underground mine openings. When present in concentrations between 5% and 15%, this gas may be ignited by open flames, electrical sparks or hot surfaces. Gas explosions may in turn produce coal-dust explosions which may propagate through large areas of a mine. Mining equipment may require special design features to prevent the generation or exposure to the ambient atmosphere of potential sources of ignition. A great variety of gas-detecting equipment has been developed through history for use in coal mining. Where gas emissions are large, the gas must be diluted to low concentrations by the ventilation system. In some areas where gas emissions are so large that conventional ventilation systems are inadequate to maintain safe levels of gas, drainage of methane through drill holes may be done before mining commences.

2. *Styles of Mining*

2.1 Underground Mining

Until the 1920s coal was mined almost exclusively using underground methods. Access to the coalbed is gained through various types of portals that may be horizontal (drifts), inclined (slopes) or vertical (shafts). Shafts and slopes and their associated hoisting systems may represent a major capital investment (especially where coalbeds are at great depth) at the beginning of a mining project, and if improperly designed may create serious bottlenecks to the passage of miners, supplies, ventilating currents and coal.

From the bottom of the shaft or slope the coalbed is developed through systems of galleries or openings (entries). Where such entries are being advanced, the coal is removed in working areas known as working faces. The entries may form various types of grid systems in which part of the coalbed is left in place (room-and-pillar mining), or may be designed to allow access to areas where large blocks of the coalbed are removed entirely (longwall methods).

With underground mining the miners attempt to remove only the coalbed, leaving the valueless roof and floor strata in place whenever possible. Mining productivity is strongly affected by the thickness of the coalbed: for a given linear advance of the face, a greater volume of coal is removed where the coalbed is thick. The efficiency of mining operations is also strongly affected by the relationship between coalbed thickness and human stature: all operations become slower and less efficient where the miners are forced to stoop or crawl.

2.2 Surface Mining

Surface mining methods have become increasingly important since their development in the late nineteenth and early twentieth centuries. In surface mining, the valueless rock (overburden) that overlies the coalbed is removed along a linear trench (pit) and cast to the side in ridges (spoil). The coal is then removed and another strip of overburden is excavated and side-cast. A critical factor is the ratio of the thickness of overburden to the thickness of the coalbed (stripping ratio): relatively thin coalbeds may be economically mined if the overburden is also relatively thin.

3. *Basic Mining Operations*

3.1 Underground Mining

Because it is a solid material, coal must be separated from the virgin coalbed by various techniques of cutting, drilling or blasting. Once the coal is separated from the coalbed, it must be loaded onto conveyances and hauled to the surface. Thus cutting, loading and hauling constitute a cyclic sequence of mining.

Until the 1950s, cutting the coalbed was usually done in conjunction with drilling and blasting (conventional mining). Picks or cutting machines (widely adopted after the 1880s) were seldom used to remove the entire thickness of the coalbeds; rather, they were used to cut slots in the coalbed (typically horizontal slots along the bottom of the coalbed). During subsequent blasting, such slots provided another free face against which the explosive wave could act. This increased the safety of blasting and produced a coarser product in an era when a premium price was paid for lump coal.

After slots were cut into the coalbed, the coal was fractured by ignition of explosives placed in drillholes (blastholes). The horizontal or inclined blastholes were drilled by hand drills, or by drilling machines powered by electricity or compressed air. Blasting presented a twofold hazard: explosives might be accidentally ignited during transportation into the mine or during their loading into blastholes, and the controlled blasting of the coal face might ignite any methane gas or dry dust present. The latter hazard was lessened by the development of so-called permissible explosives whose flame is too cool to ignite methane gas.

Other nonexplosive methods of breaking coal have been developed that utilize the introduction into the coalbed of compressed gases (air or carbon dioxide) through steel cylinders placed in drill holes: when the gas is released from the cylinder it expands and fractures the coal.

After being cut from the face, the coal was loaded onto a conveyance by hand-shoveling or by electrically powered loading machines (widely adopted after the 1920s). Some operations still use conventional mining with modern cutting and drilling equipment and blasting methods, but conventional mining methods have been widely supplanted by continuous mining and longwall methods.

Beginning in the 1950s there was widespread adoption of continuous mining methods. In these methods, electrically powered cutting machines remove the entire thickness of the coalbed at the working face, eliminating the need for drilling and blasting. Many varieties of continuous mining machines have been developed, most of which utilize numerous cutting bits mounted on moving chains (rippers), drums or milling heads. As the coal is cut and falls to the floor, it is gathered up by mechanical arms or rotating disks onto a conveyor system and discharged from the continuous mining machine onto a waiting conveyance.

The term continuous mining is something of a misnomer because the cutting and loading operations must periodically be interrupted to install roof supports in the newly mined area. Roof supports used in the face area include wooden props (timbers), steel jacks and roof bolts. Roof bolts consist essentially of steel reinforcing rods emplaced into the roof strata through vertical drill holes. The vertical holes are drilled by electrical or compressed-air machines. An important advantage of roof bolts over props and jacks is that they leave the entry fully open and present no hindrance to the movement of machinery and no resistance to ventilating currents.

The two basic principles that are believed to be involved in the support of the roof by roof bolts are the suspension of weak strata from stronger overlying strata and the clamping together of strata to form a stronger beam. Many different roof-bolting systems have been developed, but most of these involve one or both of two main types of bolt: mechanically anchored and resin-anchored bolts. Mechanically anchored bolts are anchored in rock at their upper end by various types of mechanically expanding plugs and then tensioned. With resin-anchored bolts a plastic envelope containing resin and a hardener in separate compartments is first inserted into the drill hole. The reinforcing bar is then inserted and spun, rupturing the envelope and mixing the resin and hardener. Upon curing, the hardened resin fully bonds the bolt to the hole wall. Other types of steel reinforcement have also been developed, including longitudinally slotted steel tubes (split sets) that operate on a springlike principle.

Away from the face areas, roof support may be supplemented by a variety of wooden structures, including beams supported by props (headers) and horizontally stacked timbers (cribs). Where high pressures are encountered, steel arches may be installed.

Props, roof bolts and other such mechanical means of support for the roof are subject to deterioration by decay and corrosion and are considered to be temporary. Permanent support may be provided in the form of unmined blocks (pillars) of coal that are left in place between entries. Room-and-pillar operations, of which there are many variations, make extensive use of pillars for roof support. In many mines such pillars are removed near the end of operations (pillar recovery) and the roof strata are allowed to collapse. Where pillars are left when the mine is abandoned, the pillars do not really provide permanent support but may gradually deteriorate, be crushed or sink into the soft underclay of the floor, creating instability in the roof strata that eventually results in mine subsidence. When this occurs sags or sinkholes may appear at the land surface decades or centuries after mining has ceased.

Modern longwall mining differs from room-and-pillar mining in several aspects. The property is developed (divided and subdivided into separate working areas or panels) using room-and-pillar methods, but within each longwall panel entire blocks of coal are completely extracted without any pillars being left in place, even temporarily. Coal is sliced off along a long face (hence the name longwall).

There are two general categories of longwall mining machines, both of which are electrically powered: plows and shearers. Plows are blades that are fitted with cutting bits or sawtype edges that are pulled along the working face by chains, so that the plow is powered from the end of the face. They tend to be used in relatively thin coalbeds. Shearers consist of rotating drums on which cutting bits are mounted; shearers are powered by machines mounted on the shearer itself, and they tend to be used more in relatively thick coalbeds. The coal broken from the face by the plow or shearer falls onto a face conveyor that dumps onto a conveying system that transports the coal out of the mine. Cutting of the coal may occur in only one direction along the face, or may be done in both directions.

The longwall mining machine, conveyor system and miners are protected from roof falls at the working face by a line of hydraulic self-advancing jack units of which there are several main varieties (chocks, frame-type and shield-type). As the slices are removed and the face advances, the strata are allowed to cave in behind the roof supports so that any subsidence of the surface tends to occur very soon after mining.

A variety of conveyances have been used to transport coal from the working face to the main haulage system. For such intermediate haulage, as well as main haulage, wagons pulled by humans or animals were widely supplanted by rail haulage (after the 1880s), by trackless wheeled vehicles and by conveyor systems (after 1900). Power for haulage systems has been provided by ropes, electrical trolley systems, batteries, compressed air and internal combustion engines. The selection of different systems has been governed by considerations of efficiency, safety and reliability, but electrical systems (including batteries) and diesel engines are presently the most widespread.

3.2 Surface Mining

As with underground mining, surface mining requires that the coal be separated from the virgin coalbed by techniques of cutting, drilling and blasting; but in surface mining the coalbed must first be exposed to the surface by removal of the overburden. Unless the overburden is thin and soft, drilling and blasting with explosives is usually required. After the overburden has been fragmented it is typically excavated and cast to the side (spoiled) by excavating machinery such as shovels, draglines or bucketwheel excavators.

As successive strips of overburden are spoiled, the edge of the pit (called the highwall) advances, generally from thinner overburden to thicker overburden. Surface mining first became commercially feasible on a large scale around World War I in areas where coalbeds lay very close to the surface and overburden was soft. From that time until the late 1960s, the major technological developments focused on the need to deal with ever-increasing thicknesses of overburden, so that excavating machinery, drilling equipment and haulage vehicles gradually and steadily increased in size. The development of explosives (especially cheap ammonium-nitrate–fuel-oil explosives after 1954) also played an important role. In some large surface mines with thick overburden, the excavating machinery may be very large, some individual pieces of equipment being among the largest mobile land machinery ever constructed. In the USA this trend to larger machinery ceased in the late 1960s with the opening of previously unexploited coal deposits (with relatively thin overburden) in the western states. Surface mining today is seldom conducted where the highwall exceeds 50 m in thickness. Where the overburden is relatively thin, and especially in relatively small surface mines, the fragmented overburden may be removed by bulldozers, scrapers or other construction-type equipment.

The exposed coal is then broken or blasted loose from the coalbed and loaded by hydraulic loaders into trucks (formerly railroad cars) and removed from the mine.

During the early decades of surface mining few efforts were made to reclaim the land, which was left as rough, steep-sided ridges barren of vegetation and subject to erosion, causing acidification and siltation of streams. The first attempts at reclamation involved reforestation, but only less desirable species of trees seemed capable of surviving on the spoil banks. In the USA, state legislation (after 1939) and federal legislation (after 1977) gradually imposed requirements for reclamation on mining companies. Requirements were established for regrading the spoil areas, burying acid-forming material, conserving topsoil and revegetating the land. Reclamation for agricultural use, rather than reforestation, became general. Some coal companies established their own subsidiary farming companies. Development of effective and less costly reclamation techniques continues.

4. Technological Development

Because of the widespread distribution of coalbeds and the relative ease of opening new mines, the coal-mining industry has been intensely competitive throughout much of its history. Because of this competitiveness, coal companies themselves have historically spent relatively little money on originating new technology: coal-mining technology is typically adapted from other industries, with much of the work of refinement and adaptation to mining conditions being done by equipment manufacturers and governmental agencies.

Labor, the human element in coal mining, has played a central role in the technological development of the industry. Coal miners throughout the world have long been at the forefront of industrial unionization, stimulating coal operators to adopt labor-saving devices and promoting industrial safety legislation by government.

The adoption of technology that was originally developed in other industries may be extremely costly in both money and human life and typically requires decades of refinement in response to operating experience to make it suitable for use in underground coal mines. Also, technology developed in one mine or mining district may not be applicable to other mines in other districts whose working conditions may be very different. In large underground mines that have been in operation for several decades, the introduction of new technology may be constrained by aspects of the original mine layout and equipment that remain in service as an integral part of the operation.

In the competition among coal districts and between coal and other fuels, new mining technology is often adopted first in districts where adverse geologic, labor, market and governmental factors prevail. When older, intensively exploited districts are brought into competition with less-developed districts (as by advances in coal transportation), many older districts experience disadvantages of unfavorable geologic conditions, more advanced organization of labor, and greater involvement and regulation of mining by government. Technological change is one of several strategies used by less-favored competitors to maintain competitive balance; the incentive to adopt new technology is greatest when other strategies to change labor or governmental relationships are ineffective.

Coal operators have been highly selective in introducing new technology into their operations. Many attempts to introduce new technology prematurely or to introduce inappropriate technology have failed. In the past, the coal-mining industry has often been characterized as backward, but the methods of mass production developed in manufacturing industries are often inappropriate to coal mines, and unfavorable characterizations of the coal-mining industry often reflect incomplete understanding of the unique circumstances facing coal-mine operators.

See also: Coal: Geology; Coal: World Resources

Bibliography

Given I A (ed.) 1973 *Mining Engineering Handbook* (2 vols.). American Institute of Mining Engineers, New York

Stefanko R 1983 *Coal Mining Technology, Theory and Practice.* American Institute of Mining Engineers, New York

D. Harper
[Indiana Geological Survey, Bloomington, Indiana, USA]

Coal: World Resources

Coal deposits occur as layers of variable thickness interbedded with shales, claystones, siltstones, sandstones and, less commonly, limestones. Coal seams originate as peat accumulations in swampy areas (mires). The types and quantities of plants and minerals making up a peat deposit vary greatly, depending on the environments of deposition (facies). Subsequent burial of the peat layer under hundreds to thousands of meters of younger sediments in regions of long-term subsidence (sedimentary basins) results in the conversion of peat into first lignite, then subbituminous and bituminous coal and finally—rather infrequently—into anthracite, owing to the rising overburden pressure and temperature with increasingly deeper burial. The latter sequence is the coalification series. Lignite and subbituminous, bituminous and anthracite coal are termed ranks of coal. The rising pressure and temperature bring about significant changes in the physical and chemical properties of the original plant material in the peat. Coal deposits are also subjected to whatever deformation (folding and faulting) the segment of crust they are part of is subjected to during its geological evolution.

These geological factors determine to a large degree how attractive a given coal deposit is for exploitation and thus the likelihood of it being used for future economic development. Other important factors are size of deposit, distance to potential markets, infrastructure and prevailing as well as anticipated market conditions.

1. Coal Resource and Reserve Classification and Databases

1.1 Coal Quality and Bonity

Coal resource and reserve databases, in order to be useful, should contain information on coal quality and bonity (Fettweis 1976, 1979). Coal quality depends on coal facies and rank, and controls the value of any production from a coal deposit. Bonity (derived from the German word "Bonität") encompasses factors that control the cost of mining (e.g., thickness, depth, deformation, location and size of deposit). Quality and bonity together determine the economic feasibility of mining a given deposit.

1.2 Degree of Geological Assurance

Information on quality and bonity is generated through exploration. As the amount of exploration increases (through geological mapping, drilling and trenching, as well as sampling and testing of samples) and increasing knowledge of coal quality and bonity is gained, the "degree of geological assurance" of the existence (or absence) of an economic coal deposit increases. Information on the degree of assurance is expressed through such descriptors as measured, proved, known, assured, demonstrated and discovered resources for high degrees of assurance; indicated, probable, inferred, additional and semiproved for lower degrees of assurance; and possible, undiscovered, hypothetical, speculative and prognostic for lowest degrees of assurance. These terms refer primarily to the degree of geological assurance of the existence of a deposit, rather than the degree of knowledge of its numerous quality and bonity parameters.

1.3 Economic Feasibility

Exploration also generates the information needed to assess the economic minability of coal deposits ("feasibility" to mine for profit). The degree of feasibility is expressed through such terms as economic, marginally economic, subeconomic, uneconomic, or similar terms. Economic minability includes an assessment of the technical minability. However, many coal deposits could be technically mined, yet they could not now be exploited for economic gain. On the other hand, a deposit of economic minability may not be legally minable; for instance, in areas set aside as wilderness or wetlands, or otherwise excluded from mining by law. In modern usage the term reserve always implies that economic minability has been established by considering both technical and legal constraints.

1.4 Classification Schemes

The degrees of economic minability and of geological assurance are combined to classify known and surmised coal deposits within a logical system that permits aggregation into databases for regions, nations, continents and the world. Unfortunately, up to 1988, no universally accepted standard for the estimation of amounts of coal has emerged. Consequently, compilations on a worldwide basis bring together data that are not necessarily comparable between countries (and even within countries). Examples of national classification standards for coal resources and reserves are illustrated in Figs. 1, 2 and 3.

(*a*) *World Energy Conference survey of world coal resources and reserves.* The categories of both degree of geological assurance and degree of economic minability are defined differently in various countries and make worldwide compilations of coal resource data inherently suspect. The World Energy Conference (WEC) has attempted to overcome these disparities through the adoption of precisely worded definitions (Table 1); however, much judgment is required in applying the WEC definitions to the various national databases in order to determine the amounts of coal within each of the WEC categories. The WEC compilation for 1984 shown in Table 2 follows previous surveys, in particular the one for 1974; significant earlier surveys were those for the World Power Conference in 1929 and the most comprehensive worldwide compilation of coal resource data completed for the International Geological Congress in Toronto in 1913. Coal resources, though not evenly distributed through the world (see Fig. 4, Table 2), nevertheless are sufficiently widespread, and occur in such abundance, that no shortages of coal supply are expected to occur anywhere in the foreseeable future (World Energy Conference 1986).

(*b*) *Resources versus reserves.* Coal that is currently or will be economically exploitable within a reasonable timescale (e.g., within 25 years in the Canadian scheme) is generally referred to as reserve (amount recoverable) or reserve base (amount in situ). Coal in the ground that may in the future become of economic interest, but cannot now or in the foreseeable future be economically recovered is commonly referred to as resource. However, the amount included in reserves and that allocated to resources depends very much on the background and intent of the person or group that promulgated or applied a particular classification scheme. The terminology is not uniform; for instance, the term reserve base has not been generally adopted and is considered confusing by some (van Rensburg 1980). Resources always reflect coal in the ground, but some schemes include coal, the existence of which may only be surmised by geological inference and the economic minability of which is unknown and may never materialize; for example, hypothetical and speculative resources in the US classification scheme. Other schemes (e.g., the South African scheme) reject the inclusion of any data on undiscovered resources in national inventories (van Rensburg 1980).

The term reserves has often been restricted to the amounts of coal that can be economically recovered; that is, after mining losses or after mining and beneficiation losses. To avoid confusion, the terms reserves and resources are sometimes used with qualifying adjectives such as recoverable (out of ground), exploitable (in situ), inferred, and so forth. However,

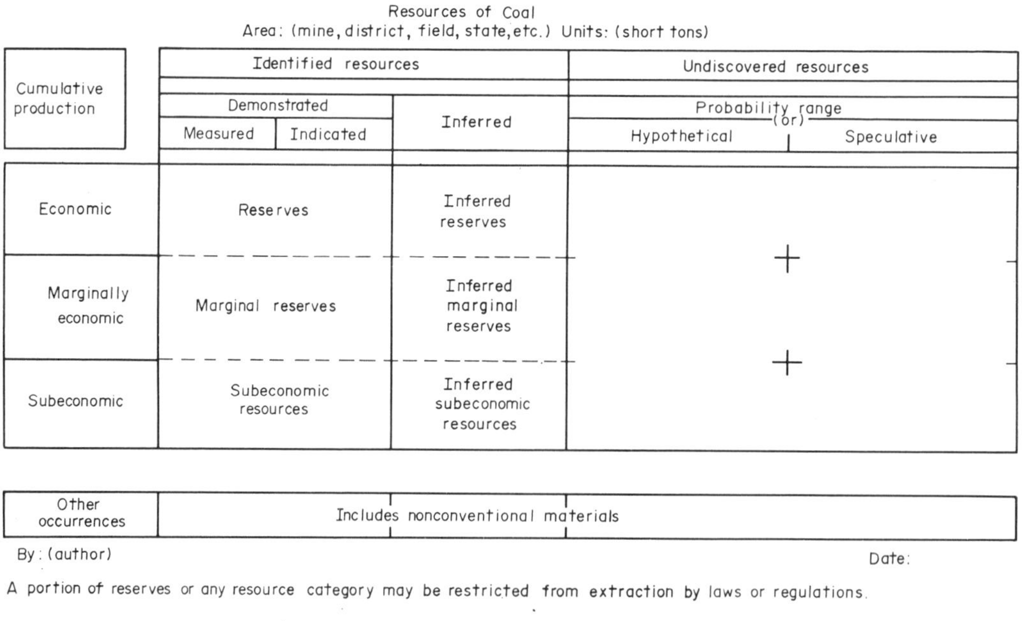

Resources of Coal

Area: (mine, district, field, state, etc.) Units: (short tons)

Cumulative production	Identified resources			Undiscovered resources	
	Demonstrated		Inferred	Probability range (or)	
	Measured	Indicated		Hypothetical	Speculative
Economic	Reserves		Inferred reserves		
Marginally economic	Marginal reserves		Inferred marginal reserves		
Subeconomic	Subeconomic resources		Inferred subeconomic resources		

Other occurrences	Includes nonconventional materials

By: (author) Date:

A portion of reserves or any resource category may be restricted from extraction by laws or regulations.

Figure 1
Classification scheme for coal resources and reserves used in the USA (after Wood et al. 1983.)

Total geological resources
Balance sheet resources
Outside balance sheet resources
Prognostic resources
A
B
C1
C2
a
b
c1
c2
d1
d2

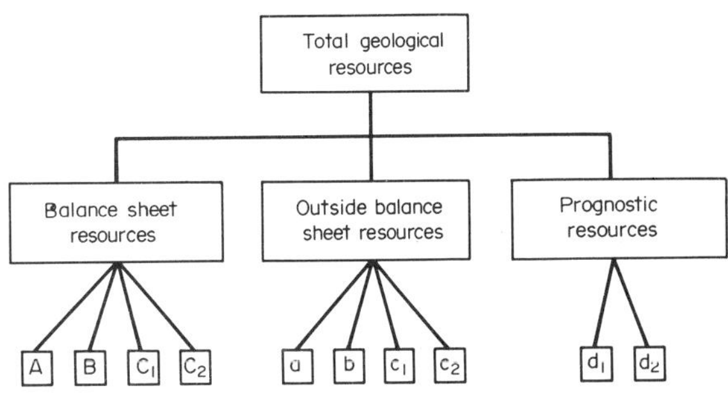

Figure 2
Classification scheme for coal resources and reserves used in the USSR and the GDR (after Fettweis 1979. © Glückauf, Essen. Reproduced with permission)

Wood et al. (1983) point out that reserves include only recoverable coal and that terms such as extractable reserves are redundant and should be avoided. Unfortunately, usage of the terms reserves and resources has been and still is inconsistent.

In the USA and Canada the classification scheme proposed by McKelvey (1973) for solid mineral resources has been widely adopted (McKelvey diagram), in particular by the agencies of the Federal governments charged with compiling national coal resource and reserve databases (Fig. 1). However, the geological surveys of individual states and provinces have not all adopted these standards. It is important to understand that resulting differences in resource estimates are a matter of preference and judgment rather than "correctness." The acceptability of estimates to the user depends on the user's requirements.

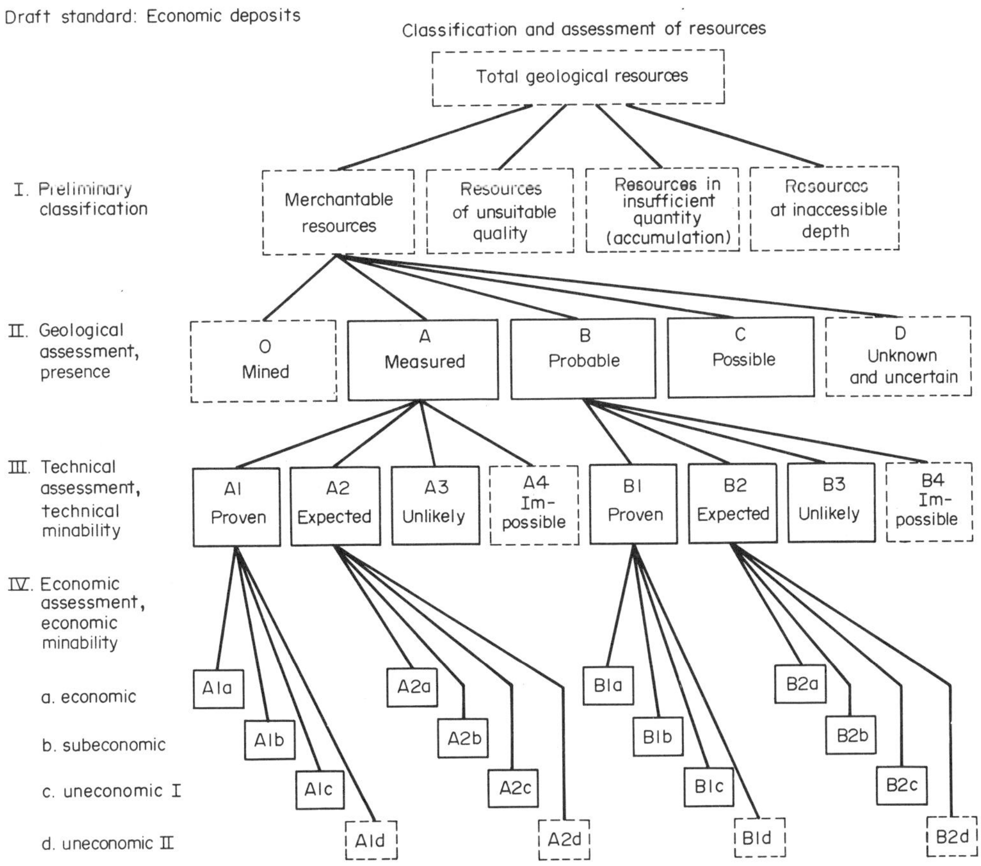

Draft standard: Economic deposits
Classification and assessment of resources
Total geological resources
I. Preliminary classification
Merchantable resources
Resources of unsuitable quality
Resources in insufficient quantity (accumulation)
Resources at inaccessible depth
II. Geological assessment, presence
O Mined
A Measured
B Probable
C Possible
D Unknown and uncertain
III. Technical assessment, technical minability
A1 Proven
A2 Expected
A3 Unlikely
A4 Im-possible
B1 Proven
B2 Expected
B3 Unlikely
B4 Im-possible
IV. Economic assessment, economic minability
a. economic
b. subeconomic
c. uneconomic I
d. uneconomic II
A1a
A1b
A1c
A1d
A2a
A2b
A2c
A2d
B1a
B1b
B1c
B1d
B2a
B2b
B2c
B2d

Figure 3
Classification scheme for coal resources and reserves used in the FRG. The dashed outlines indicate resources which are without particular importance for mining (after Fettweis 1979. © Glückauf, Essen. Reproduced with permission)

Table 1
Terminology for world resources and reserves adopted by the World Energy Conference for the 1984 survey

Term	Definition
Proved amount in place	Tonnage that has been carefully measured and is exploitable under present and expected local economic conditions with existing technology
Proved recoverable reserves[a]	Tonnage of proved amount in place that can be extracted in raw form under present and expected local economic conditions with existing technology
Estimated additional amount in place	Indicated and inferred tonnage additional to the proved amount in place, e.g., unexplored extensions of known deposits, undiscovered deposits in areas with known coal resources, amounts inferred through knowledge of favorable geological conditions. Deposits whose existence is merely speculative are not included.
Estimated additional reserves recoverable	Quantity of the estimated additional amount in place which might become recoverable within foreseeable economic and technological limits.
Accessible coal in significant coal fields	Based on figures from the International Energy Agency (IEA), Coal Information 1986; this includes amount of coal likely to be considered for extraction from significant coal fields within the next 20 years. A significant coal field is one whose collective physical characteristics render it likely either to make significant contributions to, or to enter into the detailed commercial mining and market evaluations required in order to achieve world coal supply over the next 20 years

[a] In most countries only a fraction of proved recoverable reserves is considered to be accessible coal

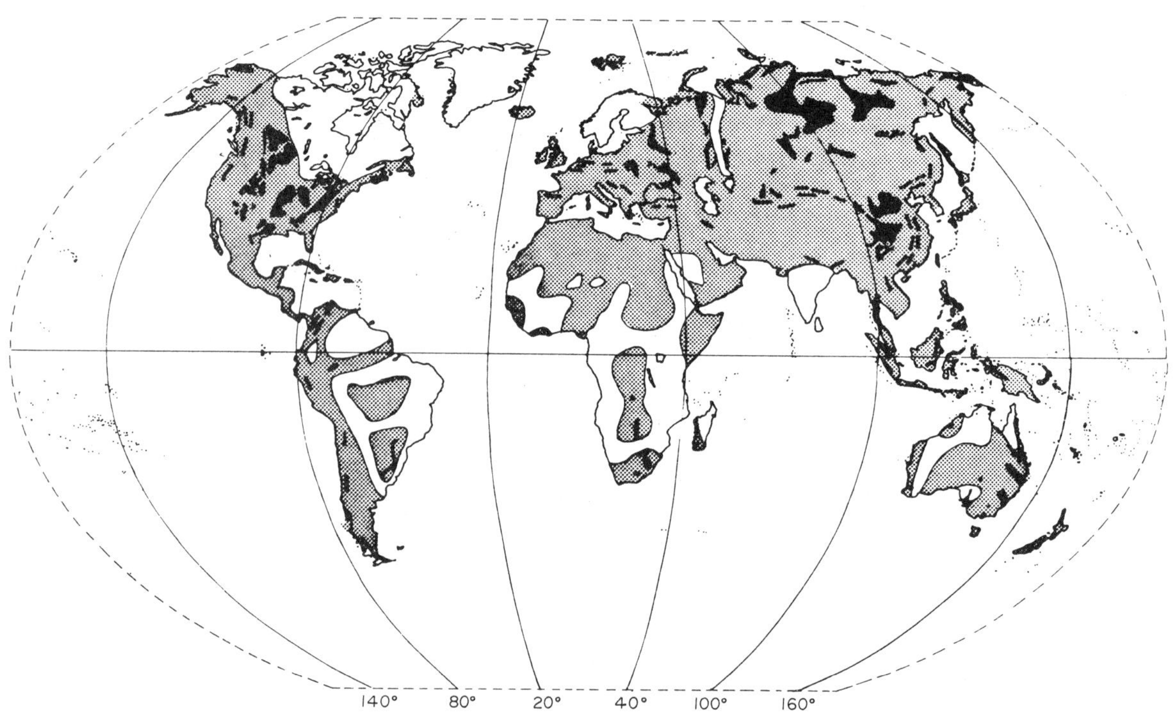

Figure 4
Important coal occurrences of the world: black areas, known and assumed coal occurrences; white areas, nearly coal-free nonsedimentary rocks; shaded areas, sedimentary rocks with potential for coal occurrences (after Fettweis 1979. © Glückauf, Essen. Reproduced with permission)

Table 2
World coal resources and reserves of the 1984 survey by the World Energy Conference (1986)

	Proved amount in place (10^9 t)				Proved recoverable reserves (10^9 t)				Accessible coal in significant coal fields (10^9 t)				Estimated additional amount in place (10^9 t)			
	BT	SB	LN	Total	BT	SB	LN	Total	BT	SB	LN	Total	BT	SB	LN	Total
Africa																
Botswana	7.0			7.0	3.5			3.5	3.0			3.0	100.0			100.0
South Africa	115.5			115.5	58.4			58.4	30.0			30.0	17.1			17.1
Zimbabwe	1.5	1.0		2.5	0.7	0		0.7	1.0			1.0	5.8	U		5.8
other	3.2	0.3	0.1	3.6	3.0	0.2	< 0.1	3.2	0.3	0.2	0.1	0.6	3.2	1.0		4.2
Total	127.2	1.3	0.1	128.6	65.6	0.2	< 0.1	65.8	34.3	0.2	0.1	34.6	126.1	1.0	0	127.1
North America																
Canada	4.9	1.3	2.8	9.0	3.5	0.9	2.4	6.8	1.4	2.2	1.3	4.9	25.7	27.8	4.2	57.7
Mexico	1.6	0.8	N	2.4	1.3	0.6	N	1.9	1.0			1.0	2.0	0.8		2.8
USA	237.6	164.4	40.9	442.9	132.0	99.2	32.7	263.9	55.0	37.0		92.0	458.1	276.2	392.8	1127.1
Total	244.1	166.5	43.7	454.3	136.8	100.7	35.1	272.6	57.4	39.2	1.3	97.9	485.8	304.8	397.0	1187.6
South and Central America																
Colombia	2.0	< 0.1	< 0.1	2.0	1.0	< 0.1	< 0.1	1.0	2.4			2.4	7.2	0.6	0.2	8.0
Venezuela	0.5			0.5	0.4			0.4	1.6		< 0.1	1.6	2.2			2.2
other	< 0.1	23.2	< 0.1	23.2	< 0.1	14.3	< 0.1	14.3	1.1	5.2		6.3	0.1	28.3	7.3	35.7
Total	2.5	23.2	< 0.1	25.7	1.4	14.3	< 0.1	15.7	5.1	5.2	< 0.1	10.3	10.5	28.9	7.5	46.9
Asia, Eastern Europe																
China	610.6		126.5	737.1	U		U	U	70.7			70.7	1700.0		300.0	2000.0
Czechoslovakia	5.8		7.2	13.0	2.7		2.9	5.6	1.0	5.0		6.0	5.5		1.6	7.2
GDR			47.0	47.0			21.0	21.0			13.0	13.0			0.0	0.0
India	26.3		1.6	27.9	U		1.6	1.6	19.0		0.5	19.5	85.6		1.9	87.5
Japan	8.5		0.2	8.7	1.0		< 0.1	1.0	0.8			0.8	U		U	U
Poland	63.0		13.2	76.2	28.3		14.4	42.7	36.5		10.9	47.4	100.5		20.4	120.9
Turkey	0.1		5.3	5.4	0.1		4.8	4.9	0.1		3.0	3.1	1.2		2.8	4.0
USSR	136.0	51.8	105.0	292.8	108.8	41.4	94.5	244.7	73.8	21.0	77.4	172.2	2163.0	2085.9	960.0	5208.9
other Asia	16.8	5.1	30.0	51.9	0.8	0.5	0.9	2.2	2.3	0.4	1.8	4.5	5.1	2.8	1.6	9.5
other Eastern Europe	1.5	4.6	25.9	32.0	0.7	2.5	21.6	24.8	0.1	0.8	0.5	1.4	1.9	2.0	4.0	7.9
Total	868.6	61.5	361.9	1292.0	142.4	44.4	161.7	348.5	204.3	27.2	107.1	338.6	4062.8	2090.7	1292.3	7445.9
Australia and New Zealand																
Australia	48.5	3.0	39.3	90.8	27.4	2.1	36.2	65.7	20.8		35.0	55.8	507.0	100.4	87.0	694.4
New Zealand	< 0.1	0.4	1.6	2.0	< 0.1	0.2	0.1	0.3		0.2		0.2	0.4	1.4	8.9	10.7
Total	48.5	3.4	40.9	92.8	27.4	2.3	36.3	66.0	20.8	0.2	35.0	56.0	507.4	101.8	95.9	705.1
Western Europe																
Belgium	0.7			0.7	0.4			0.4	0.8			0.8	1.4			1.4
France	0.9	0.2	U	1.1	0.3	< 0.1	U	0.3	0.4			0.4	NR	NR	< 0.1	< 0.1
FRG	44.0		55.0	99.0	23.9		35.2	59.1	13.4		12.0	25.4	186.3		U	186.3
UK	U		0.4	0.4	4.6		U	4.6	14.0			14.0	185.4		U	185.4
Greece			5.3	5.3			3.0	3.0			1.5	1.5			U	U
Spain	0.9	0.5	0.3	1.7	0.4	0.2	0.2	0.9	0.4		1.2	1.6	2.6	0.9	0.2	3.7
other Western Europe	1.4	0.1	0.3	1.8	0.5	0.1	0.1	0.7					0.1	1.4	3.6	5.1
Total	47.9	0.8	61.3	110.0	30.1	0.3	38.5	69.0	29.0		14.7	43.7	375.8	2.3	3.8	381.9
World total (rounded)	1340	257	508	2105	404	162	272	838	351	72	158	581	5568	2530	1797	9895

Key: BT, bituminous and anthracitic coals; SB, subbituminous coal; LN, lignite; U, unknown or not available; C, confidential; N, negligible amount; NR, not reported (*c*) *Classification standards.* In the USA the US Geological Survey (USGS) and the US Bureau of Mines (USBM) adopted a classification standard (US Geological Survey and US Bureau of Mines 1976, US Geological Survey 1980) which has become the standard reference for many Federal and State agencies, as well as the coal industry (Fig. 1). Subsequently, the USGS adopted a detailed, specific version of this system for its own coal resource studies (Wood et al. 1983) which is represented in Fig. 5 and Table 3 as an example of the implementation of a national classification scheme. In Fig. 5 original resources refers to the resources before mining, and subeconomic and inferred subeconomic resources include coal left in room

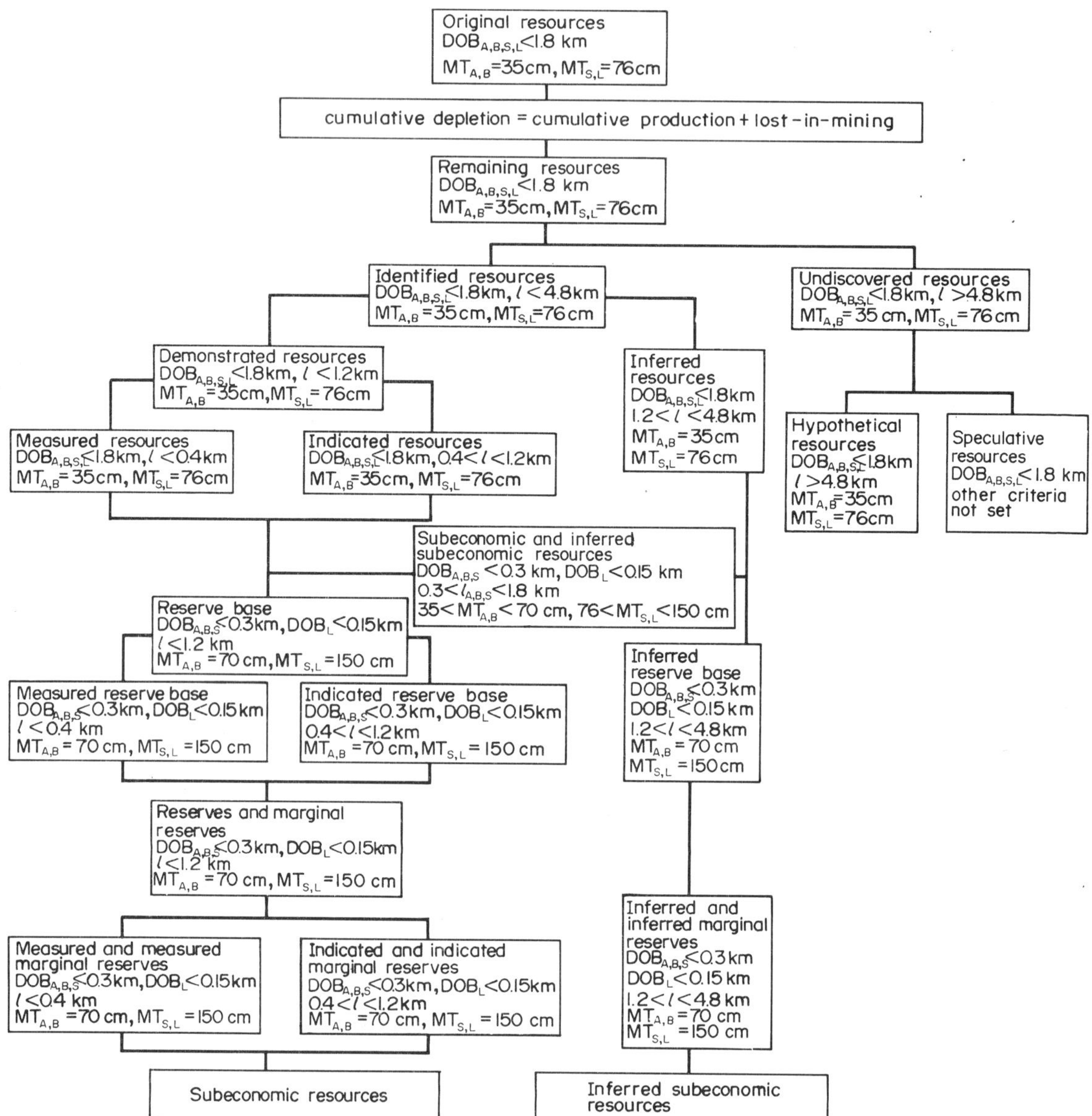

Figure 5
Classification and hierarchy of coal resources as currently used by the US Geological Survey (Wood et al. 1983). Coal resource terms are defined in Table 3. The criteria given in each box are the depth of burial (DOB), the maximum distance l from the point of measurement of coal thickness and the minimum thickness (MT) of the coal. The subscripts A, B, S and L refer to anthracite, bituminous coal, subbituminous coal and lignite

Table 3.
Abbreviated definitions of terms used by the US Geological Survey (USGS) to classify coal resources (after Wood et al. 1983)

Term	Definition
Original resources	Amount of coal in-place before production, occurring in coal seams at least 0.35 m (bituminous and anthracitic coals) or 0.75 m (subbituminous and lignitic coals) at depths to 1800 m
Remaining resources	Resources in-place after mining; does not include any amounts left in ground rendered unminable through exploitation
Identified resources	Resources whose location, rank, quality and quantity are known or estimated from specific geological evidence; they include economic, marginally economic and subeconomic components. They are subdivided into measured, indicated and inferred resources.
Measured resources	Resources having a high degree of geological assurance; specifically, they are computed by projection of thickness of coal and overburden, rank and quality data for a radius of 0.4 km from points of measurement
Indicated resources	Resources having a moderate degree of geological assurance; specifically, they lie between 0.4 and 1.2 km from points of measurement
Demonstrated resources	Measured plus indicated resources
Inferred resources	Resources having a low degree of geological assurance; specifically, they lie between 1.2 and 4.8 km from points of measurement
Undiscovered resources	Resources, the existence of which is only postulated, comprising deposits that are either separate from or are extensions of identified resources; they are subdivided into hypothetical and speculative resources
Hypothetical resources	Resources that are either similar to known coal deposits which may be reasonably expected to exist in the same coal field or region under analogous geological conditions or are an extension from inferred resources, beyond 4.8 km of points of measurement
Speculative resources	Undiscovered resources that may occur either in known types of deposits in favorable geological settings where coal deposits have not been discovered or in types of deposits as yet unrecognized for their economic potential. (Note: the USGS has not estimated the amount of speculative coal resources for the USA)
Economic resources	Implies that profitable extraction under defined investment assumptions has been established, analytically demonstrated, or assumed with reasonable certainty
Marginal economic resources	Implies that coal does not meet economic criteria under current costs and prices but may do so under reasonable future projections
Subeconomic resources	Resources which do not meet current or reasonable future criteria for profitable extraction
Reserve base	Those parts of the identified resources that meet specified minimum physical and chemical criteria related to current mining and production practices, including those for quality, depth, thickness, rank and distance from points of measurement. Specifically, the reserve base is the in-place demonstrated (measured plus indicated) resource from which reserves are estimated: coal seams thicker than 0.7 m (bituminous and anthracite) or 1.5 m (subbituminous and lignite), subbituminous coals only to a depth of 300 m and lignites only to a depth of 150 m are included.
Reserves	Virgin and (or) accessed parts of coal reserve base which could be economically extracted considering environmental, legal and technological constraints. The term reserves need not signify that extraction facilities are in-place or operative. Reserves include only recoverable coal, not coal lost in mining
Restricted reserves	Those parts of any reserve category that are restricted or prohibited by laws or regulations from extraction

and pillar mining, in property barriers, coal too thick to be recovered completely by conventional mining, and mine and preparation waste. Occurrences other than those shown in the hierarchy include those:

(a) less than minimum thickness at any depth,

(b) containing more than 33wt% ash on dry basis, or

(c) buried at depths of more than 1.8 km.

Estimated tonnage, where calculated, should be reported as "other occurrences" and not as resources, unless mined. Mined tonnage quantity is included in reserve base and reserve estimates.

(*d*) *Problems in quantifying the degree of geological assurance.* It will be noted that the degree of geological assurance of the existence of a resource in the 1983 USGS classification system is defined in terms of distance from points of measurement of thickness. Other classification schemes define assurance categories in terms of percentage certainty of the resource calculation (Fig. 6). The definitions of assurance categories in the USGS and USBM (1976) classification include a statement of percentage error margin for measured resources (less than 20%), but none for indicated resources, inferred resources, and so forth. Such restraint is not without reason. The less information available on a coal resource, the more difficult it becomes to calculate or estimate the percentage error of a computed tonnage. Tewalt et al. (1983) discussed the difficulties of estimating geological uncertainties for quantitative coal resource evaluations.

Classical statistical and geostatistical analyses can be applied to determine the uncertainties inherent in any calculated resource tonnage (or average quality), but the methods are most meaningful when a sufficiently large, closely spaced (say 0.5–1 km grid) database is available. Geostatistical methods, e.g., Kriging, should be, but are not generally, applied where the area distribution of data points is uneven, which is usually the case for exploration data on a regional scale. Generally, only deposits considered or targeted for exploitation are sufficiently explored through a regular grid of exploration wells to permit meaningful statistical analysis. Even then, simple statistical analyses (averages, standard deviations, etc.) which are in common use in the industry may be misleading because they disregard regional and local trends in quality and bonity parameters. Trend surface, regression and other geostatistical analyses produce more meaningful results, but are not yet commonly used even at the level of reserve blocks for individual mines.

Terms for presence	Assurance %
A, a, measured (certain)	100–90
B, b, probable	90–80–70
C_1, indicated C, c possible C_2, inferred	70–60–50–40–30
d_1, δ_1, uncertain	30–20–10
d_2, δ_2, unknown	10–0

Figure 6
Degrees of geological assurance used in the FRG and eastern European mineral classification schemes. 100% assurance indicates that the presence of mineral resources is totally certain and 0% assurance indicates that the presence is totally unknown. Upper-case letters refer to currently minable resources while lower-case letters refer to potentially minable resources (compare with Figs. 2 and 3)

2. *Outlook*

The introduction of computers and specialized software in the coal-mining and consulting industries and in government agencies is leading to changes in the ways coal deposits are evaluated. Large computerized coal resource databases, which incorporate not only the traditional information on amount (thickness and size), depth and rank, but also many other quality and bonity parameters, have been compiled by governments and industry, particularly since the oil embargo of 1973. These databases and software, written for their analysis, are now maturing in the USA, Canada, the FRG and other countries. They will eventually permit more flexible approaches to coal resource assessments than were possible in the past. Instead of preselecting boundary values (e.g., for thickness, depth, rank) at the beginning of a resource calculation, as was required for "manual" compilations (to limit the number of resource categories to be carried to a manageable level), the computer permits the manipulation of a vast amount of data for evaluation according to "what if" type questions. Eventually, these databases will permit re-analysis according to various national and international standards, as well as to any other standards preferred by industry, various agencies or individual researchers.

See also: Coal: Geology; Coal: Mining

Bibliography

Averitt P 1975 *Coal Resources of the United States, January 1, 1974*, US Geological Survey Bulletin 1412. US Government Printing Office, Washington, DC

Bielenstein H V, Chrismas L P, Latour B A, Tibbetts T E 1980 Coal resources and reserves of Canada. *Can. Dep. Energy, Mines Resour. Rep.* ER 79–9

Fettweis G B 1976 *Weltkohlenvorräte. Eine vergleichende Analyse ihrer Erfassung und Bewertung.* Glückauf, Essen

Fettweis G B 1979 *World Coal Resources. Methods of Assessment and Results*, Developments in Geology Vol. 10. Elsevier Scientific, Amsterdam

International Geological Congress 1913 *The Coal Resources of the World*, Vols. 1, 2, 3. 12th International Geological Congress, Toronto, Canada

McKelvey V E 1973 *Mineral Resources Estimates and Public Policy*, US Geological Survey Professional Paper 820. US Government Printing Office, Washington, DC, pp. 9–19

Tewalt S J, Bauer M A, Mathew D, Roberts M P, Ayers W B, Barnes J W, Kaiser W R 1983 Estimation of uncertainty in coal resources. *Rept. Invest. Univ. Tex. Austin, Bur. Econ. Geol.* 136

US Department of Energy 1981 *Demonstrated Reserve Base of Coal in the United States on January 1, 1979*, US DOE/EIA-0280 (79) UC-88. US Department of Energy, Washington, DC

US Geological Survey 1980 *Principles of a Resource/Reserve Classification for Minerals*, US Geological Survey Circular 831. USGS, Washington, DC

US Geological Survey and US Bureau of Mines 1976 *Coal Resource Classification System of the U.S. Bureau of Mines and U.S. Geological Survey*, US Geological Survey Bulletin 1450-B. US Government Printing Office, Washington, DC, pp. B1–B7

van Rensburg W C J 1980 The classification of coal resources and reserves. *Miner. Resour. Circ. (Univ. Tex. Austin, Bur. Econ. Geol.)* 65

Wood G H Jr, Kehn T M, Carter M D, Culbertson W C 1983 *Coal Resource Classification System of the U.S. Geological Survey*, US Geological Survey Circular 891. USGS, Washington, DC

World Energy Conference 1986 *Survey of Energy Resources. Section 1–Coal (Including Lignite)*. WEC, London, pp. 6–28.

H. H. Damberger
[Illinois State Geological Survey, Champaign, Illinois, USA]

Cobalt Resources

Cobalt is classified as a strategic and critical metal because of its applications in defense-related industries and the dependence of industrialized nations on imports of it. Moreover, because of the rapid escalation in price in 1978–79 and concerns about availability, much attention has been focused on developing new substitutes, increasing production efficiency and recycling. These concerns arose not from the physical depletion of known reserves, but rather from the possibility of political disruption in southern Africa, where about 60% of the world's cobalt is produced. Therefore, the cobalt market presents an interesting case for study of the factors that result in materials substitution, increased recycling and increased production, in response to price and supply uncertainty.

1. Applications

Cobalt has developed through the centuries from an obscure coloring additive into an essential element in many alloys and an important component of chemical compounds. Applications as chemical compounds include pigments, enamels, rubber (tires), driers (organic compounds added to paints) and catalysts (inorganic compounds, mainly those used in the desulfurization of petroleum distillates). In metallic form, cobalt is used primarily in superalloys, permanent magnets, cemented carbides and hardfacing alloys. Extrafine cobalt powder is a special form used as a matrix and serving as a binder in tungsten carbide cutting tools and drill bits. This is the "cement" of cemented carbides. Cobalt-bearing superalloys are used in the hottest sections of gas-turbine engines, where high strength at high temperatures is a requirement. Hardfacing alloys are used in mining machinery and equipment, valve seats, chain saws and other industrial applications. Permanent magnets are used mainly in loudspeakers, electric meters, relay devices, telephone receivers, line printers and other electrical applications (Kirk 1985).

2. Resources, Reserves and Production

Cobalt is a metallic element which occurs in the earth's crust almost exclusively as the stable isotope ^{59}Co. Its atomic number is 27, its atomic weight is 58.93, and its abundance in the earth's crust is estimated to be about 20 ppm. Geochemically and metallurgically, it behaves in much the same manner as nickel, partly because of its similar atomic size and weight. Of the main rock types in the earth's crust, mafic and ultramafic intrusive igneous rocks contain, on average, greater abundances of cobalt. However, the richest deposits in terms of optimum grade and tonnage are the mineralized sedimentary rocks of Zaire and Zambia. Cobalt deposits may be classified genetically as: (a) hypogene deposits associated with mafic intrusive igneous rocks; (b) contact metamorphic deposits also associated with mafic rocks; (c) lateritic (weathered) deposits; (d) massive sulfide deposits in metamorphic rocks, chiefly of volcanic origin; (e) hydrothermal deposits; (f) strata-bound deposits; and (g) deposits formed as chemical precipitates (Vhay et al. 1973). Cobalt is usually mined as a by-product or coproduct of copper and/or nickel. Table 1 contains estimated world reserves, resources and production in 1987.

2.1 The Copper Belt of Zaire and Zambia

The copper–cobalt deposits of this region are categorized as strata-bound. The copper belt stretches across Zambia and into Zaire, in the shape of a large folded arc, for a distance of 500 km and with a width of about 30 km. The mineralized beds may be 20–50 m thick and in many places are highly folded. The average grade of the various ore deposits in these beds ranges from 0.2% to 0.4% Co; copper grade averages about 3.5%. However, not all of the copper deposits contain recoverable cobalt. Cobalt-bearing minerals include carrolite ($(Co_2Cu)S_4$) and linnaeite (Co_3S_4).

The principal ore deposits are replacements of lithologic units in the Katanga System which includes the Roan Group (Bilbrey 1962). In the Shaba Province of Zaire, mineralization occurs in the Upper Roan Group, as islands on younger sedimentary rocks; mineralization in Zambia occurs in the Lower Roan Group, structurally controlled by crystalline basement

Table 1
World reserves, reserve base and mine production of cobalt in 1987 (all figures in thousand kg)

	Reserves	Reserve base	Mine production
North America			
USA	0	950000	0
Canada	50000	285000	3000
Cuba	1150000	2000000	1500
Total	1200000	3235000	4500
Europe			
Finland	25000	37500	1000
USSR	150000	250000	3000
Total	175000	287000	4000
Africa			
Zaire	1500000	2300000	20000
Zambia	400000	600000	4000
Total	1900000	2900000	24000
Oceania			
Australia	25000	100000	500
New Caledonia	250000	950000	750
Philippines	150000	440000	0
Total	425000	1490000	1250
World total (land based)	3800000	9200000	34900[a]
World total (seabed nodules)	0	205000000	0

[a] Includes Albania 650, others 500

rocks on which the sediments were deposited (White 1979).

In Zaire, the state-owned company Générale des Carrières et des Mines (GÉCAMINES) mines nearly all of the copper and all cobalt. Zambia Consolidated Copper Mines Ltd. mines cobalt and copper in Zambia. This company is also state-controlled.

2.2 The Cobalt Sulfarsenides of Morocco and the Blackbird District of Idaho

These deposits may be classified as hydrothermal. They are the only significant deposits in the world where cobalt has been mined as the principal product. In the Bou Azzer district of Morocco, ore occurs in complex sulfide lenses and veins in Precambrian gneiss exposed by the erosion of an anticline of younger rocks. The deposit is located at the eastern edge of the Atlas Mountains at an altitude of 1400 m, in a rocky, arid region. The seven ore-bearing veins range in thickness from 15 m to 36 m and are found over a 46 km, east-trending structural lineament that passes through Bou Azzer and El Graara, Morocco. The ore is a complex suite of arsenic and sulfide minerals that include skutterudite ($CoAs_3$), erythrite ($3Co.As_2O_5.8H_2O$) and safflorite ($(Co, Fe)As_2$); it is reported to contain an average of about 1.2% Co.

The Blackbird district of Idaho, near Salmon, is somewhat similar geologically to the Bou Azzer district of Morocco. The deposit consists of Precambrian, slightly metamorphosed bedded and banded gray quartzites, sandy pelites and biotitic argillites. The principal ore mineral is cobaltite ($(Co, Fe)AsS$), which occurs as large veins and in irregular bodies that may vary from tabular to podlike. The grade of ore is approximately 0.6% Co, with resources estimated at 50 000 tonnes of cobalt. The Blackbird mine was active during the 1950s.

2.3 The Sulfides of Finland and Canada

The significant sulfide ores in Finland were formed as hydrothermal Cu–Ni–Co deposits in basic and ultrabasic Precambrian metamorphic rocks in the Main Sulfide Ore Belt. This belt extends diagonally across the country and is from 40 km to 150 km wide. The principal cobalt-bearing deposits are located in a large mass of quartzite surrounded by a mica schist zone (*Mining Magazine* 1978). Principal ore minerals are linnaeite and pentlandite ($(FeNi)_9S_8$); ore grade averages about 0.2% Co.

The Sudbury district of Ontario Province, Canada, is a hypogene deposit associated with mafic intrusive igneous rocks. Nickel–copper sulfides containing cobalt occur in pods, lenses, veins, stringers and disseminated grains in a layered sequence of norite, quartz diorite, and micropegmatite of Precambrian age. The elliptical basin containing the deposit is about 60 km long and 27 km wide. Average cobalt grade is about 0.07%. The origin of the deposit has been extensively debated; many geologists believe that the deposit resulted indirectly from a meteorite impact.

2.4 The Laterites of Cuba, New Caledonia, the Phillippines and Australia

Cobaltiferous laterite deposits are scattered around the world. Laterites consist of a mantle of completely weathered mafic rocks forming a soil, from around 1 m to more than 30 m thick, mainly in tropical or subtropical climates. Cobalt content generally ranges from 0.01% to 0.3%, with 40–50% Fe and 1–2% Ni. Typically the weathering of peridotite and serpentine bodies yields, through leaching by groundwater, a laterite enriched in hydrated forms of iron, nickel, cobalt and chromium, frequently high in magnesium. Although resources of cobalt in laterite are vast, relatively little cobalt is recovered in most countries because of the relatively low content of cobalt in the ore. In the USA, laterites occur in northern California, southern Oregon and Puerto Rico.

2.5 Other Significant Deposits

Other deposits in the USA that have been mined include the hydrothermally formed deposits of the Madison Mine near Fredericktown, Missouri, where cobalt, copper, nickel and lead have been produced, and the contact metamorphic iron deposit at Cornwall, Pennsylvania, where a diabasic intrusive caused adjacent carbonate rocks to be mineralized. The Duluth gabbro in northeastern Minnesota is a layered basic intrusive in which copper, nickel and cobalt are believed to have been concentrated as sulfides in the contact zone. However, it has yet to be mined. In Canada, cobalt–silver arsenides occur in the Cobalt and Gowganda Districts, Ontario Province. Other significant deposits of various types are located in the USSR, Zimbabwe, South Africa, Botswana and Australia. Very large resources of cobalt occur in seabed manganese nodules and crusts, principally in the Pacific.

3. *Outlook*

Cumulative domestic requirements for primary cobalt from 1983 to 2000 are estimated at 176 million kg, at an average annual growth rate of 3.2%. The cumulative demand for primary cobalt in the rest of the world, at an average annual growth rate of 3.3%, is forecast at 351 million kg. Therefore, the world probable cumulative demand for primary cobalt for the forecast period is 527 million kg. The potential supply of primary cobalt in the world, from known reserves only, is 3.6 billion kg, easily adequate to meet this potential demand (Kirk 1985). However, because of its by-product character, and the control which just a few countries exercise over supply, periodic shortages of cobalt may occur. Over the long term, new markets in energy-related applications may expand demand significantly. To meet this demand, supply from non-traditional sources, such as cobalt-bearing seabed nodules and crusts, may be developed.

Bibliography

Bilbrey J H 1962 *Cobalt: A Materials Survey*, US Bureau of Mines Information Circular 8103. USBM, Washington, DC, p. 43

Kirk W S 1983 Cobalt. *Minerals Yearbook 1983*, Vol. I. US Bureau of Mines, Washington, DC, pp. 1–10

Kirk W S 1985 Cobalt. *Mineral Facts and Problems*, Bulletin 675. USBM, Washington, DC, pp. 1–13

Mining Magazine Vol. 139, 1978. Finland: Meeting the challenge. No. 6, pp. 569–91

Vhay J S, Brobst D A, Pratt W P, Heyl A V 1973 Cobalt. *United States Mineral Resources*, US Geological Survey Professional Paper 820. USGS, Washington, DC, pp. 143–57

US Bureau of Mines 1988 *Mineral Commodity Summaries 1988*. USBM, Washington, DC

White L 1979 Zambia; Zaire. *Eng. Min. J.* 180 (11): 146–206

S. F. Sibley and W. S. Kirk
[US Bureau of Mines, Washington, DC, USA]

Construction Materials: Crushed Stone

Crushed stone is a prosaic but nearly indispensable construction material. About 3 billion tonnes are produced annually worldwide, excluding the centrally-planned-economy countries, from a great variety of rock types. This immense amount of material is used for a multitude of purposes, ranging from low-cost bulk aggregate for road building and repair—its largest use—to specialty products, such as fillers and coatings, which command high prices but are a comparatively small part of the total production.

Crushed stone must have physical properties that suit the requirements for each use. For example, crushed stone for road building must be tough enough to withstand the crushing stresses and wear of modern motorized vehicles, a property found in many different rock types, and rock used for whiting must be very white and bright, properties found in only a few types of rock. Thus, for any particular use there may be many types of suitable rock; while to cover the whole spectrum of usage, many different rock types are needed.

The crushed-stone industry seems nearly ubiquitous. Large reserves of stone are widespread, and even where modern efficient machinery is used in processing and transporting the stone, bulk stone cannot be hauled more than 40 or 50 km by truck before transporting the stone becomes more costly than producing it. Numerous mines therefore have been opened within short distances of each other to serve local needs. Because most, if not all, countries are self-sufficient in crushed stone for the same reasons, there is little international trade in crushed stone for aggregate.

1. *Definitions*

Crushed stone is rock that has been mined and crushed to sizes that meet various consumer requirements. It is also commonly described as an aggregate: i.e., a mass or an assemblage of sized rock fragments that does not denote any particular rock type, but that is commonly composed of only one or two rock types. Crushed stone can be bound in cement or in asphalt, or can be used loose for many products, such as roadstone, fill, ballast, riprap and agricultural limestone. This last use and others depend on the chemical properties as well as the physical properties of crushed stone, but it needs only to be durable and reasonably inert for most physical uses.

Classifications used by industry for types of crushed stone are somewhat ambiguous geologically, and include many rock types (Table 1). Most crushed stone is classified as limestone, which includes dolomite, a closely related carbonate rock. Limestone and dolomite are commonly intermixed and are described by geologists as dolomitic limestone (limestone that contains between 10% and 50% of the mineral dolomite)

Table 1
Selected rock types used for crushed stone

Rock group	Rock type	Commercial classification
Igneous	basalt	traprock
	gabbro	
	diorite	
	andesite	
	granodiorite	granite
	quartz diorite	
	quartz monzanite	
	syenite	
	granite	
Metamorphic	gneiss	
	slate	slate
	marble	marble or limestone
	quartzite	quartzite or sandstone
Sedimentary	orthoquartzite	
	sandstone	sandstone
	graywacke	
	conglomerate	
	coral	shell
	shell	
	dolomite	limestone or marble
	limestone	
	chalk	
	marl	

or calcareous (limy) dolomite (dolomite that contains between 10% and 50% of the mineral calcite). High-calcium limestone is more than 95% calcium carbonate ($CaCO_3$), ultrahigh-calcium limestone is more than 97% $CaCO_3$, and high-magnesium dolomite is more than 95% calcium magnesium carbonate ($CaMg(CO_3)_2$).

Granite includes all coarse-grained light-colored igneous rocks. Sandstone includes impure sandstones and most siliceous rocks with a grained texture, and may include the metamorphic rock, quartzite. But the commercial term quartzite includes true quartzite, a metamorphosed sandstone, and may also include the sedimentary rock, orthoquartzite, which is sandstone that is strongly cemented with quartz. Similarly, the commercial term marble includes true marble, a metamorphosed limestone, and may include sedimentary crystalline limestone that is dense and will take a polish. All fine-grained dark igneous rocks are commonly included in the classification traprock.

2. *Rock Types and Uses*

Of the estimated 2.6 billion tonnes of crushed stone produced in developed market economy countries in 1983, 0.8 billion tonnes, or more than one quarter, was produced in the USA. Limestone dominated the crushed-stone market in the USA; it accounted for nearly three quarters of the production and was followed by granite, traprock, sandstone, shell and marble. Limestone was also a large part of the 78 million tonnes of crushed stone produced in Canada and the 247 million tonnes produced in the UK. Other large producers of stone were the Federal Republic of Germany, Australia, France and Japan.

Limestone and dolomite, which are included in the same classification, have close geologic affinities. Dolomite is commonly a replacement for limestone, or an alteration of limestone to dolomite by the addition of magnesium. This chemical reaction takes place under natural conditions through geologic time and is accompanied by a molecular reduction in volume of about 11%, which may result in a dolomitic rock that is porous or vuggy.

Much limestone that is mined was first deposited in widespread sheets in ancient marine seas, although patchy deposits may be mined in some areas. Some strata were tilted or folded by large-scale earth movements through long periods of time. High-quality limestone or dolomite strata range from a few feet to hundreds of feet in thickness, and are commonly interbedded with strata of various rock types of varying quality. Extensive layers of limestone and dolomite may provide a widespread source of crushed stone that has consistent physical characteristics.

Reefal deposits, which support many quarries in the Great Lakes area of the central USA and are a source of high-quality limestone and dolomite in other areas, also grew in some of the ancient seas and may be over a hundred meters thick. Reefs of Silurian age in the Great Lakes area range in size from 30–60 m in diameter and less than that in thickness, to more than 1.5 km in diameter and 120 m thick. These reefs are surrounded by inter-reef limestones and dolomites that are only partly suitable for aggregate (Fig. 1).

Crushed limestone and dolomite have many uses. More than 30 uses were reported in a canvass by the US Bureau of Mines (Tepordei 1987), and more than 70 were described by Lamar (1961). Secondary uses, such as those for calcined lime, increase the number of uses several fold. By far the greatest use of crushed limestone and dolomite in the USA is for roadstone and coverings, which accounted for more than 142 million tonnes of the 649 million tonnes used in 1985. It was followed in usage by concrete aggregate which exceeded 79 million tonnes, and cement which was about 84 million tonnes.

The construction of a major interstate-highway system in the USA during the 1960s and 1970s used large amounts of crushed stone, mostly limestone and dolomite. But road building has decreased, and road repair and other construction needs have increased.

Crushed limestone with less than about 5% magnesium carbonate is mined for cement making in many localities, although the restrictions on its composition make suitable deposits much less common than those for ordinary aggregate. Demand and exploration for

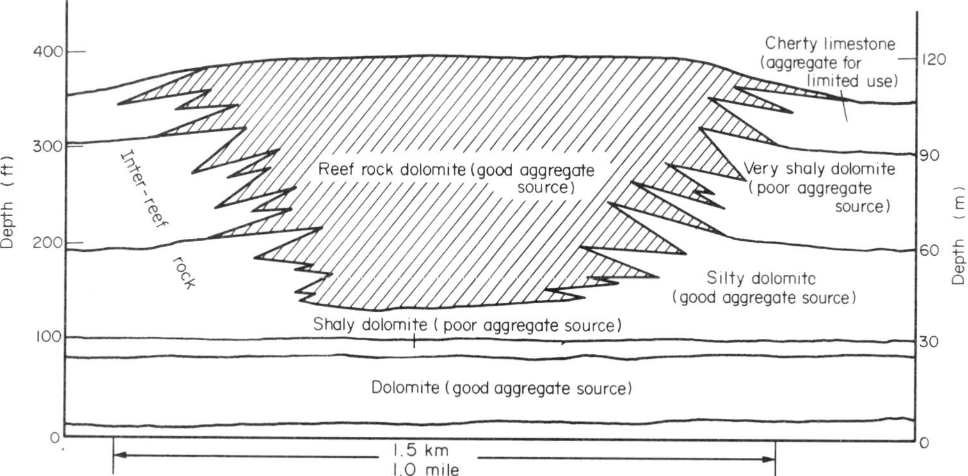

Figure 1
Cross section showing sources of crushed-stone aggregate in typical reefal and inter-reefal rocks of Silurian age in the Great Lakes area

relatively scarce new sources of high-calcium limestone, ultrahigh-calcium limestone, high-magnesium dolomite for calcined lime (includes calcined dolomite), fluxstone and many other chemical uses have increased in recent years.

Uses for crushed marble are nearly identical to those of limestone, with particular uses in terrazzo, as high-purity filler, and for decorative crushed stone because of its exceptional whiteness and brightness in some deposits. True marble is a metamorphosed limestone that is found in geologic terranes that have been subjected to great pressure and heat. The geographic distribution of marble is limited, and the amount used for crushed stone is further limited because of its greater value as dimension stone. Most marble that is crushed for aggregate comes from waste piles and discards at dimension-stone quarries and mills. Only 2.2 million tonnes was produced in the USA in 1985; that amount is less than 1% of the limestone used that year. An undetermined amount, probably quite small, is produced with and reported as limestone.

Chalk, which may also be classified as limestone, is produced in considerable amounts in the UK for cement making. A small amount from England and France is used as a mild abrasive. The high surface area and purity of chalk in some deposits make it an excellent raw material for stack-gas desulfurization processes, but its limited geographic distribution may keep it from becoming an important material for this purpose.

Marl, a calcareous material deposited especially in freshwater lakes, bogs and marshes, is an earthy substance that may be high-purity calcium carbonate, but more commonly contains considerable clay and silt impurities. It is crushed in comparatively small amounts for cement manufacture and soil conditioning.

Shell, which is produced in several southern states in the USA, is used mostly for construction and road aggregate, and in lesser amounts for cement making and lime manufacturing. Shell is derived mainly from fossil mounds of oyster shell, mostly by dredging, which limits its market area. Even so, 8.3 million tonnes was produced in the USA in 1985.

Although the commercial classification of granite includes the geologic rock types of granite gneiss, syenite, quartz monzanite, quartz diorite and diorite, most of the granite used for crushed stone has a composition in the family of granitic rocks that includes true granite and granodiorite.

Granite is a common igneous rock found in many places in folded and intruded mountain ranges and in shield areas. This distribution makes reserves of the stone practically inexhaustible in areas of widespread igneous terrane.

Fresh, unweathered granites are exceptionally durable and have low porosity, hardly any detectable loss on freeze-and-thaw tests, exceptionally high hardness, and high compressive strength. Because of the favorable physical properties of granite, nearness to market is more important for locating quarries than variation in physical properties.

In the USA, most of the granite used for crushed stone is for concrete and roadbase aggregate, which consumed nearly 45 million tonnes in 1985. Granite was also used for roadstone, concrete aggregate and bituminous aggregate. But total consumption in the USA for 1985 was about 132 million tonnes, an amount comparable to only 20% of the limestone produced.

Major physical uses of crushed granite parallel those of crushed limestone. Open-pit quarries for crushed granite are similar to limestone operations, except that equipment for producing crushed granite must take into account its high strength and abrasiveness. Most granites have preferred partings. This is of little consequence in open-pit quarries, but it is important for granite produced for dimension stone (see *Construction Materials: Dimension Stone*).

Nearly all traprock is basalt and diorite, although all dark fine-grained igneous rock falls within this classification (Table 1). Seventy-one percent of the 76 million tonnes of traprock used in the USA in 1985 was for construction aggregate.

Basalts and diorites occur mostly in flows of intrusive or extrusive sheets, sills and dikes that range in thickness from 15 m to more than 300 m. Weathered zones or vesicular layers may be undesirable, but great quantities of fresh, unweathered stone are usually available in areas where traprock is found. Traprock is the heaviest crushed stone of the above categories, which increases its cost of trasportation somewhat, but most traprock is tough and durable and makes an excellent aggregate. Its dark color may make it less desirable for aesthetic reasons.

The commercial classification sandstone includes graywacke, a tough sandstone that contains a considerable amount of impurities, and conglomerate, a poorly sorted pebbly sandstone. Since quartz is hard but brittle, sandstones tend to lack the toughness of limestone or granite, but will give satisfactory service for most aggregate uses if they are well cemented. Crushed sandstone is used for aggregate, ballast, fluxstone, and as a source of silica for cement among other uses, but in much smaller amounts than limestone; US consumption in 1985 was only 21 million tonnes. The quartz minerals in sandstone are abrasive and can cause rapid wear of processing equipment. Quartzite serves well as an aggregate and is crushed to sand sizes for high-silica uses.

3. Sampling and Testing

Physical characteristics of rock types are affected by their mineralogy, texture, cementation and fabric. Physical properties differ among rock types, and even among similarly classified rocks. Impurities or minor constituents can be a major cause of this variance. Thin shale laminae, for example, which forms only a small percentage of the rock in which they are found, can make the stone unsuitable for many purposes. Conversely, impurities may actually increase the durability and usefulness of some stone. For example, a fine-grained limestone that contains some disseminated quartz silt or sand may have more skid-resistance qualities or may be more durable than the same limestone without these "impurities."

Accurate sampling and testing when particular physical qualities are required for specific uses cannot therefore be overemphasized. Sampling and testing are inexpensive procedures when compared with the cost of opening a mine and building a crushing plant. Sampling patterns and techniques are nearly as varied as the rock types sampled, but the sample should be as representative of the rock being tested as it can be.

The number of samples needed to characterize a stone deposit adequately depends on the geologic variability and the intended use of the stone. Limestone that has physical properties near the minimum required for a particular use will require more samples to account for the variability than a limestone that is well above the minimum. If a sampling pattern for a few test holes reveals variability of physical properties that might make the deposit unsuitable for its intended use, additional holes will be needed to define the limits and location of suitable reserves.

Rock cores from drill tests will normally yield representative samples, although the core must be large enough to provide enough sample for testing, about 5 cm or more in diameter for most layered deposits. Thick, nearly homogeneous stone (e.g., some thick calcareous or dolomitic reefs) is characterized by fewer samples than thinly layered deposits, which may require many samples to detect adequately significant changes in physical properties.

Changes in the composition and gross texture of the stone obviously require additional sampling, whether from cores or from outcrops of the stone. Changes in color may not be indicative of significant changes in physical properties, but are always suspect, and differently colored rocks should be sampled separately.

Sampling stone that contains sporadic impurities presents problems. One common example is limestone or dolomite containing chert. The chert, especially if chalky and weathered, is deleterious in cement and bituminous mixtures. Chert can be present in bands or as discrete lenses or nodules. Commonly, the bands are not continuous, and a drill-hole test may penetrate between nodules or through gaps in the bands. A much better estimate of the amount of chert in a deposit can be had from quarry or outcrop exposures, although this is also not without its dangers, because the amount of chert in a limestone or dolomite may vary from one part of a quarry to another.

Because of its sporadic occurrence, then, any chert at all in a core makes that rock suspect for certain uses of aggregate, and a more precise quantitative measure of the amount of chert in stone may wait until the quarry or mine is opened and operating. A regional knowledge of the character of the rock and the amount of chert is of great help in the initial site evaluations.

Many tests for crushed-stone aggregate in the USA have been standardized by the American Association of State Highway and Transportation Officials (AASHTO) and the American Society for Testing and Materials (ASTM) (Table 2). Most states in the USA have additional specifications and tests for local needs.

Table 2
Selected tests standardized by the American Association of State Highway and Transportation Officials (AASHTO) and the American Society for Testing and Materials (ASTM) for crushed-stone aggregate

AASHTO designation	ASTM designation	Purpose	Procedure
Sampling			
T 2	D 75	Sampling stone, slag, gravel, sand and stone block	Sets forth procedures for sampling sources of aggregate at the deposit, at the plant and at the delivery point
T 248	C 702	To reduce field samples of aggregate to testing size	Describes the use of mechanical splitters and splitting by quartering
Particle size			
T 27	C 136	Determining particle-size distribution of aggregates	Crushed-stone aggregate is passed through a nest of sieves and the percentage retained on each sieve is calculated
Specific gravity and absorption			
T 84	C 128	Determining bulk and apparent specific gravity and absorption for aggregate	A weighed sample is introduced into a pycnometer, water added, and weighed. Calculations are from mass of pycnometer, amount of sample and mass of water
T 85	C 127	Determining bulk and apparent specific gravity and absorption for aggregate greater than 4.75 mm (No. 8 sieve)	Sample is dried, weighed, soaked in water and reweighed
Durability			
T 3	D 289	Testing resistance to abrasion by use of the Deval machine	Rock cubes are cut from aggregate particles, weighed and rotated in a cylinder of the Deval machine. Amount of loss of material below a 4.75 mm sieve is measured.
T 96	C 131	Testing resistance of aggregate to abrasion and wear by use of the Los Angeles machine	A weighed and sieved sample is tumbled in a cylindrical drum containing loose steel spheres. The amount of sample that is reduced below a 1.70 mm sieve is calculated
T 103		Testing resistance of aggregate to freezing and thawing in water	Sized and weighed aggregate is frozen in water and thawed for a specified number of cycles. Amount of loss of sample less than certain size is calculated
T 104	C 88	Testing resistance of aggregate to disintegration by solutions of sodium sulfate or magnesium sulfate	A sized and weighed sample is soaked in the solution, removed, and dried for a number of cycles. Loss below certain size is determined

Standards in the UK are issued by the British Standards Institution, and international standards are promoted by the International Federation for the Applications for Standards.

The most common physical tests include those for determining specific gravity, absorption, resistance to freezing and thawing (soundness), and resistance to abrasion (durability).

A knowledge of specific gravity is needed to determine the volume occupied by aggregate in concrete. Absorption measurements will indicate the tendency of crushed stone to absorb water from concrete mixes, which affects the strength of the concrete. Bonding between the particles and the cement is also affected by the water absorbed, and the amount of absorption of hydrocarbons is important in asphalt mixes. The standard AASHTO T-85 test for these properties for coarse aggregate is simple and inexpensive (Table 2). The maximum allowable absorption for many purposes is 5%.

Soundness properties are measured in part by subjecting crushed-stone particles to a series of freezing and thawing cycles to simulate seasonal changes and the action of ice and water on aggregate. Clay minerals, shaly laminae and finely fractured particles are particularly susceptible to these processes. Sedimentary stone containing abundant clay minerals or fine fractures frequently fails the freeze-and-thaw test, which has limits for most uses of about 10–20% loss on the ASSHTO T-103 test.

A more rigorous test for soundness is the sodium sulfate or magnesium sulfate test (AASHTO T-104), which simulates freezing and thawing by alternately soaking and drying the material in a saturated solution. As the particles dry, crystals of sodium or magnesium sulfate form and expand, which causes disintegration of the aggregate. This test is fast, but does not simulate actual field weathering as well as the more time-consuming freeze and thaw test. Some specifications, therefore, allow retesting of aggregate by freezing and thawing if the material fails the sodium sulfate test.

The grinding and crushing action that takes place when vehicles pass over loose aggregate or macadam roads, or when heavy trains flex railroad beds, is difficult, if not impossible, to reproduce in a laboratory. For lack of better terms, this resistance to grinding and crushing might be called durability or strength, but it is commonly by a test called resistance to abrasion—somewhat of a misnomer—that some measure of the property is taken.

AASHTO Test T-96, resistance to abrasion, subjects aggregate to at least part of this grinding and crushing action by rotating it in a cylindrical drum in which steel spheres are dropped on the aggregate every revolution. This test has been widely accepted in the USA for many years, although the relationship of the results of the test to the actual wear and tear of aggregate in roads remains unclear.

Research to understand problems with aggregate and to develop better methods of testing aggregate for roads and for many other uses is conducted in many countries, particularly Germany, Australia, the UK and the USA. This research is conducted mostly by governmental organizations, although universities and private organizations sponsor some research.

Of continuing interest is the work on antiskid aggregates from two different lines of approach: (a) determining kinds of aggregate that polish the most on road surfaces, and (b) determining rock types that have the greatest natural antiskid properties. A few standardized tests have been developed for measuring these properties. Some success has been achieved by using limestone, a rock that by itself may polish but that may contain impurities such as angular sand or chert particles, which provide an increase in friction in surface courses.

For a surface aggregate to remain rough through extensive use and to continue to provide good traction for a long time, the particles on the surface should wear at different rates and continue to present unworn particles to traffic. Such requirements may open a new market for aggregate containing impurities that have been rejected in the past, but much work must be done before practical and effective aggregates and mixes can be found or developed.

Considerable research also continues on better testing for durability and resistance to impact, and on the role of particle shape in bound or loose mixtures.

4. Mining and Processing

Sources of crushed stone have historically been so plentiful that early mining was simply a matter of removing stone from an outcrop or small open pit and crushing it by hand at the quarry or at the site of use. Only since the early and mid-1900s has widespread use been made of explosives and machinery to remove and process large amounts of stone from centralized quarries.

Although modern quarry operations are still relatively simple compared with some other industrial operations, important matters must still be considered. Stone, which has been cleared of overburden, must first be blasted from the quarry face. The method of blasting will partly determine the size and shape of aggregate particles and the amount of fines; the character of the rock itself will account for much of its shape. Sedimentary rock containing laminae of soft material, for example, may result in thin, elongate or bladed particles as compared with a homogeneous rock, which will produce stronger, more nearly cubical particles.

Stone that is reduced as near as possible to the desired product size by blasting is less expensive to produce than stone requiring undue mechanical crushing after blasting. Blasting for efficient production of rock fragments that will require minimal further processing includes judicious spacing and arrangement of blast holes, selection of correct types of explosive, designed loading of blast holes, and calculated timing of sequential blasts.

Broken rock from the quarry face is loaded into trucks by shovels or front-end loaders and transported to a primary crusher, which is usually as close as is practical to the quarry face. Crushing equipment can be selected to efficiently process the type of stone being mined, produce the size and shape of the aggregate particles needed, and minimize the amount of waste sizes. A jaw crusher, for example, may produce a slightly elongate particle but usually with less fines than an impact crusher, which may produce more equant particles and efficiently handle large amounts of some materials. Cost considerations of different types of crushers, or the ability of a crusher to efficiently process different rock types, can be more important to an operator than the slight differences in particle shapes or amounts of waste stone.

The crushed stone is screened and conveyed to stockpiles containing various sizes of aggregate, and any oversize particles that were screened out previously are crushed again in a secondary crusher. The stone is loaded into trucks, weighed and transported to the site of use. Rail, ship and barge transportation requires additional loading and unloading facilities, and is not economical unless the sources of stone are scarce and transporting the stone long distances can be justified.

Some of the largest operations are captive quarries producing stone for major industries, such as cement, steel and burned lime, but many more quarries produce stone for local construction needs, largely for road building and repair. Individual quarries may produce only a few tens of thousands of tonnes of aggregate per year, to more than a million tonnes per year near metropolitan areas.

Complicated quarrying can be caused by unusual or complex geologic structures. In areas where the stone strata are essentially flat lying, the quarry may extend over a large area if the stone section is relatively thin. Thick sections of suitable stone will encourage the quarrier to deepen his pit, especially in areas where additional land is high priced or unavailable because of environmental or other reasons.

Tilted or folded beds may necessitate production from irregular or deep, narrow quarries. Deep quarries may have greater water influx, and overburden thickness can increase dramatically in a short horizontal distance in tilted beds. Tunnel mines into the sides of quarries with thick overburden can be an economical solution.

Although stone reserves in many areas are essentially inexhaustible, lack of available land, intensive surface use of land, zoning regulations, thick overburden or environmental considerations may make underground mining attractive to crushed-stone producers. Underground mining, though commonly more expensive than open-pit mining, is becoming more common and has some advantages compared with open-pit mining. Underground mines can be operated on a year-round basis, and many environmental problems of surface mines can be eliminated or moved underground. Extensive removal of overburden is not required, and steeply dipping rock strata can be mined.

Tunnel mines into the sides of deeper quarries allow continued mining of high-quality stone in areas of high surface use, with a minimum increase in the cost of mining since in some mines much of the same equipment can be used in large underground rooms, and a new processing plant may not be needed, keeping costs down.

Combined mining of sand and gravel and crushed stone at the same mine site has become more common in the last decade. Sand and gravel pits near population centers may have bedrock floors suitable as a source of crushed stone at a depth shallow enough to be quarried after the sand and gravel have been removed. Intense use of surface lands or limited sand and gravel reserves may restrict horizontal expansion of the operation and make the bedrock the only source of additional stone reserves.

A fortunate set of geologic circumstances, however, must be present for combined mines to be practical. Sand and gravel must be sufficient to allow economical removal of all overburden from the bedrock, which then must be suitable for use as crushed-stone aggregate (see *Construction Materials: Sand and Gravel*).

5. *Environmental Concerns*

Mining is a heavy industry that will generate dust, noise, blasting vibrations and heavy traffic near surface operations. Underground mines are much less visible but also can affect the environment in significant ways.

The amount of environmental disturbance caused by a surface quarry depends mainly on its size and activity. The perception of its effect may be greatly influenced by its location: a mine near a metropolitan area is very noticeable, but a mine in sparsely settled country may attract little if any notice from nearby inhabitants. The impact of time and size on environment is illustrated by many small quarries that were opened to supply aggregate for rural-road construction during the early 1900s in the midwestern USA. Some of these quarries were active for only a few weeks and most of them have now been reclaimed by nature. They are overgrown with vegetation or filled in by stream sediments and blend well into the countryside. Although there are many more of these old quarries than there are active quarries today, they have practically no effect on the environment and in fact may be difficult to recognize in places.

The large centralized quarries of today, however, are impossible to miss, and much planning is required to make them acceptable, both as active quarries and as abandoned ones. Obvious steps can be taken to erect barriers of vegetation or fences to block the sight of unsightly equipment, and dust- and noise-suppressing equipment is helpful. Planning for eventual rehabilitation of a large quarry may be more complex, but the uses for the quarry can be numerous and range from water reservoirs in some old pits to drive-in theaters in dry pits. But quarries do have limitations for rehabilitation. Limestone quarries may not be suitable for waste disposal because of possible groundwater contamination through solution channels or porous formations. Water influx is common in quarries, and the expense of pumping the water out may make them unsuitable for many uses; but a water-filled quarry may have much utility and may increase the value of the site for industrial, residential or recreational development.

Underground mines remove much of the environmental disturbances of an open-pit quarry to underground quarters, but some surface installations still remain with some of the same environmental problems as surface mines. When an underground mine is abandoned, its utility depends on such factors as accessibility, roof stability, room size and amount of water influx. If a tunnel mine in competent rock is above the water table and has direct outlet to the surface from the side of a hill, the possibilities near an urban area for using the mine for offices, storage, underground factories, or many similar uses are excellent.

Underground mines served by a shaft may present difficulties, although a deep dry mine with no roof problems may have utility similar to that mentioned above. Mines below the water table may serve as a reservoir for water near areas of concentrated population.

Underground mines in durable materials may last indefinitely without roof collapse, and examples of such old mines can still be seen; but roof collapse, especially in shallow mines, can cause serious subsidence of surface land. The subsidence above room-and-pillar mines is variable and can be disruptive.

Planning for sequential land use solves many environmental problems. Much valuable stone that could have been mined by open-pit methods has been covered by industrial complexes or the buildings and streets of expanding cities and towns. Zoning areas for quarrying stone as the first land use and for reclamation of the sites for recreation or residential construction as the intermediate or the final use would have allowed exploitation of much stone that is now lost to man's use.

See also: Construction Materials: Fill and Soil; Construction Materials: Granules; Construction Materials: Industrial Minerals

Bibliography

Harben P W, Bates R L 1984 *Geology of the Nonmetallics.* Metal Bulletin, Worchester Park

International Road Federation 1980 *World Survey of Current Research and Development on Roads and Road Transport.* International Road Federation, Washington, DC

Lamar J E 1961 *Uses of Limestone and Dolomite,* Illinois Geological Survey Circular 321. Illinois Geological Survey, Urbana-Champaign, Illinois

Rooney L F, Carr D D 1971 *Applied Geology of Industrial Limestone and Dolomite,* Indiana Geological Survey, Bulletin 46. Indiana Geological Survey, Bloomington, Indiana

Tepordei V V 1985 Crushed stone. *Mineral Facts and Problems,* US Bureau of Mines Bulletin 675. USBM, Washington, DC, pp. 757–68

Tepordei V V 1987 Crushed Stone. *Minerals Yearbook 1985,* Vol. 1. US Bureau of Mines, Washington, DC

C. H. Ault
[Indiana Geological Survey, Bloomington, Indiana, USA]

Construction Materials: Dimension Stone

Dimension stone is a comprehensive term for rock that is cut, split or worked to form masses or units of desired shapes and sizes for specific uses, or that is selected from naturally occurring material to meet requirements of shape and size. The term does not cover crushed, broken or ground stone (see *Construction Materials: Crushed Stone*). Most dimension stone is used for building, but other uses are numerous and include monuments, markers, paving, flagging, curbing, whetstones, grindstones, surface plates, beds for billiard and pool tables, refractory units, chalk boards, electrical panels, laboratory sinks and furniture, curling stones, and a host of ornamental objects and industrial items.

1. Building Stone

Virtually every variety of rock that is firm enough to hold together until it can be put into its intended place has been used, somewhere and at some time, as building stone, but only five lithologic types of rock are produced commercially to any appreciable extent. These are granite, limestone, marble, sandstone and slate. Several of the terms as they are used in the stone industry have greater latitude than the strict petrologic definition would permit.

All of the three main generic rock types—igneous, sedimentary and metamorphic—are utilized as dimension stone. Igneous rocks are those that have cooled from a melt; they are made up largely of silicate minerals and free silica in the form of quartz. Those that cool at depth are called intrusive rocks; they range in texture from visible grain size to coarse crystals. Extrusive igneous rocks cool much more rapidly and are generally fine-grained. Some igneous rocks, called porphyries, consist of a fine-grained ground mass containing separate coarser crystals of one or more of the minerals.

Sedimentary rocks are deposited by the geological agencies of transportation (running water, marine currents, wave action or the wind) or by chemical precipitation.

Metamorphic rocks are formed from preexisting igneous rocks and sedimentary rocks, and from earlier to more advanced stages of metamorphic rocks, by pressure or heat (commonly both) within the earth's crust.

1.1 Granite

In geological terms, granite is an intrusive igneous rock composed of discrete, visible, crystals of quartz and the potassium and sodium feldspars, with a minor permitted content of biotite and muscovite mica, and of certain other accessory minerals, such as augite. Within the dimension stone industry, however, igneous rocks of wider variety are classed as granite. They include: quartz–feldspar mixtures in which calcium feldspars are fairly abundant or predominant; igneous rocks that have little or no free quartz; others that have a high, rather than accessory, content of biotite, augite and other ferromagnesian minerals; and porphyries in which the matrix is below visible grain size.

1.2 Limestone

A sedimentary rock that consists largely of the carbonate minerals calcite or dolomite, or a mixture of the two, is a limestone. The color is generally light, ranging from white or near-white through gray or buff or tan, but limestones may be dark because of mineral impurities. In texture, the limestones range from those that are so fine-grained as to lack visible particles, through the sand sizes, to coarse material; and in compactness they may range from dense to very open texture with sizable interstices, vugs and cavities. Bedding may be thin, thick or massive, and this controls the architectural use to which they may be put. Thin-bedded stone, in which nature has determined one dimension of the building unit, is usable for flagging, for vertical application and for masonry patterns emphasizing horizontal directions. Massive stone, from which sizable blocks may be quarried, is necessary for large masonry units, columns and sculpture.

1.3 Sandstone

A sedimentary rock composed of grains within the range of sand size (0.0625–2 mm) is a sandstone. Most of the sandstones used for building and other dimension purposes are composed largely of quartz grains, but the rock is still a sandstone if a high proportion of the grains are feldspar or other minerals. The cementation holding the sand grains together may be calcareous, dolomitic, siliceous or ferruginous. Siliceous sandstones, given equal degree of cementation, are harder than those that have carbonate cement, and in consequence they are more difficult to work but more durable. Sandstones range more widely in color than do the limestones. Many of those used in building are light, ranging from near-white to gray and tan, but highly colored varieties have dominated the building sandstone trade at various times and places. Examples are the "brownstone" of the Eastern Seaboard of the USA, red sandstones from the upper Midwest to the New York region of the USA, red and dark-gray sandstones in Scotland, and red sandstone in India. A wide range of colors, mostly in the warm hues, is produced in northeast-central Ohio, at Crossville, Tennessee, and at various locations in the western states. In some of these products, the same billet of finished stone contains a range of colors.

1.4 Marble

Geologically, marble is a metamorphic rock formed by the recrystallization of limestone or dolomite through some combination of heat and pressure, and many of the marbles utilized in classical structures of antiquity and up to the present day fit this definition, but by custom and precedent various rocks of other origin and somewhat different composition are classified as trade marbles. They include limestones and dolomites that will take a polish because they are microcrystalline or have recrystallized because of processes other than dynamic or thermal metamorphism, including alteration by groundwater. The term encompasses travertine, a generally banded hotspring or cavern deposit that typically has vesicular openings arranged in layers, and "onyx" marble, which is very fine-grained and is generally considered to have been formed by cold-water precipitation of calcium carbonate. Finally, the serpentine marbles are truly metamorphic and consist of that mineral, with or without accompanying carbonate minerals and chlorite or biotite; some contain flecks of magnetite. A necessary characteristic of all the trade marbles, except travertine, is that they take a polish; the solid part of the travertine structure does polish, but the voids prevent a mirror finish. Of all the building stones, the marbles present the widest variety of colors, in some instances as a single pervasive color and in others as a wide range distributed in flecks, bands, pods or aggregates. Some of these deposits of true metamorphic marbles are most elusive and difficult to explore and to develop, as color, texture and composition may change drastically in a short distance within the quarry area. Thus, an unusual marble may be literally "worked out" with little prospect of finding more—this is true to a greater extent than in the case of other main types of building stone.

1.5 Slate

The fifth major category of dimension stone is slate, which is metamorphic. It is a microcrystalline rock produced by dynamic metamorphism of clays and shales, causing the flat, micaceous particles to be oriented in planar fashion, with their long axes perpendicular to the direction of pressure. This gives rise to slaty cleavage which permits the rock to be split into sheets. The original bedding of shale may show faintly, or prominently, across the cleavage surfaces of slate, and for most dimension stone uses this is an undesirable characteristic. Slate of this kind is termed "ribbon" slate.

2. *Nature of Building Stone Use*

Within the last century, a vast change has taken place in the manner in which stone is used for building. Through most of the nineteenth century stone was used largely as a structural material to carry the weight of the building. The walls were bearing walls, and the load of the roof and floors was supported by the exterior walls, and by interior masonry walls, commonly of brick. Beams, joists and rafters were carried on the exterior and interior masonry walls. In these circumstances a great volume of stone was used. To erect a tall building it was necessary for the walls to be very thick at the base, although they could be progressively narrowed upward as the load decreased. Fenestration and entryways were awkward and cumbersome because of the wall thickness. The advent of the elevator permitted buildings to be ever higher, and it was necessary to devise new methods of building. The resulting development was the skeletal framework, in which structural steel or reinforced concrete, or a combination of the two, carried the load of the building, and stone came to be used mainly as cladding. Under these circumstances the stone could be appreciably thinner, and needed to be to reduce weight and unnecessary bulk. Originally, much of the stone cladding was still fairly thick, but it has become progressively thinner, in some instances so thin as to cause problems with anchoring the stone to the walls and with wind load. These problems have been increased by a trend towards panels of larger area, in many instances extending floor-to-floor, and even to multifloor panels. In single-story and other low-rise residential and commercial buildings, stone is used as veneer over frame and block buildings. These developments have resulted in the use of a lower volume of stone relative to the amount of space enclosed.

3. Quarrying

Dimension stone is quarried by numerous methods that range from primitive to advanced. At most sites of open-cut removal, soil or other unconsolidated material overlies the bedrock, and at some it is also necessary to remove rock overburden to gain access to the stone to be quarried. Soil cover is generally removed by bulldozers or other earth-moving machinery. Hydraulic sluicing has been used on occasion, and may still be applied in places, although environmental protection controls, where they exist, act as deterrents. Rock overburden is commonly quarried away by methods similar to those employed for the desired stone if blasting is likely to damage the material beneath, as is frequently true for limestone and marble. Explosives in drill holes are used for loosening unusable rock overburden in many granite and sandstone quarries. Even where the stone that is sought is the most shallow bedrock, the surface to some variable depth may be weathered, jointed or creviced to an extent that precludes efficient recovery and necessitates removal as waste.

Most building stone is obtained from surface quarries, but underground removal is feasible, particularly for limestone and marble. A room-and-pillar pattern is used. Recently developed belt saws and chain saws are capable of horizontal as well as vertical cutting; they are called "gallery" saws, and they hold promise for more efficient underground extraction. Subsurface quarrying already enjoys the advantages of stabilized year-long working conditions, improved ability to follow desired types of stone without removal of excessive amounts of overburden and reduced impact on the surface environment.

One method of quarrying employs a channelling machine that moves along a track, cutting a slot, typically about 5 cm wide, along one or both sides. The cutting is carried out by multiple cutter bars, sheaved together, and edged and hardened at their lower ends. These bars, actuated by a cam, move successively upward and downward, deepening the channel, as the machine moves back and forth by some amount ranging from a small fraction of a centimeter in hard stone to several centimeters in very soft stone.

A second method of separating stone from the bedrock is by use of wire saws. A power-driven, endless, helical wire is drawn initially over the surface of the ledge, and later in the slot that it has cut, carrying a mixture of water and abrasive. Quartz sand or crushed chert (flint) is sufficiently harder than limestone or marble to serve as the cutting agent, but for granite a harder abrasive, such as silicon carbide, is necessary. The wire is under tension, exerting constant pressure on the stone at the base of the slot, and feeding over sheaves that slowly descend or advance as the slot deepens. A single length of wire, typically 300 m or more, is actuated by a power unit and is so rigged over an assembly of pulleys and posts that it may perform many cuts at different places in the quarry simultaneously. Wire sawing is used principally for vertical cutting, but it can be used for undercutting in the horizontal direction.

A variant of the wire saw, generally used in shorter increments, carries "beads," impregnated or coated with diamond and spaced along the wire.

Especially in the limestone industry, either the channelling machine or the wire saw permits the separation of long "cuts" about 1.2 m in width and ranging in depth from about 2 to 4 m. At either end of such a cut, crosscuts are made by either channeller or wire saw, and a substantial tonnage of stone is thus freed from bedrock, except at the base where, unless a natural separation is present, sleeved wedges or points may be driven into closely spaced pneumatic drill holes until the cut breaks. Alternatively, the base may be freed by undercutting with a wire saw or chain saw. Then the cut, in some instances as long as 20–25 m, is tipped forward (turned down) by crane, derrick or other heavy equipment, or increasingly by distention of inflatable air bags that are lowered into the channel or slot behind the cut, onto a series of "pillows" made up of chunks of stone that crush on impact and cushion the fall, reducing or eliminating breakage. Once on its face, the cut is sectioned into mill blocks of desired size by using wedges and slips in pneumatic drill holes.

An alternative to channelling or wire sawing, but a system that still permits the turning down of cuts, is the use of a quarry bar, a massive assembly along which a drill may slide and be locked into successive positions that permit the drilling of closely spaced holes with webs of stone between them—the web then being removed with a spade-like broaching bar, also actuated by a pneumatic drill, leaving a ribbed vertical surface.

Chain saws of various types may be utilized either to cut channels or to slice a quarry floor into sections of desired width. A single chain saw, its teeth generally armored with tungsten carbide, diamond, or other hard materials, may travel on a track to cut a single channel, or multiple saws may be ganged to create thinner units and supplant a substantial part of the later mill work that is generally required. Chain saws are more effective in softer stone than in harder material. They are most commonly used for fairly soft limestones, although they will function in firm, well-consolidated stone, the speed of cutting diminishing with increased hardness. They may be used for sandstone, slate and marble.

Belt saws, which substitute tough, mechanically driven belts for toothed chains, are a relatively recent development of impressive effectiveness. The belts are cored with multiple steel wires that pass through spaced metallic segments impregnated to some depth with diamond.

For both granite and sandstone, it is possible to drill holes 30 cm or more apart in the same plane and achieve a fairly flat surface of breakage by detonating charges in them.

With granite, and no other common type of dimension stone, it is possible to cut a channel or slot by a process called jet piercing, which utilizes a very high-temperature flame attained by combustion of fuel oil or diesel oil in a high-pressure discharge of compressed air. The cutting action results from thermal shock, the coefficient of thermal expansion being different in the various minerals that constitute the granite, with the result that a shower of rock flakes is constantly driven off in advance of the flame tip.

Hydraulic piercing, achieved through use of a water jet at very high pressure is at an early stage of development but is successful in rock as obdurate as granite and may further supplant other methods for various parts of the quarrying process as the technology improves.

All of the foregoing methods apply to situations in which stone must be actually cut from a continuous body, but it should be noted that much sedimentary dimension stone is taken up in its natural bedding thickness, one of the dimensions having been determined by nature, leaving only the need to free the stone in the other two directions by one or another of the methods previously described. Similarly, marble and slate, and especially granite, may have the vertical dimension of the blocks established by sheeting, roughly following the earth's surface, occasioned by reduction of pressure as the rocks formed at depth were unloaded by erosion following uplift.

4. Milling

Building stone may be used as it comes from the quarry or in the form in which it is collected as field stone, stream shingle or boulders, but most is processed to a minor or major degree in a building stone mill.

Mill blocks of granite, marble, sandstone or limestone are generally trimmed or slabbed in a gangsaw, which slices the block to the desired thicknesses with parallel steel blades that travel back and forth, deepening the cuts with each pass. The steel blades may carry an abrasive (sand, crushed chert, steel shot or silicon carbide) in a water or slurry mixture to do the cutting, or they may have toothed diamond or tungsten carbide inserts. The various gangsawing methods yield different finishes, sand or toothed blades being the smoothest. Alternatively, the blocks may be slabbed or trimmed by wire saws, including those that use spaced diamond-impregnated beads as the cutting agents, or by circular saws.

The slabs from such sawing procedures are commonly trimmed to nearer final size by diamond-toothed circular saws. In the softer varieties of stone the principal face of the unit, the edges and moldings may be planed to the final finish and shape. Carved details and parts of the work initiated by the planer (such as the ends of rabbeted grooves) must be finished by hand. Columns, balusters and any turned work are done on a lathe. Concealed features, such as anchor slots and holes for hanging the units, and relieved areas for conduit and framework are provided by stone cutters. Stone carvers execute Ionic, Corinthian or compound capitals, decorative details, rose windows, tracery, sculpture, friezes and inscriptions.

A high proportion of building stone is sawed or otherwise cut in some way, but in much of the industry the term "cut stone" refers to units that are completely dimensioned for installation at a specific location in the finished structure. Numerous units that appear to be identical after placement are actually different in their accommodations for anchorage. Cut stone is installed by stone setters, a separate craft.

Stone supplied in modular course heights and cut to length on the job (in which case wall pattern is determined by the mason) is not, under such a restricted definition, considered to be cut stone.

Finishes range from rock face, which means a broken irregular surface, through sawed, planed, honed and polished surfaces. Broken but fairly planar surfaces result from splitting sawed slabs with a hydraulic shearing device, either straight-bladed or with articulated teeth, called a "guillotine." Flame finishes for granite are achieved through the thermal shock from a high-temperature jet flame passed over the surface, causing spalling. Grooved, ribbed and other textured surfaces are created by routing tools in planers, or by hand work. Batted or scabbled surfaces may be imparted by hand tools such as scabbling picks. Bush-hammered finishes are created by hand or mechanical work.

5. Specifications and Tests

Since rocks are inhomogeneous and naturally formed materials, varying in such aspects as mineralogy, texture, cementation and moisture, and since test specimens cannot be prepared and tested uniformly, they exhibit variations in physical properties, such as resistance to abrasion, absorption, compressive strength, rupture modulus and specific gravity.

Technical bodies and trade organizations have established standards to assure the quality of building stone, and testing procedures have been developed for determining whether materials meet stated requirements. Among the most widely recognized are those issued by the American Society for Testing and Materials (ASTM). ASTM specifications for various rock types include ANSI (American National Standards Institute)/ASTM specifications for limestone, sandstone, granite, marble and slate. Some are restricted to interior or exterior use, and some are for special uses, such as slate for backboards or roofing. Comparable specifications and tests, in general based on the same physical characteristics, exist in various other countries.

6. *Selection of Stone*

Aesthetic considerations are more important than any others in the selection of building stone by architects for most purposes. Exceptions are surfaces subject to substantial wear from foot traffic; surfaces vulnerable to attrition and impact, such as base courses; interior base courses likely to be affected by washing and chemicals; and exterior paving that may be damaged by salt or other deicing compounds.

Persons familiar with building stone are frequently asked what guidance the standards, tests and specifications offer for selection of stone for specific construction projects, and the answer is that this is not their purpose, although they may be helpful. Examples of their usefulness are to be found in the choice of stone for stair treads, flooring and paving where the stipulation of minimum resistance to abrasion may be important, depending on the amount of traffic that may be expected. Absorption is a consideration where exposure to liquids, to spillage and soiling, or to freezing are likely to be factors. For the most part, however, the physical characteristics expressed in the standards are principally assurance to the consumer that the stone purchased falls within the satisfactory range for the type of stone that has been selected. For example, minimum compressive strength is not specified for, say, the various categories of limestone or granite to guarantee that the stone will not crush in use, as most construction practice does not subject the stone to vertical pressures that will pose a problem in this respect. For example, Currier (1960) mentioned that the load on the stone at the base of the Washington Monument, a 169 m high load-bearing column, is only 4140 kPa; hollow load-bearing concrete block used in the same way requires compressive strength of only 4830–6890 kPa, and hollow block that is not load bearing (much building stone is not load bearing in use) has only 2070 kPa minimum. However, the specified minimum compressive strength for one category of limestone is 27 600 kPa, and the minimum for granite is 131 000 kPa. Compressive strength that falls below the normal range for the type of stone being tested indicates possible flaws, or at least peculiarities, that suggest the possibility of performance unsatisfactory in ways other than compressive failure.

The ultimate test of whether stone is satisfactory for any given purpose is performance, and various building stones have withstood the test of time, in many cases for centuries, in all types of buildings and in bridges and monuments distributed through a wide diversity of climates and environments. Such difficulties as have been encountered are attributable in a high proportion of the cases to placement in unsuitable situations (below grade, in constantly wet locations or subject to salt action), to faulty construction procedures (improper anchorage or lack of expansion space), and to the effects of incompatible associated materials (e.g., leaching of alkalis from mortar or concrete), rather than to any inherent frailty in the stone.

See also: Construction Materials: Crushed Stone; Construction Materials: Fill and Soil; Construction Materials: Granules; Construction Materials: Industrial Minerals; Construction Materials: Lightweight Aggregate

Bibliography

American Society for Testing and Materials 1984 *Annual Book of ASTM Standards*, Sect. 4, Vol. 4.08, *Soil and Rock: Building Stones*. American Society for Testing and Materials, Philadelphia, Pennsylvania, pp. 1–40, 78, 79

Ashurst J, Dimes F G 1977 *Stone in Building: Its Use and Potential Today*. Architectural Press, London

Bowles O 1939 *The Stone Industries: Dimension Stone, Crushed Stone, Geology, Technology, Distribution, Utilization*, 2nd edn. McGraw-Hill, New York

Bowles O 1960 Dimension stone. In: Gillson J L (ed.) 1960 *Industrial Minerals and Rocks: (Nonmetallics Other than Fuels)*, 3rd edn. American Institute of Mining, Metallurgical and Petroleum Engineers, New York, pp. 321–37

Bowles O 1960 Slate. In: Gillson J L (ed.) 1960 *Industrial Minerals and Rocks: (Nonmetallics Other than Fuels)*, 3rd edn. American Institute of Mining, Metallurgical and Petroleum Engineers, New York, pp. 791–98

British Standards Institution 1957 *Glossary of Terms for Stone Used in Building*, BS 2847, BSI, London

Currier L W 1960 *Geologic Appraisal of Dimension-Stone Deposits*, US Geological Survey Bulletin 1109. USGS, Reston, Virginia

Patton J B 1974 *Glossary of Building Stone and Masonry Terms*, Indiana Geological Survey Occasional Paper 6. Indiana Geological Survey, Bloomington, Indiana

Patton J B, Carr D D 1982 *The Salem Limestone in the Indiana Building Stone District*. Indiana Geological Survey Occasional Paper 38. Indiana Geological Survey, Bloomington, Indiana

Power W R 1983 Dimension and cut stone. In: Lefond S J (ed.) 1983 *Industrial Minerals and Rocks: (Nonmetallics Other than Fuels)*, 5th edn., Vol. 1. American Institute of Mining, Metallurgical and Petroleum Engineers, New York, pp. 161–81

Winkler E M 1973 *Stone: Properties, Durability in Man's Environment*. Springer, New York

J. B. Patton†

Construction Materials: Fill and Soil

Fill is any man-made deposit of natural soil, rock products or waste materials. Fills can be carefully constructed to form dense, stable embankments, or they can be random accumulations of soils, rocks, organic materials and even trash and garbage. To the geologist, pedologist and agronomist, soil is the very superficial layer of the earth's crust which is capable of supporting plant life. To the civil engineer, interest in the soil goes much deeper because of construction activities which must take place on or in soil, such as foundations and tunnels, or which use soil, such as dams, embankments and levees. Thus, soils in an engineering sense are the sediments and other unconsolidated accumulations of materials found above the bedrock, which are produced by the physical and chemical disintegration of rocks and which may or may not contain organic or even man-made materials.

Even though soils often possess less than desirable properties compared with other civil engineering materials such as concrete and steel, they are often used as engineering materials because they are so economical. But they must first be stabilized, reinforced or otherwise changed to improve their engineering properties.

1. Soil Formation and the Nature of Soil Constituents

Soils are primarily the products of weathering in the earth's crust. Physical (mechanical) weathering causes disintegration of rocks by alternate freezing and thawing, temperature changes, erosion, glaciation and the activities of plants and animals. Chemical weathering decomposes rock minerals by oxidation, reduction and carbonation. In general, chemical weathering is much more important than physical weathering in soil formation. Soils can be weathered in place (residual), or they may be transported by water, wind, glaciers or gravity. Soils may consist of purely mineral matter or they may also contain organic material. Some cumulose soils such as peat and muck may be almost entirely organic. Because soils consist of discrete particulate materials, they have voids and usually the voids contain both air and water.

The predominant minerals in rock and therefore in soils are silica, mica, feldspar, calcite and dolomite, ferromagnesium minerals, iron oxides and the clay minerals. The latter are basically hydrous aluminosilicates plus other metallic ions which result from chemical weathering of especially feldspars, micas and the ferromagnesium minerals. They are very small plate-shaped crystals, with a diameter of the order of 1 μm or less. The most important clay minerals in soils are kaolinite, montmorillonite, illite and chlorite. Detailed descriptions of clay minerals are given in Holtz and Kovacs (1981) and Mitchell (1976), as well as in other articles in this Encyclopedia (see *Binders: Clay Minerals*; *Fillers and Coatings: Clay Minerals*; *Filters, Sorbents and Ion Exchangers: Clay Minerals*; *Well-Drilling Materials: Clay and Nonclay Minerals*).

2. Soil Texture and Classification

The texture of a soil is related to its appearance or "feel," and it depends on the relative sizes and shapes of the particles as well as on the distribution of those sizes. Soils with particles larger than about 0.05 mm, which is about the smallest grain that is visible to the naked eye, are coarse grained and they appear coarse textured. Examples are boulders, cobbles, gravels and sands. Engineering properties of these soils are strongly dependent on grain size and distribution and grain shape. Fine-grained and fine-textured soils are silts and clays. The term clay is applied to soils which contain clay minerals in sufficient quantities to affect their engineering behavior. The behavior of fine-grained soils, especially clays, has very little relation to texture, grain size, shape or grain size distribution. Their specific surfaces are relatively large and thus their behavior is more strongly affected by electrochemical than by gravitational forces. Clays have plasticity and cohesiveness, and the presence of water significantly affects their engineering behavior. Silts, although fine grained, are granular, nonplastic and cohesionless.

Soil classification based on soil texture and plasticity is very useful for correlation with engineering properties. Many systems have been proposed for specific engineering purposes; for example, highways, earth dams and airfield construction.

3. Grain Size and Grain Size Distribution

The range of particle sizes in soils spans at least eight orders of magnitude. Sizes can range from boulders or cobbles of several centimeters in diameter down to ultrafine-grained colloidal particles. Even though the individual particles may have shapes very different from spheres, mean diameters of the particles are determined either by mechanical (sieve) analysis or some procedure using Stokes' law (American Society for Testing and Materials 1986). Typical cumulative grain size distribution curves are shown in Fig. 1.

4. Atterberg Limits of Fine-Grained Soils

The presence of water in the voids is probably the most important factor affecting the engineering behavior of fine-grained soils. Not only is the water content important but it is also useful to compare the water content with certain standards of engineering behavior. Atterberg limits indicate different boundaries of mechanical response in terms of water content, and they are very useful in the classification of fine-grained soils because they are good indices of soil behavior

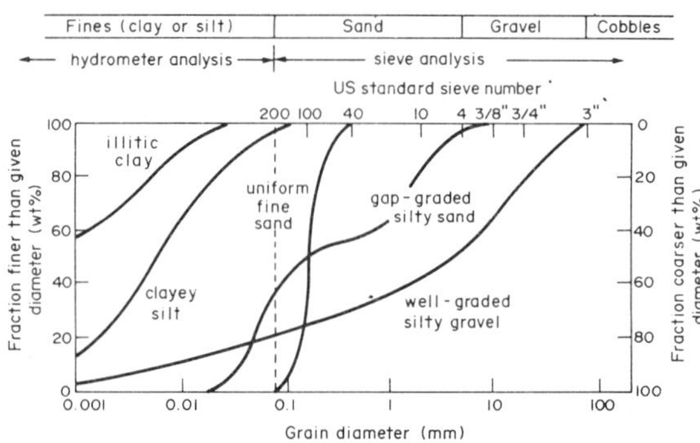

Figure 1
Typical grain size distribution curves

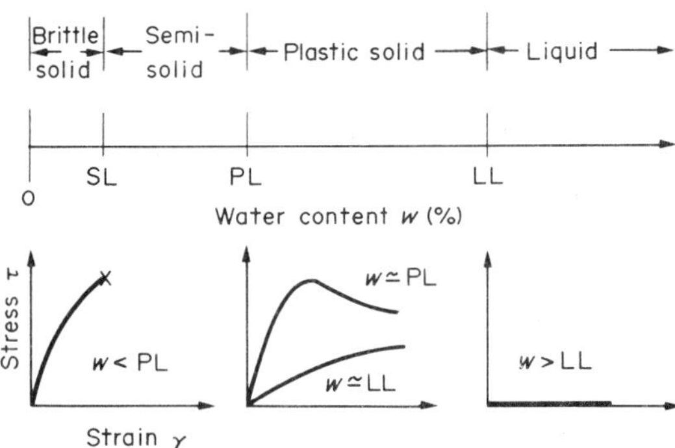

Figure 2
Behavior states and stress–strain response of soils as a function of water content

(Holtz and Kovacs 1981). It is possible for a soil to have behavior ranging all the way from a brittle solid to a liquid depending only on the water content (Fig. 2). At water contents (w) greater than the liquid limit (LL), soils can support essentially no static shear stress. For water contents less than the liquid limit but greater than the plastic limit (PL), soils behave as if they were plastic or Bingham solids. This range is called the plasticity index, or $PI = LL - PL$. At water contents below PL, soils behave as if they were semisolid or even brittle. Below the shrinkage limit (SL), no further decrease in volume occurs as drying continues. Standard tests for Atterberg limits are given by the American Society for Testing and Materials (1986).

5. *Soil Structure and Fabric*

The structure of a soil mass controls to a very large extent its engineering behavior, and this is true for both natural as well as man-made soil deposits. Soil structure includes both the geometric arrangement or fabric of the soil particles and their interparticle forces. For granular soils, since the surface activities and interparticle forces are small, structure is identical to fabric. In general, the fabric of these soils is single grained. Fabrics range from loose with a low bulk density and open structure to dense with close packing and a high density. It is also possible for the grains to have a bulked or honeycomb structure which can support some static load but collapses if it is vibrated. Large masses of loose saturated sands may liquefy if subjected to dynamic loading such as earthquakes.

With the development of the scanning electron microscope, fabrics of fine-grained soils have been found to be extremely complex. Because of their large surface activities, they always seem to be aggregated or flocculated into basic fabric units (domains) and fabric assemblages (clusters and peds) with pore spaces in between.

The macrostructure, including peds, joints, fissures, silt and sand seams, root holes and varves, often controls the engineering behavior of soil masses. For example, the strength of a soil is significantly less along a crack or fissure than through the intact material, while the permeability is greater because water flow is concentrated in joints and fissures. Engineering problems involving stability, settlement or drainage must carefully consider soil macrostructure. Although the structure of natural deposits is destroyed by excavation, soils in earth fills can have rather complex fabrics owing to the construction process (Mitchell 1976).

6. *Engineering Properties of Soils*

Important engineering properties of soils include permeability, compressibility, modulus and shear strength. For fills and embankments, the shrink–swell and frost susceptibility characteristics also must be considered (Holtz and Kovacs 1981). Permeability determines how water flows through the soil pores, and is a function of the texture, grain size and distribution, density, and soil structure. The physical and chemical interaction of the pore water and air with the solid particles in the soil structure causes both natural and man-made soil deposits to have very complex mechanical behavior. The response of soils to boundary stresses depends on the rate of load application, permeability and soil structure. For example, fine-grained saturated soils loaded rapidly will develop pressures in the pore water that strongly affect both their modulus and strength (Holtz and Kovacs 1981). The developed pore pressures can be either negative (dilatancy) or positive, depending on the density and stress history of the soil. Because soils are in general nonlinear, nonconservative, anisotropic, viscoelastic–plastic materials, it is impossible to formulate exact mathematical descriptions of their stress–strain–time behavior. Consequently, for engineering purposes great simplifications are made, and designs are often based on empirical rules of thumb, precedence and the experience of the engineer. Empirical correction of "safety" factors must be applied to account for real material response and the unknowns imposed by the geological conditions and the construction process. Therefore, the engineering of soil materials is more of an art than a science (Terzaghi and Peck 1967, Winterkorn and Fang 1975).

7. Stabilization and Compaction

Because their engineering properties are usually poor, the economic use of soils in fills requires some type of improvement or stabilization.

Stabilization can be thermal, electrical, chemical or mechanical (compaction). The first two are only occasionally used. Chemical stabilization includes the mixing or injecting of materials such as portland cement, lime, asphalt and other bituminous materials, calcium chloride, sodium chloride and even industrial wastes into soils (Winterkorn and Fang 1975, American Society of Civil Engineers 1978, 1987).

Mechanical stabilization, called compaction, is the densification of soils by the application of some type of mechanical energy. Densification and particle rearrangement occur because air is expelled from the soil voids without significant change in the water content. Soil strength is increased, compressibility and permeability decreased and detrimental effects of shrinkage, swelling and frost action often can be ameliorated. Desirable engineering properties of compacted fills can be achieved by proper consideration of the soil type and the most appropriate compaction technique and equipment (United States Bureau of Reclamation 1974).

Cohesionless soils are most efficiently compacted by vibration. Particle rearrangement and densification occur owing to cyclic deformations produced by the vibrating equipment. Equipment used for compacting granular fills includes vibrating plates, tampers, motorized rollers and large weights dropped from a crane.

With increasing cohesion the efficiency of vibration as a densification agent significantly decreases because even slight bonding or cohesion between the particles interferes with particle rearrangement. In the field, common compaction equipment includes hand-operated tampers, sheepsfoot rollers and rubber-tired rollers. Rubber-tired rollers are most effective for compacting sandy and silty soils with relatively low cohesion. Sheepsfoot rollers are better for more cohesive soils. Most efficient compaction takes place if the soils are laid down and compacted in relatively thin (15–30 cm) layers or lifts.

8. Theory of Compaction of Fine-Grained Soils

The compaction of fine-grained soils is a function of dry density, water content, soil type and compactive effort which is a measure of the mechanical energy applied to the soil mass (Holtz and Kovacs 1981). In the field, compactive effort is the number of passes or coverages of a roller of a certain type and mass. In the laboratory, a hammer is dropped several times on a soil sample of known volume, or a tamper kneads the soil by applying a given pressure for a few milliseconds which is supposed to simulate the effect of field compaction equipment.

The process of compaction can be illustrated by the common laboratory compaction or Proctor test (American Society for Testing and Materials 1986). Several samples of the same soil, but at different water contents, are compacted with the same compactive effort. When the bulk dry densities of each sample are plotted against water content, a characteristic parabolic curve is obtained (Fig. 3). The curve is unique for a given soil type, method of compaction and compactive effort. The water content corresponding to the maximum dry density is called the optimum water content. Typical values of maximum dry density are 1.6–2.0 Mg m^{-3} with an outside range of about 1.3–2.4 Mg m^{-3}. Optimum water contents range between 10 and 20% with a maximum range of 5–40%.

Even at very high water contents and high compaction energies, the compaction curve never quite reaches the 100% saturation line. However, more energy will increase the maximum dry density and decrease the optimum water content (curve B, Fig. 3).

Different soils will have different shaped compaction curves even when compacted by the same compactive effort (Fig. 4). Well-graded soils will have a higher dry density than uniform soils, and, as plasticity increases, the maximum dry density tends to decrease. Different types of compaction, that is, impact, static or kneading, produce different soil structures and therefore somewhat different shaped compaction curves (Holtz and Kovacs 1981).

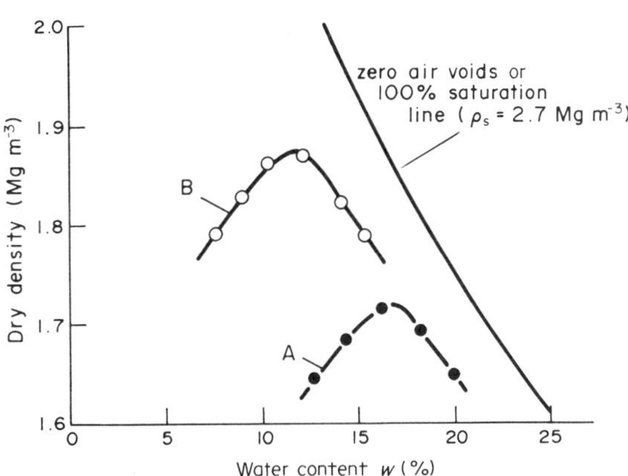

Figure 3
Typical standard Proctor (A) and modified Proctor (B) laboratory compaction curves

Table 1
Comparison of soil properties: dry of optimum versus wet of optimum compaction

Property	Comparison
1. Structure:	
A. Particle arrangement	Dry side more random
B. Water deficiency	Dry side more deficient; thus imbibes more water, swells more, has less (negative) pore pressure
C. Permanence	Dry side structure more sensitive to change
2. Permeability:	
A. Magnitude	Dry side more permeable
B. Permanence	Dry side permeability reduced much more by permeation
3. Compressibility:	
A. Magnitude	Wet side more compressible in low pressure range, dry side in high pressure range
B. Rate	Dry side consolidates more rapidly
4. Strength:	
A. As molded:	
a. Undrained	Dry side much higher
b. Drained	Dry side somewhat higher
B. After saturation:	
a. Undrained	Dry side somewhat higher if swelling is prevented; wet side can be higher if swelling is permitted
b. Drained	Dry side about the same or slightly greater
C. Pore water pressure at failure	Wet side higher
D. Stress–strain modulus	Dry side much greater
E. Sensitivity	Dry side more likely to be sensitive
5. Shrinkage	Wet side greater
6. Swell	Dry side much greater

9. Properties of Compacted Cohesive Soils

The structure, and thus the engineering properties, of compacted cohesive soils depends on the type of compaction, compactive effort, soil type and water content. When soils are compacted dry of the optimum water content, the structure is always flocculated and the engineering properties are essentially independent of the type of compaction. However, such soils are much more sensitive to changes in water content because of environmental factors. The type of compaction has a significant effect on soils compacted wet of optimum. As the water content increases, the soil fabric becomes increasingly more oriented as clusters of particles tend to align themselves along zones of

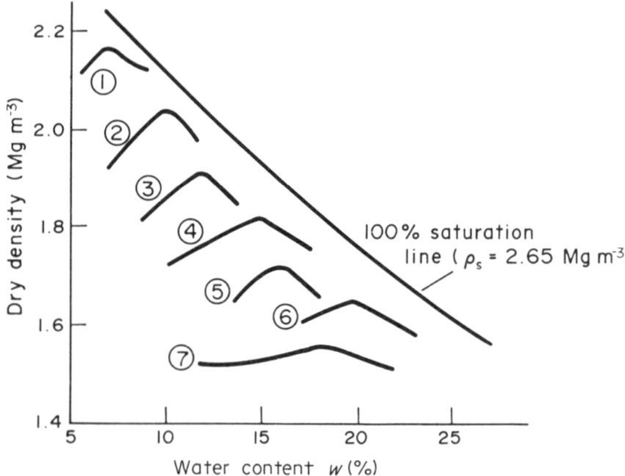

Figure 4
Water content/dry density relationships for seven soils compacted by the standard Proctor method: (1) well-graded silty sand; (2) well-graded clayey sand; (3) medium-graded clayey sand; (4) sandy silty clay; (5) clayey silt; (6) fat clay; (7) poor graded sand

movement caused by the compacting device. Table 1 compares different soil properties wet and dry of optimum.

See also: Construction Materials: Crushed Stone; Construction Materials: Dimension Stone; Construction Materials: Granules; Construction Materials: Industrial Minerals; Construction Materials; Lightweight Aggregate

Bibliography

American Society for Testing and Materials 1986 *Annual Book of ASTM Standards*, Vol. 04.08, *Soil and Rock; Building Stones.* American Society for Testing and Materials, Philadelphia, Pennsylvania

American Society of Civil Engineers 1978 *Soil Improvement: History, Capabilities, and Outlook.* Committee on Placement and Improvement of Soils, Geotechnical Engineering Division, American Society of Civil Engineers, New York

American Society of Civil Engineers 1987 *Placement and Improvement of Soils: A Ten Year Update.* Committee on Placement and Improvement of Soils, Geotechnical Engineering Division, American Society of Civil Engineers, New York

Harr M E 1977 *Mechanics of Particulate Media: A Probabilistic Approach.* McGraw-Hill, New York

Holtz R D, Kovacs W D 1981 *An Introduction to Geotechnical Engineering.* Prentice-Hall, Englewood Cliffs, New Jersey

Mitchell J K 1976 *Soil Behavior.* Wiley, New York

Perloff W H, Barron W 1976 *Soil Mechanics.* Ronald Press, New York

Rowe P W 1972 The relevance of soil fabric to site investigation practice. *Géotechnique* 22(2): 195–300

Scott R F 1963 *Principles of Soil Mechanics.* Addison-Wesley, Reading, Massachusetts

Terzaghi K, Peck R B 1967 *Soil Mechanics in Engineering Practice*, 2nd edn. Wiley, New York

US Bureau of Reclamation 1974 *Earth Manual*, 2nd edn. US Bureau of Reclamation, Engineering and Research Center, Denver, Colorado
Winterkorn H F, Fang H-Y 1975 *Foundation Engineering Handbook*. Van Nostrand Reinhold, New York

R. D. Holtz
[Purdue University, West Lafayette, Indiana, USA]

Construction Materials: Granules

Granules as used in construction (specifically roofing materials) are defined as natural or synthetic mineral grains smaller than gravel and larger than sand, used primarily to form a protective and decorative coating on the weather-exposed surface of composition roofing.

1. Historical Development

Roofing granules in today's market are the products of a long technological evolution. Primitive constructions of burlap saturated with tar appeared as early as 1780. Dusting of talc and mica on the coated surfaces to keep the roofing from sticking to itself in storage, represents an early improvement to the original construction product. Sometimes pigmentation was added with the dusting material for initial color. Exposure to the sun caused the asphalt to become brittle and it eventually cracked and leaked. This led to the first important use for crushed mineral products on asphalt-coated shingles and provided the impetus for the subsequent growth of the asphalt roofing industry. When a layer of crushed mineral was applied on the exposed surface the asphalt coating was protected from ultraviolet light for long periods of time.

Screened waste from slate quarries was the most common granule used in the early asphalt roofing products after the significant technological advances of the 1900s. Records show that coarse slate screenings were used for asphalt roofing as early as 1906. As the asphalt roofing industry further developed, other naturally colored rocks were used in roofing surfaces. These included sand, quartz, mica, feldspar, talc, glass, slag, basalt, granite, rhyolite and oyster shells. In addition, a number of manufactured granular mineral materials, such as crushed brick and tile, fired clay and white porcelain, were used. By 1910, mineral-surfaced asphalt roofing was becoming better known as an economical, long-lasting serviceable product.

The first artificially colored roofing granule was produced in 1914, by running molten slag into a solution of alkaline metal silicate which was absorbed by the porous slag. Various colors were obtained by adding pigments to the silicate coating and by subsequently heat treating the granules. In 1922 another technique was developed in which green slate granules were saturated with copper sulfate solution and heated to 760 °C to form copper oxide. Many other granule coloring processes were patented by 1970 and probably the most significant was the sodium silicate coating introduced about 1975.

2. Specifications and Manufacture

A rock deposit suitable for roofing granules must meet the following requirements.

(a) Resist weathering: the weathering properties of a roofing granule are of primary importance. If it does not adequately withstand the elements of nature, the granule will not provide the necessary protection to the asphalt matrix, nor will it give the color performance to the roofing shingle.

(b) Accept coloring: after crushing, many mineral granules do not have the surface characteristics necessary for granule base—for example, quartz with its conchoidal fracture is very difficult to coat. Also, the mineral constituents must be chemically inert to processing chemicals. Many rocks spall in the heating process because of their thermal expansion characteristics and the presence of molecular water. This condition alters the grade and leaves uncolored surfaces on the granule. To maintain acceptable standards of quality, it is of the utmost importance that the rock base be of uniform quality and homogeneous in color as well as in other physical characteristics.

(c) Low porosity: high porosity allows the penetration of moisture and allows the seepage of water into the asphalt shingle, causing it to blister and/or crack. It must be completely opaque to ultraviolet light. The actinic rays of the sun rapidly deteriorate asphalt roofing. The quality of toughness is especially important during the manufacturing cycle where unavoidable attrition by conveyors, chutes, screens and elevators could severely reduce the size of the original grains.

The production of granules involves exploration, plant site location, quarrying, crushing, screening and coloring. Detailed geological exploration, laboratory evaluations and marketing considerations determine the plant location. Accessibility to mass transportation routes, water, power and labor sources are important economic factors to be considered. Quarrying operations are similar to those of rock aggregates. Roofing granules are produced principally in open-pit operations. After removal of overburden, drilling and blasting are followed by hauling the rock to the primary crusher. Crushing is carried out by either jaw or gyratory crushers. Depending on the quarry and weather conditions, a washing operation might be necessary. Secondary crushing is usually performed by cones, rolls or hammer mills in circuit with vibrating screens. In the modern granule plant, as illustrated in Figs. 1 and 2, elaborate dust-control systems are

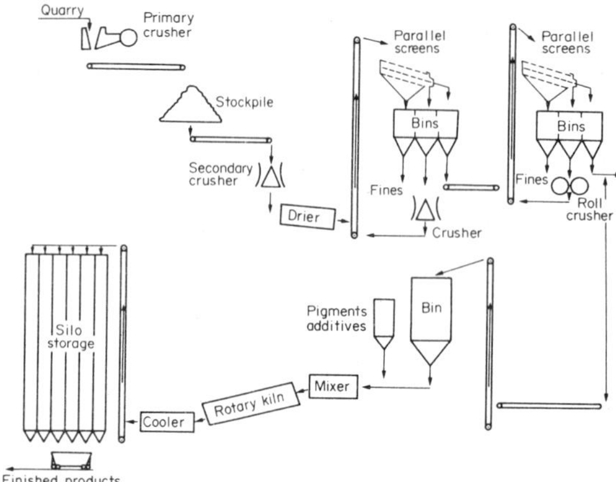

Figure 1
Flowsheet for a typical roofing granule plant (after Jewett et al. 1975)

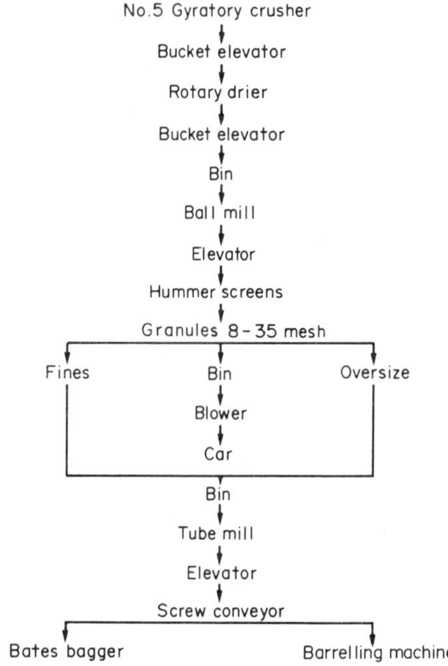

Figure 2
Flowsheet of a mill for manufacturing slate granules

necessary, not only to protect employees' health but also to comply with local, State and Federal laws (in the USA). The crushed and screened granules are transported to the coloring section of the plant, where they are mixed with pigments and other coating materials in rotary mixers. The coated granules are then fed into a rotary kiln for the final heat treatment to mature the bond. Finally, the granules are discharged from the kiln into a rotary cooler or similar device. Rescreening on vibratory screens is usually necessary for quality control. All modern granule plants are supported by elaborate quality control departments in order to meet the stringent requirements imposed by rigid product specifications.

Recent advances have been made in synthetic granules technology. Methods have been developed for the production of granules, organic and inorganic, for use in chemical and pollution technologies with controlled morphology, particle size, particle shape and flow characteristics, and extreme inertness to various liquid and gaseous media in purification applications.

3. Uses

The principal use of granules is in asphalt roofing production. With the increasing importance of the quality and color of roofing materials, ceramic-coated artificially colored granules have become the dominant mineral surfacing on asphalt roofing products.

Today the utilization range of synthetic granules, organic and inorganic, goes far beyond roofing shingles, covering areas of pollution control, industrial filtration and the production of high-technology ceramics. An example of the use of granules in pollution control technology involves the various filtering devices such as granular bed filters used in industrial cleaning processes. Mathematical models have been developed describing the filtration efficiency and pressure drop across granular bed filters. Clarification of mine waters by filtration through granular charges is well known.

Many ceramic processes employ granulated raw materials and raw material mixes in the production of final ceramic ware. Such is the case in the refractories industry. Sintered granular materials enable the ceramic technologist to produce refractory wares of high density, which is necessary for improved high-temperature mechanical properties and resistance to slag attack.

See also: Construction Materials: Crushed Stone; Construction Materials: Dimension Stone; Construction Materials: Fill and Soil; Construction Materials: Industrial Minerals; Construction Materials: Lightweight Aggregate

Bibliography

Bowles O 1922 *The Technology of Slate*, US Bureau of Mines Bulletin No. 218. US Government Printing Office, Washington, DC, pp. 87–103

Bowles O 1934 *The Stone Industries*, 1st edn. McGraw-Hill, New York, pp. 461–64

Brown J A 1960 Granules. In: Gillson J L (ed.) 1960 *Industrial Minerals and Rocks (Nonmetallics Other Than Fuels)*, 3rd edn. American Institute of Mining, Metallurgical and Petroleum Engineers, New York, pp. 443–54

Henry E C 1948 Utilization of slate waste. *J. Can. Ceram. Soc.* 17: 27–31

Jewett C L, Collins R C, Weaver L W, McSheer T 1983 Roofing granules. In: Lefond S J (ed.) 1983 *Industrial Minerals and Rocks (Nonmetallics Other Than Fuels)*, 5th edn., Vol. 1. American Institute of Mining, Metallurgical and Petroleum Engineers, New York, pp. 203–12
Johnstone S J 1954 *Minerals for the Chemical and Allied Industries*. Wiley, New York, pp. 461–64
Josephson G W 1949 Granules. In: Dolbear S H (ed.) 1949 *Industrial Minerals and Rocks (Nonmetallics Other Than Fuels)*, 2nd edn. American Institute of Mining, Metallurgical and Petroleum Engineers. New York, pp. 404–16
Larabee D M 1968 Roofing granules. *Mineral Resources of the Appalachian Region*, US Geological Survey Professional Paper 580. US Government Printing Office, Washington, DC, pp. 252–54
Saxena S C 1980 *Mechanistic Modeling of Granular Bed Filters for the Removal of Particulate Matter from High-Temperature High-Pressure Gas Streams: A Critical Assessment*, Special Publication DOE/METC/SP-80/24. US Department of Energy, Washington, DC
Sherrington P J, Oliver R 1981 *Granulation*. Heyden, New York
Subramaniam A F 1967 Charnockites and the granulites of southern India: A review. *Dan. Geol. Foren.* 17: 473–93
Tohata H 1979 Development of granulation technology. *Kemikaro Enjiniyaringu* 24(1): 76–81

R. S. Kalyoncu
[US Bureau of Mines, Tuscaloosa, Alabama, USA]

Construction Materials: Industrial Minerals

Natural industrial minerals used for construction are crushed stone, sand and gravel, dimension stone, fill and soil, and granules. All except granules are used either alone or with a paste or a binder, such as cement or asphalt, in the construction of, for example, highways, airports, railroads, dams, homes and public buildings. Granules have a specialized use in making asphalt roof materials. Construction materials have so many applications, although often disguised, that the average person fails to recognize their true importance. In 1988, the USA, the leading consumer of construction materials, produced about 2 billion tonnes of sand and gravel and crushed stone—about 9 tonnes per person. Except for dimension stone and granules, construction materials are generally produced near the places where they are used. In 1988, sand and gravel and fill and soil were produced in all 50 states; crushed stone was produced in 49. Because dimension stone and granules require special properties not found widely distributed, they have more restricted production.

The natural industrial minerals used for construction are characterized by low unit cost and high unit weight. Because they are generally consumed in highly populated areas, places inhospitable to mining operations, they must sometimes be transported from surrounding areas. In some places the cost of transportation to market exceeds the cost of production. Processing natural construction materials is generally simple and, if needed at all, consists of breaking, crushing, washing or screening to size. In general, as processing increases for a product, the cost increases. Dimension stone, for example, may have a high unit cost when it requires cutting, carving or polishing to meet specifications.

Although chemical properties of a rock may be important for some construction applications, physical properties are far more important. In some construction projects, the material is used to occupy space or to provide the base on which other materials are used. The material must therefore be competent yet chemically inert in the environment in which it is used. For example, when rock is crushed for aggregate it should not split along cleavage planes or thin partings but form equidimensional fragments; the aggregate should be hard, low in porosity, strong and free of reactive or nondurable materials. Government agencies that consume large amounts of materials generally set specifications for use, but these agencies may instead follow the recommendations of an association or a society that has established specifications for using particular products. Some users have their own specifications, which vary according to availability and cost of the materials.

See also: Construction Materials: Crushed Stone; Construction Materials: Dimension Stone; Construction Materials: Fill and Soil; Construction Materials: Granules; Construction Materials: Lightweight Aggregate; Construction Materials: Sand and Gravel

Bibliography

Brobst D A, Pratt W P (eds.) 1973 *United States Mineral Resources*, US Geological Survey Professional Paper 820. USGS, Washington, DC
Lefond S J (ed.) 1975 *Industrial Minerals and Rocks*, 4th edn. American Institute of Mining. Metallurgical and Petroleum Engineers, New York
US Bureau of Mines 1980 *Mineral Facts and Problems*. US Government Printing Office, Washington, DC

D. D. Carr
[Indiana Geological Survey, Bloomington, Indiana, USA]

Construction Materials: Lightweight Aggregate

Lightweight aggregates are classified as those minerals, natural rock materials, rock-like products and byproducts of manufacturing processes which are used as bulk fillers in concrete building blocks, lightweight structural concrete, precast concrete structural units, road surfacing materials, plaster aggregates and insulating fill. Other uses include soil conditioners, architectural wall covers and suspended ceilings.

Lightweight aggregates for concrete masonry units, according to the American Society for Testing and Materials (1986), consist of "aggregates of low density used to produce lightweight concrete, including: pumice, scoria, volcanic cinders, tuff and diatomite; expanded or sintered clay, shale, slate, diatomaceous shale, perlite, vermiculite or slag; and end products of coal or coke combustion."

Materials used for lightweight aggregate may be classified into four general groups:

(a) natural lightweight aggregate materials such as pumice, scoria, tuff, breccia and volcanic cinders;

(b) manufactured structural lightweight aggregates prepared by pyroprocessing clay, shale or slate in rotary kilns or on travelling-grate sintering machines;

(c) manufactured insulating ultralightweight aggregates prepared by pyroprocessing perlite, vermiculite and diatomite; and

(d) by-product lightweight aggregate prepared by crushing and sizing foamed and granulated slag, cinders and coke breeze.

The last group of lightweight materials, with the exception of slag, has become less significant with time and will not be addressed in this article.

The principal properties sought in materials which are used as lightweight aggregates vary according to the end use. Lightweight aggregates are normally valued for the following characteristics:

(a) lightweight, thermal and acoustical insulation properties;

(b) high fire resistance, toughness and substantial compressive strength;

(c) low water absorption and good resistance to freezing and thawing;

(d) low shrinkage on drying and minimal thermal expansion;

(e) good bonding properties with cement; and

(f) chemical inertness, good elastic properties and good abrasion resistance.

The list of the varieties and the uses of materials as lightweight aggregates is extensive. Most lightweight aggregates and aggregate materials are bulky and low in unit value. Usually they cannot be transported great distances in their final form yet still remain competitive with alternative building materials. Some, which have an economical transportation means available to them, service international markets. Special materials, such as perlite, vermiculite and diatomite, service both national and international markets. This is owing to their relatively high unit value. They are normally shipped prior to pyroprocessing, which expands these materials to a much larger size.

Pumice, scoria and expanded slag may be used extensively and in substantial quantities in some areas. These may not be available economically in other locations. The more specialized the end use, the greater the market area served. The greater the unit value of these materials, the greater the geographical distribution and use.

1. Naturally Occurring Lightweight Aggregates

Naturally occurring lightweight aggregates, such as pumice, scoria, tuff and volcanic cinders, are products of explosive volcanic eruptions. Pumice and pumicite are produced by the violent expansion of dissolved gases in a viscous silicic lava such as rhyolite or dacite. The relatively high viscosity and the volatile content of this type of lava are an explosive combination. During formation the pressure of dissolved gases is suddenly released and the expansion of the gases generates frothy masses of expelled lava. These masses cool very quickly in the atmosphere to form fragments of glass containing innumerable bubble-shaped cavities. Each bubble cavity or cell is sealed off from its neighboring cell by a thin membrane of glass, making the permeability of pumice very low. Pumicite consists of grains, flakes, threads and shards of glass finer than 4 mm in diameter.

Scoria, the mafic counterpart of pumice, is a rusty-red to black pyroclastic material, generally of andesitic to basaltic composition that accumulates in volcanic cinder cones. The vesicles of scoria tend to be coarser and more variable than those of pumice. The cell walls are thicker and the texture is finer grained and more compact. Scoriaceous material less than 25 mm in particle diameter is termed volcanic cinders.

The cellular nature of pumice, pumicite, scoria and volcanic cinders provides these materials with two main properties of value: they are lightweight and have insulating ability. These materials are used primarily in the construction industry as aggregates in lightweight structural concrete, as a plaster aggregate and as loose-fill insulation in structures. Coarse pumice aggregate is used extensively in the manufacture of concrete blocks and masonry products where its low specific gravity reduces the weight of concrete blocks, which makes handling easier. Pumice also minimizes the bulk of foundations and supporting columns in multistory construction, reducing the reinforcement required in the suspended floors. It has high insulation and good acoustical properties, is fireproof and resists condensation, mildew and vermin. Block pumice ranges in size from 5 cm to more than 1.5 m in diameter. Larger blocks may be sawed and used architecturally for wall coverings and suspended ceilings. Scoria is used as a decorative stone in landscaping. Other uses include material for road surfacing, ballast and roofing granules. In 1985 crushed volcanic cinder and scoria sold or used by producers in the USA totalled 2.4 million tonnes for all uses. Use as lightweight aggregates consumed 371 870 t.

Commercial production of volcanic lightweight materials tends to be concentrated in two main zones—the circum-Pacific belt and a belt that spreads from the Mediterrean Sea, along the Himalayas and into the East Indies. The 1985 production in the USA of pumice and pumicite was 368 242 t; it was concentrated in the western USA with 22 mines operated by 21 companies in eight states.

In 1985 world production of pumice and related volcanic materials declined by 4% from the 1984 production level to 11 million tonnes. The combined output of Greece and Italy accounted for 65% of the total world production. France, Spain, the USA and Yugoslavia produced 63% of the remaining 35%.

These materials are normally mined by open-pit methods. The exact method of extraction depends upon the degree of consolidation of the raw materials. In 1984, US pumice was priced at US\$8.98 ton^{-1}. Imported pumice from Greece was quoted at US\$6.47 ton^{-1} for use in concrete masonry products.

2. *Manufactured Lightweight Aggregates*

Manufactured lightweight aggregates are produced by either the rotary-kiln or sintering-grate processes. Clays, shales and slate used in the manufacture of brick, tile, sewer pipe and other structural clay products are frequently satisfactory sources for manufactured lightweight aggregates. These materials must contain certain impurities, such as iron oxides, carbon, sulfides, alkalis or alkaline earths, to be suitable for consideration as a lightweight aggregate. When the shale, clay or slate is heated to incipient fusion the impurities decompose and generate gases, which are trapped in the viscous glassy exterior of the particle. The rotary-kiln process produces rounded particles which have a glassy impervious coating and a fine internal pore structure with unconnected pores. This process produces an aggregate that is very low in weight and also low in absorption.

The sintering process uses similar materials that have been mixed with a proper amount of combustible materials, such as coke, coal or petroleum residue. In this process the material is fired on a sintering grate, where it partially melts; the added combustibles burn out leaving voids. Some sintered materials will expand slightly. The product from this process is then crushed to specification size. The exposed pore structures of the angular particles make the absorption higher than that of rotary-kiln fired aggregates. When used in concrete, sintered aggregate requires more cement to obtain a specific compressive strength than other aggregates because the capillary action of the pore structure tends to draw additional amounts of the cement into the void.

In 1985, consumption of clay, shale and slate in the manufacture of lightweight aggregate increased by 8% from that consumed in 1984. The largest category of use was in the production of concrete block (2.2 million tonnes), and the second largest use was in structural concrete (1.2 million tonnes). Highway surfacing, recreational and horticultural uses consumed the balance (544 200 t). In the USA the average price for all uses in 1985 was US\$6.11 ton^{-1}.

3. *Manufactured Ultralightweight Aggregates*

3.1 Perlite

Perlite is a glassy volcanic rock with a rhyolitic composition containing 2–5% combined water. True perlite is characterized by a system of macroscopic and microscopic concentric onion-like structures, produced as a result of shrinkage during cooling. Commercially, the term includes any volcanic glass that will expand or "pop" when heated quickly, forming a lightweight frothy material.

Perlite occurs naturally as steep-sided block-type lava domes restricted in area which are formed by the extrusion of highly viscous magmas. Perlite deposits are restricted to Tertiary age or younger deposits of rhyolitic composition. Rhyolitic glasses are unstable and devitrify with age into a microcrystalline aggregate of quartz and feldspar. They are also susceptible to hydrothermal alteration, forming clay minerals and zeolites.

Perlite differs from other volcanic glasses principally by its combined water content. Perlite is formed by the secondary hydration of obsidian after its implacement. The combined water exists in two forms—molecular and hydroxyl. The ratio of these two kinds of water is different for perlite of different origins. It is the volatilization of the combined water that provides the energy to "pop" the perlite into a frothy material.

When grains of crushed perlite are brought to the point of incipient fusion, the contained water is converted to steam and the grains swell into light, fluffy cellular particles. Expansion takes place in either a stationary vertical furnace or a horizontal rotary furnace. Some preheating may be utilized to drive off excessive surface water on particles. The temperature at which expansion takes place lies in the range 760–1090 °C. A volume increase of 10–20 times the original value is normal.

Expanded perlite has a low thermal conductivity, high sound absorption and a high resistance to heat. More than half of the perlite produced goes into the construction industry as aggregate in insulation board, plaster and concrete. It is also used as a rooting medium and soil conditioner.

In 1985, crude perlite produced in the USA was mined by 12 operations in six western states. New Mexico was the major producer with 86% (557 800 t) of the total production. Other sources were in Arizona, California, Colorado, Idaho and Nevada. In 1985, expanded perlite was produced by 64 plants in 31 states. Perlite, because of its lightweight characteristics and limited distribution, can be shipped long distances prior to expansion and still remain competitive with other materials. The average price of processed perlite in 1985 was US\$33.85 ton^{-1}. Greece and the USSR are other major producers of perlite. The major deposits in Greece are located on the islands of Milos and Kos. The deposits in the USSR are located in Armenia, Kazakhatan and Buryat. Smaller deposits are located in Turkey, Italy, Japan, Australia, New Zealand and South Africa.

3.2 Vermiculite

The term vermiculite is used to classify or define a small group of minerals that have chemical affinities with montmorillonite, chlorite and biotite. Physically, vermiculites are similar to the micas, having the typical cleavage that allows splitting into thin flexible flakes. The mineral is an alteration product of biotite and other micas. The three broad types of deposits are:

(a) those formed within large ultramafic intrusive masses, such as pyroxenitic plutons cut by syenite or alkalic granite and by carbonatitic rocks and pegmatites;

(b) those associated with ultramafic intrusive bodies, such as dunite and unzoned pyroxenite, and peridotite cut by pegmatite and syenitic or granitic intrusives; and

(c) those formed from ultramafic metamorphic rocks such as amphibole schist in contact with pyroxenite or peridotite and cut by pegmatite.

Vermiculite has a hardness of 2.1–2.8, a specific gravity of 2.5, and a bronze to yellowish-brown or black color. The unit cell of vermiculite contains a layer of water. In rapid heating (to above 870 °C), the water flashes into steam, forcing the layers apart at right angles to the plane of cleavage. Some of the resulting particles have a worm-like appearance; hence, the name "vermiculite," which is the Latin for "little worm." Pure vermiculite can expand to 30 times its original volume, but generally averages 8–12 times.

In 1985, 284 800 t of vermiculite concentrate valued at \$32.4 million was produced in the USA. In the same year, world production of vermiculite concentrate was estimated to be more than 498 850 t. More than 90% of world production comes from Libby, Montana and Palabora, South Africa. Other noted deposits occur in the Blue Ridge Mountains area of North Carolina, the Green Springs area of Louisa County, Virginia, and the Enoree district of South Carolina. Other countries with significant vermiculite production are Japan, Brazil and Argentina.

See also: Construction Materials: Crushed Stone; Construction Materials: Dimension Stone; Construction Materials: Fill and Soil; Construction Materials: Granules; Construction Materials: Industrial Minerals; Construction Materials: Sand and Gravel

Bibilography

American Society for Testing and Materials 1986 *Annual Book of ASTM Standards*, Part 16, *Lightweight Aggregates for Concrete Masonry Units*. ASTM, Philadelphia, Pennsylvania

Harben P W, Bates R L 1984 *Geology of the Nonmetallics*. Metal Bulletin, New York, pp. 64–76, 282–84

Kadey F L Jr 1983 Perlite. In: Lefond S J (ed.) 1983 *Industrial Minerals and Rocks*, 5th edn. American Institute of Mining, Metallurgical and Petroleum Engineers, New York, pp. 997–1015

McCarl H N 1983 Lightweight aggregates. In: Lefond S J (ed.) 1983 *Industrial Minerals and Rocks*, 5th edn. American Institute of Mining, Metallurgical and Petroleum Engineers, New York, pp. 81–95

Peterson N V, Mason R S 1983 Pumice, Pumicite, and volcanic cinders. In: Lefond S J (ed.) 1983 *Industrial Minerals and Rocks*, 5th edn. American Institute of Mining, Metallurgical and Petroleum Engineers, New York, pp. 1079–84

Strand P R, Steward O F 1983 Vermiculite. In: Lefond S J (ed.) 1983 *Industrial Minerals and Rocks*, 5th edn. American Institute of Mining, Metallurgical and Petroleum Engineers, New York, pp. 1375–81

US Bureau of Mines 1987 *Minerals Yearbook 1985*. USBM, Washington, DC

B. H. Mason
[The France Stone Co., Waterville, Ohio, USA]

Construction Materials: Sand and Gravel

Sand and gravel are among the more important materials of the construction industry, and therefore are essential resources among the heavily populated, industrialized nations of the world. Sand and gravel are used for road base, fill, road surfacing, and for aggregate in both bituminous and portland cement mixes. Sand is used alone in mortar and plaster.

Sand and gravel are widely distributed throughout the world, but because they are high-bulk, low-volume materials they are almost invariably mined near places where they are used, and except in border areas they are unimportant in international trade.

World production of sand and gravel was estimated by the US Bureau of Mines in 1987 to be about 9000 Mt per year; most of that production was concentrated in Europe and North America. Heavily industrialized nations with large populations, like the USA and the USSR, produce as much as 1000 Mt of sand and gravel per year. Other industrialized nations with smaller populations, such as the UK, France and the Federal Republic of Germany, produce 100–200 Mt per year, and the lesser industrialized countries in Europe and South America may produce 5–20 Mt per year. Nonindustrialized nations generally have negligible production of sand and gravel for construction.

1. Definitions

The definitions of terms relating to sand and gravel that are used throughout the industry in the USA are those proposed by the American Society for Testing and Materials (ASTM C125-76). Sand is defined as fine aggregate resulting from natural disintegration and abrasion of rock or processing of completely friable sandstone. As fine aggregate the material must pass the $\frac{3}{8}$ in. (9.5 mm) sieve and almost entirely pass the No. 4 (4.75 mm) sieve. Manufactured sand is the product resulting from crushing rock, gravel or sand.

Gravel is coarse aggregate resulting from natural disintegration and abrasion of rock or processing of weakly bound conglomerate. Coarse aggregate must be predominantly retained on the No. 4 sieve. Crushed gravel is the product resulting from artificially crushing gravel and having substantially all fragments with at least one face resulting from fracture.

2. Composition

Gravel may be composed of many rock types that differ in their physical properties. Igneous rocks form from molten materials. Those that are intruded deep within the earth's crust cool slowly and form a coarsely crystalline texture. Gravels derived from these rocks tend to be durable, tough and chemically inert. Molten material extruded onto the earth's surface forms finely crystalline igneous rocks. Some of these rocks may contain minerals that can produce undesirable chemical reactions within concrete, but most of them are acceptable as aggregate. Certain porous volcanic types are finding increased use as lightweight aggregate.

Metamorphic rocks have been altered from their original state by high temperature and pressure well below the earth's surface. Many metamorphic rocks produce high-quality aggregate, but schists and gneisses with high mica content deteriorate rapidly in gravel deposits, and slates contain cleavage planes that reduce the durability of the rock.

Sedimentary rocks are deposited on the earth's surface by water, wind, chemical action or organic activity. Chemical rocks include salt, gypsum and anhydrite; all are unsuitable for use as aggregate because of their water solubility. Chert is a special type of chemical rock that forms as an inclusion in a host rock. Certain types of chert have long been known to produce severe problems in aggregate because of their potential to cause deleterious chemical reactions with the alkalis in portland cement. Other types of chert are not susceptible to dissolution in portland cement, and because chert has a high resistance to abrasion, these types may be considered high-quality aggregate.

Organic rocks include coal, which is a deleterious component of aggregate because of its low strength and limestone and dolomite, which, except for certain argillaceous and siliceous varieties, are acceptable for use as aggregate.

Sedimentary rocks deposited by water or wind include conglomerates, sandstone, siltstone and shale. Highly cemented varieties of these rocks are usually acceptable, but many of them are too porous or their individual grains are not sufficiently cemented together to produce durable rock.

Sand is produced when rocks are more completely broken down, many of them to their individual mineral components. Sands tend to be composed predominantly of quartz, because it is the most abundant mineral in many rock types and is more resistant to chemical and mechanical breakdown than most other minerals. Quartz is a desirable component of fine aggregate because of its hardness, durability and relative insolubility. Besides quartz, sands may contain labile minerals, mineral aggregates, chert, coal or mica, which reduce the quality of the aggregate.

3. Origin and Occurrence

3.1 Introduction

Sand and gravel in lower latitudes accumulate in rivers, on beaches or as residuum. The deposits tend to be lithologically simple, reflecting the sorting and abrading action of transport mechanisms and the limited size of the source area from which the materials were derived. Besides, in humid low-latitude regions chemical weathering may be so intense that all but the most resistant rocks are destroyed.

Higher-latitude deposits originated directly or indirectly from the glaciers that covered large portions of the mid- and upper latitudes during the Pleistocene epoch (Ice Age). Glaciers originated in subpolar regions, moved south hundreds of kilometers, and incorporated into their sediment load all rock types represented in the bedrock over which they passed. Sorting action is minimal, and chemical weathering does not take place during glacial transport. Crushing and plucking rather than abrasion are the dominant modes of mechanical breakdown. For these reasons the sediment load carried by glaciers ranges in size from clay to boulders, and tends to consist of angular clasts that may include rock types that are soft or susceptible to rapid chemical degradation.

Fortunately, glacial sand and gravel deposits are almost invariably the result of the concentration of the coarse-grained fraction of the glacial load by running water. Depending on the mode of origin, sorting and abrasion have altered the original material to varying degrees. Fluvial transport in ice-contact deposits such as kames and eskers is limited, but where sediments are carried away from the glacier, as in outwash fans and valley trains, the resulting deposits have the characteristics of fluvial sediments.

The geological setting and mode of origin control not only the mineralogy of sand and gravel deposits but also the size and shape of the deposits and the amount and distribution of waste materials. These are the dominant factors affecting the quality and quantity of the aggregate and the cost of extracting it.

3.2 Nonglacial Deposits

Sand and gravel suitable for aggregate are found in many active and inactive stream channels. Although local bedrock outcrop may contribute deleterious materials to nearby deposits, the washing and abrading action of streams tends to produce clean deposits of durable material. The deposits may occur in channels or bars of the present river or on adjacent terraces that formed when the river flowed at a higher level (Fig. 1).

Alluvial fans occur in mountainous regions, where streams abruptly decrease in gradient as they emerge from mountains onto basins or plains. Coarse material is deposited near the mountain front, while finer-grained sediments are carried out onto the basin or plain. The composition of the deposit reflects the bedrock composition in the adjacent uplands. Poor-quality aggregate can result, because transport distances in many places are too short to allow abrasion of deleterious material. And as the fan builds, stream positions shift rapidly, causing a complex arrangement of sand and gravel and fine-grained waste material (Fig. 1).

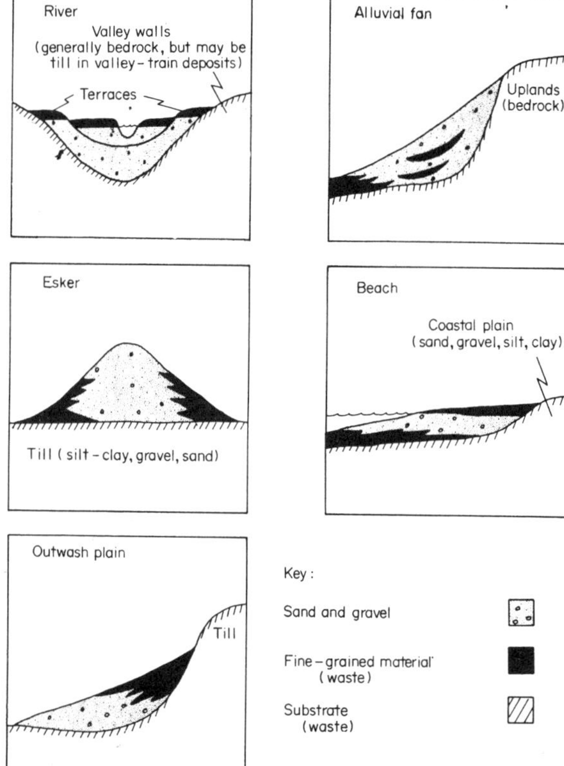

Figure 1
Cross sections of selected types of sand and gravel deposits showing their shapes and the relationships between waste and minable material

Beach deposits consist of clean sand and gravel that tend to be relatively free of deleterious materials owing to the abrading action of waves. The deposits tend to be thin, but many of them extend along the coast for great distances, and they may extend both landward and seaward of the actual shoreline (Fig. 1). In nonglaciated areas with wide coastal plains they may be an important source of aggregate.

Residual deposits result from the concentration of the more durable products of deeply weathered rock. These deposits may show all stages of transport, from the near in situ concentration at the site of weathering to some stream transport. Extensive deposits of late Tertiary age cover wide areas. But many of these deposits consist of chert that may be generally unsuitable for use as aggregate without beneficiation, and many of these clasts have well-developed iron oxide coatings that can adversely affect the cement–aggregate bond. In some areas, such as the southeastern USA, residual deposits may be the only local source of aggregate.

3.3 Glacial Deposits

Ice-contact deposits are usually small, irregularly shaped, and vertically and laterally inhomogeneous. Wide variations in grain size within a deposit make it difficult to produce a uniform product from them, and interbedded fine-grained waste material can seriously impede mining operations (Fig. 1). Because of short transport distances, gravels tend to be angular, and in many places consist of labile clasts that adversely affect the quality of the aggregate.

Deposits that are carried beyond the glacial margin are more laterally and vertically uniform, and are therefore much easier to exploit than ice-contact deposits. Outwash plains are broad, thick deposits similar to alluvial fans in that local accumulations of fine-grained waste material can hinder mining operations (Fig. 1). Valley-train deposits occur along the rivers that drained glaciers, and are similar in many ways to nonglacial fluvial sediments. They differ from other fluvial deposits in the wide variety of rock types that may be included. Use of depositional models to predict the location and characteristics of sand and gravel deposits in glacial terrains has increased recently owing to advances in understanding how glaciers deposit these materials (Eyles 1983).

4. Mining and Processing

Mining operations range in size from those producing a few thousand tonnes per year of pit-run (unprocessed) material, to those producing several million tonnes per year and consisting of large, complex plants able to process a wide variety of construction products. Production has tended to be concentrated in these large plants, but because of steadily mounting transportation costs and improvements in the design of small plants, especially wheel-mounted portable processors, smaller plants nearer places of use may become more common.

Sand and gravel may be dredged from natural bodies of water, such as lakes and rivers. On land they may be dredged from wet pits where the deposit extends below the water table, or they may be excavated from dry pits where the water table is below the pit floor or where dry conditions are maintained by pumping.

There are two steps in producing sand and gravel for aggregate. The material is excavated, and then it is processed according to the character of the deposit and the intended use of the material produced (Fig. 2).

In rivers, lakes or oceans, sand and gravel are excavated by floating dredges, some of which have complete processing plants aboard. The material may be excavated by dragline, or by suction, bucket-ladder or cutterhead dredges. It is then either processed aboard or transferred to barges for transport ashore. Excavation is similar for wet pits on land, except that draglines operate from the shore and dredges are usually tethered to the shore, and the sand and gravel are immediately transferred to land-based processing plants by floating pipelines. In a dry pit, sand and gravel may be mined by front-end loaders, draglines or power shovels.

After excavation, the sand and gravel are usually run through a coarse screen to remove particles too large to be used in the plant. The resultant material can then be stockpiled as pit-run gravel to be used for fill, or it can be sent on to the processing plant by truck or rail. Where material is transported to the plant by conveyor belt, the coarse screen is usually combined with a primary crusher that reduces boulders to sizes that can be carried on the belt.

At the plant the aggregate is first screened to separate sand from gravel. Sand is fed to the sand plant, where secondary screening may remove pea gravel, and finer sand is size graded, usually in wet classifiers. Very fine sand is trapped in cyclones, and material which passes a No. 200 screen (0.074 mm) is passed in slurry form to a settling basin, where the water is reclaimed or drained away.

Gravel-size material can enter one of two lines. In one line the material is fed into a crusher and then screened to produce a size-graded material suitable for use as road base or fill. In the other line, the material is

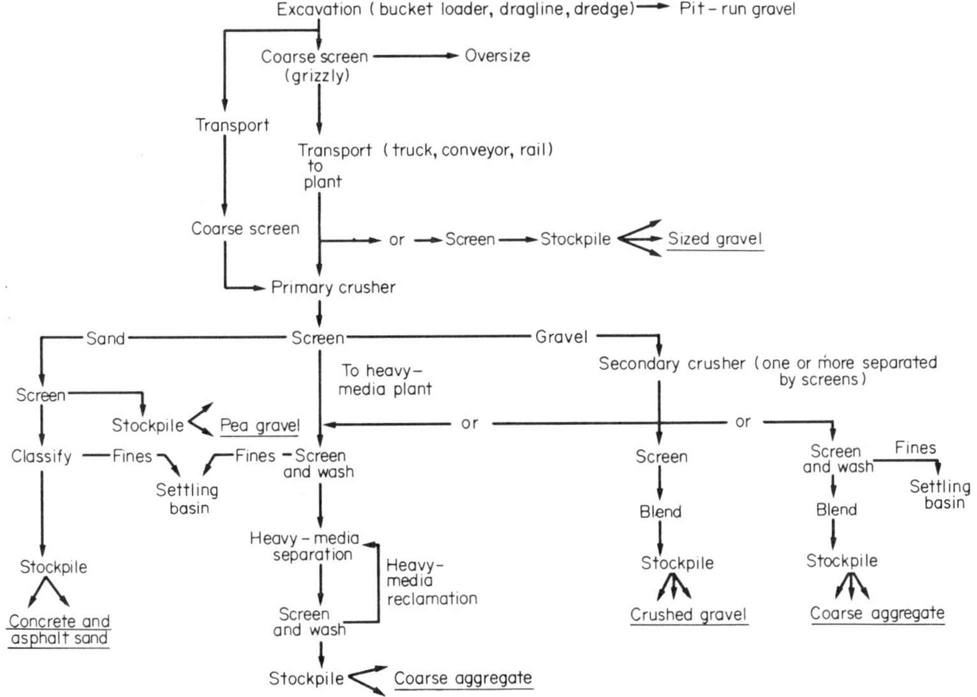

Figure 2
Flow chart showing the various stages in mining and processing sand and gravel. The processed material becomes more valuable as it moves through the system increasing its capability to meet more stringent specifications

crushed, screened and washed, producing a material that can be used for concrete or bituminous aggregate. After screening, the material can be stockpiled, or the various sizes can be blended to produce the different classes of aggregate.

In some plants, beneficiation of the raw material must take place. The most common type of beneficiation is heavy-media separation, whereby lightweight deleterious materials, such as coal, shale and chert, are floated off in mixtures of water, magnetite and ferrosilicon. Other types of beneficiation include specialized screens to remove flat particles, differential impactors to reduce soft particles to fines that may be washed away, and elastic fractionation, which relies on the difference in elasticity between hard particles and soft deleterious particles to effect separation.

Sand and gravel operations, like all extractive industries, place a heavy burden on the environment. They release large amounts of fine particulate matter into the air (or water in dredging operations), and noise levels near pits can be high. The industry uses different measures to ameliorate this burden. Pit roads can be periodically sprayed with water or chemical dust suppressants. Cyclones and settling ponds can be used to trap fine-grained waste material, and if water supply is critical, wash plants can be fitted with settling tanks, where effluent from the plant is treated with flocculants that cause fine particles to settle out rapidly so that wash water can be quickly recycled. Fences and berms can be used to cut noise levels, and landscaping can be used imaginatively not only to reduce noise levels but also to improve the appearance of the operation.

Many abandoned pits are near expanding population centers, and the large areas they occupy have considerable value in land-short urban areas as potential residential or commercial developments or as recreational areas. The industry is becoming increasingly aware of the value of successfully reclaimed sand and gravel pits, and some firms now routinely plan their mining strategy to minimize the adverse environmental aspects of the operation and to maximize the potential of the reclaimed land.

5. *Quality*

For sand and gravel to be suitable aggregate, they must be capable of being bound together by cementing agents to produce a durable material able to withsand stress for a long time. Asphalt or portland cement are the most common cement materials, but certain sulfur compounds are potential asphalt substitutes. The aggregate must be clean, durable and chemically inert, and should have the proper shape. The aggregate and the cement should also have similar thermal characteristics. A series of rather stringent specifications, based on the performance records of various constituents of aggregate, has been written by the American Society for Testing and Materials (ASTM) to assess the probable performance of untested aggregate from new sources (Table 1). Most other specifications written by government agencies and private firms are based on these tests. Other countries have their own specifications and tests, but most of these regulations differ only slightly from those written by the ASTM.

Besides promoting uniformity in testing and specifications, the ASTM also acts as a clearing house for information dealing with the physical properties of materials. The organization in Europe that performs similar functions is the Réunion Internationale des Laboratoires d'Essais et de Recherches sur les Matériaux et les Constructions (RILEM).

The aggregate must first be clean—i.e., free of very fine particles, mica and organic matter. There also should be no coatings by such materials as clay, calcium carbonate, iron oxide or manganese that might inhibit the bonding of cement to particle. Fines can be determined by sieving the potential aggregate on a No. 200 (0.074 mm) sieve. In general, no more than 5% of a −200 mesh material will be allowed in aggregate. Excessive amounts of organic matter can be determined by a chemical colorimetric test, and mica and surface coatings can be determined by petrographic examination.

The aggregate must also be durable and free of deleterious particles that may affect the strength of the concrete. Deleterious materials, such as coal and lignite, can be identified by petrographic analysis, and soft or poorly lithified particles can be identified by the scratch-hardness test. The amount of these deleterious materials allowable in an aggregate is strictly limited. Among the tests for measuring the durability of an aggregate are the Los Angeles or Deval Abrasion tests, which determine the resistance of an aggregate to abrasion. The resistance to weathering is determined by the freeze–thaw test, or by the magnesium or sodium-sulfate soundness test. In both of these tests, particles are subjected to cycles of saturation and either freezing or drying. Crystals of ice or sodium or magnesium sulfate grow in the particle and cause accelerated splitting, flaking or crumbling. Many factors, including alumina (clay) content, carbonate mineralogy, specific gravity and hardness, contribute to the performance of rocks in the test, but the two with the greatest effect are the presence of fractures and laminae, and the absorption of water.

Another property of aggregate which is receiving increasing attention is its ability to withstand polishing under road conditions. It is desirable for aggregate, especially in bituminous mixes, to present a rough surface to increase the antiskid properties of the road. Such aggregate may contain hard minerals that resist polishing, such as quartz or chert, or softer rocks, such as siliceous limestone, containing harder impurities that continue to present fresh, rough surfaces as they wear down. Two tests currently in use measure the skid-resistance properties of aggregate under simulated and actual road conditions.

Table 1
Quality-control tests frequently performed on sand and gravel used for aggregate

ASTM[a] designation	Test	Purpose
		Sampling
D75-71	sampling stone, slag, gravel, sand and stone block	describes the methods used to take representative samples at the deposit or plant or on delivery
C702-75	reducing field samples of aggregate to testing size	describes methods used to reduce field samples to an appropriate size for various tests
		Deleterious materials
C117-76	amount of material finer than 0.074 mm (No. 200 screen) in aggregate	outlines the procedure for determining the total quantity of material finer than a standard No. 200 sieve
C40-73	organic impurities in sands for concrete	describes the method used to determine injurious organic compounds through a colorimetric test, and warns whether further tests may be necessary
C87-69	effect of organic impurities in fine aggregate on strength of mortar	determines the effect of organic impurities on mortar strength by means of a compression test on mortar of plastic consistency
C142-71	clay lumps and friable particles in aggregates	determines approximate amount of clay lumps and friable particles in natural aggregates through disaggregation and washing
C123-69	lightweight pieces in aggregate	determines the approximate percentage of lightweight pieces in aggregate by means of heavy-liquid separation
C586-69, C289-71 and C227-71	potential alkali reactivity of concrete aggregates (rock-cylinder method (C586-69), chemical method (C289-71), or mortar-bar method (C227-71))	determines the potential reactivity of aggregate with alkalis in portland cement by rock-cylinder expansion, by chemical reactivity, or mortar-bar expansion
		Durability
C131-76	abrasion of small-size coarse aggregate by use of the Los Angeles machine	describes procedure used in testing resistance to abrasion of coarse aggregate smaller than 37.5 mm ($1\frac{1}{2}$ in.) in the Los Angeles machine
C535-69	abrasion of large-size coarse aggregate by use of the Los Angeles machine	describes procedure used in testing resistance to abrasion of coarse aggregate larger than 37.5 mm ($1\frac{1}{2}$ in.) in the Los Angeles machine
C88-76	soundness of aggregates by sodium sulfate or magnesium sulfate	determines the resistance of aggregates to disintegration by saturated solutions of sodium or magnesium sulfate to judge soundness of aggregate subjected to weathering under freeze–thaw conditions
C235-68	scratch hardness of coarse aggregate particles	identifies and determines the quantity of soft or poorly lithified particles in coarse aggregate on basis of scratch hardness
C295-65	petrographic examination of aggregates for concrete	outlines procedures to be used in examining aggregate by optical microscopes to determine physical and chemical properties and to describe and classify constituents, especially those that bear on the quality of the material
E274 and E303	skid resistance of paved surfaces by using a full-scale tire (E274) and measuring surface frictional properties using the British portable tester (E303)	determines the skid resistance of aggregate in pavement under actual or simulated road conditions
		Other properties
C136-76	sieve analysis of fine and coarse aggregates	determines the particle-size distribution of fine and coarse aggregate by using sieves with square openings
C128-73	specific gravity and absorption of fine aggregate	determines bulk and apparent specific gravity (to calculate aggregate displacement in concrete) and absorption of fine aggregate
C127-77	specific gravity and absorption of coarse aggregate	determines the bulk and apparent specific gravity (to calculate aggregate displacement in concrete) and absorption of coarse aggregate
C70-73	surface moisture in fine aggregate	determines in the field the surface moisture in fine aggregate by water displacement; may also be used for coarse aggregate with appropriate changes in method
C29-76	unit weight of aggregate	determines the unit weight of fine, coarse or mixed aggregates

[a] American Society for Testing and Materials 1979 *Annual Book of Standards*, Pt. 14. ASTM, Philadelphia, Pennsylvania

Other tests measure the properties of the aggregate which affect the behavior of the concrete during its preparation and while it is in use. The specific gravity is measured to calculate the volume occupied by the aggregate in portland cement, and bulk specific gravity is required to exceed 2.50 for concrete aggregate in most specifications. Water absorption is measured, because the particle–cement bond is adversely affected when water is absorbed by aggregate particles in concrete. A limit of 5% water absorption is usually placed on concrete aggregate. The moisture content of the aggregate must also be known for proper preparation of concrete, and the total evaporable water and surface moisture, especially of fine aggregate, are routinely measured.

The shape and grain-size distribution of the aggregate are carefully controlled because these properties affect the void ratio and flowability of the concrete. Shape, as determined during petrographic examination, is especially important in determining the quality of aggregate. In concrete, rounded equant clasts are the most desirable because highly angular or flat particles reduce the workability of poured concrete, especially when it is used in forms for walls. In asphalt, however, crushed gravel and manufactured sand are preferred, because the asphalt bonds better to angular surfaces and angular clasts pack together better. This combination produces a road surface less subject to flow under heavy loads.

The potential of chemical reactivity between aggregate particles and portland cement must be considered. Most silicate and carbonate rocks are relatively inert. Some silicate rocks, chiefly opaline and chalcedonic chert, some volcanic and zeolitic rocks, and even some siliceous carbonate rocks can react with alkalis in cement and form alkali–silica gels. These gels can absorb water and expand, and therefore cause expansion failures in concrete. Among the properties of the deleterious rock types that determine the potential reactivity are a low degree of crystallinity and a large surface area owing to high microporosity. Other factors also affect the reactivity, and the degree to which these factors combine can be determined by measuring the expansion of rock cylinders or mortar bars that have been saturated in alkaline solutions, or by measuring the degree to which siliceous minerals are dissolved in alkaline solutions.

The tests described above do not measure the actual behavior of the aggregate in cement, but instead measure those properties that are believed to influence performance. Because the behavior of aggregate in cement is a function of the complex interaction of many physical and chemical properties, the tests may be misleading for some aggregates. The only true test of the quality of aggregate is its actual performance in use.

See also: Construction Materials: Industrial Minerals

Bibliography

American Society for Testing and Materials 1978 *Significance of Tests and Properties of Concrete and Concrete-Making Materials.* ASTM Special Technical Publication 169B. American Society for Testing and Materials, Philadelphia, Pennsylvania

Bates R L 1969 *Geology of the Industrial Rocks and Minerals.* Dover Publications, New York

Dunn J R 1983 Aggregates—Sand and gravel. In: Lefond S J (ed.) 1983 *Industrial Minerals and Rocks*, 5th edn., Vol. 1. American Institute of Mining, Metallurgical and Petroleum Engineers, New York, pp. 96–110

Evans J R 1978 *Sand and Gravel*, Mineral Commodities Profiles MCP-23. US Department of the Interior, Washington, DC

Eyles N (ed.) 1983 *Glacial Geology: An Introduction For Engineers And Earth Scientists.* Pergamon, Oxford

Goldman H B, Reining D 1975 Sand and gravel. In: Lefond S J (ed.) 1975 *Industrial Minerals and Rocks*, 4th edn. American Institute of Mining, Metallurgical and Petroleum Engineers, New York, pp. 1027–42

Lenhart W B 1960 Sand and gravel. In: Gillson J L (ed.) 1960 *Industrial Minerals and Rocks*, 3rd edn. American Institute of Mining, Metallurgical and Petroleum Engineers, New York, pp. 733–58

Neville A 1973 *Properties of Concrete.* Wiley, New York

Orchard D F 1962 *Concrete Technology*, Vol. 1, *Properties of Materials.* Contractors Record, London

Schellie K L 1977 *Sand and Gravel Operations—A Transitional Land Use.* National Sand and Gravel Association, Washington, DC

G. S. Fraser
[Indiana Geological Survey, Bloomington, Indiana, USA]

Copper Resources

Copper was one of the earliest metals used by man and still is one of the most useful. The reasons for the early popularity of the metal are that it is one of the more abundant heavy metals, occurring in mineral deposits in many parts of the world; it occurs as the native metal and in minerals that were easily smelted to obtain the metal; it has an attractive color and good resistance to corrosion; and it can be easily worked and alloyed with other metals to satisfy a variety of needs. In the modern industrial era, the high electrical and thermal conductivity of copper are its most important properties.

Copper minerals are concentrated and deposited in the earth's crust by a variety of geological processes in a wide range of geological environments. The discovery, development and mining of copper deposits, and the treatment of ores and refining of the metal are the main steps in the long and complex process of transforming resources into copper supply.

1. Properties

Copper (atomic number $Z = 29$) occupies subgroup Ib of the periodic table, above silver and gold. The copper atom contains 2, 8 and 18 electrons in its inner shells and 1 electron in its outer shell. It forms the cuprous ion (Cu^+) by loss of the outermost electron and the cupric ion (Cu^{2+}) by loss of a second electron. Copper has an atomic weight of 63.54, and has two stable isotopes, ^{63}Cu and ^{65}Cu.

Copper is the only metallic element except gold to have a natural color other than gray. The highly reflective, reddish surface of the clean metal tarnishes to brown or green by formation of copper carbonates. Like the other elements in subgroup Ib, copper oxidizes slowly in air and in fresh and salt water. It has little corrosion resistance to sulfur, however. Copper is relatively soft (Mohs hardness 2.5–3.0), malleable and ductile; it has a specific gravity of 8.89 and a melting point of 1083 °C. The electrical conductivity of copper by cross section (diameter of conducting wire) is exceeded only by that of silver, and its electrical conductivity by weight is exceeded only by that of aluminum. Only silver possesses a thermal conductivity greater than that of copper.

Copper may be alloyed with tin to form bronze and with zinc to form brass; both these alloys exceed copper in hardness and strength, and have improved mechanical properties. Some copper compounds are highly toxic to fungi but, in general, copper has a low toxicity to most of the higher plants and animals.

2. *Uses*

Copper is one of the few elements that is used mainly in its pure form rather than as an alloy or compound. The electrical industry consumes about 50% of the supply of copper to produce wire for motors, generators, and power distribution and control equipment. Copper is preferred for underground and household power transmission, but it has been supplanted by aluminum for long-distance overhead transmission. Copper is also being supplanted by photooptical fibers which are gaining rapid acceptance in the transmission of telephone and television signals.

The mechanical properties and corrosion resistance of copper and its alloys make it useful in plumbing and roofing as well as for small parts for the automotive industry, shell casings for ordnance use and marine equipment.

High thermal conductivity and ease of brazing and soldering make copper useful for radiators, heaters and solar collectors. Minor uses include coins, jewelry, cooking utensils and decorative items. Copper compounds are used in agriculture for fungus control.

In addition to the substitution of aluminum and silicon for copper in the applications mentioned above, steel and plastics may be used in place of copper in applications where lower cost and lighter weight are important.

3. *Supply*

Copper ores are extracted from open-pit mines in the case of large near-surface deposits, and from underground mines in deposits where stripping of the overburden is not feasible. In some mines, copper is leached by the introduction of leaching solutions into rocks that have been fractured in place by explosives. Mined copper ores are crushed and ground, and the copper minerals are either concentrated by froth flotation or they are leached and the leachate stripped of copper by electrolysis. In the former process, which accounts for the bulk of copper production, organic collectors and frothing agents are introduced into a suspension of ground ore in water, and bubbles are created by air jets. The collectors cause the copper minerals to adhere to the surface of the bubbles and float to the surface, where they are removed mechanically.

Copper is recovered from these sulfide mineral concentrates by smelting, a process which involves oxidation of the sulfide minerals and removal of sulfur as SO_2 gas and reduction of the copper oxide to molten metallic copper which is then cast in the form of anodes to be electrolytically refined. To a limited extent, these pyrometallurgical smelting methods are being replaced by hydrometallurgical processes which employ dissolution and electrolysis. Sulfur dioxide and other undesirable emissions that result from pyrometallurgical techniques are thus eliminated, but at the cost of a large expenditure of electrical energy.

During 1987, the world total production of primary copper was 8.34 Mt. Of this, 79–80% came from eleven countries led by Chile, the USA and Canada (Table 1). In 1982 and 1983 Chile moved from second to first place, and is expected to retain this position because of its high-quality ore reserves. The remainder came from 45 countries including the People's Republic of China, South Africa, Zaire, Zambia, Zimbabwe, Mexico, Bulgaria, Yugoslavia, Malaysia and Indonesia. New production is expected in the future from Colombia, Argentina, Brazil, Panama and Fiji, and a significant increase in production is expected from Australia, the People's Republic of China and Iran.

The USA is the most important copper-producing country, with 17% of smelter production and 21% of refinery production. Japan, the Federal Republic of Germany and Belgium account for nearly one fifth of the world's refinery production although they have insignificant mine production. Japan imports most of the copper concentrates produced in the southwest Pacific Islands, and receives shipments from many other countries, including Canada and the USA. Belgium refines copper ore imported mainly from Zaire. This pattern may be changed in the future as developing countries seek to industrialize and to retain the value added in copper smelting and refining.

Statistics for the world supply of copper from scrap (secondary copper) are not readily available. In the

Table 1
Copper reserve base and share of annual world production by country in 1987

Country	Deposit types	Share of national reserve base (%)	Reserve base (Mt)	Share of world production (%)
Chile	mainly prophyry		120	17.0
USA	porphyry	91	90	15.2
	sedimentary	7		
	massive sulfide	2		
Canada	porphyry	18	23	8.9
	massive sulfide	62		
	magmatic	20		
USSR	sedimentary	32	54	7.7
	magmatic	28		
	porphyry	19		
	massive sulfide	21		
Zaire	sedimentary		30	6.7
Peru	mainly porphyry		32	5.4
Poland	sedimentary		15	5.2
Zambia	sedimentary		34	5.2
Papua New Guinea	porphyry		14	3.0
Australia	mainly sedimentary		41	2.8
Philippines	mainly porphyry		18	2.6
Other			109	20.3
Total			580	100.0

USA, recycling of old scrap provided 19–31% of total consumption during the period 1976–1980.

4. Abundance and Distribution

In cosmic abundance, copper occupies a position—along with zinc ($Z = 30$), germanium ($Z = 32$) and arsenic ($Z = 33$)—between the lighter elements, whose abundance increases exponentially with decreasing atomic number, and the heavier elements, whose abundance decreases gradually with increasing atomic number. Within the earth's continental crust, copper is estimated to have an abundance of about 55 ppm, being less abundant than chromium, nickel and zinc, and more abundant than tungsten, tin and cobalt.

Within the igneous rocks—those formed by crystallization from a silicate melt or magma—copper is most abundant in the magnesium- and iron-rich (mafic) rocks (40–160 ppm), and least abundant in the silica-rich rocks (1–40 ppm). Within the sedimentary rocks—those formed by accumulation and cementation of mineral fragments and by chemical precipitation of minerals—copper is most abundant in shale (70 ppm) and least abundant in limestone (6 ppm).

5. Copper Minerals

Copper forms a variety of sulfide, sulfosalt, oxide, carbonate, silicate and sulfate minerals. Chalcopyrite ($CuFeS_2$), a soft, metallic yellow mineral, is the most abundant, being widely dispersed in rocks and concentrated in the largest ore deposits. Chalcocite (Cu_2S) and minerals with similar composition are concentrated in certain sedimentary environments having low oxygen activity.

During weathering and oxidation of copper deposits, chalcopyrite and other copper minerals are dissolved, and copper may be fixed in the oxidized zone as hydrous carbonates, silicates, sulfates, chlorides, oxides and native metal. Of these minerals, bright green malachite ($Cu_2(OH)_2CO_3$) and brilliant blue azurite ($Cu_3(OH)_2(CO_3)_2$) are the most common and their colors have been a prospecting guide for thousands of years. Oxidation of copper deposits that contain abundant iron sulfide minerals produces ground water of very low pH, and dissolved copper is kept in solution. This dissolved copper may be deposited as copper sulfides on contact with reducing ground water at depth.

6. Copper Ore Deposits

An ore deposit is a concentration of minerals from which one or more useful products can be extracted at a profit. The concentration necessary for a copper ore deposit is at least a hundred times the average crustal abundance. Geological conditions that produce such concentrations are not common; only about one millionth of the copper contained in the crust is concentrated in ore deposits.

Most copper ore deposits can be classified according to form, mineralogy and geologic environment (Table 2). Porphyry, and some vein and replacement ore deposits are believed to form within and beneath

Table 2
Classification of copper deposits

Type	Form	Geologic process	Typical ore tonnages and grades	Coproduct or by-product
Porphyry	disseminated chalcopyrite in and near igneous intrusions	intrusion of granitic magma, fracturing, and flow of hot saline fluids	2–2000 Mt at 0.4–1.1% Cu	Mo, Au, Ag
Sedimentary	beds of sandstone or shale impregnated by Cu_2S minerals or chalcopyrite	flow of saline waters through reactive and/or permeable sedimentary rocks	1–1000 Mt at 0.7–10% Cu	Ag, Co
Massive sulfide	pods and layers of metal sulfides containing chalcopyrite	submarine volcanism and outflow of metal-bearing saline fluids onto the sea floor	0.5–100 Mt at 0.5–5% Cu	Zn, Pb, Au, Ag
Magmatic	layers or lenses of massive or disseminated sulfide in mafic igneous rock	gravitational settling of sulfides directly from a molten rock	variable tonnage, 0.2–0.5% Cu	Ni, Co
Native copper	irregular masses and disseminations of native copper filling voids in lavas and sedimentary rock	flow of waters through fragmental, mafic–volcanic rocks and sediments	small tonnage, variable grade	Ag
Vein and replacement	quartz-filled fractures containing chalcopyrite and other copper sulfides and sulfosalts; deposits may replace layers of sedimentary rocks	flow of hot saline fluids through fractured and/or reactive rocks	small tonnage, variable grade	Ag, Au, Pb, Zn, As, Sb, Ba, Sn, Mn

volcanoes of the kind that are common in the Andes. Sedimentary copper deposits form in continental rifting environments in arid regions; the metal-rich brines lying in pools on the floor of the Red Sea may be genetically related to deposits of this type. Massive sulfide deposits form during volcanism on the sea floor and may be related to modern metal-sulfide-rich hot springs on the East Pacific Rise.

There is much evidence to suggest that copper deposits (with the exception of the magmatic deposits) form in response to the flow of chloride-rich solutions through fractured and permeable rocks. Transport of copper as a chloride complex ion is believed to be an important common factor in the origin of the various deposit types. Fluid flow is accelerated by heat from igneous heat sources during the formation of porphyry, massive sulfide and vein deposits. Fluid pressures caused by the compaction of sediments may be important in the flow of the solutions that form sedimentary deposits.

Much remains to be learned about the uneven distribution of copper deposits around the world. For example, Andean and island-arc volcanoes, believed to be favorable environments for the formation of porphyry copper deposits, have been active around the Pacific rim for at least 100 million years, yet important porphyry deposits are known only on the eastern side (Chile to British Colombia) and the southwestern side (Fiji to Philippines and eastern China). Many other such examples of favorable geologic environments lack significant copper deposits, and there is increasing speculation that copper sources beneath the crust may be important in controlling the distribution of copper provinces.

7. *Reserve Base*

The reserve base is the measured in-place tonnage of copper in deposits that either are currently minable or have a reasonable potential for becoming economically available in the future. In 1987, the US Department of Interior estimated the world reserve base to be 570 Mt. Approximately 81% of this reserve is distributed among eleven countries in three deposit types as shown in Table 1. Chilean and US porphyry deposits lead the list with 36% of the reserve base. Porphyry copper deposits constitute over 50% of the world total, whereas sedimentary deposits constitute about 30%.

The uneven distribution of reserves reflects the fact that tonnage of contained copper in deposits is distributed lognormally. Because large deposits contain an order of magnitude more copper than average size deposits, the distribution of the large deposits strongly affects the ranking of countries according to copper reserves.

8. Additional Resources

Copper in deposits or parts of deposits that have been incompletely studied, or that have a low potential for becoming available in the future, is classified as additional resources.

Among these resources, perhaps the largest category is that of copper contained in the deep, lower grade, or otherwise unavailable parts of the world's largest deposits. Copper from these resources may become part of supply when currently developed reserves in the deposits become depleted.

Another major category of additional resources is represented by recently discovered deposits that are large but not well enough measured to be part of the reserve base. Examples, each with several billion tons of ore containing about 1% copper, are in the Roxby Downs area of South Australia and the Carajas area in the Amazon basin of Brazil.

A third category of additional resources includes copper contained in unconventional deposits for which economic mining methods have not been established. The largest example of this type of resource is the copper- and nickel-bearing manganese nodules that are found in deep ocean basins.

Bibliography

Cox D P 1979 The distribution of copper in common rocks and ore deposits. In: Nriagu J O (ed.) *Copper in the Environment*. Pt. 1: *Ecological Cycling*. Wiley-Interscience, New York, pp. 19–42

Cox D P, Wright N A, Coakley G J 1981 *The Nature and Use of Copper Reserve and Resource Data*, US Geological Survey Professional Paper 907-F. US Government Printing Office, Washington, DC

Joralemon I B 1973 *Copper, the Encompassing Story of Mankind's First Metal.* Howell North, Berkeley, California

McMahon A D 1965 *Copper Materials Survey*, US Bureau of Mines Information Circular 8225. USBM, Washington, DC

Schroeder H J 1977 *Copper*, US Bureau of Mines Mineral Commodity Profile 3. USBM, Washington, DC

US Bureau of Mines 1988 *Mineral Commodity Summaries 1988*. USBM, Washington, DC, pp. 44–45

D. P. Cox
[US Geological Survey, Menlo Park, California, USA]

Critical and Strategic Materials

The terms critical materials and strategic materials are often used jointly or interchangeably. In general, a material is considered critical by a country if future events involving it threaten to inflict serious damage on the nation. A material that is needed for military purposes is considered strategic, although a strategic material need not be critical, nor a critical material strategic. Assessments of whether materials should be considered critical and/or strategic can vary from country to country. Metals like platinum, manganese and chromium are often considered critical in the USA, but in South Africa, a major source of all three, they are not. This article focuses on the US perspective concerning critical and strategic materials, and it will use the terms critical and strategic jointly. It also concentrates on metals and minerals rather than organic materials such as natural rubber and opium.

US legislation concerning stockpiles of critical and strategic materials defines them as those that: (a) would be needed to supply the military, industrial, and essential civilian needs of the USA during a national emergency; and (b) are not found or produced in the USA in sufficient quantities to meet such a need. This definition is fairly precise, but it leaves considerable room for disagreement on practical grounds. Some materials (such as nickel) are defined as critical and strategic, even though all US requirements during a national emergency could be satisfied by imports from a relatively secure source of supply, such as Canada. Although technically a critical and strategic material, nickel is seldom considered a serious problem for the USA. Criteria are needed for measuring criticality and for comparing it across different materials.

1. Criteria for Evaluating Criticality

Many criteria have been proposed for analyzing the degree of criticality of materials. They can generally be grouped into one of three categories: availability of foreign supplies, availability of domestic supplies, and means for reducing domestic consumption.

1.1 Foreign Supplies

Several issues can be raised concerning foreign supplies of a material that might be considered critical or strategic: how likely is a disruption, how severe might a disruption be, and how long might a disruption last. These general issues are usually analyzed by considering a number of factors.

A major concern is whether production is concentrated in a small number of countries. If production is dominated by a small number of suppliers, the possibility exists for cartelization of the market. This risk is considered quite small for most mineral markets, particularly in the long term. Of greater practical concern is the diversification of supplies. Materials-processing operations can experience disruptions caused by labor unrest, fires, floods, earthquakes and local political or economic problems. If a market is supplied by a large number of small producers, one or more is likely to have production problems at any given time, but the variance of total production will be relatively low. If the market is dominated by just a few producers, disruptions will be more unusual events, but the variance of total production will be relatively high. Markets dominated by a small number of supply sources are therefore of relatively high risk.

The location of these supply sources is also of concern. Some, such as Canada, might be considered relatively safe on the grounds of both political orientation and security of supply routes. A country like Australia might be considered secure politically, but the length of the supply route from Australia to the USA raises some concerns. Supplies of materials from many African countries might be considered insecure on both grounds.

The speed with which foreign producers could increase production during a shortage is also of interest. Most minerals projects have long lead times, in the range of five to ten years for new properties. If adequate reserves are available, existing capacity can be expanded much more quickly, sometimes in as little as one to two years.

1.2 Domestic Supplies

Issues concerning domestic supplies are somewhat different. The availability of domestic raw materials and processing capacity are of major importance, but these two factors must be considered separately. In the USA, some materials (e.g., copper) are primarily mined and processed domestically. Others, like aluminum, are primarily processed domestically from foreign raw materials. Still others, such as cobalt, are primarily imported in metallic form. The ability of the domestic industry to expand during an emergency will depend on the availability of raw materials, the availability of domestic processing capacity, and the speed with which both can be expanded. Expanding domestic processing capabilities is of little benefit, however, if additional raw materials would not be available during a crisis. The costs of new or additional production are important from a policy standpoint, because the costs of producing domestically during an emergency must be compared with the costs of other policies, such as stockpiling.

1.3 Domestic Demand

Demand issues include substitution, conservation, and recycling possibilities, the normal business cycle, and effects of new technologies. In some instances, substitute materials that provide equivalent properties may be readily available with little or no increase in costs. However, this is seldom the case. Potential substitutes are typically either more expensive or provide somewhat inferior properties, and they almost always require time to develop and/or implement. These lags can be as long as ten years, depending on the amount of development required. Substitutions can be either direct or functional. Consider, for example, the use of cobalt as a binder in tungsten carbide cutting materials. A direct substitution might involve replacing some of the cobalt with nickel, while a functional substitution might involve replacing a machined part with a forged part.

Conservation possibilities also tend to take time to explore and implement, and usually entail either inferior properties or higher costs. For example, the amount of manganese used in carbon steel might be reduced in several ways. One approach would involve more precise control of the manganese ferroalloy additions to the steel furnace, which would permit the steelmaker to aim for the low end of the acceptable manganese range, rather than the midpoint. This increases the costs of making the steel but does not affect the steel's properties, and could be accomplished rather quickly. Another approach would be to lower the manganese specification for steels in which the manganese is used primarily as a desulfurization rather than alloying agent. This would affect the properties of the steel, but not unacceptably so in many applications. Educating steel consumers to use lower quality steel where acceptable could take some time. A much more expensive, time-consuming approach to conserving manganese might involve recycling steel furnace slags. This approach conserves manganese by changing processing methods, rather than by just using less with the same processing methods. For some materials, the conservation allowed by more material-efficient (but more expensive) processing and fabrication methods can be substantial.

Recycling, if technically and economically feasible, can be an effective way to reduce domestic demand during an emergency. Recycling of some materials, like platinum, is extensive even during normal market conditions. The situation can be very different for other materials, however. For example, prior to 1978 virtually no superalloy scrap, except some of that generated by the superalloy producers themselves, was recycled. Technically, recycling mixed superalloy scrap is at best marginally feasible, and in the past the costs of segregating and storing scrap by alloy type exceeded the value of the scrap. When cobalt prices rose to very high levels in 1978 and 1979, segregating the scrap by type of superalloy became economically feasible, and recycling of scrap generated in fabrication became quite common. Recycling of old jet engine parts made of superalloys, however, was still minimal because of technical and economic problems.

Variations in demand over the business cycle can also cause problems for consumers of materials. For example, the aerospace industry has occasionally experienced large surges in demand. This has led to corresponding increases in the demand for titanium castings and forgings, sometimes outstripping the capacity of the industry. Analysis of issues concerning the demand for critical and strategic materials must cover the possibility that supply emergencies could coincide with upswings in the business cycle.

Another potential problem with demand is the emergence of new technologies that require large amounts of materials. If these technologies emerge suddenly, shortages can develop. Usually, however, firms assure themselves of a steady supply of raw materials before they commit themselves to new products. One example of this is the use of platinum in automobile catalytic converters in the USA. Before the US automobile manufacturers committed themselves to this product, they negotiated contracts with the South African platinum producers to assure themselves of continued supplies, and even promoted capacity expansions by the smaller producers to diversify their supplies somewhat. Catalytic converters are now the largest single use of new platinum in the USA.

2. Methods for Ranking Criticality

After all the factors important in analyzing criticality have been analyzed, methods for ranking the relative importance of different materials are needed. A number of approaches have been suggested, all of which have drawbacks.

The simplest approach is to rank materials by import dependence, the percent of normal demand that is met through imports. Although it is relatively easy to obtain data and use import dependence as a measure of criticality, the approach ignores so many important issues that it is generally useful only for rough comparisons across materials. For instance, this method does not account for scrap as a source of supply or changes in government stockpiles in estimating import dependence.

Purely judgmental approaches have been used at times to rank materials by their degree of criticality. The US Congress Office of Technology Assessment (1976) used such an approach in a study of stockpiling policies. The OTA study relied on a consensus of expert opinion to compare materials. Judgmental methods may be acceptable for simply ranking materials by their perceived degree of criticality, but they can be criticized for not providing numerical estimates of the degree to which some materials can be considered more critical than others.

Modified judgmental methods have been proposed to produce numeric indices of criticality. In these studies, such as those by King and Cameron (1974), King (1977), and Hughes (1975) a number of important factors were identified that affect criticality, and for each material, numeric scores (on arbitrary scales) were assigned to each factor. Summing the scores across factors yields a criticality index for each material. This approach is an improvement on the simple judgmental methods, but the arbitrary numeric indices are difficult to estimate and interpret.

One method for measuring the criticality of materials in monetary terms involves using input–output models of a national economy. An input–output model can be used to estimate the change in an aggregate measure of the economy, such as gross national product (GNP), that arises when imports of a material deviate from some normal or base level (Levine and Yabroff 1975).

Multiplying the monetary measure of the costs of a disruption in supplies by the probability that such a disruption will occur yields an estimate of the expected costs of a disruption for a material. This expected disruption cost can be used as a criticality index. This approach requires much more information than the judgmental and relative weighting approaches, but it makes explicit, visible assumptions concerning important issues, and it leads to monetary measures of criticality that can be used directly in performing policy analyses. The input–output approach is probably best suited for studying short-run criticality issues, because input–output models are ill-suited for reflecting inventory, supply and demand responses to shortages.

The market model approach to measuring criticality overcomes these objections to the input–output approach, but it requires even more information. A market model for a material can be constructed using engineering and economic information on domestic and foreign supply, demand and inventory behavior. The model can be used to produce forecasts of market conditions during both normal times and a variety of market emergencies. Measures of economic losses during these emergencies can be constructed, usually relying on the concept of economic surplus (consumer surplus and producer surplus). As with the input–output approach, the measures of economic loss are weighted by the disruption probabilities to estimate a monetary index of criticality. The major drawback of this approach is its large information requirements. Detailed market models are needed for each of the materials to be studied.

3. Policy Options

Many policies have been implemented or proposed to deal with perceived problems of critical and strategic materials. Stockpiling has been the most commonly followed practice. Other policies have generally been designed to reduce domestic consumption or increase domestic production.

In most cases stockpiling has proved to be the most cost-effective policy in providing supplies for potential emergencies. The carrying costs for large stockpiles can be quite high, however, even if storage costs are negligible, so other options have been investigated regularly.

The other policies most commonly implemented in the USA have been designed to encourage domestic production. Such policies have included price guarantees for firms considering whether to develop domestic capacity, guaranteed loans, and direct capital subsidies. Federally funded research programs to investigate methods for making use of low-grade domestic resources have been quite common.

Policies directed at reducing consumption or increasing recycling have primarily involved research and development, and information programs. Although never implemented, direct taxes on primary consumption would be one method to encourage recycling and discourage primary consumption.

Some policies, such as import tariffs and import quotas, affect both consumption and production. Their effects are similar: both raise the domestic price of a material, relative to the world price. This encourages recycling and domestic production and discourages domestic consumption. Import tariffs and quotas have often been imposed on industries, but usually for purely protective reasons. The US oil import quota in the 1960s was, however, called a strategic policy.

4. Markets for Chromium, Manganese and Cobalt

The markets for chromium, manganese and cobalt are of concern to policymakers in some developed countries because they are considered to be essential materials for some defense applications and because supplies are highly concentrated in a few potentially unstable or hostile countries. Moreover, most developed countries import more than 95% of their supplies of these materials.

Before discussing these markets, however, it is useful to consider briefly what is meant by the terms requirements and demand for materials.

The term demand refers to the quantity of consumption that exists at varying levels of price, other factors being constant. It is usually expected that the demand for a material is inversely correlated with its price and directly proportional to the level of economic activity in the industrial sectors that use the material. If prices rise, demand is expected to decrease if economic activity and other factors remain unchanged. Requirements refer to the level of consumption of a material by domestic customers that is expected to occur based on technological and market growth assumptions. The concept of requirements thus implicitly assumes that the price elasticity of demand is small enough to be neglected.

Much of the debate concerning critical and strategic materials has focused on projections of the requirements for these materials and not on an analysis of demand. However, it is the demand analysis that is most useful for evaluating future materials consumption. It is desirable to evaluate the demand for these materials this year and in the future if prices rise (e.g., because of a supply disruption).

4.1 Chromium

Chromium production is highly concentrated in a few countries.The most important supplier to the market economies of the world is South Africa, which is of concern to some because of the potential for political disruptions in the future. South Africa accounted for about 32% of total world production in 1987. The second most important supplier to the West in the 1980s was the USSR, which accounted for 29% of total world production in 1987. Other important western producers have been Zimbabwe, the Philippines and Turkey, all of which have experienced severe financial or political difficulties in recent years. This suggests that a moderate disruption in chromium supplies is likely to occur within the next decade or so, and a severe disruption is also possible.

Metallurgical grade chromite ores are found primarily in the USSR, Turkey and Zimbabwe. Although South African production is of a lower grade, most useful for chemical uses, technological developments have made it possible to use these ores in metallurgical applications. In the 1970s, South Africa gradually assumed an increasingly important position in the world market. The chromite ores found in the Philippines are primarily useful for refractory applications.

In the developed world chromium is used for three main applications: production of metallic alloys, chemicals and refractories. In the USA, about two thirds of the chromium consumed in a typical year is in the form of ferrochromium for metallurgical applications. In 1987, 79% of this was used for the production of stainless steels, and 18% was consumed in alloy steels; 3% was used in superalloys. Most of the chromium used by the steel industry is in the form of high-carbon ferrochromium, which cannot be used in superalloys because of requirements for high-purity materials.

It is often stated that there is no substitute for chromium in the production of stainless steels and superalloys. For instance, it is estimated that only 10% of current superalloy usage, which accounts for 1–2% of total chromium consumption in the USA, could be substituted for, even at very high prices (e.g., 50 times current chromite prices). Moreover, since most of these materials are used in aircraft and industrial gas turbine applications, which require long testing and qualification programs before new alloys can be used, there is a 3–5 year lag time before a significant amount of the substitutions could be accomplished. Moreover, only a limited amount of substitution for stainless steels by other materials would occur, even at higher prices. More importantly, however, there is opportunity for considerable substitution among grades of stainless steels (e.g., ferritic grades for austenitic grades) and for conservation in the production processes. Tables 1–3 summarize the results of a study of the potential for various substitution and conservation options at prices that are various multiples of a base price. The prices are expressed in terms of the form of the material that is actually consumed in one application, and therefore the price elasticity of demand cannot be estimated in terms of equivalent prices. However, it has been shown that a twofold increase in the price of chromite would lead to an increase of

~6% in the cost of producing an austenitic stainless steel. There is considerably more elasticity of demand with respect to price in refractory applications than in stainless steel applications.

4.2 Manganese

Manganese is used as a bulk ferroalloy even more than chromium. In the USA over 90% of the manganese consumed is used by the metallurgical industry, mostly in carbon steels. Alloy steels, particularly the Hadfield steels, use more manganese per ton of steel than do carbon steels, but the much higher production levels for carbon steels make them the most important market for manganese. Manganese is an essential input in modern steelmaking, for desulfurization, which is why manganese is of vital concern to all steel producers.

Most manganese is consumed in the form of standard high-carbon ferromanganese, which contains about 78% manganese and 5% carbon. Standard ferromanganese is well suited for use in producing carbon steel, but alloy steels and other specialized metallurgical products often require more refined forms. Medium- and low-carbon ferromanganese directly follow standard ferromanganese in importance, and electrolytic manganese has only a very small market. Manganese ore is used directly by the chemical industry, particularly for use in dry cell batteries.

The world's largest producer of manganese is the USSR, but East–West trade is minimal. The most important producer in the noncommunist world, from the standpoint of both current output and reserves, is South Africa. In 1987 it accounted for 26% of noncommunist supply, and 62% of noncommunist reserves. Brazil, Australia and Gabon are all important, accounting for 54% of noncommunist supply. Other important producers include Mexico and India. The USA currently produces no manganese.

Manganese reserves (1987) are estimated to be of the order of 1.0×10^9 t, with South Africa and the USSR accounting for about 40 and 32% of the total, respectively. There are considerable additional tonnages of land-based manganese deposits that could potentially be recovered at higher prices, but these are also located primarily in South Africa and the USSR.

There is not much concern about short-term problems with manganese supplies because the leading free-market producer—South Africa—accounts for only 13.5% of total production. Moreover, the other major producers (Gabon, Australia, Brazil and India) account for >10% of noncommunist output each. Therefore, a supply disruption in any individual country would not be of great consequence. The major concern is over the potential for long-term problems. Of particular concern is that developed market economy countries may be dependent upon only two major suppliers of manganese (South Africa and the USSR) by the year 2000. The potential situation may not be as troublesome as some fear, however; there are conservation and substitution options available to consumers in developed countries if manganese prices escalate. For instance, it has been estimated that if the price of manganese ore doubles, consumption of manganese by the carbon steel industry in the USA would decline by about 10% in one year and by approximately 24% at the end of five years (Table 4). A tenfold increase in the price of ore would result in a 15% decrease in consumption in the first year and a decline of almost 50% by the end of seven years. Similar decreases in the consumption of manganese in other industrial sectors and in other geographical regions would be expected at higher prices.

Table 1
Stainless steel consumption relative to base consumption at different stainless steel price levels

Price relative to base price	Years after price increase				
	0	1	2	3	4
1.00	1.00	1.00	1.00	1.00	1.00
1.10	0.97	0.96	0.95	0.94	0.93
1.25	0.92	0.88	0.86	0.84	0.83
1.50	0.83	0.76	0.70	0.68	0.66
2.00	0.75	0.66	0.60	0.55	0.51
3.00	0.60	0.51	0.45	0.40	0.37

Table 2
Chromium consumption in alloy steels, relative to base consumption at different ferrochromium price levels

Price relative to base price	Years after price increase								
	0	1	2	3	4	5	6	7	8
1.00	1.00	1.00	1.00	1.00	1.00	1.00	1.00	1.00	1.00
1.67	0.92	0.92	0.92	0.92	0.92	0.92	0.92	0.92	0.92
2.67	0.80	0.79	0.77	0.76	0.74	0.72	0.70	0.69	0.67
4.33	0.75	0.72	0.70	0.67	0.63	0.60	0.55	0.51	0.47
9.33	0.72	0.66	0.60	0.55	0.50	0.45	0.36	0.29	0.23
17.67	0.72	0.65	0.58	0.52	0.47	0.42	0.31	0.23	0.17

Table 3
US chromite consumption in refractories relative to base consumption at different chromite price levels

Price relative to base price	Years after price increase 0	1	2	3
1	1.00	1.00	1.00	1.00
2	0.70	0.59	0.48	0.35
5	0.49	0.44	0.36	0.24
10	0.46	0.38	0.29	0.19
25	0.45	0.35	0.24	0.13
50	0.45	0.33	0.22	0.10

Table 4
Manganese demand in carbon steel production

Price factor[a]	Time lag (years) 0	2	4	7
USA (Base steel production = 127 630 000 short tons)				
1 ×	810450			
2 ×	750000	725000	700000	619000
5 ×	680000	640000	595000	505000
10 ×	635000	585000	515000	420000
25 ×	615000	500000	440000	330000
50 ×	600000	490000	420000	325000
Japan (Base steel production = 110 135 000 short tons)				
1 ×	610147			
2 ×	585000	574000	558000	544000
5 ×	572000	512000	480000	462000
10 ×	556000	482000	426000	405000
25 ×	550000	465000	410000	263000
50 ×	544866	461465	406398	257715
EEC (Base steel production = 155 058 000 short tons)				
1 ×	1052840			
2 ×	981000	954000	924000	883000
5 ×	929000	852000	815000	766000
10 ×	906000	810000	751000	688000
25 ×	890000	770000	695000	485000
50 ×	882800	765987	688458	479129
Other (Base steel production = 92 734 000 short tons)				
1 ×	634000			
2 ×	612000	594000	578000	532000
5 ×	577000	524000	495000	440000
10 ×	558000	480000	444000	398000
25 ×	540000	450000	415000	330000
50 ×	532293	439559	397829	263365

[a] Base manganese price = 55 US cents kg^{-1}

4.3 Cobalt

Cobalt is considered by many to be both a strategic and a critical material for the USA; strategic because of its use as a high-temperature material in jet engines and industrial gas turbines, and as a catalyst for petroleum desulfurization; critical, because the USA imports virtually all of its supply of primary cobalt, and a large percentage of these imports comes from countries with potentially unstable political environments.

Zaire is the dominant producer in the cobalt market, accounting for 50–60% of world production in most years. Cobalt is a coproduct with copper in Zaire, and serious declines in production occurred from 1975 through 1978. In Zambia, cobalt is a by-product of copper and production declined after 1975 as copper prices fell. In Canada, the Philippines, Australia, Indonesia, New Caledonia and Botswana, cobalt is recovered as a by-product of nickel mining and processing operations. Disposals of cobalt from the US strategic stockpile played an important role in the cobalt market in the early 1970s, but sales of cobalt ended in 1976.

Zaire has experienced severe problems in supplying cobalt in recent years. In 1975, the Benguela railroad from Zaire through Angola to the Atlantic was cut off due to the civil war in Angola. In 1977, the Shaba province of Zaire, where cobalt is produced, was invaded. Cobalt production was not affected that year, but the invasion in 1978 did affect production. The damage to the facilities in 1978 was slight, but virtually all of the foreign workers fled, and many facilities were in a state of poor repair. Low copper prices have left Zaire almost bankrupt, unable to improve its domestic transportation network or to maintain the copper and cobalt facilities. The International Monetary Fund and the World Bank have been attempting to remedy Zaire's financial difficulties, but it remains to be seen if their attempts will prove successful. It is possible that Zaire could just manage for the next several years until the copper market recovers and the efforts by the IMF and World Bank bear fruit. Despite attempts by Zaire to improve its economy and mining industry, another cutoff in supplies, although unlikely, could happen at any time.

The three most important uses of cobalt on a tonnage basis in the USA are in superalloys, as an alloying element in permanent magnets and in chemicals. Other important uses include cemented carbides, hard-facing and welding materials and as an alloying element in steels.

A recent study of the world cobalt market found that there is considerable price elasticity in the demand for cobalt, because, as the price rises, consumers shift their preferences to alternative materials and designs. Furthermore, the increase in price encourages conservation of cobalt through increased recycling and improvements in processing techniques. For instance, the study showed that if the price of cobalt were to stabilize at 55 US dollars per kg (1978) the demand for cobalt would decline to less than half of its initial value within five years. The short-term response would be less, but there would still be a decrease of approximately 15% in the total demand for cobalt within one year if the price were to increase from 22 to 55 US dollars per kg. The results of the substitution analysis, which was developed in 1978, have undergone the test of time well, and were used as part of an overall supply–demand framework to predict, with considerable accuracy, world cobalt prices over the period 1978–1982.

The results of the substitution analysis are crucial to an understanding of the economic and national security consequences to the USA of a supply disruption in the world cobalt market. It is conceivable, even likely, that another supply disruption will occur because of the political instability in southern and central Africa. In such an event, it is clear that the price of cobalt will again rise as the result of psychological factors and actual production curtailments. The magnitude of the price increase is uncertain, but it is quite unlikely to increase above 110 US dollars per kg for any appreciable period because of the downward pressure from substitution and additional supplies from other nations. The results of the analysis may be interpreted as follows. If the price were to increase by a factor of 2.5, and other parameters (such as industrial activity and prices of substitutes) were to remain unchanged, the total demand for cobalt in the USA would decrease by 15% in a one-year period. A price increase by a factor of five would bring a demand decrease of 20% in the first year. Currently, about 15% of cobalt supplies in the USA come from recycled (secondary) sources. Thus even if supplies were restricted by 50% in the first year of the disruption only up to 20% of demand would have to be met by releases from the government stockpile. The greater the price, the less the demand and the greater the contribution from recycled material.

If the supply restriction were to last for more than a year, demand would have more time to adjust and there would be more substitution. Thus the overall demand for cobalt in the USA would decrease by 50% in years two through seven, with a price increase factor of 2.5, if other factors were to remain unchanged. In the event of larger price increases, substitution would be greater. If the disruption were to continue, there would also be time for additional sources of supply to respond to the higher price.

Indeed, such responses have been shown by both the demand and supply sides of the market to the 1978 disruption. Following the disruption, the producer price was set at 55 US dollars per kg and an attempt was made to maintain the price at that level. However, because of substitution and increased supply from other sources outside Zaire and Zambia, large stocks were built up by producers, and the free market price declined to less than 22 US dollars per kg in 1982. Consequently, extreme pressure was placed on the price, which gradually decreased to about 11 US dollars per kg in early 1983, before rebounding to about 25 US dollars per kg in late 1984. The price declined to 11.40 US dollars per kg in 1987.

Two conclusions may be drawn from this analysis. Firstly, a disruption in cobalt supplies from Zambia and/or Zaire would not have a measurable effect on the national security of the USA. The increased prices would result in substantial substitution away from cobalt in nonessential uses. However, there would be a time delay, and it might be necessary to release some cobalt from the government stockpile. Secondly, there would be an economic cost to the USA resulting from a supply disruption because of the loss in consumer surplus due to higher market prices and reduction in quantities consumed. The costs of such a disruption could be substantial but would be short lived because of the downward pressure on the price.

See also: Stockpiling

Bibliography

Charles River Associates 1978 *Measuring Materials Criticality for National Policy Analysis*, prepared for the US General Accounting Office. Charles River Associates, Boston, Massachusetts

Charles River Associates 1981 *The Effects of Supply Restrictions on the Demand for Manganese, Chromium, Cobalt and Platinum*, report to the US Department of the Interior. Charles River Associates, Boston, Massachusetts

Hughes E E 1975 *Strategic Resources and National Security: An Initial Assessment*, prepared for the Defense Advanced Research Projects Agency. Stanford Research Institute, Stanford, California

King A H 1977 *Materials Vulnerability of the United States—An Update*. US Army War College, Strategic Studies Institute, Carlisle Barracks, Pennsylvania

King A H, Cameron J R 1974 *Materials and the New Dimensions of Conflict*. US Army War College, Strategic Studies Institute, Carlisle Barracks, Pennsylvania

Levine M D, Yabroff I W 1975 *Department of Defense Materials Consumption and the Impact of Material and Energy Resource Shortages*, prepared for the Advanced Research Projects Agency, Stanford Research Institute, Stanford, California

US Bureau of Mines 1988 *Mineral Commodity Summaries 1988*. USBM, Washington, DC

US Congress, Office of Technology Assessment 1976 *An Assessment of Alternative Economic Stockpiling Policies*. US Government Printing Office, Washington, DC

J. P. Clark
[Massachusetts Institute of Technology, Cambridge, Massachusetts, USA]

B. Reddy
[Charles River Associates, Boston, Massachusetts, USA]

E

Economic Theory of Mineral Resources and Markets

Theoretical economics provides analytical frameworks to aid the understanding of how nonrenewable resources are efficiently discovered, developed and valued over time through market operations. Theories permit empirical tests of hypotheses; for example, interpretations of data on mineral reserves, the rates of depletion, technical change and the trends in prices or production. They also offer current insights into the uncertain future values of resources.

1. Equilibrium and Capital Theory

Because mineral resource information is costly to collect, the efficiency of a given theoretical framework and its correlates, the information's structure and the data are always important considerations to both industry and public policymakers. The efficiency of an economic information system is defined by its ability to reach equilibrium solutions to problems and by the degree to which a particular theory models well important aspects of the market, that is, with parsimony, special insights and some measure of confidence. How well a theory models reality determines its validity. Measures of confidence depend on how well the important relations of the market, such as supply and demand, are specified and identified, and on how robust the results are under changing conditions. The sensitivity of an application is examined by describing the nature of the equilibrium and its stability under the specified market adjustment mechanism.

A materials industry's market structure is defined by the behavior of its buyers and sellers and varies by industry and over time. For example, the dominance of a few sellers in a national market has characterized particular nonfuels at times. The behavior of a firm in such a market will be different from that of one among many sellers; namely, prices will be higher and levels of output lower than in the more competitive case. Furthermore, both the industry's technologies and information on the relevant economic geology of deposits change. Thus, a single static or dynamic model is not likely to accommodate all the interesting questions addressable to resource information, requiring that the objectives of different mineral economics applications be properly distinguished. The building of an appropriate theoretical model can assist an industry firm in negotiating contract terms, in predicting the movements of spot market prices in the short run, or in critically assessing the longer-run policy decisions of agencies concerned with mineral conservation, regulation or price stability.

Most fundamentally, mineral deposits are a form of capital and their transformation from the asset side of the industry's balance sheet into flows of materials from producers to consumers is a major problem of mineral economics analyses. This article is limited to three principal issues of applied capital theory under perfectly competitive market conditions:

(a) the valuation of specific depletable resources;

(b) the dynamics of capital investment's growth or decline under conditions of risk and uncertainty; and

(c) the problems of inter-regional trade, including transfers of mineral wealth.

This division emphasizes the analysis of mineral and energy resources as nonrenewable capital stocks and flows and hence the inevitability of exhaustion, the uncertainties and high risks of exploration, environmental damage and other economic externalities, and the prominence of spatial equilibrium and economic rents. A glossary of convenient terms describing mineral markets is provided in Table 1.

2. Mineral Resource Valuation

In the long run, future expected prices $\hat{p}_t$ form the basis for measuring the extent of resources and their potential for transformation to reserves over time. This transformation takes place largely through investment in exploration and mine development. Reserves are valued by finding the present discounted worth of their expected net income stream, given the rate of interest r_t and the risk premium, just as in the case of renewable capital assets. However, the expected price path will rise, due to depletion as deposits are mined to exhaustion. It is optimal for each firm to equate in all time periods the present discounted net return received over its deposit life. Assuming constant deposit quality and zero extraction costs for simplicity, this means that in the time interval $(t, t+\theta)$ the return to conserving the resource is $(p_{t+\theta}-p_t)/p_t$. The optimal rate of depletion will thus equal the return for holding a bond of equivalent risk the same amount of time, $r_t\theta$ (Hotelling 1931). This "pure theory of exhaustion," therefore, produces the basic arbitrage equation forecasting an upward path for mineral prices:

$$\dot{p}/p=r_t \tag{1}$$

If the interval describes the life of a single mine, the typical grade–tonnage relation will reinforce Hotelling's result, and there will always be an economic rent or "user cost" assignable to the deposit above monopoly or other rents and costs. If the

Table 1
Glossary of terms and symbols

Symbol	Term
a	vector of supply entry costs in consuming center
b	capacity and demand constraints vector
d	own-price slopes of demand vector
e	variable extraction costs vector
f	demand intercept vector
g	transportation costs
q_t^i	quantity supplied at time t from the ith source
p_t^i	supply price $=f(q_t^i)$ at time t at the ith source
q_t^j	quantity demanded at time t at the jth sink
p_t^j	demand price $=f(q_t^j)$ at time t at the jth sink
$\bar{q}_t$	initial quantity vector $(\bar{q}^j, \bar{q}^i)$
$\bar{p}_t$	initial price vector $(\bar{p}^j, \bar{p}^i)$
q^i-q^j	excess demand
$\dot{p}$, dp/dt	mechanism of price adjustment
$\hat{p}$	price expectations $(\hat{p}^j, \hat{p}^i)$
$\hat{q}$	quantity expectations $(\hat{q}^j, \hat{q}^i)$
r	"the" rate of interest
t	time
u	vector of errors in observed price
v	vector of errors in observed lags
x	vector of aggregated regional outputs of i regions
y	vector of individual deposit outputs
z	inter-regional flow vector
G	matrix of zeros and vectors routing flows
Q	matrix of intercepts, slopes and bounds
R	matrix of constraints
S	matrix of own-price slopes of supply
β	parameter for estimation
θ	time difference lag

interval describes long-run industry supply, then offsetting factors such as technological change, substitution or capital stock effects can subvert the effects of depletion. However, in the very long run, resource depletion impacts are inevitable and imply difficult choices for society in the determination of levels of conservation.

In the more immediate run, current market prices p_t determine the decision of both the level of producers' output and its allocation over existing capacities. Figure 1 depicts how conventional theories of market equilibrium and growth portray a typical materials market by a system of equations and an adjustment mechanism. These can be used to explain the pattern of observed price variations in specific investigations. At point $A(\bar{p}, \bar{q})$, an initial equilibrium, excess demand is zero for both short- and long-run supply and demand. Assuming a shift in long-run "permanent" demand from some initial level OA to a higher level Of, due, say, to materials-intensive innovation, positive excess demand $(\hat{q}-\bar{q})$ results. Such shortages are removed in the short run by firms increasing outputs from existing supply capacities to the expected point $\hat{B}(\hat{p}, \hat{q})$.

Emphasis is placed on the mechanism of such adjustments, which typically relate the magnitude of excess demand to the speed with which changes in price occur from A to $\hat{B}$ and to the initial variable values:

$$\dot{p}=f[(q_t^j-q_t^i), \bar{p}_t, \bar{q}_t] \tag{2}$$

Here i identifies the ith commodity source and j the jth consuming center. In the longer run, capacity changes are indicated by the long-run supply function AC. New capital in the form of exploration or development will increase reserve response to the higher price expectations, forcing a limit to profits at $\hat{p}$ and eventually a decline, following the intermediate or "lagged" supply schedule AB. However, the ultimate long-run equilibrium price and quantity after all such adjustments at $C(p', q')$ are expected to be higher than at A, supporting the solid-line pattern of price and output changes over time in Fig. 2 forecast by depletion theory. Such forecasts assume that technological change does not offset depletion entirely by shifting AC downward and to the right.

Economic geologists, in collaboration with mineral economists, have recently addressed the problems of integrating geological information more explicitly into the modelling of changes in the resource–reserve base

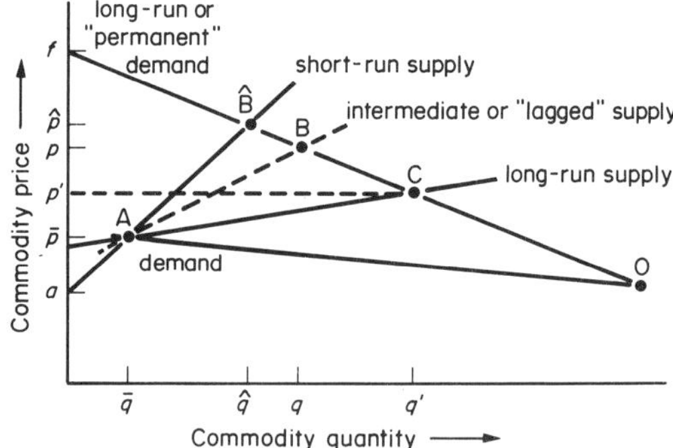

Figure 1
Short- and long-run supply adjustments to a change in "permanent" demand

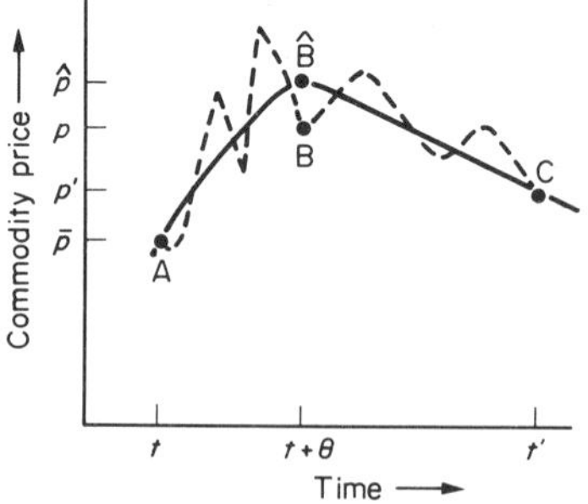

Figure 2
Expected (——) and observed (- - -) price paths under dynamic demand and uncertain conditions

of large regions. This broadens market theory applications to include the assessment of a region's stock of deposits or endowment and the partitioning of the set of potential and known deposits into economic and uneconomic subsets. The discovery of these reserves is modelled, ordering them by their net present values. Such a stocks schedule is a "potential supply function." The forward prices indicated by such appraisals supplement the results of the conventional economic models described above.

3. *Dynamic Equilibrium Under Uncertain Conditions*

While benefits can be derived from the insights of conventional geostatistical and economic theory applications as described, the equilibrium conditions of the competitive models remain essentially "stationary," and the pretense is required that the futures they forecast are "certain." Even with such simplifications, the mathematical requirements to ensure modelling solutions are quite severe. For instance, production must be "well behaved," in the sense that no interdependencies can exist among agents or activities; externalities, such as pollution, are not easily accommodated; no economies are permitted such as those of scale; no costs of information, transport or transactions are included; and so on. The mineral industries offer abundant evidence that these restrictions do not portray realistically the dynamic and uncertain decisions of actual markets. For instance, investment choices require the undertaking of unknown risks of exploration, based on expectations of prices and profits as far removed as 30 years.

The question of how conventional theory might be rendered more dynamic and econometrically relevant has occupied economists since Jevons (1865), who analyzed Britain's coal industry. A step toward realism taken by more recent theorists to reflect uncertainties relies on the observation that the price expectations of suppliers depend on assumptions about past observed prices. This leads to difference equation adjustment mechanisms which produce fluctuating price paths because of lags:

$$p_t = \beta_1 + \beta_2 p_{t-\theta} \quad (3)$$

The constants β of this equation are reductions, being functions of the slope and intercept parameters of the structural supply and demand equations depicted in Fig. 1. Depending on the β values, distributed lag models produce oscillating price behavior more suggestive of the observed short-term cyclical variations of mineral prices about an expected trend. A case is shown by the dotted path of prices over the interval (t, t') in Fig. 2. In this formulation, the stability conditions are open to empirical inquiry, while explicit tests of the inflexibilities and lags in supply or demand associated with observed market fluctuations can be econometrically attempted.

The excess demand differences can similarly be employed to measure the extent to which intermediate price expectations on the way to longer-run target outputs fail to be met in actuality $(\hat{p}-p)$. This can be the result of inflexibilities in capacity adjustments in capital-intensive segments of a minerals industry. In such a case, the estimation of the distributed lags can be based on the relationships observed between ongoing investment and the previous output prices. When the shifts of supply or demand can be exactly identified, the estimation of the β coefficients in Eqn. (3) permits the representation of actual price oscillations p_t as the addition to expected prices $\hat{p}$ of disturbed departures of observed prices from that expectation, p_t^*, plus an error term. In a simple example by Nerlove (1972):

$$p_t = \hat{p} + p_t^* + u_t \quad (4)$$

where

$$p_t^* = \beta p_{t-\theta}^* + v_1 \quad (5)$$

Under conditions where the error terms u and v are normal and independent with zero means and constant variances, $(\hat{p}+p_t^*)$ may be interpreted as the long-run normal or trend from initial price A $(\bar{p}, \bar{q})$ to target price C(p', q') in Fig. 2. Then, the mean square error forecasts of $(p_{t+\theta}-\hat{p})$, that is, $(\hat{\mathrm{B}}-\mathrm{B})$, will be the conditional expectations of the price deviations at $t+\theta$ through time t. Under these simple conditions, and with target prices fixed, well-behaved trends and measures of confidence can be stated for the statistical estimates of the distributed lags obtaining for actual materials market price data.

More-realistic complex versions can be formulated, which link measures of resource depletion to shifts in supply and demand or which incorporate the uncertain impacts of environmental regulations, technological change and the like into the study of materials price trends or cycles. Such theoretical models have led to valuable insights into decisions under these stylized conditions, termed "rational expectations hypotheses," and into the dynamic investment behavior of the markets for minerals or fuels. They have also formed the basis for the econometric analysis of materials prices. Nevertheless, few such studies have produced entirely convincing results, and much pioneering work remains to be formulated.

Despite early notice of the need for dynamic models, theorists have only recently attempted empirical applications explaining the dynamic underpinnings of distributed lag models fitted to observed data. However, factors other than past prices cannot readily be incorporated into the discussion of the dynamic stability of these systems. Thus, the models assume fixed targets and certain long-run equilibrium prices for tractability. Although attempts to identify statistically the disequilibrium elements associated with observed market shifts are conceivable, the mathematics of dealing with and defining optimal investment behavior under realistic long-run dynamic conditions remain complex. As more adequate mathematical insights are obtained, more econometrically relevant analyses become feasible.

An alternative is the adaptation of control theory and probability, combined with simulations in which exploration and the other important empirical relationships underlying long-run potential supply might be identified. To this end, stock supply functions derived from cost engineering data and the information of economic geology have become increasingly useful. The results are more-tractable models in which greater attention can be paid to the rigidities, interdependences and externalities typical of real investment growth or decline in the materials markets. Clearly, too, regional geology and transportation can take more-prominent roles in such simulations.

4. Spatial and Temporal Market Equilibrium

Empirically useful valuations of regional materials flows often require a spatial framework for the description of market equilibrium. Following Samuelson (1952), multimarket models offer a mathematical representation of allocative efficiency that is consistent with activity analysis (input–output) approaches as well as the maximization of surplus (rents) in interregional trade. These approaches reduce multicommodity and multimarket equilibria to the joint minimization of total delivered resource costs, and recognize economic rents as an essential part of the allocation problem. Rents can be computed by region, as in the seminal programming study by Henderson (1958).

Programming theory presents an ideal model of linked markets as multiple sources i and sinks j interconnected by transport costs g, in which both aggregated (x) and disaggregated (y) supplies are defined in optimal inter-regional flow patterns G. In his simple linear program, Henderson fixed demands. Then the matrices of parameters for supply singly (S) and jointly with other constraints (R) are easily identified and the net social payoff is equivalent to economic rent. The surplus is maximized by finding $(y, z) > 0$, such that

$$\min_{y,z} e^{\mathrm{T}}y + g^{\mathrm{T}}z$$

$$\text{subject to } Ry + Gz \geqslant b \ (y, z > 0) \qquad (6)$$

The equivalent quadratic programming objective function is

$$\min_{y,z} a^{\mathrm{T}}x + \tfrac{1}{2}x^{\mathrm{T}}Sx + g^{\mathrm{T}}z$$

$$\text{subject to } Qx + Gz \geqslant b \ (x, z > 0) \qquad (7)$$

where T indicates the vector transpose.

As reformulated, Eqn. (7) has been increasingly employed to present an efficient spatial informational system for comparative studies of the reassignment of mineral rents; for example, through taxation or changes in the transportation system. An example is provided by current oil markets where such rents, represented by the area $\bar{p}Cp'$ in long-run competitive equilibrium, can be significant. Clearly they are most impressive in the short run, represented by the area $a\hat{B}f$ in Fig. 1.

See also: Prices of Metals: History; Resource Appraisal

Bibliography

Harris D 1982 Mineral endowment, geostatistical theory and methods for appraisal. In: Newcomb R (ed.) 1982 *Future Resources: Their Geostatistical Appraisal.* West Virginia University Press, Morgantown, pp. 83–135

Henderson J M 1958 *The Efficiency of the Coal Industry.* Harvard University Press, Cambridge, Massachusetts

Hotelling H 1931 The economics of exhaustible resources. *J. Polit. Econ.* 3: 137–75

Jevons J S 1865 *The Coal Question.* Macmillan, London

Nerlove M 1972 Lags in economic behavior. *Econometrica* 40: 221–50

Newcomb R 1967 Measuring technical progress in the resource industries. *Council of Economics Proceedings,* American Institute of Mining, Metallurgical and Petroleum Engineers, New York, pp. 53–68

Samuelson P A 1947 *Foundations of Economic Analysis.* Harvard University Press, Cambridge, Massachusetts

Samuelson P A 1952 Spatial price equilibrium and linear programming. *Am. Econ. Rev.* 52: 283–303

Samuelson P A 1978 The canonical classical model of political economy. *J. Econ. Lit.* 16(4): 1415–28

R. T. Newcomb
[University of Arizona, Tucson, Arizona, USA]

Electronic and Optical Minerals

Minerals have been an essential component of human society since the discovery that, when properly struck, flint could be used to start a fire; however, the explosion in high-technology uses for minerals probably can be traced to early twentieth century experiments with crystal oscillators that were the forerunners of the modern communications industry. Since those pioneering uses of quartz, galena and germanium crystals, technological uses for minerals have multiplied rapidly. Minerals from altaite (lead telluride) to zincite (zinc oxide) have been used to exploit their special optical or electronic properties.

It is ironic that the same developments that led to so many applications for minerals also led to requirements of size, shape and purity such that only a few minerals are suitable for use as they occur in nature. For this reason, synthetic crystal growing techniques now provide most of the highly pure, sometimes exotic, minerals used in photoconducting devices (cadmium sulfide, gallium arsenide), infrared spectrophotometers (sodium chloride, zinc selenide), lasers (aluminum oxide) and similar instruments.

The following discussion will emphasize those commercially important minerals that can be used as they occur naturally. Quartz is included because of its technical importance, and because natural quartz crystal continues to be the feedstock for synthetic crystal growth.

1. Quartz

1.1 Definitions

Quartz, in both natural and synthetic forms, is a crystalline material with a chemical composition very close to 100% SiO_2. Several different grades and types are defined. *Electronic-grade quartz* is quartz that, although it may be colored, is sufficiently free of impurities to be suitable for electronic applications. For such applications, the quartz must also be free from electrical twinning. *Optical-grade quartz* is quartz that is high in optical clarity and free from optical twinning. *Lasca* is a term of Brazilian origin denoting undersize fragments of high-purity quartz used as feedstock for synthetic quartz crystal growth. *Fused silica* is a material prepared by direct melting of quartz crystal (often lasca).

1.2 Supply

The greatest known natural quartz deposits occur in Brazil and are distributed over a large area, mainly in the States of Minas Gerais, Santa Catarina, Sao Paulo, Paraiba, Parana, Bahia and Rio de Janeiro. Other exporters of natural quartz include Canada, Spain, Mexico, and the Netherlands. Although the USA is the world leader in synthetic quartz production, no significant quantity of natural electronic grade quartz is presently produced.

The trend to synthetic quartz use is evident from the fact that, whereas natural quartz requirements in the USA were formerly met by imports, in 1970 synthetic quartz production exceeded imports of natural quartz, and in 1971 synthetic quartz consumption exceeded that of natural quartz. Between 1969 and 1985, imports of natural quartz (including lasca) fell from an estimated 563 000 kg to about 78 000 kg. Since the 1970s supplies of natural electronic grade quartz to US manufacturers have been provided by sales from the US Government stockpile, which had surpassed its inventory goal. The importance of quartz to industry can be inferred from the fact that quartz crystal is one of 93 commodities designated by the US Government as a basic stockpile material.

1.3 Mineralogy and Physical Properties

Quartz is one of the most abundant minerals in the earth's crust and occurs in rocks of igneous, metamorphic and sedimentary origin. It exists in two distinct forms:

(a) α-Quartz (low quartz). This is the thermodynamically stable form below 573 °C. α-Quartz has trigonal symmetry and belongs to the enantiomorphous crystal class 32. The crystal consists of an array of SiO_4 tetrahedra connected at each corner to other tetrahedra so that each silicon atom has four oxygen atoms and each oxygen atom has two silicon atoms as nearest neighbors. The linked tetrahedra form helices.

(b) β-Quartz (high quartz). Above 573 °C α-quartz is transformed to β-quartz, which is stable to 870 °C, by minor atomic movements that do not involve Si–O bond breaking or atom interchange. The structure of β-quartz is similar to that of α-quartz except that the SiO_4 tetrahedra are more regularly arranged. β-Quartz exhibits hexagonal symmetry and belongs to enantiomorphous crystal class 622.

Because quartz is enantiomorphous, single crystals are distinctly right- or left-handed and will rotate plane-polarized light. The extent of this rotation is expressed as the rotatory power α, which for the NaD line is 27.71° mm^{-1}. Quartz is weakly birefringent, with indices of refraction, also for the NaD line, $N_\omega = 1.5442$ and $N_\varepsilon = 1.5533$. It melts at 1610 °C, has a specific gravity of 2.65 and a hardness of 7 on the Mohs scale. The principal value of quartz derives from its piezoelectric property and high acoustic strength Q, which for natural quartz varies from about 0.1×10^6 to 3×10^6. Q is a dimensionless measure of energy loss in an oscillator attributable to internal friction and will be high for oscillators with low internal friction. Advances in technology for growing quartz crystal have led to synthetic crystals with Q approaching that of natural quartz.

1.4 Mining

Because of the need to preserve the crystals, and also because of the irregular distribution of the mineral, conventional mining methods such as blasting are avoided, and quartz is mined and sorted almost exclusively by hand.

1.5 Technology and Uses

Optical uses for quartz, which account for a minor fraction of consumption, include prisms and cells for spectrometers. Because of its transparency in the uv region, quartz is required as a window material for uv spectrometer sample cells. Figure 1 depicts the transmission characteristics of representative cells and demonstrates the superiority of quartz to optical glass, particularly for wavelengths shorter than 300 nm. Spectrometer prisms for use at wavelengths shorter than about 350 nm must be made from quartz or fused silica.

Most quartz is consumed in electronic applications such as frequency controls for radio transmitters, radiofrequency telephone circuits, filters in receiving circuits, transducers for interconversion of electronic and mechanical forces, and oscillators for watches and other timing devices.

Figure 2 shows an autoclave for single-crystal synthetic quartz production. Lasca is charged into the quartz supply region after which crystal seed racks are lowered into the autoclave barrel. After adding sodium hydroxide or sodium carbonate solution, the autoclave is sealed and heating is commenced. The autoclave is typically operated at 250–450 °C and a pressure of 3.4×10^7 to 1.4×10^8 Pa. Convection currents induced by the temperature difference between the heated lower chamber and the upper seed chamber carry saturated solution from the warm supply region to the seed chamber, where silica crystallizes. Losses in processing synthetic quartz may exceed 50%; however, losses in processing natural quartz for the same end use are usually greater. It is estimated that 1 kg of synthetic crystal is equivalent to 3–10 kg of natural quartz of the size and quality most commonly used in industry.

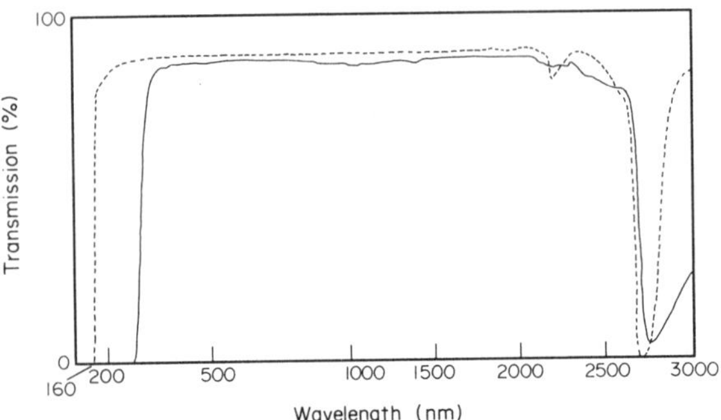

Figure 1
Transmission characteristics of silica (----) and optical (——) glass sample cells (courtesy of Beckman Instruments, Fullerton, California)

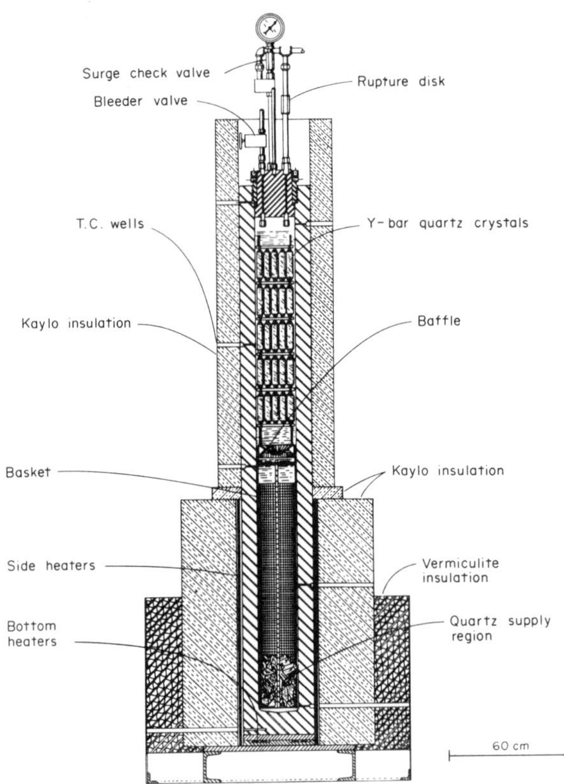

Figure 2
Autoclave for single-crystal synthetic quartz production

2. *Asbestos*

Asbestos is a general term applied to certain minerals that are classified under crystal-structure-based groups, when these minerals crystallize as the asbestiform (fibrous) variety. The principal mineral of commerce is chrysotile ($Mg_6Si_4O_{10}(OH)_8$), a member of the serpentine group; however the term asbestos is also applied to several minerals of the amphibole group, such as crocidolite ($Na_2Fe_2^{3+}Si_8O_{22}(OH,F)_2$).

The principal producer of asbestos, mainly chrysotile, is the USSR, with 58% of world production, followed by Canada (18%), Zimbabwe (4%), the People's Republic of China (3.5%) and the USA (about 1.4% of the world total, mainly in California).

Asbestos is recovered in both open-pit and underground mines and is milled in a complex process designed to separate fiber from rock by crushing and impacting, and to classify the fiber according to length. Classification is accomplished by multiple screens, trommels and air classifiers.

Chrysotile, the most commonly used asbestiform mineral, is essentially a hydrated magnesium silicate. It has a specific gravity of 2.4–2.6 and a hardness of 2.5–4.0 on the Mohs scale. Precise mineralogical definitions and characterization criteria for asbestiform minerals have been discussed by Campbell et al. (1977).

A small quantity of asbestos is used in electronic applications for thermal and/or electrical insulation. It deserves mention because of continuing concern about health hazards possibly associated with exposure to this mineral. No firm data are available on the extent to which asbestos is presently used in electronic applications; however, the heat and flame resistance, high tensile strength, and adaptability of asbestos to widely varying physical requirements will continue to make it attractive as an insulator, especially as questions regarding safety in handling are resolved.

3. *Calcite*

Calcite is crystalline $CaCO_3$, with a chemical composition very close to that of the pure compound. Although it is among the most abundant minerals in the earth's crust, calcite crystals of a size and quality suitable for optical uses are rare. For many years the best-known source of optical calcite was Iceland, which accounts for the common name, Iceland spar. Other sources of optical calcite have included the UK (Cornwall), Spain, South Africa and the Harz Mountains in Germany. No economical method for producing synthetic calcite crystal has been developed, and much of the optical calcite presently used is natural crystal obtained from Mexico. Production and consumption data for optical calcite are unavailable.

Because end-use requirements for calcite crystal are similar to those for quartz, mining and processing methods for calcite crystal are similar to those for quartz and consist essentially of hand collection and sorting.

Calcite is ubiquitous and occurs in rocks of sedimentary, metamorphic and igneous origin. Optical-quality calcite is water-clear and colorless, although natural variants occur having yellow, bluish, red or green color owing to the presence of impurities. Some specimens are luminescent or thermoluminescent, and some exhibit fluorescence under uv irradiation. Calcite exhibits trigonal symmetry and crystallizes in the hexagonal system. Unlike quartz, calcite has a center of symmetry and is not piezoelectric. The pure mineral has a specific gravity of 2.71 and a hardness on the Mohs scale of 3.0. In cold, dilute hydrochloric acid calcite dissolves readily, with vigorous effervescence caused by liberation of carbon dioxide.

The principal use for optical-quality calcite crystal is in polarizing instruments, such as microscopes, polarimeters and dichroscopes. For this application, two calcite prisms are cemented together with the cemented interface forming an angle such that the ordinary ray is totally reflected out of the prism and the extraordinary ray is transmitted through it. The most common conformations are the Nicol and Glan–Thompson prisms, of which the Glan–Thompson is preferred even though its construction requires larger pieces of crystal than does the Nicol prism.

4. Mica

Like the term asbestos, mica is a generic name that refers to a group of minerals with similar properties. A generalized formula describing the composition of micas is $X_2Y_{4-6}Z_8O_{20}(OH,F)_4$, where X is mainly potassium, sodium or calcium; Y is mainly aluminum, magnesium or iron; and Z is mainly silicon or aluminum. Electronic applications require sheet mica, which is distinguished from scrap and flake mica by its size and suitability for certain uses rather than by its composition.

The world's most productive sheet mica districts are Bihar, Madras and Rajas in India, and these large reserves have been the primary world source for years. Large sheet mica reserves are also located in Brazil, Western Africa and the USSR. Smaller, but still substantial, deposits are found in Argentina, Australia and Zimbabwe.

Although mica minerals vary considerably in their physical and chemical properties, they are characterized by a common platy morphology and perfect basal cleavage attributable to their layered atomic structure. Among the micas, the most important commercially are:

(a) muscovite, $K_2Al_4[Si_6Al_2O_{20}](OH,F)_4$; and

(b) phlogopite, $K_2(Mg,Fe^{2+})_6[Si_6Al_2O_{20}](OH,F)_4$.

Muscovite is the principal mica used because it is found in greater abundance and has superior electrical properties. Phlogopite remains stable at high temperatures and finds application where a combination of electrical properties and high heat stability is required. Table 1 lists selected properties of natural mica.

Where practical, mica is mined hydraulically because subsequent screening and washing operations require large quantities of water. In general, owing to the irregular distribution of the mineral, sheet mica mining involves small operations that use very little

Table 1
Selected properties of natural mica

Property	Muscovite	Phlogopite
Specific gravity	2.6–3.2	2.6–3.2
Density (kg m^{-3})	2560–3200	2560–3200
Mohs hardness	2.0–3.2	2.5–3.0
Crystal structure	monoclinic	monoclinic
Optical axial angle (2V) (degrees)	38–47	0–10
Temperature at decomposition (°C)	400–500	850–1000
Thermal conductivity (W m^{-3} K^{-1})	1392.5	1392.5
Modulus of elasticity (N m^{-2})	1.7×10^{-11}	1.7×10^{-11}
Dielectric constant	6.9–9.0	5.0–6.0
Resistivity (ohm cm)	10^{12}–10^{15}	10^{10}–10^{13}

mechanical equipment. Mining is done by both underground and open-pit techniques that share a common concern for preserving the crystals. Great care is exercised to avoid drilling through crystals, and only small charges of low-velocity explosive are used. After mining, crystals are hand separated from waste rock, and further processing also involves considerable hand labor.

Sheet mica is used principally in the electronic and electrical industries. Its usefulness in these applications stems from its unique electrical and thermal insulating properties and its mechanical properties which allow it to be cut, punched and stamped to close tolerances.

The largest use of block mica is in vacuum tubes, where fabricated mica spacers position, insulate and support the tube elements. High-quality block mica is also used to line the gauge glasses of high-pressure steam boilers. Transparency, flexibility and resistance to heat and chemical attack are the properties that make mica suitable for this use. Other uses include optical filters, retardation plates in helium–neon lasers, pyrometers and thermal regulators. Block mica is also fabricated into washers which act as insulators in electronic equipment. Washers have also been used extensively in the computer industry as gap separators in recording heads.

The emergence of the burgeoning solid-state electronics industry has greatly reduced mica demand for electronic applications and consumption of sheet mica has decreased from 6.7×10^6 kg in 1953 to 1.7×10^6 kg in 1983.

Bibliography

Campbell W J, Blake R L, Brown L L, Cather E E, Sjoberg J J 1977 *Selected Silicate Minerals and their Asbestiform Varieties, Mineralogical Definition and Identification-Characterization*, Information Circular 8751. US Bureau of Mines, Washington, DC

Dana E S 1966 *A Textbook of Mineralogy*, 4th edn. Wiley, London

Deer W A, Howie R A, Zussman J 1966 *An Introduction to the Rock Forming Minerals*. Wiley, London

Hale D R, Blair R E 1983 Electronic and optical uses. In: Lefond S J (ed.) 1983 *Industrial Minerals and Rocks*, Vol. 1, 5th edn. American Institute of Mining, Metallurgical and Petroleum Engineers, New York, pp. 213–31

Thrush P W 1968 *A Dictionary of Mining, Mineral, and Related Terms*. US Bureau of Mines, Washington, DC

US Bureau of Mines 1985 *Mineral Facts and Problems*, Bulletin 675. USBM, Washington, DC

W. N. Marchant
[US Department of the Interior, Washington, DC, USA]

D. L. Barna
[US Bureau of Mines, Washington, DC, USA]

F

Fillers and Coatings: Carbonate Minerals

The uses of carbonate minerals in fillers and coatings closely parallel those of clay, siliceous and sulfate minerals. Utilization varies primarily with performance and cost, though it is affected by a myriad of chemical or physical characteristics. Some factors which give the carbonate minerals their popularity include availability, geographically widespread production locations, lack of abrasiveness, whiteness (when required) and low cost.

In the USA, relatively pure (95% or greater $CaCO_3$) limestones are generally the industry standard product, though some dolomites are marketed. Dolomite is more frequently utilized as a filler in Europe.

1. Sources

The carbonate-filler industry should be considered in two parts:

(a) those utilizations requiring a relatively white product such as paints, paper, and in some instances plastics and rubber; and

(b) those utilizations where color may not be a factor such as asphalts and carpet backings.

The white carbonate fillers are almost invariably obtained from deposits of white marble, which are found where extensive metamorphic activity has occurred. In the eastern USA the producers are in or along the Appalachian Mountain chain from Vermont to Alabama. In the far west the deposits are associated with the Rocky Mountain chain.

Limestone with proper characteristics for many filler applications is available throughout the USA. Production tends to come from specialized operations that may also produce limestone for its chemical characteristics. As a result these "filler limestones" can be chemically pure although this purity may not be necessary for the filler application.

Some of the ultrafine-ground limestones may be chemically bleached to improve whiteness, permitting their utilization in applications previously restricted to white marble products.

2. Mineralogy and Physical Properties

This analysis, like the sources, must be divided into the two categories based on color. The marbles, which are most desirable in terms of whiteness, are crystalline with visible crystals varying from about 0.2 to 4 mm. Although calcite is white, most deposits have impurities that include pyrite, graphite and a large number of silicate minerals. The eastern (Appalachian) deposits are generally bedded, extensively folded, and extend laterally, in some instances, for several kilometers (though the purity may vary substantially from point to point). The western deposits tend to be more recent "pendants" of marble with recrystallization primarily a result of igneous intrusives.

The unmetamorphosed limestones are generally not as white as marble and vary texturally from granular to crystalline, depending on original composition of the limestone and subsequent diagenesis. The chemical purity can vary widely, but most filler-producing facilities operate from relatively pure deposits to satisfy the requirements of other markets. High silica contents (more than 2%) are generally considered detrimental since silica tends to increase machinery wear for both the producer and the user.

Most limestone deposits are distinctly bedded with significant fluctuations in quality of the stone within a formation. In many cases only a single bed will be of acceptable quality. In recent years it has been found that limestones originating in high-energy environments, such as reefal and oolitic deposits, are a basis for relatively pure lighter colored limestones that are desirable for a broad range of filler applications.

3. Mining and Production Technology

Carbonates for filler production are mined by underground and open-pit methods, with systems similar to those used for other minerals. The specific system will depend on the geology, overburden, structure and other characteristics of the individual deposit. The mining method will frequently provide for additional selectivity to produce a whiter, or more chemically uniform, product.

Production technology varies depending on end use. Most marble is crushed, then washed to remove mud or other color-degrading material. In some instances it may then be sorted (by hand or by machine) to remove pieces of darker colored rock. Subsequent crushing steps produce a feed for grinding circuits, and may produce very coarse filler products of size 20–200 mesh (0.84–0.074 mm).

Many grinding circuits incorporate a color beneficiation step, usually selective flotation, which rejects undesirable minerals. The finely pulverized fillers are produced by wet or dry grinding, and in most instances the grinding is performed in closed circuit with a particle-size classifier. The coarser sizes are produced in roller mills, impact mills or hammermills whereas the finer sizes are generally ground in some form of pebble mill. The most common grinding medium is high-density alumina. If color degradation is to be minimized, silica or alumina mill linings are used. Wet grinding circuits utilize centrifuges for particle classification, whereas dry circuits use an air separator specifically designed for low micrometer-size separations. Ultrafine grinding is accomplished in attrition mills.

Finished products are shipped in bags or bulk, with the utilization of bulk increasing in response to the high costs of bags and bag handling. A new family of products is appearing which are chemically coated to improve their performance, primarily in plastics systems. There has been increased interest in calcium carbonates ground to 1.5 μm or less. The primary market for this product group is paper filling, paper coating and plastics. In addition to meeting a new market demand, it may also substitute for chemically precipitated calcium carbonate.

Ground carbonate testing procedures include sieve analysis, usually as fine as 325 mesh (0.044 mm); sedimentation procedures to measure particle sizes down to 0.3 μm; color, measured as reflectance from a smooth cake of the ground limestone; chemical composition, determined by the usual analytical procedures; and a number of specialized tests that may be required for a particular using industry. Most calcium carbonate filler producers have extensive laboratory facilities for testing their products, assisting customers in solving problems and developing new product uses.

4. Uses

4.1 Bituminous Products

Carbonate minerals are the primary filler in asphaltic type products. Typical end uses include roofing shingles, roll roofing, asphaltic coatings and sealing compounds, and asphalt-based patching materials. The filler elevates the softening point, increases viscosity and improves weather resistance. Color is not usually a factor considered. Relatively coarse ground products (50% finer than 200 mesh (0.074 mm)) are generally used for fillers, whereas somewhat finer grinds (50% finer than 325 mesh (0.044 mm)) are used for bitumastic coatings. Since most fillers are mixed at elevated temperatures, the filler must be dry.

4.2 Plastics

The carbonates are the most commonly used filler for plastics compounding. They enjoy this popularity because of their low abrasion, high brightness, ease of dispersion and relatively low cost. Upper particle size limit and average particle size will vary with the specific application but fall into the 0.1–20 μm range. End uses include automotive parts, pipe and plumbing, boats, and floorings. The cultured marble used for bathroom lavatories and table tops is heavily filled with carbonate fillers that are in some instances quite coarse (50 mesh (0.297 mm) and finer).

4.3 Rubber

Carbonate fillers are used in a wide range of rubber products where their use tends to maintain the qualities of softness, resilience, and elongation, even at relatively heavy loadings. They can provide whiteness (e.g., in athletic footwear), but in many instances, such as with automotive floor mats, color is not critical and ground limestones that are not white can be used. In some formulations magnesium ions have a chelating effect on the compounds, changing their processing or final characteristics; hence, a high calcium limestone filler rather than a dolomitic one may be specified.

4.4 Protective and Decorative Coatings

Although a number of minerals are used in this group, calcium carbonate is again the leading one. The high brightness and broad range of particle sizes available promote its utilization. The coarser grinds are used in flat (matte) paints, exterior paints and traffic paint, while the finer grinds are used in those products where a glossy finish is desired. They are specified with equal effectiveness in the traditional oil-based coatings as well as the more recently developed emulsion paints.

4.5 Paper

Some calcium carbonate has traditionally been utilized directly in the paper industry, but most was chemically precipitated calcium carbonate. In recent years the calcium carbonate industry has developed ultrafine ground products (finer than 1 μm) that are receiving increased acceptance. Calcium carbonate is included in filling the paper stock when an alkaline binder is employed in paper manufacturing, but is not generally compatible with acidic binders. In coatings, which are applied after the base sheet is formed and filled, the addition of ultrafine calcium carbonate to the traditional kaolin filler improves brightness, opacity and ink receptivity, and may assist in higher production rates on the coating machines.

4.6 Other Applications

Carbonate fillers are common in a host of other applications. Some of these that require substantial quantities include: the latex on the back of tufted carpet and the carpet cushion, whether it be attached to the carpet or a separate underlay; adhesives, mastics and putty that are used in many industries; dry-wall joint cement that is used to give the smooth final surface to gypsum-board walls; and the soundproofing compounds used in the automotive industry. The versatility of the carbonate fillers is illustrated by their inclusion in chewing gum, concrete curing compounds, children's crayons and oil-well drilling mud.

See also: Fillers and Coatings: Clay Minerals; Fillers and Coatings: Industrial Minerals; Fillers and Coatings: Siliceous Minerals; Fillers and Coatings: Sulfate Minerals

Bibliography

Carr D D, Rooney L F 1983 Limestone and dolomite. In: Lefond S J (ed.) 1983 *Industrial Minerals and Rocks: (Nonmetallics Other Than Fuels)*, 5th edn., Vol. 2. American Institute of Mining, Metallurgical and Petroleum Engineers, New York, pp. 833–68

Galvanek E 1979 *Extender and Filler Pigments*. Fairfield, New Jersey, pp. 85–136

Hall R F 1964 Calcium carbonate pigments. *Off. Dig. Fed. Soc. Paint Technol.* 183–92

Power W R 1983 Dimension and cut stone. In: Lefond S J (ed.) 1983 *Industrial Minerals and Rocks: (Nonmetallics Other Than Fuels)*, 5th edn., Vol. 1. American Institute of Mining, Metallurgical and Petroleum Engineers, New York, pp. 161–81

Severinghaus N Jr 1983 Fillers, filters, and absorbents. In: Lefond S J (ed.) 1983 *Industrial Minerals and Rocks: (Nonmetallics Other Than Fuels)*, 5th edn., Vol. 1. American Institute of Mining, Metallurgical and Petroleum Engineers, New York, pp. 243–57

N. Severinghaus Jr.
[Franklin Limestone Co., Nashville, Tennessee, USA]

Fillers and Coatings: Clay Minerals

One of the most widely used minerals for fillers and coatings is kaolinite. Kaolinite is a hydrated aluminum silicate ($Al_2O_3.2SiO_2.2H_2O$) and is the major clay mineral constituent in the rock kaolin. The paper industry is by far the largest user of kaolin, both as a filler and as a coating; it uses over 50% of the total kaolin produced in the world. A second clay that is employed as a filler and coating material is smectite. Smectite is a group name for hydrated sodium, calcium, iron, magnesium and aluminum silicates known as montmorillonites. Industrially the term bentonite is used to describe these clays.

Some physical and chemical properties of a Georgia kaolinite and a Wyoming bentonite are shown in Table 1. Each of these two clay types are discussed separately below as they have very different properties and are used for different purposes.

Table 1
Chemical analysis and physical properties of a Georgia kaolinite and Wyoming bentonite

Property	Georgia kaolinite	Wyoming bentonite
Specific gravity	2.6	2.0
Hardness (Mohs scale)	1.5–2.0	1.5–2.0
Refractive index	1.57	1.64
Bulk density (kg m^{-3})	320–800	640–960
Oil absorption (cm^{-3} g)	0.25–0.60	0.20–0.30
Color	off white	brown, tan, cream
Base-exchange capacity (Meq/100 g)	1–2	75–100
Brightness (457 μm%)	78–90	60–75
Chemical composition (%)		
SiO_2	45.30	54.62
Al_2O_3	38.38	20.08
Fe_2O_3	0.30	3.39
TiO_2	1.44	0.11
MgO	0.25	2.15
CaO	0.05	0.40
Na_2O	0.27	2.26
K_2O	0.04	0.35
Loss on ignition (%)	13.97	16.36

1. Kaolin

Over 18 Mt of kaolin were mined throughout the world in 1980 and over half came from two areas; 7 Mt from Georgia and South Carolina in the USA and 3.5 Mt from Cornwall, UK. Other areas producing significant quantities of kaolin are Brazil, Czechoslovakia, the German Democratic Republic, France, Russia, Spain and the Federal Republic of Germany. Kaolinite is a unique industrial mineral because, except for catalytic activity in some organic systems, it is chemically inert over a relatively wide pH range; is white in color; has good covering power when used as a pigment or extender in coating and filling applications; is soft and nonabrasive; has low conductivity of heat and electricity; is very fine in particle size and thus disperses readily; and is lower in cost than most materials with which it competes. Many grades of kaolin are especially designed for specific end uses, particularly those used in paper, paint, rubber, ceramics, plastics and ink.

The term kaolin in addition to being a rock name is a group name for the minerals kaolinite, nacrite, dickite and halloysite. Kaolinite is by far the most common mineral of the group. With the exception of the hydrated form of halloysite, which has two more water molecules per unit cell, all these minerals have essentially the same composition. Halloysite has an elongated tubular shape, whereas the other kaolin minerals are pseudohexagonal (Fig. 1).

Kaolins are classed as primary or secondary deposits. Primary kaolins are those that have formed by the alteration of crystalline rocks such as granite, and remain in the place where they were formed. The English china clay deposits are an example of this type. Secondary deposits of kaolin are sedimentary, and have been transported from their place of origin and deposited in beds or lenses associated with other sedimentary rocks, such as sands. The Georgia–South Carolina deposits are examples of secondary or sedimentary deposits. Most kaolins are processed to some extent because none are naturally pure. Beneficiation is accomplished using either a dry or a wet process. All kaolins used for coating paper are wet processed, because the product is more uniform, is relatively free from impurities and has a better color.

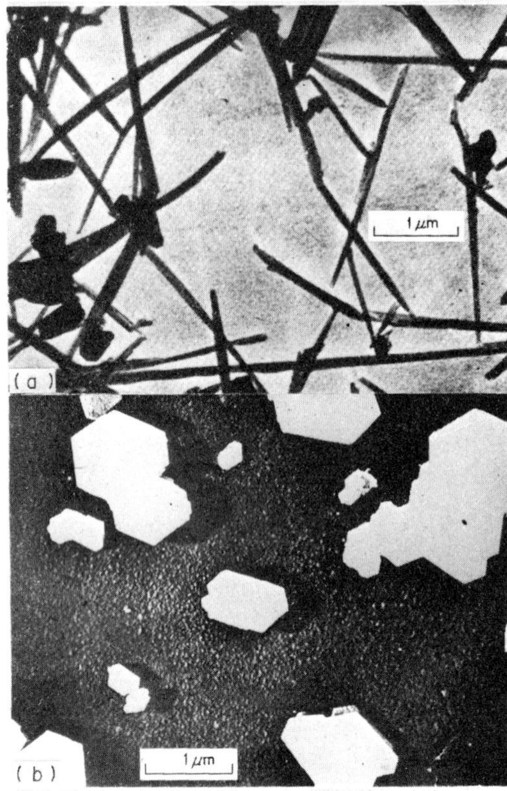

Figure 1
Electron micrographs of kaolin minerals: (a) halloysite, (b) kaolinite

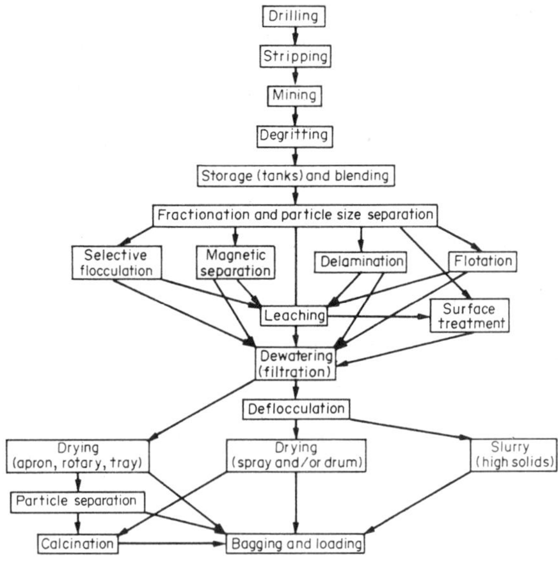

Figure 2
Generalized flow sheet for wet processing of kaolin

The dry process is simple in that the kaolin is crushed to approximately egg size, dried, disintegrated, and air floated to remove the grit particles. This process yields a lower cost and lower quality product than the wet process. In the wet process, the kaolin can be beneficiated to produce a relatively pure product with strict specifications as to particle size, brightness, grit content and other properties. A general flowsheet for wet processing is shown in Fig. 2. Leaching is to remove iron coloration, delamination is a process to reduce coarse kaolin stacks to thin plates and calcination is thermal treatment of the kaolin to develop certain properties for special uses. With the exception of high-intensity wet magnetic separation, the other processes are widely used in mineral processing. The major impurities that are detrimental to color are iron and titanium oxides and hydroxides, and a major portion of these fine mineral particles can be extracted using high-intensity magnetic separation.

1.1 Major Uses

Kaolin used in paper fills the interstices of the sheet, producing better color, higher opacity, better printability and lower costs. As a coating on paper, kaolin gives the sheet better smoothness, gloss, brightness, opacity and printability, in that colored inks have much more fidelity when printed on kaolin-coated sheets. The low viscosity of kaolin at both high and low shear rates is a very important property, because at today's very high-speed production rates, the coating must be applied at a high solids content, and still give the correct coating thickness and smoothness to the paper. High-quality paper-coating kaolin flows readily when applied with very high-speed coating equipment, giving the paper a smooth and even-coated film.

The rubber industry uses large amounts of kaolin as a filler or extender in both natural and synthetic rubber. Kaolin is incorporated in the rubber mix to lower costs and improve properties such as strength, abrasion resistance and rigidity. Through the years, the terms hard and soft kaolins have become commonplace in describing kaolins used by the rubber industry. Hard kaolin has a very fine particle size that improves the tensile strength of the rubber and its resistance to tear and abrasion. Soft kaolin is coarser in particle size, and is used to lower elasticity and improve abrasion resistance.

Kaolin is used as an extender in paints, because it is chemically inert, has a high covering power, gives desirable flow properties, is low in cost, is white and reduces the amount of expensive pigment required. In addition, kaolin has excellent suspension properties and is available in a wide range of particle sizes which can be used in many types of paints. Coarse-particle kaolins are used in paints where a dull or flat finish is required, and fine-particle kaolins are used in high-gloss paints. Large quantities of calcined and water-beneficiated kaolins are used in interior wall paints and metal primers. Calcined kaolin is particularly suited for use in paint as an extender for TiO_2, because of its resistance to abrasion and dry covering properties.

Kaolin is used extensively as a filler in plastics. The desirable characteristics that can be obtained with kaolin include smoother surfaces, more attractive finishes, dimensional stability and resistance to chemical attacks. Manufacturers to poly(vinyl chloride) (PVC) use kaolin as a reinforcing agent, and it makes the plastic more durable. Calcined and partially calcined kaolins are used as a filler in PVC wire insulation to improve electrical resistivity. In manufacturing glass-reinforced polyesters, kaolin has helped eliminate flow problems which hindered production of large products such as boat hulls. The use of kaolin actually results in a stronger and more uniform body.

Kaolin has many other important filler applications; these include inks, adhesives, insecticides, medicines, catalyst preparations, fertilizers, detergents, pastes, linoleum, textiles and cosmetics.

1.2 Specifications and Grades

For paper filler application, coarse, medium and fine particle sizes are used. The coarse fillers have particle sizes of 20–40% <2 μm, the intermediate fillers have particle sizes of 40–60% <2 μm and the fine filler kaolins have particle sizes in the range of 60–80% <2 μm. In addition to particle size, a brightness of 80–85% is specified by the Technical Association of the Pulp and Paper Industry (TAPPI). Another important specification is screen residue, which is called grit. This is material held on a 325 mesh (44 μm) screen. Because of abrasivity and scratching, only a very small percentage of grit (about 0.01%) can be tolerated. Another property that is important for a paper filler is its abrasion index. For clays this is usually not a problem, although some primary kaolins contain fine quartz and mica that cause abrasion problems. The test used in the paper industry is the Valley Abrasion test which again is described by TAPPI.

The most common coating clay grades are shown in Table 2. The #3 and #2 coating grades are generally used in high-solids formulations, for publication-grade paper and medium-finish lower cost enamels. The #1 grade is used to coat high-quality boxboard and enamel papers. This grade gives excellent gloss and brightness and relatively uniform ink receptivity and good smoothness. The fine #1 grade is used where maximum gloss and very good ink holdout are desired.

The regular-delaminated grade is used to coat lightweight paper and paper that has a rough surface, where good printability must be maintained. This clay makes up some of the base-sheet deficiencies and gives coatings that are exceptionally smooth with excellent ink and varnish holdout. The fine-delaminated grade is used to achieve maximum gloss, good runnability and high opacity.

Table 2
Particle size and brightness of some coating clay grades

Clay grade	Particle size (% <2 μm)	Brightness (%)
Regular		
#3	73	85–86.5
#2	80–82	85.5–87.0
#1	90–92	87.0–88.0
fine #1	95	86.0 87.5
Delaminated		
regular	80	88.0–90.0
fine	95	87.0–88.0
High brightness		
#2	80	89–91
#1	92	89–91
Calcined		
dehydroxylated	60	85.5–88
fully calcined	90	90–92

The high-brightness grades are very bright and have a blue-white tint. The coatings using the #1 high-brightness grade are exceptionally smooth, and give excellent gloss, ink holdout and good printability. These high-brightness grades are used to coat premium grade paper.

The dehydroxylated or partially calcined kaolin is used as a paper coating additive to enhance resiliency and opacity in low basis-weight sheets. The addition of dehydroxylated kaolin to a coating formulation provides increased thickness to the coating, increased opacity, a significant reduction in loss of brightness and opacity on calendering, improved ink receptivity and more uniform color printing. The fully calcined kaolin coating clay increases light scattering of the film in proportion to its content. This enhanced light scattering makes it possible to achieve opacity and brightness specifications at lower levels of titanium dioxide, which reduces the pigment coat required.

A very critical specification for coating kaolins is viscosity. The flow properties of kaolin are very important to the paper coater because they affect the functioning of the coating operation, as well as the final quality of the paper product. Two viscosity tests are used, one to measure high-shear viscosity and the other to measure low-shear viscosity. The measurements are normally made on slurry suspensions at 71% solids. A Brookfield four-speed viscometer is used to measure low-shear viscosity, and a Hercules viscometer is used to measure high-shear viscosity.

In addition to the particle size, brightness, pH, moisture and grit specifications, kaolins used in paints and plastics must meet certain electrical resistivity specifications. The resistivity test is designed to give an indication of the amount of residual salts that are contained in the clay; high resistivity values reflect a low soluble salt content. For certain paint applications, the Hegman fineness-of-grind test specification is important. This test is intended to measure the degree of dispersion, and involves the use of a fineness gauge and a scraper. The gauge is calibrated from 0 to 8, and the fineness is reported by number. The higher the number, the finer the particles and the better the dispersion. Some paint manufacturers specify that a pigment or extender must be finer than a particular Hegman number.

2. Bentonite

The definition of bentonite proposed by Grim at the International Clay Conference at Madrid, Spain, in 1972 is used here. According to Grim, bentonite is a clay consisting essentially of smectite minerals regardless of origin or occurrence.

One way of classifying bentonite is based on its swelling capacities when wet, or added to water. Bentonite containing sodium as the dominant exchangeable ion, typically has very high swelling capacities and forms gel-like masses when added to water. This type of bentonite is produced in Wyoming, Montana and South Dakota, and is commonly called Wyoming or Western bentonite. Bentonite in which the major exchangeable cation is calcium has a much lower swelling capacity, and is commonly referred to as subbentonite or Southern bentonite, because it occurs in Texas, Louisiana and Mississippi.

Bentonite production in the USA in 1980 was slightly over 4 Mt, and the world production was estimated to be about 8 Mt. The processing of bentonite is rather unusual, in that the mining is openpit using shovels or payloaders. The bentonite is crushed to egg size or smaller, dried in rotary flash or fluidized bed driers, and screened and/or further ground to meet the size specifications required. A very small proportion of the bentonite is wet processed using centrifuges, filter presses or rotary vacuum filters for dewatering, and flash drying to produce very pure special grades. Also some bentonite is surface coated with organic compounds such as amines, for use in paints, greases and oil-based drilling muds.

Bentonite is produced in many other countries of the world including Mexico, Canada, Argentina, Brazil, Peru, Cyprus, Czechoslovakia, France, Greece, Hungary, Italy, Poland, Romania, Spain, the UK, Russia, the Federal Republic of Germany, Yugoslavia, Algeria, Morocco, Mozambique, South Africa, Japan, Pakistan, Turkey and New Zealand.

Although the major uses of bentonites are in drilling muds (see *Well-Drilling Materials: Clay and Nonclay Minerals*), foundry bonding agents and bondants for metal ore pellets, some bentonites are used in cosmetics, pharmaceuticals, paints, detergents, grease and paper. Near Gonzales, Texas, and on the Island of Ponza near Naples, Italy, white bentonites are mined that are particularly good for applications where color is important (e.g., pharmaceuticals, paints, detergents and paper).

A special product, an organic-treated bentonite, is used to make thick nondrip paints, to gel organic liquids and to produce greases having superior adherence to metal, ability to repel water and resistance to high temperatures. This organic-clad bentonite is processed in such a way that the original inorganic exchangeable ion on the smectite is replaced by an alkali amine organic cation. This reaction produces a hydrophobic clay, because the inorganic ions that can be hydrated are removed, and a large part of the mineral surface formerly capable of absorbing water is coated by hydrocarbon chains. These special products are made in the USA, the UK, France, the Federal Republic of Germany and Japan. The processing is done in water after a large proportion of the nonsmectite minerals have been removed by centrifuging or sedimentation. The drying is accomplished using special flash dryers.

Another special smectite product is used to coat paper for making multiple copies without using carbon paper. This product is wet-processed and treated with acid to activate the smectite. The activator removes the sodium, calcium and some aluminum and/or iron, making the clay surface highly charged with hydrogen ions. This special smectite clay is used to coat the receiving surface of the copy sheet; it contains about 25% smectite, with starch, dextrin and butyl rubber as the binder. The transfer surface is coated with starch containing minute encapsulated droplets of colorless dyestuffs, such as triethynyl methane derivatives with a lactone structure. Writing or typing breaks the dyestuff globules, so that they penetrate the smectite-coated layer, which catalyzes their conversion into a colored letter or mark. This type of clay is produced in the USA, Mexico, Japan and the Federal Republic of Germany.

See also: Fillers and Coatings: Carbonate Minerals; Fillers and Coatings: Industrial Minerals; Fillers and Coatings: Siliceous Minerals; Fillers and Coatings: Sulfate Minerals

Bibliography

Murray H H 1984 Clay. In: Hagemeyer R W (ed.) 1984 *Pigments for Paper*. Technical Association of The Pulp and Paper Industry, Atlanta, Georgia, pp. 95–193

Patterson S H, Murray H H 1983 Clays. In: Lefond S J (ed.) 1983 *Industrial Minerals and Rocks*, 5th edn., Vol. 1. American Institute of Mining, Metallurgical and Petroleum Engineers, New York, pp. 585–651

H. H. Murray
[Indiana University, Bloomington, Indiana, USA]

Fillers and Coatings: Industrial Minerals

The use of finely powdered minerals to load or fill such materials as paper has been common practice for thousands of years, and for other materials such as rubber and plastics a relatively few tens of years. In addition to minerals being incorporated in the body of the paper itself, minerals are also applied as a thin film or coating on the surface. The choice of mineral fillers and coaters is wide, and the selection is based upon the properties desired in the finished product, and the cost.

Mineral fillers are defined as inert materials that are incorporated in a composition for some useful purpose. Included are a broad group of minerals: asbestos, barite, bentonite, calcite, diatomite, feldspar, kaolin, micas, nepheline syenite, perlite, silica, talc, titanium dioxide, tripoli and wollastonite. Mineral fillers can be included in compounds to attain a variety of resultant properties. Hardness, brittleness, impact strength, softening point, fire resistance, surface smoothness, color, opacity, electrical conductivity and a multitude of other physical characteristics may be modified by the inclusion of fillers.

The characteristics of the product are the result of the physical and chemical properties of the fillers or coaters, including chemical activity, hardness, particle size, shape and distribution, color, density, surface area, structure, chemistry, refractive index, and dispersibility. In addition to the physical and chemical characteristics, a most important consideration is cost. Certainly, the inclusion of a filler that costs 10¢ kg^{-1} will reduce the cost of a plastic whose resin costs \$1 kg^{-1}, or pulp for paper that costs 55¢ kg^{-1}. The decision on the use of a particular filler, however, is not as simple as just reducing the cost. For example, a certain filler may react with a particular ingredient, making the compound exceedingly viscous such that it will not flow into a mold as an injection molding compound, although it may be very useful in certain caulking compounds to prevent sagging. Thus, it is very important to evaluate a filler in any compound with which it is to be used, in order to determine fully the resultant properties.

Fillers can be used to modify cost, flow characteristics and other properties, but seldom is the desired modification achieved without some sacrifice of other qualities. For example, some plastics may become excessively brittle when cost-reducing fillers are incorporated, or a paint may tend to settle out in the can when antichalking fillers are added. Therefore, some properties are usually compromised in order to achieve a degree of modification that is acceptable. In many cases, a combination of fillers is used to attain a number of desirable modifications; for example, a paint may contain kaolin, calcite, talc, bentonite and perhaps other fillers. One filler may be substituted for another in some uses, but as technology advances and specifications become tighter, substitution becomes more difficult. Replacement or partial substitution may require total reformulation, which the consumer is reluctant to undertake unless significant cost and/or performance improvements are likely to be forthcoming.

The physical characteristics of a filler or coater are the result of the inherent characteristics of the mineral and the features imparted by processing. Inherent characteristics include hardness, particle size and shape, color, refractive index, chemical composition and chemical reactivity. The presence of trace quantities of other minerals may preclude the use of a filler or coater in certain applications if that particular trace mineral cannot be removed during processing. Hardness of a mineral affects its utilization as a filler or coater. Silica, for example, has a hardness of 7 on the Mohs scale, and will score and wear blades, knives and rolls, meaning that in most instances a softer mineral such as calcite, kaolin or talc must be used. Particle size and particle size distribution are probably the most important physical characteristics of a filler or a coating material. The properties affected by particle size and size distribution include smoothness, viscosity, binder demand, film integrity, density, gloss, color and, in the case of coated paper, the printing characteristics and ink receptivity. Particle shape affects utilization, in that acicular particles such as asbestos provide reinforcement and strength, while plates of kaolin and mica provide a tough, smooth surface film. Color is very important in many products, so the color of the filler or coating pigment becomes an important characteristic. Most mineral fillers and coaters are more desirable the whiter they are. Index of refraction is closely related to color and is very important in some applications. If the index of refraction of the mineral filler is very close to that of the material it is filling, the color of the filler will be very faint and the product will be translucent, whereas if there is a large difference the product will become opaque. Chemical composition is important and affects the use of many fillers. Calcium carbonate cannot be used in acid systems because it is soluble, so a nonreactive filler such as kaolin must be used. Many other physical characteristics are important and influence the intended use of the filler or coater mineral.

Fillers are used extensively in many applications. They are utilized in bituminous compositions, including asphalt shingles and roll roofing, road-paving asphalt, flooring materials, waterproof coatings and other compositions where up to 50% loadings are not uncommon. Both the thermoplastic and thermosetting branches of the plastics industry use fillers, not just to gain cost savings, but to perform a function in determining the properties of the final product—in some cases up to 80% loadings are used. The rubber industry uses fillers to achieve hardness, abrasion resistance, tensile strength and to modify several other properties. Large quantities of fillers and coaters are used by the paper industry (estimated to be over 3 million tonnes in the USA in 1983) to improve opacity, smoothness, ink receptivity, body, feel and printing fidelity. The paint industry utilizes functional fillers to control flow characteristics, to improve sheen or control gloss or flatness, to enhance film strength, scrubability and weather resistance, to reduce film tack, and to influence many other properties. Other areas where fillers are employed include fertilizers, adhesives, textiles, cement, caulking, inks, lubricants, food, cosmetics, pharmaceuticals and insecticides.

Coater minerals are used primarily by the paper industry, and to some extent by the wire and cable industry. The coater minerals are suspended in an adhesive–water mixture such as starch or synthetic latex. Normally, the adhesive content is 12–18% based upon the dry weight of the coater minerals. The purpose of coating paper is to improve the printing characteristics, including brightness, opacity, gloss, smoothness and ink receptivity. In wire and cable coatings, minerals are used to improve the electrical characteristics and must have a high dielectric constant and low heat conductivity.

The filler and coater minerals are a necessity in many industries because they are not just inert cost-saving space fillers, but perform significant functions in controlling the physical and chemical properties of the finished product.

See also: Fillers and Coatings: Carbonate Minerals; Fillers and Coatings: Clay Minerals; Fillers and Coatings: Siliceous Minerals; Fillers and Coatings: Sulfate Minerals

Bibliography

Deanin R D, Schott N R (eds.) 1974 *Fillers and Reinforcements for Plastics*, Advances in Chemistry Series No. 134. American Chemical Society, Washington, DC

Hagemeyer R W 1976 *Paper Coating Pigments*, TAPPI Monograph Series No. 38. Technical Association of the Pulp and Paper Industry, Atlanta, Georgia

Patton T C 1973 *Pigment Handbook*, Vol. I, *Properties and Economics*. Wiley, New York

Technical Association of the Pulp and Paper Industry 1958 *Paper Loading Materials*, TAPPI Monograph Series No. 19. Technical Association of the Pulp and Paper Industry, Atlanta, Georgia

H. H. Murray
[Indiana University, Bloomington, Indiana, USA]

Fillers and Coatings: Siliceous Minerals

A large number of siliceous minerals are used as fillers and coatings. This article concentrates on the following: asbestos, diatomite, perlite, pumice, pyrophillite, silica and talc.

Industrial minerals in this category have some common characteristics: all have a refractive index in the range 1.4–1.7, a specific gravity of 2.0–2.9 and a pH of 6–10. They all receive processing that enhances their natural beneficial properties and which allows them to be used in a variety of applications. Some receive additional processing: for example, perlite is thermally expanded to a material of low density similar to pumice, and diatomite is flux calcined to improve color. Generally, they all contribute economic benefits in their applications. They supply bulk, porosity, selective abrasiveness (hardness), strength and/or reactivity.

1. Sources

Table 1 gives an abbreviated summary of the principal states or countries where there are operating deposits of siliceous minerals. The sources are listed in an approximate order of decreasing annual production. However, since these minerals are used in a broad variety of applications that include nonfiller applications, it is difficult to assign an order of filler production.

2. Mineralogy

Typical mineral compositions of the siliceous fillers discussed in this article are shown in Table 2. Actual chemical compositions of commercial products vary from these values owing to dilution by impurities. Generally, small amounts of impurities do not adversely affect the utility of the product for filler uses, although they may affect color or brightness which is important in some applications, determining a premium price for purer deposits. Impurities may affect expansion characteristics of perlite, and some applications of silica such as ceramics that require high purity for color.

Asbestos. Asbestos is a name which applies to a group of natural fibrous minerals, of which chrysotile (a type of serpentine) and the amphiboles have the greatest commercial importance, with chrysotile accounting for over 90% of commercial asbestos. Chrysotile probably results from two geologic steps: metamorphism of volcanic ultrabasic rock to serpentine, followed by solution recrystallization of serpentine in cracks and fissures as chrysotile.

Diatomite. Diatomite or kieselguhr is a sedimentary material formed by centuries of life cycles of aquatic diatoms, a simple plant in the algae family with an opaline silica cell wall. Thousands of species of diatoms have flourished and continue to do so in both marine and lacustrine environments. Fossilized skeletal remains of diatoms in commercial quantities are found in many parts of the world.

Perlite. Perlite is believed to result from hydration of volcanic glass or obsidian. Generally, hydration is about 2–5%; this water content is important to the expansibility of the perlite, influencing melting point and supplying expansion steam.

Pumice. The rapid expansion of dissolved gases in silica lavas during volcanic eruptions produces the

Table 1
Main sources of supply of fillers

Asbestos	Diatomite	Perlite	Pumice	Pyrophillite	Silica	Talc
Canada	USA	USA	USA	USA	USA	USA
Quebec	California	New Mexico	Arizona	North Carolina	Illinois	California
Newfoundland	Nevada	Arizona	California	Japan	Missouri	Texas
Ontario	Washington	California	New Mexico	South Korea	Arkanas	Nebraska
British Columbia	Oregon	Idaho	Idaho	Australia	Oklahoma	Vermont
Yukon	Arizona	Colorado	Colorado	Canada	Tennessee	Montana
USSR	France	Nevada	Oregon	India	Georgia	New York
Central Urals	Denmark	Hungary	Nevada	Thailand	Pennsylvania	Belgium
Kazakhstan	USSR	Greece	Iceland	South Africa	New Jersey	Italy
Tuva		USSR	Hungary		West Virginia	Canada
South Africa		Italy	Italy		Texas	France
Zimbabwe		Japan	New Zealand		Canada	Finland
China		China	Greece		Ontario	Japan
Italy		Bulgaria			Quebec	
Swaziland		Turkey				
		Mexico				
		Australia				

Table 2
Typical mineral compositions (%) of siliceous fillers

Mineral	MgO	SiO_2	H_2O	FeO
Asbestos				
chrysotile	40	40	12	
amphiboles				
amosite	5	50	4	40
anthophyllite	30	56	4	10
crocidolite[a]		50	4	15
Diatomite		86–92	0–4	
Perlite[b]		74	4	
Pyrophillite[c]		67	5	
Silica		97–99+		
Talc				
theoretical	32	64	5	
typical	26–33	35–62	5	

[a] Plus 5% NaO, 18% Fe_2O_3 [b] Plus 4% Na_2O, 4% K_2O, 12% Al_2O_3 [c] Plus 28% Al_2O_3

light density pumice or pumicite. The finer pumicite particles are transported by wind away from the source volcano, whereas pumice accumulates closer to the vent.

Pyrophillite. The hydrous aluminum silicate, pyrophillite, is formed by hydrothermal metamorphism of acid tuffs or breccias.

Silica. Silica sand is frequently obtained from the weathering of quartz-containing rock. Decomposition and disintegration of the rock with decomposition of other minerals leaves a primarily quartz sand that has been concentrated by water movement. Induration of sands to sandstone results in another source for silica sand. Amorphous silica, or more properly cryptocrystalline or microcrystalline silica, is formed by the slow leaching of siliceous limestone or calcareous chert.

Talc. Talc is formed by the metamorphic (hydrothermal) alteration of magnesium silicates such as serpentine, pyroxene or dolomite.

3. Physical Properties

These siliceous fillers are generally inert in most applications as shown by pH values in the range 6–10. Asbestos, diatomite and expanded perlite can contribute higher oil absorptions owing to their structure or porosity. The other materials have lower oil absorptions of less than 50 g per 100 g. Some grades of talc, silica and diatomite can offer high brightness or white color that enhances certain applications. All of these fillers have refractive indices around 1.5 which is similar to many application liquids, giving them little pigment value. They can, however, contribute by acting as pigment extenders, spacing the individual higher-refractive-index pigment particles so that they contribute to the maximum extent. The porosity of diatomite and expanded perlite can give some hiding power when formulated above the critical pigment volume concentration.

Typical filler physical properties are shown in Table 3.

4. Mining and Production Techniques

Although a variety of mining and production methods are used with these minerals, they are generally all mined by open-pit techniques; asbestos, silica and talc, however, are sometimes obtained by underground mining. The harder minerals like perlite, asbestos, pyrophillite, silica and talc usually require drilling and

Table 3
Typical physical properties

	pH	Oil absorption (g per 100 g)	Brightness (%)[a]	Specific gravity	Refractive index
Asbestos					
chrysotile	9–10	50–180	<80	2.56	1.53–1.56
Diatomite	6–9.6	90–300	65–90	2.0–2.3	1.40–1.49
Perlite (expanded)	7.6–9.0	50–275	80–90	2.2–2.6	1.48–1.49
Pumice	7–9	30–40		2.2–2.6	1.49–1.50
Pyrophillite	6–8	25–70	80	2.8–2.9	1.57–1.59
Silica					
cryptocrystalline	6.8–7.2	15–30	81–90	2.63–2.65	1.54–1.55
quartz	7.0	13–50	53–89	2.65	1.55
Talc	8.1–9.5	20–52	72–94	2.6–2.9	1.54–1.59

[a] using MgO standard

Table 4
Typical mining and production techniques

Mining and production	Asbestos	Diatomite	Perlite	Pumice	Pyrophillite	Silica	Talc
Mine type							
open pit	×	×	×	×	×	×	×
underground	×					×	×
Ore mining method							
drilling and blasting	×		×		×	×	×
hydraulic jetting						×	
ripping		×	×				
bulldozing		×	×		×		×
dredging						×	
drag line on shovel	×	×		×	×	×	×
trucks	×	×	×	×	×		×
Ore processing methods							
crude blending		×			×		×
primary crushing	×	×	×		×		×
drying	×	×	×	×	×	×	×
crushing/milling	×	×	×	×	×	×	×
classification	×	×	×	×	×		×
benefication		×					×
screening	×		×	×	×	×	×
wet classification						×	×
flash drying							×
air classification	×	×	×	×	×	×	×
fluid energy milling						×	×
calcination		×					
thermal expansion			×				
post calcination/thermal expansion/classification		×	×				
Product size							
coarse	×	×	×	×	×	×	×
medium	×	×	×	×	×	×	×
fine	×	×	×	×	×	×	×
ultrafine		×				×	×
Typical % yields (dry basis)[a]	10	80–90	80–90	(90)	(90)	(90)	(90)
World production (10^3 t)	5700	1700	1800	16000	1700		5000
US production (10^3 t)	93	650	860	4000	127	30000	1200

[a] Estimated values are given in parentheses

blasting, although occasionally the minerals can be ripped or bulldozed. All the minerals undergo some drying, crushing, and/or milling and air classification and/or screening. They are marketed in a series of particle size ranges from coarse to fine or ultrafine.

Specialized processing techniques, such as wet classification by froth flotation, hydrocyclones and magnetic separation, are used to achieve color improvements in silica and talc. Fluidized energy milling is employed to achieve extrafine size reduction in silica and talc. Diatomite often receives flux calcination to improve its color for filler applications. Perlite is always thermally expanded or "popped" like popcorn to a much lighter density material.

Table 4 summarizes and compares the variety of techniques used to mine and produce these fillers. It also shows overall production tonnages. Production includes uses other than fillers and coatings, since it is not possible to develop figures by individual market areas.

Most of the products typically result in over 80% yield on a dry basis. Some loss results from size-reduction processes which grind material into unwanted fines. Loss also occurs owing to beneficiation or classification designed to remove some of the impurities included during mining or interlayered in the deposit in a way that selective mining would not avoid. The lowest recovery is asbestos which is found in thin layers or bands in serpentine rock. A great amount of rock must be mined and processed to recover the asbestos with yields of about 10%.

5. *Safety*

Nearly all these minerals have some hazards in their use if only as "nuisance dusts" that have low levels of hazards, but usually require dust control, ventilation, and use of dust masks. Asbestos, silica (quartz) and flux-calcined diatomite (cristobalite), require more stringent control to ensure that safe dust counts are achieved in dry use areas (see *Asbestos Hazards*).

Table 5
Typical uses of mineral filler

Mineral	Uses
Asbestos	roofing cement, caulking, paints, plastics, cement products
Diatomite	paints, abrasives, plastics, fertilizers, paper, insecticides, insulation, filter aids
Expanded perlite	acoustical, insulating and horticultural products, concrete, plaster, abrasives, paints, filter aids
Pumice	concrete, abrasive, roofing, paints, insecticides, building blocks
Pyrophillite	refractories, insecticides, ceramics, asphalt, floor tiles, paints, crayons
Silica	cement, ceramics, paints, plastics, adhesives, abrasives, caulking, plastics, rubber, filter aids
Talc	paints, paper, cosmetics, plastics, rubber, ceramics, insecticides, roofing

6. *Industrial Uses*

In coating applications, the properties of these fillers are used to achieve a variety of functions. They can control gloss and sheen, modify consistency and flow, influence film permeability, provide "tooth," improve solvent release and improve sanding resistance. They also can control polishing of flat paints, transparency in clear finishes, corrosion resistance, tint retention, improved touch-up or lapping, suspension, stain resistance, weathering, dispersion and color uniformity. Generally, fillers or pigment extenders are used in combination, in an attempt to achieve the best balance in properties.

In caulking, plastics, asphalt, rubber and adhesives, they can supply viscosity, surface appearance, strength, water resistance, antiblock, bulking, resiliency and higher softening point.

Other applications of siliceous fillers and coatings include:

(a) roofing, where they can supply weathering resistance and strength;

(b) paper, where they can supply "tooth," pitch control, porosity and color, and where they can influence printing characteristics, strike-through and opacity;

(c) insecticides, where they are used to provide chemical inertness and controlled dilution for even distribution;

(d) insulation, concrete, plastic, acoustical products, and refractory building blocks, where they supply light density, thermal stability, low heat conductivity and reaction with cement; and

(e) abrasives, where they supply controlled particle size, friability and various hardnesses (see *Abrasives: Siliceous Minerals*).

The wide variety of uses for these minerals is outlined in Table 5.

See also: Fillers and Coatings: Carbonate Minerals; Fillers and Coatings: Clay Minerals; Fillers and Coatings: Industrial Minerals; Fillers and Coatings: Sulfate Minerals

Bibliography

Patton T C (ed.) 1973 *Pigment Handbook*, Vol. 1: *Properties and Economics*. Wiley, New York

Severinghaus N Jr 1983 Fillers, filters and absorbents. In: Lefond S J (ed.) 1983 *Industrial Minerals and Rocks: Nonmetallics other than Fuels*, 5th edn. American Institute of Mining, Metallurgical and Petroleum Engineers, New York, pp. 243–57

Speil S, Leineweber J P 1969 Asbestos minerals in modern technology. *Environ. Res.* 2: 166–208

G. Coombs
[Johns-Manville Sales Corporation, Denver, Colorado, USA]

Fillers and Coatings: Sulfate Minerals

Sulfate minerals, and in particular gypsum and barite, are used on a limited scale as mineral fillers and coatings, chiefly in paints, rubber, plastics, paper and textiles. These sulfate minerals are generally of high purity, are easily ground and classified, are white to near-white in color, and are chemically inert. Barite is especially useful because of its high density. Commercial deposits of gypsum generally contain large reserves and are widespread, whereas barite resources are more limited both in distribution and reserves.

1. Sources

Gypsum reserves are large and commercial deposits are widely distributed in the USA and throughout the world. The mineral has a low unit value, one of the chief attributes of using gypsum as a filler, and transportation costs are kept at a minimum by mining deposits close to industrial and population centers.

Major gypsum-producing districts in the USA and Canada are: the Permian Basin region of New Mexico, Texas, Oklahoma and Kansas; the Michigan Basin and surrounding areas in Michigan, Ohio, New York and Ontario; and the maritime provinces of Nova Scotia and Newfoundland. Other important districts are in southern Indiana, Iowa, southern California, Nevada, Wyoming, Arizona and Colorado. Other North American districts include British Columbia, San Marcos Island and the Caribbean nations of Jamaica and the Dominican Republic. Major gypsum-producing countries on other continents include Japan, France, Spain, Iran, the People's Republic of China and the USSR.

US consumption of gypsum in 1985 was about 22.4 Mt. Most of the gypsum was used in the construction industry (wallboard, plasters and cement retarders) or in agriculture, and less than 2% was used as fillers and coatings.

Barite reserves are somewhat limited in the USA and on a worldwide basis. The decreased demand in recent years for barite in oil-field drilling muds has lowered US consumption to about 2.5 Mt, of which about 3% is now used as fillers and coatings.

The major barite deposits in the USA are in Nevada, which yielded about 80% of the total US output in 1985. Other major world producers include the People's Republic of China, India, the USSR, Mexico and Morocco.

2. Mineralogy and Physical Properties

Gypsum is the name given to the mineral that consists of hydrous calcium sulfate ($CaSO_4 . 2H_2O$), and also to the sedimentary rock that consists primarily of this mineral. In its pure state, gypsum contains 32.6% lime (CaO), 46.5% sulfur trioxide (SO_3), and 20.9% water. Single crystals and rock masses that approach this theoretical purity are generally colorless to white, but in practice, the presence of impurities such as clay, dolomite, silica and iron imparts a gray, brown, red or pink color to the rock.

There are three common varieties of gypsum: selenite, which occurs as transparent or translucent crystals or plates; satin spar, which occurs as thin veins (typically white) of fibrous gypsum crystals; and alabaster, which is compact, fine-grained gypsum that has a smooth, even-textured appearance. Most deposits of rock gypsum that are suitable for industrial purposes are aggregates of fine to coarse gypsum crystals that have intergrown to produce a thick, massive sedimentary rock unit that is 90–98% gypsum. Alabaster is highly prized because of its uniformly fine particle size, but the more common deposits of rock gypsum consisting of coarser-grained selenite can generally be crushed and ground to produce a suitable filler and coating material.

Gypsum has a hardness of 2 on the Mohs scale, and can be scratched with the fingernail. Large rock masses are easily crushed and ground to a fine powder. The specific gravity of pure gypsum is about 2.31, and the refractive index is about 1.53. Gypsum is slightly soluble in water, but it is an inert substance that resists chemical change. The oil-absorption capacity of gypsum is fairly low ($0.17–0.25\ cm^3\ g^{-1}$).

Raw or crude gypsum is one of the forms used as fillers and coatings, but for some purposes calcined or deadburned gypsum is desired. In calcining, the gypsum is heated to about 120–160 °C to drive off free water and partially remove the water of crystallization. The calcined material, or stucco, has a chemical composition of $CaSO_4 . \frac{1}{2}H_2O$, and it readily takes up water. Calcination at higher temperatures (500–725 °C) results in a product called deadburned gypsum, which has a composition of $CaSO_4$.

Anhydrite, a sulfate mineral and rock that is closely associated with gypsum in nature and has minor uses as a filler, is anhydrous calcium sulfate ($CaSO_4$) containing 41.2% CaO and 58.8% SO_3. It is typically fine grained (like alabaster), and occurs in thick, massive sedimentary rock units. Anhydrite usually is white or bluish gray when pure, but it may be discolored by impurities. Anhydrite has a hardness of 3.5, a specific gravity of 2.98, and a refractive index of 1.57–1.61.

Barite, also called barytes or heavy spar, is a mineral that consists of barium sulfate ($BaSO_4$). Pure barite, which is 65.7% BaO and 34.3% SO_3, is white, but many commercial deposits are discolored by small amounts of impurities.

Commonly, barite is well crystallized and occurs as thin to thick tabular crystals that have intergrown as aggregates. Commercial deposits of barite occur (a) as irregular masses, concretions, or nodules in vein and cavity fillings; (b) as residual masses produced by weathering of preexisting rocks; and (c) as fine-grained, massive rock layers in bedded deposits several centimeters to more than 15 m thick. Any of these forms of barite is readily crushed and ground to produce fillers and coating material, although the vein or residual types are more commonly used for this purpose because of their white or off-white color.

Barite has a hardness of 2.5 to 3.5, and therefore cannot be scratched with the fingernail. The calculated specific gravity of pure barite is 4.5, but inclusions and chemical impurities in commercial deposits cause this value to range from 4.3 to 4.6. Barite is thus the highest density mineral among those commonly used as mineral fillers. The refractive index of barite is about 1.64. The mineral is relatively insoluble in water and acids, and can thus be used as a chemically inert material. The oil-absorption capacity of barite is low, with values of 0.06–0.10 $cm^3 g^{-1}$.

Celestite is a high-density strontium sulfate ($SrSO_4$) mineral consisting of 56.4% SrO and 43.6% SO_3. With a density of 3.96, celestite may be a suitable substitute for barite in paints and other uses where extra weight is needed. Celestite has a hardness of 3 to 3.5, a refractive index of 1.62, is chemically inert, and typically is white with a faint blue or red tint. Celestite is commonly a secondary mineral associated with some deposits of limestone, gypsum, or rock salt. Principal production of celestite comes from the UK, Mexico, Spain, and Canada; no deposits in the USA are believed to be commercial at this time.

3. *Mining and Production Methods*

Gypsum is produced both from surface pits and from underground mines. Open-pit mining is most desired where the gypsum bed is within 15 m of the surface and/or where the overburden thickness is only one to three times greater than that of the gypsum bed itself. Draglines, scrapers, or bulldozers are used to remove the overburden, and power shovels or front-end loaders are used to load the gypsum. Explosives are used to break up rock overburden and also are used to fragment the gypsum.

In underground mines, gypsum is fragmented with explosives and is loaded on self-loading dump units or on conveyor belts, rail cars or diesel-powered trucks by front-end loaders or gathering-arm loaders. Square or rectangular gypsum pillars 6–9 m wide are left in place to support the mine roof, whereas the mined-out areas surrounding the pillars commonly are 9–12 m across. Such "room-and-pillar" mining methods enable recovery of up to 75% of the gypsum, and are being employed at depths commonly ranging from 15–180 m below ground.

Individual gypsum mines normally produce 0.1–0.6 Mt annually. The mining process is generally highly mechanized, and working conditions are good because of the lack of explosive gases or harmful dust.

Processing of raw or crude gypsum consists merely of crushing and grinding the mine-run rock to specified sizes. Processing of calcined or deadburned gypsum after heating involves only grinding the material and classifying it by particle size.

Barite is also mined by surface or underground methods, depending upon the depth of the ore body and whether the deposit is residual, vein-type or bedded. Residual barite deposits are normally poorly consolidated and just below the topsoil. They are generally mined by open-pit methods. After removal of topsoil, the barite-bearing ore body (consisting of clay, rock and about 10–25% barite) is removed by power shovel, dragline or front-end loader, and is hauled on dump trucks to a washer plant. where the barite is separated from waste material.

Bedded and vein deposits of barite are mined either by open-pit or underground methods. Open-pit mining of shallow deposits is preferred, because of the lower costs and higher recovery, but underground mines can also be operated because of the integrity of adjacent roof and wall rock in most of these types of deposits. Ore from these deposits must normally be crushed, ground and beneficiated by jigging or by flotation of barite with anionic reagents.

Barite is processed either from crude lumps, or from a jigged or flotation concentrate. The barite is crushed and ground either wet or dry. For some filler uses, bleaching with sulfuric acid is necessary to remove objectionable iron stains.

4. *Uses*

Gypsum is not a major mineral filler or coating material. It is, however, used on a modest scale in paints, paper, cloth, pesticides, pharmaceuticals (such as aspirin) and toothpaste. Properties that make it especially useful as a filler include its widespread distribution, relatively low unit cost, high purity, white to near-white color, softness, freedom from grit, fine particle size, low solubility and chemical inertness. Gypsum may be used raw or in the calcined or deadburned state. Anhydrite also is used sparingly as a higher density filler in paints.

Barite is more widely used than gypsum as a mineral filler or coating material, with much of its demand resulting from its high density. Barite is used in rubber, paints, plastics, paper and textiles. Barite may be used either raw or after being bleached in sulfuric acid to remove iron impurities. The high density of barite is its main value as a filler, but other desirable properties include its relatively low unit cost, high purity, white to off-white color, moderate hardness, freedom from grit, fine particle size, low solubility and chemical inertness. A specialized use of barite, based on its high density and its absorption of γ radiation, is as a shielding material in nuclear reactors to reduce the amount of expensive lead shielding that would otherwise be needed.

See also: Fillers and Coatings: Carbonate Minerals; Fillers and Coatings: Clay Minerals; Fillers and Coatings: Industrial Minerals; Fillers and Coatings: Siliceous Minerals

Bibliography

Brobst D A, Pratt W P (eds.) 1973 *United States Mineral Resources*, US Geological Survey Professional Paper 820. US Government Printing Office, Washington, DC

Severinghaus N Jr 1983 Fillers, filters and absorbents. In: Lefond S J (ed.) 1983 *Industrial Minerals and Rocks*, 5th edn., Vol. 1. American Institute of Mining, Metallurgical and Petroleum Engineers, New York, pp. 243–57

Sibbing E 1976 Barytes: Raw material for fillers and chemicals. *Ind. Miner. (London)* 101: 33–43

US Bureau of Mines 1985 *Mineral Facts and Problems*. US Bureau of Mines Bulletin 675. US Government Printing Office, Washington, DC

K. S. Johnson
[Oklahoma Geological Survey, Norman, Oklahoma, USA]

Filters, Sorbents and Ion Exchangers

Mineral filters, sorbents and ion exchangers are employed in a wide variety of special applications. Mineral filters are generally used to remove solid particles from liquid media, and in particular for applications requiring the filtration of high volumes of liquids. Mineral sorbents are employed in the purification by selective absorption of industrial items such as edible oils and fats, and also as carriers to absorb insecticides, herbicides and other agricultural chemicals. These mineral sorbents are also used as sweeping compounds to absorb oil spills and other industrial chemicals. The largest use of mineral absorbents is for pet litter. Ion-exchange minerals are employed mainly in the conditioning of process waters and in the treatment of wastewaters and effluents.

This article serves as a general introduction to filters, sorbents and ion exchangers. Detailed information on siliceous filters such as sand, diatomite, perlite and asbestos, clay sorbents and filters, and minerals with ion-exchange properties can be found elsewhere (see *Filters, Sorbents and Ion Exchangers: Clay Minerals*; *Filters, Sorbents and Ion Exchangers: Siliceous Minerals*; *Filters, Sorbents and Ion Exchangers: Zeolite Minerals*).

For industrial purposes, a good filter aid must possess several characteristics. These include the ability to form a very porous cake, a low surface area, correct particle size distribution and nonuniform sizing, and availability in a number of grades. The performance of a mineral filter aid is measured by the clarity of the filtrate, the rate of flow of the liquid through the filter bed and the length of the filter cycle. Mineral filter aids are used in food products, metallurgical processes, petroleum refining and processing, drug and pharmaceutical manufacture, and in many other industrial manufacturing processes. They also find use in the dry-cleaning process, in waste disposal and for water filtration of all types. The most widely used mineral filter aid is diatomaceous earth, and the largest producer is the USA with about 0.7 Mt per year. World production is estimated at about 2 Mt. Diatomaceous earth has a very low bulk density, so freight costs are a very important factor in marketing. Exploration for new deposits has intensified recently because of the high freight cost.

Mineral sorbents are highly absorbent and adsorbent in their natural or modified form. The most important are attapulgite (palygorskite), sepiolite, smectites (bentonites) and diatomite. A high surface area is the leading factor in the absorptive capacity of these minerals, although charged sites on the surface are also a factor for the clays. Absorption is readily reversible by simple low-temperature heating, and most sorbents can be restored to their original absorptive character by such processing. Adsorption, on the other hand, is a selective process and is not readily reversible by low-temperature heating.

The terminology of the clays used as mineral sorbents needs some clarification. The term Fuller's earth normally signifies clay or other fine-grained earthy material suitable for bleaching and absorbent uses. It has no compositional or mineralogical meaning. The term was first applied to material used in cleansing and fulling wool, thereby removing the lanolin and dirt from it. When it was found that some of the earths used for fulling wool would also serve to decolorize and purify mineral, vegetable and animal oils, and as a general sorbent material, the term Fuller's earth was modified to include this usage. The terms bleaching clay and bleaching earth are applied to both naturally active and activated clay, including activated bauxite. Activated clays is a term used for clays which are treated with acid or otherwise altered to improve their desirable properties. Bleaching earths then refers to clays that in their natural state, or after chemical or physical activation, have the capacity to adsorb coloring matter from oils. The term absorbent clay is applied to Fuller's earth used for a wide variety of absorbing purposes different from those of processing oils.

Sorbent clays are produced in the USA, the UK, Germany, Spain, India, Japan, the USSR, Hungary, Rumania, Italy, Morocco, Syria, Argentina, Mexico and Australia. The USA is the largest producer followed by Germany and the UK. The product is either granular or powdered dependent on the use requirements. In many instances, the granules are hardened by thermal treatment so that they can withstand attrition without readily breaking down into fines. The size of the granules range from 30 to 60 mesh (9.8–23.2 μm). In addition to the size of the granules, the amount of oil and water that is absorbed per gram of sample is important.

Two processes are used to decolorize various types of oil, namely percolation and contact. In the percolation process the bleaching clay is placed in a filter bed and the oil to be processed percolates through the bed. In the contact process, the clay is mixed with the oil to be decolorized, and then the clay is separated out in filter presses or some other filtration device.

The future outlook for mineral filters, sorbents and ion exchangers indicates a slow to moderate growth in usage. Since the late 1960s, most companies producing mineral filters expanded their capacity to meet the increased demands. In the late 1970s, the growth rate slowed considerably, but it is anticipated that as developing countries progress, new markets for mineral filter aids will open up. In the case of sorbents much the same market pattern exists. The tremendous growth of absorbent clays as pet litter took place in the 1960s and 1970s, and future growth will be slow but steady. Agricultural uses of absorbents particularly as carriers of pesticides could increase dramatically as the developing countries modernize their agriculture. The use of bleaching clays could increase significantly as more used oils are filtered and decolorized to be reused. The high prices of essential petroleum products will mandate that more used oil will be cleaned, which means that bleaching clays should have a promising future.

Bibliography

Lefond S J (ed.) 1983 *Industrial Minerals and Rocks*, 5th edn. American Institute for Mining, Metallurgical and Petroleum Engineers, New York

Mumpton F A 1983 Commercial utilization of natural zeolites. In: Lefond 1983, pp. 1418–31

Patterson S H, Murray H H 1983 Clays. In: Lefond 1983, pp. 585–651

Severinghaus N Jr 1975 Fillers, filters, and absorbents. In: Lefond 1983, pp. 243–57

H. H. Murray

[Indiana University, Bloomington, Indiana, USA]

Filters, Sorbents and Ion Exchangers: Clay Minerals

Clays find widespread use as sorbents, and are also employed in minor quantities as filter aids. The specific clay minerals used as mineral sorbents are members of the hormite and the smectite groups. The major clay mineral used as a filter aid is kaolinite.

Sorbent clays are employed in decolorizing, clarifying and absorbing various types of oil, and in pesticide and chemical carriers, pet litters and carriers for microencapsulated ink in no-carbon copy paper. They are also used in clarification of sugar cane juice, beers and wine. The largest use is as pet litter.

Clay minerals are ideal sorbents for many industrial processes because of their fine particle size, high surface area and surface charge sites. The absorbent markets are thought to tie in closely with the gross national product and sales fluctuate with the economy. The granular pesticide carrier markets have only been fully developed in the USA and Japan, so there is a high potential market in Europe and the rest of the world.

1. Mineralogy

1.1 Hormite Group

The hormite group of clay minerals are hydrated magnesium aluminum silicates. Palygorskite (or attapulgite) has a fibrous, lath-like form, which along with its crystal structure results in unusual colloidal and sorptive properties. Sepiolite is structurally similar to attapulgite, but has a slightly larger unit cell and the individual laths are about one-third wider. Sepiolite in its compacted form is known as meerschaum.

1.2 Smectite Group

The best sorbent amongst the smectite minerals is calcium montmorillonite, sometimes called a subbentonite or nonswelling bentonite. Calcium montmorillonite is a hydrated calcium aluminum silicate that normally has considerable iron and magnesium substitution of aluminum and calcium. The mineral consists of thin sheet-like irregular particles that are extremely fine in particle size ($\leqslant 2$ μm).

2. Sources and Production Techniques

Attapulgite is not as common as many other commercial clays. It is produced in Georgia and Florida in the USA, and in the USSR and Senegal. By far the largest production is from the USA, although Senegal has the most extensive reserves.

Sepiolite is also a relatively rare clay mineral in comparison with other commercial clays such as kaolins and bentonites. It is produced in Spain and Turkey, and in Nevada in the USA. The largest production is from Spain.

Attapulgite and sepiolite are mined by open-pit stripping methods. Typically, the overburden is removed in panels, the clay mined by loading trucks with shovels or endloaders, and the overburden from an adjacent panel shifted to the mined-out area. Most of the hormite group minerals are selectively mined because their properties vary considerably in most deposits. Processing involves simple milling techniques that involve removal of water and grinding to suitable sizes. Granular products are screened, and the granules hardened by heating to 650 °C, which drives out all the absorbed water.

Calcium montmorillonites used as sorbent or bleaching clays are much more widely distributed than the hormite group clays. Production is from the USA, Mexico, Italy, Spain, the UK, the Federal Republic of Germany, South Africa, India, Japan, Turkey and Canada. The largest production is from the USA; the Federal Republic of Germany and the UK are also large producers.

Calcium montmorillonites are mined and processed in a similar fashion to attapulgite and sepiolite. Activated clays for use in decolorizing, for paper coating and other specialized uses require special processing. The calcium montmorillonite is slurried in water at about 10–20% solids and the sand is settled or screened out. The slurry is heated and sulfuric and/or hydrochloric acid is added to make a hydrogen clay; this treatment also removes a small amount of aluminum from the clay mineral structure. The acidified clay slurry is kept hot using steam, and after the appropriate time the slurry is filtered, washed and then dried. Almost all the product is bagged for shipment, although small quantities are now shipped in bulk.

3. *Physical and Chemical Properties*

Attapulgite and sepiolite absorbent clays are used for many miscellaneous uses, with more than 90 different grades of attapulgite being produced. In the USA, most absorbent granules are prepared to fulfill the requirements outlined in Federal Specification P-A-1056A. A similar set of specifications is outlined in the American Society for Testing and Materials Standard C431-65.

The US Federal specifications require that absorbent granules must be clean, uniform and free of lumps or foreign matter, and that no more than 10% of the granules can pass through an 80 mesh (0.177 mm) sieve in an attrition resistance test. This test is carried out by shaking the granules with steel balls on a screen, according to a specified procedure.

Test methods for evaluating bentonite and other types of Fuller's earth for bleaching soybean and cottonseed oils are outlined by the American Oil Chemists Society. Their specifications contain instructions on bench-type tests, including stirring time, heating rates and temperatures; approved equipment; quantities of raw oil and clay required; and color determination methods. They also require the comparison of the material tested with an official natural bleaching earth. Purchase specifications are based on comparison of the color of oils bleached by the official earth with those bleached by the test clay.

For pet litter the granules must be absorbent and dust free so that the cat does not track the dust, and so that the dust does not fly when the litter box is filled with fresh litter. Also the granules must be hard enough so that they do not break up as they are handled, or when the cat scratches them around in the litter box.

When granules are used as an insecticide carrier, in addition to meeting the absorbent specifications outlined above, specific tests using the particular insecticide chemical must be performed to make sure that the chemical will be released and will not be polymerized or cracked into other compounds. This can only be determined by making absorption tests under controlled conditions.

4. *Industrial Applications*

The largest market, and perhaps the area of highest competition, is in domestic pet litter, agricultural carrier, and oil spillage applications. In the pet litter markets, the utility of a particular clay is dependent not only on its sorptive characteristics, but also on balling properties and bulk density, whereas in the oil spillage markets the ability of the clay to resist slaking and its wear properties are of importance along with bulk density. Bulk density in both of these applications governs the quantity of material needed in a particular situation and consequently the overall costs. As an agricultural carrier the percentage of chemical that can be absorbed and still flow readily is very important.

Another large market is in the refining of liquids—mainly oils, but also sugar cane juice, beers and wines. Large amounts of naturally active calcium bentonites are used for this purpose, but the acid-activated clays, which have superior absorption properties, are gradually replacing the naturally active clays. The acid-activated clays are more efficient decolorizers because of their larger surface area. However, in the treatment of some very light oils which are particularly sensitive, a natural bleaching clay is usually prescribed. Attapulgite also has a share of the bleaching and decolorizing market, particularly in the USA, and a significant proportion of the sepiolite production from Turkey is used for the extraction of sulfur from light paraffin oils. The refining of glyceride oils of both animal and vegetable origin is a significant market for calcium montmorillonites. Oil recycling has been stimulated by the antidumping laws, particularly in the USA, and also by the high cost of new oils. Many types of industrial oils and lubricants are purified and decolorized using granular attapulgite and activated calcium bentonite. The principal types of oil refined in the mineral-oil field are naphthas, fuel oils, lubricating oils, waxes and greases. The more important vegetable and animal oils refined with bleaching clays are cottonseed, soybean, linseed, coconut, palm, tallow and grease. The principal object of applying bleaching earths is to remove color, but there are other important uses in all fields of refining; for example, in the treatment of naphtha, reduction of the amount of gum and stability are most important factors. In the refining of lubricating oil, improved sludge content, carbon content, oxidation test, acidity, emulsion test and viscosity index are also important. In processing of vegetable oils, stabilization and removal of residual soap, gums, odor and taste are also important improvements after treatment using bleaching earths.

See also: Filters, Sorbents and Ion Exchangers; Filters, Sorbents and Ion Exchangers: Siliceous Minerals; Filters, Sorbents and Ion Exchangers: Zeolite Minerals

Bibliography

Oultin T D 1965 *Mining, Production and Uses of Attapulgite Clay Products*, SME Preprint 65H39. American Institute of Mining, Metallurgical and Petroleum Engineers, New York

Patterson S M, Murray H H 1983 Clays. In: Lefond S J (ed.) 1983 *Industrial Minerals and Rocks*, 5th edn. American Institute of Mining, Metallurgical and Petroleum Engineers, New York, pp. 585–651

Rich A D 1960 Bleaching clay. In: Gillson J L (ed.) 1960 *Industrial Minerals and Rocks*, 3rd edn. American Institute of Mining, Metallurgical and Petroleum Engineers, New York, pp. 93–101

H. H. Murray
[Indiana University, Bloomington, Indiana, USA]

Filters, Sorbents and Ion Exchangers: Siliceous Minerals

Siliceous filtration materials function in several different ways. Diatomite and perlite are used as filtration aids to remove suspended particles from liquids by straining or trapping the particles on a layer of filter aid supported by a filter septum or screen as the liquid passes through the filter aid. This is accomplished in a variety of units; for example, pressure filters, vacuum filters, plate and frame filters, and rotary precoat filters.

Body addition or controlled addition of filter aid to the "dirty" liquid can help to improve filter cake porosity by spacing the dirt particles with the more porous filter-aid particles. The flow rate and filtration ability of the filter aids are normally a function of crude quality and processing.

The sorptive properties of asbestos can be used to improve some filtrations of very fine particles by attracting them to the asbestos surface and allowing filtration of the sorbed dirt with the asbestos.

Silica is used in sand filters to form a porous bed that traps the dirt particles. Filtration in sand beds is often aided by chemical coagulants or flocculents to make dirt more readily trapped by the relatively porous sand.

Garnet is also used in sand filters to form a porous bed that tends to stay at the bottom of the sand bed owing to its specific gravity of 4.2.

1. Sources of Supply

Sources of asbestos, diatomite and perlite are discussed elsewhere (see *Fillers and Coatings: Siliceous Minerals*). Sources of quartz filtration sand are extensive. A good filtration sand should be a clean river sand, free of organics, fines, clay and friable particles, as well as low in acid solubles. Rounded sand grains (river sand) work best in filter beds since they require the least backwashing pressure, scour best during backwash, and fluidize more easily during cleaning cycles. The hardness of silica is needed to resist the intense scouring. However, the prime requirement is very tight size gradation, often in the 0.35–2.0 mm size range.

In addition to quartz filtration sands, garnet sands are used owing to their much higher specific gravity which allows maintenance of a layered bed after backwashing. Garnet sands are angular since they need crushing to generate the required sizes. Cost and availability considerations restrict the use of garnet in filtration.

2. Mineralogy

2.1 Asbestos

Chrysotile forms by the solution recrystallization of serpentine in cracks and fissures of the host serpentine. The chrysotile fibrils have been shown to be hollow cylinders formed from a magnesium silicate layer. These fibrils occur in fiber bundles that must be opened by careful milling to maximize the usefulness of the asbestos.

2.2 Diatomite

Natural diatomite comprises the opaline silica skeletons of the various species of diatoms that flourished in marine or freshwater environments. Typical commercial processing of diatomite includes either calcination or flux calcination to modify the flow rates of filter aids. Calcination "burns out" some of the finest tertiary silica structure and helps sinter fines together. Flux calcination accentuates these changes and may also improve color. Calcination dehydrates the opaline silica and causes an increase in specific gravity and some conversion to cristobalite. Flux calcination increases the specific gravity to 2.3 and results in as much as 40–60% conversion to cristobalite.

2.3 Perlite

Perlite is typically a sodium, potassium or aluminum silicate glass hydrated with about 2–5% water. It is believed that perlite results from hydration of volcanic glass or obsidian. Size-graded perlite, typically in the range 2.00–0.15 mm, expands up to 30 times its original size when heated to 760–1090 °C in either a horizontal or vertical furnace. The expanded perlite has the same composition as the ore, but without the water of hydration.

2.4 Silica

Filtration sand is typically quartz or quartzite sand, preferably obtained from river transported sources so that grains are rounded and uniform.

2.5 Garnet

A variety of types of mineral garnets exist, including grossularite, pyrope, almandite, spessartite and andradite. These are all aluminum silicates with calcium, magnesium, iron or manganese (see *Abrasives: Precious and Semiprecious Minerals*).

3. Physical and Chemical Properties

In filtration applications, diatomite, expanded perlite, silica and garnet normally do not react with the liquids being filtered. They can show trace solubilities with acids, and with strong alkalis (generally with pH values greater than 10 or 11) siliceous minerals will show solubility, particularly at increased temperatures and concentrations.

Particle shape and porosity are important properties of diatomite and expanded perlite when used as filter aids. The pore-size distribution of these materials and the particle-size distribution of turbidities can often be related to assist in selecting suitable filter aids.

Sand and garnet filtrations are generally thicker beds than diatomite or perlite. They trap the turbidities in the relatively larger pores between sand grains. The proper use of chemical flocculents can greatly benefit sand filtration.

Chrysotile asbestos contributes to filtration in two ways. Because of its fibrous nature, it can be used to form a strong precoat over the filter septum. It also has a positive zeta potential which aids filtration by attracting generally negatively charged turbidity particles to its surface.

Physical and chemical properties of siliceous minerals used as filter aids are shown in Table 1.

4. Mining and Production Techniques

4.1 Asbestos

Asbestos is recovered from the crushed serpentine host rock through a variety of screening and air-clarification techniques. Generally, the shorter fibers are chosen for filtration applications, and preferably, the fiber is well opened to maximize surface area.

4.2 Diatomite

The cyclone fraction of dried, milled and air-classified natural diatomite is used for maximum clarity low-flow-rate filtrations. Straight calcination at 760–1090 °C agglomerates some fines and melts some of the tertiary diatomite structure. After milling and air classification to remove fines, the cyclone fraction has faster flow rate but lower clarity than the natural filter aids. Straight or nonflux calcined diatomite is typically pink owing to oxidation of small amounts of iron. Still faster flow rates can be attained by flux calcination with 2–8% soda ash, sodium chloride, or similar fluxes, and calcination at 760–1090 °C. Flux calcination results in substantial agglomeration of fines and almost complete melting of tertiary structure. Generally, white colors are obtained as a result of the flux reacting with the color-causing impurities like iron and aluminum.

4.3 Perlite

Perlite ore is dried of free surface moisture, crushed and size graded to various mesh-sized ores that may

Table 1
Some physical and chemical properties of siliceous minerals used as filter aids

Mineral	Specific gravity	pH	Bulk density ($g\ cm^{-3}$)	Surface area ($m^2\ g^{-1}$)	Particle size (mm)	Flow rate ($m^3\ m^{-2}\ h^{-1}$)
Asbestos						
chrysotile	2.56	9–10	0.8–1.2	4–50	fibrous	0.10–2.0
Diatomite						
natural	2.0	6.0–7.5	0.25–0.37	10–30	0.01–0.05	0.16–2.0
calcined	2.1–2.2	5.0–7.5	0.25–0.33	3–7	0.01–0.10	0.04–2.0
flux calcined	2.3	8.5–10.5	0.29–0.37	0.5–5	0.01–0.50	0.04–4.9
Perlite						
milled, expanded	22–23	6.6–8.0	0.13–0.28	1	0.01–0.50	0.05–4.0
Silica sand						
fine	2.6	7.0	1.0–1.3	0.10–0.02	0.26–1.50	5
medium	2.6	7.0	1.0–1.3	0.01–0.02	0.34–1.80	12
coarse	2.6	7.0	1.0–1.3	0.01–0.02	0.41–2.00	50
Garnet	3.9–4.1		1.6–2.0		0.21–2.00	5–50

be used as they are or blended together to give a selected distribution of particles. The sized ore is expanded in either horizontal or vertical furnaces. In vertical furnaces, the ore is fed into the flame and is concurrently carried to collection cyclones and/or baghouses. The ore may be preheated in the jacket of the furnace to aid expansion. In horizontal furnaces, the ore may also be preheated in the jacket of the rotating tube. The ore drops across the flame and is carried out of the tube by rotation and gas velocity. The perlite is generally expanded to 0.03–0.10 $g\,cm^{-3}$ for filter aid production. Filter aids are milled to reduce floaters or bubbles, and classified to give various flow rates.

4.4 Silica Sand

Natural bank sand usually needs to be modified to achieve the necessary sizing and uniformity to accuracies as close as economically possible. Generally, oversize is screened out and fines are washed out. This can be done by the sand supplier or by the sand user by running their own sand washers and sand separators or by shovelling off fines washed to the top of the bed after backwashing.

4.5 Garnet Sand

The production of garnet sand is discussed elsewhere (see *Abrasives: Precious and Semiprecious Minerals*).

5. *Safety*

These minerals are all possible sources of dust that at least have to be treated as a nuisance and require adequate dust control. Asbestos, silica (quartz) and flux-calcined diatomite (cristobalite) require a higher degree of dust control. Generally, the coarser particles used in filtration are not in the respirable size range. However, safe dust levels must be maintained (see *Asbestos Hazards*).

6. *Industrial Uses*

Filter-aid filtration is used for a wide variety of applications. Beverages such as beer, wine, fruit and vegetable juices, and spirits are frequently filtered. Cane, beet and corn sugars, animal and vegetable oils, antibiotics, waxes, gelatin, inorganic chemicals, acids, lube oils, rolling and cutting oils, varnishes, lacquers, jet fuels, and organic chemicals use filter-aid filtration. Other uses are in process, potable and waste waters.

Turbidities removed can vary in characteristics from noncompressible particles to compressible, gelatinous or slimy materials. They also can vary widely in particle size and concentration. In addition to turbidities, filtration is used to remove or recover adsorbents, catalysts and other solids added to a liquid as part of a chemical processing step. The selection of the proper filter aid and type of filter to use strongly influences filtration rate and clarity of filtrate.

Asbestos is used to give a high degree of clarity or polish to critical filtrations. It is also used to form a precoat layer over the filter septum that has greater stability against loss of filter cake owing to pressure or flow-rate fluctuations.

Sand filtration is used primarily for water filtration; for example, in producing potable water and in swimming pool systems. Generally, low levels of turbidities are involved and filtration is aided by flocculating with, for example, alum, organic polymers and polyelectrolytes.

See also: Filters, Sorbents and Ion Exchangers; Filters, Sorbents and Ion Exchangers: Clay Minerals; Filters, Sorbents and Ion Exchangers: Zeolite Minerals

Bibliography

Fair G M, Geyer J C, Okun D A 1968 *Water and Wastewater Engineering*, Vol. 2, *Water Purification and Wastewater Treatment and Disposal.* Wiley, New York

Manual of Instruction for Water Treatment Plant Operators. New York State Department of Health, Health Education Service, Albany, New York

Severinghaus N Jr 1983 Fillers, fitters, and absorbents. In: Lefond S J (ed.) 1983 *Industrial Minerals and Rocks*, 5th edn. American Institute of Mining, Metallurgical and Petroleum Engineers, New York, pp. 243–57

Speil S, Leinweber J P 1969 Asbestos minerals in modern technology. *Environ. Res.* 2: 166–208

G. Coombs
[Johns-Manville Sales Corporation, Denver, Colorado, USA]

Filters, Sorbents and Ion Exchangers: Zeolite Minerals

Zeolites are a group of 34 separate mineral species which are crystalline, hydrated aluminosilicates containing exchangeable alkali or alkaline-earth cations. Extensive deposits of high-purity zeolite minerals occur in Cenozoic-age volcanics and sedimentary rocks of volcanic origin. The natural zeolite minerals and their synthetic counterparts, which are known as molecular sieves, are used principally as filters, sorbents and ion exchangers, and as such they find application in pollution control, radioactive-waste management, petroleum refining and gas purification processes. Natural zeolite minerals are also used in construction materials, carriers, fillers and coatings, aquaculture and agriculture. Major future uses will include wastewater treatment, solar energy and synthetic fuel production.

1. *History*

Zeolite minerals were discovered in 1756 by Freiherr Axel Fredrick Cronstedt, a Swedish mineralogist who named them from two Greek words, *zein* and *lithos*, meaning boiling stones. The name described the peculiar frothing or boiling characteristics the minerals exhibited when heated in a blowpipe flame. For nearly 200 years after their discovery, the unique dehydration, adsorption and ion-exchange properties of the zeolite minerals were studied by chemists and mineralogists.

By the 1920s, researchers discovered that zeolite minerals adsorbed some substances and not others. In describing these unique properties, J. W. McBain of Stanford University first referred to zeolite minerals as molecular sieves. By 1948 research was underway to determine whether zeolites might be used in the separation of atmospheric gases. Utilization was impeded by an almost universally accepted belief that zeolite minerals occurred in limited quantities only as vug or cavity fillings in basalt or trap rocks. Although an adequate supply of zeolite minerals was available from these crystal-lined vugs for research purposes, it was much too limited to be considered as a potential commercial source. Therefore, researchers were forced to turn to synthesis as a means of providing a dependable supply of zeolite minerals.

In 1949 R. M. Milton of the Linde Division of Union Carbide Corporation began experiments in zeolite synthesis which resulted in the discovery of the Linde Type A zeolite. Synthetic zeolite minerals or molecular sieves were in commercial production in the 1950s, delaying utilization of the natural zeolite minerals in the USA, for example, by nearly a decade after the first industrial application of the synthetic ones. In fact, it was not until the discovery of large deposits of zeolitically altered tuffs, vitrophyres and sediments in Japan in 1949, and the USA in 1957, that their potential as an important industrial mineral was recognized. Since then, natural zeolites have supplied a small but growing segment of the market for ion-exchange and adsorbent products.

2. *Mineralogy*

Chemically the 34 minerals that have been identified in the zeolite group are crystalline, hydrated aluminosilicates of alkali and alkaline-earth elements, usually sodium, potassium, magnesium, calcium, strontium and barium. Structurally the minerals are framework aluminosilicates consisting of infinitely extending three-dimensional networks of AlO_4 and SiO_4 tetrahedra linked to each other by the sharing of all oxygen atoms. The minerals might best be described as three-dimensional-framework aluminosilicates of alkali and alkaline-earth cations, predominantly sodium and calcium, which contain highly variable amounts of water within the voids of the framework. Though similar in chemical composition to the feldspars and feldspathoids, the zeolite minerals are dissimilar in crystal structure. Whereas the feldspars and feldspathoids have tight and compact structures, the zeolite minerals have open structures containing large cavities filled with water molecules. The cavities are interconnected by pores or channels. When the zeolite mineral is activated by driving off the water molecules (by heating to a temperature over 100 °C), the crystal structure remains intact and the cations become coordinated with the oxygen along the inner surfaces of the cavities. Thus, the zeolite crystal becomes a porous solid consisting of up to 50% void space, permeated with cavities which are interconnected by channels having diameters ranging from 2 to 7 Å (see Fig. 1). The result is a molecular sieve which, depending upon the size of the channels, can adsorb gases and effectively separate gaseous mixtures.

Water molecules can pass through the channels and pores, allowing cations present in solution to be exchanged for cations in the structure. This process is known as ion exchange, and each zeolite mineral has a characteristic ion-exchange selectivity and capacity. Nevertheless, many factors, including pH, solution strength, temperature and the presence of other cations or anions in solution, can profoundly affect both the ion-exchange selectivity and capacity of a specific zeolite mineral.

As catalysts, zeolites promote the molecular rearrangements known as catalytic reactions which, like adsorption, occur within the cavities or pores. The zeolite minerals best suited for catalysts possess large, three-dimensional pore systems and a high silica-to-alumina ratio. Among natural zeolite minerals faujusite has been the subject of more research than any other. Faujusite, unfortunately, is not known to occur in minable deposits. Though small quantities of modified clinoptilolite and mordenite are used as catalysts in both Bulgaria and Hungary, the catalyst market is dominated by synthetic zeolites.

3. *Geology*

At least 20 zeolite minerals have been reported from deposits in zeolitically altered rocks. However, only the following nine minerals are known to occur in deposits large enough to mine:

Analcime $NaAlSi_2O_6 \cdot H_2O$

Chabazite $(Ca,Na_2)Al_2Si_4O_{12} \cdot 6H_2O$

Clinoptilolite $(Na_2,K_2,Ca)_3Al_6Si_{30}O_{72} \cdot 24H_2O$

Erionite $(Na_2,K_2,Ca)_{4.5}Al_9Si_{27} \cdot 27H_2O$

Ferrierite $(K,Na)_2(Mg,Ca)_2Al_6Si_{30}O_{72} \cdot 18H_2O$

Heulandite $(Ca,Na_2)_4Al_8Si_{28}O_{72} \cdot 24\,H_2O$

Laumontite $Ca_4Al_8Si_{16}O_{48} \cdot 16H_2O$

Mordenite $(Na_2,K_2,Ca)Al_2Si_{10}O_{24} \cdot 7H_2O$

Phillipsite $(K_2,Na_2,Ca)_2Al_4Si_{12}O_{32} \cdot 12H_2O$

Only clinoptilolite, mordenite, chabazite, erionite and ferrierite have been used in commercial applications.

Figure 1
Scanning electron micrograph of clinoptilolite from Buckhorn, New Mexico. Scale is 31 μm between white marks (courtesy of R. A. Sheppard, US Geological Survey)

Zeolite deposits can be classified into six types based on origin:

(a) closed system or saline–alkaline lake deposits,

(b) open-system deposits,

(c) burial diagenetic deposits,

(d) soil and land-surface deposits,

(e) hydrothermal deposits, and

(f) marine sedimentary deposits.

These zeolite deposits are all the result of the alteration of volcanic glass into zeolite minerals. This alteration process occurs from the chemical reaction between saline or alkaline surface, ground or hydrothermal waters and the volcanic glass in the tuff, vitrophyre or sediment. The two most important types of zeolite deposits are formed in either the closed or the open hydrologic systems. Zeolite deposits of the closed-system type result from the reaction of volcanic glass with saline–alkaline water in a closed hydrologic system. These zeolite deposits are usually interbedded with lake sediments. Open-system zeolite deposits result from the reaction of volcanic glass with subsurface meteoric water. These zeolite deposits are usually intercalated with lava flows and tuffaceous rocks in thick volcanic piles. In Japan, however, most of the zeolite deposits are of a third type known as burial diagenetic deposits. These deposits are formed at elevated temperatures while buried within great thicknesses of volcanic sediments.

4. *Resources*

Over 1000 occurrences of zeolite minerals in volcanic and sedimentary rocks of volcanic origin have been reported from over 40 countries. More than 300 of these occurrences are in the USA. Nevertheless, large deposits of high-purity zeolite minerals possessing the physical and chemical properties required for specific industrial applications are uncommon. The quantity of zeolite resources occurring in near-surface deposits in the Basin and Range Province of the western USA is estimated to exceed 200 million tonnes of the minerals clinoptilolite, chabazite, erionite, mordenite and phillipsite. Though this is a conservative estimate of the zeolite resource available, the known reserves of high-purity zeolite minerals covered by shallow overburden are much smaller. It is impossible to make a meaningful estimate of the worldwide zeolite resources other than to conclude that the resources available are many times larger than those in the USA.

5. *Major Deposits*

Deposits of zeolite minerals are mined in Bulgaria, China, Cuba, Hungary, Italy, Japan, Korea, South Africa, Yugoslavia, the USSR and the USA. These countries account for virtually all of the known world production of natural zeolites. In Bulgaria, clinoptilolite and mordenite are produced in the northeastern part of the Rhodopes Mountains. These extensive zeolite deposits occur in Oligocene-age marine sediments of volcanic origin. In Hungary, large deposits of clinoptilolite and mordenite occur in vitroclastic tuff units within the Miocene-age Samarian formation which forms the core of the Tokaji Mountains. Clinoptilolite and mordenite are produced in Japan from deposits in Miocene-age sediments of volcanic origin on the islands of Honshu and Hokkaido. In the USA, the most important zeolite deposit is near Bowie in southeastern Arizona, where high-purity chabazite is produced from a selectively mined zeolitic tuff bed which ranges from 10 to 20 cm thick. The other productive zeolite deposits in the USA, ranked in approximate order of tonnage produced, are listed in Table 1.

6. *Production*

Zeolite minerals are produced from deposits by selective open-pit or strip mining methods. The mined product is then crushed and screened to the desired size. Zeolite minerals, usually clinoptilolite, used in ion-exchange applications are sold in bag or bulk as screened products, usually in the −10 to +50 mesh (2000–297 μm) size range. In Hungary, catalysts are produced from clinoptilolite and mordenite. The zeolites are ground to a size of 1000–1600 μm. These granules are modified by ion exchange with either ammonia or hydrogen ions, depending on the specific

Table 1
Principal zeolite deposits in the USA

Location	Zeolite produced
Chrisman Hill, Idaho	clinoptilolite
Succor Creek, Oregon	clinoptilolite
Ash Meadows, California	clinoptilolite
Mud Hills, California	clinoptilolite
Tilden, Texas	clinoptilolite
Washakie Basin, Wyoming	clinoptilolite
Eastgate, Nevada	mordenite
Union Pass, Arizona	mordenite
Buckhorn, New Mexico	clinoptilolite
Castle Creek, Idaho	clinoptilolite
Hector, California	clinoptilolite
Jersey Valley, Nevada	erionite
Pine Valley, Nevada	erionite

catalyst application. Zeolite minerals such as chabazite and mordenite used as specialty adsorbent, ion-exchange or catalyst products are ground to −200 mesh (74 μm) then mixed with water and a clay binder, usually attapulgite, plus wetting agents and lubricants. The wet zeolite mixture is formed into extrudates and beads which are rolled to improve attrition resistance and then dried and activated by heating for an hour at 427 °C. The activated extrudates and beads are sold in sealed drums.

Reliable production statistics for natural zeolites are unavailable. In 1986 the world production of zeolite minerals was estimated to be about 300 000 t valued at US$33 million. Annual production in the USA is estimated to be 10 000 t. The most important zeolite deposit in the USA near Bowie, Arizona since 1961 produced 12 000 t of high-purity chabazite valued at US$40 million. The highest annual production rate of 2100 t was attained in both 1978 and 1979. The largest producer at the deposit is the Linde Division of Union Carbide Corporation. The other producers include The Norton Company, NRG Inc., GSA Resources Inc., and Letcher and Associates. The other important producers of zeolites in the USA include Teague Mineral Products which produces clinoptilolite from the Chrisman Hill, Idaho and Succor Creek, Oregon deposits. East West Minerals Inc. produces clinoptilolite from the Ash Meadows, California deposit and mordenite from the Eastgate, Nevada deposit. Tenneco Minerals produces clinoptilolite from the Mud Hills deposit in California and Zeotech Corporation produced clinoptilolite from a deposit near Tilden, Texas. About 40–50 kt of clinoptilolite are produced in Hungary annually by National Ore and Mineral Mines from the Hegyalja Works in Mad. In Japan, 14 companies produce zeolites; however, only two companies produce more than 10 000 t per year. These are Zeiklite Chemical Mining Company Ltd, which produces clinoptilolite in the Itaya area of Yamagata Prefecture and Sun Zeolite Industries Company, Ltd, which produces clinoptilolite from a deposit at Futasui-Machi, Yamamota-Gun in Akita Prefecture. Other Japanese producers include Shin-shin Development Company Ltd and Asahi Kasei Kogyo Ltd, which both produce mordenite. In Italy, Euro-Zeolite, a subsidiary of Eurimport Srl, produces natural zeolites from a deposit in southern Italy. Burgaria is also an important producer of clinoptilolite. Smaller quantities of zeolites are produced in China, Cuba, Korea, South Africa, Yugoslavia and the USSR.

7. *Uses*

Worldwide, the building and construction industry is the largest consumer of natural zeolite minerals. The principal uses include lightweight aggregates, pozzolans and building stone. The paper industry in Japan uses zeolites in fillers and coatings. Zeolites are also used in agriculture as soil conditioners and animal-feed supplements. Collectively, these uses consume between 80 and 90% of production worldwide. The remaining 10–20% of production is consumed in high-unit-value industrial applications which utilize the ion-exchange, adsorption and catalytic properties of natural zeolites. The ion-exchange properties of zeolite minerals are utilized principally in the conditioning of process waters and the treatment of municipal, agricultural and industrial wastewaters and effluents. Both of the advanced wastewater treatment (AWT) plants which treat municipal wastewater at Lake Tahoe, California, and Manassas, Virginia, use clinoptilolite in the CH_2M Hill process to denitrify or remove the ammonia nitrogen from sewage effluent. A smaller plant which also uses clinoptilolite in a similar process to denitrify wastewater is in operation at Rosemount, Minnesota. Phillipsite may have applications in wastewater treatment, particularly if the effluent contains a high percentage of competing cations such as magnesium.

In Japan, clinoptilolite has been used successfully for years in the treatment of poultry, swine and cattle manure. Several proprietary processes use clinoptilolite to convert the manure into a dry, odor-free organic fertilizer: in these processes, the clinoptilolite is used to remove the ammonia nitrogen and water from the manure.

In Hungary, both clinoptilolite and mordenite are used as fodder additives for swine, poultry and cattle. Other zeolite products have been developed specifically for use in intensive fish-breeding to remove ammonia nitrogen from the water. ATEC Inc. in Riverton, Wyoming, manufactures moving-bed ion-exchange equipment which uses clinoptilolite to remove ammonia nitrogen from tanks at large fish hatcheries. The process allows recycling of the hatchery water which in turn stabilizes water temperatures and increases both the number of fish hatched and their survival rate.

Clinoptilolite and chabazite have been used to remove radioactive cesium (^{137}Cs) and strontium (^{90}Sr) from high-level radioactive processing-wastes at the Hanford Atomic Energy Project. Additionally, several million gallons of low-level ^{137}Cs have been processed through zeolite ion exchangers at Hanford since about 1960. Chabazite from the Bowie deposit sold by the Linde Division of Union Carbide Corporation was used in the submerged demineralizer system in the containment building in the crippled Three Mile Island Nuclear Power Station Unit 2 to recover ^{137}Cs and ^{90}Sr from high-activity-level water expelled from the primary coolant system. Clinoptilolite from Tenneco Minerals Mud Hills deposit is used in British Nuclear Fuels Ltd's Sellafield reprocessing plant in the UK to reduce ^{137}Cs and ^{90}Sr discharges from the plant into the Irish Sea. Several similar processes utilizing zeolite minerals for removal and fixation of long-lived fission products have been used in Bulgaria, Canada, France, Germany, the UK, Hungary, Japan, Mexico and the USSR. Other potential applications which utilize the ion-exchange properties of zeolite minerals include the removal of heavy metals from metallurgical effluents, and the removal of ammonia nitrogen from fish hatchery tanks.

The adsorption properties of zeolite minerals are utilized to remove water and carbon dioxide from methane, particularly in the presence of high concentrations of hydrogen sulfide; the treated methane must contain less than 4 ppm hydrogen sulfide. The Bowie chabazite is stable at a pH as low as 2.5, which makes it suitable for removing hydrogen chloride from hydrogen streams, water from chlorine and sulfur dioxide from stack gases. Zeolite minerals have been used to remove water from ethanol in the production of gasohol. Chabazite from Bowie is used in a gas treatment plant in the Salt Lake field near Los Angeles, California. The methane casing-head gas, which contains 16–20% CO_2, has a heating value of 850–950 Btu ft^{-3} (approximately 32–35.5 MJ m^{-3}). After passing through a column of activated chabazite which removes the CO_2, the heating value of the treated gas is increased to over 1100 Btu ft^{-3} (approximately 41 MJ m^{-3}). A similar methane purification plant using the NRG Nufuel process is in operation at the Palos Verde landfill near Los Angeles. Chabazite is used to remove water, oil and other contaminants from compressed air and freon. The Linde Division of Union Carbide Corporation developed its new product Absorbite, which is based on Bowie chabazite, to adsorb the highly corrosive liquid which collects in automotive mufflers. The addition of Absorbite to a muffler reduces corrosion and significantly lengthens the service life of the muffler. Chabazite is also being used to remove water and CO_2 from liquified natural gas (methane) and liquified petroleum gases (propane and butane). Mordenite is used in pressure-swing adsorption units to produce enriched air containing about 70% oxygen. Nitrogen can also be produced from similar pressure-swing adsorption units. Mordenite from the Idado mine in Akita Prefecture, Japan, is used in the oxygen plant of the Takaokakogyo Company at Toyohashi City, Japan, and several other installations. Both clinoptilolite and chabazite have been used successfully in solar-powered refrigerators. Other potential applications utilizing their adsorption properties include the removal of sulfur dioxide, nitrogen oxides and mercury from stack gases. Both mordenite and erionite have been used in hydrocracking applications. The hydrogen form of mordenite has been used as a catalyst in nitrogen oxide reduction processes.

The zeolite minerals clinoptilolite and mordenite are used as catalysts in Hungary. The impurities such as iron oxides and quartz do not appear to interfere with xylene isomerization; consequently, the natural minerals are satisfactory even though their capacities are somewhat lower than synthetic zeolites.

8. Health Hazards

The zeolite mineral erionite may be associated with mesothelioma, a rare form of lung cancer, previously reported only in workers exposed to asbestos particles. An investigation of an area in central Turkey underlain by zeolitic tuffs, where the disease is prevalent, has failed to establish a clear-cut correlation between the occurrence of erionite or any other zeolite and the incidence of malignant disease. However, as a safety precaution, an approved respirator should be used by workers exposed to zeolite or other mineral dusts.

See also: Filters, Sorbents and Ion Exchangers; Filters, Sorbents and Ion Exchangers: Clay Minerals; Filters, Sorbents and Ion Exchangers: Siliceous Minerals

Bibliography

Breck D W, Mumpton F A, Olson R H, Sheppard R A 1983 Zeolites. In: Lefond S J (ed.) 1983 *Industrial Minerals and Rocks*, 5th edn., Vol. 2. American Institute of Mining, Metallurgical and Petroleum Engineers, New York, pp. 391–431.

Drzaj B, Hocevar S, Pejovnik S (eds.) 1983 *Zeolites Synthesis, Structure, Technology and Application*, Studies in Surface Science and Catalysis Vol. 24. Elsevier, Amsterdam

Gottardi G, Galli E 1985 *Natural Zeolites*. Springer, Berlin

Sand L B, Mumpton F A (eds.) *Natural Zeolites, Occurrence, Properties, Use*. Pergamon, Oxford

Sheppard R A 1973 Zeolites in sedimentary rocks. In: Brobst D A, Pratt W P (eds.) 1973 *United States Mineral Resources*, US Geological Survey Professional Paper 820. US Government Printing Office, Washington, DC, pp. 689–95

T. H. Eyde
[GSA Resources, Cortaro, Arizona, USA]

Foundry Sands: Classification and Sources

Sands are defined as granular particles of diameter 0.05–2 mm that result from the disintegration or crushing of rocks. The most widely used foundry sand, silica sand, is composed mainly of quartz (SiO_2) grains; however, sands based on zircon, chromite, olivine and ground ceramic minerals are also employed in the production of castings (see *Foundry Sands: Silica Sand and Nonsilica Minerals*). In foundry practice, a wide range of sands is used which vary in purity, structure, refractoriness and also in grain size, shape and distribution.

1. Classification

Foundry molding sands are generally classified into two broad categories: naturally bonded sands and unbonded sands. The naturally bonded molding sands are water- or wind-deposited sands which are naturally combined with clay minerals. They occur with a wide range of grain fineness and clay contents depending on their source and origin. The processing of this type of sand consists primarily of screening to remove oversize particles, roots and other deleterious material. For some sands, this treatment is all that is required to meet specifications. Other sands may be dried to approximately 6% moisture and then passed through mullers, cage mills or dry pans to mill the sand and distribute the natural bond. Sands from different pits or different strata in the same pit, having different compositions, are often blended with or without clay additions to produce a range of sands having consistent uniform properties. Sand with a wide range of properties can be produced; specifications are negotiated between producer and consumer.

Many unbonded sands are referred to as "washed-and-dried" silica sands. These sands are prepared for use in casting production by being bonded with the required amount of binders, additives and "temper water." The total sand mixture is prepared in a sand muller in order to gain uniform distribution and dispersion of the clay and additives around the individual sand grains.

2. Mode of Occurrence

Foundry-sand deposits occur naturally as a result of the geological sedimentary processes of weathering, erosion, transportation by wind, water and ice, and deposition. The composition of the deposits depends on the nature of the materials eroded and the manner in which they were deposited. For example, some silica sands were deposited by rivers, whereas others were deposited along coasts of ancient seas.

2.1 Consolidated Deposits

In many places, individual grains of sand have been cemented into a consolidated rock known as sandstone, and these deposits have been covered by other layers of rock. The normal processes of uplift, faulting and erosion eventually cause these sandstones to be exposed at the surface. Where erosion of the material covering the sandstone has been extensive, only a thin layer of overburden will remain, making the sandstone much easier to mine.

2.2 Unconsolidated Deposits

Unconsolidated deposits with a low clay content are generally referred to as "bank sands," although they are usually subjected to considerable processing and blending before shipment. Bank sands are a product of the disintegration of sandstone or other rocks by weathering and erosion. These sands were transported by streams or blown by wind over vast areas and heaped into small banks. Bank sands vary in purity, depending on the foreign materials and minerals with which they are mixed. In many areas they are of high purity and suitable for foundry use. In the processing of bank sands, after stripping of overburden, the sand is hauled to the plant. A front-end loader dumps the sand into a hopper, from where it is conveyed to a rotary screen with 13 mm openings to remove roots. Sand for use as a system sand addition and as a synthetic sand base is dried and shipped without further processing. Where a cleaner and more consistent product is required, the screened sand is washed. The sand to be washed is dropped into a surge bin to assure a uniform feed and then conveyed to a tank where it is mixed with water before being pumped to a scalping screen where fine roots and organic materials are removed. The sand is then pumped to a deslimer to remove the clay and silt. Fresh water is added, and the sand slurry is either pumped to a second deslimer and then to ground storage, or it may be fed to banks of screens to separate it into fine and coarse particle sizes. Each sand grade is collected in a sump, mixed with water, pumped to a deslimer and dewatering cyclone, and piled for drainage and storage. The sand is then dried, cooled and passed over a final scalping screen before being conveyed to storage bins ready for shipment.

Unconsolidated deposits composed of sands produced by the erosion of rock along the shores of lakes are known as "lake sands." (This term also includes dune sands formed from surface sand by wind action.) In a typical lake-sand processing operation, the overburden is first removed and then a front-end loader is used to mine the sand and feed a hopper which discharges to a portable screen in order to remove deleterious material. Conveyor belts transport the sand to the plant site for further processing, and the hopper and screen can be moved along these as required.

During the natural deposition process, classification by size, composition and specific gravity can take place. In general, the degree of purity of a deposit is dependent on the purity of the parent material and the history of its deposition. Based on their origin, sands vary in grain size, shape, composition, surface texture and distribution of sizes. These physical properties, in addition to chemical composition, sintering point and expansion characteristics, are important considerations when using sands as the base molding or core aggregate in metal casting.

See also: Foundry Sands: Silica Sand and Nonsilica Minerals

Bibliography

Ammen C W 1954 Preparation of synthetic sand. *Foundry* 82: 202–8

Goettman F P 1947 Preparation of foundry sands for market. *Trans. Am. Foundrymen's Soc.* 55: 490–92

Goettman F P 1975 Production of sands for the foundry industry. *Trans. Am. Foundrymen's Soc.* 83: 15–24

Hardy T H 1954 Natural sands of mid-America. *Foundry* 82: 114–15

Wilborg H E, Henderson G V 1983 Foundry sand. In: Lefond S J (ed.) 1983 *Industrial Minerals and Rocks*, 5th edn., Vol. 1. American Institute of Mining, Metallurgical, and Petroleum Engineers, New York, pp. 271–78

Winter W P 1960 Molding sand. *Foundry* 8: 103–5

E. L. Kotzin
[American Foundrymen's Society, Des Plaines, Illinois, USA]

Foundry Sands: Silica Sand and Nonsilica Minerals

Refractory materials to make molds and cores have been very important to foundrymen since metal casting began hundreds of years ago. Early requirements were simple, but as core and mold sand technology has become more sophisticated and binder technology continues to be developed, the physical properties of foundry sands have become more critical.

1. Silica Sand

Over the years, the most commonly used foundry sand has been silica sand. This is not surprising because it is the most abundant and one of the most easily mined minerals at the earth's surface. Silica sand and other foundry materials are produced from rocks which formed under a variety of geologic conditions and environments. Silica is the common term applied to SiO_2 in the mineral form of quartz. In addition to foundry sands, quartz is used in a number of industrial applications, including abrasives, aggregates, sand blasting abrasives, fillers, filters, well-drilling materials, electronic and optical minerals and refractories (see *Abrasives: Siliceous Minerals*; *Sand and Gravel*; *Fillers and Coatings: Siliceous Minerals*; *Filters, Sorbents and Ion Exchangers: Siliceous Minerals*; *Well-Drilling Materials: Industrial Minerals*; *Electronic and Optical Minerals*).

Although silica sand has the advantages of abundance, ease of bonding with organic or inorganic binders, low cost and ability to be reclaimed for reuse by wet, dry or thermal methods, it also possesses certain disadvantages when used in the production of metal castings. Undoubtedly the major disadvantage of silica sand is its high thermal expansion. It is this expansion behavior which causes casting quality problems (e.g., rat-tails, buckles and scabs), as well as contributing to other expansion-type defects. Figure 1 illustrates the typical thermal expansion of various silica materials, and depicts α-quartz—a polymorph of quartz—expanding at a constant rate until a temperature of nearly 600 °C is reached. As the temperature increases above this point, a sudden expansion takes place due to the change in crystal form from α-quartz to β-quartz. This high thermal expansion requires carefully controlled additions of cushioning materials (e.g., cellulose additives) to minimize the deformation and rupture of mold surfaces in contact with molten metal. In addition, silica sand is unable to resist metal penetration and reaction when in contact with casting surfaces where there are re-entrant angles in hot-spot areas of large iron and steel castings, or when the steel contains high amounts of manganese (e.g., Hadfield's austenitic manganese steel castings) which wet and attack the silica sand mold surfaces. In addition, considerations of occupational health place increased stress on silica sand's potential as a cause of silicosis (see *Silicon Dioxide Hazards*).

The use of silica sands for metal casting has been the subject of studies by many investigators. However, the continued demands of casting purchasers for improved cast surfaces and closer dimensional tolerances have also emphasized some of the physical and chemical limitations of silica sands and their inability to provide all of the properties required to meet these quality levels.

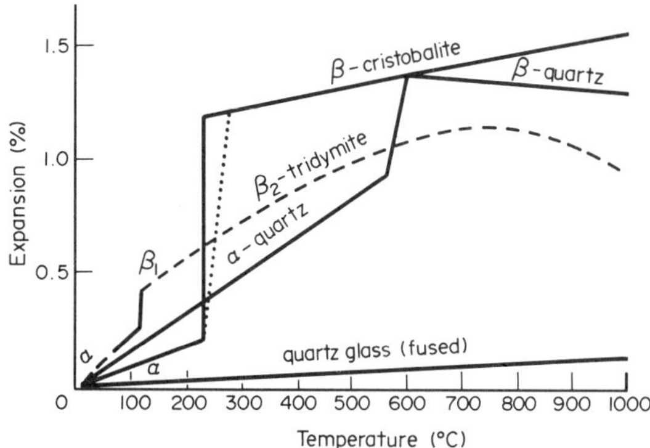

Figure 1
Thermal expansion properties of silica materials

2. Nonsilica Minerals

A number of specialty sands of material other than silica possess properties which make them superior to silica sand for certain applications. The major nonsilica sands, zircon, chromite and olivine, have gained wide recognition and use throughout the metal-casting world for their individual properties. Although other nonsilica molding media such as aluminum silicate sand and staurolite sand also possess unique properties, their use in casting production has been on a much smaller scale. Figure 2 shows the typical thermal expansion data for the nonsilica materials as compared with silica sand. It is readily apparent that the nonsilica aggregates have a much lower rate of thermal expansion than silica sand, and none of the nonsilica minerals undergoes the phase change which is characteristic for silica sand. These data and experience in the use of the major nonsilica minerals indicate that expansion-type defects are greatly reduced and, in many cases, eliminated. Table 1 shows some of the typical properties of the nonsilica sands in comparison with those of silica sand.

2.1 Zircon Sand

The major sources of zircon sand are found in Australia (east and west coasts), the USA (north central Florida), Sri Lanka, Malaysia and South Africa. Zircon and other heavy minerals, such as tourmaline, spinel, kyanite, sillimanite, corundum, topaz and staurolite, are recovered as by-products in mining of the titanium-bearing minerals, ilmenite, leucoxene and rutile.

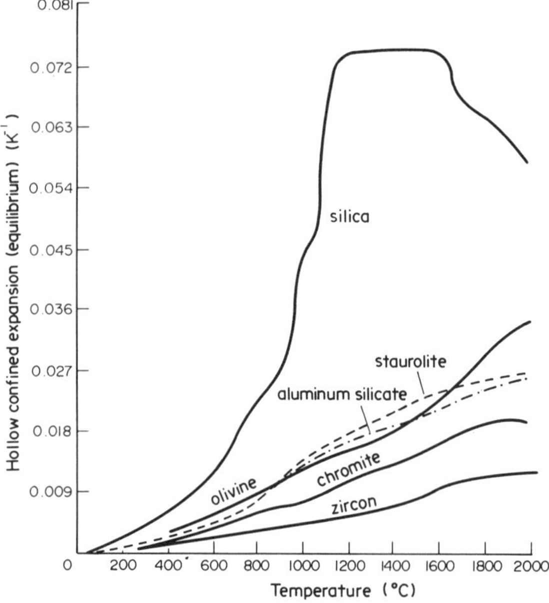

Figure 2
Comparison of thermal expansion properties of silica and nonsilica sands

After stripping of overburden, a typical sand mining operation uses a suction-type dredge capable of producing up to 1100 t of sand per hour. The sand, containing about 4% heavy minerals, is pumped to a wet mill on a floating barge. The wet mill produces a concentrate averaging 85% heavy minerals with a recovery of 80%. Three stages of gravity concentration using spiral classifiers are carried out. The concentrate is then pumped to a land-based dry mill. The scrubbed ore is dried and subjected to various magnetic and electrostatic separations to remove titanium-bearing minerals. The tailings from the titanium separation are fed to high-intensity magnets, which produce sand suitable for foundry use consisting of zircon and staurolite with small percentages of tourmaline, spinel, kyanite and sillimanite.

Zircon sand is chemically and thermally stable and possesses most of the desirable properties for a foundry sand. Its major advantages are as follows:

(a) it possesses the lowest thermal expansion of any foundry sand;

(b) it has a high thermal conductivity and bulk density, giving approximately four times the cooling rate of quartz;

(c) it is completely unwetted by molten metal;

(d) it is chemically unreactive with molten metal;

(e) it requires less binder than any of the other sands when using chemical binders;

(f) it is compatible with all known binder systems (organic or inorganic);

(g) it possesses excellent dimensional and thermal stability characteristics at elevated temperature; and

(h) it has a pH that is neutral or slightly acid.

2.2 Chromite Sand

As indicated in Table 1, the unusual properties of chromite sand lend themselves ideally to fulfilling some of the extreme thermal requirements of heavy-sectioned ferrous castings. In addition, the basic nature of chromite sand also makes its use highly desirable when producing castings of Hadfield's austenitic manganese steel.

Most of the world production of chromite comes from South Africa (Steelport and Rustenburg areas of the Transvaal), the USSR and Zimbabwe, but Albania, Turkey, Finland and the Philippines also produce small amounts (see *Chromite Resources*). Although the USA has a small amount of chromite reserves, over 92% of the material has a high iron content, which is undesirable for foundry purposes. About 60% of the world's production of chromite is consumed in metal

Table 1
Comparison of silica and nonsilica sand properties

Property	Silica	Olivine	Chromite	Zircon	Zircon/aluminum silicate	Staurolite
Origin	USA	USA (Washington, N. Carolina), Norway	Republic of South Africa	USA, Australia	USA (Florida)	USA (Florida)
Color	white–light brown	greenish gray	black	white–brown	black and white	dark brown
Hardness	6.0–7.0	6.5–7.0	5.5–7.0	7.0–7.5	6.5–7.0	6.5–7.0
Dry bulk density ($g\ cm^{-3}$)	1.36–1.60	1.60–2.00	2.48–2.64	2.56–2.96	2.48–2.69	2.29–2.34
Specific gravity	2.2–2.6	3.2–3.6	4.3–4.5	4.4–4.7	3.2–4.0	3.1–3.8
Grain shape	angular/rounded	angular	angular	rounded/angular	rounded	rounded
Thermal expansion	0.018	0.0083	0.005	0.003	0.005[b]	0.007[b]
Apparent heat transfer	average	low	very high	high	high	high
Fusion point (°C)	1427–1760	1538–1760	1760–1982	2038–2204	1815–1982	1371–1538
High-temperature reaction	acid	basic	basic	acid	slightly acid	slightly acid
Wettability with molten metal	easily	not generally	resistant	resistant	resistant	resistant
Chemical reaction	acid-neutral	basic	neutral-basic	acid-neutral	neutral	neutral
Grain distribution[a]	2–5 screens	3–4 screens	4–5 screens	2–3 screens	3 screens	3–4 screens
AFS grain fineness number ranges	25–180	40–160	50–90	95–160	≈80	≈70

[a] ⩾10% retained on USA standard sieve [b] Clay-bonded sand mixture

production and processing, 20% in refractories, 12% in chemicals and 8% in foundry sand.

Chromite deposits are associated with the basic igneous peridotites and pyroxenes, or with serpentine resulting from the metamorphic alteration of these rocks. Most chromite is mined by underground methods. The ore is drilled, blasted, loaded onto conveyors or mine cars, and conveyed to the processing plant. Hand sorting and/or mechanical screening is used to separate lumps and chips from the sand, followed in some cases by bar or rod milling to reduce the sand to American Foundrymen's Society grain fineness number (AFS GFN) 80 and release the gangue minerals (primarily serpentine and talc). A natural grain size of approximately AFS GNF 55 is generally preferred for metal-casting purposes. The chromite sand is further beneficiated by the use of auger-type washers to remove the slimes, and then gravity concentration using riffling tables or spiral concentrators. The sand is then stored on concrete pads to drain. At this point, the sand is assayed to ensure that it contains a minimum of 45% Cr_2O_3. A turbidity test has been developed to control the working process and assure the production of clean sand. The sand is then transported to a finishing plant where it is dried, cooled and screened to produce grades ranging from AFS GFN 55 to 80. The sand may also be further washed to below 150 ppm turbidity for certain core operations. Quality control tests include the turbidity test, acid-demand values, and screen and chemical analyses.

Chromite sand was introduced to the US foundry industry in the early 1960s, when zircon sand was in short supply. Later, in the face of rising zircon sand prices, foundrymen adopted chromite sand as a substitute for zircon sand. Chromite has been successful as a nonsilica foundry sand because it possesses excellent thermal stability and good heat diffusivity characteristics, and is not easily wetted by molten metals, is highly refractory and chemically unreactive with molten metals. One of the major disadvantages in using chromite sands is the presence of hydrous impurities which can contribute to pinhole formation and blows.

2.3 Olivine Sand

Olivine is a constituent of many basic igneous rocks, such as basalt, dunite, gabbro and peridotite, but most commercial production comes from dunite. Large deposits of olivine-bearing dunite are found in Norway, Sweden, the USSR, Austria, Zimbabwe, South Africa and the USA. Norway supplies much of the olivine sand consumed in Europe, while dunite deposits in the states of Washington and North Carolina furnish the major amounts of olivine sand for foundry use in the USA.

In a typical mining operation, rock is broken by blasting and a drop ball to produce fragments less than 1 m in diameter. Ore is passed through a jaw crusher, reduced to <15 cm and stored for feed for the processing plant. At the mill, rock is further reduced to particles <1.5 cm by jaw and gyratory crushers. It is then passed through a screw classifier to wash out clays and slimes and then to a rod mill, from which it is pumped to scalping screens. Oversized material is returned to the rod mill, and undersized material is fed to riffling tables to make a gravity separation of the olivine from the gangue, which is predominantly serpentine and talc. Riffling tables are fed a water slurry containing 25% solids. The amount of water and the tilt and reciprocating speed of the tables control the quality of the product. The olivine concentrate, representing 60% of the table feed, is pumped to a dewatering screen and then to drainage tanks. The sand is dried before screening to produce four basic grades.

Olivine sand is less stable under thermal shock than zircon or chromite sand, but since the thermal expansion of olivine is much less than silica, it has been very successful in nonferrous as well as ferrous casting production where high thermal requirements are not necessary. The basic character of olivine also makes it an ideal aggregate for the production of castings of Hadfield's austenitic manganese steel. The presence of quantities of hydrous magnesium silicates, such as serpentine, may contribute to a pinholing or pockmarking condition when the uncalcined sand is used in the production of low-carbon-steel castings. Because of mineralogical characteristics, olivine sand is less durable than other nonsilica sands.

2.4 Aluminum Silicate Sand

Aluminum silicate (Al_2OSiO_4) is trimorphous and exists in nature as the minerals andalusite, kyanite and sillimanite. Along with zircon, aluminum silicate minerals are produced as by-products of mining of titanium-rich minerals. The three minerals are thermally stable up to a temperature at which they break down to yield 88% mullite and 12% silica (cristobalite). Mullite is stable to 1810 °C and imparts the properties of high refractoriness, low thermal expansion, resistance to thermal shock, intermediate thermal conductivity and resistance to chemical erosion.

The aluminum silicate minerals are processed by means of first scrubbing and washing the sand to remove contaminants and fines to yield a rounded grain sand much like zircon sand that is associated with it. Aluminum silicate sands, being neutral, have been successfully used in the production of Hadfield's austenitic manganese steel castings as well as production of other ferrous metals. These nonsilica sands are also compatible with all known binder systems.

2.5 Staurolite Sand

Staurolite sand is a naturally occurring mineral with the general formula $(Fe, Mg)_2Al_2SiO_4O_{23}(OH)$, and it is produced as a by-product of the same heavy mineral separation which yields zircon and aluminum silicate sands. Staurolite sand consists primarily of the mineral staurolite with small amounts of tourmaline, kyanite, zircon and other silicate minerals. In a typical processing operation, staurolite and the other heavy minerals are removed from the ore body by use of a floating dredge, then concentrated in three stages of spiral separators on a floating wet mill. The heavy mineral concentrates are pumped from a wet mill to a conditioning unit where they are scrubbed with caustic to remove organic and clay coatings from the grain surfaces. After scrubbing, the minerals are rinsed and dried, and the staurolite is removed by means of electric and magnetic separation.

Like other nonsilica minerals, staurolite sand is used in casting production because of its low thermal expansion, refractoriness, durability and compatibility with the various binder systems. Although the thermal expansion of staurolite sand is slightly greater than that exhibited by zircon or chromite sands, it is approximately equal to the thermal expansion of aluminum silicate and olivine sands. Staurolite sands have been used singly as a base aggregate or blended with silica sands in the production of nonferrous castings and ferrous castings of light- to medium-section thickness, although, because of its lower melting point (≈ 1538 °C), it is not recommended for use in steel-casting production.

See also: Foundry Sands: Classification and Sources

Bibliography

American Foundrymen's Society 1978 *Mold and Core Test Handbook*, 1st edn. American Foundrymen's Society, Des Plaines, Illinois

Asanti P 1963 The wetting of quartz and olivine sand by iron at high temperature. *International Foundry Congress, Congress Papers*, Vol. 30. International Committee of Foundry Technical Associations, Zurich

Asanti P 1966 On the interface reactions of chromite, olivine and quartz sands with molten steel. *International Foundry Congress, Congress Papers*, Vol. 33. International Committee of Foundry Technical Associations, Zurich

Beckius K 1967 Chromite sand in the steel foundry. *Foundry Trade J.* 10: 562–66

Beckius K 1973 Olivine sand: Recent research and experience. *International Foundry Congress, Congress Papers*, Vol. 40. International Committee of Foundry Technical Associations, Zurich

Garnar T E Jr 1977 Mineralogy of foundry sands and its effect on performance and properties. *Trans. Am. Foundrymen's Soc.* 85: 399–416

Locke C, Briggs C W, Ashbrook R L 1954 Heat transfer of various molding materials for steel castings. *Trans. Am. Foundrymen's Soc.* 62: 589–99

Marais J J 1966 Foundry techniques using chromite sands. *Mod. Cast.* 50(4): 45–47

Middleton J M, Bownes F F 1970 Properties of chromite and its application in the foundry. *Br. Foundryman* 63: 248–60

Petro A, Flinn R A 1978 Mold–metal interactions between chromite sand and cast steel. *Trans. Am. Foundrymen's Soc.* 86: 357–64

Rassenfoss J A 1977 Mold materials for ferrous castings. *Trans. Am. Foundrymen's Soc.* 85: 583–96

Schaller G S, Snyder W S 1954 Olivine–silica molding sands. *Trans. Am. Foundrymen's Soc.* 62: 413–19

E. L. Kotzin
[American Foundrymen's Society, Des Plaines, Illinois, USA]

G

Gallium, Germanium and Indium Resources

Gallium, germanium and indium are dispersed in trace amounts in many rocks and minerals. Commercial production of these important elements, used in solid-state semiconductor electronics, is limited to minor by-product recovery in refining zinc and aluminum. Approximate world annual production ranges are (1987): gallium, 50–55 t; germanium, 80–90 t; and indium, 45–60 t (US Bureau of Mines 1988).

Bauxite, the major aluminum ore mineral, is the principal source of gallium; all three metals are recovered in smelting sphalerite (ZnS_2), the major zinc ore mineral. Caustic leaching of bauxite concentrates gallium in a sodium aluminate solution, with recovery accomplished by fractional carbonation and electrolysis. In roasting zinc sulfide, gallium and germanium are reduced with the zinc metal, which is distilled off, and the germanium is then separated from the gallium by oxidation, chloridization and fractional distillation. Indium is recovered from various smelter dusts, anode slimes, and bullion and cadmium recovery wastes.

Crustal abundance estimates ($g\,t^{-1}$) are: gallium, 18; germanium, 1.5; and indium, 0.14. None occur as native elements, and they are essential elements in only a few rare minerals. Initial discoveries in the latter part of the nineteenth century were all made in spectrographic studies of sphalerites.

Gallium is coordinated with aluminum, and is thus present in most aluminosilicate minerals, especially bauxite and clay minerals. Germanium is coordinated with silicon and tin, and is present in many silicate minerals, especially late magmatic and hydrothermal minerals including fluorine-rich pegmatites, greisens and veins. Indium is slightly coordinated with tin. All three enter the crystal lattice of sulfide minerals, especially sphalerite. Gallium and germanium occur in plant ash, and are also present in coal and petroleum ashes. No specific biological function has been reported.

Gallium is more abundant in bauxites derived from granitic and alkalic rocks, and the Ga:Al ratio increases with weathering intensity. In zinc ores, the gallium and germanium contents are higher in replacement deposits formed at lower temperatures, and the indium content is higher in deposits formed at intermediate temperatures. The Tri-State and Upper Mississippi Valley districts have been the source of US production of gallium and germanium, and the western mining districts have provided most of the US and Canadian indium.

The Tsumeb deposit in Namibia, a pipelike replacement body of copper–lead–zinc sulfides, is unique in its production of a germanite–renierite concentrate, which has been the dominant source for European refining of gallium and germanium. This deposit has been the source for most of the other reported germanium and gallium minerals from oxidized ores.

World production potential from zinc and aluminum resources appears to be adequate for projected demand. Additional resources of all three elements may be available in other smelter by-products, notably in the refining of copper and tin. Germanium and gallium may also be recoverable from combustion wastes at major power installations.

See also: Aluminum Resources; Zinc Resources.

Bibliography

Shaw D M 1957 The geochemistry of gallium, indium, thallium—A review. In: Ahrens L H, Press F, Rankama K, Runcorn S K (eds.) 1957 *Physics and Chemistry of the Earth*, Vol. 2. McGraw-Hill, New York, pp. 164–211

US Bureau of Mines 1988 *Mineral Commodity Summaries 1988*. USBM, Washington, DC, pp. 54–55, 60–61, 74–75

Weber J N (ed.) 1973 *Geochemistry of Germanium*, Benchmark Papers in Geology Vol. 11. Dowden, Hutchinson and Ross, Stroudsburg, Pennsylvania

Weeks R A 1973 Gallium, germanium, and indium. In: Brobst D A, Pratt W P (eds.) 1973 *United States Mineral Resources*, US Geological Survey Professional Paper 820. US Government Printing Office, Washington, DC, pp. 237–46

R. A. Weeks
[St. Cloud, Florida, USA; retired,
US Geological Survey]

Geochemical Distribution of the Elements

The manner of occurrence, distribution and abundance of almost all the naturally occurring chemical elements in the accessible parts of the Earth are well known. This has come about largely as a result of the researches initiated by Victor Moritz Goldschmidt, summarized in his posthumous text (Goldschmidt 1954) and further updated in the *Handbook of Geochemistry* (Wedepohl 1969–79). The latter allocates a chapter to each element or to coherent groups of elements (e.g., lanthanides) and should be consulted for details beyond the scope of this article.

Goldschmidt's work, and that of his colleagues and students, was influenced profoundly by earlier and contemporary research by F W Clarke in the USA and V I Vernadsky in the USSR. More-recent work has established with some confidence the abundances of many elements in the solar photosphere, other stellar bodies, meteorites, the Moon, other planetary bodies and interstellar gas and dust, but only terrestrial occurrences are considered here.

Following geochemical practice, abundances of the elements are given in weight percent or more generally in weight fractions. It is convenient to divide the periodic table into:

(a) major elements, in the general range 1–100%;

(b) minor elements, in the range 0.1–1%; and

(c) trace elements, at lower concentrations, often expressed as parts per million or grams per tonne, but here (Table 1) expressed for convenience in floating-point fractional notation.

Questions of analytical accuracy and precision are not considered here, but the number of digits quoted in the tables gives an indication of the confidence to be

Table 1
Element abundances in different Earth units (weight fractions)[a]

Atomic number	Symbol	Atmosphere[b]	Sea[c]	Crust[d]	Mantle[e]
1	H		0.1082	15E-4	
2	He	73E-8	72E-13		
3	Li		18E-8	22E-6	21E-7
4	Be		60E-13	28E-7	1E-7
5	B		46E-7	1E-5	33E-8
6	C	46E-5	29E-6	52E-4	
7	N	0.755	15E-6	2E-5	
8	O	0.232	0.859	0.4688	0.439
9	F		13E-7	62E-5	163E-7
10	Ne	127E-7	12E-11		
11	Na		1076E-5	213E-4	25E-4
12	Mg		1294E-6	233E-4	0.246
13	Al		2E-9	808E-4	18E-3
14	Si		29E-7	0.2686	0.208
15	P		6E-8	66E-5	105E-6
16	S		904E-6	5E-4	319E-6
17	Cl		1935E-5	18E-5	23E-6
18	Ar	13E-3	4E-9		
19	K		399E-6	255E-4	252E-6
20	Ca		412E-6	499E-4	184E-4
21	Sc		6E-13	7E-6	11E-6
22	Ti		1E-9	50E-4	15E-4
23	V		25E-10	53E-6	84E-6
24	Cr		3E-10	35E-6	31E-4
25	Mn		2E-10	53E-5	10E-4
26	Fe		2E-9	509E-4	62E-3
27	Co		5E-11	12E-6	1E-4
28	Ni		6E-10	19E-6	24E-4
29	Cu		3E-10	14E-6	1E-5
30	Zn		3E-9	71E-6	62E-6
31	Ga		3E-11	15E-6	28E-7
32	Ge		6E-11	15E-7	85E-8
33	As		3E-9	18E-7	1E-7
34	Se		9E-11	5E-8	41E-9
35	Br		67E-6	25E-7	11E-9
36	Kr	327E-8	21E-11		
37	Rb		12E-8	11E-5	86E-8
38	Sr		79E-7	316E-6	21E-6
39	Y		1E-12	21E-6	29E-7
40	Zr		3E-11	24E-5	78E-7
41	Nb		15E-12	26E-6	6E-7
42	Mo		1E-8	15E-7	
44	Ru				6E-10
45	Rh				2E-9
46	Pd				43E-10
47	Ag		8E-12	7E-8	19E-9
48	Cd		1E-10	98E-9	4E-8
49	In		1E-13	8E-9	18E-9
50	Sn		1E-11	2E-6	5E-7
51	Sb		33E-11	2E-7	25E-9
52	Te				22E-9
53	I		6E-8	5E-7	1E-8
54	Xe	403E-9	47E-12		
55	Cs		4E-10	12E-7	19E-9
56	Ba		2E-8	107E-5	49E-7
57	La		34E-13	32E-6	89E-8
58	Ce		1E-12	66E-6	225E-8
59	Pr		6E-13		19E-8
60	Nd		28E-13	26E-6	97E-8
62	Sm		4E-13	45E-7	32E-8
63	Eu		1E-13	94E-8	17E-8
64	Gd		7E-13	28E-7	42E-8
65	Tb		1E-13	48E-8	79E-9
66	Dy		9E-13	3E-6	52E-8
67	Ho		2E-13	62E-8	13E-8
68	Er		9E-13	28E-7	34E-8
69	Tm		2E-13		48E-9
70	Yb		8E-13	15E-7	33E-8
71	Lu		1E-13	23E-8	52E-9
72	Hf			58E-7	23E-8
73	Ta			2E-6	4E-8
74	W		4E-12	15E-7	31E-8
75	Re				4E-10
76	Os				51E-10
77	Ir			24E-12	36E-10
78	Pt				7E-8
79	Au		5E-12	18E-10	27E-10
80	Hg		2E-11	9E-8	8E-8
81	Tl		1E-11	75E-8	11E-9
82	Pb		3E-11	21E-6	2E-7
83	Bi		2E-11	13E-8	1E-8
90	Th		15E-13	1E-5	8E-8
92	U		33E-10	26E-7	2E-8

[a] Concentrations exceeding 10wt% are given as weight fractions; remainder are in floating-point notation (e.g., 13E-7 denotes 13×10^{-7} weight fraction) [b] Holland (1978, Chap. 6), Smith (1977, Table 2) [c] Holland (1978, Chap. 5) except H, O (Smith 1977, Table 2) [d] Heinrichs et al. (1980), Shaw (1980, Table 1), Smith (1977, Table 2) [e] Heinrichs et al. (1980), Morgan et al. (1980a, 1980b), Jagoutz et al. (1979), O'Nions et al. (1979), Shaw (1980, Table V), Smith (1977, Table 2), Taylor (1980, Table 1), Wood (1979)

attributed to the values; in some cases (Table 2), average values are taken from the original references without rounding. In general, the least confidence is to be attributed to abundance values for trace elements, but this is not universally the case: some rare elements (e.g., U) are well characterized and some commoner ones (S, Cl) are not.

Familiarity of an element is of course no guide to its abundance. Some of the rarest elements are familiar to all (Au, Hg, Pt). Some of the commonest are unknown except to scientists and engineers (Rb, V, La); such elements largely owe their unfamiliarity to their manner of occurrence as dispersed elements, forming few natural chemical compounds in which they are enriched. Many natural compounds are familiar to the public without any reference to the elements composing them (emerald, asbestos, fuller's earth).

1. Distribution of the Elements

Certain elements are abundant in the atmosphere, in natural waters and in biological materials. The principal constituents of the atmosphere and of normal salinity ocean water are reported in Table 1, omitting reference to complexes and organic compounds. Biologically major elements are few in number and are not further considered here, although many trace elements have important roles, both essential and pathological, in biological systems.

Only a minuscule fraction of the solid Earth is accessible for sampling and direct study. Geophysical measurements nevertheless indicate the Earth to be made up of roughly concentric units (Table 3).

The nature of the core has been controversial, although its high density suggests a similar composition to the nickel-bearing iron phase of meteorites. Although an inner solid core exists (radius 1217 km), the outer core is in a liquid state. Other proposals for the possible composition of the core include condensed solar matter (mostly hydrogen) and mantle silicates transformed to the metallic state at high pressure (the pressure is ~1.3 Mbar at the mantle–core interface).

Recent measurements of bulk sound velocities in a variety of materials at ultrahigh pressures, compared with seismic information from the core, indicate the average atomic number of the outer-core material to be in the range 24–26 (see Ringwood 1979, Chap. 3). Adjacent elements (atomic numbers from 23 to 29) are V, Cr, Mn, Fe, Co, Ni and Cu, of which Fe has by far the greatest abundance in stellar photospheres and meteorites and is therefore the most likely major element in the outer core. The density of the outer core is nevertheless about 11.5% lower than that of pure Fe. This and other measurements indicate that the core also contains some 5–15% of elements of lower atomic weight than Fe. The most likely candidates are S, Si and O.

Table 2
Major element composition of important rocks, expressed mostly as oxides (wt%)

	GR[a]	GD[b]	AND[c]	MORB[d]	SH[e]	SST[f]	LST[g]	UCC[h]	UM[i]
SiO_2	72.08	66.88	59.5	49.14	58.9	78.7	6.9	64.93	44.66
TiO_2	0.37	0.57	0.70	1.17	0.78	0.25	0.05	0.52	0.19
Al_2O_3	13.86	15.66	17.2	15.64	16.7	4.8	1.7	14.63	2.80
Fe_2O_3	0.86	1.33		2.64	2.8	1.1	0.98	1.36	
FeO	1.67	2.59	6.10	6.66	3.7	0.3	1.3	2.75	7.92
ΣFeO				9.14					
MnO	0.06	0.07	0.12	0.16	0.09	0.03	0.08	0.068	0.12
MgO	0.52	1.57	3.42	8.22	2.6	1.2	0.97	2.24	40.97
CaO	1.33	3.56	7.03	11.84	2.2	5.5	47.6	4.12	2.59
Na_2O	3.08	3.84	3.68	2.40	1.6	0.45	0.08	3.46	0.30
K_2O	5.46	3.07	1.60	0.20	3.6	1.3	0.57	3.10	0.12
H_2O+	0.53	0.65		0.75	5.0	1.3	0.84	0.79	
H_2O-				0.34				0.13	
P_2O_5	0.18	0.21		0.12	0.16	0.08	0.16	0.15	
S				0.04	0.24		0.11	0.06	
Cr_2O_3				0.04					0.45
CO_2					1.3	5.0	38.3	1.28	
NiO									0.31
F								0.05	
Total	100.00	100.00	99.35	99.36	99.67	100.01	99.64	99.588	100.43

[a] GR, average of 72 granites (Nockolds 1954) [b] GD, average of 137 granodiorites (Nockolds 1954) [c] AND, average of andesite (Taylor 1968) [d] MORB, average of 387 mid-ocean-ridge basalts (Wedepohl 1969) [e] SH, average of 277 shales (Wedepohl 1969) [f] SST, average of 253 sandstones (Clark in Wedepohl 1969) [g] LST, average of 93 limestones (Wedepohl 1969) [h] UCC, upper continental crust (Shaw et al. 1967) [i] UM, upper mantle, as average of 20 garnet peridotites (Carswell 1968)

Table 3
Structural makeup of the Earth, according to geophysical evidence

	Radius/ thickness (km)	Mass (10^{25} g)	Mean density (g cm^{-3})
Core	3471	192	11.0
Mantle	2880	403	4.5
Crust	20	2.5	2.8
Oceans	3.8	0.14	1.03
Whole Earth	6371	597.6	5.52

The mantle of the Earth comprises material above the liquid outer core and below the seismic discontinuity at the base of the crust. Its density and seismic velocities match rather well with those of compressed iron–magnesium silicates having a ratio MgO/(MgO + FeO) of ~0.9 and containing minor amounts of other components. In detail, the seismic structure of the mantle is complex and additional discontinuities occur at various depths: the lower mantle is taken as lying between depths of 2900 and 1000 km, overlain by a transition zone of some 600 km and the upper mantle, about 400 km thick. The composition of the mantle may also be expected to vary and there is much indirect evidence to suggest that the upper mantle may be rather heterogeneous in composition, though probably not so much so as the crust.

Geophysical evidence as to the nature of the mantle is supplemented by information from samples of it found with surface rocks, including:

(a) small xenoliths or inclusions brought to the surface:
 (i) by erupting basalitic lavas, or
 (ii) in diamondiferous kimberlite pipes; and

(b) larger masses outcropping over many km^2 and believed to represent wedges or slices of upper mantle, intruded in mountain belts (Alpine peridotites).

The principal mineral components of these rocks are olivine, orthopyroxene, clinopyroxene and in some cases garnet, spinel and other minerals. When fresh, such rocks are called peridotite; hydrous alteration converts them to serpentine. Their major elements are O, Si, Mg, Fe, Al, Ca and their minor elements are Cr, Ni, Na, Ti, Mn, K, both groups in order of decreasing abundance.

Since the mantle constitutes > 67% of the mass of the Earth, it is the most important reservoir of most of the metallic elements and many of the nonmetallic ones. It is probably the source from which the crust of the Earth formed early in its history and from which additions to the crust are still provided, in the form of volcanic magmas which generate lava and ash deposits. These magmas form by partial melting, in part of mantle rocks and in part of recycled crustal materials, carried back down into the mantle by plate tectonic processes. For these reasons it is important to attempt estimates of elemental abundances in the mantle. Such estimates are obviously of limited value, being made on sparse sampling of those surface rocks which are believed to be mantle-derived, but they are better than nothing. The values provided in Table 1 should be considered appropriate to the upper mantle, since even less is known of the lower mantle. The number of digits in these values relates to precision of analysis more than to the range in composition between mantle samples of different mineralogy.

As regards the crust of the Earth, it is evident that there is a great diversity in the rocks exposed at the surface, probably of much greater degree than within the upper mantle. This arises from the fact that crustal rocks include a variety of low-temperature products, formed by interaction of igneous rocks with the atmosphere and hydrosphere. Such sedimentary rocks may persist for long periods or, in certain cases, are subjected to alteration produced by high temperature and pressure. This alteration, or metamorphism, may also affect igneous rocks produced from magma, giving rise to metamorphic rocks. In extreme cases metamorphism may lead to partial melting, giving rise to regenerated magma of various kinds.

Among the great diversity of crustal rocks there are a number of commoner varieties. These include granite, granodiorite, andesite, basalt, shale, sandstone and limestone. The average abundances of major and minor elements in these rocks are given in Table 2, mostly expressed as oxides. In addition there is an estimate of the upper mantle and an estimate of the average surface composition of a large area of the Canadian Precambrian shield, which has been shown to be remarkably close to similar measurements for several other Precambrian shields. Such data are commonly taken as representing the composition of the continental crust, though it is acknowledged that they are of course surface estimates and do not take account of possible composition changes with depth. It is inferred that composition does change with depth for some elements, both from studies of rocks believed to represent lower crustal varieties and by comparing surface heat flow with surface abundances of U, Th and K, which generate heat by radioactive decay.

The crustal abundances of trace elements in Table 1 are representative only of the continental crust. The oceanic crust is both thinner and less variable in composition than the continental crust. Apart from a thin veneer of sediment at the surface, the bulk of the ocean crust is constituted by basaltic rocks (column headed MORB, Table 2) of fresh or altered character.

By far the major parts of both continental and oceanic crust are composed of the rocks whose composition is shown in Table 2 and by related varieties of intermediate composition. Economic mineral and fuel deposits are of negligible dimensions and have been produced by special conditions working at locally favorable situations, which resulted in abnormally high enrichment processes.

2. *Principles of Element Distribution in the Solid Earth*

The major achievement of Goldschmidt (1954) and his successors was to demonstrate that element distributions exhibit regularities based for the most part on simple chemical and crystallographic properties.

The principal rock types composing the crust and mantle are aggregates of silicate minerals, with smaller amounts of oxides, carbonates, phosphates and other compounds. The major rock-forming silicate minerals are listed in Table 4, with their general formulae. X-ray diffraction and crystal chemistry have demonstrated that minerals of this kind form extensive solid-solution series and are largely indifferent to the chemical nature of the cation components, so long as these ions are of the right size for a particular crystallographic site and provided that charge differences can be compensated so that the structure is electrically neutral. The right size is determined by the number of coordinating oxygen ions, but a purely ionic model is too simple and has to be modified to some degree for bonding polarization effects.

This behavior explains the main principles of cationic element distribution. The bulk composition of a particular chemical system which crystallizes to form an igneous, sedimentary or metamorphic rock determines the distribution of the elements within the rock. The particular mineral phases to crystallize are determined by the major elements present and by the phase relations appropriate to the system, but external factors largely control the crystallization process (equilibrium or fractionation, partial or complete, for example). The trace elements (or non-structure-determining elements) are picked up and held in appropriate crystal sites under the thermodynamic and kinetic control of solution chemistry, according to ionic size (Table 5). Since fractional crystallization is common in natural processes, minerals formed at different stages (early, middle, residual) may differ in kind or, for a given species, in chemical composition.

An example of this process is the typical crystallization of basaltic magma, widely documented and illustrated by the typical trace element abundances in Table 6. The early products of fractional crystallization are peridotitic, consisting largely of olivine, pyroxenes and perhaps some plagioclase, removing preferentially the major elements Mg and Ca. Crystal sites in olivine and pyroxenes favor sixfold coordinated cations such as Cr, Ni and Co, which therefore accumulate in such rocks. The main phase of crystallization typically shows pyroxenes and plagioclase precipitating, depleting the magma in Mg, Ca and Fe but enriching it in Na, K and Al. Trace components such as the ferrides continue to be accepted in the minerals, but at lower concentrations.

The later stages of crystallization are characterized by sodic plagioclase, alkali feldspar, amphibole, micas and quartz. These minerals, except the last, have larger cation sites and so accommodate the larger cations such as La, Ba, Rb and Cs which have been excluded (by their size) from the earlier mineral phases. In addition, small amounts of accessory minerals rich in some minor or trace element (e.g., zircon, $ZrSiO_4$) also now begin to form. Throughout the crystallization process Fe–Ti oxides and phosphate may also crystallize; the former picks up other ferrides, while the latter takes in the rare earth elements and Sr.

Table 4
Structural formulae of the common silicates (generalized and simplified)[a]

Olivine	Y_2ZO_4	Z mostly Si
Clinopyroxene	XYZ_2O_6	
Rhombic pyroxene	$Y_2Z_2O_6$	Z mostly Si
Clinoamphibole	$(OH,F)_2X_{2-3}Y_5(Z_4O_{11})_2$	
Mica	$(OH,F)_2Y_{2-3}Z_4O_{10}\cdot W$	W mostly K
Chlorite	$(OH)_8Y_6Z_4O_{10}$	
Chloritoid	$(OH)_4Y_2Z_4O_{10}$	Y mostly Fe, Al
Staurolite	$(OH)Y_{11}Z_4O_{23}$	Y approximately Fe_2Al_9
Garnet	$X_3Y_2Z_3O_{12}$	garnet is unusual in that Fe, Mg and Mn occur in octahedral (X) sites
Cordierite	$Y_2Z'_3Z''_6O_{18}$	Z is Al; Z″ is mainly Si
Alkali feldspar	$(X,W)Z_4O_8$	Z_4 usually $AlSi_3$
Plagioclase feldspar	XZ_4O_8	mostly a mixture of $NaAlSi_3O_8$ and $CaAl_2Si_2O_8$
Epidote	$(OH,F)X_2Y_3Z_3O_{12}$	Y is normally restricted to Al and Fe
Nepheline	$WX_3Z_8O_{16}$	W mostly K; X mostly Na
Scapolite	$W_4Z_{12}O_{24}(Cl_2,CO_3,SO_4)$	
Sillimanite, kyanite, andalusite	Y_2ZO_5	Y is all Al except in andalusite, which may be ferrian

[a] W, X, Y, Z denote cations in coordination >8, 8, 6, 4, respectively

Table 5
Some ionic radii in groups of similar size

Symbol	Coordination	Size (Å)	Ion[a]
Z	4	0.55–0.64	Si^{4+}, (Al^{3+}), (Ti^{4+}), Ge^{4+}
Y	6	0.65–0.90	Mg^{2+}, Fe^{2+}, Fe^{3+}, Al^{3+}, Li^{+}, Ti^{4+}, Mn^{3+}, Ni^{2+}, Co^{2+}, Cr^{3+}, V^{3+}, Zn^{2+}, Zr^{4+}, Sc^{3+}, Sn^{4+}, Ga^{3+}, Mo^{4+}, Nb^{5+}, Ta^{5+}, (Mn^{2+})
X	8	0.90–1.22	Ca^{2+}, Na^{+}, Mn^{2+}, (K^{-}), (Sr^{2+}), (Pb^{2+}), La^{3+}, Y^{3+}, $Ce^{3+} \rightarrow Lu^{3+}$
W	8–12	1.20–1.65	K^{+}, (Na^{+}), Rb^{+}, Cs^{+}, Ba^{2+}, Sr^{2+}, Pb^{2+}

[a] Elements in brackets are transitional and may be found in two coordinations

Table 6
Typical trace element[a] abundances (ppm) in the products of igneous fractional crystallization

	Peridotite	Basalt	Granite
Cr	3100	120	10
Ni	2400	50	5
Co	150	50	5
Zr	8	100	300
U	0.1	1	5
La	3	15	100
Ba	0	680	1300
Rb	1	30	250

[a] The elements are arranged in sequence of increasing size from top to bottom

The remark made earlier, that the chemical nature of a cation plays no role in the process just described, is of course an oversimplification. When other anions are present in the system, such as S, certain metals preferentially bond with them rather than with O. Some transition metals, such as Fe, Ni, Co, Cu, Zn and Pb, characteristically form sulfides, and such processes dominate their occurrence as economic deposits (except for Fe). Both in these cases and in ordinary magmatic crystallization, the redox potential of the system greatly affects the behavior of polyvalent metals.

The distribution of elements may also follow different patterns in aqueous environments, in minerals formed from the sea, from aqueous alteration of preexisting rocks, from hot springs and from hydrothermal waters percolating through rock pores and fissures. In addition to redox potential, pH is of great importance here, and in sedimentation processes so also is adsorption of metals (or their complexes) on fine-grained particles. In products of chemical sedimentation, such as carbonates, clays, phosphorites, bauxites and Fe oxides, these minerals also take up trace cations of appropriate size and charge to substitute for major cations, just as in silicates.

In general, therefore, the distribution of elements in rocks is controlled by the principles:

(a) that major elements determine the mineral structures that can form; and

(b) that every natural mineral is impure, but the impurities are determined by thermodynamic properties, kinetic limitations and availability of the trace impurities.

3. Economic Deposits

It is appropriate to consider the relationship between the crustal abundance of an element and the concentration in a host rock at which the rock would be commercially exploitable as an ore deposit. For certain metals (Au, Hg, Cd, Bi, Ag, Sb, Mo, As, Pb, Cu, Zn, P, K, Fe), it has been shown by McKelvey (1974) that the domestic reserves in the USA in tonnes are in fact in the range 10^9–10^{10} times the percentage abundance in the crust.

This apparent relationship suggests, however, a more direct expression of economic value than in fact exists, and suggests also that the supply of every element is virtually unlimited, so that if demand increases it can always be met by exploiting a lower-grade source. Such a cornucopian view of resources has been subjected to close analysis by Cloud (1975), who has identified the flaws inherent in this view of mineral wealth and has shown how energy expenditures, population pressure and social attitudes affect the expression of "need" for raw materials, and also that sources are not in fact unlimited in practical terms.

It is consequently not easy to indicate how much of a mineral resource is necessary to make extraction worthwhile. First of all the word "worthwhile" relates to the need of a particular society at a particular time. For example, uranium was long a valueless by-product from the extraction of radium. Secondly, the costs of extraction are closely related to geography, energy availability, transportation costs, political stability and so forth, as well as to technical questions of metallurgical treatment. In Arctic Canada, for example, a lead–zinc mine must be of higher grade, or larger in tonnage, than one close to energy and markets. Thirdly, the chemical and geological nature of a metallic deposit influence its value. Some metals occur in minerals of which they are major components (e.g., Pb or Zn in simple sulfides), whereas others occur in problematical compounds at low concentrations even though in vast tonnages (e.g., U in phosphorites, Mo in black shales). Some elements form few or no minerals of their own (e.g., Rb, Hf, Br) and can be obtained only as by-products from other extractions.

But beyond these economic and technical considerations, the prime influence is the needs of society for a particular resource, and these needs continually wax and wane as a function of the current technology; for example, Li, Zr and B are currently in high demand, whereas Tl and In are not. Similarly, biogenic fuels such as hydrocarbons and coal are now of immense importance as energy sources but will eventually be largely replaced by fission and fusion power, if mankind survives.

It appears, therefore, inappropriate to try to relate elemental natural abundances to commercially exploitable ore grades.

See also: Mineral Resources: Definitions, Uses, Classification and Future Availability; Ore Minerals

Bibliography

Carswell D A 1968 Possible primary upper mantle peridotite in Norwegian basal gneiss. *Lithos.* 1: 322–55

Cloud P E 1975 Mineral resources today and tomorrow. In: Murdoch W W (ed.) 1975 *Environment*, 2nd edn. Sinauer Publishers, Sunderland, Massachusetts, pp. 97–120

Goldschmidt V W 1954 *Geochemistry.* Oxford University Press, Oxford

Heinrichs H, Schulz-Dobrick B, Wedepohl K H 1980 Terrestrial geochemistry of Cd, Bi, Tl, Pb, Zn and Rb. *Geochim. Cosmochim. Acta.* 44: 1519–34

Holland H D 1978 *The Chemistry of the Atmosphere and Oceans.* Wiley, New York

Jagoutz E, Palme H, Baddenhausen H, Blum K, Cendales M, Dreibus G, Spettel B, Lorenz V, Wanke H 1979 The abundances of major, minor and trace elements in the earth's mantle as derived from primitive ultramafic nodules. *Proc. 10th Lunar Planet. Sci. Conf.* Pergamon, Oxford, pp. 2031–50

McKelvey V E 1974 Mineral resource estimates and public policy. In: Utgard R O, McKenzie G D (eds) 1974 *Man's Finite Earth*, 1st edn. Burgess, Minneapolis, Minnesota, pp. 246–65

Morgan J W, Wandless G A, Petrie R K, Irving A J 1980a Composition of the earth's upper mantle: II Volatile trace elements in ultramafic xenoliths. *Proc. 11th Lunar Planet. Sci. Conf.* Pergamon, Oxford, pp. 213–34

Morgan J W, Wandless G A, Petrie R K, Irving A J 1980b Earth's upper mantle: Volatile element distribution and origin of siderophile element content. *Proc. 11th Lunar Planet. Sci. Conf.* Pergamon, Oxford, pp. 740–42

O'Nions R K, Evensen N M, Hamilton P J 1979 Geochemical modeling of mantle differentiation and crustal growth. *J. Geophys. Res.* 84: 6091–101

Nockolds S R 1954 Average chemical compositions of some igneous rocks. *Bull. Geol. Soc. Am.* 65: 1007–32

Ringwood A E 1979 *Origin of the Earth and Moon.* Springer, Heidelberg

Shaw D M 1980 Development of the early continental crust. Part III. Depletion of incompatible elements in the mantle. *Precambrian Res.* 10: 281–99

Shaw D M, Reilly G A, Muysson J R, Pattenden G E, Campbell F E 1967 The chemical composition of the Canadian precambrian shield. *Can. J. Earth Sci.* 4: 829–54

Smith J V 1977 Possible controls on the bulk composition of the earth: Implications for the origin of the earth and moon. *Proc. 8th Lunar Planet. Sci. Conf.* Pergamon, Oxford, pp. 333–69

Taylor S R 1968 Geochemistry of andesites. In: Ahrens L E (ed.) 1968 *Origin and Distribution of the Elements.* Pergamon, Oxford, pp. 559–83

Taylor S R 1980 Refractory and moderately volatile element abundances in the earth, moon and meteorites. *Proc. 11th Lunar Planet. Sci. Conf.* Pergamon, Oxford, pp. 333–48

Wedepohl K H 1969 Composition and abundance of common sedimentary rocks. *Handbook of Geochemistry*, Vol. 1. Springer, Heidelberg, pp. 250–71

Wedepohl K H (ed.) 1969–1979 *Handbook of Geochemistry.* Springer, Heidelberg

Wood D A 1979 A variety veined suboceanic upper mantle—Genetic significance for mid-ocean ridge basalts from geochemical evidence. *Geology* 7: 499–503

D. M. Shaw
[McMaster University, Hamilton, Canada]

Geochemical Exploration for Mineral Deposits

Geochemical exploration is the systematic sampling and analysis of natural materials to detect anomalous concentrations of chemical elements derived from mineral deposits. The materials commonly sampled are rocks, soils, stream sediments, vegetation and water. The chemical elements analyzed may be the ore metals or other elements that are associated with a mineral deposit. Element concentrations are plotted on maps and contoured at various intervals to delineate anomalous areas that are favorable for ore deposits. Geochemical prospecting can thus be thought of as a means of extending our ability to detect mineral deposits by using modern analytical techniques. As geochemical exploration methods are nondestructive to the environment, they are ideally suited for use in modern mineral exploration programs.

1. Geochemical Anomalies

1.1 Primary Processes and Anomalies

Primary geochemical anomalies are those formed when an ore deposit is emplaced. The ore metals or their compounds are transported in hydrothermal solutions, until changing physicochemical conditions result in the precipitation of new minerals that form the deposit. These may be deposited in veins, stock works, open fillings or replacements, or may be disseminated through large volumes of rock. Although there are many other types of ore deposits (e.g, laterites, sedimentary deposits, pegmatites and magmatic segregations), hydrothermal deposits are the most common, most widely distributed and most amenable to discovery by geochemical exploration methods.

Although an ore deposit is limited to a volume of rock from which metals can be economically extracted, lesser concentrations of the metals or associated elements are transported for considerable distances into the surrounding rocks, forming halos around the deposit. These halos or geochemical anomalies are commonly much larger than the deposit, and extend both laterally and vertically. This is illustrated by a geochemical study done around a porphyry copper deposit at Ely, Nevada, USA. The main ore metal, copper, is concentrated in and near the quartz–monzonite intrusive rock, and forms the core of the deposit. Molybdenum also occurs in the core rocks in much smaller amounts. Gold, in small amounts, is found just outside the central core, while zinc and lead extend even further out. More volatile (or mobile) elements, such as As, Ag, Sb, Hg, Tl, Te and Mn, are enriched in halos further out from the core, resulting in zonal patterns. These zones are not sharply defined, but overlap.

1.2 Secondary Processes and Anomalies

After erosional processes have removed most of the overlying rock from an ore deposit, weathering processes begin to alter its physical and chemical makeup. Ore deposits exposed at the surface are dispersed by weathering, and the constituent minerals and trace elements are transported downstream to produce anomalies that can be detected in the sediments for many kilometers. Transportation can occur either through suspension of fine grains, or through dissolution of the mineral. Thus, sampling stream sediments can effectively indicate mineralized areas upstream.

In some cases, non-ore elements associated with the ore are more mobile than the ore elements, and produce more extensive, broader anomalies, which are more useful in exploration. Geochemical anomalies in soils result from weathering and disintegration of the bedrock. Where the bedrock is mineralized, anomalous concentrations of the ore elements are commonly found in the overlying soil.

Stream or lake sediment, water and vegetation anomalies are also secondary. They are usually weaker in amplitude than bedrock anomalies, owing to dispersion and dilution. Most geochemical surveys conducted today are based on the detection of secondary anomalies.

2. Geochemical Surveys

Geochemical exploration surveys are used for both regional and local prospecting. Regional or reconnaissance prospecting is conducted to detect mineral provinces or deposits, whereas local geochemical surveys are conducted when a single deposit or mineralized area is sought and usually cover only a few square kilometers.

Stream sediments are effective for regional exploration, because the entire area of a drainage basin can be assessed with carefully selected samples. In the USSR, geochemical sampling of soil is done routinely in conjunction with other geological studies to prospect regions covering hundreds of thousands of square kilometers.

In local geochemical surveys, rock or soil samples are preferred to other media, and sample spacing may be only a few meters. The disseminated gold deposit at Cortez, Nevada, USA was discovered by sampling surface rocks, and soil sampling has been successfully used for exploration in all parts of the world.

Large disseminated deposits such as molybdenum or porphyry copper deposits may be detected by sample spacings as great as 1 km, but more commonly samples are taken at intervals of ten to a few hundred meters. When vein deposits are suspected, smaller sample intervals may be required. Traverses are usually made normal to the strike of the vein or tabular bodies, or on a grid. The sample interval can be determined by doing an orientation study using close sample spacing, and plotting the analytical data to show the extent of the geochemical halo. In general, the smaller the body of exposed ore, the shorter the sample interval required.

Rock samples have the distinct advantage of providing geochemical data on primary dispersion halos. Sampling quartz or other material in veins or fractures will show the extent of the hydrothermal "plumbing" system.

Geochemical anomalies in soils are often broader than the underlying mineralized area. Where the overburden is transported (e.g., glacial till or loess), surface soil sampling may be of limited usefulness. In Scandinavia, mineralized boulders in glacial till have been successfully used to find ore in bedrock.

Not uncommonly, the principal ore metal will have a less extensive or more subdued geochemical anomaly than other, associated trace elements. This pattern is exemplified by a geochemical survey for cobalt deposits in northern Idaho, USA. The cobalt occurred as cobaltite (CoAsS), but the ore also contained abundant pyrite. Soil samples were collected and analyzed for cobalt and arsenic. The cobalt anomalies were less useful in outlining the deposits than were the arsenic anomalies, because the acid solutions generated by the dissolution of pyrite during weathering leached the cobalt out of the surface soils, leaving the arsenic behind. In addition to cobalt, copper and zinc tend to be depleted in surface material during weathering.

Where soils are on flat terrain, geochemical anomalies will be found directly over the ore in the bedrock. Where the terrain is hilly, geochemical anomalies will be transported downslope by soil creep. Soils developed on transported overburden, such as alluvium or glacial till, may give false anomalies that reflect mineralized rock in the overburden, but not in the bedrock.

2.1 Coordination with Other Methods of Exploration

Other exploration methods are commonly used in conjunction with geochemical methods. Airborne magnetic surveys are often used in the early stages of exploration. Large tracts of land can be evaluated rapidly, and favorable target areas can be identified for later detailed ground exploration. Widely spaced geochemical sampling accomplishes the same purpose. Sophisticated techniques for analyzing satellite imagery reveal areas of subtle alteration, which may be mineralized.

2.2 Climate in Study Area

Climate influences the dispersion of trace elements and determines the type of sampling and the trace elements that are likely to be most useful in tracing mineralization. In arid climates, dispersion of trace elements will be largely by physical transport, whereas in areas of considerable precipitation these elements will be dissolved and transported in water. If a stream sediment survey is done in an arid environment, the trace elements will be largely confined to discrete mineral grains in the sediment. Orientation studies of stream sediments in arid environments have shown that a coarse fraction (e.g., 600–180 μm) is often most useful in showing anomalies.

In wet environments where chemical weathering and dispersion predominate, the fine fraction of the stream sediment (e.g., <180 μm) usually produces the best anomalies, because fine particles and clays have a greater surface area and readily adsorb metal ions or compounds from water. The adsorption of iron and manganese on fine particles results in enrichment of their oxides with concomitant adsorption of many trace elements, and frequently results in enhanced anomalies. Although these rules-of-thumb are helpful, an orientation study should be conducted to determine the best size fraction for the survey. Water sampling may also be done in these areas.

2.3 Surface Exposures

Rocks or minerals in mineralized districts or provinces are often enriched in trace elements and can be used to delineate these areas. Analyses for trace metals in specific minerals (e.g., biotite) in intrusive rocks have been used to indicate the favorability for porphyry copper provinces in regional exploration.

2.4 Planning a Survey

All geochemical surveys involve sampling and chemical analysis. Sample locations are plotted on topographic maps or on aerial photographs. In reconnaissance (regional) prospecting, stream sediments are the most widely used sample medium, because the mineral potential of large areas can be assessed with few samples. Although less used, lake-bottom sediments have the same advantage. In reconnaissance surveys with widely spaced sample sites (one sample per 5–20 km^2), about 25 samples per day may be collected by one sampler or sampling team. In local surveys as many as 100 samples per day may be collected.

Perhaps the most important aspect in planning geochemical sampling is to be consistent, as some elements are more concentrated in one rock type or soil horizon than in another. Therefore, it is necessary to sample consistently to obtain meaningful anomalies.

2.5 Rock and Minerals

Although nonconsolidated media (e.g., residual soil) may reflect primary halos, weathering removes the more soluble elements and may obscure primary zoning patterns. Selective altered or mineralized material reflects the hydrothermal plumbing system. Vein material, fracture coating and jasperoid are examples of this type of material. Individual minerals in rocks are sometimes found to contain anomalous concentrations of trace elements in the vicinity of ore deposits. In the southwestern USA porphyry copper province, biotite in ore-related intrusives contains several hundred ppm Cu or more, whereas whole-rock analysis reveals no such anomaly. Mineral analysis may be most useful in identifying large regions favorable for mineralization. Manganese and iron oxides tend to coprecipitate or adsorb many trace metals. As sample media, these take advantage of a natural process of concentration and may reveal ore-related anomalies that whole-rock analysis fails to detect.

2.6 Soil

Soil samples are used more for local than for regional exploration. Residual soils reflect the underlying bedrock, and the mineral deposits it contains. Some ore elements are readily leached from soils by the action of surface water, and therefore analyzing the soils for indicator elements is often more appropriate for locating mineral deposits. Chemical analysis shows the best horizon to sample, namely the one where the anomaly-to-background ratio is greatest. The element or elements best correlating with mineral deposits are then selected for the soil survey. Transported soils, such as those in glaciated areas, may be of little use.

2.7 Water

Water sampling is used for both regional and local exploration programs. Analysis of stream waters, like that of stream sediments, can indicate the mineral deposits of an entire drainage basin. However, trace-element concentration in water is usually very low—in the ppb range—and dilution limits the distance an anomaly may be detected downstream. Natural springs issuing from favorable stratigraphic horizons have been used to detect uranium and other metals. Well samples have been successfully used to locate molybdenum deposits in arid environments.

2.8 Vegetation

Plants are used in exploration in several ways. Essential nutrients for plants include trace elements derived from the soil and rock on which they grow. Hence, the presence of certain plant species indicates that specific trace elements occur in the soil. This characteristic has been used in the Colorado Plateaus province in the USA to prospect for selenium-rich uranium ores, by mapping the distribution of the plant *Astragalus*.

More commonly, plants are sampled and chemically analyzed for trace elements like other sample media. Widespread root systems may extend to considerable depth, and elements dissolved in ground water are taken up and deposited in the plants. Thus, geochemical anomalies in plants reflect mineralized areas in the same manner as soils, rocks or water.

Another manner in which plants are used for geochemical exploration is to examine changes in plants that are caused by abnormal concentrations of metals and other elements. Changes in color, such as yellowing of leaves or foliage deformities, can sometimes be seen in aerial photographs or satellite imagery, and can indicate mineralized areas.

2.9 Other Media

Airborne dust from minerals exposed at the surface has been collected and analyzed in South Africa, and geochemical anomalies in these particulates are found to correlate with base metal deposits. Other particulates are organic, and are given off from the surface of vegetation. These also contain anomalous concentrations of trace metals, which correlate with exposed and buried mineral deposits.

Anomalous concentrations of He, CO_2, SO_2, H_2S, COS, CH_4 and other gases are found in soil over buried deposits. Gas chromatographic or mass spectrometric techniques are used to analyze constituent gases. A soil gas survey over a massive sulfide deposit in glaciated terrain of the northern USA showed anomalous concentrations of CO_2 and hydrocarbon gases over the deposit; these anomalies were detected through as much as 60 m of transported glacial overburden.

3. Sample Analysis

Analytical methods useful in geochemical exploration must (a) have sufficient sensitivity to obtain positive values on most samples, (b) be rapid, and (c) be sufficiently reproducible to distinguish relatively small differences in trace element concentration. The methods in common use are semiquantitative, and give results that are $\pm 30\%$ of the true value. The absolute concentration of an element is usually of less importance than the differences between samples. The element or elements selected for analysis will depend on the type of mineral deposit sought, the sample media and the local weathering environment. Analyzing the samples for a single element may define a deposit, but in many cases, analyses for several elements more fully define the type and extent of the deposit. An example is the disseminated gold deposits in the Basin and Range province of the western USA, where the elements Au, As, Sb, Hg and W are characteristic.

3.1 Colorimetric Methods

Colorimetric methods were among those first used in geochemical exploration. The trace element concentration of an element is determined by observing the change in color of an organic reagent, the change being proportional to the trace element content of the sample. Colorimetric methods have been used for the following individual elements: Cu, Pb, Zn, Mo, W, V, Mn, Fe, Co, Ni, Ge, As, Se, Nb, Ag, Sn, Sb, Te, Au, Bi and U.

Some of these methods have been adapted to the analysis of stream sediment or soil without a pretreatment or sample digestion step; these are called cold-extraction methods. They require little equipment and few reagents; they are convenient and find extensive use in on-site soil or stream sediment analyses.

3.2 Emission Spectrographic Methods

These methods can be used for the simultaneous determination of 30 or more elements, and are often used when multielement analysis is sought. The limit of sensitivity is in the low ppm range for many elements, including Cu, Pb, Ag, Mn, B, Ba, Be, Bi, Co, Cr, Mo, Ni, Sc, V, Y and Zr. The advantage of emission spectrographic methods is that many elements can be determined simultaneously; however, the instruments are costly and require skilled operators. As many as 50 samples can be analyzed per day by one spectrographer. It is necessary to have a finely divided, homogeneous sample, since a very small sample (10 mg) is used for the analysis.

3.3 Atomic Absorption Methods

These methods have largely superseded colorimetric methods of analysis because of their greater sensitivity and specificity. Most atomic absorption instruments determine one element at a time, although a single sample solution can be used to determine several elements. The sample is usually digested in acid, and is introduced into a flame that produces a fine spray with large numbers of atoms in the atomic state. A second method used to atomize samples is direct evaporation of the element at elevated temperatures. These methods are termed flame and flameless atomic absorption, respectively.

The advantages of these methods are their great sensitivity, freedom from interferences and relative simplicity of operation. Instrument costs are moderate to high, but highly trained personnel are not required. They are currently the preferred methods of analysis in most geochemical laboratories.

3.4 Other Methods

In addition to the commonly applied methods given above, others used include x-ray fluorescence spectrometry, partial extractions, and several methods with limited applications. Radiation detection instruments, such as Geiger counters, are used in exploration for U and Th. Portable instruments containing radioactive sources have been used to detect Be and Ag. Specialized instruments are also used to detect volatile elements (e.g., Hg) or compounds in air or soil gas. These include gas chromatographs, mass spectrometers and mercury detectors.

4. Data Handling and Interpretation

Geochemical data are best shown by plotting the analytical data for each element on a map of the area, and drawing contours to show the geochemical anomalies. Contours are drawn on the map to enclose different ranges of element concentration, thus delineating the geochemical anomalies.

The values or ranges to be contoured are selected by examination of all of the analytical data. Statistical and graphical techniques are used to determine the range and distribution of each element. The background or threshold value for an element is the concentration that marks the dividing line between normal and anomalous geochemical values in a group of samples. This value varies from area to area and largely depends upon lithology.

Geochemical exploration programs, either local or regional, frequently involve the collection of hundreds or even thousands of samples, and analysis for several elements results in a large amount of data. Computers are being increasingly used to store and analyze the data, and even to plot the geochemical maps. The use of computers not only saves a great deal of time, but provides the geologist with powerful programs which greatly facilitate interpretation.

5. Costs of Geochemical Surveys

The costs of geochemical surveys are lower than those of many other exploration methods. Commonly the cost of sample collection is about half of the total, the other half being for analysis and interpretation of data. In countries where the cost of trained personnel is high, sophisticated machines and instruments are employed to minimize the number of skilled people utilized. In countries where the cost of manpower is low, more people are employed to conduct geochemical programs. Thus the local availability and cost of manpower and equipment should be considered in planning exploration programs.

Sample preparation and analysis can also be done with emphasis on greater use of manpower or instrumentation. The decision to build and equip a geochemical laboratory will depend on the size and extent of the exploration program and on the availability of skilled manpower and equipment. In areas where little or no prospecting has been done and mineralized outcrops exist, simple methods of sampling and analysis may be used to discover ore deposits.

6. Environmental Impact

There is an ever-increasing concern about the effect of exploration on the environment. In the past, exploration programs often involved building new roads and clearing brush from lines with a bulldozer. These practices are now being restricted in many countries. One example is new guidelines for exploration enacted by Botswana in 1977. At least 85% of the land surface in Botswana is dune sand, a surface that is particularly vulnerable to surface disturbances which accelerate erosion. The new act specifies exploration methods that may be used, and geochemical exploration is one of the accepted (specified) methods.

7. Other Uses

Although the primary use of trace element geochemical data is for mineral exploration, these data have also found application in other fields. A correlation has been noted between incidence of certain diseases in man and animals and the trace element content in foodstuffs and water. It is also becoming increasingly evident that many trace elements are essential to human health. Thus, in addition to exploration for minerals, geochemical data can be used to evaluate the effects of geology on human health.

See also: Geochemical Distribution of the Elements; Geological Exploration for Mineral Deposits; Geophysical Exploration for Mineral Deposits; Remote Sensing in the Search for Mineral Deposits

Bibliography

Beus A A, Grigorian S V 1977 *Geochemical Exploration Methods for Mineral Deposits.* Applied Publishing, Wilmette, Illinois

Brooks R R 1972 *Geobotany and Biogeochemistry in Mineral Exploration.* Harper and Row, New York

Levinson A A 1974 *Introduction to Exploration Geochemistry.* Applied Publishing, Calgary, Canada

McCarthy J H Jr, Reimer G M 1986 Advances in soil gas geochemical exploration for natural resources: Some current examples and practices. *JGR, J. Geophys. Res.* 91: 12 327–8.

Rose A W, Hawkes H E, Webb J S 1979 *Geochemistry in Mineral Exploration*, 2nd edn. Academic Press, New York

Siegel F R 1974 *Applied Geochemistry.* Wiley-Interscience, New York

J. H. McCarthy Jr.
[US Geological Survey, Denver, Colorado, USA]

Geological Exploration for Mineral Deposits

Geological exploration is the fundamental method of searching for ore deposits. Based on geological concepts and field observations, it provides the context for supporting work in remote sensing, geophysics and geochemistry (see *Remote Sensing in the Search for Mineral Deposits*; *Geochemical Exploration for Mineral Deposits*; *Geophysical Exploration for Mineral Deposits*).

Exploration may begin with the reconnaissance of a metallogenic province, or favorable region for a particular kind of orebody, and may involve examination of a single prospect or of a few known occurrences of ore minerals. Exploration may also be initiated within a mining district where the characteristics of recently discovered or nearly depleted orebodies provide incentives and geological guidelines for finding additional ore. Regardless of the initial step, an exploration effort is predicated on a conceptual model or image of the orebody that is being sought.

Based on genetic and environmental characteristics of well-studied orebodies in geologically similar terrain, an exploration model comprises an expected association of minerals in a particular setting of rock type, stratigraphy and geological structure. Relevant geophysical and geochemical patterns or signatures are expected; these are part of the exploration model, and they furnish means for recognizing the presence of ore mineralization.

In field exploration work, an array of signals or observations is obtained that can be related either to ordinary geological terrain or to an anomaly—a condition that may possibly be associated with an orebody. An anomaly or a group of anomalies that seem to fit a particular exploration model may derive from an orebody or from a halo of related geological features surrounding an orebody. In a succession of field investigations, anomalies are tested in relation to the exploration model, and because no two orebodies are exactly alike, the model itself is tested against new field data. The model and its working hypotheses are revised if need be, and the anomalies are evaluated by further testing until they are verified by an ore discovery, or until the search in that locality is abandoned.

1. Conceptual Models in the Subsurface Environment

Pure conceptual models of ore deposits are images of undisturbed mineral associations at a certain depth below the influence of the zone of weathering. The geological evidence for an orebody obtained from exploration at the earth's surface is more likely to have been affected by surface processes, such that all that is exposed is a weathered derivative of the deeper ore. A complete exploration model pertains to the entire system of an orebody, with its leached surface expression, with a zone of secondary or supergene mineralization resulting from near-surface processes and with the deeper primary or hypogene zone that represents the basic conceptual model.

Figure 1 shows a representative conceptual model in cross section. In this instance, the expected orebody is a hydrothermal vein-type copper deposit that has occupied a fault zone in a limestone host rock; ore has replaced some of the walls of the fault zone and also a thin zone of host rock beneath a relatively impermeable shale cover. If normal faulting is envisioned, steeper portions of the fault zone should be associated with greater widths of ore. Satellitic orebodies and ore-filled gash fractures are expected to occur in the vicinity of the main orebody.

In this model, zonation of metals and minerals is anticipated, with manganese minerals dominating the outer limits, lead and zinc minerals characterizing an intermediate zone and copper minerals predominant in the central part of the orebody. Wall rock alteration is expected to occur in a halo around the orebody, possibly in a zonal pattern with an inner zone of

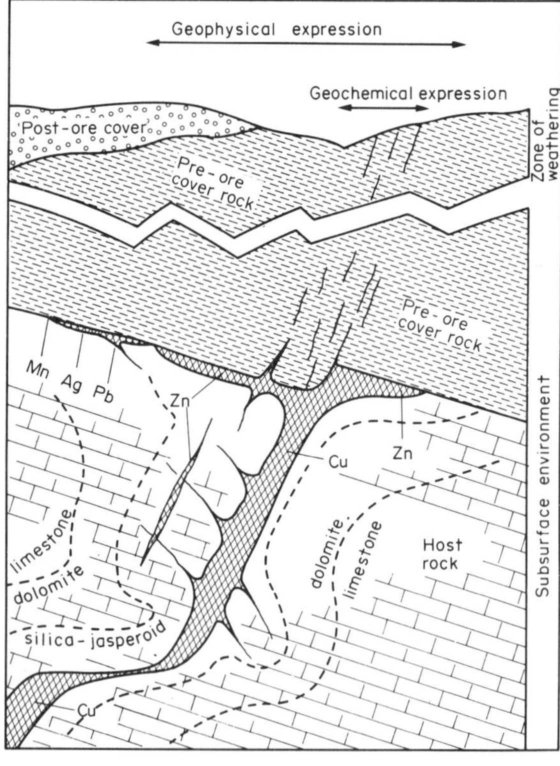

Figure 1
Representative conceptual model of a copper deposit, as used in the design of an exploration program (cross section, not to scale)

silica–jasperoid rock and an outer zone of recrystallized dolomite or dolomitic marble. Leakage of hydrothermal ore-forming solutions through the controlling fault zone is expected to have extended into less-favorable overlying rock for some distance from the orebody, possibly to the level of the present-day surface, where it may have formed thin veinlets and weak alteration halos.

In an exploration model based on the particular conceptual model shown, a geochemical expression would be expected with the thin leakage veinlets. If a geophysical expression were to be detected at the surface, it would represent as much of the orebody, the satellitic features and the extension of the ore to depth as would come within the range of the particular geophysical method.

In a regional exploration program where several kinds of geological terrain are considered, there may be several conceptual models of ore deposits. In addition, conceptual models need not be based solely on the most thoroughly established and accepted principles of ore genesis; new theories of ore deposition can provide models for innovative exploration programs in new terrain.

2. *Zone of Weathering*

As a result of continual erosion, orebodies that were in equilibrium with a deeper geological environment are exposed to the surface zone of weathering and to an environment characterized by meteoric water, atmospheric pressure and atmospheric temperature. Some ore minerals such as gold and cassiterite (SnO_2), and some of the associated gangue minerals such as quartz and tourmaline, are relatively stable in the zone of weathering; most, however, are unstable and are destroyed or changed into minerals that can remain in equilibrium with the new environment.

Oxidation, the breakdown of subsurface ore minerals, takes place from the surface down to the water table, as shown in Fig. 2, which represents the conceptual model of Fig. 1 as it would appear in the zone of weathering. The depth of oxidation may extend to 30 m in temperate humid climates, and to 200 m or more in deserts. In the upper or leached part of the zone of oxidation, a cellular accumulation of iron oxide and quartz is left behind by the weathering of a sulfide ore-bearing vein. The remaining gossan does not normally contain ore minerals of the base metals, but it may contain a residual accumulation of native gold and also silver halide minerals.

In the oxidized ore zone of Fig. 2, gold, silver and base-metal mineralization occurs in the form of oxides, silicates, carbonates, elements and other minerals that have formed in situ or have been formed by supergene processes from solutions descending through the entire zone of leaching. Beginning at the water table and extending downward from a few to several hundred meters is an interval of supergene or secondary sulfide ore. Supergene ore, both in oxidized and sulfide form, commonly provides an enriched zone; in early mines this was often the only zone that was mined, because of a sharp decrease in ore grade at the top of the hypogene or primary zone.

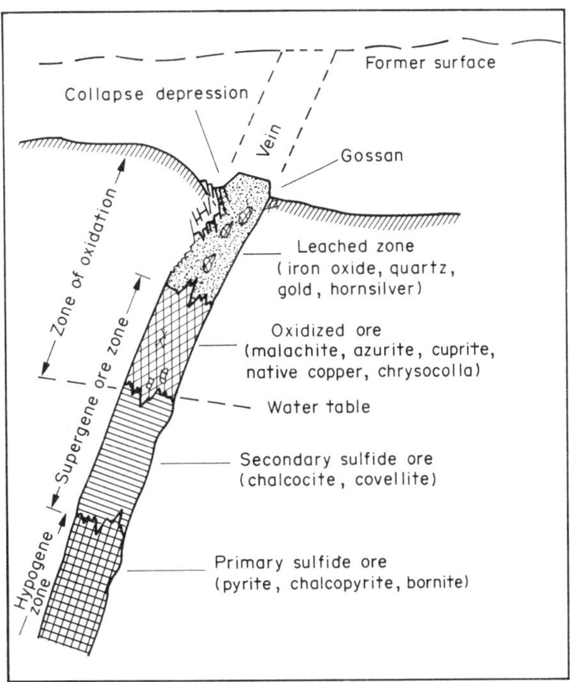

Figure 2
Diagrammatic representation of the oxidation and supergene enrichment of a copper–quartz vein with gold–silver values

3. *Nature of Exploration*

The time-honored approach to geological exploration has been to examine ore occurrences or suspected gossan reported in the literature, or prospects brought to the attention of a mining firm by prospectors or promoters. A geologist or mining engineer would investigate previous findings, project the geological conditions into a potentially larger or deeper orebody, acquire the mineral rights for his or her firm, and test the projections in new mine workings or drill-holes. The examination of prospects still has a fundamental place in mineral exploration, but it is being supplanted by, and combined with, systematic programs of stepwise investigation in broader regions and districts.

Emphasis on concept-oriented regional exploration has increased because of the rarity of ore outcrops. A regional search for hidden and blind orebodies relies heavily on geophysical and geochemical techniques and on drill-hole investigation, with the programming of the techniques and the interpretation of the data done in the context of geology and a currently accepted exploration model.

Although artisanal mining and individual prospecting are still important elements of mineral exploration in many of the developing countries, exploration in the industrialized countries has become an effort that is more commonly carried out by large organizations. The acquisition of mineral rights from prospectors and small-scale miners continues to be a factor in all exploration, but the role of the individual prospector as a discoverer has shrunk with the depletion of near-surface high-grade ore.

4. Objectives of Mineral Exploration

An exploration program is related to objectives which are in turn related to the specific needs of an organization and to the financial risk the organization is willing to accept in searching for an orebody. The need may be for a mining opportunity in a specific region, or it may be for a particular commodity at any location in the world. A depleting orebody or an inadequate supply of mineral material for an established industrial complex may define an objective in terms of both commodity and region.

The financial risk or uncertainty inherent in exploration has several facets. It may be largely a geological risk, the 1000:1 or 10:1 probability that the anticipated orebody will not be found. It may also involve the risk that the orebody will be diminished in value or made uneconomic by low mineral prices, unfavorable political climate or overwhelming technological problems.

The decision to explore in a particular place for a particular mineral commodity is sensitive to trends in mineral markets, access to geographical regions and new concepts in science and technology.

5. Exploration Programs and Patterns

Figure 3 shows the stages involved in a regional exploration program. In its preliminary or design phase, an exploration program is organized and budgeted on the basis of an objective, a region and an exploration model. During this phase, metallogenic maps provide indices for the selection of a favorable region, and orientation surveys provide information for use in the planning of appropriate geochemical and geophysical techniques (see *Metallogenic Provinces*).

The initial field phase in a regional exploration program is reconnaissance, a screening of the entire region to define smaller areas of specific interest. Geological mapping is the main method used to gather reconnaissance information, with a strong dependence on airborne or satellite imagery, geochemical sampling in major drainages, and airborne geophysics.

Geological field work in the reconnaissance phase begins with the preparation of a small-scale synoptic

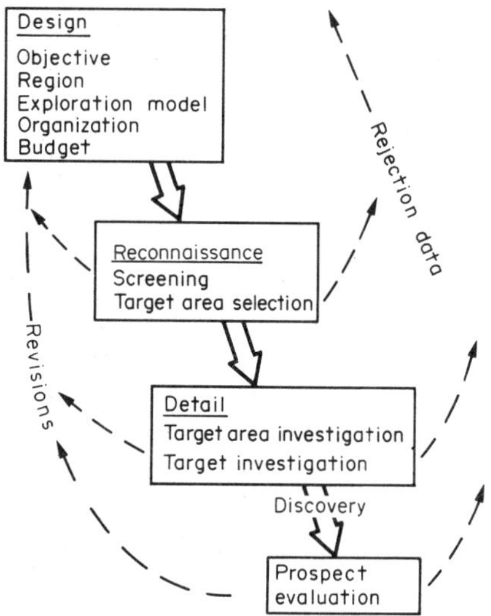

Figure 3
Stages involved in a regional exploration program

map of the region, so that major elements in geomorphology, structure and lithology can be evaluated in relation to the exploration model. A mapping scale of about 1:250 000 is common in the initial stage of reconnaissance. At this scale, field geological observation sites within a spacing of 1 km can be plotted on topographic maps and on satellite imagery, and can be related to principal topographic and cultural features. Patterns of aeromagnetic, airborne radiometric and reconnaissance geochemical data serve as guides to the outlining and mapping of major geological domains in the field.

Using the information obtained from a synoptic map, the region is screened to provide areas of specific interest for further reconnaissance mapping at a scale in the range of 1:50 000–1:25 000. At this scale, geological observations at intervals of 50–100 m can be plotted on maps, and aerial photographs, geochemical stream sediment sampling sites and airborne geophysical flight lines can be shown. Other information such as individual mine workings, stratigraphic boundaries, local structural conditions and major areas of rock alteration can also be discriminated.

A further appraisal of areas and further reconnaissance geological mapping at 1:10 000 scale is sometimes needed in order to provide a sufficient resolution of geological features for designating target areas.

Geological mapping for target area investigation is commonly done at a scale in the range of 1:5000–1:1000, as shown in Fig. 4. This permits the boundaries of a fault zone or a dike a few meters in width to be shown without exaggeration, and also permits the plotting of pits or trenches and of individual stations in

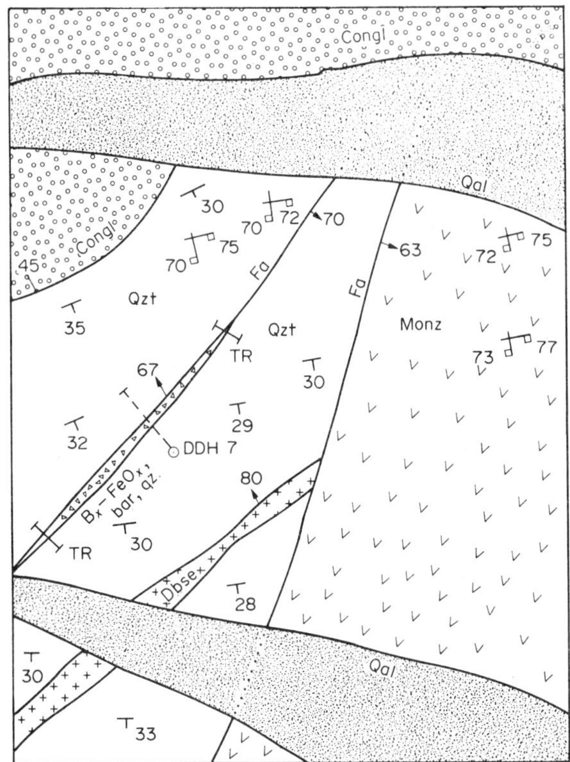

Figure 4
Part of a geological map of an exploration target area

detailed geophysical and geochemical work. Geological observations are most commonly recorded in the field on aerial photographs, and then transferred to large-scale topographic maps. Considerable detail is recorded, with emphasis given to significant evidence of mineralization, zones of rock alteration, and the attitudes of bedding, jointing, rock cleavage and schistosity. As a result of detailed investigations within target areas, one or more specific targets are expected to be located.

Geological information from drill holes, though sometimes used in reconnaissance investigations where regional changes in stratigraphy are significant, is most commonly obtained during target investigations. Detailed geological mapping and geophysical and geochemical work provide the outline of a target that can be related to the exploration model and ultimately to an expected orebody at a particular depth. Drill-holes are then used to evaluate the target. If ore mineralization is discovered, drilling continues until a significant body of ore can be outlined. With the recognition of significant ore mineralization, the target becomes a prospect.

In the evaluation of a prospect; a close pattern of drill-holes is used to outline and sample the ore, and to determine whether the ore mineralization can be designated as an orebody. In addition to the close pattern of drilling, shafts or adits and underground mine workings are often required for an adequate evaluation.

During evaluation, geological mapping of the immediate area may be done at a larger scale than during target investigation. Surface geological maps with scales of about 1:500 are commonly used to show the information in drill holes and trenches. Geological maps and sections of underground mine workings are commonly made at a scale of 1:250 or 1:500.

Prospect evaluation (see *Mineral Deposits: Examination and Evaluation*) involves economic and engineering considerations as well as geology. The final step in a successful exploration program and in a favorable prospect evaluation is marked by a decision to enter into a preproduction program to develop a mine.

Each stage in an exploration program, from reconnaissance to preproduction, is related to a series of decisions. On this basis the program is continued, revised or halted. Revisions, based on the accumulation of new information, may make it necessary to reexamine a target area that was once rejected. With a decision to halt or abandon an exploration program, data are accumulated for future use in the formulation of exploration models and the design of new exploration programs.

See also: Geochemical Distribution of the Elements; Ore Minerals

Bibliography

Edwards R, Atkinson K 1986 *Ore Deposits Geology.* Chapman and Hall, London

Kužvart M, Böhmer M 1986 *Prospecting and Exploration of Mineral Deposits*, 2nd edn. Elsevier, Amsterdam

Lacy W C 1983 *Mineral Exploration.* Van Nostrand Reinhold, New York

Peters W C 1987 *Exploration and Mining Geology*, 2nd edn. Wiley, New York

Reedman J H 1979 *Techniques in Mineral Exploration.* Applied Science, London

W. C. Peters
[Mining and Geological Consultant, Tucson, Arizona, USA]

Geophysical Exploration for Mineral Deposits

Ground, airborne, well-logging and marine methods are commonly used in geophysical prospecting. All of the methods and their survey applications are complex, but a simple approach to a prospecting problem usually consists of the following stages:

(a) flying over the area and utilizing available airborne techniques;

(b) carefully evaluating the data before sending in ground exploration teams and applying ground techniques; and

(c) collecting subsurface information by monitoring geophysical instruments or probes placed in wells (well logging).

General textbooks on the subject are included in the Bibliography.

1. Electrical Techniques

Electrical techniques can be divided into two basic categories: (a) those which depend on naturally occurring influence fields, such as self-potential, telluric-current and magnetotelluric methods; and (b) those based on introducing fields into the earth artificially, such as electromagnetic, resistivity and induced-polarization methods. Electrical techniques are used in metals and minerals exploration, in shallow exploration (<500 m) and sometimes in oil exploration. They also aid in geothermal exploration, engineering geology, hydrology and archaeological studies.

The three basic electrical properties of rocks on which these techniques are based are resistivity, electrochemical activity (including the chemical composition and electrolyte concentration) and the dielectric constant.

Electrical methods are summarized below, starting with methods based on introducing fields into the earth artificially and concluding with methods based on naturally occurring influence fields.

1.1 Electromagnetic (EM) Methods

Electromagnetic methods are based on the induction of electric currents in conductors buried in the earth. These produce a magnetic field which induces a current in a pick-up coil. This induction, a function of the resistivity of the buried conductor, results from magnetic components of electromagnetic waves generated at the earth's surface or in the air, and which originate from alternating field currents of varying frequency. Optimum depth versus frequency relationships exist, such as $h(f/\rho_a)^{1/2}=10$, where h is the depth, f is the frequency (Hz) and ρ_a is the apparent resistivity (Ω m).

A number of specific electromagnetic techniques have been developed for specialized applications. Examples of some of these are summarized below.

(a) *Audio-frequency magnetic field (AFMAG) method.* Variations (inclination and phase changes) in the natural horizontal electromagnetic field due to buried conductors are measured. This natural electromagnetic field is the result of electrical storms in the atmosphere and has a frequency of 30–20 000 Hz, in the audio range.

(b) *Very low frequency (VLF) method.* A strong electromagnetic field is produced by employing vertical antennas. The operation range is 15–30 kHz, and variations in the field due to near-surface buried conductors are measured.

(c) *Sequential electromagnetic signals (SES) method.* Induced currents produced at two or three frequencies are used to give interfering electromagnetic fields. The resultant field induces a current in a pick-up coil, which is then measured.

(d) *Induced-pulse transient (INPUT) method.* Short ac wave-crest pulses, separated by 2 ms pauses, induce a field in the earth. This field produces charged conductors which act like capacitors. Interrupting the current allows these capacitors to discharge, thus producing secondary electromagnetic fields, which are then measured. A good conductor has a long current-decay curve.

(e) *Turam method.* A straight conductor produces electromagnetic waves using two or three ac frequencies. Measurements consist of determining the phase differences and amplitude ratios of the resultant polarized electromagnetic waves.

(f) *Tilt-angle method.* A stationary transmitter produces a primary electromagnetic field, which in turn induces current flow in a buried conductor. A mobile pick-up coil is used to measure the angle between the ground surface and the resultant vector (tilt angle) arising from interaction between the primary and secondary fields. The minimum tilt angle is observed immediately above the buried conductor.

1.2 Resistivity Methods

Resistivity methods are based on applying a current to the ground through electrodes and measuring the change in electrical potential at the surface. Basically, the method yields a measure of the resistance to current flow through the ground. Plots of apparent resistivity ρ_a are developed from the fundamental resistivity equation:

$$\rho_a=(2\pi V/I)D \tag{1}$$

where V is the potential difference between two surfaces of constant potential, I is the current in the conducting body and D is some function of the electrode spacing which varies depending on the exploration technique employed.

Field procedures involve profiling or sounding techniques. Electrical profiling, in which the electrode spacing is fixed, provides a rapid ground surveying technique. It is generally used in locating sand and gravel, ore bodies, faults, and steeply dipping contacts between different rock bodies. The fixed spacing of electrodes yields a reading which represents a weighted average of all resistivities present. In electrical sounding, the center of a spread is fixed and the electrode spacing is increased. This technique is generally used to monitor ρ_a variability with depth, to differentiate a sequence of zones with high and low ρ_a, to determine the depth of a layer, and to estimate the thickness of a layer. This method assumes that all ρ_a variability is vertical.

1.3 Induced-Polarization (IP) Methods

Induced-polarization methods are the most widely used ground-based geophysical techniques in the mining industry, and are generally used in searching for disseminated sulfide ores and in ground water exploration (controlled by clay dispersal). Principally, the technique involves passing a variable or constant frequency, ac or dc current into the ground, which gives rise to exchange-ions at surface contacts between metallic substances and electrolytes in pore-space fluids. This electrochemical exchange produces a voltage which opposes the current flow through the metallic substance. A voltage, termed the overvoltage, is used to overcome this opposition. After stability is reached, the applied current is removed, and the electrochemical voltages at metallic grain surfaces allowed to decay. The decaying voltages are measured, and the ratio of the overvoltage amplitudes after and before shut-off gives an indication of the concentration of the subsurface metallic substance. Electrode configuration in the field is similar to that used for resistivity measurements.

1.4 Self-Potential (SP) Method

The self-potential or spontaneous polarization method is based on the existence of electric potentials which naturally develop within the earth as a result of electrochemical action. Ore bodies are good conductors which carry currents from oxidizing electrolytes above a water table to reducing electrolytes below a water-table surface. Potential anomalies thus produced over graphite, magnetite and sulfide (pyrrhotite and pyrite) ore bodies are negative. Field detection is based on profiling and sounding techniques using two electrodes and a potentiometer. Applications involve shallow-depth prospecting, and require terrain corrections which may indicate tilting of the water table.

1.5 Telluric-Current (TC) Method

The telluric-current method is based on the presence of natural, large, sheet-like currents which flow within the earth in a fixed pattern with respect to the location of the sun. These currents are induced below the earth's surface by ionospheric currents which correlate with diurnal changes in the earth's magnetic field. Potential differences, controlled by variations in subsurface resistivity, are measured employing self-potential methods at two differing sites. At each site, the amplitudes of simultaneous oscillations of the sheet-like currents are compared, and the potential difference is measured in two orthogonal directions.

1.6 Magnetotelluric (MT) Method

The magnetotelluric method is based on the simultaneous measurement at the same location of an east–west and north–south component of the oscillating electric (sheet-like) field within the earth, and an east–west, north–south and vertical component of the magnetic field at the site. Measurements are made with nonpolarizing electrodes and buried induction coils. The method is generally used for large area reconnaissance (rapid basin delineation), and provides greater depth information than any other electrical technique.

2. Magnetic Techniques

Magnetic techniques are based on the fact that the earth has a magnetic field which is variable and easily measured at or above the surface. Contributing to this field are the main field (approximating to a dipole and originating within the outer (liquid) core of the earth) and the nondipole field, which includes nondipole components of the total field. These components generally result from (a) secular variations in the field due to long-term interaction of the outer core fluid with the lower mantle interface, (b) micropulsations and diurnal variations in the field due to short-term magnetic field characteristics arising from currents induced within the earth by ionospheric currents, and (c) field components produced by variations in the natural remanent magnetic properties of the earth's crust.

The natural induced and remanent magnetization of rocks generally resides in a suite of important iron-bearing minerals, including iron oxides (magnetite, hematite), iron–titanium oxides (ilmenite), iron oxyhydroxides (goethite), iron carbonates (siderite) and iron sulfides (pyrrhotite, greigite). Magnetic methods are particularly well suited to detection of concentrations of, and variations in, these minerals.

Two types of magnetic methods are employed in exploration for deposits. By far the most important of these is measurement of local variations in the magnetic field, at or above the ground surface, which result from buried magnetic bodies. The magnetic properties of rocks can also be directly measured, but are not often used in exploration for deposits. The three types of magnetometers in general use in exploration geophysics are discussed below.

2.1 Flux-Gate Magnetometers

These magnetometers are composed of two balanced alternating current (primary) coils wound in opposite directions around bars of high-permeability magnetic material, all surrounded by a single direct current (secondary) coil. A net imbalance between the primary coils induced by the earth's ambient field is measured using the secondary coil. The component of the field which is measured in this way is that which is parallel with the axis of the cores. In order to measure the total field, three orthogonal elements must be measured. This can be accomplished remotely using servo motors. The method is not absolute and has a sensitivity of about 1 nT.

2.2 Proton Magnetometers

A bottle of fluid is surrounded by a coil through which a current is passed. Protons in the fluid are aligned with the magnetic field thus produced, and begin to precess when the current is removed. This precession is proportional to the total magnetic field, and can be measured using the energizing coil. The method yields an absolute estimate of the total field, requires periodic current flow and decay cycles, and can have a sensitivity of 50 pT.

2.3 Optical Pumping Magnetometers

These instruments yield an absolute, continuous reading of the total magnetic field and have a sensitivity of 1–10 pT. They are, however, extremely expensive and cumbersome. Measurement is based on the resonance frequency of rubidium, cesium or helium vapors being proportional to the ambient magnetic field.

3. Gravimetric Techniques

Variations in the near-surface gravity field are a function of horizontal and vertical variations in density. Therefore, gravimetric techniques have application where such variations are geologically important. Generally, only few applications exist in ore exploration, but gravimetric techniques are important in oil exploration, where variations in density can help to delineate salt domes and other structures.

Gravimeters can be airborne, carried at sea or positioned on the ground, and gravity interpretation usually entails producing contour maps of density variability. The data are often modelled in cross section, but some knowledge of the geology must exist to make meaningful interpretations, since many structures that lie at different depths and have different densities can have similar gravity anomaly patterns. By utilizing filtering and smoothing techniques, a regional (long wavelength: deep) anomaly pattern is defined from which local (short wavelength: shallow) anomalies can be subtracted. The resulting residual maps are useful in exploration.

Great care must be taken while conducting gravity surveys to determine station location, elevation, local variations in terrain, and estimates of density, based on rock type. For exploration purposes, two types of gravity anomaly are calculated: (a) a free-air anomaly, where the observed gravity is corrected for the decrease in gravity due to elevation above sea level, with $C_F = g_o - (g_\phi - f_g)$, where g_o is the observed gravity value, g_ϕ is the theoretical gravity value at the site latitude, and f_g is the free-air correction; and (b) a Bouguer anomaly, where the observed gravity is corrected for the attraction of crustal material (a slab) between a base level and the station elevation h above the base level. It is given as $C_B = g_o - (g_\phi - f_g + S - T)$, where S is the slab correction and T is a terrain correction.

Generally, negative Bouguer anomalies are observed over sedimentary basins, salt domes, granites and grabens. Positive anomalies can be correlated with uplifts (horsts, reverse faults) and mafic rocks.

4. Seismic Techniques

Seismic methods are sea- or ground-based exploration techniques which can give an accurate indication of subsurface depth of rock units, as well as giving a reasonably comprehensive picture of subsurface structure. These exploration techniques are based on the premise that contrasts in the seismic impedence of rocks reflect and refract elastic waves within the earth. These waves are termed body (P and S) and surface (Rayleigh and Love) waves. The velocity of propagation of these waves is a function of the elastic properties and densities of the subsurface rocks. These are controlled by such factors as compaction, cementation and metamorphic grade.

Principal applications in industry are associated with exploration for oil. The method involves generating seismic waves at the surface of the earth by using a source (e.g., explosives), recording the arrival times of these waves at receivers (geophones, or at sea by hydrophones, usually set out in a straight line around the source) and then processing the data. The time it takes a wave to travel the distance to the receiver is the basic information available for analysis. Since the earth is not homogeneous and propagating waves tend to move with varying speeds in different directions, and since different types of elastic waves move at different velocities, such analyses can become complex.

5. Radiometric Methods

Certain radioactive elements in near-surface rocks give off α and β particles and γ rays during radioactive disintegration. Important examples are ^{238}U, ^{235}U, ^{232}Th and ^{40}K. While the α and β particles are readily absorbed by very thin layers of sediment, γ rays can be easily detected. This is done by using a device such as a scintillation or a Geiger counter. Certain rocks (granite) or sediments (beach sand) intrinsically contain or accumulate greater concentrations of uranium, thorium and radioactive potassium than do other rocks, and γ-ray emission from these units is often readily detectable. These data can then be compiled into maps which provide geological information on near-surface rock types and structure. Such maps are also useful in uranium-ore exploration.

6. Thermal Methods

Thermal methods are based on the generation of heat within the earth as a result of radioactive decay. The flow of heat Q out of the earth can be measured from the relation:

$$Q = -K \operatorname{grad} T \tag{2}$$

where K is the thermal conductivity of the rock and grad $T = dT/dZ$, the rate of temperature increase with depth ($W\ m^{-2}\ s^{-1}$).

Local variations in Q may be the result of (a) high concentrations of radioactive heat sources, (b) variations in thermal conductivity of the rocks, (c) subsurface geothermal sources, or (d) subsurface chemical reactions. Thermal exploration is currently being employed for shallow-level prospecting, and is useful in delineating subsurface aquifers, salt domes, fault contacts, thermal reservoirs, granites, sulfide ore bodies and other geological features.

7. *Spectral Techniques*

Spectral methods involve detection of electromagnetic waves, including visible light, ultraviolet and infrared rays and high-frequency radio waves. Methods involve simple photography (black and white, color and ordinary infrared), optical–mechanical scanning (thermal infrared), side-scan and direct-scan radar (radio detection and ranging), satellite imagery, and spectral reflectance measurements.

Air and satellite images and radar techniques provide extremely important geological data, which can give rapid estimates of structure and lithologies. These are based on information such as linear trends and affinities of some vegetation types for soil chemicals present in certain lithologic units. Combined with thermal infrared data, areas of minor heat emissions can be identified, which arise from subsurface chemical reactions such as those known to occur in association with porphyry copper deposits. Other uses exist, such as remote identification of hydrothermal sources and of variations in water-table level.

Spectral reflectance measurements can be used to detect changes in rocks and soils caused by mineralization, metal stresses in plants and ground-water-related phenomena. Spectral reflectance information is possible because certain minerals when excited by sunlight emit a characteristic wavelength of visible light. This data has proven useful in directly delineating zones of hydrothermal alteration where distinct iron and clay spectral bands are emitted. It has been demonstrated that specific mineral identification is possible using such methods, especially when dealing with clay minerals such as kaolinite and montmorillonite. Alunite and iron minerals are also easily identified.

8. *Well Logging*

Well logging is a prospecting tool generally used in oil exploration and prospect evaluation. It is not extensively applied to metallic minerals exploration, probably due to the greater structural complexity in locations where ore deposits are found. Well-logging techniques are designed to provide direct information from the monitoring of instruments lowered into drill holes, and in conjunction with instruments distributed around the drill holes, to evaluate the subsurface geology. Generally, the rock properties evaluated using electrical and acoustic well-logging methods are porosity, permeability and the nature of fluids within the rock. Using magnetic susceptibility, radiometric and heat-flow methods, mineral constituents of the rock can be determined, and using gravimetric methods, density is estimated. Instruments designed to be inserted into drill holes generally provide continuous output graphs as they are lowered into the hole. These graphs, which show the variations of physical properties with depth, are usually interpreted as stratigraphic logs. Such interpretations are qualitative and rely on independent checks on the data, such as rock cuttings from cores and data from nearby wells.

See also: Geochemical Distribution of the Elements; Geochemical Exploration for Mineral Deposits; Geological Exploration for Mineral Deposits; Remote Sensing in the Search for Mineral Deposits

Bibliography

Dobrin M B 1976 *Introduction to Geophysical Prospecting*. McGraw-Hill, New York

Kazmitcheff A 1974 *Modern Mineral Prospecting*. Casterman, Tournai, Belgium

Kearey P, Brooks M 1984 *An Introduction to Geophysical Exploration*. Blackwell Scientific, London

Keller G V, Frischknecht F C 1977 *Electrical Methods in Geophysical Prospecting*, International Series of Monographs on Electromagnetic Waves, Vol. 10. Pergamon, Oxford

Nettleton L L 1976 *Gravity and Magnetics in Oil Prospecting*. McGraw-Hill, New York

Parasnis D S 1979 *Principles of Applied Geophysics*, 3rd edn. Chapman and Hall, London

Sharma P V 1986 *Geophysical Methods in Geology*, 2nd edn. Elsevier, Amsterdam

Telford W M, Geldart L P, Sheriff R E, Keys D A 1976 *Applied Geophysics*. Cambridge University Press, Cambridge

B. B. Ellwood
[University of Texas, Arlington, Texas, USA]

Glass: An Overview

Glass is an essential material in modern technological society. In addition to traditional uses such as windows and containers, glass is used in a host of specialized ways in lamps, optics, composites and electronics. Recent developments of fiber-optic waveguides, laser optics for initiating fusion reactions, and repositories for radioactive waste—all of glass—demonstrate the versatility, importance and promise of glass for future applications. This article discusses the history, definition, uses and main types of glass.

1. History

Archaeological evidence shows that natural glasses such as obsidian have been used by man from the earliest times. Synthetic glass and glazes were made far back in human history. The earliest known glaze (coating of glass) dated from about 12 000 BC, and the earliest solid glass from about 7000 BC; both were found in Egypt, which continued to be a center of glass and glaze manufacture throughout ancient history. At first, glass was used only for decorative purposes, but later it was molded or pressed into vessels. The invention of glass blowing in about the first century BC greatly increased the use of glass for practical purposes in Roman times, mainly for vessels but later for windows also. In Europe, glassmaking dispersed to isolated sites after the fall of the Roman Empire, but was continued in the Byzantine Empire (Syria especially), and then by the Arabs in Syria, Egypt and Persia. Venice became the center of a resurgent glass industry in Europe after about 1300. Artistic development paralleled that of the Renaissance in the fifteenth century, the most significant advance in Venice being the manufacture of a clear, colorless glass called cristallo. With this glass Venice became the premier center of glass manufacture and export until the seventeenth century, when centers sprang up throughout Europe, especially in Bohemia, Nuremberg and the UK.

Of special interest in this period was the publication of *L'Arte Vetraria* by Neri in Florence in 1612; this work released the art and technology of glassmaking from the secret hold of the Venetian artisans and made it available to the world.

After 1600 the development of glassmaking was rapid. Cut glass was first made about 1600, and the use of coal instead of wood for fuel gave higher temperatures and a more reliable fire. Late in the seventeenth century, lead oxide was added to glass in England, resulting in lower melting temperatures, a wider temperature range for working the glass, and "sparkle" from a higher refractive index. Better selection and some purification of raw materials in the eighteenth century gave clearer glasses and better control of melting and working conditions and of the color of the glass.

In the nineteenth century, the quality of glass produced in Europe increased tremendously in response to the increased affluence of industrial society. At the end of the nineteenth century, automatic methods of making glass were developed, such as the bottle-blowing machine.

Traditionally, window glass was made by hand by either the crown or cylinder process. In the crown process a glob of glass was blown out and one side of the resulting globe flattened. A solid iron rod was attached to the flat part and the blowing pipe detached. Then the globe was reheated and rotated until it formed a flat disk about three feet in diameter. Panes of glass were cut from the disk after it was slowly cooled. The part attached to the rod was the "bull's-eye"—still seen in some older windows. In the cylinder process the blower made a large cylinder, which was then split open and flattened. Cylinder glass was also made by machine.

In the early part of the twentieth century, processes for drawing sheet glass directly from the glass melt were developed (the Fourcault and Colburn processes). When linked with a continuous-glass-melting furnace, this process is capable of producing large quantities of flat glass of reasonable quality. Plate glass of the highest quality was, until recently, made by flowing the glass from the furnace through rollers, which gives a rough surface, and then grinding and polishing it by large automatic machines. This process requires a large capital investment, but is fairly economical because of the large quantity of glass that is drawn directly from the melt. The sheet-glass surface has a fine fire-polished finish, but shows some surface distortion because of variations in processing conditions as the glass is drawn from the melt.

In the 1950s, an ingenious new method of making relatively inexpensive flat glass of high quality was developed by Alistair Pilkington of the Pilkington Glass Company of England. In this float process, a continuous strip of glass from the melting furnace floats onto the surface of a molten metal, usually tin, at a carefully controlled temperature. The flat surface of the molten metal gives a smooth, undistorted surface to the glass as it cools. After sufficient cooling, the glass is rigid and can be handled on rollers without damaging the surface finish. High forming speeds are possible, and the cost is much less than for similar quality glass made by grinding and polishing. As a result, one glass manufacturer after another has converted to the float process and today most flat glass is made by this process.

Glass containers such as bottles and jars are made by blowing hot glass into a mold on a continuous machine. Lamp bulbs are made in a similar way by blowing hot glass into a mold; however, in this case the glass is fed from the melting furnace as a ribbon, rather than blowing out an individual glob of glass as in the container machines. A special nozzle blows glass in the ribbon into a mold. The high-speed "ribbon machine" for lamp bulbs is a spectacular sight as lamp bulbs pop out from it at a rate of 2000 per minute.

Certain glass objects such as plates, tumblers and vases can be made inexpensively by pressing the hot glass in a mold. This pressed glass was especially popular in the USA in the nineteenth century because it was much cheaper than cut-crystal glass imported from Europe. In this method, a glob of hot glass is placed into a metallic mold, and a metallic plunger is forced into the mold to shape the glass in the desired way. Patterns in the mold surface are imprinted in the glass and make it look like an expensive article. Pressed glass can also be made continuously using automatic feeders for molds on a rotating bed.

The continuous manufacture of glass in a large furnace, called a "tank," was developed in response to the demand for greater, cheaper and more uniform production of glass. This furnace requires improved refractory materials to line it, and is fuelled with natural gas or fuel oil.

A variety of new methods for making special types and forms of glass have been developed. Glass fibers as fine as 10 μm in diameter are drawn and coated at high speeds. Fused-silica tubing is drawn at temperatures above 1800 °C. Highly pure and clear glass is made for a number of optical purposes. A large variety of glass compositions have been developed for diverse applications. The infinite variability of glass compositions leads to a great assortment of possible properties which are only beginning to be exploited.

2. *Definition*

Glass is an amorphous solid. A material is amorphous when it has no long-range order, that is, when there is no regularity in the arrangement of its molecular constituents on a scale larger than a few times the size of these groups. For example, the average distance between silicon atoms in vitreous silica (SiO_2) is about 3.6 Å, and there is no order between these atoms at distances above about 10 Å. A solid is a rigid material: it does not flow when it is subjected to moderate forces. More quantitatively, a solid can be defined as a material with a viscosity of more than about 10^{15} P (poise).

Many glass technologists object to the above definition of a glass. These workers prepare a glass by cooling a liquid in such a way that it does not crystallize, and feel that this process is an essential characteristic of a glass. Many earlier writers insist on this criterion: "glass is an inorganic product of fusion which has been cooled to a rigid condition without crystallization" as taken from the ASTM standards for glass. The difficulty with this view is that glasses can be prepared without cooling from the liquid state. Glass coatings are deposited from the vapor or liquid solution, sometimes with chemical reactions. Thus sodium-silicate glass can be made by evaporating an aqueous solution of sodium silicate (water glass) and baking the deposit to remove water. The product of this process is indistinguishable from sodium-silicate glass of the same composition made by cooling from the liquid. It seems wise to use the same name for materials with the same molecular structure and properties no matter how they are made; consequently, the broader definition given here is preferred.

3. *Uses*

The original use of glass for decorative and artistic objects continues to this day. One need only visit the Stueben exhibit in New York City, the Corning museum in Corning, New York, or glass exhibits in Czechoslovakia to sense the originality and excitement of present-day artistic work in glass.

The second use of glass was for containers, for food, perfume, oils and many other substances; this application still uses the largest quantity of glass today. The production of flat glass, mainly for windows in buildings and vehicles, is now the second largest item of glass manufacture. Lamp envelopes and seals are another major area of use. There are many special applications of glass, some in small quantity but of high value, most of which have been developed in the last few decades. Some of these special applications are glass ceramics and surface-strengthened glass for higher strength and chemical durability; lightweight composites of fiberglass in polymer matrices; glass for laser hosts and optic waveguides for long-distance communication; fused silica for melting semiconductors, telescope mirrors and arc lamps; amorphous silica layers on silicon in electronic devices; encapsulation of these devices; and as a medium for consolidating radioactive wastes. New compositions of silicate glasses, and even nonoxide glasses, are being rapidly developed for a variety of new applications.

These uses of glass are based on a variety of desirable properties, such as ease of forming into many different shapes, cheap and widely available raw materials, chemical durability, transparency, high-temperature durability, wide "solubility" of constituent oxides and low electrical conductivity.

4. *Main Types of Glass*

The possibility of incorporating a large number of different oxides in a silicate glass has led to a wide variety of commercial glasses. Nevertheless, there are a relatively small number of glasses that make up the large majority of glass production. Some of these glasses are listed in Table 1, and their approximate compositions given in Table 2. The softening temperature in Table 1 is either the temperature at which a rod 25 cm long and 0.76 mm in diameter elongates by 1 mm min^{-1}, or the temperature above which the glass can be easily deformed and worked.

By far the most common glass is based on the soda-lime–silicate (sodium calcium silicate) system. All ancient glasses contained the oxides of sodium, calcium and silicon. These glasses are cheap, chemically durable and relatively easily melted and formed. Many minor additions to the basic composition (not far from 70% silica, 15% soda, 10% calcium oxide and magnesium oxide, and 5% other oxides) are made to improve properties of melting and forming; alumina for improved chemical durability and reduced devitrification

Table 1
Some types of silicate glasses and their uses

Glass	Use	Approximate softening temperature (°C)	Coefficient of thermal expansion (per °C × 10^6)
Soda lime	containers, windows, lamp bulbs	700	9.0
Pyrex borosilicate	headlamps, cookware, laboratory ware	830	3.2
Vitreous silica	semiconductor crucibles, lamps, optical components, fiber optics	1600	0.5
Alkali lead	lamp tubing, sealing	620	9.3
"E" lime aluminosilicate	fibers	830	6.0
Lime–magnesia aluminosilicate	high temperatures, cookware	900	4.6

Table 2
Approximate compositions (wt%) of the glasses in Table 1

Glass	SiO_2	B_2O_3	Al_2O_3	CaO	MgO	PbO	Na_2O	K_2O
Soda lime	72	1	2	5	4		15	
Pyrex borosilicate	81	13	2				4	
Vitreous silica	100							
Alkali lead	77			1		8	9	5
"E" lime aluminosilicate	55	7	15	21			1	
Lime–magnesia aluminosilicate	64	5	10	9	10		1	1

(crystallization), borates for easier working and lower thermal expansion, zinc oxide for lower melting temperatures, and arsenic and antimony oxides for fining (removal of bubbles). Soda-lime glass is often termed "soft" glass because of its relatively low softening temperatures.

Pyrex-borosilicate glass was developed by the Corning Glass Works to be more resistant to thermal shock and more chemically durable than soda-lime glass, yet still melt at a similar temperature. The borate in this glass reduces its viscosity and the coefficient of thermal expansion, and allows low sodium content, which increases chemical durability. Pyrex-borosilicate glass is often called "hard" glass because of its higher softening temperature compared to soda-lime glass, and is somewhat more expensive than soda-lime glass because of its higher melting temperature and more expensive borate raw material. The mirror for the Mount Palomar telescope was made of Pyrex-borosilicate glass because of its low thermal expansion; nevertheless, it is necessary to correct minute distortions of the mirror surface caused by temperature differences.

Vitreous silica is made of pure silica, SiO_2, giving it excellent high-temperature stability, optical properties and thermal-shock resistance.

A variety of lead glasses are important as low-melting reading and solder glasses with a wide working-temperature range. Lead glass for fine "crystal" contains much more lead than these glasses.

Some nonoxide glasses find special applications and many others are being studied for new uses. Fluoride glasses based on beryllium fluoride have low values of refractive index, Abbé number and nonlinear index, and therefore are the best glass compositions for high-power laser optics. However, beryllium fluoride is highly toxic and the glasses are sensitive to moisture.

A new series of fluoride glasses based on zirconium fluoride and other heavy metal fluorides are optically transparent in the infrared below about 8 μm. Thus they have potential for application as optical fibers for transmission of information and as infrared optical components and sensors.

Calcogenide glasses, based on sulfur, selenium and tellurium compounds rather than oxides, also transmit in the infrared and are also electronic conductors (semiconductors). Various kinds of switches and devices can be made of these glasses.

See also: Traditional Ceramics: An Overview

Bibliography

History of glass
Charleston R J 1980 *Masterpieces of Glass: A World History from the Corning Museum of Glass*. Abrams, New York

Douglas R W, Frank S 1972 *A History of Glass Making*. Foulis, Henley-on-Thames, UK
Morey G W 1954 *The Properties of Glass*, 2nd edn. Reinhold, New York, Chap. 1
Neri A 1612 *L'Arte Vetraria*. Florence, Italy, and many subsequent editions and translations
Polak A B 1975 *Glass: Its Tradition and Its Makers*. Putnam, New York
Zerwick C 1980 *A Short History of Glass*. Corning Museum of Glass, Corning, New York

Monographs on glass
Babcock C L 1977 *Silicate Glass Technology Methods*. Wiley, New York
Doremus R H 1973 *Glass Science*. Wiley, New York
Dunken H H 1981 *Physikalische Chemie der Glasoberfläche*. VEB Deutscher Verlag für Grundstoffindustrie, Leipzig
Holloway D G 1973 *The Physical Properties of Glass*. Wykeham, London
Izumitani T S 1984 *Optical Glass*. Kyoritsu Shuppan, Tokyo, (Translation to English in preparation)
McMillan P W 1979 *Glass–Ceramics*, 2nd edn. Academic Press, New York
Morey G W 1954 *The Properties of Glass*, 2nd edn. Reinhold, New York
Mott N F, Davis E A 1979 *Electronic Processes in Non-Crystalline Materials*, 2nd edn. Clarendon, Oxford
Scholze H 1965 *Glas: Natur, Struktur, und Eigenschaften*. Vieweg, Braunschweig
Vogel W 1979 *Glaschemie*. VEB Deutscher Verlag für Grundstoffindustrie, Leipzig
Wong J, Angell C A 1976 *Glass Structure by Spectroscopy*. Dekker, New York
Zallen R 1983 *The Physics of Amorphous Solids*. Wiley, New York
Zarzycki J 1982 *Les Verres et L'Etat Vitreux*. Masson, Paris

Collections of review articles
Mackenzie J D (ed.) 1960, 1962, 1964 *Modern Aspects of The Vitreous State*, Vols. 1–3. Butterworth, London
Pye L D, Stevens H J, LaCourse W C (eds.) 1972 *Introduction to Glass Science*. Plenum, New York
Tomozawa M, Doremus R H (eds.) 1977, 1979, 1982, 1985 *Treatise on Materials Science and Technology*, Vols. 12, 17, 23, 26, *Glass I, II, III, IV*. Academic Press, New York
Uhlmann D R, Kreidl N J (eds.) 1980–1987 *Glass: Science and Technology*, Vols. 1–5. Academic Press, New York

Reference works
Bansal N P, Doremus R H 1986 *Handbook of Glass Properties*. Academic Press, Orlando, Florida
Mazurin O V, Streltsina M V, Shvaiko-Shvaikovskaya T P 1983, 1985. *Handbook of Glass Data*, Parts A, B. Elsevier, Amsterdam

Journals on glass
Glass Technology and Physics and Chemistry of Glasses. Society of Glass Technology, Sheffield
Glastechnische Berichte (in German). Deutsche Glastechniche Gesellschaft, Frankfurt Main
Journal of the American Ceramic Society. American Ceramic Society, Columbus, Ohio
Journal of Noncrystalline Solids. North-Holland, Amsterdam
Soviet Journal of Glass Physics and Chemistry (in Russian, also translation into English). Consultants Bureau, New York
Yogyo Kyokaishi, Journal of the Ceramic Association of Japan (mostly in Japanese, with English abstracts). Ceramic Association of Japan, Tokyo

R. H. Doremus
[Rensselaer Polytechnic Institute, Troy, New York, USA]

Gold Resources

Gold is a soft yellow metal with a melting point of 1063.0 °C. Its discovery and first use by people are shrouded in the mists of antiquity, but by 1000 BC it had attained wide usage as a metal for currency. Since that time gold has played a significant role in the history of many individuals and nations. Today the importance of gold continues undiminished, for it not only provides a valuable standard against which various currencies can be measured, and a metal of superb beauty for decorative purposes, but its unique properties of malleability, conductivity and resistance to corrosion are finding increasing utilization in modern technology. These factors, combined with worldwide inflationary pressures, have led to sharp increases in the price of gold in recent years.

1. Crustal Distribution and Geochemistry

The average gold content of the earth's upper lithosphere is approximately 5 ppb. Rocks of basaltic composition contain an average 7 ppb, those of andesitic composition 5 ppb, and those of rhyolitic composition 3 ppb (Boyle 1979).

Sedimentary rocks understandably exhibit a broader range of gold content, but on average conglomerates and sandstones contain 30 ppb, shales 4 ppb and limestones 3 ppb. However, much higher values, up to 2 ppm, are found in certain graphitic and sulfidic schists, iron-rich chemical sediments, and some sandstones and conglomerates. Native gold is extremely insoluble in virtually all surficial environments, and hence occurs in only very small amounts in natural fresh waters (~0.03 ppb) and seawater (~0.012 ppb) (Boyle 1979).

The concentration of gold to levels at which it can be economically mined takes place by either mechanical or chemical means, but even in the former case a preconcentration of gold into particulate form by chemical means is required. For this reason considerable attention has been focused on gold solubility in aqueous solutions. Gold occurs mainly in two ionic forms, Au^+ (aurous) and Au^{3+} (auric), and can form a series of complex ions. The most important of these in terms of transport of gold in natural environments appears to be $AuCl_2^-$ (Helgeson and Garrels 1968), but $AuCl_4^-$, $Au(OH)_4^-$, AuS and $Au(HS)_2^-$ may also be significant. These complex ions dominate the aqueous chemistry of gold because Au^+ and Au^{3+} are unstable in water owing to their high oxidation potentials. On the basis of natural geochemical associations it seems probable that thioarsenito and thioantimonito complexes, such as $[Au(AsS_3)]^{2-}$ and $[Au(SbS_3)]^{2-}$ are also locally important in gold transport.

The atomic radius of gold (1.44 Å) is identical to that of silver, and as a consequence native gold inevitably contains at least a few percent silver, and gold–silver alloys (electrum) are not uncommon. Gold also forms alloys with copper and platinoid metals. The only other important group of gold minerals are the gold tellurides (e.g., calaverite, $AuTe_2$) and a series of gold–silver tellurides (e.g., petzite, Ag_3AuTe_2). Finally, the mineral aurostibite ($AuSb_2$) has been reported from a number of gold deposits that contain antimony.

2. *Principal Types of Gold Deposits*

It is estimated that up to 1988 approximately 3 billion ounces of gold have been extracted from the earth's crust. A small percentage of that amount has come as a by-product of the mining of other metalliferous ores, especially of base metals and nickel. The bulk of the remainder has come from placer deposits, of both modern and fossil type. The word "placer" means pleasure in Spanish, an allusion to the ease of mining the unconsolidated auriferous materials that constitute modern placer deposits.

The development of gold placers requires the presence of particulate gold in veins or disseminations in preexisting rocks, followed by a period of deep weathering and subsequent concentration of the free gold by some agency, generally running water. Modern placer deposits occur in many parts of the world, but most are now mined out. Those in California yielded about 42 million ounces, those in the Lena-Amur area, USSR, 40 million troy ounces, and those in Colombia about 32 million ounces of gold. The rich gold placers along the Klondike in the Yukon yielded 9 million ounces of gold. As noted by Henley and Adams (1979), these giant placers and many others appear to be related to young uplifts around the Pacific rim.

2.1 Fossil Placer Gold Deposits

Quartz pebble conglomerates of lower Proterozoic age are the most important sources of gold in the world. Auriferous ancient conglomerates of this age are known in South Africa, Ghana and Brazil. Although of similar type, the deposits of Jacobina in Brazil and Tarkwa in Ghana are dwarfed by the gold-rich conglomerates of the Witwatersrand Basin in South Africa. The Witwatersrand Supergroup actually was deposited 2.5–2.8 billion years ago, but is of typical lower Proterozoic style in terms of its clastic sediments. Altogether 14 000 m of clastic sediments and volcanics are represented, but virtually all the gold mineralization is associated with six large fluvial fan systems, the sites of which were controlled by structural domes to the north and west of the basin (Pretorius 1981a). Maximum gold concentrations are in medium–small pebble conglomerates in the mid fan facies, but gold values also occur in fanhead and fanbase facies. Such conditions prevailed a number of times during basin-filling as gold-bearing "reefs" or bankets occur at a number of stratigraphic levels. Recent work has shown that algal mats along the fanbase facies were important traps for the finest gold (and uranium) particles, and possibly some gold in solution. Recent speculations on the tectonic setting of the Witwatersrand Basin (Burke et al. 1985, Winter 1986) broadly favor the concept of a foreland basin with an orogenic system somewhere to the north and west.

2.2 Archean Lode Gold Deposits

A significant amount of the remaining (i.e., nonplacer) gold production has come from numerous quartz-rich lode deposits that occur in many of the late Archean greenstone belts in Canada, Australia and Africa. Two of the largest such systems, each with production of close to 10^9 grams of gold are the Golden Mile, western Australia (Phillips 1986), and the Hollinger–McIntyre vein complex in Ontario (Spooner et al. 1985).

Late Archean gold lodes occur within a wide variety of the rock types in greenstone belts, and display a range of geometries, but most exhibit a spatial relationship to major crustal linears in the form of faults or shear zones (Colvine et al. 1984). Iron-rich rocks, especially oxide facies banded iron formation, and mafic and ultramafic igneous rocks provide favorable chemical environments for gold deposition (MacDonald 1983). Conceptual models for the source, transport and deposition of the gold range from magmatic (Spooner et al. 1985) to metamorphogenic (Kerrich and Fyfe 1981), but the genesis of these important deposits remains unclear. Why the late Archean was so productive of major gold deposits is not understood, but may relate to some unique tectonomagmatic event (Hodgson 1986).

2.3 Homestake-Type Gold Deposits

The Proterozoic era was not particularly productive in terms of the generation of gold deposits, but the Homestake Mine, Lead, South Dakota (Slaughter 1968) is a major exception. It is of interest because the gold orebodies are confined to a narrow stratigraphic interval within a thick metasedimentary sequence of early Proterozoic age. Sawkins and Rye (1974) suggested that it represents the type example of a subcategory of gold deposits that are syngenetic in origin, and associated with iron-rich sediments. The important gold deposits at Morro Velho (Brazil), Kolar (India) and Lupin (northern Canada) are probably of this type.

An important characteristic of these deposits is the continuity of the auriferous horizons. Morro Velho and the Kolar goldfields, for example, have been exploited to depths of over 3000 m.

2.4 Phanerozoic Gold Deposits

Considerable diversity is exhibited by gold deposits of Phanerozoic age, but some valid subtypes can be recognized on the basis of lithologic setting. Most important, both in terms of total production and geographic extent, are volcanic-hosted deposits of Tertiary age. Bonham (1986) divides volcanic-hosted deposits into three major divisions:

(a) Silver-rich, low total sulfur deposits associated mainly with rhyolites (e.g., Tonapah, Nevada and Pachuca-Real del Monte, Mexico);

(b) high total sulfur deposits characterized by presence of the copper antimonide enargite, and related to intermediate calcalkalic igneous rocks (e.g., El Indio, Chile; Goldfield, Nevada; and Pueblo Viejo, Dominican Republic); and

(c) gold telluride-bearing deposits that have an association with alkalic igneous rocks (e.g., Cripple Creek, Colorado and Vatakoula, Fiji).

The young age of all these deposits relates to the fact that they form at shallow depths ($\leqslant 1$ km) in subduction-related orogenic belts and are thus highly prone to removal by erosion.

Sediment-hosted gold deposits of Phanerozoic age are of two quite disparate types. Turbidite-hosted gold deposits (Keppie et al. 1986) occur most notably within thick deformed turbidite sequences of Paleozoic age. The saddle reef gold–quartz orebodies of the Ballarat and Rendigo goldfields, Victoria, Australia (Sandiford and Keays 1986) are famous examples.

Finally, there are the replacement-type sediment-hosted gold deposits that are known mainly from Nevada, USA and are often called Carlin-type deposits after the first gold deposit of the category to be discovered (Radtke et al. 1980). Such deposits have represented the focus of very active exploration activity during the past decade and numerous new orebodies have been found.

3. Production and Reserves

Total world gold production in 1984 was 46.4 million ounces. As can be seen from Table 1, South Africa produced slightly less than one half of this amount. Close to 30% of the remainder comes from the USSR. South Africa has dominated world production of gold for many decades, and has produced about 1 billion ounces of the yellow metal. Given maintenance of the current gold price, there should still be tonnages of gold ore that will allow South Africa to produce an additional 500 million ounces or more in the future.

Table 1
World gold production in 1984 (US Minerals Yearbook)

Country	Production	
	10^6 oz	Percent
South Africa	21.9	47.2
USSR	8.6	18.5
Canada	2.6	5.6
USA	2.1	4.5
China	1.9 (est.)	4.1
Brazil	1.8	3.9
Australia	1.3	2.8
Papua New Guinea	0.84	1.8
Colombia	0.80	1.7
Philippines	0.79	1.7
Zimbabwe	0.48	1.0
Dominican Republic	0.34	0.7
Ghana	0.29	0.6
Mexico	0.27	0.6
Peru	0.20	0.4
Zaire	0.12	0.3
Japan	0.10	0.3
India	0.08	0.2
Nicaragua	0.04	0.1
World total	46.41	

In 1987 world gold production was 54×10^6 oz (US Bureau of Mines 1988). Leading producers (in 10^6 oz) were: South Africa 20 (37.0%), the USSR 8.9 (16.5%), the USA 4.9 (9.1%), Canada 3.8 (7.0%), and Australia 3.2 (5.9%). Other significant producers include China, Brazil, Australia, Papua New Guinea, Colombia and the Philippines. It is noteworthy that virtually all gold production from Papua New Guinea and most of that from the Philippines comes from gold-rich porphyry copper deposits.

4. Future Production and Global Resources

Although it appears that future production of gold from Canada, the USA, Brazil, Australia and the southwest Pacific will be enhanced, a downward trend in total gold production in the western world is predicted by Pretorius (1981b). South African production is expected to decline slightly through the 1980s and then decline to about 50% of its present value by the year 2000.

Between 1979 and 1984, production of gold in the USA more than doubled, from 1.0 to 2.1 million ounces. Other major increases in output over the same period were achieved by Canada (from 1.7 to 2.6 million ounces), Brazil (from 0.9 to 1.8 million ounces) and Australia (from 0.65 to 1.3 million ounces). Production in all of these countries is continuing to rise, spurred on by major new discoveries (e.g., the Hemlo deposit, Ontario) and advances in recovery techniques of gold from low-grade ores. Definitely planned new mines in Canada and the USA will add over 100 million tons to reserves, but the average grade of these in the USA is only 0.08 oz ton^{-1}. Kavanagh (1980) estimated that by 1985 Canadian gold production should reach 2.5 million ounces, obtained from ore with an average grade of 0.2 oz ton^{-1}.

The main problem with future gold supply, especially in the western world, is the indicated falloff of South African gold production during the 1990s. If this situation is to be averted it is important that another major early Proterozoic fossil placer basin, similar to the Witwatersrand, be discovered. It is extremely unlikely that increased gold production from porphyry, Carlin-type deposits, or from conventional vein deposits could make up for a serious decrease in South African gold production.

Global resources of gold, assuming an average price of US $600 per ounce over the next two decades, are estimated at 2118 million ounces (Pretorius 1981b). This will represent a distinct shortfall with respect to anticipated demand. If energy and labor costs continue to rise and world demand for gold remains strong, a major increase in the price of gold within the next decade appears likely.

Bibliography

Bonham H F Jr 1986 Models for volcanic-hosted epithermal precious metal deposits: A review. In: Brathwaite R L, Brown P R L, Roberts P J (eds.) 1986 *Proc. Symp. 5: Volcanism, Hydrothermal Systems and Related Mineralization*, International Volcanological Congress, Auckland, New Zealand. Australasian Institute of Mining and Metallurgy, New Zealand Branch, pp. 13–18

Boyle R W 1979 *The Geochemistry of Gold and its Deposits*, Bulletin 280. Geological Survey of Canada, Ottawa

Burke K, Kidd W S F, Kusky T 1985 Is the Ventersdorp rift system of southern-Africa related to a continental collision between the Kaapvaal and Zimbabwe cratons at 2.64 Ga ago? *Tectonophysics* 115 (1–2): 1–24

Burrows D R, Spooner E T C 1986 McIntyre Cu–(Au–Mo) depsoit, Timmins, Ontario, Canada. In: MacDonald A J (ed.) 1986 *Proc. Gold 1986*, Toronto, pp. 23–39

Colvine A C, Fyon J A, Marmont S, Smith P N, Troop D J 1988 *Archean Lode Gold Deposits in Ontario*, Miscellaneous Paper 139. Ontario Geological Survey, Ontario

Helgeson H C, Garrels R M 1968 Hydrothermal transport and deposition of gold. *Econ. Geol.* 63: 622–35

Henley R W, Adams J 1979 On the evolution of giant gold placers. *Trans. Inst. Min. Metall. Sect. B* 88: 41–50

Hodgson C J 1986 Place of gold ore formation in the geological development of Abitibi greenstone belt, Ontario, Canada. *Trans. Inst. Min. Metall.* 95: 185–94

Kavanagh P M 1980 The world's gold reserves and production capacity. *Geosci. Can.* 7: 73–81

Keppie J D, Boyle R W, Haynes S J (eds.) 1986 Turbidite-hosted gold deposits. *Spec. Pap. —Geol. Assoc. Can.* 32

Kerrich R, Fyfe W S 1981 The gold-carbonate association: Source of CO_2 and CO_2 fixation reactions in Archean lode deposits. *Chem. Geol.* 33: 265–94

McDonald J A 1983 The iron formation-gold association: Evidence from Geraldton area. In: Colvine A C (ed.) 1983 The geology of gold in Ontario, *Ont. Geol. Surv. Misc. Pap.* 110: 75–83

Phillips G N 1986 Geology and alteration in the Golden Mile, Kalgoorlie. *Econ. Geol.* 81: 778–808

Pretorius D A 1981a *Gold and Uranium in Quartz-Pebble Conglomerates*, Information Circular 151. University of the Witwatersrand, Johannesburg

Pretorius D A 1981b *Gold, Geld, Gilt: Future Supply and Demand*, Information Circular 152. University of the Witwatersrand, Johannesburg

Radtke A S, Rye R O, Dickson F W 1980 Geology and stable isotope studies of the Carlin gold deposit, Nevada. *Econ. Geol.* 75: 641–72

Sandiford M, Keays R R 1986 Structural and tectonic constraints on the origin of gold deposits in the Ballarat State Belt, Victoria. In: Keppie J D et al. (eds.) 1986 Turbidite-hosted gold deposits, *Spec. Pap. —Geol. Assoc. Can.* 32: 15–24

Sawkins F J, Rye D M 1974 Relationship of Homestake-type gold deposits to iron-rich Precambrian sedimentary rocks. *Trans. Inst. Min. Metall. B.* 83(810): 56–59

Slaughter A L 1968 The Homestake Mine. In: Ridge J D (ed.) 1968 *Ore Deposits of the United States, 1933–1967.* (Graton Sales Vol.). American Institute of Mining, Metallurgical and Petroleum Engineers, New York, pp. 1436–59

US Bureau of Mines 1988 *Mineral Commodity Summaries* 1988. USBM, Washington, DC, pp. 62–63

Winter D H de la R 1986 Cratonic foreland model for Witwatersrand basin development in a continental back-arc plate tectonic setting. *Geol. Soc. S. Afr. Geocongress* 86 (Abstract Vol.): 75–80

F. J. Sawkins
[University of Minnesota, Minneapolis, Minnesota, USA]

Gypsum and Anhydrite

Gypsum, the dihydrate form of calcium sulfate ($CaSO_4.2H_2O$), and anhydrite, the anhydrous form ($CaSO_4$), are found in close association in many parts of the earth's surface. Anhydrite has minor economic importance, but gypsum is a widely used industrial mineral owing to its unique ability to give up and retain water of crystallization at moderate temperatures. To understand the position of gypsum and anhydrite in the present and future world economy, it is appropriate to examine their uses, the structure of the industries and the properties of the materials.

1. Uses

The majority of uses of gypsum are based on the reversible reaction

$$CaSO_4.2H_2O \rightleftharpoons CaSO_4.\tfrac{1}{2}H_2O + 1\tfrac{1}{2}H_2O$$

and gypsum's modest solubility of 0.24 g per 100 cm^3 water at 0 °C. Anhydrite reacts exothermically with water to form gypsum, but, except for a few specialized uses, the reaction rate is too slow to be of practical value.

Stucco, the hemihydrate of calcium sulfate ($CaSO_4 \cdot \frac{1}{2}H_2O$), is also known as plaster of paris. When it is mixed with water and various additives it forms a fluid that can be molded or shaped into various products. With time the stucco and water recrystallize to form gypsum, a process known as setting. When gypsum is heated, the water of crystallization that is driven off provides fireproofing, a desirable property for construction uses. Building materials is the main area of application of gypsum.

Gypsum panels and plaster are major products. Paper-faced gypsum panels or wallboard, invented about 1890, is the main gypsum-based product in countries where building systems are based on lumber or formed-metal alternatives. Major plaster usage dates from the 1870s when set-control additives were discovered. In countries where building partitions are primarily masonry, plaster is the common gypsum-based building material, competing with cement. Plaster is applied directly to masonry walls, but is also used for decorative ceiling and wall panels.

In cement, gypsum and anhydrite act as set-controlling agents; about 5% of the finished weight of cement is gypsum or anhydrite ground with the cement clinker.

Landplaster is natural gypsum used as a soil amendment, supplying calcium and sulfur. Calcium replaces sodium in bentonitic soils, permanently swelling the clay mineral lattice and making the soil amenable to agriculture.

Although specialty plasters and fillers are a minor use of gypsum in terms of volume, they are an important application. Specialty plasters include molding plasters and castings, including medical and dental plasters, oil-well cements and absorbents. High-purity gypsum is used as a calcium supplement in food and for fillers, filters, glass manufacture and coal-mine rock dust.

In European coal mines, some use is made of anhydrite to build mine pillars with the help of accelerators to convert anhydrite to gypsum. Small amounts of alabaster gypsum are used for artistic carvings and other *objets d'art*. Gypsum or anhydrite can also be used as a raw material for the manufacture of lime, sulfur, sulfuric acid and cement. These processes have high energy requirements and this factor together with wide variations in the sulfur market have resulted in a decrease of this use for gypsum and anhydrite.

2. *Extraction Methods*

Most of the world's gypsum is produced by quarrying, or surface mining, owing to the fact that most gypsum deposits are the result of near-surface hydration of anhydrite. Underground mines are developed in regions where groundwater has dissolved outcrop gypsum or glaciation has covered it and quarrying is not economic. Most underground mines are room-and-pillar construction with extraction ratios between 70% and 85% and pillar sizes dependent on overburden load and the strength of the roof. Local geology dictates mining height. Roof bolting is commonly practised and both quarries and mines are highly mechanized in Europe, Canada, the USA and other developed countries.

In areas where labor is cheap and abundant, contrasting extraction styles are common; local markets are usually served by small operations with minimal equipment and export markets are supplied by highly mechanized quarries. Production varies widely between 1 and 100 t per person-day.

Gypsum rock is crushed and generally screened to a size suitable for calcining. Depending on the calcining method, 90% under No. 100 screen to 15 cm maximum size is appropriate. Ground gypsum, ready for calcining, is generally called landplaster. After crushing, impurities are sometimes removed by screening or hand picking, but mechanical beneficiation to upgrade gypsum is generally not practised in preference to selective mining or quarrying.

3. *Manufacturing Technology*

At atmospheric pressure the calcining of gypsum begins at 43°C; formation of the beta hemihydrate, stucco, is essentially complete at about 120 °C. Calcining with steam produces alpha hemihydrate. These hemihydrates are only two phases of the system $CaO–SO_3–H_2O$. Other phases, soluble anhydrite and insoluble anhydrite, along with alpha hemihydrate, are used for specialty products and plasters where a high-strength, hard, dense material is desirable.

Uncalcined gypsum uses are based on gypsum purity, lack of deleterious impurities and particle size. Processing is usually limited to crushing, grinding, screening or air separation.

Calcined gypsum is made by several methods. The most common method is in a kettle, an enclosed pot-like vessel with a central shaft for stirring the landplaster while the water boils off. Other calcining methods include: hollow flights, a hollow-core auger-like device that has hot oil circulating through the core; immersed combustion conical kettles; and rotary, beehive and Moorish kilns.

Moorish kilns remain unchanged from when they were first used in the 1400s or earlier. They are simple half ellipsoidal (long axis being vertical) hillside excavations lined with coarse noncalcining rocks. An oval doorway allows entry to stack coarse rock, introduce fuel and remove calcined rock, which is then ground into plaster. Stucco produced from kettles generally does not need regrinding before use in plaster or wallboard; however, regrinding stucco for plaster will produce a denser plaster.

Setting or rehydration time of plaster and stucco is controlled by additives. Organic compounds and mineral salts will increase and decrease setting times, respectively. In less technologically developed countries where setting control additives are generally not employed, only as much plaster as can be used within the natural setting time of 15 to 30 min is mixed at any one time.

Wallboard is made by creating a sandwich of two continuous sheets of paper with a layer of stucco–water slurry between, laid out continuously on a conveyor belt; the wallboard thickness is controlled by using a master roll or forming plate. The stucco slurry rehydrates as the newly formed board travels on the belt line away from the stucco–water slurry mixer. Retarders and accelerators are used so that the setting time is closely controlled and generally ranges between three and seven minutes to accommodate boardline speeds in the range 20–100 m min^{-1}. Wallboard weight is an important market factor related to gypsum purity.

4. Geology

4.1 Mineralogy

The mineralogy of the natural calcium-sulfate minerals is relatively simple, consisting of three minerals: anhydrite, bassanite and gypsum. The mineral species that exists at any particular location is dependent on pressure and availability of water, which is itself related to a wide range of variables, including pressure, temperature and other constituent mineral salts.

Gypsum is the common calcium-sulfate mineral in the near-surface environment. The common rock gypsum is usually easily distinguished from other minerals by its low hardness of 2 on Mohs scale, low specific gravity of 2.2–2.4, and its ability to give up water easily. Selenite is the very coarsely crystalline variety; alabaster is compact and very finely crystalline; and satin spar is a variety that crystallizes parallel to an elongated *c* axis and is generally found as vein-type fracture filling. Gypsum is commonly white, pink or tan. Gypsite is an impure variety of gypsum deposit which is gray or brown in color, depending, in part on impurities. Anhydrite is harder, rating 3 to 3.5 on Mohs scale, and heavier than gypsum with a specific gravity of 2.7–3.0, and it is commonly brown, blue, gray or tan. Bassanite, $CaSO_4 \cdot \frac{1}{2}H_2O$, is distinguishable only by x-ray diffraction and occurs in nature only in trace amounts.

4.2 Deposits

In the classical evaporite precipitation sequence, gypsum precipitates from saturated brines after calcium carbonate and before halite. Owing to the fact that concentrated brines are widely distributed on the earth's surface, depositional environments for gypsum are numerous and consist of playa lakes, brine pans, sabkhas and enclosed or restricted marine basins. Most gypsum deposits are depositionally associated with a basin or basin margin where circulation of seawater is restricted. Studies of modern deposits of anhydrite and gypsum on sabkhas along the Trucial Coast of the Sheikdom of Abu Dhabi have been useful in the understanding of ancient depositional environments. An example of ancient sabkha deposits in which anhydrite and gypsum, but few other evaporites, occur is the Illinois Basin. In contrast, the Michigan Basin contains a nearly complete suite of evaporites and has anhydrite commingled with halite. Lithologic textures and sedimentary structures of sabkhas, typified by nodular structures, allow definition of the depositional environment of many gypsum deposits.

A class of deposits not extensively used for the gypsum itself, but economically important, is salt-dome caps. As salt goes into solution in groundwater near the top of the dome, residual anhydrite collects and is converted to gypsum. Bacterial activity based on hydrocarbons results in the gypsum being converted to calcium carbonate and native sulfur. The sulfur-reducing bacteria utilize the lighter hydrocarbons, preferentially resulting in heavy oil and asphaltic accumulations being associated with salt domes that contain sulfur deposits. Gypsum deposits are associated with hydrocarbons in many deposits.

Similarly, dissolution of salt in bedded deposits can result in gypsum deposits that have characteristics of both shallow- and deep-water deposits, making the original depositional environment difficult to determine.

Solution of existing gypsum deposits by groundwater and subsequent deposition by evaporation results in gypsite deposits. Gypsite usually contains considerable quantities, up to 50%, of the original near-surface materials.

Commercial gypsum deposits are found in sediments of Ordovician age and younger. Owing to the relatively easy destruction of gypsum deposits by solution, Cambrian and older rocks contain few deposits and no known commercially exploited ones. The geological age and location of commercial gypsum deposits are given in Table 1.

Regardless of the original deposition, most actively exploited gypsum deposits are the result of hydration of anhydrite as it becomes exposed to groundwater. Exceptions are found in arid tropical or subtropical climates where Tertiary-age gypsum has not been buried deeply enough to be converted to anhydrite.

As hydration of anhydrite occurs, impurities that are contained in the anhydrite lattice and associated with anhydrite surfaces are rejected and not included in the gypsum crystal lattice. The first product of hydration of anhydrite is likely to be bassanite, which is metastable, and hydrates to gypsum if water is available.

In many places gypsum associated with anhydrite is very hard and dense, but recrystallization appears to continue in many deposits, resulting in coarsely crystallized, nearly selenitic, gypsum. Regardless of whether or not the originally deposited calcium-sulfate mineral was anhydrite or gypsum, as gypsum

Table 1
Geological age and location of main commercial sources of gypsum deposits

Geological age	Area
Tertiary	California, Caribbean, Europe, Mexico, Nevada
Jurassic	Arkansas, Colorado, Europe, Iowa, Nevada, Utah
Permian	Kansas, New Mexico, Oklahoma, Texas
Carboniferous	Europe, Indiana, Nova Scotia, Virginia
Devonian	British Columbia, Iowa, Manitoba
Silurian	Manitoba, New York, Ohio, Ontario

becomes buried, it dehydrates to anhydrite at pressures and temperatures approximating those at 1300 m below the land surface.

4.3 Impurities

Impurities are important considerations in the utilization of gypsum deposits. Deposits that originate in low-energy basins are usually very pure, having less than 5% carbonate and fine clastics as impurities. Deposits that originate in sabkha environments also can be pure, but they commonly contain materials that were part of the original sediment of the coastal plain. Carbonates, mainly dolomite, are very common impurities. Clastic impurities are generally little affected by brines, with the exception of smectites which may be altered from sodium-based clays to calcium-based clays.

Soluble salts that may be deposited in small amounts include other sulfates, especially those of potassium, sodium and magnesium, and chlorides. Soluble salts are especially important in gypsum and anhydrite manufacturing technology because they affect water vapor pressure, important in calcining, and paper bonding to the gypsum core in wallboard manufacture.

4.4 Prospecting

Gypsum deposits are relatively easy to find when they crop out on the earth's surface. Blind deposits, those that do not crop out, are generally discovered by core drilling in favorable areas. Geological and hydrological studies, literature searches and examination of existing drill cuttings and cores in repositories are among the methods used to find gypsum deposits. Geophysical methods have little applicability. Gypsum is a low-cost bulk material that has place value; however, a more general rule of gypsum value is that the lowest production cost in the market area has the greatest value. Thus, the value of a gypsum deposit is a combination of place value and other market factors.

4.5 Synthetic Gypsum

Synthetic gypsum is a term applied to gypsum that is the by-product of another process. Several chemical processes result in synthetic gypsum, the most common ones being the manufacture of phosphate fertilizers, acetic acid and titanium dioxide pigments. Gypsum is also a common product of acid neutralization; for example, pickling liquor. Flue-gas desulfurization at power plants produces large quantities of gypsum, second only to phosphogypsum resulting from fertilizer manufacture. Desulfurization gypsum generally contains soluble salts from the original coal: fly-ash, if not collected separately; calcium sulfite, a result of incomplete oxidation; and unreacted calcium carbonate or other desulfurizing agent. Phosphogypsum commonly is very fine, contains a high moisture content and, owing to decay of the radium in the original phosphate rock, gives up radon daughter products.

The amount of synthetic gypsum produced worldwide is approximately twice the amount of natural gypsum that is mined or quarried. Usage of synthetic gypsum, however, has been slow to increase for the reasons below.

(a) Most synthetic gypsums are of lower quality or contain deleterious impurities that limit their utility. The cost to process synthetic gypsum, especially phosphogypsum and desulfurization gypsum, is usually prohibitive.

(b) Sources of synthetic gypsum are not often close to preexisting manufacturing facilities that use gypsum. The manufacturing facilities usually are capital intensive and not often constructed or moved.

(c) Most gypsum-producing mines and quarries are captive operations owned by the manufacturing companies which are sensitive to raw-material supply.

(d) Many synthetic gypsums are not amenable to the material handling systems that are present at existing operations, resulting in prohibitive cost penalties if used. High moisture contents of up to 25% and fine particle size are examples of this problem.

5. Industry

5.1 Structure

In most western countries the individual markets for manufactured gypsum products are generally dominated by three or four companies. One example is Canada were over 95% of all gypsum products are manufactured by Domtar Gypsum Company, Canadian Gypsum Company and Westroc Gypsum Company.

In the USA the two largest manufacturers, United States Gypsum Company and National Gypsum Company, satisfy over half of the market demand, and the next three largest competitors, Domtar Gypsum Celotex and Georgia–Pacific, supply approximately 35% more. The remaining 10–15% of the manufactured product market is split among ten, one- and two-plant, operating companies. The exception to a very competitive industry is in the UK where British Plasterboard Ltd is a regulated monopoly. In countries with a gypsum industry based on plaster usage, industry is structured to localized supplies and usage, with two or more producers commonly sharing the local market.

Gypsum products are heavy and can normally stand long-distance transportation costs only in times of good markets. Recent building markets in the USA have brought about some changes in product distribution that would have been uneconomic only a few years ago. Examples of such changes are wallboard manufactured in Montana being shipped by truck to the East-Coast markets and Norwegian wallboard being imported into Florida.

Most leading North American companies maintain product development facilities to improve product quality and manufacturing techniques, and have technical agreements with European counterparts.

Raw gypsum markets in the USA are very different from the markets of the manufactured products. The major producers of agricultural gypsum and cement rock include several producers which have large local markets and no facilities for making wallboard.

5.2 Distribution of Operations

Manufactured gypsum products are heavy and so many gypsum-product operations are located at deposits that are controlled by the producing companies. Historic major gypsum districts in the USA and Canada are in upstate New York, northern Ohio, Michigan, Iowa, Ontario and Nova Scotia. Of these, Michigan, Iowa, Ontario and Nova Scotia are still leading producing areas. Major producing states in the USA are Texas, California, Michigan, Iowa, Nevada, Indiana, Oklahoma, Arizona, Ohio, New York, Colorado, Arkansas, Kansas and Virginia. There are no major gypsum deposits on any coasts of the USA. The three coasts, Atlantic, Gulf and Pacific are supplied to some extent from interior deposits, but most of the gypsum in coastal manufacturing facilities is imported from Canada or Mexico.

In Canada gypsum is produced in Newfoundland, Nova Scotia, Ontario, Manitoba and British Columbia. Nova Scotia is the major producer, supplying both internal Canadian markets and much of the East Coast of the USA.

In Mexico, gypsum is common and occurs in many states. Markets in the gulf coastal plain, which lacks gypsum deposits, are supplied by internal deposits.

Most European countries have a good distribution of deposits with the exception of the Scandinavian countries, which have none. The gypsum deposits of the Paris Basin and the Midlands of England are well-known historic producers that continue to be major producing areas. Spain is believed to have the largest reserves of high-purity gypsum in Europe.

The Middle Eastern countries have enormous reserves, but they are mostly undeveloped. Australia has large reserves as does Thailand. Japan's gypsum industry is based on synthetic gypsum as it has no natural gypsum deposits.

5.3 World Production and Trade

The USA is the largest producer and consumer of gypsum, producing about 15 million tonnes and consuming an additional 7 to 8 million tonnes of imported rock. Of the gypsum imported into the USA, 4 to 5 million tonnes is supplied by Canada, mostly from Nova Scotia, about 2 million tonnes is from Mexico and approximately one million tonnes is from Spain. Major producers of gypsum are shown in Table 2.

Major importers are the USA, Japan, the Scandinavian countries and Taiwan. The main exporters are Canada, Spain, Australia and Thailand. Many countries export smaller amounts, a few hundred thousand tonnes or less, and many countries import roughly the same quantities. Most of this smaller quantity trade is for portland cement and plaster use.

An important change in the world trade of gypsum has come about since the late 1970s with the availability of low-cost high-purity gypsum supplies from Spain. This gypsum enjoys a very low-cost freight rate, returning to North America in ships that had carried grain or other bulk cargo to Europe. Spanish gypsum

Table 2
Major producers of gypsum in 1986

Country	Amount produced (10^6 t)
Australia	1.2
Canada	8.9
People's Republic of China	5.0
UK	3.1
France	5.5
FRG	2.3
Italy	1.4
Iran	4.8
Japan	6.2
Mexico	3.2
Poland	1.3
Romania	1.6
USSR	4.9
Spain	5.4
Thailand	1.4
USA	14.8

Source: US Bureau of Mines

now dominates the cement-rock business on the East and Gulf Coasts of the USA, and a wallboard plant being constructed in the New York area will use Spanish gypsum as a raw material.

Monetary devaluation is also a strong factor in gypsum trade. In the late 1980s, gypsum from Mexico has been delivered to ports in the USA at a lower dollar cost per tonne than at any time since 1960, and during that time devaluation of the Spanish peseta has roughly paralleled the devaluation of the dollar.

5.4 Industry Changes

Over the long term, gypsum usage is likely to continue growing at a few percent annually. Business cycles that affect the gypsum industries result in increases of as much as 50% of the low production figures. Because of the large investment required to build most gypsum facilities, and other factors, the break-even capacity of many gypsum operations is 75 to 80% of capacity; profitability is achieved only when operating at relatively high operating capacities.

In the USA, gypsum usage has historically been related to new home construction. Commercial construction, whose building cycle is different from the housing cycle, an enlarging repair and remodelling market and the proliferation of home mortgage financing has greatly changed the historic building cycle, resulting in levelling out of gypsum usage. Future annual changes in gypsum usage are likely to be smaller than in the past as a result of these moderating factors.

Manufacturing technology will continue to improve, resulting in lighter-weight wallboard and higher boardline speeds. Board weights of 1750 lb MSF^{-1} (pounds per thousand square feet of 1/2″ thick board) (8.54 kg m^{-2} per 1.27 cm thick board) are now common compared with 1900–2000 lb MSF^{-1} (9.3–9.8 kg m^{-2} per 1.27 cm thick board) in the early 1960s. Wallboard imports into the USA are dependent on monetary exchange rates and are not likely to grow significantly. Wallboard usage is likely to grow slowly in countries now using gypsum plaster as acceptance of metal framing and labor costs increases. Use of synthetic gypsum will grow and displace some natural gypsums as better quality synthetic gypsum becomes available, historic gypsum deposits become depleted and disposal of synthetic gypsum becomes increasingly problematic.

Bibliography

Appleyard F C 1983 Construction materials, gypsum and anhydrite. In: LeFond S J (ed.) 1983 *Industrial Minerals and Rocks*, 5th edn. American Institute of Mining, Metallurgical and Petroleum Engineers, New York, pp. 183–97

Appleyard F C 1983 Gypsum and Anhydrite. In: LeFond S J (ed.) 1983 *Industrial Minerals and Rocks*, 5th edn. American Institute of Mining, Metallurgical and Petroleum Engineers, New York, pp. 775–92

Kelly K K et al. 1941 *Thermodynamic Properties of Gypsum and its Dehydration Products*, Technical Paper 625. US Bureau of Mines, Washington, DC

Kinsman D J J 1969 Modes of formation, sedimentary associations and diagenetic features of shallow water and supratidal evaporites. *Am. Assoc. Pet. Geol. Bull.* 53: 830–40

MacDonald G F J 1953 Anhydrite–gypsum equilibrium relations. *Am. J. Sci.* 251: 883–98

US Bureau of Mines 1985 *Mineral Facts and Problems*, US Bureau of Mines Bulletin 675. USBM, Washington, DC

D. B. Jorgensen
[US Gypsum Co., Chicago, Illinois, USA]

H

Hydrocarbons: Origin, Migration and Accumulation

1. Origin of Hydrocarbons

Modern consensus on the origin of hydrocarbons states that formation takes place under exclusively organic (biogenic) conditions. Modern fauna and flora, in their nanomicroscopic to ultramicroscopic elements, reveal an actual production of hydrocarbons during the life cycle of many living taxa. Proponents of an organic origin therefore assume that hydrocarbons were present also in paleotaxa, and that these paleohydrocarbons contributed at least in part to the formation of petroleum.

Although this consensus has been strong for more than a half century, however, it has not gone unchallenged. Adherents of an inorganic (abiogenic) origin of hydrocarbons continue to present their views, thus keeping alive a theory more than a hundred years old. In the 1980s these adherents have turned actively to seismology, to a carbon budget of the crust, to assumptions of massive outgassing from the earth's crust and mantle, and even to the drilling of a deep borehole in search of evidence for their theory.

1.1 Inorganic (Abiogenic) Origin

Proponents of the inorganic theory of the origin of hydrocarbons accentuate the widely accepted presumption that hydrogen and carbon are present through most or all levels of the earth's crust. The presence of these same two elements in the underlying mantle also seems possible. Generation of the hydrogen may come from chemical reaction between superheated steam and mineral oxides, especially iron oxide. Catalysts to combine carbon and hydrogen deep within the crust, on the other hand, are unlikely to exist. Proponents of the inorganic theory of origin believe, however, that heat and pressure at depths many tens of kilometers below the surface of the earth may substitute for catalysts, and that the resulting hydrocarbons begin their existence at depths far below sediments, whether such sediments lie as thin columns on the craton or whether they lie as thick columns in structural foredeeps. The theory that heat and pressure can accomplish an inorganic production receives support from the presence of microscopic inclusions of hydrocarbon in diamond mined from pipes of kimberlite.

With regard to heat and pressure in the earth's crust, some new parameters which contribute to the inorganic view have appeared with the drilling of SG-3 on the Baltic Shield of the Kola Peninsula, USSR. This tremendous borehole, already more than 14 years in the drilling and by far the deepest borehole in the world, had reached a depth of some 12 000 m by 1984. Reports from the USSR mentioned gases and inflows of strongly mineralized waters at more than 11 000 m below the earth's surface. These came into the borehole despite pressures greater than 3000 bars and temperatures greater than 180 °C. Included in the gases were hydrocarbons.

Those who support the inorganic theory of origin might argue that the discovery of hydrocarbons by a single borehole in deeply buried Archean gneiss bodes well for the presence of widespread inorganic hydrocarbons in the deep crust. This borehole, however, contains 7000 m of Proterozoic sedimentary cover above the gneiss. With dolomites, siltstones and phyllites present, a generation of organic (biogenic) hydrocarbons is possible within this cover despite no more than its "cryptozoic" level of organic material. Such organic hydrocarbons may have migrated downward from the Proterozoic sedimentary beds through fractures to lodge in Archean gneiss. Such a possibility clouds the status of the "superdeep" well SG-3 as a rigid test of the inorganic (abiogenic) origin of hydrocarbons.

A deep borehole drilled near Siljan in central Sweden differs from SG-3 on the Kola Peninsula in having, as its sole objective, a definitive proof of the inorganic origin of hydrocarbons. Suspended at 6300 m, the borehole at Siljan drilled fractured granite from top to bottom, and reportedly recovered "significant" (though probably small) quantities of hydrogen and helium some 200 m above the bottom.

Opponents of inorganic theories of origin, however, note that the Siljan granite, more than 35 km wide, is almost ringed by fossiliferous Ordovician shales with seeps of gaseous, liquid and bituminous petroleum. Opponents believe, accordingly, that if the operators continue drilling below 6300 m and discover methane, they will have proved only that organic (biogenic) methane had migrated through fractures into the granite from peripheral Ordovician source rocks. As with SG-3 on the Kola Peninsula, the borehole at Siljan in central Sweden, if deepened, may ignite considerable debate concerning its trustworthiness as a test of the origin of hydrocarbons. Insofar as the operators of Siljan claim an inorganic origin for the hydrogen and the helium in their borehole, they appear justified. The presence of these two gases through 6300 vertical meters of uninterrupted granite also justifies their assertion that these gases (but only these gases) were present in an early inorganic phase of the earth's history.

1.2 Organic (Biogenic) Origin

Proponents of the organic theory of the origin of hydrocarbons agree that carbon and hydrogen were present in the primordial earth. They maintain, however, that only in an organic regime could these two elements combine to form recognizable molecules of hydrocarbons, and eventually of petroleum. For supporting evidence they cite the exclusive association of important deposits of petroleum with sedimentary rocks. These sedimentary rocks, whether formed under marine or continental conditions, contain organic material in an original or transformed state. Up to 1988 no important deposits of petroleum have appeared in terranes of purely igneous or purely metamorphic lithofacies.

Whereas the primordial earth was cooling from a high-temperature state, the geochemical evidence from petroleum itself is for an origin of hydrocarbons at relatively moderate temperatures. Petroleum rotates the plane of polarized light; this indicates that it contains molecules almost certainly of organic origin, formed at no more than moderate temperatures. Some of these molecules belong to the porphyrins, formed either from the chlorophyll of plants or the hemin of blood. Other compounds present in petroleum also indicate derivation from living matter; for example, cholesterol, carotene and terpenoids are derived from plants.

Modern geochemists suggest an origin of oil from organic matter in two ways:

(a) directly from the hydrocarbons that many taxa, especially marine taxa, contain in their living cells—the contribution from this source may amount to about 10% of the total; and

(b) from decay and chemical alteration of buried organic matter—the contribution from this source, much larger than that from the first source, may amount to the remaining 90% of the total.

Evidence for the smaller contribution comes from a similarity of the hydrocarbons in petroleum to those found in plankton, kelp and in taxa even as advanced as fish. For the larger contribution, geochemists assert that burial, decay and chemical alteration produce homologues and polymers of methane. Present are all hydrocarbons that contain up to ten carbon atoms, and possibly half of all hydrocarbons that contain more than ten carbon atoms.

Before burial, photosynthesis of plants and algae converts water and carbon dioxide into oxygen and the organic saccharide glucose. Polysaccharides and a variety of complex organic compounds follow. At the end of the food chain, oxidation completes a cycle by breaking down complex organic molecules back to their original carbon dioxide and water. When burial takes place, however, the complex organic compounds escape a complete impact of the cycle, and it is these, as mentioned above, that furnish many of the hydrocarbons with ten or more carbon atoms.

In areas of incremental deposition, the depth of burial increases continually, and organic material in the buried sediment must experience continual increases of temperature and pressure. In response to these conditions the organic material must pass through a series of changes that geochemists have divided into the following three stages.

(a) *Diagenesis.* This occurs not far below the earth's surface, and with only a small increase of temperature and pressure. In this stage, reactions of organic character include decay through attack by bacteria (reactions of inorganic character also occur), the content of oxygen declines, and after a loss of methane, carbon dioxide and water the residue consists of kerogen, a solid and complex hydrocarbon.

(b) *Catagenesis.* Increasing temperature and pressure act on the residual kerogen, whose transformation during this stage, however, appears to be largely a thermal process. In addition, as in the preceding stage of diagenesis, the ratio of hydrogen to carbon declines, and theory supposes a release of oil and gas from kerogen in this stage.

(c) *Metagenesis.* Temperatures and pressures are considerably greater than in the second stage of catagenesis and are sufficient, ultimately, to expel all hydrocarbons and to convert the organic carbon to inorganic graphite.

Kerogen, described as the organic matrix of rocks, contains chemical compounds derived from carbohydrates, proteins and lipids, all three of which yield hydrocarbons under appropriate thermal conditions. Chemically, however, lipids bear the closest relations to hydrocarbons, for which reason lipids represent the most likely precursors of petroleum. In kerogen itself, slight differences in its percentages by weight of carbon, hydrogen and oxygen lead to a threefold classification.

(a) Kerogen type I—derived for the most part from algae. In this group, lipids are dominant, derivatives of oils, fats and waxes are present, there are many aliphatic chains, and there is the highest ratio of hydrogen to carbon.

(b) Kerogen type II—characterized by liptinite (exinite). Like Kerogen type I, this class contains aliphatic chains, but has derivatives of algal detritus accompanied by derivatives of zooplankton and phytoplankton, and a lesser ratio of hydrogen to carbon.

(c) Kerogen type III—this is the humic type, derived from the lignin of terrestrial woody plants, with few aliphatic chains and many aromatics with least ratio of hydrogen to carbon, and believed therefore to produce more gas than oil.

Kerogen is said to mature. Many variables, however, enter into its assumed maturation and attempts to quantify the process have taken two main forms. The first employs an index of thermal maturity (ITT) which takes into account the temperature and time of burial in each interval of temperature through each increment of 10 °C. The second approach utilizes an organic index (LOM) which is based on an assumption that maturation doubles for each increment of 10 °C. These quantifications, on the whole, correspond largely to the formation of oil and gas in catagenesis, and to the formation mainly of gas in earliest metagenesis.

2. *Migration of Hydrocarbons*

The rocks that serve as reservoirs for large accumulations of hydrocarbons consist for the most part of sediments deposited in or not far below the photic zone, where light, aeration and conditions of high energy would inhibit the formation of much organic material. For this reason and others, explorationists assume that oil and gas do not form in the reservoir itself, but that they migrate towards and into it from another rock, the source rock. The migration from source to reservoir can be effectively continuous, but mainly it resolves into a primary migration and a secondary migration.

2.1 Primary Migration

In the simplest form of this concept, water carries either petroleum or its precursor, in extremely low concentrations, out of the source rock as compaction drives fluid from the latter. If the source rock consists of a shale or a siltstone, its initial condition would have been that of a mud containing 70% or more of water. A constantly increasing overburden of uninterrupted sedimentation squeezes water continuously out of the compacted mud, and if organic material has survived in the original mud, some of this organic material presumably moves out together with the escaping water. Fluids migrating from the source rock may move vertically in the initial stages, but as compaction proceeds, the flat clay minerals present in the source rock assume a horizontal position, and they may impose a largely lateral motion upon migrating fluids.

The mechanism by which water carries hydrocarbons towards eventual accumulation remains largely conceptual, to a degree that the subject of primary migration has attained the status of "the last great mystery of petroleum geology." Petrophysicists believe that oil is not capable of passing through the pores of clayey mud either as small drops or as colloids. It is reasoned, consequently, that oil within the range of low molecular weights migrates in solution, and that its precursors migrate also in solution, but within the range of high molecular weights. The weak point of this model appears to be its need to convert the precursors to complete hydrocarbons by "mild cracking" after they enter the reservoir rock, presumably by "coagulation" in a changed "physical–chemical environment."

The time needed for the formation of hydrocarbons by transformation of kerogen appears to conflict, at least in part, with the model of a relatively early expulsion of hydrocarbons together with water under the impetus of simple compaction. Some petrophysicists, for this reason, have proposed that there is an increase in specific volumes and pressure within the source rock as the presumed transformation of kerogen to hydrocarbons of low molecular weight takes place. This internal pressure, developed later than the pressure of initial compaction, may drive additional specific hydrocarbons out of the source rock into a primary migration. This model, while reasonable, may be insufficient to cover the whole spectrum of primary migration. A repetition or repetitions of increase in specific volumes, however, could improve considerably the expulsion of hydrocarbons and precursors from the source rock even during the comparatively late phases of its compaction. Another possibility in the late stages of compaction, especially if the pressure of the fluids in pores rises above rock pressure, is the development of microfractures. Should these microfractures connect with each other, they would offer an important pathway to primary migration.

With respect to the distance covered by hydrocarbons in primary migration, the petrophysical consensus is to the effect that the distance is short. Examples of statements concerning this factor are "only as far as needed to reach a permeable bed" or "distances covered by primary migration are commonly in the order of meters or tens of meters." The latter statement implies a strong vertical component in primary migration; that is, migration to permeable rock immediately above or below the source rock.

2.2 Secondary Migration

This phase of the movement of hydrocarbons is one that virtually by definition follows the phase of their expulsion from source rocks. The permeable beds in which secondary migration takes place are the carrier beds, precisely because they carry hydrocarbons to their ultimate accumulations. If a carrier bed remains porous without interruption in a way that does not deter the impetus of buoyancy, the length of movement of hydrocarbons through that carrier bed becomes almost limitless. Where explorationists see indications of broad and largely undeformed structures they invoke secondary migrations as long as hundreds of kilometers. It should be noted that this concept of long migration is not new. Explorationists more than a half century ago referred to long "gathering grounds" as a favorable condition for prolific final entrapment of hydrocarbons.

3. *Accumulation of Hydrocarbons*

With the buoyancy of migrating hydrocarbons as the main motive force to propel them through carrier beds, any impediment to buoyancy terminates migration and brings about accumulation. In two dimensions, a reversal in dip of the carrier bed to form an anticline is sufficient to provide a trap in which gas can accumulate at the top and oil can accumulate below the gas. Under real subsurface conditions in three dimensions, however, the trap can accumulate and retain its hydrocarbons only if a double plunge of the structure provides complete closure. A double plunge is also necessary to retain hydrocarbons when accumulation begins with truncation or wedging at an unconformity. Faulting may replace plunge partly or wholly as an agent of closure. In trapping that is wholly stratigraphic, the agent of closure varies. Where accumulation takes place in reefs or bioherms, envelopment of these carbonate buildups by shales provides closure. Where salt penetrates carrier beds, the flanks of the salt itself convert carrier beds into reservoirs with accumulated hydrocarbons. Where a paleotopography develops on the upper surfaces of unconformities, buoyancy sends oil and gas into the high areas of this paleotopography, and accumulation occurs there under the cover of cap rock.

4. *Conclusions*

An inorganic (abiogenic) theory of the origin of hydrocarbons, now at least a century old, continues to find adherents, especially during the last decade, despite an almost overpowering consensus for an organic origin that has gathered strength since the 1940s. Adherents to the inorganic theory maintain that hydrocarbons, especially methane, escape from the mantle and the crust in an outgassing that has continued from the time of the primordial earth to the present.

Proponents of an organic theory of the origin of hydrocarbons cite an almost exclusive association of important deposits of hydrocarbons with sedimentary rocks. They cite also the ability of oil to rotate the plane of polarized light and the presence in oil of porphyrins as indications of organic (biogenic) character. Additional evidence cited in support of this theory includes the presence of microscopic quantities of actual hydrocarbons in the cells of living taxa and the conversion of inorganic carbon dioxide and water by photosynthesis into organic glucose, the precursor of polysaccharides and other organic complex molecules.

Under incremental deposition, organic material must experience continual increases of temperature and pressure, loss of oxygen, carbon dioxide and water, and the formation of kerogen, the "organic matrix" of rocks. Kerogen itself presumably "matures," a process that theorists have quantified, and which corresponds for the most part to the release of oil and gas during the theoretical stage of "catagenesis."

The almost exclusive association of large, medium and small deposits of hydrocarbons with sedimentary rocks is certain. That such an association renders an inorganic theory of the origin of hydrocarbons untenable, is less certain. Possibly overlooked is the circumstance that for the tens of thousands of wells drilled specifically for hydrocarbons in sedimentary rocks, only one has drilled a thick body of granite from surface to total depth. The ratio of drilling in igneous rocks specifically for hydrocarbons to similar drilling in sedimentary rocks should therefore increase to even a minimum value before opponents can put forward a reasoned devaluation of the inorganic theory. An important amount of drilling in granites and gneisses, for example, might show the presence of residual methane. The polymerization of this simple aliphatic hydrocarbon to hydrocarbons of high molecular weight might seem unlikely at first glance. In theory, however, methane from the mantle and the crust could enter the sedimentary realm and become the beneficiary of pressure owing to overburden as well as the beneficiary of catalysis by clay minerals. In such a model the origin of hydrocarbons in sediments could be part inorganic and part organic.

In the organic model of recent years, details of primary migration remain purely speculative. Uncertainty exists over the relationship of the expulsion of hydrocarbons to compaction of the source rock, and over the nature of the hydrocarbons themselves at the time of expulsion. Precursors of oil and gas (protopetroleum) may be present as fluid leaves the source rock. Alternatively, oil and gas already formed may leave the source rock together with precursors. In either case, the distance to a porous carrier bed may amount to no more than a few meters. The carrier bed, by implication, is thus likely to lie immediately above the source rock. Once in the carrier bed, buoyancy becomes the motive force for the propulsion of hydrocarbons and precursors, and secondary migration has begun.

On broad epeirogenically simple forelands, secondary migration may continue along the carrier bed for distances of well over a hundred kilometers. Only when tectonic modification occurs, and when the tilt of the carrier bed reverses itself, do conditions for accumulation arise. For accumulation and consequent permanent retention of hydrocarbons, structural closure is necessary. Among such types of closure are doubly plunging anticlines, truncation or wedging, faulting, envelopment of reefs or bioherms by shale, truncation against salt and crestal paleotopography.

See also: Petroleum: Oil and Gas Fields; Petroleum: World Resources

Bibliography

Anon 1987 Swedish test of Gold's gas theory inconclusive. *Pet. Rev.* October 1987: 40

Atwater G I 1980 Petroleum. In: Goetz P W (ed.) 1980 *The*

New Encylopaedia Brittanica, Macropaedia, Vol. 14. Benton, Chicago, Illinois, pp. 164–75
Giardini A A, Melton C E 1983 A scientific explanation for the origin and location of petroleum accumulations. *J. Pet. Geol.* 6(2): 117–38
Giardini A A, Melton C E, Mitchell R S 1982 The nature of the upper 400 km of the Earth and its potential as the source for non-biogenic petroleum. *J. Pet. Geol.* 5(2): 173–90
Gold T 1979 Terrestrial sources of carbon and earth outgassing. *J. Pet. Geol.* 1(3): 3–19
Hobson G D (ed.) 1977 *Developments in Petroleum Geology—1*. Applied Science, London
Hunt J M 1979 *Petroleum Geochemistry and Geology*. Freeman, San Francisco, California
Hunt J M 1982 Petroleum, origin of. In: Parker S P (ed.) 1982 *McGraw-Hill Encyclopedia of Science and Technology*, Vol. 10. McGraw-Hill, New York, pp. 77–78
Landes K K 1959 *Petroleum Geology*. Wiley, New York
Levorsen A I 1954 *Geology of Petroleum*. Freeman, San Francisco, California
Selley R C 1985 *Elements of Petroleum Geology*. Freeman, New York
Shirley K 1987 Siljan project stays in cross fire. *AAPG Explorer* January 1987: 1, 12, 13
Tissot B P, Welte D H 1984 *Petroleum Formation and Occurrence*. Springer, New York
Whiteman A 1987 Gold's hypothesis. *Pet. Rev.* November 1987: 8

M. Kamen-Kaye
[Cambridge, Massachusetts, USA]

I

Igneous Ore Deposits: Formation Processes

Today, mineral resources have become almost synonymous with industrial power, and industrial power is in turn dependent upon ownership of, or access to, large quantities of mineral resources. It is startling to realize that the rise of the industrial age has so accelerated the demand for minerals that the world has dug up and consumed more of its mineral resources since the beginning of World War II than in all preceding periods. This insatiable demand for minerals to feed industry has made some sources of supply that we used to think were adequate now look rather small, and sources capable of meeting large demands are being depleted rapidly.

More than a century and a half ago, the rate of increase in the consumption of copper and iron was directly proportional to the increase in population. If the population doubled, the consumption of these metals doubled. In contrast, during the first hundred years of the Industrial Revolution, roughly from 1812 to 1912, the population increased fourfold whereas the consumption of copper increased 80 times and that of iron 100 times.

The quest for the wealth of minerals wrested from the earth for man's vanity, necessities or comforts has always been a powerful incentive to discover, explore and trade. Geologists study the occurrences of mineral deposits, their mode of formation, the types of rocks in which they occur, and use geophysical tools and geochemical processes as aids in attempts to locate deposits.

Geologists have been aided in the search for mineral deposits by devising several means of classifying the variety of mineral and rock concentrations. Some classifications refer to the geometry of a deposit, e.g., tabular or veined, flat or stratigraphic, irregular or "bunched," and numerous fine veins or fissures. Another classification is by time of formation. Deposits formed at the same time as the host rocks are called syngenetic, and if the mineralization invaded pre-existing host rocks, epigenetic.

Classifications based on temperatures, pressures, reduction–oxidation conditions and partial pressures of fluids provide much information on conditions of formation of mineral deposits. Of course, these factors are not easily determined but relating the results of laboratory experiments to detailed observations of field relationships of minerals and rocks has provided much understanding of the conditions of formation of specific types of deposits.

Lastly, as many mineral deposits form only in specific kinds of rocks, a classification based on the mode of formation and associated rock types can aid the exploration geologist in the search for new deposits. Rocks are separated into three groups based on their mode of origin. Igneous rocks are formed from molten magma, plus its dissolved fluids, either intrusive (formed below the surface of the earth) or extrusive (formed on the surface of the earth); sedimentary rocks are formed by deposition in seas or lakes, or are deposited as thin layers on the surface of the earth; and metamorphic rocks are formed from pre-existing rocks in environments of high temperatures and pressures existing below the surface of the earth.

Igneous rocks are aggregates of minerals that form by precipitation or crystallization from a melt or magma, i.e., molten rock plus its dissolved fluids (gas and liquid). Evidence from plate tectonics suggests that the provenance of most magmas is at plate margins, a result of sea-floor spreading or subducting plates. At depths from some tens to several hundred kilometers, magmas are formed by partial melting of specific constituents which then rise from the mantle or lower crust as somewhat plastic masses and may adsorb water when reaching upper crustal, water-bearing sediments. Igneous rocks are classified by: (a) texture, or the size, shape and arrangement of the mineral grains in the rocks; and (b) the composition of the minerals in the rocks.

1. Intrusive Rocks

The felsic rocks generally contain about 70–80 wt% SiO_2 with a high content of quartz and alkali feldspars ($KAlSi_3O_8$ and $NaAlSi_3O_8$). Since Si, Al and alkalis are important determinants of viscosity, the magma that forms with these minerals has a high viscosity. In contrast, mafic magmas containing higher charged ions such as Ca, Fe or Mg, and Si-poor or late stage differentiates of felsic magmas which are OH-rich, are less viscous or moderately fluid. Rhyolite magma, high in SiO_2, is highly viscous and even when cooled slowly forms a glass which leaves the silica tetrahedrons $(SiO_4)^{4-}$ in random positions. In contrast mafic magma may be cooled at the same rate or even more rapidly but because of its high fluidity, mineral crystals have sufficient time to form. Because of this high fluidity, mafic glasses are rare.

2. Magmatic Deposits and Processes

Magmatic deposits result from simple crystallization or from concentration by differentiation of mafic or ultramafic magmas. Magmatic deposits are characterized by their close relationship with ultramafic deep-seated igneous rocks commonly derived from the upper mantle and include diamonds, nickel, platinum group minerals, titanium–iron, chromite and other valuable ores.

The valuable metals deposited are intermixed or juxtaposed with the igneous body. They may form only during each crystallization in the igneous rock, or they may form in a continuous manner as the magma slowly crystallizes, or they may remain in the liquidus until the body has almost completely crystallized. The times of formation will result in different concentrations of the metals because of varying solubilities of the minerals in the solution melt. A diagram showing this process and the effects of gravity settling and forming a liquid deposit is shown in Fig. 1.

Some deposits consist of disseminated ore minerals. If such crystals are valuable and concentrated enough, the result is a magmatic mineral deposit. Diamonds disseminated in the kimberlite pipes of South Africa are examples. Other bodies form as either early or late

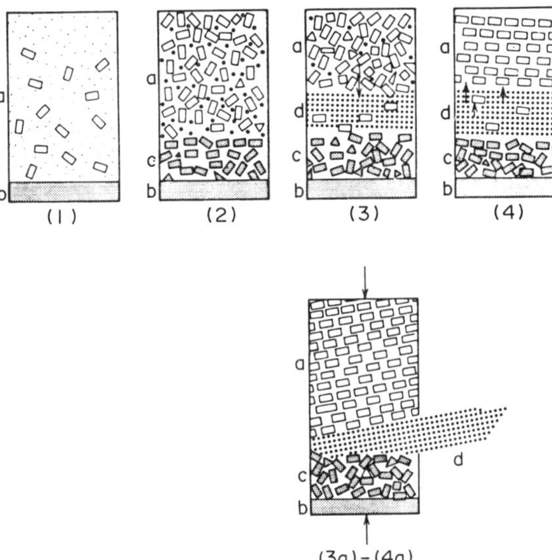

Figure 1
Idealized diagrammatic representation of late gravitational liquid accumulation. (1) Early stage of crystallization of basic magma a, after formation of chill zone b; (2) layer of sunken early-formed ferromagnesian crystals c, resting on chill zone b, with mesh of later silicate crystals above, whose interstices are occupied by residual magma enriched in ore oxides; (3) mobile, oxide-rich, residual liquid draining down to layer d, and, floating up, later silicate crystals; (4) formation of concordant oxide ore body in which a few late silicate crystals are trapped as mobile, enriched gravitational accumulations are squeezed out or decanted to form the magmatic injections (after Jensen and Bateman 1980)

segregations. The Bushveld Igneous Complex (BIC) with chromite and platinum minerals, and Allard Lake, Quebec, with titaniferous hematite are examples. The Muskox intrusion, Northwest Territories, Canada, the Duluth Complex, Minnesota, the Stillwater Complex, Montana, and the Skaergaard intrusion, Greenland, are additional examples with specific economic materials. The Skaergaard and the Muskox intrusions have occurrences of sulfides but not in economic concentrations, and the Stillwater complex is being explored for economic platinum–palladium deposits. There are many other ultramafic igneous bodies throughout the world which are of interest because of their size and, therefore, their potential for economic mineral production.

3. *Hydrothermal Deposits and Processes*

The most common intrusive rocks are granite, granodiorite, diorite and monzonite. They generally form nearer the surface than ultramafic rocks and are associated with subducting plate boundaries. They generally form as a magma at some depth and rise until they begin to crystallize in the overlying crustal rocks. Those portions of the magma that reach the earth's surface are extrusive igneous rocks. They cool rapidly and develop either a glassy or fine-grained texture. Felsic magmas become enriched in water as the magma moves upward. This water is gained either from the formations that are intruded or partially melted or it is magmatic water, residually concentrated by the crystallization of other components. Water content may increase from less than 1% in the upper mantle to 10% in the upper crust. This water acts as a fluidizer and mineralizer and allows the magma to crystallize at relatively low temperatures, 625 °C and less. In the sequence of magmatic crystallization, the nonvolatile minerals crystallize first with anorthite-rich plagioclase, pyroxenes and amphiboles. Sodium-rich plagioclase (albite), orthoclase, micas and quartz form late with volatile-rich aqueous solutions the last to crystallize.

As the volatile content increases in amount, the metals begin to form complexes such as $CuCl_4^{2-}$, $PbCl^+$ and $PbCl_3^-$. Laboratory research has indicated that complexes of metal chlorides, especially lead, silver and many other metals, are relatively soluble at temperatures of a few hundred °C. In contrast, these metals alone or as sulfides are essentially insoluble in aqueous solutions at many hundreds of °C. The phenomenon of metal transport into veins, fractures, or pores of rocks has been a major problem to geologists. Certainly strong acids could dissolve some of the metals and metal sulfides but these veinlet paths are commonly seen crossing limestone beds with no reaction.

In the last few decades, deep wells in the Salton Sea, brines of the Red Sea and fluid inclusions have shown that the concentrations of the metals in these solutions are surprisingly high. For more than a century, fluid inclusions (Fig. 2) have been studied in thin slices of minerals with a polarizing microscope. These fluid-filled cavities are microsamples of the solution in which the mineral was growing. As the fluid was trapped by crystal growth, it was retained and is available now for analysis. The inclusions contain predominantly water and/or liquid carbon dioxide but brine solutions in the inclusion sometimes make up from a few percent to 30–40% NaCl.

Recently, laboratory and computer studies have indicated that hydrochloric acid (HCl) does not dissociate into H^+ and Cl^- ions at temperatures of a few hundred °C. Metals can be transported as relatively soluble chloride complexes in only slightly acid solutions because the HCl with its low dissociation constant is not an acid at all at the mineralizing temperatures. As the temperature decreases, the HCl begins to dissociate into H^+ and Cl^- ions, with a resulting decrease in pH; the metal chlorides react with H_2S resulting in the deposition of metal sulfides.

4. Hydrothermal Alteration

Released H^+ ions of the hydrothermal solution react with existing minerals which results in chemical changes and the formation of new minerals. As the mineralizing fluids are much more abundant and more widespread than the smaller hidden ore deposit, recognition of hydrothermal alteration of surface bed rock is a prime tool that may lead to a buried economically viable ore deposit.

Feldspar minerals, such as orthoclase, are prime examples of minerals that undergo hydrothermal alteration. The supposed reaction is:

$$3KAlSi_3O_8(\text{orthoclase}) + 2H^+ \rightarrow KAl_2AlSi_2O_{10}(OH)_2(\text{sericite}) + 2K^+ + 6SiO_2$$

Note that only the H^+ ion is needed for the reaction. Because of this, some types of hydrothermal alteration are referred to as hydrogen ion metasomatism. Sericite is a fine grained, white mica. It is soft and is easily distinguished from unaltered orthoclase which is hard and cannot be scratched with a knife.

Silicification is often associated with hydrothermal alteration or mineralization. Note in the reaction above that silica (SiO_2) is released during the alteration of orthoclase to sericite. Thus silicification commonly occurs with mineralization.

Argillic alteration results from hydrothermal alteration which leaches lime and forms clay minerals. Commonly, kaolinite, montmorillonite, dickite, halloysite, and illite develop in altered rocks with the argillic alteration varying from weak to advanced. This type of hydrothermal alteration is common in porphyry copper and molybdenum deposits.

Stable isotopic studies are now applied to problems of sources of fluids, especially water (D/H and $^{18}O/^{16}O$ ratios) and are used to determine temperatures of mineral formation and hydrothermal alteration. Because of isotopic exchange equilibrium reactions, H_2O may exchange oxygen isotopes with minerals forming the wall rocks. Figure 3 illustrates the isopleths or equal temperature zones, determined by isotopic measurements with the highest temperature near the ore where the lowest $^{18}O/^{16}O$ ratios were found.

Other types of wall-rock or hydrothermal alteration exist; for example, pyritization, aluminization jasperoid and many others.

5. Replacement and Vein Deposits

The large replacement lead–zinc deposits of Leadville and Gilman, Colorado, were some of the first deposits to be recognized as replacement of limestone by galena

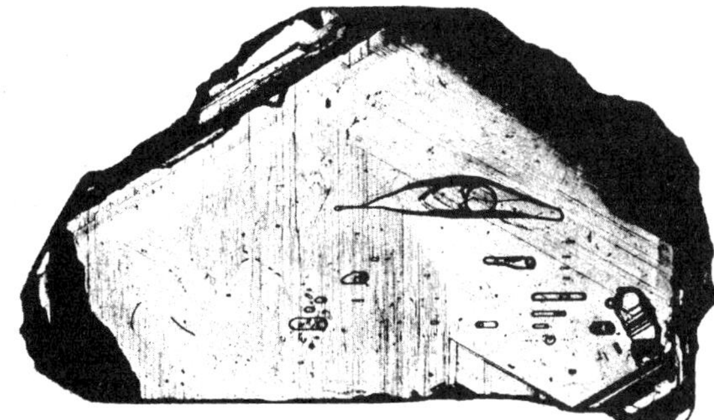

Figure 2
Large primary inclusions just under striated surface of clean, zoned sphalerite crystal from Cripple Creek, Colorado (after Roedder, US Geological Survey Professional Paper 440JJ)

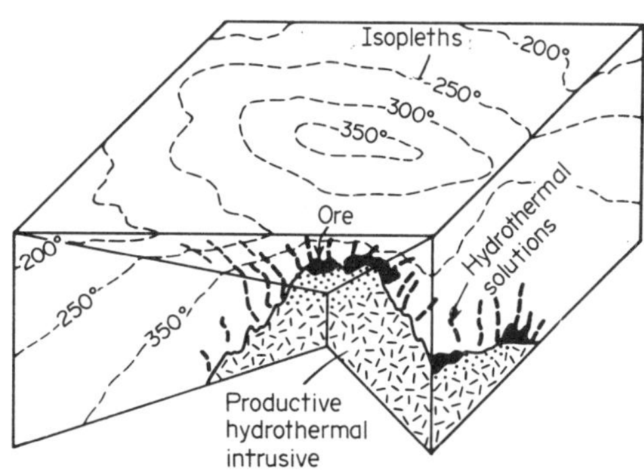

Figure 3
Idealized potential application of ^{18}O studies to hydrothermally altered contact zones near a mineralizing fluid source of water permeating carbonate rocks (after Jensen and Bateman 1980)

(PbS) and sphalerite (ZnS). The process is not fully understood, but the evidence is conclusive that pre-existing limestone and dolomite formations had been selectively removed and replaced by sulfide minerals. Some original fine textures are retained indicating that the process must have been a removal of minute portions of the carbonate rocks occupied almost simultaneously by the sulfide minerals. If the process continued for long periods (for example, hundreds of thousands of years), large deposits resulted.

In the Tintic mining district, Juab County, Utah, silver–lead–mineralization exists in fissure veins and selective replacement deposits. The replacement deposits formed predominantly in the Devonian Bluebell dolomite that is mineralized over a horizontal distance in excess of 3 km. Other formations are also replaced but of the total US\$425 000 000 gross value of ore mined in the district, US\$110 000 000 was from the Bluebird dolomite.

Hydrothermal replacement and vein deposits are not limited to sulfide ore bodies. Gold exists as hydrothermal replacement occurrences in the Homestake, South Dakota mine and in the Carlin Mine, Nevada. The Mother lode of California is a hydrothermal lode (closely spaced veins) system, as is the famous Virginia City silver vein, Nevada. Tin occurs in hydrothermal vein systems; the Llallagna deposit in Bolivia is a good example. Iron ore replacement deposits are abundant, as in the Iron Springs magnetite deposits in southern Utah where magnetite has formed contract metasomatic, hydrothermal replacement deposits in the host Paleozoic limestone formation.

6. *Porphyry Deposits*

These deposits, whether porphyry copper, molybdenum, or porphyry tin, are closely associated with intrusions of felsic magma, usually quartz monzonite or quartz diorite, generally porphyritic and almost all Mesozoic or Tertiary in age. Porphyritic igneous rocks have large phenocrysts (crystals) surrounded by a finer grained groundmass. The New Guinea highland deposits are comparably much younger being mid-to-late Miocene in age with some ore deposits believed to be Pleistocene.

Mineralization occurs as disseminations of sulfide minerals within or near the igneous body, the parent source of the ore minerals, but formed in the last phase of the crystallization of the intrusion. In some cases, the volatile pressure is so high and with a thin overburden, that shattering of the cupola of the intrusion results. This provides innumerable (planar) fractures that provide the "plumbing system" for the alteration fluids and ore-bearing fluids to permeate the cupola and form disseminated economic ore deposits. The grade rarely exceeds a few percent sulfides, predominantly pyrite, and ore sulfides of less to slightly more than 1%. A large porphyry copper deposit should contain no less than 0.4 to 0.6% copper to be profitable to mine.

Many of the deposits have undergone leaching from meteoric waters that react with pyrite to produce ferric sulfate $Fe_2(SO_4)_3$ and sulfuric acid (H_2SO_4) that dissolve copper sulfides (Cu_2S, $CuFeS_2$ and CuS) and form soluble sulfates. These percolate downward until the oxidation–reduction boundary is reached; there, the metal-carrying solutions replace pyrite and commonly form chalcocite pseudomorphs after pyrite. These are supergene deposits containing several percent copper and are zones of secondarily enriched, high-grade material.

About 80% of US copper is derived from porphyry copper deposits. Their origin, especially the origin of the felsic intrusions that gave rise to the deposits, is intimately related to plate tectonics and subduction zones that were active at the edges of continental plates. The deposits are numerous in the Andes along the western portion of South America; from there, they extend into North America in New Mexico, Arizona, Utah and Nevada, and into the cordillera of British Columbia.

Subducted oceanic plates in the east Pacific have resulted in numerous deposits in the Philippine Islands, Bougainville and Indonesia. Deposits also occur along the Himalaya–Alpine chain, in Afghanistan, Iran, Turkey and the Balkans.

The search continues for these large volume, multimillion tonne, low-grade porphyry deposits. Some contain copper ore that may have a gross value of about US\$10 per tonne but where 100 000 tonnes per day may be mined by comparatively low-cost, open-pit methods, they may provide a gross production of about US\$1 000 000 per day.

7. *Extrusive Igneous Rocks*

Many mineral deposits are related to extrusive igneous rocks, specifically volcanic rocks. Many epithermal deposits (that is, low-temperature and near-surface deposits) are often enclosed in altered volcanic rocks of felsic composition, generally latites, dacites and andesites. Many of the silver and gold deposits of Nevada are replacement fissure vein or disseminated deposits in volcanic rocks. Examples in Nevada are Carlin, Goldfield, Tonopah, Virginia City, Rhyolite, Golden Arrow, Reveille and the Divide district, as well as many other smaller deposits. Portions of the Ruth porphyry copper deposit, Nevada, involve pre-ore volcanic rocks that are hydrothermally altered and mineralized.

The Frisco mine in Millard County, Utah, formed along a vertical fault separating limestone from felsic volcanic rocks. The limestone was a much less favorable host replacement rock than the extensively replaced and mineralized volcanic rock.

8. Exhalative Submarine Volcanogenic Ores

Ore deposits of this genesis are being recognized in large numbers. This is not only because of new discoveries but also as a result of reclassifying some deposits formerly referred to as massive sulfide deposits.

The fairly recently described process of volcanogenic ore formation applies to deposits formed in shallow seas by submarine volcanism accompanied by mineralizing fluids that result in stratabound, massive sulfide deposits. Generally, the ore zones are sandwiched between overlying felsic volcanic rocks and underlying mafic volcanic rocks. As this process may extend laterally, associated with volcanic ores during a volcanogenic epoch, the search and recognition in continents for such deposits can be concentrated in former volcanic arcs along the contacts of andesitic and rhyolitic rocks.

In the Mattagami district, Quebec, Canada, such contacts can be traced along a westward plunging anticline for more than 35 km. Several ore bodies, such as Bell Channel, Radiore "A," Bell Allard, Orchan, New Hosco and Mallard Lake occur between the two volcanic groups of the younger Wabasse group (andesite, basalt, dacite and rhyodacite) that form the outer limbs of the anticline with the older Watson Lake group, predominantly rhyolite. The ore bodies of massive Fe–Zn–Cu sulfides are stratabound, of Archean age, and highly metamorphosed.

Other exhalative submarine volcanogenic deposits include many formerly classified as massive sulfide deposits, such as the deposits of the Noranda area, the Bathurst–Newcastle area, New Brunswick and Japanese Koroko deposits. The Flin Flon district, Saskatchewan and Manitoba, Kidd Creek, Timmons district, Ontario and Mount Isa, Queensland, Australia are also considered to be volcanogenic.

See also: Metamorphic Ore Deposits: Formation Processes; Sedimentary Ore Deposits: Formation Processes

Bibliography

Barnes H L (ed.) 1967 *Geochemistry of Hydrothermal Ore Deposits.* Holt, Rinehart and Winston, New York

Gilbert J M, Park C F Jr 1986 *The Geology of Oil Deposits.* Freeman, New York

Jensen M L, Bateman A M 1980 *Economic Mineral Deposits,* 3rd edn. Wiley, New York

Lindgren W 1933 *Mineral Deposits,* 4th edn. McGraw-Hill, New York

Park C F, McDiarmid R A 1975 *Ore Deposits,* 3rd edn. McGraw-Hill, New York

Peters W C 1978 *Exploration Mining and Geology.* Wiley, New York

Ridge J D 1968 *Ore Deposits of the United States 1933–1967.* American Institute of Mining and Metallurgical Engineers, New York

Stanton R L 1972 *Ore Petrology.* McGraw-Hill, New York

M. L. Jensen
[University of Utah, Salt Lake City, Utah, USA]

Insulation Raw Materials

Insulation materials are used to control or slow heat flow. Heat flow can be reduced over a wide range of temperatures from cryogenics at -273 K to well above 2000 K. Insulations are made in a broad variety of types, from organics to minerals. Some minerals are used essentially as mined, others are significantly modified or composited to give improved qualities. Minerals used as insulation raw materials can be categorized by the degree of processing they require in order to be converted into useful insulation products.

The first group of minerals are used with only minor processing which could include mining, drying, size gradation, some beneficiation, and either expansion or higher-temperature sintering. Examples are perlite, vermiculite and diatomite which are all used as loose-fill insulations, cavity fill and block fill, or may also be consolidated into structural boards, bricks, or other shapes by use of binders such as cement, sodium silicate, clays, and so forth.

The second group of minerals requires more complex chemical and/or thermal processing before they are converted into a form that can be used as insulations. These insulation raw materials are mined and frequently receive drying, size gradation and beneficiation but they then receive chemical or thermal processing. Table 1 gives, in a broad review, examples of these raw minerals, of the processing they receive and of the types of insulation usually produced. In Table 1 and throughout this article, those minerals that undergo minor processing are referred to as group I minerals and those that are chemically or thermally processed as group II minerals.

1. Sources

An abbreviated summary of the principal states in the USA and the world countries that have operating deposits of the raw materials used in insulation products is given in Table 2. The sources are listed in an approximate order of decreasing annual production. These minerals are widely available and, therefore, only the larger producing states and countries are listed.

2. Mineralogy

As a wide variety of mineral raw materials are used in producing insulation products, the mineralogy of these materials will not be discussed here. The reader is referred to the appropriate articles on specific materials.

3. Physical Properties

Important physical properties of insulating products are summarized in Table 3. Insulating materials must endure a wide range of operating conditions. Since no

Table 1
Examples of insulation raw materials, processing and products

Raw materials	Processing	Products
I Minor processing diatomite, perlite vermiculite	mining, milling, drying, sizing, expansion or sintering	loose-fill powders, granules
II Chemical or thermal processing a. bauxite	mining, milling, solubilization, precipitation, calcining	powders, bricks, cements, refractories
b. silica, limestone, soda ash, feldspar	mining, milling, drying, blending, melting, fiberization	glass wools, rock wools, refractory wools, microspheres
c. silica sand	mining, milling, drying, reaction with NaOH or Cl purification, reaction or precipitation	precipitated silicas, silica gels, aerogels, fumed silicas
d. colemanite, ulexite-probertite, kernite brines	mining, milling, washing, flotation, calcining, evaporation	borates, borax
e. limestone, silica, alumina, gypsum, iron oxide, Wollastonite	mining, milling, drying, blending, calcining, hydrothermal processing	calcium silicates, cements, bricks, blocks, pipe covering

one material is likely to have all the correct properties, including installation cost, the selection of a suitable insulation takes into account such properties as the melting point or maximum operating temperature, thermal conductivity at the operating condition and the necessary strength properties for the particular use. In addition, the insulation often must be chemically compatible with its environment, which may require that it have a selected chemical composition. For example, contact with stainless steel or electrical equipment often requires low sodium and low chloride content to avoid corrosion.

4. Mining and Production Techniques

The generalized production techniques for each group of minerals are given in Table 1.

The minerals in group I, the minor-processing category, are commonly open-pit mined using ripping, bulldozing and/or drilling and blasting, if necessary. The mined materials are dried, milled and classified with some beneficiation to remove tramp contaminants. Perlite and vermiculite can be "expanded" to a significantly lower density by rapid heating that converts combined water to steam and "pops" these minerals like popcorn. Diatomites are usually sintered, often with a sodium flux to improve color. The sintering process bonds smaller particles together to improve handling, increase density and remove combined water.

The minerals in group II, the chemically or thermally processed minerals, are also commonly open-pit mined with drilling and blasting. The mined materials are dried, milled and beneficiated prior to chemical and thermal processing. Since this processing varies with the subgroups under group II in Table 1, they will be discussed individually.

(a) Group IIa. Bauxite is solubilized in caustic, precipitated as aluminum hydroxide, then calcined to aluminum oxide. Alumina is used with binders, clays, bauxitic clays, etc., to make pressed, extruded, cast materials, and the like, that are recalcined to give various refractory insulations. Alumina is also used with hydraulic cements and other materials to make refractory cements to bond bricks, blocks, and so on. Aluminum metal, reduced from aluminum oxide, is used as thin foils for reflective vapor barrier insulations, often in combination with other insulations. In multiple layers and with the interlayer space evacuated, aluminum foils become effective insulations without other interlayered materials.

(b) Group IIb. Silica, soda ash, limestone, feldspar, borax and a variety of other glass-forming minerals are combined and melted into various melting point, solubility and viscosity glasses that are fiberized by drawing, flame attenuation and/or rotary spinning methods. Used as bulk fibers or formed into boards or blankets with various binders, such as phenolic resins, these fibers become effective insulating materials. Slags from other thermal-processing or various natural minerals can be fiberized, as are the synthetic glasses, to produce rock wool or mineral wool. Refractory fibers are made from even higher melting point combinations of alumina, silica and selected metal oxides.

Glass-forming minerals can also be formed into hollow bubbles or microspheres and used as insulating ingredients in composite materials or as loose-fill insulations.

Table 2
Main sources of insulation raw materials

Raw material	Main sources
Diatomite	California, Nevada, Washington, Oregon, Arizona, France, Denmark, USSR, Iceland, Spain, Mexico
Perlite	New Mexico, Arizona, California, Idaho, Colorado, Nevada, Hungary, Greece, USSR, Italy, Japan, the People's Republic of China, Bulgaria, Turkey, Mexico, Australia
Vermiculite	Montana, South Carolina, Virginia, South Africa, Japan, Brazil
Bauxite	Arkanas, Alaska, Georgia, Australia, Guinea, Jamaica, Surinam, USSR, Brazil, Ghana, India, Guyana
Silica	Illinois, Missouri, Arkanas, Oklahoma, Tennessee, Georgia, Pennsylvania, New Jersey, West Virginia, Texas, Ontario, Quebec
Limestone	Pennsylvania, Ohio, Indiana, Michigan, Texas, New York, USSR, Japan, FRG, Brazil, Mexico
Soda ash	Wyoming, California, USSR, the People's Republic of China, Bulgaria, France, UK, FRG, Japan, GDR, Romania, Poland, India
Felspar	Italy, North Carolina, Connecticut, Georgia, California, Oklahoma, South Dakota, FRG, USSR, France, Brazil, Mexico, Finland, Norway, Sweden
Borates	California, Turkey, USSR, Argentina
Gypsum	Texas, Oklahoma, Michigan, Iowa, California, Nevada, Canada, Japan, Spain, France, Iran, USSR, China, UK, Mexico
Wollastonite	New York, California, the People's Republic of China, Finland, India

(c) Group IIc. Silica gels, aerogels, precipitated silicas and fumed silicas are special categories of processed minerals that are used in specialized insulations. These are generally formed by converting silica sand to sodium silicate or to silica tetrachloride by high-temperature reaction with sodium hydroxide or with chlorine, respectively. Sodium silicate is neutralized with acid to precipitate silica which is recovered and dried to give silica gels and precipitated silicas, or dried by removal of liquid above the triple point to give aerogels. Silica tetrachloride is purified by distillation and then reacted with, or burned in, an oxygen–hydrogen flame to produce fumed silica. These synthetic silicas are used in composites as thickening agents, fillers, pozzolans, etc.; they contribute insulation characteristics owing to their low bulk density.

(d) Group IId. Boron chemicals are produced from colemanites or ulexite–probertite by mining, grinding, washing and calcining. Flotation processing is also used. Lake brines are evaporated to collect boric acids and borax. Boron minerals are used in glass-fiber compositions to produce insulations. Borates are used as fireproofing agents in organic insulations like cellulose.

(e) Group IIe. Limestone is converted to lime by calcination. Cement is produced by calcination of mixtures of limestone, silica, alumina, gypsum, iron oxide, etc. Lime or cement may be combined with silica, expanded perlite, vermiculite, wollastonite and other minerals and fibers, and made into various insulating materials by either hydrothermal reaction of calcium hydroxide with silica or by hydration of cement with silica ingredients. Calcium silicates are used as pipe and block insulations. Cement or concrete bricks, blocks and coverings are used in less-demanding insulation areas or as structural components. Insulation efficiency can be increased by incorporation of air bubbles, making lightweight concrete.

5. Safety

Essentially all of the raw materials for insulations have some hazards in their use, even if it is only as nuisance dusts that have low levels of hazards but require dust control, ventilation and the use of dust masks. The same would apply to the insulation products made from these materials, particularly when handled in ways to produce dust such as sawing or sanding. Insulation raw materials that contain crystalline silica above 1% need to be treated with proper care.

Fibrous insulation materials are being evaluated for specific safety hazards. Generally, safety relates to the ability of the human body to solubilize fibers. Durable fibers are more likely to cause health problems. It should be noted that asbestos has not been included in this discussion of insulation raw materials. The reason for this is that although it has been used in many forms and combinations because of its excellent insulation characteristics, the overwhelming health hazard has significantly reduced asbestos usage in the USA and is decreasing in use on a worldwide basis (see *Asbestos: Alternatives*).

High-temperature insulation uses cause particular safety problems with some materials. Long-term use at elevated temperatures can result in conversion of less-hazardous minerals to more-hazardous materials, such as cristobalite and tridymite.

Table 3
Physical property ranges for typical insulation product types

Property	Diatomite, calcined	Perlite, expanded	Vermiculite, expanded	Bauxite: Refractory brick	Bauxite: Castable cements	Aluminum foil	Silica: Glass fibers	Silica: Refractory fibers	Silica: Rock wool	Silica: Silicas	Limestone calcium silicate
Density (kg m^3)	128–320	32–160	48–280	560–1040	850–1250	12	46–176	64–385	16–320	10–160	192–736
Maximum operating temperature (°C)	800–900	800–900	700–1100	1000–1760	1150–1420	500	450–650	870–1540	540–980	1000	650–730
Thermal conductivity (10^{-2} W m^{-1} °C^{-1})											
−100 °C	1.1–1.7	1.1–2.3				0.0057[a]			1.7	2.3[b]	0.72[b]
0 °C		4–6	6.6				3.2	3.5	3.7–4.0	2.0	
150 °C	6.9					4.3	5.0		6.8	2.3–8.7	5.2–12.0
200 °C	7.2–10.0	7.9	10.0	13–56	21–36	6.6	6.6–7.9	7.2	7.6		5.6–13.0
400 °C	7.5–12.0	10.0	15		24–40		10	10–12	10		7.6–13.0
550 °C	8.1–13.0	11–16	15–19	16–59	26–42			13–16	13		8.9–13.0
750 °C	8.9				42–46			18–27			
1000 °C								26–39			
1100 °C				22–62				32–49			
Composition—ignited											
SiO_2	80–90	72–74	34–40	30–62	35–44		50–60	20–70	30–65	98–100	50–60
Na_2O+K_2O	0.8–2.0	7–10	2–8	0.1–1.0			5–20		0–20		−2
Al_2O_3	1–6	12–13	12–14	31–78	34–51	(188)	5–15	30–60	0–40		1–5
B_2O_3							5–20				
CaO	0.5–3.0	0.4–0.9	1–2	0–12	4.7–17		10–20		0–40		20–40
Cr_2O_3								0–5	0–5		
Fe_2O_3	1–3	0.7–2.0	6–13	−1–2	0.5–7.0			0–1	0–15		−2
MgO	0.5–1.0		18–24				2–6		0–40		−2
MnO								0–1	0–40		
TiO_2								0–1.5	0–10		
ZrO_2								0–20	0–10		

[a] Foils with separators [b] Vacuum

6. Industrial Uses

As mentioned earlier, insulations are used to retard the flow of heat over a wide range of temperatures, from cryogenics and the very low temperatures of space to the very high temperatures of furnaces, plasmas, and so forth. Insulations are selected to minimize heat transfer at a suitable cost while giving the necessary strength and system compatibility.

Some insulations are used under high vacuums, others are used at normal pressures. Insulation characteristics involve the various methods of heat transfer: conduction, convection and radiation. Gas conduction and gas convection are influenced by the concentration of gas molecules and are therefore decreased by reductions in pressure (such as in a vacuum). Solid conduction is influenced by the nature of the particle shape, size, porosity, and so on. Radiation is influenced by opacity, emissivity, adsorption and scattering. All mechanisms are usually involved and the optimum insulation must minimize all of them, particularly the mechanisms that contribute the largest percentages of the total heat transfer.

Bibliography

Dillon J B 1978 *Thermal Insulation, Recent Developments.* Noyes, Park Ridge, New Jersey

Kirk R E, Othmer D F 1978–1980 *Encyclopedia of Chemical Technology*, 3rd edn. Wiley, New York

Kujawa R J 1983 Insulating materials—Thermal and sound. In: Lefond S J (ed.) 1983 *Industrial Minerals and Rocks*, 5th edn. American Institute of Mining, Metallurgical and Petroleum Engineers, New York

Malloy J F 1969 *Thermal Insulation.* Van Nostrand Reinhold, New York

US Bureau of Mines 1984 Metals and minerals. *US Bureau of Mines Minerals Yearbook.* US Bureau of Mines, Washington, DC

G. Coombs
[Johns–Manville Sales Corporation, Denver, Colorado, USA]

Iron Resources

Iron is the fourth most abundant element in the earth's crust, representing 5.06% of the lithosphere, and is an important constituent element in several hundred minerals. Iron readily combines with carbon dioxide, oxygen, sulfur or silica to form carbonates, oxides, sulfides or silicates, and only rarely occurs as the native element in meteorites and in basalts as at Ovifak, Disko Island, Greenland, where it is found as small grains through to masses up to 20 t in weight. Iron is an important element in sedimentary, igneous and metamorphic rocks and is particularly stable at and near the earth's surface under oxidizing conditions as ferric iron oxide or hydrated ferric iron oxide, which gives many surface materials a red, red–brown, brown or yellow color.

About 98% of all iron ore produced is used in the manufacture of iron and steel. Lesser amounts are used in the manufacture of portland cement, as an aggregate material to add weight to concrete, as a heavy media material in mineral or coal processing, as a source of iron supplement in animal feed and in some fertilizers, as a paint pigment, and in the production of ferroalloys, ferrites and abrasive rouge.

1. Iron Ores

Iron-bearing materials in deposits that are of sufficient size and iron content to be either used or potentially used as commercial sources of iron are termed iron ores. The vast majority of iron ore bodies are composed of iron oxides in the minerals goethite ($Fe_2O_3.H_2O$), hematite (Fe_2O_3) or magnetite (Fe_3O_4) with lesser amounts of other iron minerals. In a few regions, as in the Michipicoten district, Ontario, Canada, siderite ($FeCO_3$) deposits are mined as iron ore. Commercial iron ores may be grouped into two major types:

(a) "merchantable ores" or "natural ores" or "direct ores" are ores that can be shipped to the steel works, as mined, after minor processing that is chiefly crushing and sizing; and

(b) concentrating-grade ores that require considerable processing which may include fine grinding to yield a concentrate.

Current iron ore markets require that most iron ore be beneficiated in some way before shipment as there are considerable cost and energy savings at the blast furnace stage that result from the use of high-grade, low-impurity, sized and/or agglomerated iron ores.

Natural, merchantable iron ores commonly contain 55–68% Fe on a dry-analysis basis with low phosphorus and other impurity levels. They are produced on a large scale in Australia, Brazil, India, Liberia, Mauritania, South Africa, USSR and Venezuela. A merchantable iron ore with an iron content of 32–38% is mined in France and Luxembourg. These ores are economic because of their location near the steel works.

Concentrating-grade iron ores include a wide range of iron ore types that are processed using a variety of procedures. Process technologies have been developed that can concentrate a wide range of crude ores containing goethite, hematite or magnetite that have favorable grades and textures. These technologies have transformed many iron-rich deposits formerly of no economic value into iron ores or potential iron ores. Most iron ores currently mined in the USA and Canada and major deposits in China, Norway, Sweden and the USSR are classed as concentrating-grade ores.

1.1 Iron Ore Production

Iron ores are mainly produced at open-pit mines that use large materials-handling equipment. Important amounts of ore are mined by underground methods in France, Luxembourg and Sweden, with relatively few underground mines elsewhere.

Processing of off-grade natural, merchantable ores is commonly by crushing, sizing, washing and gravity methods that include heavy media and spirals. The sizing yields shipping-grade coarse ore material at about minus 4 in. (10.2 cm) and plus $\frac{1}{4}$ to $\frac{3}{8}$ in. (6.37–9.55 mm). Fine ore is commonly sintered or pelletized.

Concentrating-grade ores are processed in complex plants using gravity, magnetic, flotation or selective flocculation–flotation systems. The process used is designed to yield a uniform, high-quality iron ore pellet or sinter with a maximum recovery of the contained iron.

1.2 Iron Ore Markets

Iron ores are a large-volume, relatively low-value commodity and are commonly sold in an oxide form. The ore markets are located at the various iron and steel plants throughout the world. The iron ores are prepared for market by a variety of sizing and concentration procedures but they still contain varying quantities of impurities that affect iron and steel production and metal quality. Silica, alumina, lime and magnesia determine the volume of slag produced whereas impurities such as Cr, Mn, P, S and Ti influence metal quality. Phosphorus is of particular importance and may have a considerable effect on ore value and acceptability. The commonly quoted iron ore price is at Lake Erie ports for a base-grade ore of 51.5% Fe content as Bessemer grade if less than 0.045% phosphorus (dry) and non-Bessemer if the phosphorus content is 0.045–0.18% (dry). The world trade in iron ore exceeds 335 million tonnes per year and is growing as high-grade iron ores and pellets replace lower-grade, local ore sources. Most world trade ore contains in excess of 63% Fe (dry) with low impurities.

There is a growing market for prereduced pellets which have been treated by a direct reduction process to yield a pellet containing 87–92% Fe. Prereduced pellets can be used in an electric furnace to make steel.

2. Geology of Iron Ore Deposits

Iron ores occur in a wide range of geological environments in deposits that formed over a period that exceeds 3 billion years. Important deposits were formed under various sedimentary, igneous, metamorphic and surficial conditions that can be grouped as shown in Table 1.

Table 1
A simplified classification of iron deposits

Bedded sedimentary and metasedimentary deposits
banded cherty iron-formations
ironstones
river terrace
black sands
Igneous deposits
magmatic segregations
titaniferous magnetites
Hydrothermal deposits
contact zone replacements
replacements in iron-formations
Residual deposits
Lake Superior type ores
brown ores
laterites

2.1 Banded Cherty Iron-Formations

A sedimentary, bedded rock termed iron-formation occurs in extensive formations in early and middle Precambrian terrains throughout the world. Iron-formations are less commonly found in areas underlain by rocks younger than about 1.9 billion years, while a few minor occurrences may be of Cambrian age. Iron-formations are banded, layered sedimentary rocks of possible chemical or biochemical origin that contain $\geqslant 15\%$ iron and typically consist of iron-rich layers alternating with silica-rich layers. The iron-rich layers may be composed of siderite, iron oxides or iron silicates, or mixtures of these materials with admixed silica as chert or quartz. The silica-rich layers are composed of silica as chert or quartz with varying amounts of iron minerals. In some iron-formations, as in the Cuyuna District, Minnesota, manganese may reach several percent. The iron content varies considerably but averages about 30%.

Iron-formations in early Precambrian terrains (+3.2–2.5 billion years) are associated with volcanic rocks and volcanogenic sediments. In middle Precambrian terrains (2.5–1.7 billion years), iron-formations are most commonly associated with metasedimentary rocks and may occur as extensive stratigraphic units. Major iron-formations of this age occur in the Lake Superior region in the USA, Labrador–Quebec region, Canada, Minas Gerais and Carajas regions in Brazil, the Hamersley region, Australia, Krivoi Rog and Kursk regions in Russia, and in China, India, Africa and Venezuela. These iron-formations are often from 100 to several hundred meters thick and extend for tens or hundreds of kilometers. In the Hamersley region, Australia, iron-formations have an aggregate thickness of about 1100 m and extend along strike for several hundred kilometers. The tonnages of iron-formations are enormous.

In the Lake Superior and Labrador–Quebec regions of North America the following types of iron-formation ores are being mined.

(a) Magnetite taconite is mined extensively in Minnesota along the Mesabi Range with a 1987 production of about 30.6 million tonnes of pellets (US Bureau of Mines 1988). Taconite-type ores are concentrated by magnetic methods after fine grinding. This type of ore is also being produced in Michigan, Wisconsin and Wyoming in the USA, in Ontario, Canada, in the Kursk region, Russia, in Norway, and in Tasmania, Australia.

(b) Jaspilite ore is mined in Michigan on the Marquette Range. This ore is a layered metasediment with alternating layers of jasper and finely crystalline hematite. This ore type is concentrated by flotation.

(c) Itabirite-type ores are being processed extensively in the Labrador–Quebec region of Canada. These ores represent metamorphosed iron-formations that consist of alternating layers of quartz and crystalline hematite with more or less magnetite and iron silicate minerals. These ores are concentrated by gravity and magnetic methods. Extensive deposits of the itabirite type occur in Brazil, Liberia and Venezuela.

(d) Oxidized banded iron-formation ores at the Tilden Mine, Marquette Range, Michigan, are being processed after very fine grinding, using a selective flocculation–flotation method. This type of iron-formation has been oxidized and partially leached by weathering processes, with most or all magnetite, siderite and iron silicates converted to hematite with more or less associated goethite. Extensive deposits of oxidized iron-formation are known in Michigan, Minnesota and Wisconsin, in the Laborador–Quebec region, Canada, in South Africa and elsewhere in the world, but their possible concentratability is not known.

(e) Siderite ores are mined in the Michipicoten District, Canada, from siderite bodies which are associated with volcanogenic sediments, tuffs and volcanics of Archean age and have an overlying low-iron, banded silica formation. These commercial deposits, unique to this district, are sintered to yield a commercial iron ore.

2.2 Ironstone Ores

Ironstones are a group of sedimentary rocks containing 20–40% Fe that typically consist of oolitic or pelletal grains of goethite, hematite or chamosite ($(Fe,Mg,Al)_6(Si,Al)_4O_{14}.(OH)_8$) in a matrix that contains varying amounts of iron oxide, chamosite, siderite, calcite or dolomite, with common, variable amounts of clastic quartz. Ironstones are associated with sandstones, shales and limestones, are often lenticular, and commonly range in thickness from about 30 cm to a maximum of about 10 m. These iron deposits range in age from late Precambrian to Tertiary. Ironstone ores are relatively high in phosphorus, which, coupled with a common 30–38% Fe content, has resulted in a worldwide decline in use. France and Luxembourg are the major producers of ironstone ores.

2.3 River-Terrace Ores

River-terrace iron ore deposits with reserves of hundreds of millions of tonnes occur in some ancient river valleys in the Hamersley region in western Australia. The iron ore beds consist of oolitic or pisolitic grains, locally with some interbedded clastic grains and pebbles of goethite or hematite, in a goethitic matrix. These deposits are from about 1 m to about 70 m in thickness and form mesas and dissected river terraces that are often several miles long. The ores contain from 50 to 60% Fe with a low phosphorus content. They are being mined on a large scale in the Robe River area. These iron ores appear to have formed in a geological environment of solution and redeposition of iron similar to conditions that result in bog iron deposits.

2.4 Black Sands

Accumulations of magnetite and ilmenite in beach sands are common. Many such deposits have been studied as potential sources of iron ore, but few are of commercial value. Sands containing magnetite are mined and concentrated on the west coast of North Island, New Zealand. About 3 million tonnes of iron ore were exported to Japan in 1979.

2.5 Igneous Deposits

Two major classes of iron ores are igneous in origin: (a) magmatic segregations that occur as intrusive bodies, veins or extrusive deposits; and (b) titaniferous magnetite segregations in basic igneous rocks.

Iron ore deposits in the Kiruna region in northern Sweden and deposits in Missouri in the USA occur as large, tabular, rather massive bodies of magnetite with lesser hematite in rhyolite, quartz porphyry or syenite porphyry. These deposits commonly contain from 1 to 3% phosphorus. A unique iron deposit that appears to be of similar origin occurs in the high Andes Mountains in Chile at El Laco where shallow intrusives and lava flows are composed of almost pure magnetite–hematite with associated apatite. The iron deposits in northern Sweden produce 20 million or more tonnes of iron ore per year, which are largely exported.

Titaniferous magnetite bodies occur as segregations in gabbro, pyroxenite and anorthosite, but have limited potential as iron ore sources because of the relatively high titanium content.

2.6 Hydrothermal Deposits

Iron ores that were deposited from hydrothermal solutions may be grouped into two classes: (a) contact zone replacements, sometimes termed "contact metamorphic" or "pyrometasomatic" deposits, where iron oxides replace the intruded rock sequence generally near the contact with an intrusive igneous body; and

(b) replacements in Precambrian iron-formations. The geological processes active in the two classes appear to be similar.

Contact zone replacement deposits are chiefly composed of magnetite with lesser hematite and minor amounts of carbonates, pyrite, chalcopyrite and pyrrhotite. The most common host rock is limestone. Skarn minerals such as garnets, pyroxenes and amphiboles are often associated with these ores. The deposits vary in shape from tabular bodies to irregular to vein-like. In size, they range up to several hundred million tonnes. Major deposits were mined in the Cornwall area, Pennsylvania, and are mined in the Iron Springs District, Utah, at Eagle Mountain, California, at Mount Magnitnaya, USSR, and at Marcona, Peru.

Replacement iron deposits in Precambrian iron-formations are known in many parts of the world where hydrothermal solutions have replaced silica with iron oxide, commonly hematite with lesser magnetite. These deposits are often high grade, some containing about 70% Fe. The ore bodies are often tabular in shape and retain the layered appearance of the original cherty iron-formation. Deposits occur in Brazil, India, Liberia, Mauritania and Venezuela.

2.7 Residual Deposits

Weathering of iron-bearing and iron-rich rocks has been an important iron ore forming process as iron oxides are relatively insoluble compared to silica and most other materials in the surficial geological environment. Oxidation converts iron to an insoluble, ferric oxide and leaching removes the more soluble elements by the water system. This oxidation–leaching system may be most active in warm and moist climates with wet and dry seasons. The zone most affected is from the surface to depths of about 100 m, but locally under very favorable conditions, the oxidation and leaching can extend to depths of 1000 m or more. As an example, soft, Lake Superior type iron ores were mined to depths of about 1000 m at the Mather Mine, Marquette Range, Michigan, with ore known to occur to depths in excess of 1500 m. The residual ores are commonly grouped as: (a) Lake Superior type or soft ores, (b) brown iron ores, and (c) laterites.

(*a*) *Lake Superior type.* These iron ores were formed by the oxidation of the iron minerals and the leaching of silica from a Precambrian, banded cherty iron-formation, leaving a residual accumulation of hematite and goethite with associated varying amounts of silica and clay. These ores are one of the most important iron ore types in the world. The time of ore formation is often unknown, as some ore bodies could have formed in Precambrian time and other bodies during later epochs that may extend to quaternary time. The ore bodies have a wide range in size and shape that often are influenced by the structural setting and stratigraphy of the iron-formation. Iron deposits of this association have yielded over 3.5 billion tonnes of shipping-grade ore in the Lake Superior region of the USA but are now largely depleted. Major production from Lake Superior type residual ore deposits is continuing from deposits in Australia, Brazil, India, Liberia, Mauritania, South Africa, the USSR and Venezuela.

(*b*) *Brown ores.* Weathering of Paleozoic limestones and dolomites in the southern Appalachian region of the USA may result in the development of irregular bodies of clays that contain lumps and nodules of iron oxide that can be shipped as iron ore after separation from the clay. In northeastern Texas the weathering of sedimentary beds containing siderite and glauconite has formed a series of iron ore deposits. These surficial deposits and others that are largely nodules and lumps of goethite and/or hematite with clay and sands that can be beneficiated by simple methods are collectively termed brown ores. Deposits are usually small.

(*c*) *Laterites.* Weathering of ultramafic rocks such as serpentinized peridotites, pyroxenite and dunites under tropical conditions often yields a surficial accumulation of iron oxide as goethite and hematite that is termed lateritic iron ore. This residual mantle is locally extensive and often ranges in thickness from about 1 m to 15 m or more. These deposits commonly contain from 40 to 55% Fe with varying amounts of alumina, chromium, nickel and cobalt, which makes them generally unsatisfactory as iron ore.

3. Iron Ore Reserves and Resources

Iron ore reserves as commercial tonnages that include all ore types and grades are very large. Iron ore resources in known potential iron ore areas are enormous. A summation of iron ore reserves and resources is shown in Table 2.

Table 2
World iron ore resources 1987 (Mt iron content) (USBM 1988)

Country	Production	Reserves	Reserve base
USA	43	3700	5900
Canada	33	4500	9800
Brazil	134	10800	11300
Venezuela	21	1200	1200
Australia	88	10100	20200
Liberia	15	500	800
South Africa	24	2900	6600
France	12	900	900
Sweden	20	1600	2400
India	54	4800	7500
USSR	248	25000	25000
China	96	3500	3500
World total	871	72000	10200

Bibliography

Klemic H, James H L, Eberlein G D 1973 Iron. In: Brobst D A, Pratt W P (eds.) 1973 *United States Mineral Resources*, US Geological Survey Professional Paper 820. US Government Printing Office, Washington, DC, pp. 291–306

McGannon H E (ed.) *The Making, Shaping, and Treating of Steel*, 9th edn. United States Steel Corporation, Pittsburgh, Pennsylvania, pp. 1–35, 178–239

Peterson E C 1980 Iron ore. *Mineral Facts and Problems*, US Bureau of Mines Bulletin 671. USBM, Washington, DC, pp. 433–53

Pounds N J G 1971 *The Geography of Iron and Steel*, 5th edn. Hutchinson University Library, London, Chap. 2

United Nations 1970 *Survey of World Iron Ore Resources*, 5th edn. United Nations, New York.

US Bureau of Mines 1988 *Mineral Commodity Summaries 1988*. USBM, Washington, DC, pp. 78–79

R. W. Marsden[†]

L

Lead Resources

The annual world usage of lead was about 3.4 Mt in 1987, making it the fifth most widely used metal after iron, aluminum, copper and zinc. Lead resources and mines are scattered throughout the world, the major producing countries being the USA, the USSR, Australia, Canada, Peru, Mexico, Yugoslavia and the People's Republic of China.

As lead is relatively inert and has a low melting point, over 50% of the lead used by industry can be recycled from scrap. Major uses are in storage batteries, antiknock additives for gasoline, construction materials, ammunition, paints and glass. The usefulness of lead is tempered somewhat by recent awareness of its toxic nature in paints, fumes and dust. Forecasts indicate the demand for lead will grow slowly at a rate of 2–3% annually. World reserves supplemented by probable new discoveries and improved recycling should be adequate to meet future demand for at least twenty years and probably well into the twenty-first century.

1. History

Lead figurines dating from 3800–3000 BC have been found in upper Egypt and western Asia Minor, proving that man had a knowledge of lead and its smelting before being able to make iron.

The Romans used lead extensively (e.g., for water pipes and wine vats), and worked lead mines in Spain, Cyprus and Greece. Silver and lead mining at Laurium, Greece, continued until recent times.

Charlemagne established a provisional capital at Goslar in Germany because of the need to control and supervise the production of silver and lead in the nearby Rammelsberg mine. The mine is still producing mixed lead, zinc, iron and silver ores.

The earliest recorded production of lead in the USA was in Virginia in 1621. Later, a traffic in galena for use as an ore of lead was started with the Indians by Perrot in 1690 near the present city of Dubuque, Iowa.

2. Geochemistry

Lead (atomic number 82) has various isotopes some of which are the product of atomic decomposition of uranium or thorium. Common lead has an atomic weight of 207.2. Lead is rarely found in nature as the metallic element; it most commonly occurs as the sulfide in the divalent state. It has a crustal abundance of about 15 ppm; the average varies from 7–20 ppm in sedimentary rocks, and in igneous rocks the range is from 5 ppm in gabbro to 20 ppm in granite. Lead ores require a minimum average grade of about 3% to be economic, so a natural concentration of about 2000 times the average crustal composition is required to form an orebody.

3. Mineralogy

In primary ore deposits the dominant mineral is galena (PbS), a mineral with prominent cubic cleavage, a specific gravity of 7.5, a bright metallic luster and a lead-gray color. The metallic minerals commonly associated with galena are pyrite, sphalerite, chalcopyrite, and silver minerals included as blebs within the galena. The near-surface part of some galena orebodies is altered to cerussite ($PbCO_3$), anglesite ($PbSO_4$), or other oxidized lead minerals, but galena is resistant to weathering and generally persists over long periods of exposure in outcrops. The primary gangue minerals of lead deposits are quartz, calcite, dolomite, barite and fluorite.

4. Geology

The most convenient classification for metal deposits is that proposed by the US Geological Survey on the basis of geological occurrence and suggested origin. The survey lists (a) stratabound deposits of syngenetic origin, (b) stratabound or stratiform deposits of epigenetic origin, (c) volcanic deposits conformable to enclosing sediments or metavolcanics, (d) replacement deposits, (e) veins and (f) contact metamorphic deposits. As the origin and time of formation of stratabound deposits are the subject of much uncertainty among geologists, the first two classifications are here combined as stratabound or stratiform deposits.

Stratabound deposits (i.e., those restricted to one sedimentary formation or unit) are commonly found in host rocks of limestone or dolomite and are the most productive lead deposits. If the ore deposit was formed contemporaneously with the enclosing sediment, it is termed syngenetic, whereas if it was formed shortly after deposition as a result of dolomitization or some other alteration, it is diagenetic. Stratabound deposits which appear to be formed long after the host rocks are termed epigenetic and account for many of the lead reserves in the world today. Sedimentary structural features, such as zones of jointing, reefs, facies changes, collapse breccias, and sinks associated with ancient karst drainage, serve as loci for orebodies within the favorable formation. Many of these mineralized areas appear to be near or under unconformities within a sedimentary series adjacent to domal or basin structures.

An example of a stratabound deposit apparently of syngenetic origin is the "Kupferschiefer" bed now being mined in Germany and Poland. Although mined chiefly for copper, it contains enough zinc, lead and silver to make these metals attractive by-products. Examples of stratabound deposits considered by many to be of epigenetic origin are those in southeast Missouri, the Missouri–Oklahoma–Kansas district, the Upper Mississippi Valley district, the Metaline area in Washington, the Pine Point deposits in Canada, and deposits in southeastern Morocco. Studies of Mississippi Valley type ore deposits have indicated that they were deposited from metal-rich brines at temperatures of 70–150 °C. The source of the metals may have been by solution from the rocks of sedimentary basins or from the crystalline basement underlying the ore-bearing beds. Others think the source of the metals may have been volcanic beds or volcanic exhalations.

Volcanic sedimentary deposits contain massive sulfide bodies interlayered conformably with volcanic or sedimentary rocks. Many such deposits are found in older severely folded and metamorphosed geological belts. The larger bodies contain some of the world's more productive mines, such as the Kidd Creek deposit near Timmins, Ontario; the Sullivan deposit at Kimberly, British Columbia; Mount Isa and Broken Hill in Australia; Rammelsberg in the Federal Republic of Germany; and Jerome, Arizona. An unmetamorphosed example of this type of orebody is the Kuroko deposits of Japan with massive pods and layers composed of sphalerite, galena, and accessory sulfosalt minerals in tuffs and other clastic rocks of relatively recent age.

Hydrothermal replacement deposits are the third most productive sources of primary lead. They are commonly irregular replacements in carbonate rocks, but some are found in quartzite or metamorphic rocks. The form and size of the orebody is determined by structural and stratigraphic controls that localize the replacement of the country rock by ore-bearing solutions. Examples include tabular or cylindrical flat-lying layers called mantos, pipelike structures that cross the bedding and irregular branching deposits forming small veins in host-rock sediments. Examples of some of the larger replacement ore deposits are at Tintic in the USA, Trepça in Yugoslavia and Tsumeb in Namibia.

Vein deposits, because they outcrop, were easily found and exploited by early miners. They are commonly situated in faults, joints or formational contacts. They range in width from narrow veinlets to bodies that are $\geqslant 15$ m wide. Persistent veins may follow faults or other structures for 5–15 km and extend down dip as much as 3 km. They contain a mixture of barren or gangue minerals with local concentration of ore minerals in pod-shaped layers or shoots 10–15 m long horizontally, and which extend hundreds of meters vertically. Only a small part of US lead comes from veins, but in Europe, Central America and South America, many highly productive mines are located on vein-type deposits. The silver–lead deposits of the Coeur d'Alene district in Idaho are in extensive vein systems.

Contact metamorphic deposits form near igneous intrusions that have either provided the solutions or emanations creating the deposit, or altered and rearranged a mineral deposit existing prior to the intrusion. Only a few contact deposits have produced lead in the USA, but many are mined for other metals. An example of this type of deposit is in the Hanover district in New Mexico, USA, where zinc is the major product and a small percentage of lead is mined as a by-product.

5. Resources

The term resources as used here includes known reserves that can be mined economically under current conditions, plus other resources that may have future economic value either in extensions of known mineral districts ("hypothetical" resources) or in as yet undiscovered deposits that can probably be found by diligent exploration ("speculative" resources).

Data on world lead resources are given in Table 1. Quantities indicated in the "reserves" column include minable reserves commonly referred to as "proven, probable and possible" ores. The values under "other resources" are subjective estimates dependent on the long-term outlook for lead demand and prices, and the continued development of technology for finding and opening new and deeper mines.

US lead reserves more than doubled during the period 1955–1970 as a result of the discovery and development of a new portion of the southeast Missouri lead district now known as the Virburnum Trend. The dominance of Missouri as a lead-producing state is shown in Table 2.

Table 1
World lead resources (US Bureau of Mines 1987)

Location	Reserves (Mt)	Other resources (Mt)	Total (Mt)
Australia	16	12	28
USA	11	11	22
Canada	8	15	23
South Africa	4	2	6
Mexico	3	1	4
Peru	2	1	3
Yugoslavia	2	1	3
Other market economy economies	9	5	14
Centrally planned economies	20	10	30
World total	75	50	125

Table 2
US production and reserves of lead

State	Mine production of recoverable lead (kt) 1980	1981	Reserves (Mt)
Missouri	497.9	389.7	34.1
Idaho	39.9	38.4	1.6
Colorado	9.7	11.4	0.9
Virginia	1.8	1.6	0.2
Other states[a]	1.8	4.4	3.2
Total	551.1	445.5	40.0

[a] Arizona, California, Illinois, Maine, Montana, Nevada, New Mexico, New York, Utah, Washington and Wisconsin

In recent years, more than half of the annual lead supply of the USA has come from secondary sources, mainly from disused storage batteries. In Europe, secondary sources provide about 40% of the lead supply, with a small but significant proportion coming from recycled architectural lead and cable sheathing.

6. Exploration, Development and Mining

Successful prospecting requires coordination of regional and detailed geological mapping, aerial photography, remote sensing from satellites or high-altitude flights, low-altitude aerial geophysical surveys, surface geophysical and geochemical surveys, and finally testing by borehole drilling. Such programs require a large and steady investment over a period of ten or twelve years.

Development requires many drill holes to outline the shape and size of the ore mass, exploratory shafts to determine the physical character of the deposit, metallurgical tests of large samples to determine methods of beneficiation and, in North America and Europe, mine permits regulating procedures for overcoming environmental hazards. The development of new primary sources of lead in recent years has been encouraging and should be sufficient to supply a normal demand for several decades.

Most lead or lead–zinc mines are underground operations entered by shafts or tunnels. A few, such as the Kidd Creek Mine in Canada, start as open pits with later development underground. The most common mining method is a room-and-pillar system as used in southeast Missouri. Such a system, with access by vertical shafts and/or inclined tunnels, can run with low costs by using rubber-tired mining and hauling equipment within the orebody with a minimum of development workings in waste rock. Many mines in western USA are in steeply dipping veins or beds, and so must use more costly mining systems, such as shrinkage, cut-and-fill, or timbered stoping methods.

Lead ores are commonly concentrated in a mill or concentrator located at or near the mine. Mine-run ore containing 3–8% lead is treated to produce a concentrate containing 50–60% lead. Flotation is the principal method of beneficiation used for lead or mixed lead–zinc–silver ores. Concentrates are shipped by truck or rail to smelters that produce lead metal or compounds.

7. Smelting

Conventional lead smelters convert lead sulfide concentrates into lead bullion in a pyrometallurgical operation built around a lead blast furnace. The ores or concentrates are first sintered or roasted to remove the sulfur as sulfur dioxide gas, which is then converted to sulfuric acid. The disposal or marketing of the sulfuric acid is one of the major economic considerations in building or modernizing a smelter. The older practice of discharging the sulfur dioxide to the atmosphere through a tall stack is no longer acceptable in most industrialized countries because of the undesirable environmental effects.

Primary smelters (operating mainly on products from mines) are situated in 35 countries with 90 units, including 12 Imperial Smelting furnaces (developed in the UK for lead–zinc concentrates), providing a combined maximum annual production capacity of about 4.7 Mt. Major producing countries are the USSR (7 plants), the USA (6 plants), Australia (5 plants), the Federal Republic of Germany (6 plants) and Japan (8 plants).

Secondary smelters (fed mainly with recycled battery lead) are generally smaller than primary smelters, but the larger number of secondary plants contributes in the aggregate a large share of world lead supply. Many secondary plants contain a reverberatory furnace and/or a small blast furnace. In the USA, over 100 plants produce recycled lead or lead alloys with 15 companies operating 28 plants that put out 80% of the secondary lead supply. In some European countries statistics on secondary production are not separated from primary ones, but the secondary share is known to be a major part of the total. Estimated world secondary production capacity exceeded 2 Mt.

Actual world annual production is invariably less than rated capacity, probably about 3.2 Mt primary and 1.7 Mt secondary, giving a total of 4.9 Mt in 1980.

8. Trade and Uses

Lead is a commodity in international trade. Prices are most often set by quotations on the London Metal Exchange (LME). Transactions in the USA are governed by a US producer price, commonly a few cents higher than the London price. The average price on the LME in 1987 was 26 US cents per pound. If lead prices are adjusted for inflation the price in 1900 would be almost equal to the 1981 price.

The relationship between domestic and imported sources of lead in the USA is shown in Table 3. Net import reliance was only 15% of US consumption in 1987. Import reliance for lead in the 1970s ranged from 30% down to 8% of consumption.

The distribution of US demand for lead by broad industrial categories is outlined in Table 4. Demand patterns in Europe show a smaller proportion of use in automotive storage batteries and more in metal for architectural or construction purposes and for radiation shielding.

9. *Environmental Aspects*

Lead is present only in very small concentrations in natural environments because of its low solubility, and hence only small amounts normally enter the human body. The major source of lead released into the atmosphere is the exhaust fumes of automobiles using leaded gasoline. In areas of high population density and heavy automobile traffic, these fumes can produce higher than normal concentration of lead in the blood of persons continuously exposed for long periods, and although no cases of lead poisoning from this source have been documented, many industrialized countries have adopted programs to reduce the use of lead in gasoline. In the USA, regulations of the Environmental Protection Agency required the average lead in gasoline to be no greater than 0.18 $g\,l^{-1}$ in 1980, about one half the concentration used in 1978.

Fumes from smelters, coking coal plants, or power plants using coal with a relatively high lead content may cause toxic effects on plants, animals or humans exposed for a long time in downwind areas. Current US laws forbid the use of leaded paints in nursery furniture or interiors. Occupational lead ingestion in storage battery manufacturing plants, primary and secondary lead smelters, and plants using soldering operations has been reduced in the USA in recent years by improved ventilation and safer working practices. In 1978 the US Occupational Safety and Health Administration issued standards limiting workplace exposure to lead to 50 $\mu g\,m^{-3}$ of air, averaged over an 8-hour shift. This rule was contested in the courts, and partial stays were granted in 1980.

Table 3
US lead supply (kt)

Supply	1978	1983	1987
US refinery production			
from domestic ore	530	460	330
from imported ore	66	55	30
from secondary sources	769	452	630
Stock as of December 31	134	159	114
Apparent consumption	1350	1136	1130

Table 4
US and world demand for lead (kt)

Demand	1978	1979	1980	1981
US demand				
storage batteries	880	820	645	769
metal products	244	237	201	185
gasoline additives	178	187	128	111
pigments	92	91	78	80
miscellaneous	39	69	18	22
total	1433	1404	1070	1167
World demand				
primary	3194	3300	3200	3300
secondary	1600	2000	1700	1700
total	4794	5300	4900	5000

10. *Future Outlook*

The potential increase in demand for lead in storage batteries in electric vehicles and for electric utility load levelling systems may justify development of new mines and refineries during the period up to the year 2000. The decrease in demand for lead in gasoline, as environmental regulations take effect, will tend to reduce such growth. Demand for metal products, ammunition, construction and pigments will show small changes dependent on population growth. A forecast made by the US Bureau of Mines predicts an annual lead demand growth rate of almost 2% in the USA and 4% in the rest of the world. The greater growth will be in the developing countries as they become industrialized.

To meet the predicted world annual primary demand of 6 Mt in 2000, a cumulative primary tonnage of 90 Mt would be required in the 1981–2000 period. World reserves are adequate to support the increased mine production required. Secondary recovery and refining is expected to grow at a higher rate as recycling techniques are improved and more secondary plants are built in developing countries.

In the USA, the lead industry has an opportunity to build for an increased production from domestic mines in order to maintain the present level of self-sufficiency. Environmental restrictions and the high cost of investment capital can be outweighed by advances in technology, favorable location of ore reserves, low-cost mining methods, and revised government policies allowing reasonable development of natural resources.

Bibliography

Fine P, Rasher H W, Wakesberg S (eds.) 1973 *Operations in the Nonferrous Scrap Metal Industry Today*. National Association Secondary Metal Industries, New York, pp. 49–54

Bureau of Mines Bulletin 671. USBM, Washington, DC,

International Lead and Zinc Study Group 1980 *World Directory: Lead and Zinc Mines and Metallurgical Works*. International Lead and Zinc Study Group, London

International Lead Zinc Research Organization 1976 *Lead Research Digest 1976*. International Lead Zinc Research Organization, New York

Lovering T G 1976 *Lead in the Environment*, US Geological Survey Professional Paper 957. US Government Printing Office, Washington, DC

Morris H T, Heyl A V, Hall R B 1973 Lead. *Mineral Resources*, US Geological Survey Professional Paper 820. US Government Printing Office, Washington, DC, pp. 313–32

Rathjen J A 1980 Lead. *Mineral Facts and Problems*, US Bureau of Mines Bulletin 671. USBM, Washington, DC, pp. 1–19

Rathjen J A, Rowland T J 1978–9 Lead. *Minerals Yearbook*. US Government Printing Office, Washington, DC, pp. 507–38

US Bureau Mines and US Geological Survey 1980 Principles of a resource/reserve classification for minerals. *Geol. Surv. Circ. (US)* 831: 1–5

US Bureau of Mines Mineral Commodity Summaries 1988. US Bureau of Mines, Washington, DC, pp. 88–89

J. M. Hague
[Platteville, Wisconsin, USA]

Lime

Lime is a manufactured chemical which is a calcined or burned form of limestone or dolomite. The primary lime products include quicklime, pebble lime, hard-burn dolomite or, when water is added, calcium hydroxide or slaked lime. The rather loose use of the term "lime" had led to much confusion and misunderstanding. The term is frequently, albeit erroneously, used to denote almost any kind of calcareous material or finely ground form of limestone or dolomite, as well as the true products of calcination.

Lime is usually made from high-calcium or high-magnesium limestone, generally having a minimum of 90% combined carbonate content. Normally, high-calcium limes contain less than 5% MgO. When the lime is produced from a high-magnesium limestone, the product is referred to as dolomitic lime.

1. Calcination

Calcination refers to a reaction wherein limestone or dolomite is heated to less than its melting point with a resultant change in chemistry and a weight loss. The basic chemical reaction is as follows:

$$CaCO_3 \text{ (limestone)} + \text{heat } (1000\text{–}1300\,°C) \rightleftharpoons CaO \text{ (quicklime)} + CO_2\uparrow$$

or

$$CaCO_3.MgCO_3 \text{ (limestone)} + \text{heat } (900\text{–}1200\,°C) \rightleftharpoons CaO.MgO \text{ (quicklime)} + 2CO_2\uparrow$$

Calcination is strongly time-variant with the different limestones. In a very broad sense this relates to the fact that the calcination reaction starts on the exterior surfaces of the limestone and then proceeds toward the center. As the calcination reaction takes place the CO_2 released at the reaction interface must make its way through the lime to the exterior surface; therefore, because calcination is limited by gas defusion to the surface of the partially calcined limestone, natural impurities in the stone, differences in crystallinity, grain-boundary chemistry, density variations and imperfections in the atomic lattice all play a significant role in the calcination rate. The suitability of a given limestone as a source material for lime production can be determined only after completion of adequate burn tests, designed to evaluate the various limiting parameters. When a coal-fired kiln system is considered, the entire process of calcination is made even more complex.

In the calcination reaction CO_2 is released. The result is a 44% weight loss during the complete calcination of a high-calcium limestone or a 48% weight loss for a highly dolomitic limestone. The trade term for this weight loss is "loss on ignition" (LOI).

2. Lime Production

The production (calcination) of lime requires the use of kilns which will facilitate the heating of the raw limestone such that the calcination reaction takes place. Although lime calcination is chemically a relatively simple operation, many kiln systems are used. These include vertical, rotary, rotary with preheaters, multiple-shaft regenerative and a variety of other types of kilns.

Selection of a lime kiln depends upon several factors, the most important ones being burning characteristics of the stone, fuel consumption and capital equipment cost. Other factors include customer specifications, the type of fuel to be utilized, environmental constraints, and so forth. As a result of severe energy-cost fluctuations since the early 1970s, there have been major changes in kiln selection. In the USA and Canada the emphasis has been on rotary preheater kilns. Concurrently, developments in calcining technology were generating interest in multiple-shaft regenerative kilns.

2.1 Vertical Kilns

Vertical kilns traditionally have burned only larger stone (75–300 mm) with a size ratio of approximately 1:2. However, the new designs can handle stone sizes as small as 20 mm and a widening size ratio to as much as 1:4. Fuel consumption has been a major advantage of vertical or shaft kilns, normally requiring less than $5.2\,GJ\,t^{-1}$ of lime produced. Power requirements for these kilns will vary with stone size but at nominal capacity will be in the range of $54\text{–}90\,MJ\,t^{-1}$. A disadvantage of traditional vertical kilns in today's energy environment is their fuel requirement of oil, natural gas or coke; however, several recent developments in burner technology have come to the point where pulverized coal can now be utilized in the modern shaft kiln.

Vertical kilns may be of stone-masonry, reinforced-concrete or boiler-plate construction. The most widely used kiln has a refractory-lined steel shell and is usually circular in cross section. These kilns may be 2.7–7.3 m in diameter and 15–4.8 m high. Capacities vary from as low as 9.5 $t\,d^{-1}$ to an excess of 700 $t\,d^{-1}$, with the larger tonnages being restricted to the newer generation of kilns.

2.2 *Multiple-Shaft Kilns*

Developed to burn stone of small diameter, this vertical kiln utilizes the parallel-flow calcining principle in double and triple shaft units. The shafts are interconnected in the burning zone and while one shaft is being fired the other is preheated. Fuel and combustion air are supplied to the burning shaft from above, ignite at the upper end of the burning zone, and calcine the lime in uniflow. The exhaust gases then pass into the second shaft, preheating the stone in counterflow. After 10–15 min, the shaft firing is reversed. Cooling air is blown into both shafts simultaneously. The regenerative system has experienced fuel consumption of 3.3–3.9 $GJ\,t^{-1}$ of lime produced. Kilns vary in capacity from 90–550 $t\,d^{-1}$ and while nominally gas or oil fired, current technology has now allowed the use of pulverized coal. These kilns have found acceptance in Europe and the USA and are now being installed in Japan.

2.3 *Annular-Shaft Kilns*

The burning process in this kiln is based upon counterflow for preheating, and calcining and uniflow for residual calcination. Partitions provided by an inner cylinder and by staggered bridges in the burning zone permit an even distribution of heat and a uniform flow of material down the kiln. Stone as small as 25 × 75 mm can be calcined to produce soft-burned lime using either oil or gas firing. Capacities vary from 90–270 $t\,d^{-1}$ with a fuel consumption of 6.2 $GJ\,t^{-1}$. This type of kiln is not currently popular anywhere in the world.

2.4 *Rotary Kilns*

Unlike vertical kilns which operate fully charged, the rotary kiln has about 90% of its volume filled with flame and hot gases. As the kiln slowly rotates, new surfaces of stone are exposed to the hot gases but there is little passage of gases through the solids. Hence, radiation interchanged between gas, solids and refractory wall plays an important part in the overall heat transfer. Because the area of solids exposed is relatively small, a rotary kiln is less efficient than a shaft kiln. However, the rotary has the advantage of burning stone from as small as 6.5 mm to as large as 57 mm. Generally the size ratio is 1:3 in order to minimize segregation.

Rotary kilns vary greatly in size ranging from 2 × 25 m to 5.5 × 190 m with capacities of 45–945 $t\,d^{-1}$. While rotary kilns can burn a wide range of fuels, the major drawback of these kilns has been their fuel requirement of 18.6 $GJ\,t^{-1}$.

2.5 *Rotary Preheater Kilns*

Of the several advances which have been made in improving the heat efficiency of rotary kilns, probably the most pronounced has been the trend away from longer kilns to medium-length kilns with external preheaters. The preheater is a refractory-lined chamber located below the raw-feed bin and ahead of the kiln. Exhaust gases from the kiln are drawn countercurrent to the stone preheating it to 650–900 °C. Retention time within the preheater is 1–1.5 h depending upon the system design, burning characteristics of the stone, and so on. During preheating, approximately 30% calcination is achieved before the stone is discharged to the kiln. Following preheating, the partially calcined stone is fed to the kiln at a predetermined rate by means of hydraulically actuated plungers. This type of preheater has been adapted to the new large-volume kilns and typically may be 5.2 m in diameter by 62.5 m long and have a capacity of 900–1000 $t\,d^{-1}$. This and other types of preheater kilns, including the travelling-grate variety, plus the use of heat exchangers and coolers which recuperate waste heat, have improved fuel efficiency to 6.9–8.5 GJ per tonne of lime.

2.6 *Calcimatic Kiln*

The rotary hearth calciner, the calcimatic kiln, consists of a preheater, circular hearth and cooler, all refractory lined. Like the preheater, this kiln burns small stone which is typically sized in a 1:3 ratio. The stone is carried on the hearth in a thin layer and one revolution of the hearth constitutes the calcining cycle. Numerous burners located inside and outside the hearth are used for firing, utilizing gas, fuel oil and, most recently, pulverized coal. Fuel requirements approximate 7.7 $GJ\,t^{-1}$ with a wide range of burns possible because of the ease of operator control and calcining. One of the primary advantages of this type of kiln is the fact that attrition loss is negligible. This calcining system can utilize very soft limestones which would otherwise be subject to high mechanical breakage. However, their capacity is generally restricted to only 90–300 $t\,d^{-1}$.

2.7 *Flash Calciners*

The flash calciner is one of the most recent developments in calcination and in reality is not a kiln. This system is designed to calcine the undersized material or fines remaining after the primary kiln feed has been sized. The system includes a preheater, flash calciner and cooler. Preheating is achieved in a series of cyclones in which the heat of rising hot gases is absorbed in the counterflowing material which is in the top and heated to 980 °C. As the preheated feed drops into the furnace calciner, a forced air vortex flow pattern pulls the particles into the center of the furnace. Within the furnace the temperature is 1200 °C. The retention time is only a few seconds. The system is designed in two modes depending upon the size of feed to be used. Where stone fines are to be calcined, 60–80% of the calcination occurs in the furnace. The final calcination takes place in a fluid bed which is located between the furnace and the cooler. In those systems where −28 mesh material is fed to the calciner, no fluid-bed reactor is necessary. This system can utilize gas, oil or coal and will normally require 6.2–8.5 GJ t^{-1} of product.

3. Lime Hydration

Hydrated lime is obtained by adding water to quicklime to produce a dry, fine powder. The affinity of quicklime for moisture is then satisfied although it still retains a strong affinity for CO_2.

Hydration is the combining of calcium oxide with water in a reversible reaction to form calcium hydroxide:

$$CaO + H_2O \rightarrow Ca(OH)_2$$

As a standard condition, the reaction is exothermic for the formation of hydrate with the accepted value for the heat of hydration being 15.456 kcal g-mol^{-1} of CaO. This is a three-step reaction with the first being the disassociation of CaO into a Ca^{2+} ion and an O^{2-} ion:

$$CaO \rightarrow Ca^{2+} + O^{2-}$$

Next the water ionizes into two hydroxal ions, owing to the strong attraction that an oxygen ion exerts on a water molecule:

$$H_2O + O^{2-} \rightarrow 2(OH)^-$$

Finally the calcium and hydroxal ions combine to form calcium hydroxide:

$$Ca^{2+} + 2(OH)^- \rightleftharpoons Ca(OH_2) + \text{heat}$$

As noted, this is a reversible reaction and hydrated lime can therefore, when heated, revert to the original oxide form.

Hydrated lime is available in two principal forms: standard, where 85% passes a No. 200 sieve and superfine or spray hydrate where 98% or more passes a No. 325 sieve. These in turn are manufactured in two types, normal (N) and special (S), each including high-calcium or dolomitic hydrate.

Type S limes are highly hydrated, generally by processes involving pressure hydration. They contain less than 8% unhydrated oxides and develop high plasticity and high water retention. Since World War II, the demand for type S lime has grown rapidly and currently it is the chief lime for structural uses. Some lime manufacturers also add entraining catalysts to the lime to increase plasticity and durability.

4. Lime Markets

Lime has possibly the greatest number of diverse uses of any chemical or mineral commodity. Over 90% is utilized by the chemical and metallurgical process industries where it may be used as a flux, acid neutralizer, causticizing agent, flocculent, hydrolyzer, bonding agent, absorbent and/or raw material. An allied product, dead-burned dolomite, is used as a refractory material. Table 1 shows the approximate percentage uses of lime resources by industry in the USA.

Of these various markets, steel is the largest consumer industry. In this application, lime acts as a scavenger in purifying steel by fluxing out, into the molten slag, acid oxide impurities such as silica, alumina, phosphorus and sulfur.

Most steel companies use pebble or pelletized lime as flux in the basic oxygen furnace (BOF). Recent European and Japanese technology has modified the basic oxygen process. This modified process, known as Q-BOP, utilizes pulverized lime.

The treatment of municipal water and industrial process water with lime is also an important market. Lime is required to remove the temporary bicarbonate hardness from the water. The high pH of 11.5 induced by the lime is of value as the secondary sterilization agent to chlorine because retention for 3–10 h at this pH will kill over 99% of the bacteria and most viruses. By introducing CO_2 into the lime-treated water, the pH is lowered to acceptable levels and most of the lime in solution is then precipitated as a carbonate sludge.

Lime is an important chemical in a number of environmental applications. The newest of these is as a primary reagent material in flue-gas desulfurization (FGD). Initially it was used in wet lime scrubbing, but more recently, attention is focusing on the so-called dry scrubber or spray dryer. FGD markets may grow

Table 1
Industrial usage of lime in the USA in 1986 as a percentage of total lime resources

Industrial use	Percentage
Steel	34.4
Water treatment and purification	10.3
Soil stabilization (includes all construction uses)	10.1
Flue-gas desulfurization	8.4
Chemical manufacturing	8.2
Pulp and paper production	7.2
Sewage treatment	6.9
Sugar refining	2.9
Refractories (dolomitic lime)	2.1
Metallurgical processing (other than steel)	2.0
Other	7.5

Source: US Bureau of Mines

significantly in the 1990s in the USA if the US Congress enacts more stringent air pollution control standards. At the other end of the environmental control spectrum, lime is utilized in acid-water treatment, sewage treatment and toxic-chemical cleanup. Overall, environmental applications may represent the largest potential growth markets for lime.

Lime is also used in a number of chemical manufacturing processes. These include the production of caustic soda, calcium carbide, bleaches and various inorganic chemicals. In addition, lime is required in the production of several important organic chemicals. Chief among these are ethylene and propylene glycols, and calcium-based organic salts such as calcium stearate, acetate, lactate and lignosulfonate.

Another primary market for lime has been in the manufacture of pulp and paper where it is universally used to causticize waste sodium carbonate. In this process it is used to regenerate sodium hydroxide; in spite of the fact that the pulp mills recover 90 to 96% of the lime employed by recalcining the dewatered calcium carbonate sludge precipitate, a substantial quantity of make-up lime continues to be used. Secondary uses of lime by the paper industry include making calcium hypochlorite bleach and the clarification of process water and color removal of wastewater effluents.

The construction industry is also a major user of lime. While finding a variety of applications, the primary consumption is the area of soil stabilization. This is particularly important in areas where the subgrade soils have very high clay contents. The lime reacts with the silica and alumina in the clay soils to form complex cementing compounds which bind the soil into hard stable masses that are not sensitive to water saturation. The plasticity of the soil, as well as shrink and swell, are markedly reduced. The amount of lime used generally ranges between 3 and 5% of the dry weight of the soil. Over 80% of the lime used in this application is hydrate which is disked into the clay soils.

5. *Specifications and Analysis*

Owing to the number of limes available, their many uses and their wide variation in physical and chemical characteristics, there is no overall lime specification. Numerous specifications and buyer requirements for lime exist, not only for most of the major uses but even among individual companies or plants.

In the USA, the National Lime Association (NLA), with its member lime companies, and a majority of the lime consumers, subscribe to the American Society for Testing and Materials (ASTM) specifications on lime which are promulgated by Committee C-7 on Lime. These specifications cover many uses in testing procedures, including chemical analysis, sampling, inspection, packing and marking of quicklime and lime products, and physical testing. The details of these specifications are found in ASTM Standard Section 4, volume 04.03.

Another important lime specification in the USA is the American Water Works Association (AWWA) Standard for Quick Lime and Hydrated Lime (B202-77).

See also: Calcium Aluminate Cements; Construction Materials: Crushed Stone; Construction Materials: Dimension Stone; Construction Materials: Fill and Soil; Construction Materials: Granules; Construction Materials: Industrial Minerals; Construction Materials: Lightweight Aggregate; Portland Cement Raw Materials; Portland Cements, Blended Cements and Mortars

Bibliography

American Society for Testing and Materials 1986 ASTM Standard, Sect. 4, Vol. 04. 03. ASTM, Philadelphia, Pennsylvania

American Water Works Association 1977 *Standard for Quick Lime and Hydrated Lime*, Standard (B202-77). AWWA, Denver, Colorado

Boynton R S 1980 *Chemistry and Technology of Lime and Limestone.* Wiley, New York

Dixon T 1982 North American lime. *Ind. Miner. (London)* 177:51–63

Freas R C, Thompson J L 1983 Lime. In: Lefond S J (ed.) 1983 *Industrial Minerals and Rocks*, 5th edn. American Institute of Mining, Metallurgical and Petroleum Engineers, New York, pp. 809–31

Gutshick K A 1987 *Lime for Industrial Uses*, STP 931. American Society for Testing and Materials, Philadelphia, Pennsylvania

Mullins R C, Hatfield J D 1969 *Effects of Calcination Conditions on Properties of Lime*, STP 462. American Society for Testing and Materials, Philadelphia, Pennsylvania, pp. 117–31

National Lime Association 1965 *Chemical Treatment of Sewage and Industrial Wastes*, Bulletin 215. NLA, Washington, DC

National Lime Association 1966 *Specifications for Lime and its Uses in Building*, Bulletin 322. NLA, Washington, DC

National Lime Association 1972 *Lime Stabilization Construction Manual*, Bulletin 326, 5th edn. NLA, Washington, DC

National Lime Association 1976 *Acid Neutralization with Lime*, Bulletin 216. NLA, Washington, DC

National Lime Association 1976 *Lime Handling Application and Storage in Treatment Processes*, Bulletin 213. NLA, Washington, DC

National Lime Association 1976 *Water Supply and Treatment*, Bulletin 211, 11th edn. NLA, Washington, DC

National Lime Association 1980 *Lime in Municipal Sludge Processing*, Bulletin 217. NLA, Washington, DC

National Lime Association 1981 *Chemical Lime Facts*, Bulletin 214, 4th edn. NLA, Washington, DC

Pressler J W 1986 Lime. *Minerals Yearbook 1986*, Vol. 1. US Bureau of Mines, Washington, DC

R. C. Freas
[Franklin Limestone Company, Nashville, Tennessee, USA]

Lithium, Cesium and Rubidium Resources

Lithium, cesium and rubidium are useful because these rare alkali metals and their compounds have unique properties. Lithium is the most available of the group. It is recovered from brines and from minerals in granite pegmatites, and in the future may also be recovered from clays. Domestic lithium reserves are adequate for conventional uses at least until the end of the century.

1. Uses

Lithium carbonates and fluorides are used in aluminum reduction plants to lower the melting temperature and power required to reduce aluminum oxide to metal. When added to glass and ceramics, lithium improves their heat-resistance properties. The use of lithium soaps in lubricants also improves their resistance to temperature changes. Cesium and rubidium are used in photosensitive electronic devices, and all three rare alkali metals are used in research and in medicine. Its light weight and high electrical potential gives lithium an advantage in batteries for electric vehicles, for storage of off-peak power, for intermittent power supplies and for fuel cells. Lithium is also an essential source for the tritium required in various thermonuclear (fusion) power reactor designs such as the tokamak where it also serves as a neutron absorber and heat exchanger.

2. Geochemistry and Occurrence

Lithium, rubidium and cesium tend to be dispersed in common rocks at the 1–100 ppm level. The disparity in abundance compared to sodium and potassium (i.e., the ratio of Li:Na for seawater is 0.000 016) means that concentration levels required to form separate minerals of these metals are rarely attained. Although nearly 40 lithium minerals have been described, only the silicates—spodumene, lepidolite and hectorite—occur in significant quantity. Pollucite is the only known cesium mineral, and no rubidium minerals are known.

2.1 Pegmatites

Spodumene, lepidolite and pollucite occur in the late stage, coarsely crystalline phase of some granite intrusions known as pegmatites. These form dikes peripheral to the main intrusion by crystallization of the residual fluids after most of the sodium and potassium have been depleted. As high temperatures and pressures prevail at the 5–10 km depth where pegmatites form, they are exposed at the earth's surface only after several kilometers of rock have been removed by erosion. Such extensive uplift and erosion is restricted to ancient continental shield terrains in Canada, Australia and Africa, and locally in such mountains as the southern Appalachians and the Black Hills of South Dakota. There is virtually no chance of discovering pegmatites in most of the USA. Zoned pegmatites in the Black Hills, containing individual spodumene crystals as much as 2 m long, or large masses of lepidolite that could be separated by hand were the chief sources of domestic lithium produced through the early 1950s. An increase in demand encouraged recovery by mechanical beneficiation of spodumene from previously known unzoned pegmatites in North Carolina in which the crystals average only about 0.05 m in length but comprise a nearly uniform 20% of dikes 15–100 m wide (average 0.7% Li). This source continues to supply about two thirds of domestic needs.

2.2 Brines

A brine containing 300 ppm Li, which occurs in the subsurface of Clayton Valley playa in Esmeralda County, Nevada, was developed as a source of lithium beginning in 1967. The brine is pumped into solar evaporating ponds and the lithium concentrated to more than 6000 ppm Li at which point it can be precipitated as the carbonate. Other brines containing more than 100 ppm Li are associated with the Smackover oil-field formation-waters of Texas and with geothermal waters in the Imperial Valley of California. Potentially commercial lithium brines are also known in the playas and salt lakes of Chile, Bolivia, Argentina, Israel and the Tibetan plateau.

2.3 Clays

Domestic supplies of lithium could be augmented by extraction of lithium from lithium-rich clays. Hectorite is a lithium clay, which occurs associated with altered volcanic sediments in certain basins in the western USA. In the future hectorite may be developed as a source of lithium.

3. Resources and Industry Outlook

Of the 4000–5000 t of lithium produced each year in the USA, more than half comes from two pegmatite mines in North Carolina and the balance from the Nevada brine field. This production has been adequate for the needs of the USA and has provided limited amounts for export. Cesium and rubidium are produced as minor by-products of the recovery of lithium from previously imported lepidolite. A reserve of pollucite from which cesium could be obtained if the demand warranted is known to occur in Manitoba, Canada.

Domestic reserves and world resources of lithium (Table 1) are regarded as adequate for conventional uses at least until the year 2000. An increase in demand for lithium resulting from new energy-related applications will probably be met by production from the large brine fields of South America. Further pressure on demand sufficient to increase the price may see a return to production from domestic lithium clays and

Table 1
In-place (geologic) reserves and resources of lithium and probable yield (in Mt)[a]

Resource class	Lithium ore and brine	Lithium in place	Probable yield	
			Ultimate	By year 2010
Pegmatites				
reserves	310	2.0	1.4	0.58
resources	1200	8.0	3.0	0.00
Brines				
reserves	3600	3.1	1.0	0.17
resources	17000	4.0	1.0	0.04
Clays				
resources	1000	3.0	1.0	0.03
Totals	23000	20.0	7.0	0.8

[a] Estimate for the world exclusive of China and USSR (adapted from Vine 1980 p. 90)

from numerous small properties in various parts of Canada.

Bibliography

Penner S S (ed.) 1978 Lithium–Needs and resources. *Energy* 3 (3): 235–413

Vine J D (ed.) 1976 *Lithium Resources and Requirements by the Year 2000*, US Geological Survey Professional Paper 1005. US Government Printing Office, Washington, DC

Vine J D 1980 *Where on Earth is All the Lithium?*, US Geological Survey Open-file Report 80–1234. US Government Printing Office, Washington, DC

J. D. Vine
[Golden, Colorado, USA]

M

Magnesium Resources

Magnesium is the eighth most abundant element in the earth's crust. It does not occur in nature as an element, but is extracted commercially from seawater, brine, salt beds and naturally occurring ores such as magnesite and dolomite. It is used in industrial applications as a metal but finds its greatest use in the form of an oxide for refractory purposes.

1. Crustal Distribution and Tenor

In its many forms magnesium makes up about 2% of the earth's crust. It occurs in over 60 different materials but only four—dolomite, magnesite, brucite and olivine—are used commercially to produce magnesium compounds.

Magnesium is also widely distributed in salt forms in seawater and brines. Seawater contains about 1.3 g Mg^{2+} per liter (0.2wt% MgO), whereas brines (either naturally occurring or obtained by solution mining of solid salt beds) contain the equivalent of 4–20% $MgCl_2$.

Natural magnesite ($MgCO_3$) occurs in two forms, crystalline and cryptocrystalline. The crystalline form occurs in relatively few but very large deposits, whereas the cryptocrystalline form occurs widely in many small deposits. The crystalline deposits have a much higher ore content (70–90%) than the cryptocrystalline deposits, which can have ore contents as low as 12% of the rock handled in mining. Both types of ore are beneficiated.

2. Origin and Occurrence

2.1 Crystalline Magnesite

Crystalline deposits of magnesite are usually associated with dolomite. All deposits have similar relationships suggesting a common mechanism of emplacement (Rosenberg and Mills 1966). The dolomite beds are widely viewed as being sedimentary and some authors have suggested a sedimentary origin for the magnesite emplacement. The consensus, however, favors hydrothermal replacement of preexisting dolomite by magnesite on a volume-for-volume basis. The coarse granularity of some magnesite deposits can be attributed to later recrystallization during regional metamorphism.

2.2 Cryptocrystalline Magnesite

Cryptocrystalline or "amorphous" magnesite is an alteration product of serpentine or allied magnesium-bearing rocks. The magnesite occurs in veins varying in width from a few millimeters to several meters, and formed by the action of carbon-dioxide-charged water percolating down or rising up through serpentinized fissures and converting serpentine to magnesite. Most of these deposits are found near the surface and are of limited depth.

3. Global Resources

Dolomite, seawater, brines, salt beds and magnesite—the main sources for most of the magnesium and magnesium compounds—are widely distributed throughout the world. While the quantities are nearly inexhaustable, there are a limited number of deposits which can be exploited economically.

The major deposits of crystalline magnesite are located in the USA, Canada, the USSR, North Korea, the People's Republic of China, Greece, Czechoslovakia, Australia, Austria, Brazil, India and Nepal. The total reserves are estimated to be about 2800 Mt.

The major cryptocrystalline deposits are located in Greece, Turkey, Australia and India. The main brine deposits are located in the USA, Mexico, and the evaporite-salts basin stretching from Scotland to the south of Poland.

4. Production

The total production in 1980 of all grades of dead-burned and calcined magnesite was about 4.5 Mt representing about 11 Mt of crude ore; about 80% of this production was in central Europe and Asia. Magnesia produced by chemical methods totalled some 2.5 Mt in 1980. Production was located mainly in the USA, Israel, Ireland, Mexico and Japan.

Magnesium metal production in 1987 was about 368 kt with production being concentrated in the USA, Canada, France, Italy, Norway, the USSR and Japan.

5. Uses and Future Outlook

About 10% of the magnesium produced is used in metallic form with 90% used in nonmetallic applications, principally for the production of high-temperature refractories. Over 80% of all refractory magnesia (MgO) is used by the steel industry. In 1958 magnesia usage was 5.25 kg per tonne of steel produced; by 1980 this had diminished to 2.27 kg per tonne.

The production of magnesium metal is expected to increase at a modest rate. The use of magnesia in cattle feed, pharmaceuticals, dyes, sizings, paper manufacture, fertilizers, filters and other chemical applications is also expected to increase, but at a slow rate. The use of magnesia for refractory applications will probably become more specialized, with an increased demand for higher-purity magnesia.

Bibliography

Duncan L R, McCracken W H 1980 The impact of energy on refractory raw materials. *Refract. Prod. Acero Cobre, Congr. Refract. ILAFA–ALAFAR*. ILAFA, Santiago, Chile

Kramer D 1985 *Magnesium Metal, Magnesium Compounds*, US Bureau of Mines Mineral Commodity Summary. USBM, Washington, DC, pp. 92–95

Rosenberg P E, Mills J W 1966 A mechanism for the emplacement of magnesite in dolomite. *Econ. Geol.* 61: 582–86

US Bureau of Mines 1988 *Mineral Commodity Summaries 1988*. USBM, Washington, DC

Wicken O M, Duncan L R 1983 Magnesite and related minerals. In: Lefond S J (ed.) 1983 *Industrial Minerals and Rocks: (Nonmetallics Other Than Fuels)*, 5th edn. American Institute of Mining, Metallurgical and Petroleum Engineers, New York

L. R. Duncan
[Harbison-Walker Refractories, Pittsburgh, Pennsylvania, USA]

Manganese Resources

Manganese is an essential and irreplaceable element in steel, and is mined from sources with natural concentrations ranging from about 15% to more than 50% manganese. From a geological standpoint, known world resources are adequate to meet foreseeable demands, but technological, economic and political considerations complicate the supply picture. South Africa is increasingly likely to be the source of the remaining high-grade ores, essential for current practices in manganese use, and could become the main supplier of manganese to market-economy countries by the turn of the century.

1. Manganese Ore Deposits

Manganese is the twelfth most abundant element in the earth's crust and constitutes slightly more than 0.1% of the crust. The economically minable manganese deposits are in natural concentrations of 150–500 times the average crustal abundance. Manganese is a mobile element, and a variety of geological processes have created natural concentrations of minable grade and size.

Deposits of principal present and future economic importance are of three types, each formed by a distinct set of geological processes

(a) Residual deposits formed by surface and near-surface chemical breakdown and the leaching of impurities from manganiferous rocks. Such deposits are restricted to humid tropical regions where intense tropical weathering has leached many elements from the rocks, forming a residual concentration of manganiferous material. Deposits of this type are mined extensively in Brazil, India and Gabon.

(b) Sedimentary deposits, formed by chemical processes that segregated manganese during deposition of sediments. Some sedimentary deposits contain vast quantities of ore because the manganiferous beds are very extensive. The world's largest manganese deposits, including those of the Nikopol and Chiatura districts of the USSR and the Kalahari district in South Africa, are of sedimentary origin. Residual concentration may have improved the grade of parts of these deposits.

(c) Seafloor nodules, containing copper, nickel, cobalt and manganese are abundant over vast areas of deep ocean basins. They can be considered as a type of sedimentary deposit, but their polymetallic nature and submarine setting, and the technical mining problems created by that setting, warrant their classification as a distinct deposit type (see *Ocean Resources and Mining*). Nodules of especially high metal content and great abundance are common in much of the northern equatorial Pacific Ocean. The better grades of nodules contain 30–35% manganese.

The mining of nodules could begin as early as 1990 if legal, political and some remaining technological problems are solved. The manganese obtained could provide a significant portion of world industrial demand by the turn of the century, although most early mining and processing of nodules may recover only copper, nickel and cobalt and, at least temporarily, discard manganese for economic reasons.

2. Uses

About 90% of manganese consumption is by the steel industry. Manganese is a quantitatively minor but essential ingredient in the process of converting iron ore to steel. It serves as a desulfurizing, deoxidizing and conditioning agent during the smelting of iron ore, and as an alloying element in steel it increases toughness, strength and hardness. About 7 kg of manganese is used in producing a tonne of steel. Manganese is used in steelmaking both as ore and as ferromanganese, a processed derivative of ore that contains 68–78% manganese.

Other lesser uses of manganese are as a depolarizing agent in dry-cell batteries and in a wide variety of applications in the chemical industry. Manganese ores for chemical and battery purposes are generally of higher grade, or have special properties not common to ores used in metallurgical applications.

3. Resources and Supply Adequacy

Worldwide production of manganese in 1987 was about 25.9 Mt of metal, contained in ores ranging from about 15 to 50% manganese. Demand for manganese grows in direct proportion to growth in steel production; common estimates are that growth worldwide will be about 3% per year to the year 2000 and that the demand for manganese will approximately double between 1980 and 2000.

Identified land-based resources of manganese are greater than 6×10^9 t, but they include material that is highly variable in grade and quality and that presents various technological problems for mining and utilization. These resources contain manganese sufficient to meet anticipated demand for many centuries if the contained manganese can be obtained economically. However, all resources are not likely to be economically minable. Present practices in the steel industry require large quantities of high-grade ore that is used either directly or blended with lower-grade ores. The utilization of some low-grade ores is contingent on the continued supply of high-grade ores with which they can be blended to produce an acceptable average grade. The critical problem for the near future, therefore, is the continued supply of higher-grade (>35% manganese) ores. Traditional suppliers of high-grade ore, such as Brazil, India, Australia and Gabon, have deposits of large but limited size; exhaustion of some of these deposits in the next few decades is foreseeable. Intense exploration has failed to find additional, economically important orebodies in these countries and other tropical regions.

Large quantities of high-grade ores are known in South Africa. About 75% of the better grades of manganese resources used by market-economy countries are located there. If the trend in recent years of depletion without new discoveries in other countries continues, South Africa will dominate the manganese supply of market-economy countries by the turn of the century.

The USSR has very large deposits of manganese, totalling nearly 3×10^9 t in two principal districts: Nikopol and Chiatura. Much of the material is of low grade and otherwise poor quality, and mining costs are high.

Improvements in world manganese supply could result from: (a) discoveries of new high-grade ore deposits; (b) development of cost-effective techniques to upgrade low-grade deposits to make synthetic high-grade ore; (c) changes in steel- and ferromanganese-making technology to use ores of lower grade than those now used; and (d) initiation of manganese recovery from seafloor nodules.

Bibliography

DeHuff G L, Jones T S 1977 *Manganese—1977*, US Bureau of Mines Mineral Commodity Profiles MCP-7. USBM, Washington, DC

DeHuff G L, Jones T S 1980 Manganese. *Mineral Facts and Problems*, US Bureau of Mines Bulletin 671. USBM, Washington, DC, pp. 549–62

Dorr J V N, Crittenden M D, Worl R G 1973 Manganese. In: Brobst D B, Pratt W P (eds.) 1973 *United States Mineral Resources*, US Geological Survey Professional Paper 820. US Government Printing Office, Washington, DC, pp. 385–99

National Materials Advisory Board 1981 *Manganese Reserves and Resources of the World and Their Industrial Implications*, NMAB-374. National Materials Advisory Board, Washington, DC

US Bureau of Mines 1988 *Mineral Commodity Summaries 1988*. USBM, Washington, DC

W. F. Cannon
[US Geological Survey, Reston, Virginia, USA]

Mercury Resources

Mercury is one of the few metals that at ordinary temperatures is a fluid. This property combined with others, such as high density, high surface tension, linear thermal expansion, good electrical conductivity, ability to alloy readily and a tendency to form very toxic compounds, has given mercury thousands of uses, for many of which no substitution is practical or desirable. Most uses, however, require only small amounts, and the annual world consumption of mercury is only about 6000 t. The hazards of mercury poisoning and environmental contamination were popularized in the 1960s and led many countries to place restrictions on its use resulting in a decline in consumption since 1973.

World production of mercury during this century has ranged from 65 000 flasks (2300 t) in 1933 to a high of 290 000 flasks (10 000 t) in 1969. For 1987, production has been estimated at 174 000 flasks (6010 t). This estimate does not include secondary recovery, which in recent years has increased to about 10 000 flasks (350 t) annually.

Six mines or districts have yielded three quarters of world production, and several thousand others have some production. Historically, the largest producers were the Almaden mine in Spain, the Idria mine in Yugoslavia, the Monte Amiata district in Italy, the Santa Barbara mine in Peru, and the New Almaden and New Idria mines in the USA. Of these, only the Almaden mine was operating in 1980, as a major part of the world demand was being supplied by more recently discovered deposits in Algeria, the USSR and the USA.

Mercury deposits lie in two nearly connected belts of Tertiary to Recent orogeny and volcanism. One is the circum-Pacific belt, which follows the western part of South, Central and North America, extends across the Aleutian Islands, runs southward through Japan and the Philippines and terminates in New Zealand. The other belt begins close to the Pacific belt in northeastern Siberia, extends southwestward through the eastern part of the USSR, Mongolia and China, west through the Himalayas and from there through Turkey, Yugoslavia, Italy, Algeria and Spain. In terms of modern plate tectonic theory, the belts, in general, coincide with zones of subduction where continental plates have overridden oceanic plates.

Although mercury deposits occur in belts of volcanism, many are not closely related to volcanic rocks, and, in fact, they occur in nearly every common kind of rock. Major deposits have been found in quartzite, limestone, sandstone, volcanic rocks of various compositions and altered serpentine. Some are formed by replacement of the host rock, whereas others are the result of ore minerals filling openings in the rock, such as pore space or open fractures. Some ores are localized beneath well-formed structural traps with "caps" of less permeable material, but for other major deposits no structural control is obvious. Despite the wide diversity of occurrence, deposits in similar kinds of rock exhibit similarities in size, grade, mineralogy and structural environment that aid in predicting their ultimate potential.

All major mercury deposits are mined for their mercury content alone, and all contain cinnabar, the red mercuric sulfide (HgS), as the chief ore mineral. The black mercuric sulfide, metacinnabar, and liquid mercury metal are locally important, and in single deposits a mercury sulfide–chloride, corderoite, and a mercury–antimony sulfide, livingstonite, are significant. Some polymetallic deposits containing mercurian tetrahedrite have been mined, and a very small amount of mercury has been recovered as a by-product of gold, iron and lead–zinc mining. Nonmercurial minerals commonly found in mercury ores are the silica minerals (quartz, chalcedony and opal), the carbonate minerals (calcite, dolomite and magnesite), the iron sulfides (pyrite and marcasite) and a few antimony and arsenic compounds. Gold, silver or base metals are generally present in trace amounts. Small amounts of hydrocarbons, such as oils and tars, also accompany some mercury ores.

Mercury deposits are epithermal, meaning they are formed at shallow depths and at relatively low temperatures (50–200 °C). Virtually all are hydrothermal; that is, deposited from hot water. Hot springs in the USA, New Zealand and the USSR are now depositing a little cinnabar at the earth's surface, but all commercial mercury deposits were formed a short distance below the surface. The deepest mined ore body, at the New Almaden mine in California, USA, had its lower limit of deposition at 800 m, and the deepest known occurrence of cinnabar is at about 1800 m.

Unlike most metals, marketable mercury is generally recovered at the mine. Mercury with a purity of 99% or more is extracted directly from cinnabar ore containing 0.3–3.0% mercury by heating to volatilize the mercury and then condensing the vapor. Two general types of equipment are used: externally heated retorts in which there is no mixing of fuel gases with mercury vapor, and internally fired furnaces where mixing occurs. Major mines generally use furnaces rated at 100–200 t of ore per day. The ore is treated as mined or reduced to 2–3 cm in size before furnacing. In recent years, to avoid the high cost of heating all the ore, flotation has been used to prepare a 75% cinnabar concentrate before roasting. The condensing system used with furnaces yields mostly "soot," which consists of finely divided mercury and carbon. The soot is purified by stirring with lime, which causes the mercury droplets to coalesce. The mercury is then bottled in steel vessels containing 34.5 kg. This flask is the standard unit of international trade.

Most uses of mercury are in technologically advanced applications. A large amount is used in electrical apparatus; smaller amounts are used in batteries, control instruments, fungicides, bacteriacides, dental preparations, pharmaceuticals and chemical laboratories.

Total reserves at most large active mercury mines are unknown because exploration is not carried beyond what the operator expects to mine in the next year. Broader world resource estimates have generally been tied to prices or economics. Those made by various experts have differed by several hundred percent but fall between 0.6 and 3.2 million tonnes of mercury in ore containing 0.1–0.2%. Thus, despite the uncertainty about the quantity, resources are clearly adequate for many decades at current rates of consumption. More than a third of the estimated resource is in the Almaden district in Spain, and a sizeable part lies in the USSR.

Bibliography

Bailey E H, Clark A L, Smith R M 1973 Mercury. In: Brobst D A, Pratt W P (eds.) 1973 *US Geological Survey Professional Paper 820*. US Government Printing Office, Washington, DC, pp. 401–14

Roskill Information Services 1978 *The Economics of Mercury*, 4th edn. Roskill Information Services, London

Smirnov B I, Kuznetsov V A, Fedorchuk V P 1975 *The Metallogeny of Mercury*. National Translations Center, Chicago, Illinois

US Bureau of Mines 1965 *Mercury Potential of the United States*, US Bureau of Mines Information Circular 8252. USBM, Washington, DC

US Bureau of Mines 1988 *Mineral Commodity Summaries 1988*. USBM, Washington, DC

Wambeke L van 1976 Le Mercure: Production, consommation et demande future de mercure dans le monde et dans la communauté européenne. *I Congreso Internacional del Mercurio, Barcelona*. Fabrica Nacional de Moneda y Timbre, Madrid, pp. 65–92

E. H. Bailey†

Metallogenic Provinces

The distribution of metals in the earth's crust is characterized by inhomogeneity. To understand this distribution, geologists have repeatedly tried to define certain crustal regions as uniquely suited to the concentration of one or more metals. These regions, metallogenic provinces, are thus areas of the crust where, during some portion of geological time, the conditions were appropriate for the concentration of metals to produce ore deposits (i.e., metal concentrations of economic significance). Such metallogenic provinces provide significant insight into the regional distribution, the genesis, and the associations of metals and metallic mineral deposits. Metallogenic maps, based on these provinces, are a common tool of exploration geologists seeking new mineral deposits.

1. Crustal Distribution of Metals in Space and Time

The metals of interest to the materials scientist range in crustal abundance from common to extremely rare. Table 1 lists the estimated abundance of 20 economically significant metals. Such estimates are, by their nature, subject to modification based on the assumptions made. The significant point is that the estimated concentration factor required to generate an economically viable metal deposit can vary from 3.4 to over 50 000 $g\,t^{-1}$, depending on the metal. Thus, to generate an economic mineral concentration, some geological process is required to upgrade average crustal rocks to the minimum cutoff grade.

Table 1
Crustal abundance, minimum grade and concentration factors of metals

Metal	Abundance in igneous rocks of upper continental crust ($g\,t^{-1}$)	Cutoff grade[a] ($g\,t^{-1}$)	Economic concentration factor
Al	78300	265000	3.4
Fe	35400	250000	7.1
Ti	4700	100000	21
Mn	690	70000	100
V	95	3000	32
Cr	70	68400	980
Zn	60	40000	670
Ni	44	3000	68
Cu	30	3000	100
Li	30	4600	150
Pb	15	40000	2700
Co	12	3000	250
U	3.5	300	86
Sn	3	2000	670
W	1.3	2000	1500
Mo	1	1000	1000
Ag	0.06	100	1700
Hg	0.03	2000	67000
Pt	0.005	3	600
Au	0.004	5	1250

[a] Minimum concentration required to generate an economically viable ore deposit based on a rock source (not brine) and with indicated metal being the sole source of revenue

These processes may be chemical, physical or both. As often as not, the processes are multistage, and may involve geological events widely spaced in time. For example, the typical aluminum deposit requires the occurrence of an effective igneous differentiation process to produce magmas of granitic to syenitic composition, followed by the appropriate combination of uplift, geochemical leaching and erosion to produce a bauxite occurrence of economic size and grade. The exploration geologist is therefore engaged first and foremost in an effort to determine the nature and sequence of geological events, to allow the prediction of where metal concentrations may be sought. This determination presumes a fairly complete understanding of the chemical and physical processes that generate ore deposits.

More commonly, the geologist relies on his technical capability to locate specific metal concentrations within a region that has been empirically determined to be an attractive exploration target. It is this empirical determination which leads to the concept of metallogenic province: "... strongly mineralized areas or regions containing ore deposits of a specific type, or groups of deposits that possess features suggesting a genetic relationship" (Turneaure 1955). Thus, a metallogenic province may be viewed as a geographically defined entity, wherein all the mineral deposits share what J E Spurr in the 1920s referred to as "a blood relationship." That such consanguinity is difficult if not impossible to prove in detail, has led to the common practice of delineating metallogenic provinces on the basis of deposits and prospects that contain the same metals or metals known to behave in a geochemically similar manner, and for which the best evidence suggests a common origin. Thus, the great Sn belt of the southwestern Pacific is characterized by Sn-bearing granites, and Sn placers derived from these granites. On the other hand, the Colorado Mineral Belt is a metallogenic province, even though the deposits found vary from Mo–W to Pb–Zn–Ag to Au. The issue is ultimately to establish their common origin.

There are a number of issues that need consideration in discussing metallogenic provinces. First, the definition of a metallogenic province requires only the recognition that similar mineralization is present over a significant area. Petraschek (1965) emphasized that at least one axis of such a province should be 1000 km long. While such a fixed dimension is undoubtedly too rigid, the 1000 km figure does indicate that to be a useful concept, a metallogenic province must be at least regional in scale. Turneaure argues on the other hand that a metallogenic province could be as small as a given mineral district. While valid philosophically, this is operationally unwieldy, and most modern students would favor the larger scale criteria.

Secondly, ore deposits occur as events within a space–time continuum. The metallogenic province concept addresses the spatial distribution of mineral occurrences only. Within a given province, there is the possibility of multiple metallogenic events producing similar or different deposits of differing geological age. Thus, the concept of metallogenic epoch has been developed. In parallel with the metallogenic province idea, a metallogenic epoch is an interval, "... of geological time favorable for the deposition of a definite mineral" (Bilibin 1968). Clearly, this definition requires expansion to include the deposition of more than a definite mineral, and may include any number of cogenetic mineral phases. The concept of the metallogenic epoch is easily applied within a given metallogenic province, and is useful in delineating the metallogenic history of a region. For example, in the metallogenic province that Turneaure called the Copper Province of Arizona and New Mexico, disseminated copper mineralization is present in Laramide age plutonic rocks (~65 million years old) over nearly a quarter of a million square kilometers. Thus, one would say that the Laramide orogeny represents a metallogenic epoch within the Copper Province of Arizona and New Mexico. Within the same province, one finds Precambrian mineralization (>600 million years) in the uplifted and exposed cores of many mountain ranges, and Eocene–Oligocene mineralization (30–40 million years) associated with the collapse of volcanic structures, especially in southwestern New Mexico. Defining metallogenic epochs as periods of similar mineralization of similar genesis (parallel to the definition of metallogenic provinces), the Copper Province of Arizona and New Mexico witnessed at least three metallogenic epochs, albeit the Laramide metallogenic epoch is the most economically significant.

To complicate matters, metallogenic epochs are often viewed as being periods in the earth's history when similar mineralization occurred on a continental or worldwide scale. Thus, metallogenic epochs need not be restricted to a given metallogenic province. The Laramide metallogenic epoch mentioned previously is known to have produced disseminated copper mineralization in the eastern Alps, from Yugoslavia into Turkey and Iran. The Precambrian shields of Canada, Brazil, Sierra Leone and Australia are all characterized by significant economic concentrations of banded iron formation, which clearly define a metallogenic epoch of a worldwide nature.

Finally, the fact that metallogenic provinces and epochs are real features of the otherwise irregular distribution of metals in the earth's crust clearly relates their metallogenesis to broader features of crustal evolution. Stanton (1972) has emphasized this by arguing that as the earth's crust evolves, the metal concentrations will also evolve. Thus, metallogenic provinces have been correlated in a direct manner with given tectonic provinces. Examples of such correlations are shown in Table 2, where three vastly different tectonic classifications have been correlated with types of ore deposits which in many parts of the world are grouped into metallogenic provinces. The table serves

Table 2
Metallogenic provinces associated with different tectonic provinces, based on different tectonic classification schemes

Source	Tectonic province	Typical metallic ore deposit association
Blondel (1936)	Shield areas	Archean Au, massive sulfices, U-veins, Co–Cu in mafic rocks, ultramafic Ni–Cu–Pt, chromite, banded iron formation
	Orogenic belts	Au-quartz veins, massive sulfides, W-deposits, disseminated Cu–Mo, epithermal Au–Ag
	Stable regions	Carbonate-hosted Pb–Zn–barite–fluorite, U–V red-bed deposits, Cu-shales
Bilibin (1968)	Ancient platforms	Pt, chromite, diamonds
	Initial geosynclinal stage	Cu–Ni, disseminated Cu, Pt
	Early geosynclinal stage	Epithermal Au–Ag, Hg, disseminated Cu–Mo
	Intermediate folded belt	Sn–W–Bi–Mo, pegmatites
	Late folded belt	Epithermal Au–Ag, disseminated Cu–Mo
	Shelf	Fluorite, barite
	Young platform	Hg
Guild (1972).	Divergent plate margins	Rift-related brines, Cyprus type Cu, podiform chromite
	Transforms	Boléo type Cu–Mn
	Convergent plate margins	Kuroko, magnetite–chalcopyrite skarn, disseminated Cu, W–Sn–Hg–Sb
	Within oceanic plates	Deep-sea nodules, evaporites
	Trailing continental margins	Black sands, shelf phosphorites
	Within continental margins	Au–U conglomerates, banded iron formation, evaporites, Cu-shales, red-bed deposits, kimberlites, carbonate-hosted Pb–Zn

to emphasize the direct correlation of metallic mineral deposits with tectonic provinces—regardless of which tectonic scheme one chooses.

Thus, metallogenic provinces and/or epochs are to be viewed in the context of the tectonic history of their host rocks. To a first approximation, one can assume that metals, being ultimately controlled by their geochemical and petrologic behavior, will be present in a given tectonic setting in a predictable and systematic manner. The metallogenic province is then a reflection of the fact that metals do behave in such a manner. This makes an Au-province, for example, as geologically significant as a granite province. Both reflect the history of the region under consideration.

The possibility that broader features of earth evolution than simple crustal tectonics may control metallogenesis is never easy to evaluate. The likelihood that mantle inhomogeneity, in terms of metal content, can effect the location of metallogenic provinces in space cannot be discounted. The time equivalent of mantle inhomogeneity (i.e., that mantle and crustal evolution would have irreversible effects on metallogenesis) has also been suggested. Sullivan (1948) suggests that the concentration of siderophile and chalcophile elements in the upper mantle through geological time would result in a change in composition in the source rocks for crustal evolution, and thus metallogenic epochs may reflect time inhomogeneity in crustal evolution. Finally, the influence of geologically long-lived crustal features, such as basement lineaments, is equally unknown. There is strong circumstantial evidence that certain areas of the earth's crust, such as the Colorado Mineral Belt, have been sites of metallogenesis through successive tectonic events, and thus imply some metallogenic link to deeper, more fundamental upper mantle or lower crustal features.

2. *Metallogenic Provinces of the Earth*

Generalized maps of the metallogenic provinces of North America, South America, Africa, Eurasia and the Western Pacific, and Australia are shown in Fig. 1. For reasons of simplicity, only provinces roughly fitting Petraschek's 1000 km restriction have been considered. All these provinces include at least one major historically producing deposit. This last criterion thus excludes metallogenic provinces that may in fact be important in the future, but which are based on speculative resource estimates or economically undeveloped reserves. In many instances, reference to tectonic maps (see *Plate Tectonic Settings for Metallic Ore Resources*) can permit significant correlations between tectonic environments and ore genesis.

The quality of these maps varies greatly from locale to locale. Clearly, areas such as North America, the Andes, Western Europe and Australia have been most actively explored for mineral resources. Thus, the maps of these areas are detailed. Regions of the world with adverse physical conditions for exploration, such as equatorial South America and Africa or Antarctica, areas with histories of significant political unrest and/or poorly developed technological infrastructures during the past several decades, and areas about which significant mineral resource data are not readily available in western literature have been mapped in significantly lesser detail.

Furthermore, the 1000 km size restriction eliminates many famous districts and ore deposits, because they happen to be singular occurrences or, at least, not presently known to represent a major crustal anomaly. It is for this reason that no Hg deposits are shown, as more than 75% of world production has come from six mines, one in Spain, two in Italy, one in Peru and two in California. None of these districts is large enough to constitute a province in the true sense of the word. Furthermore, for minor metals, such as Li or Be for example, and for such alkaline earths as As, Sb or Bi, there is insufficient industrial demand for there to be significant primary extractive industry built about these commodities. Many, thus, are produced from a few, small mines and districts or are by-products of other metallic mines. This is not to imply that there may not be large crustal anomalies which fulfill the definition of a metallogenic province, it simply means that in practice the data are not available and the impetus for deriving such models is not present.

Finally, the application of the criterion that the deposits should be of the same metal and of common origin masks the fundamental observation that there are tectonically defined regions of the earth's crust which, at some period in earth history, have been remarkably metal-rich. This is true in spite of the fact that the deposits may vary widely in their metal content and origin. Thus, for example, the South African shield is immensely rich in Au- and U-bearing conglomerates, in Cr- and Pt-bearing ultramafites, and in Mn, Ni and Co, even though the mechanisms whereby these deposits originated are extremely different. An additional example is the American southwest, with its sedimentary-hosted U–V deposits, plutonic-generated Cu–Mo, and volcanically derived Pb–Ag–Au. Such regions lead to the concept of the metallotect, a set of geological processes which broadly define an area of the crust favorable to ore deposition. Typical metallotects may be sedimentary basins on continental cratons, island arc systems, Archean shields or continental rift systems. Thus, a metallotect becomes the bridge between the empirically derived metallogenic province, and the theoretically defined correlations of ore genesis and tectonism.

3. *Metallogenic Maps*

One by-product of the study of the global distribution of ore deposits is the metallogenic map. Such maps attempt to show the size, form, metallic content, location and origin of mineral deposits. National, regional, and continental maps of varying scales and

Figure 1
Metallogenic provinces of major continental areas of the world. Provinces are outlined to include major districts which meet criteria discussed in the text. Principal economic metals are given for each province using chemical symbols except for the Colorado Mineral Belt, an area of multielemental concentrations of extreme complexity, and for regions of pegmatites, host rocks that often produce a host of nonmetals and metals, as well as gems. The reader is referred to Bilibin (1968) and Turneaure (1955) for more complete presentations on many of the specific provinces.

employing variously ambitious legends have been produced. At present, the Metallogenic Subcommission of the Commission for the Geological Map of the World, which is a UNESCO-sponsored organization, is attempting to produce a series of continental maps which may eventually be combined into a world map. These maps do not expressly define metallogenic provinces, but they represent the most readily available data source for the delineation of such areas. The only current map with a global emphasis is that produced by the Ministerstvo Geologii SSSR (1972).

See also: Geochemical Distribution of the Elements; Igneous Ore Deposits: Formation Processes; Metamorphic Ore Deposits: Formation Processes; Sedimentary Ore Deposits: Formation Processes

Bibliography

Bilibin Y A 1968 Metallogenic provinces and metallogenic epochs. In: Walker W (ed.) 1968 *Metallogeny and Global Tectonics*, Benchmark Papers in Geology, Vol. 29. Dowden, Hutchinson and Ross, New York, pp. 13–50

Blondel F 1936 *La Géologie et les Mines de Vieilles Platformes.* Société d'Editions Géographiques, Maritimes et Coloniales, Paris

Derry D R 1980 *World Atlas of Geology and Mineral Deposits.* Wiley, New York

Guilbert J M, Park C F Jr 1986 *The Geology of Ore Deposits.* Freeman, New York
Guild P W 1972 Distribution of metallogenic provinces in relation to major earth features. In: Petraschek W E (ed.) 1972 *Metallogenetische und Geochemische Provinzen.* Springer, Vienna, pp. 10–24
Ministerstvo Geologii SSSR 1972 *Karta Poleznykh Iskopaemykh Kontinentov.* Mira, Moscow
Noble J A 1970 Metal provinces of the western United States. *Geol. Soc. Am. Bull.* 81: 1607–24
Nockolds S R 1954 Average chemical composition of some igneous rocks. *Geol. Soc. Am. Bull.* 65: 1007–32
Peters W C 1978 *Exploration and Mining Geology.* Wiley, New York
Petraschek W E 1965 Typical features of metallogenic provinces. *Econ. Geol. Bull. Soc. Econ. Geol.* 60: 1620–34
Stanton R L 1972 *Ore Petrology.* McGraw-Hill, New York
Sullivan C J 1948 Ore and granitization. *Econ. Geol. Bull. Soc. Econ. Geol.* 43: 471–98; 44: 336–45
Turneaure F S 1955 Metallogenetic provinces and epochs. In: Bateman A M (ed.) *Economic Geology*, Fiftieth Anniversary Volume, pp. 38–98
United States Bureau of Mines 1970 *Mineral Facts and Problems*, US Bureau of Mines Bulletin 650. USBM, Washington, DC
Wedepohl K H 1971 *Geochemistry.* Holt, Rinehart and Winston, New York

P. G. Feiss
[University of North Carolina, Chapel Hill, North Carolina, USA]

Metamorphic Ore Deposits: Formation Processes

Metamorphic rocks are the least abundant of the three different types of rocks which also include igneous and sedimentary rocks. Mineral deposits are associated with each type; many nonmetallic mineral deposits, especially, are associated with metamorphic rocks. Metamorphic minerals generally form under different *P–T* (pressure–temperature) conditions to those under which the original rock formed. If the chemical composition is also changed, the term metasomatism is used. Metamorphism (changed form) and metasomatism (changed composition) processes profoundly alter preexisting rocks and minerals and form new ones. Some preexisting mineral deposits have been metamorphosed, but the majority of mineral deposits formed by metamorphism are nonmetallic (e.g., asbestos, talc, graphite, garnet and corundum).

Metamorphism varies in its *P–T* conditions. Specific minerals form at low, intermediate or high *P–T* environments, which are referred to as ranks or grades of metamorphism. The formation of certain index minerals in an assemblage of rocks and minerals demonstrates that chemical equilibrium has been reached during metamorphism within any *P–T* range. In an area of varying ranks of metamorphism, each zone is referred to as a specific metamorphic facies; metamorphic facies grade into each other.

1. Pressure and Temperature

Metamorphic processes occur in adjustment to the chemical potential of any system of minerals and rocks to *P–T* changes and over geological periods of time. An increase in pressure will cause a reaction to move in a direction in which the total volume of the system decreases. For example, increasing pressure results in a reduction in the total molar volume.

olivine + anorthite → garnet

nephelene + albite → glaucophane or jadeite

andalusite → sillimanite or kyanite

An increase in temperature may result in endothermic reactions unless the system is not in equilibrium conditions. An example is the conversion of pyroxene to hornblende during the metamorphism of diabase to amphibolite. Metamorphic reactions contrast with low-temperature surficial weathering processes of oxidation and hydration that also convert minerals into others. As an example, the chemical weathering of orthoclase to form clay, potassium carbonate and silica occurs at the surface of the earth; the reverse reaction occurs during metamorphism.

2. Deposits and Minerals Formed by Metamorphism

Several kinds of mineral deposits are formed as a result of regional metamorphism. The source materials are rock constituents that have undergone recrystallization or recombination or both with the major chemical change in water or carbon dioxide content.

2.1 Sillimanite Group

The four minerals, kyanite, andalusite, sillimanite and dumortierite withstand high temperature and convert to mullite and quartz. The first three have identical compositions ($Al_2O_3 . SiO_2$) but differ in crystal form. Andalusite and sillimanite form in the orthorhombic system and kyanite is triclinic. Dumortierite, much rarer than the other three minerals, is an orthorhombic basic aluminium borosilicate.

Mullite and vitreous silica are produced commercially from the four minerals at temperatures of 1100–1650 °C. The former material, rare in nature, remains stable up to 1810 °C. It is heat and shock resistant and is a good thermal insulator.

2.2 Asbestos

This is a fibrous or needle-like variety of silicate minerals that can be woven to form heat-resistant and low thermal conductivity gloves, clothes, mats, vehicular brake band linings, insulation and many other similar materials. It can be mechanically separated into fine filaments of considerable tensile strength and flexibility, and can withstand temperatures up to 2570 °C.

The two main varieties of asbestos are serpentine and amphibole. Serpentine asbestos is a hydrous magnesium silicate and includes the minerals chrysotile and picrolite; fine, silky chrysotile is the most valuable as it provides the spinning fiber. The asbestos amphiboles are hydrous silicates of calcium, magnesium, iron, sodium and aluminium, and include the minerals amosite, crocidolite, tremolite, actinolite and anthophyllite (Jensen and Bateman 1980).

Chrysotile asbestos occurs in serpentinite. The rock is derived from the metamorphic alteration of ultrabasic igneous rocks such as dunite, or magnesium limestones, or dolomite. The serpentization of dunite is of prime importance as it provides 93% of the world's asbestos supply.

Dunite is a dense greenish to gray–green rock, with a specific gravity of 3.3, and is composed almost entirely of olivine. It crystallizes early from a magma and may be partially converted at shallow depths to serpentinite. Some is emplaced at continental margins, derived from slabs of the upper mantle, dragged or rafted along by a moving oceanic plate, scraped off at the continental margin and structurally emplaced as the remainder of the plate is subducted.

Asbestos fiber occurs in lens-like veinlets with three modes of occurrence in serpentinized ultrabasic rocks. They are: (a) crossfiber, i.e., fibers normal to the walls; (b) slip fiber, parallel or oblique to the walls; and (c) mass fiber, variably oriented generally short fibers. All may occur in the same deposit.

Cross-fiber chrysotile is the most desired asbestos as it occurs in veins up to 10–12 cm in width but generally much less. It usually makes up only a few percent of the rock, rarely in excess of 10–20%.

The origin and emplacement of asbestos are unresolved problems. One idea is that hot water moved along fractures or minute openings in serpentine and formed veinlets of asbestos. In some cases, asbestos apparently replaced serpentine.

2.3 Graphite

This is a form of carbon that occurs in two varieties: crystalline, consisting of thin, nearly pure, black flakes; and amorphous, a noncrystalline impure variety. It is soft and black, has a greasy feel, and marks paper, hence the term graphite (from the Greek "write"). The mineral is soft because carbon layers of the structure are held together by weak Van der Waals bonds.

Grapite forms in metamorphic rocks during either regional or contact metamorphism. It occurs in marble, gneiss, schist, quartzite and, in Rhode Island, in metamorphosed coal beds.

Graphite originates by (a) regional metamorphism, (b) original crystallization, (c) contact metamorphism, and (d) introduction of hydrothermal solutions. The first process is the most common whereby hydrocarbons or organic matter are reduced to graphite, or carbonate rocks are broken down by metamorphism.

2.4 Talc, Soapstone and Pyrophyllite

Talc is a product of metamorphism; it is a hydrous magnesium silicate, $H_2Mg_3(SiO_3)_4$, and is the source of talcum powder and other cosmetic products. Soapstone is a soft rock comprised of fine talc but contains other mineral impurities including chlorite, sericite, magnesite and others. Pyrophyllite, sometimes mistaken for soapstone, is a hydrous aluminum silicate. These minerals are formed only as the result of metamorphism of other rocks or minerals. A typical reaction that forms talc from pyroxene (enstatite) is:

$$4MgSiO_3 + CO_2 + H_2O \rightarrow H_2Mg_3Si_4O_{12} + MgCO_3$$

Commercial talc and soapstone deposits occur in dolomitic (Ca, Mg)CO_3 limestones or in altered ultramafic intrusives. The deposits are lens shaped and reach widths of 40 m. Their shape suggests simple dynamic metamorphism in the presence of water and carbon dioxide.

3. Other Metamorphic Minerals and Deposits

Garnet forms as the result of metamorphism but is also common in the higher temperature zone of contact metasomatic deposits. Its prime use is as a grinding agent. Emery is a mixture of corundum and magnetite with hematite or spinel that forms by contact metamorphism. It is also a prime grinding agent.

See also: Igneous Ore Deposits: Formation Processes; Sedimentary Ore Deposits: Formation Processes

Bibliography

Guilbert J M, Park C F Jr 1986 *The Geology of Ore Deposits.* Freeman, New York.

Jensen M L, Bateman A M 1980 *Economic Mineral Deposits,* 3rd edn. Wiley, New York

Probst D A, Pratt W D 1973 *United States Mineral Resources,* US Geological Survey Professional Paper 820. US Government Printing Office, Washington, DC

M. L. Jensen
[University of Utah, Salt Lake City, Utah, USA]

Mineral Deposits: Examination and Evaluation

Mineral deposits must be examined in the course of exploration, so that provisional geological guidelines can be established for prospecting at the same site, or in the same region. The evaluation of a mineral deposit is more conclusive, and it involves engineering and economic considerations as well as geology; a prospect or a mine must be examined and a potential value placed on it.

Prospects—deposits that have not yet been mined—and mines that have had some degree of production are evaluated for various purposes. Potentially significant ore mineralization encountered during an exploration program constitutes a new prospect that needs to be evaluated as a justification for more thorough investigation. A prospect or mine in the ownership of an individual, an organization or a government constitutes a mineral property to be evaluated as a potential source of revenue and for potential acquisition, investment or joint venture. Long-term purchasers of minerals for use in manufacturing may require an evaluation of the source or sources of the particular minerals.

Mineral-deposit evaluation is commonly made in stages, with a brief preliminary evaluation serving as the justification and guide for follow-up evaluations, and for an eventual comprehensive evaluation and feasibility study in which the decision is made to develop a mine. Where a mineral property cannot be seen to warrant further investigation, the search for an economic ore deposit is directed elsewhere. While most mineral property evaluations are not carried beyond the preliminary stage, properties are commonly examined and evaluated time after time by different organizations under the influence of changing mineral prices, technological innovations and new economic conditions. Geological theories of ore genesis and the capabilities of the tools of exploration will often provide a basis for reevaluating prospects that were rejected at an earlier time as uneconomic.

Table 1 is a listing of principal considerations in the evaluation of mineral deposit, whether it be a raw prospect, a mine that has ceased production or a mine that offers a potential for increased production.

Geographical setting is a first consideration. In a regional sense, an iron ore deposit in a remote inland part of Alaska would not be subject to the same economic determinants as an iron ore deposit in an industrialized area. Both orebodies might be attractive targets for development, but the Alaskan deposit would need to be larger, much higher in grade and much better suited to low-cost mining. In a consideration of local geography, site conditions might permit or rule out otherwise appropriate methods of development, mining and processing.

The history of a mine provides important information for assessing the potential value of an expanded or extended operation. Zonal characteristics in mineralogy, the pattern of orebodies and variations in geotechnical conditions are reflected in the statistics of ore production and profitability. Ownership of the mineral body and its possible extensions, and ownership of the necessary surface and water rights can be an overriding factor in the evaluation of a prospect or mine for acquisition.

The geology of the orebody and its immediate environment is basic; regional geology provides the context. In all mineral deposit evaluations, consideration is given to the next-indicated steps in an investigation. Further geological work may be in order, and the type of work may be clearly indicated, or geological work may be too costly or inconclusive to justify. Geophysical and geochemical work may be appropriate and there may be specific sites to test; on the other hand, the contrast in character between ore and rock may not be sufficient, and there may be conditions that could cause difficulty with interpretation.

Table 1
Principal factors affecting mineral deposit evaluation

Geography	location, access, topography, climate, environmental character
History	production and profitability, at the deposit and in the district
Ownership	mineral, water and surface rights, land-use restrictions, and owner's terms for agreement
Geology	regional and local rock units, structural and metallogenic setting; type of mineralization, ore controls, geometric character; response to geological, geochemical, geophysical investigations
Sampling	procedures, patterns, analytical methods
Reserves	tonnage, grade, cutoff grade, and ore mineralogy
Geotechnics	geomechanics, hydrology
Mining and processing	past and proposed methods
Marketing	proximity, specifications, prices
Capital requirement	plant, preproduction work, infrastructure
Production cost	total cost (labor, operating, ownership)
Profitability	potential return on investment, present value

1. Sampling of Mineral Deposits

Samples from an orebody should represent the orebody, yet the amount of material taken in sampling is an infinitesimally small part of the total tonnage. In most mineral deposit evaluations, samples cannot be directly related to an entire orebody because they represent only the accessible part of the deposit. Geological projection and reasoning must be called upon. Where sampling is done at the surface, the effects of weathering (leaching) and of residual accumulation or secondary enrichment mask the true nature of the underlying orebody. Where samples are taken from existing underground mineworkings, they represent only the low-grade ore, thin ore and wallrock left behind, and the sample sites are generally contaminated by dust and efflorescences of metal salts.

Orebodies are most effectively sampled in new mine workings and in drill holes that are designed to penetrate all parts of the orebody at close intervals. The cost of the new mine workings and of the drilling program will have been justified by preliminary evaluations and preliminary sampling.

Drill-hole samples are taken by diamond coring and by rotary or percussion drilling from locations at the surface and in underground workings. Diamond-core drilling provides the best geological data and generally yields a near-complete sample for laboratory analysis. In solid ore and rock, full core recovery is expected; in fractured or soft material, some of the sample is commonly lost. Rotary and percussion drilling is done at a lower cost than diamond drilling; the geological data obtained from finely broken chips (cuttings) are less satisfactory, but a sample is obtained for laboratory analysis. In all types of drilling, the loss of even a small amount of material may bias the results. If the missing material is recognized to constitute soft or friable ore minerals, an appropriate factor is taken into account.

The basic calculation is dealing with sampled slots (channels) in mine workings and with material from drill holes is to find a composite value or weighted average for the series of samples taken within designated widths of ore and waste rock. Analyses (assays) from samples of various widths are averaged by the following formula:

$$\text{average assay} = \frac{\Sigma \text{ width} \times \text{assay}}{\Sigma \text{ width}}$$

The average grade of three samples taken from a length of drill hole intersecting a vein of lead–silver ore would be calculated as below with the data from Table 2:

$$11.6/1.8 = 6.4\% \text{ Pb}$$

$$53.1/1.8 = 29.5 \text{ g t}^{-1} \text{ Ag}$$

The pattern of sampling is a function of the objective of the evaluation, the sampling method, the accessibility of sampling sites and the geological situation. Where possible, a uniform grid of sample locations is attempted. Sample locations may be spaced 200 m or more apart in a large and relatively uniform deposit, or they may be spaced as closely as 2–3 m in a small and irregular deposit. In the latter situation, "scout" samples are often taken first at greater intervals, with the more detailed sampling done in critical zones.

Table 2
Data for average assay of mineral deposit samples

Width of deposit sampled (m)	Assay (% Pb)	$W \times A$	Assay (Ag, g t^{-1})	$W \times A$
0.3	3.0	0.9	27	8.1
1.0	9.7	9.7	35	35.0
0.5	2.0	1.0	20	10.0
1.8		11.6		53.1

2. Ore Reserves

A block of ore is described by its tonnage and average grade. Other descriptive characteristics are the range in grade and the degree of certainty that the block actually exists. Information comprises assay results from sampling, dimensional measurements in geological plans and sections, and density measurements. In addition, there must be a provision for establishing a cutoff grade (the lowest grade of ore that can be mined economically), a provision for estimating the potential recovery of ore in mining as a percentage of the total ore block, and a provision for estimating the potential dilution in mining as a percentage increase in tonnage with little or no increase in the total content of ore minerals.

The part of an orebody so thoroughly sampled that its outline, tonnage and grade are estimated with a high degree of certainty is designated as proved ore (Fig. 1). Another part of the orebody in which sampling may not have been so thorough but in which there is enough geological information for a reasonably secure estimate of the tonnage and grade is designated as probable ore. The fringes of the orebody, where the tonnages and grade are estimated on the

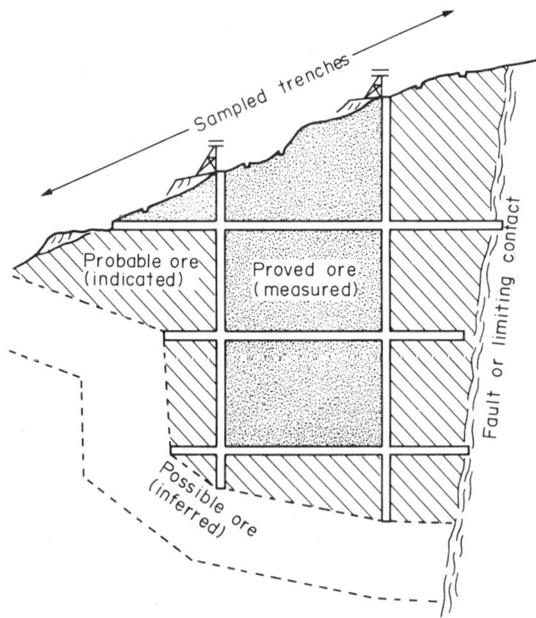

Figure 1
Ore reserve designations

basis of a few samples, but where there is supporting evidence of ore mineralization, is designated as having possible ore.

The precise meanings of ore reserve designations vary from one type of deposit and one organization to another, but in the sense of classical vein deposits, proved ore has been blocked out (sampled) on four sides, probable ore has been blocked out on two or three sides, and possible ore has either been exposed on one side or projected from better-known blocks. Mining companies usually refer to proved and probable ore as their actual reserves. In a preliminary evaluation of a mineral deposit, there may only be possible ore, with probable and proved ore to be designated in later evaluations.

The terms measured, indicated and inferred, used by several government organizations, are similar to the designations of proved, probable and possible, but they may refer to identified resources of paramarginal and submarginal nature as well as to ore reserves.

3. Technical Considerations

The geotechnical characteristics of a mineral deposit, and the attendant suitability of the deposit for mining and processing by various methods, must be estimated during an evaluation. In a preliminary evaluation, order-of-magnitude estimates are made on the basis of analogous conditions in better-known mines and deposits. Follow-up evaluations are accompanied by more detailed study, by preliminary estimates, and finally by definitive estimates based on a thorough study of the accumulated data. At each stage in mineral property evaluation, estimates of technical parameters are of critical importance in dealing with the economic considerations in marketing, capital requirements, production cost and profitability.

4. Economic Considerations: Expressions of Value

An ultimate statement in prospect evaluation is to place an expression of profitability on the deposit as a potential mining venture. Working estimates are prepared for the geological, economic and technological characteristics of the deposit. Market prices for the mineral commodities, an optimum level of production, and a yearly revenue from sale of the product are estimated. The information is used in a financial analysis to determine an expected annual cash flow.

With an estimate of cash flow and an estimate of the capital investment, the indexed expressions of payback period, discounted cash flow return on investment (DCFROI), and net present value are obtained. The expressions are quantitative, but they are derived from a series of estimated sizes, shapes, values, rates and proportions that cannot be closely measured. Thus, there is a large element of risk to be taken into account in expressing the potential value of a mineral deposit.

All mining investments must be made with risk as a consideration. There are geological risks that the expected orebody will be smaller, lower in grade or less uniform than anticipated. There are technological risks that the ore will be more difficult to develop, mine and process than indicated from evaluations. Economic risks, that mineral markets, taxation and land tenure will take an unfavorable turn, are to be taken into account. In response to the risk element in mining, the minimum acceptable rate of return on a potential investment must be stated at a somewhat higher level than that used for safer investments. For each dozen or so prospects that warrant high levels of expenditure in evaluation, one or two will actually become a mine, and each mine that enters production still has a risk element. The only measurements and analyses that can be made without some degree of risk are those made on a completed mining operation with a mined-out orebody.

See also: Geological Exploration for Mineral Deposits; Resource Appraisal

Bibliography

Baxter C H, Parks R D 1957 *Examination and Valuation of Mineral Property*, 4th edn. Addison–Wesley, Reading, Massachusetts

Forrester J D 1973 Examinations, valuation and reports. In: Given I A (ed.) 1973 *SME Mining Engineering Handbook*, Vol. 2. Society of Mining Engineers, American Institute of Mining, Metallurgical and Petroleum Engineers, New York, pp. 2–56

Peters W C 1987 *Exploration and Mining Geology*, 2nd edn Wiley, New York

Raymond L C 1976 Valuation of a mineral property. In: Vogely W A (ed.) 1976 *Economics of the Mineral Industries*, 3rd edn. American Institute of Mining, Metallurgical and Petroleum Engineers, New York, pp. 433–60

Reedman J H 1979 *Techniques in Mineral Exploration*. Applied Science, London

Van Landingham S L 1983 *Economic Evaluation of Mineral Property*. Hutchinson Ross, Stroudsburg, Pennsylvania

W. C. Peters
[Tucson, Arizona, USA]

Mineral Resources: Definitions, Uses, Classification and Future Availability

Minerals are directly or indirectly the source of most of the materials used by industrial society. Their increasing use has made possible the rise in standard of living in both developed and developing countries. With increased use, concern and controversy have developed about the adequacy of resources to meet future needs. Much of both the concern and controversy arises from lack of understanding of what mineral resources are and of how supplies are developed.

1. Definitions

The term mineral has several connotations. Used by the mineralogist, it may refer to a naturally occurring solid inorganic substance with distinctive physical properties and a composition that can be described by a chemical formula. Used by the mineral economist, it may refer to a commercially traded product derived from a naturally occurring, nonliving, organic or inorganic, solid, liquid or gaseous substance. Used to describe resources, it refers to naturally occurring, nonliving, organic or inorganic, solid, liquid or gaseous substances that have a known or potential use. Reference to the mineral resources of a country or region, for example, would be likely to be all-encompassing and include its water resources, its energy resources, such as coal, gas, petroleum, uranium and geothermal energy, and its potentially usable deposits of metallic and nonmetallic minerals, such as deposits of ferrous, base and precious metals, phosphate rock, saline minerals, sand, gravel and building stone.

In reporting specific aspects about mineral resources, the scope of the subject is generally limited to substances that are commercially traded. Water resources are thus treated separately, except for brines containing commercially recoverable dissolved salts. Air is not commonly thought of as a mineral resource, but gases recovered commercially from the atmosphere—argon, nitrogen and oxygen, for example—are counted as minerals in production statistics.

The term ore refers to solid substances currently recoverable at a profit. Generally it applies to explored and developed deposits of metallic minerals but is sometimes used to describe producible deposits of a few nonmetallic minerals such as phosphates or potash.

An important part of the concept of mineral resources is that they have a known or potential use. They thus do not represent permanently fixed quantities but tend to be extended over time as new uses are found for naturally occurring substances and as technologies are developed to recover valuable products from deposits once considered too lean, refractory or inaccessible to have any potential use. Mineral resources to prehistoric man would have thus been limited to flint, salt, water and rocks that could be used for tools, weapons or crude construction. By the end of the nineteenth century about 30 chemical elements were in commercial use; aluminum was known only in the laboratory and deposits rich in that element would not have been considered a resource. Today about 80 of the elements and 135 mineral commodities are in commercial use in the USA and mineral deposits, such as taconite (a hard, highly siliceous iron-bearing rock), that were once considered worthless are now in commercial production.

Mineral resources (excluding water) have been described as nonrenewable resources, in contrast to renewable resources such as surface water and timber, which are or can be replenished naturally or artificially. The geological processes by which most mineral deposits are formed take so long for completion that they cannot be thought to replenish deposits extracted from the ground and dispersed by use. However, it is important to recognize that mineral resources are extendable as the result of advancing technology that develops uses for substances not usable before, permits concealed deposits to be discovered, improves recovery efficiency or makes it possible to mine and process deposits once considered worthless. The life of mineral resources can also be extended by increased efficiency of use. For example, the card-size calculator weighing 35 g and costing about US$5 in 1984 will perform more functions than the calculating machine of 1950 weighing 15 kg and costing US$500 or more.

2. Uses

Some mineral products are used in the form in which they are mined, with only shaping, crushing, cleaning or other treatment that does not change their composition. Examples include sand, gravel, crushed stone, metal, building stones, coal in most of its uses, natural gas, gemstones and some abrasives. Most minerals are processed, however, to yield an artificial product—a metal, chemical or other material from which most of the things used by industrial society are made. Minerals provide the materials from which labor-saving machines, lighting, heating and cooling systems and transportation and communication networks are built and powered. Processed and utilized with the aid of inanimate energy, minerals are the source of most chemicals, drugs, paints, ceramics, synthetic textiles, plastics and metal products. They are the raw materials from which most homes, buildings, dams and other engineering works are constructed. Also, converted to fertilizers, insecticides, farm machinery and inanimate energy, they are key elements in the high productivity of modern agriculture.

It is the extensive use of mineral resources, in fact, that forms the physical basis of the high standard of living enjoyed by industrial societies. Strong positive correlation is found, for example, between per capita national income and per capita consumption of energy and key mineral commodities in a given country over time and among the countries of the world today. Key though they are to affluence, their value as raw materials is relatively small in the gross national product of industrialized countries—for example, only ~ 6.3% of it in the USA in 1987.

Along with the affluence developing from the high use of mineral resources have come two other phenomena. One is growth in world population, made possible mainly by increased food production, improved sanitation and control of infectious diseases. The other is environmental strain related mainly to air, water and land pollution arising from industrial society's high use of chemicals and fuels and, to a lesser extent, to mining. If unmitigated, these phenomena could cancel the social gains from the advance of mineral technology. Two of the most important challenges of the future are to achieve population stability and to avoid environmental pollution that would reduce the carrying capacity of the land or affect the long-term stability of the global ecological system (Brooks and Andrews 1974).

3. Classification

For many commercial and governmental purposes it is desirable to know the magnitude of specific mineral resources—at a specific mine site, in a region, in a country and in the world at large. Over the years, diverse terms and classifications have been introduced in recognition that estimates of mineral resources must be labelled to indicate the degree of certainty involved. None of these terms or classifications has achieved universality in use, but the terms *proved*, *probable* and *possible*, and *measured*, *indicated* and *inferred* are illustrative of attempts to indicate degrees of certainty in resource estimates. For many years, such estimates were made only of known deposits believed to be recoverable at a profit, but, as advancing technology has lowered cutoff grades and shown the power of exploration in discovering new deposits, and as mineral shortages have developed or threatened to do so, interest has grown in the quantitative assessment of deposits that are as yet undiscovered or that are not profitably producible now but that may come within economic reach in the future.

Figure 1 shows a classification based on the degree of certainty of the estimate (increasing from right to left) and the economic feasibility of recovery (increasing from bottom to top). A similar system was adopted by the US Bureau of Mines and the US Geological Survey in 1976 and modified somewhat in 1980 (see also Wood et al. 1983, McKelvey and Masters 1984). A group of experts convened by the Secretary General of the United Nations in 1979 recommended an almost identical system, using letters and numbers for the various categories instead of names that may not be translatable into certain other languages. Although there is not yet universal agreement on a classification system for mineral resources, publication of diagrams similar to Fig. 1, drawn from earlier versions of it (e.g., McKelvey 1972, Schanz 1975) in many other countries indicates that its principles are widely accepted, even though the system has some shortcomings (Zwartendyk 1981, Harris and Skinner 1982). The following discussion relates mainly to Fig. 1, but it is the principles of classification that are being emphasized, rather than the terminology or the exact definition of the boundaries of individual categories.

The most important feature of the classification is to distinguish between identified deposits that are recoverable with existing technology and at current prices from those that are as yet undiscovered or, if known, are not yet within economic reach. In this

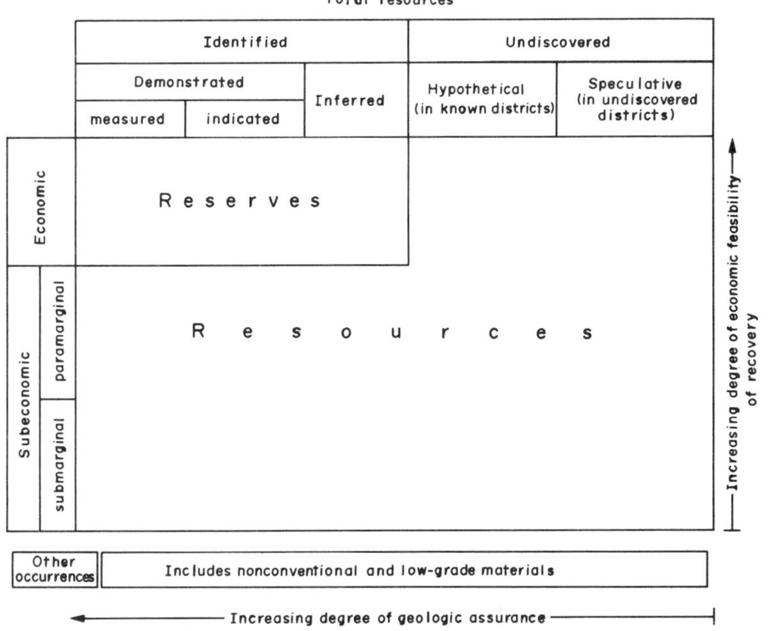

Figure 1
Classification of mineral resources

classification the term reserves is used only for identified deposits that could be produced commercially at the time the estimate is made; the term resources is used for subeconomic and undiscovered categories and for the totality of all categories for either one or all minerals. Although resources may include undiscovered deposits that, even if found, are far from being economically recoverable, they include only those deposits conceived to be recoverable at some time in the future (in the United Nations recommendations the future is defined as within the next two or three decades). Estimates of deposits not conceived to have such a potential would be classed as *other occurrences.*

Another important distinction for some minerals, not self-evident in Fig. 1 but stressed in the United Nations report and to a lesser extent in the 1980 US Bureau of Mines–Geological Survey system, is between recoverable and in situ quantities. In Fig. 1, reserves would include only currently recoverable amounts, but some of the remainder might be shown in one or more of the subeconomic categories. Until recently, for example, crude oil recovery averaged only ~30% of that originally in place; with higher prices and advancing technology it is reasonable to expect that the percentage would increase, perhaps to 60% or so, and that part of the oil originally in place could be classed as a paramarginal and part as a subeconomic resource. In the system recommended by the United Nations expert group, estimates of in situ versus recoverable would be reported separately, as they would to some extent in the 1980 US Geological Survey scheme. However the distinction is made, it is an important one for those minerals for which current extraction practice leaves a substantial portion in the ground. At least some of the remainder is a resource, and an estimate of its amount establishes a target for advance in extractive technology or an indication of amounts that might become available with increased prices.

A good indication of the uncertainty that exists in estimates, even where discoveries have been made and some exploratory drilling completed, is seen in the report of the National Research Council (NRC) (1976) on gas reserve estimation of offshore producible shut-in leases in the Gulf of Mexico; that is, leases on which a commercial discovery of oil or gas has been made but that are not yet in production. Because of concern that outer continental shelf Federal leases were deliberately shut in to raise gas prices, in 1974 the Federal Power Commission (FPC) estimated the proved and probable reserves of the 168 leases then classified as producible shut-in. The results, showing proved reserves of 130×10^9 m^3 and probable reserves of 93×10^9 m^3 of gas, were challenged as being much too conservative.

Accordingly, the NRC was asked to make an independent estimate, using the same data from US Geological Survey files used by the FPC. The NRC selected 33 as a representative sample of the 168 leases and commissioned two consulting firms to make independent estimates of the gas reserves. Their results indicate that rather than being too low, the FPC estimates may have been much too high. In the sample, the FPC's estimate of both proved and probable reserves was 42×10^9 m^3, compared with "best" estimates of the two consultants of 18×10^9 and 13×10^9 m^3 (the consultants also made high and low estimates which were 18–28% lower and 32–45% higher than their best estimates). Estimates of reserves in some individual reservoirs differed by more than an order of magnitude and, with few exceptions, the FPC was generally the highest, consultant A generally the lowest and consultant B generally in between.

Because many of the reservoirs are structural traps associated with salt domes, differences in interpretation of the extent of reservoirs at this early stage of exploration are understandable, but that such differences were not random indicates that differences in philosophy—possibly born from experience or the lack of it—played a significant part. In any case, the wide range in estimates of reserves in discovered and partly explored fields in the Gulf of Mexico made by experts using the same data is indicative of the great uncertainty that necessarily accompanies estimates of reserves, to say nothing of undiscovered resources. They are fittingly described as estimates, with the lack of precision implied by that term.

In the light of such uncertainty, it may be asked if such estimates serve any useful purpose. Obviously they must be used with the understanding that they are only estimates, subject to a wide margin of error, and that they need to be continually updated as new information is acquired. But taken with that understanding, they show with respect to reserves that there are, or are not, known deposits that can be mined with existing technology at current prices; they form the basis for commercial sale and purchase of mineral properties, guide plans for mineral development and production and provide governments with information bearing on the design of near-term supply policies. Estimates of identified subeconomic resources show whether there are known deposits large enough to justify technological research and development—valuable information for both industry and government. Estimates of undiscovered resources show whether there are targets for additional exploration—information as essential to government in formulating policy as it is to industry in laying plans to develop future supply.

Estimates in the framework of the classification illustrated in Fig. 1 help to emphasize that the development of mineral resources is a dynamic process (Schanz 1977) and that the magnitude of usable resources, although limited by a finite earth (a point that some would not concede, now that space exploration has begun), is not fixed but increases with advancing technology and exploration and, of course, decreases with increasing production. The mineral deposits being produced today were once in the category of undiscovered subeconomic resources. For most minerals, the measured reserves of tomorrow will come from the now subeconomic or undiscovered categories; for some they may come from today's "other occurrence" category and for still others they may come from unknown types of deposits that would not even be listed in 1984 in the "other occurrence" category (DeYoung 1978).

4. *Future Availability*

Concern about the adequacy of mineral resources to meet future national and world needs has mounted from time to time, receiving initial prominence in the USA with the Governors' conference on conservation convened by President Theodore Roosevelt in 1908. In his opening address to the conference (see Nolan 1958) he said:

> I have asked you to come together now because the enormous consumption of these resources, and the threat of imminent exhaustion of some of them, due to reckless and wasteful use, once more calls for common effort, common action. We want to take action that will prevent the advent of a woodless age and defer as long as possible the advent of an ironless age.

Many others since then have expressed similar concerns. In the widely publicized study sponsored by the Club of Rome (Meadows et al. 1972) it was concluded that world resources would be exhausted in less than 100 years if growth in consumption were to continue at current rates. Skinner (1976), foreseeing the dawn of a second iron age to be "much closer than most of us suspect" concluded that by the end of the next century "mining of scarce metals will be increasingly a memory." Concerned about the prospects of shortages of several metals, some believe that a "resource war" designed to capture supplies of metals that come largely from a few African countries is imminent if not already under way (Council on Economics and National Security 1980).

Others, however, see no signs of impending mineral scarcity. Barnett and Morse (1963), for example, found that the labor cost per unit of output in the USA declined rather steadily from 1870 to 1957, as did the unit costs of minerals compared with other extractive goods, and Barnett (1979) reports that these trends continued through 1970 and seem to apply throughout most of the world. Simon (1981), expressing views with which many disagree, recently stated that "over the course of history, up to this very moment, copper and other minerals have been becoming not more scarce but more abundant" and concluded that "with knowledge, imagination, and enterprise, we and our descendants can muster from the earth all the mineral raw materials that we need and desire, at prices that grow smaller relative to other prices and to our total income." Table 1 shows declining prices and average grade and increasing reserves for copper in the USA from 1900 to 1977. Similar trends have been shown by many other minerals as advancing technology and exploration have transformed undiscovered and submarginal resources into measured reserves. But what about the future?

Table 1
Changes in copper price, grade of ore mined, and US ore reserve estimates, 1900–1977[a]

Year	Average annual refinery price (¢ kg^{-1}): Quoted refinery	Based on constant 1970 dollars	Average grade of ore mined (% Cu)	US reserves 10^6 t Cu)
1900	36.69	140.0	4.0	NA
1910	28.06	95.9	1.88	NA
1920	38.47	66.1	1.63	NA
1930	28.62	71.6	1.43	16.8
1940	24.89	65·0	1.20	21.5
1950	47.00	66.1	0.89	26.5
1955	82.65	90.4	0.83	22.7
1960	70.66	80.5	0.73	29.5
1965	77.21	88.0	0.70	29.5
1970	128.31	128.3	0.60	77.5
1974	170.42	134.3	0.49	NA
1976	153.44	104.5	0.51	84.4
1977	147.00	94.8	0.52	97.1

[a] US Bureau of Mines (1979)

In thinking about future mineral availability it is useful to differentiate several kinds of availability (Sheldon 1977, Brobst 1979). One is the geological availability of the mineral or the mineral resource as it is currently conceived. Another is technological availability. For example, the availability of energy from fusion of deuterium is currently limited by technology. If there should be a technological breakthrough, economic availability may become limiting, at least at first. In the 1980 revised mineral classification of the US Bureau of Mines and US Geological Survey still another kind of availability—legal—is recognized; restricted reserves or resources are defined as deposits that meet all the requirements of the definitions of reserves and resources but are restricted from extraction by laws, regulations or governmental policy.

Also, of course, there is marketplace availability, partly a function of the other limiting factors but also of the extent to which exploration and development have resulted in productive capacity adequate to meet demand. The USA was not short of economically producible petroleum resources during the Arab embargo of 1973, but it was short of the productive capacity to meet all its own needs. Availability limited by technology, economics, production capacity or legal restriction is subject to change over time, but geological availability is finite, even though the magnitude of the deposits of a mineral or of its resources are unknown or at least uncertain.

In examining the future availability of minerals, the rate of consumption as well as the magnitude of unmined mineral resources must be considered. Meadows et al. (1972) assumed a fixed 250-year supply of minerals if used at present levels; their estimate of a 100-year lifetime of reserves if used at exponentially increasing rates of consumption thus has no merit as a prediction (which they did not claim it to be). But they were right in emphasizing the impossibility of continued exponential growth in mineral consumption. For example, a billion-year supply of anything if used at current levels would be exhausted in only 584 years if used at an exponentially increasing rate of 3% a year (which is a modest rate of increase compared with that of many minerals in recent decades). A 2-billion-year supply would last only one more doubling time—24 years— at the 3% annual rate of increase. Described in another equally disturbing way, no matter how many doubling times it takes to use, at an exponentially increasing rate, 25% of a fixed supply, the remainder would be exhausted in only two more doubling times. Exponential growth would never continue unabated to the point of exhaustion, of course, but by the time market forces signalled an impending scarcity, maneuvering room for adjustment might be uncomfortably small.

There appears, however, to be a tendency for the rate of increase in consumption to slacken, once initial uses are satisfied and recycling begins. In the USA, for example, the rate of annual increase in primary mineral demand for 24 commodities has been declining since the 1960s and for 33 more it averaged less in the 1970–76 period than in the 1960s. Nevertheless, exponential growth in consumption is continuing for most minerals.

Mineral resources are irregularly distributed over the earth's surface and no country has domestic supplies of all the minerals it needs. The situation and outlook with respect to mineral availability thus varies from country to country and commodity to commodity. The discussion that follows focuses on mineral trends in the USA, many of which are representative of those in other industrialized countries, and in the world at large, the mineral trends of which are strongly influenced by the industrialized countries on the demand side and by the developing countries on the supply side.

As shown by data assembled on the nonfuel minerals by the US Bureau of Mines (1979, 1981), US production was greater in 1979 than in 1958 for 42 mineral commodities and less for 18 others, reflecting declining availability of competitive sources for some, such as cobalt, fluorite, gold, manganese and platinum, and increasing costs for others, such as dimension stone. Taking price in constant dollars as a related measure of availability, 1979 prices were lower than those in 1958 for 42 commodities and higher for 40 others. Because the improved technology resulting in decreasing prices has been related to the increased use of low-cost energy in the production process for many minerals, the large increase in energy costs since 1973 led to some reversals in a previously declining trend. Thus, the price was lower in 1973 than in 1958 for 20 of the 40 commodities for which 1979 prices were higher than 1958 prices. For some 18 commodities, however, the price increase over the period was relatively continuous, reflecting for the most part diminishing availability coupled with increasing demand.

World production of all mineral commodities but six has been steadily increasing and imports have thus been offsetting diminishing domestic supplies in the industrialized countries. The six for which 1979 production was less than in 1968 are arsenic, beryllium, gold, indium, mercury and thallium. Gold production peaked in 1970, when its price in 1978 dollars was US$60.59 a troy ounce, and the continued increase in its price subsequently did not reverse the decline in world production through 1979. Gold production did increase, however, in both the USA and the world in 1980–83 (US Bureau of Mines 1983, 1984) and world production of beryllium and mercury also increased in that period.

Attempting to assess mineral availability for the remainder of the century, the US Bureau of Mines (1979) forecast a decline in US demand during the period 1976–2000 for only mercury, thallium, sheet mica and barite and an annual increase of 2% or more per year for 64 of the 84 commodities projected. It estimated that primary US production will supply 90% or more of the cumulative demand for 32 commodities and $<10\%$ for 20 others. In addition to its estimate of year 2000 production, the Bureau calculated what 2000 production would be if based on a projection of the past 20-year trend; for 23 commodities this "projected" 2000 production would be less than half of the Bureau's "estimated" year 2000 production and for 9 others it would be $<100\%$ of the estimated production, indicating anticipation of developments that will accelerate domestic production of many minerals over what would be expected from the trend of the last 20 years.

The Bureau also compared these demand estimates with US reserves and identified resources. For the USA, the ratio of reserves to 1976–2000 cumulative demand is 0.1 or less for 18 commodities, between 0.1 and 1 for 15 others, >1 for 36 and "adequate" for 4 more. For the 38 commodities for which there are estimates of identified resources, the ratio to cumulative demand is >1 for all but 10. Presently undiscovered resources will surely be found for many minerals and will add to their future supply.

For the world as a whole, the Bureau of Mines estimates that the ratio of reserves to 1976–2000 cumulative demand is 0.1 or less for only fluorine,

0.1–1 for 12 commodities, >1 for 61 others and "adequate" for 5 more. The ratio of world identified resources to cumulative demand is <1 for only 2 commodities: fluorine and asbestos.

Turning to the future availability of energy resources, the changing situation and prospects for future supplies of energy are illustrative of future availability problems and possible solutions for mineral resources in general. The shortage of crude oil in 1973 as a result of the Arab embargo on exports to a world that had become dependent on them to a significant extent, followed by US shortages of natural gas during the winter of 1976–77, were strong shocks that have had a wide range of consequences. World oil prices, controlled largely by the Organization of Petroleum Exporting Countries cartel, skyrocketed to levels many times those prevailing in 1973, resulting in sharp increases in the prices of consumer products such as gasoline.

Widened recognition of the industrial society's dependence on energy and broadened realization that low-cost conventional reserves of oil and gas are limited led the US government to develop policies that would encourage conservation of energy and its more efficient use on the one hand and encourage exploration and the development of alternate sources of energy on the other. Crude-oil prices established after the embargo were decontrolled, and decontrol of natural-gas prices is in process. Several significant trends are developing as a result of these changes. Energy consumption per unit of gross national product was 20% less on a constant-dollar basis in 1982 than in 1973 (US Department of Energy 1984a). Industry consumed 28% less energy in 1982 per unit of output than in 1973, but its output increased; its energy consumption per unit of output has decreased by ~3.5% a year since 1970. Consumption of both oil and gas began to decline as people drove at slower speeds, shifted to improved-gas-mileage automobiles, added insulation to their homes and took other conservation measures, and in 1978 oil imports also began to decline. Oil consumption decreased 21% in the period 1979–83 (US Department of Energy 1984b), and in 1983 oil imports were down to 36% of the 1973 level.

Higher prices have also stimulated oil and gas exploration. More than twice as many rotary rigs were active in the USA in 1979 than in 1971, completing 1.88 times as many wells, drilling 1.90 as much footage and finding 1.45 times as much liquid hydrocarbons and 2.38 times as much natural gas as in the low-discovery years of the early 1970s. Whereas production of crude oil had begun to decline after apparently peaking in 1970, it began to increase again in 1977. Even though the increases are modest, that trend is still continuing—a reversal similar to that which began in 1963 for the production of lead in the USA (Harris 1977). Research and development to improve the percentage of oil recovered from the ground has been stepped up, leading to a reversal in the production decline in California in 1976 and promising to make available extensive heavy crude oil resources elsewhere.

Production of energy in the USA has also begun to increase from other sources—coal, hydropower, geothermal energy, oil shale, biomass and solar. Plans have been announced for the development of tar sand production in Texas and construction of a peat-supplied 120 MW power plant in North Carolina. Research and development are under way on the recovery of natural gas as well as geothermal energy from the extensive geopressured reservoirs of the Gulf Coast and on the recovery of methane occluded in coal. Except for coal and hydropower, these sources in 1981 were not making significant contributions nationally, but they have the potential for doing so.

In spite of public uncertainty about its safety, even nuclear energy has yielded significantly increased production in recent years. As of 1973, 40 plants with a total of 19 843 MW capacity had been licensed by the Nuclear Regulatory Commission; as of the end of 1983, 81 plants were licensed with a total capacity of 64 058 MW (US Department of Energy 1984a,b). According to the Office of Technology Assessment, however, as reported by *Chemical and Engineering News* on 13 February, 1984, "Without significant changes in technology, management and level of public acceptance, nuclear power in the US is unlikely to be expanded beyond the reactors already under construction."

The recent developments in the energy field illustrate how mineral resources can be created and extended not only by advancing technology and exploration but also by political, social and economic ingenuity. As shortages develop, price increases (if permitted) encourage exploration, improved recovery technology, technically more efficient use, conservation in the sense of curtailing unnecessary use and avoiding waste, recycling and substitution by a more abundant material—any or all of which extend supplies. The limits of these processes cannot be thought to have been reached for any mineral resource. Even for fluorine, for which the US Bureau of Mines estimated the ratio of world identified resources to world cumulative demand to the year 2000 to be only 0.8, some relief is already in sight, for the Bureau did not include in its estimates fluorine in identified resources of phosphate rock, in which it is present in amounts of 3–4% and from which its by-product recovery has begun in recent years.

Because of the near impossibility of predicting the result of the interplay of the factors that extend resources or accelerate their decline, it is impossible to be definitive about the lifetime of mineral resources beyond the certainty that exponential growth in consumption cannot continue indefinitely. Reserves for all minerals can be extended for a time, perhaps a very long time for some minerals, particularly if some of the very large potential resources of energy—such as dissolved gas in geopressured reservoirs, tar sands, deuterium in natural waters for fusion—can be brought within technological and economic reach and if the environmental degradation that might deter their development can be avoided. Still uncertain, for example, is whether or not continued combustion of fossil fuels will raise the atmospheric temperature as a result of the greenhouse effect caused by increased CO_2 content. The climatic effects of such a development might be so adverse as to necessitate curtailment rather than expansion in the use of fossil fuels.

Even though economic tests do not indicate developing scarcity of many minerals on a world scale, there are many signs of regional and national depletion, especially in industrialized countries. The USA, for example, has shifted from being a net exporter to a net importer of many minerals, including oil, not necessarily because all ultimately producible deposits of those minerals have been exhausted but because some of the remaining deposits are economically not competitive with foreign sources or because they are not adequate to meet expanded needs. Producible deposits of nearly all minerals are becoming more difficult to find; this is especially true for some of those—titanium, for example—for which production has fallen in spite of rising prices. New discoveries are still being made, however, and ultimately producible resources should not be thought to be limited to those now identified.

5. *Summary*

Mineral resources are naturally occurring solid, liquid and gaseous substances useful or potentially useful to man. They are the source of most of the materials and energy used by the industrial society. Historically, the per capita gross national product of an industrial society has increased in nearly direct proportion to increase in their consumption. Mineral resources include identified deposits producible under current economic and technological conditions (termed *reserves*), undiscovered deposits of the same quality (*undiscovered economic resources*) and known and unknown deposits that cannot yet be produced profitably (*identified* and *undiscovered subeconomic resources*).

Estimates of the magnitude of mineral resources have low precision and may be expected to change over time as the result of further geological exploration and technological advance that transform undiscovered and subeconomic resources into reserves. In assessing the future availability of mineral resources, account must be taken of rate of consumption. Over the past century, mineral resource output in the USA has increased at an annual rate of $\sim 3.7\%$ per year, and in its forecasts of demand to the year 2000 the US Bureau of Mines has assumed growth rates for most nonfuel minerals in the range 2–6%.One of the few certainties about the future availability of mineral resources is that none is adequate for long-continued, exponential growth in consumption. In the USA, higher prices for crude oil began in the late 1970s to reduce consumption through more efficient and conservative use while at the same time stimulating actions that enlarge supply.

The USA has already become a net importer of many minerals. Estimated reserves are lower than the forecast cumulative demand to the year 2000 for nearly half of the commodities for which estimates are available, and estimated identified resources are insufficient for about a quarter of them. For the world at large, estimates of reserves and identified resources are larger than the forecast cumulative demand to the year 2000 for nearly all commodities. Production of a few has begun to decline in spite of rising prices, indicating that profitably producible deposits are becoming extremely difficult to find. Even so, new discoveries, as of gold are still being made and the limits of efforts to extend supply through exploration, improvement in extractive technology, increase in efficiency of use, reduction of waste and conservative use, recycling and substitution have not been reached for any resource. Geological availability—unknown and perhaps unknowable in any but the most approximate sense—of a mineral is finite, but mineral resources are created by technological ingenuity and their life can be much extended by scientific, technological, political, economic and sociological creativity.

See also: Mineral Deposits: Examination and Evaluation; Ores of the Future

Bibliography

Barnett H J 1979 Scarcity and growth revisited. In: Smith V K (ed.) 1979 *Scarcity and Growth Reconsidered, Resources for the Future.* Johns Hopkins, Baltimore, Maryland, pp. 163–217

Barnett H J, Morse C 1963 *Scarcity and Growth, Resources for the Future.* Johns Hopkins, Baltimore, Maryland, pp. 1–288.

Brobst D A 1979 Fundamental concepts for the analysis of resource availability. In: Smith V K (ed.) 1979 *Scarcity and Growth Reconsidered, Resources for the Future.* Johns Hopkins, Baltimore, Maryland, pp. 106–42

Brooks D B, Andrews P W 1974 Mineral resources, economic growth, and world population. *Science* 185: 13–19

Council on Economics and National Security 1980 *The Resource War and the U.S. Business Community: The Case for a Council on Economics and National Security.* National Strategy Information Center, Washington, pp. 1–42.

DeYoung J H Jr 1978 *Mineral Resources Map of the Chandalar Quadrangle, Alaska,* US Geological Survey Map MF 878-B. USGS, Arlington, Virginia

Group of Experts on Definitions and Terminology for Mineral Resources 1979 *The International Classification of Mineral Resources,* Secretariat Economic Report No. 1. United Nations, New York, pp. 1–8

Harris D P 1977 Conventional crude oil resources of the United States: Recent estimates, methods for estimation and policy considerations. *Mater. Soc.* 1: 263–86

Harris D P, Skinner B J 1982 The assessment of long-term supplies of minerals. In: Smith V K, Krutilla J V (eds.) 1982 *Explorations in Natural Resource Economics*. Johns Hopkins, Baltimore, Maryland, pp. 247–326
McKelvey V E 1972 Mineral resource estimates and public policy. *Am. Sci.* 60: 32–40
McKelvey V E, Masters C D 1984 Undiscovered oil and gas resources: Procedures and problems of estimation. *Proc. 27th Int. Geological Congr.*, Vol. 13. VNU Science Press, pp. 335–52
Meadows D H, Meadows D L, Randers J, Behrens W W III 1972 *The Limits to Growth*. Universe Books, New York, pp. 1–205
Miller B M, Thomsen H L, Dalton G L, Coury A B, Hendricks T A, Lennartz F E, Powers R B, Sable E G, Varnes K L 1975 *Geological Estimates of Undiscovered Recoverable Oil and Gas Resources in the United States*, US Geological Survey Circular 725. USGS, Arlington, Virginia, pp. 1–78
National Research Council, Panel on Gas Reserve Estimation, Board on Mineral Resources 1976 *Gas Reserves Estimation of Offshore Producible Shut-in Leases in the Gulf of Mexico*. National Academy of Sciences, Washington, DC, pp. 1–170
Nolan T B 1958 The inexhaustible resource of technology. *Perspectives on Conservation, Resources for the Future*. Johns Hopkins, Baltimore, Maryland, pp. 49–66
Schanz J J Jr 1975 Resource Terminology: An examination of concepts and terms and recommendations for improvement. *Resources for the Future*. Johns Hopkins, Baltimore, Maryland, pp. 1–116
Schanz J J Jr 1977 United States minerals—A perspective. *Min. Cong. J.* 63(2): 26–29
Sheldon R P 1977 Factors limiting the future primary mineral supply system. In: Marios M (ed.) 1977 *Towards a Plan of Action for Mankind*. Vol. 1, *Long Range Mineral Resources and Growth*. Pergamon, Oxford, pp. 63–66
Simon J L 1981 The scarcity of raw materials. *Atl. Mon.* June: 33–41
Skinner B J 1976 A second iron age ahead? *Am. Sci.* 64: 258–69
US Bureau of Mines 1979 *Mineral Trends and Forecasts*. US Government Printing Office, Washington, DC, pp. 1–25
US Bureau of Mines 1981 *Mineral Facts and Problems*, US Bureau of Mines Bulletin 671. USBM, Washington, DC
US Bureau of Mines 1983 *Minerals Yearbook*, Vol. 1. USBM, Washington, DC
US Bureau of Mines 1984 *Mineral Commodity Summaries*, USBM, Washington, DC
US Bureau of Mines, US Geological Survey 1976 Coal resource classification system of the US Bureau of Mines and US Geological Survey. *Geol. Surv. Bull. (US)* 1450-B: 1–7
US Bureau of Mines, US Geological Survey 1980 Principles of a reserve/resource classification for minerals. *Geol. Surv. Circ. (US)* 831: 1–5
US Department of Energy 1984a *Annual Report of Energy Conservation Indicators for 1982*, DOE/EIA-0441 (82). US Government Printing Office, Washington, DC
US Department of Energy 1984b *Monthly Energy Review* for December 1983, US Government Printing Office, Washington, DC
Wood G H Jr, Kehn T M, Carter M D, Culbertson W C 1983 Coal resource classification of the US Geological Survey. *Geol. Surv. Circ. (US)* 891: 1–65
Zwartendyk J 1981 Economic issues in mineral resource adequacy and in the long-term supply of minerals. *Econ. Geol.* 76: 999–1006

V. E. McKelvey[†]

Mining and Processing: Environmental Problems

Environmental quality is a new parameter both for industry and for the regulatory agencies created to oversee environmental control. A certain price in environmental change must be paid to obtain the minerals and energy that industrial societies consume, but it is now recognized that a balance must be achieved between use and protection that requires compromise and sensible control policies. Thus, it is essential that advance-planning procedures based on sound research be adopted by industry and governments to minimize the costs of meeting prescribed environmental standards.

The US National Environmental Policy Act of 1969 assured attention to environmental protection within each Federal agency. In 1970, the US Environmental Protection Agency (EPA) was formed to monitor and control air and water pollution, solid waste disposal, and use of pesticides and environmental radiation. The Clean Air Acts mandated the EPA to set national air-quality standards for gaseous and particulate emissions, as well as standards for facilities emitting hazardous substances. The Water Quality Act of 1972 established waste-quality and treatment standards, and a permit system for discharge of pollution into waterways. Many other national governments have established similar protections and control agencies, although different standards are frequently adopted.

Mining, milling and smelting in various degrees cause damage to the environment and present hazards to animal and plant life. Chief among the environmental problems associated with mining are the disposal of spoil piles from strip mining, tailings from milling, airborne contamination from smelters, water pollution, and surface subsidence from underground mining.

1. Coal and Metal Mining Problems

Air and water pollution, subsidence and land reclamation are problems common to both coal and metal mining; these problems tend to receive more attention in areas of high population density than in more sparsely populated regions.

1.1 Air Pollution

Air-pollution prevention requires proper ventilation and extraction of rock dust from the air in mines.

Among health problems associated with air pollution, silicosis is a major consideration in mining of all kinds, although problems have been much lessened since the early 1900s. Coal worker's pneumoconiosis in bituminous-coal miners and asbestotic pneumoconiosis are specifically associated with the chemistry and particle size of the substances being mined.

1.2 Water Pollution

Water pollution traceable to mining activities is a well-recognized environmental problem, resulting in accelerated sedimentation, acidification and metal contamination. Sources include active or abandoned surface or underground mines, processing plants, waste-disposal areas, haulage roads or tailings ponds.

Acid mine drainage is one of the most serious aspects of water pollution. Where pyrite (a common accessory mineral in sulfide ores) comes into contact with air and water, ferrous hydroxide and sulfuric acid are formed. The ferrous ions are then changed to the ferric state and form the relatively insoluble ferric hydroxide, which precipitates on the stream bottom; other metals may be adsorbed on or coprecipitated with the hydroxide. The second product of the weathering of pyrite, sulfuric acid, increases the acidity of the water. Acid drainage increases treatment costs, destroys living organisms in the stream, corrodes culverts, piers, boat hulls, pumps and other metal equipment in contact with the acid waters and renders the water unacceptable for recreational use. Reports from the US Fish and Wildlife Service indicate that acid mine drainage adversely affects animal life in about 10 000 km of streams and 6000 ha of impoundments in the USA. The metal content and acidity gradually decrease below mines owing to dilution, chemical precipitation and adsorption on organic matter. Studies in Colorado showed that no benthic organisms or fish could exist for about 30 km below mine tailings. This distance would be reduced if sufficient algae were present to concentrate the metals.

Water infiltration into mines is caused by natural movement of underground water into mine openings or by the infiltration of surface water through boreholes drilled during exploration. Water from surface mines or strip mines often penetrates underground workings; effluent from strip mines may enter the groundwater system. Infiltration of water into mines can be controlled by increasing surface water runoff through backfilling, grading or sealing with impermeable materials to decrease the permeability of rock strata. Regrading requires revegetation, for which irrigation may be necessary in arid regions. Various kinds of regrading may be used over surface mines, depending upon the topography and available backfill.

Drainage from mines may be controlled at the source by preventing or restricting the pollutants from reaching the exterior drainage or collecting the mine waters for treatment. Water handling may include the development of evaporation ponds in semiarid areas, alkaline regrading, controlled-release holding ponds, connector wells and slurry trenching (in which the sides of the trench are maintained by a water–clay slurry that forms a groundwater dam). Slurry trenching is an effective way of inundating abandoned mines to reduce acid drainage where the down-dip outcrop has been strip mined. Clay sealing of abandoned mine shafts and drainage openings, reclamation of surface mines and revegetation have been shown to be feasible but initially costly; the resulting stability is optimal and maintenance costs low. Several other methods to control the quality of water discharged from abandoned mines are possible. Backfilling, a method of disposing of mine and milling wastes, has been used for more than a century to control mine fires, and to protect the ground surface from subsidence. The method is to inject solids or slurry through boreholes. The degree of pollution control by backfilling has not been demonstrated, although the oxidation of sulfides is definitely inhibited. Experimental programs have also shown the feasibility of pressurizing abandoned underground mines with inert gas to reduce the oxidation of sulfides by limiting free oxygen. Laboratory and field tests also indicate the feasibility of underground precipitation of sludge by injection of alkaline slurries to produce a treated effluent, while filling the abandoned mine with sludge that would otherwise cause a surface disposal problem.

1.3 Subsidence

Subsidence of the ground surface over mines was not a serious problem when full-recovery mining was largely under agricultural lands, but with the rapid expansion of suburban housing over both old and active mines, difficulties with subsidence have multiplied. Until recently in the USA, a large percentage of mining rights included a waiver of surface damage or the right to subside the land. Later, owners were permitted to purchase an agreement with the company to leave supportive pillars in the ground, with no guarantee against subsidence. The cost of the agreement is based on the tonnes of coal required to support the surface structures. More recently, certain coal companies have revised the agreements to guarantee restitution for damage. Poor legislation and poor public relations forced the coal companies to cooperate with legislators for the enactment of the Bituminous Mine Subsidence and Land Conservation Act of 1966. The Act requires strict policing and enforcement and protects the home owner. Similar regulatory control is mandatory in the metal-mining industry.

1.4 Land Reclamation

Land reclamation is an important environmental aspect in all surface mining and milling operations. In area strip mining on a relatively flat terrain, the overburden from each successive trench is deposited in the cut previously excavated. These spoil piles create a washboard appearance on the landscape. The overburden may be graded, revegetated and the acid-producing materials buried to reduce environmental damage. The last cut may be filled with water to prevent oxidation of sulfides. A certain amount of acid mine drainage does, however, occur.

Contour stripping on steep hillsides creates a shelf and an unstabilized spoil bank, which is more vulnerable to erosion and landslides than the undisturbed slope. In some US states, the permitted width of the bench and the steepness of the slope are defined by law. Grading toward the bench wall may cause acid water to accumulate; grading away from the bench wall may cause serious erosion unless diversion ditches are dug. Accidental breaching of spoil piles may release sediment which can cover highways, kill crops and clog water courses, causing floods. Toxic materials in the lower part of the overburden will be brought by natural processes of leaching to the surface unless, at a relatively small cost, the top soil is set aside to be used as a cover and to be revegetated. Experience in the UK indicates that there should be constant maintenance for at least 5 years after the operation has ceased.

Concurrent contour mining and reclamation can be provided by several techniques of backfilling, spoil-bank shaping, slope reduction and revegetation. Preplanning, proper mining practices and reclamation help to minimize environmental damage. Included in the planning prior to mining should be studies of the chemistry of the overburden, geology, hydrology and biology of the watershed, local use of water and land, siting of haulage roads and topography. If old underground mines will be breached, plans to seal the openings by clay compaction or other means need to be made. Concurrently or after mining, the area should be fertilized and revegetated with quick-growing cover for effective erosion control. The introduced plants must be sufficiently hardy to provide permanent growth under difficult conditions. Poplar, crown vetch and grasses have proved satisfactory in the US eastern coal lands; saltbush and native grasses have proved best in the western strip-mining areas. Revegetation, however, requires research into the texture, acidity and chemistry of the overburden, particularly the contents of the toxic sulfides, and of the potash and phosphate nutrients. Lakes formed in the final cuts may become valuable recreation areas.

2. *Anticipated Problems in Oil-Shale Development*

Oil shale is present in many areas of the world. There are estimated reserves of 2×10^{12} barrels of oil in the shales of the USA, but probably no more than 60% can be extracted. Oil shale is commonly interbedded with saline deposits of sodium bicarbonate, sodium aluminum carbonate and sodium chloride. Shale oil may be produced by strip mining or underground mining, in which the shale is crushed and heated to 480 °C in above-ground retorts, or it may be produced in situ by passing retorting gases and produced liquids through fractured shale. The latter requires mining the shale in the lower part of a room to provide space when the shale above is collapsed and burned. By either method, the oil extraction process increases the bulk volume of the shale by 20–30%. Whether the spent shale is disposed of at or below the surface, some surface disposal will be required because of this increase in volume.

In the expected production of 10^6 barrels of shale oil per day in the western USA, 1.3 Mt of spent shale per day must be disposed of. Probably much of the spent shale will be put into canyons and each layer compacted to minimize the volume. Since snow destroys compaction of the top two feet of the spent shale, channelling of the local runoff away from the canyons will be necessary to reduce erosion. The spent shale not put into canyons must be contoured, covered with topsoil and revegetated. Runoff water from areas of spent shale potentially may contain high concentrations of Na, Mg and SO_4, but this hazard can be minimized by proper compaction and grading. Mines in some areas will require dewatering. Excessively saline water will require reinjection, evaporation or transportation from the area; the latter process constitutes a real hazard through the addition of salts to rivers, by which they would be transported to irrigated agricultural land.

The piles of spent shale can be stabilized chemically by adding a reagent to react with the residue to form a hard crust. Also, spent shale will support native grasses and shrubs if it is leached, mulched to a depth of 10 cm, fertilized and irrigated. However, considerable water is required initially for leaching and the establishment of vegetative growth; continuous long-term irrigation is required for maintenance. Over a long period, oil-shale operations affect fish and wildlife habitats and grazing and agricultural activities over a wide area (at least 33 000 ha in the USA). The large reserves of shale oil cannot be extracted without a high environmental cost, unless technological innovation of the highest order is fostered.

3. *Specific Problems in Metal Mining*

Environmental problems related to metal or mineral mining, other than those discussed under coal and oil shale, include the disposal or stabilization of tailings and the control of emissions from smelters. Certain problems also arise in the mining and processing of any given material.

3.1 Tailings Disposal

Tailings disposal problems have become aggravated by the finer grinding used today to separate low-grade-ore minerals from their matrix, and the increasing use of toxic materials in processing. Tailings ponds operate as sedimentation catchment basins and as biological treatment devices in mineral mining. Water pollution is the greatest environmental problem associated with tailings disposal, and the purposes of impoundment are to effect pollution control, reuse of water, and to permit reworking for recovery of lost metals. Impoundment is expensive because it affords a risk of structural failure, and the impoundment is a source of air and water pollutants. Water treatment may be necessary, and sulfides in the tailings may affect revegetation efforts and produce a vegetative cover that is toxic to animals. Wholly impervious impoundment dams on impervious foundations must be designed for maintenance-free abandonment as well as to withstand events such as earthquakes or floods. The fine particle size of mill tailings makes the surface of tailings ponds susceptible to wind erosion. To prevent deflation and the movement of dust into the air, the tailings must be compacted and the surface stabilized chemically or with vegetation. Tailings ponds may be the source of acid effluent bearing toxic heavy metals, dissolved salts, and suspended particles which enter and pollute the natural drainage system. The effluent can be collected by a second interception dam and either returned to the main impoundment or treated separately. Treatment methods include sedimentation, flocculation, filtration, acid neutralization or precipitation using chemical reagents. Few mineral mines outside arid regions use recirculated water to fill more than 50% of their water requirements, because the costs of pumps and pipes are high, and pollutants such as sulfuric acid may interfere with the flotation process.

Reclamation of tailings impoundment to prevent a safety hazard or other environmental damage is possible if the tailings are stabilized by physical or chemical means. Visual improvement and land use are then possible with revegetation. Nonacid tailings in temperate climates can be readily revegetated by adding organic matter and fertilizer before seeding. In arid regions, irrigation systems are necessary. The vegetation on heavily polluted soils in mining districts commonly consists of resistant forms of only a few plant species. In the UK, the most resistant species have been found to be bent grass (*Agrostis*), fescue (*Festuca*), sorrel (*Rumex acetosa*) and plantain (*Plantago lanceata*). Culture-solution experiments have shown that tolerant gene-controlled races of these species have evolved that can withstand highly metalliferous low-moisture soils. These races of plants are dwarfed and cannot grow if transplanted to normal soils. They are also metal-specific and cannot grow on soils containing a different assemblage of heavy metals. A technique involving fertilization and seeding with seed collected from races of *Agrostis* and *Festuca* growing on soils of similar heavy-metal content has proved successful on old mine tailings, and nullifies the necessity for covering the tailings with topsoil.

Where possible, marine disposal should be considered. Operating costs are low, the volume of storage area is great and impoundment failure or slumping is minimized; however, there is little possibility for future reworking of the tailings for metal recovery. Environmental advantages are several. The alkaline seawater and low oxygen content at depth prevent oxidation of pyritic tailings; maximum dilution of tailings and mixing of toxic elements takes place; and no reclamation or revegetation is necessary. Studies of tailings disposal at sea from several copper and molybdenum mines in different parts of the world show that effects on marine life are negligible. On the other hand, effluent carried down rivers to the sea may seriously damage seaboard industry (e.g., in the UK, lead-mining effluent seriously affected the shell-fishing industry). Marine disposal will no doubt receive greater attention in the future, but terrestrial impoundment is likely to continue to be the predominant method of disposal.

3.2 Smelter Emissions

Emissions from smelter stacks are a mixture of particulate matter and gases, including sulfur dioxide, carbon monoxide and carbon dioxide. At some installations, these gases are discharged directly to the atmosphere; elsewhere, a part is converted to sulfuric acid. Dust generated in the smelting process is generally collected by baghouse filters, electrostatic precipitators or scrubbing devices, but these facilities are never completely effective. During processing, sulfur oxides and sulfates are released in such quantities as may destroy the nearby vegetation and affect the biosphere. Correlations have been found between sulfur dioxide levels and chronic respiratory problems, but other oxides and sulfates (sulfuric acid mist) may act synergistically to increase the hazards.

3.3 Zinc, Lead and Cadmium

Zinc, lead and cadmium commonly occur together in base-metal sulfide ore deposits. Environmental studies before and for three years after commencement of production from the world's largest lead deposit in Missouri showed only minor increases in dissolved heavy metals in drainage below the mine, but showed fine suspended solids to be the major transport mechanism. Therefore, aquatic life is more affected by the silting and by the release of toxic milling reagents such as copper and zinc cyanide and xanthate than by lead. At one mine, a series of shallow meanders was installed below the final tailings dam. Algae growing in the stream trapped the metals and rock flour. A final sedimentation basin contained floating algal mats. The creek below the treatment system now exceeds the allowable concentrations for heavy metals only in times of floods. Two lead smelters located in the oak forests of the area, however, have raised the lead content of the leaf litter and soils for as much as 40 km, and of copper, zinc and cadmium for 25 km. Lead contents of tree leaves increased many-fold for at least 10 km around smelters.

Atmospheric lead released from smelters may cause health problems. Zinc, on the other hand, rarely reaches toxic levels although it is easily dispersed and zinc concentrations are raised in mining areas and in the vicinity of smelters. The chief health concern in zinc districts relates to the accompanying cadmium.

The major pollution sources of cadmium are from particulate emissions during mining of zinc, and the leaching of cadmium from the overburden especially in acid waters. Typically, 15–20% of the initial cadmium accumulates in the dust during the sintering process and is vented to the atmosphere. Emissions from sinters consist mainly of cadmium oxides, whereas emissions from roasters are mainly sulfates. Of the cadmium deposited near smelters, the sulfates may be leached from the soils by rainfall, but the oxides will accumulate. Analyses of soils and vegetation near smelters show cadmium accumulations in soils, but only a small percentage is taken up by vegetation. Of particular importance is the increased cadmium to zinc ratio, which has been found to be more harmful to health than their actual quantities.

3.4 *Copper and Molybdenum*

Copper and molybdenum are commonly associated in sulfide ore deposits, although each may occur and be mined separately. Increased concentrations of these elements occur in soils around mines and smelters. Soluble molybdenum compounds become available to vegetation under alkaline conditions, and copper becomes soluble in acid soils, creating deficiencies and excesses that affect grazing stock.

3.5 *Nickel*

Nickel is not readily available to the food chain, even in contaminated areas. It is accumulated in large quantities by certain plant species that have developed a tolerance to lateritic soils and soils derived from ultrabasic rocks; locally, these may be so rich in nickel as to be minable. Nickel is very toxic, however, to most normal plant species; vegetables and forage are not grown in high-nickel areas. Nickel is usually not detected in groundwater unless it is present as a result of industrial pollution.

3.6 *Iron Ore*

Massive iron ore deposits occur in the Lake Superior region of North America, and are mined by room-and-pillar, caving and open-pit methods. Subsidence is a major problem, and turbid alkaline minewaters of a red–orange color pollute the environment; they are commonly high in iron, manganese and silica. These waters affect fish and wildlife, and also aesthetically spoil recreational areas and lakeshore property. Mine drainage waters from taconite iron ores can contaminate the drainage system with asbestos-like fibers.

3.7 *Asbestos*

Asbestos in asbestos mines and asbestos-like fibers in iron and talc mines are released during removal of the overburden in preparation for open-pit mining, and during drying, crushing, grinding and screening of the ore. Dust-control measures are necessary in the mines and open pits, and also on waste dumps and during ore haulage. Asbestosis (asbestotic pneumoconiosis) and calcified pleural plaques are common in workers exposed to asbestos. Recent information supports a relationship between asbestos and malignant mesothelioma (see *Asbestos Hazards*).

3.8 *Precious Metals*

The mining and processing of gold, silver and other precious metals has little effect on health other than the poisoning of areas near the mines with chloride and mercury, and the release of arsenic to the watershed and vegetation. The distribution of arsenic in gold-mining districts is more widespread than is generally realized, but whether or not the arsenic becomes reduced from arsenate to the toxic arsenite by plants and animals that are consumed by man is not known. Cases of arsenic poisoning have been reported. Evidence is mounting that metallic mercury can be mobilized by biomethylation to produce absorbable mercury compounds of increased toxicity that can enter the food chain through air or water. This natural conversion has been observed in stream sediments below mercury mines.

3.9 *Fluoride*

Fluoride in locally hazardous concentrations may be released to the atmosphere and to drainage basins by industrial operations connected with the manufacture of phosphate fertilizer, aluminum and steel. The health of grazing animals has been affected near phosphate plants and financial settlements have been made.

3.10 *Uranium and Selenium*

Environmental problems related to uranium mining include the release of radioactive dust, mine drainage and tailings disposal. Ventilation in active mines is presently monitored to prevent excess accumulation of ^{222}Rn which can cause lung cancer. The disposition of tailings from present uranium mills and the removal of old tailings previously dumped into rivers is now subject to monitoring and legislative control. The sale of radioactive materials for housing landfill is now prohibited in the USA (EPA 1975). The long-term effects of low-level radiation requires extended epidemiological study. An accompanying element, selenium, which is toxic to grazing animals, occurs in selenium-accumulating plants that grow in certain uranium districts. These areas must be restricted from grazing.

4. Conclusion

Little is known about the effects on animal and plant life of low-level concentrations of many of the elements and compounds that have been discussed. Questions that need to be answered are:

(a) What happens to metals and fossil fuels after they are released to the environment?

(b) What effects do they have at low levels over long periods of time?

(c) Do they interact with additive, synergistic or neutralizing effects?

Bibliography

Bradshaw A 1970 Pollution and plant evolution. *New Sci.* 48: 497–500

Down C G, Stocks J 1977 Environmental problems of tailings disposal. *Min. Mag. London* 137: 25–33

Environmental Protection Agency 1975 *Inactive and Abandoned Underground Mines: Water Pollution Prevention and Control*, EPA Report No. 440/9-75-007. Environmental Protection Agency, Washington, DC

Fleischer M, Sarofim A F, Fassett D W, Hammond P, Shacklette H T, Nisbett A C T, Epstein S 1974 Environmental impact of cadmium: A review by the Panel on Hazardous Trace Substances. *Environ. Health Perspect.* 7: 253–323

Geological Society of America 1974 Development of oil shale in the Green River formation. *Report of the Committee on Environmental and Public Planning* 9(4)(suppl.): 1–8

Hill R D 1971 Restoration of a territorial environment—The surface mine. *ASB Bull.* 18: 107–16

Jennett J C, Wixson B G, Lowsley I H 1977 Control of industrial emissions of lead to the environment. In: Bogges W R (ed.) 1977 *Lead in the Environment*, National Science Foundation Report NSF/RA-770214. US Government Printing Office, Washington, DC, pp. 1–272

Metz W D 1975 Oil shale: A huge resource of low-grade fuel. In: Van Tassel A J (ed.) 1975 *Environmental Price of Energy*. Lexington Books, Lexington, Massachusetts, pp. 3–326

National Committee on Materials Policy 1973 *Material Needs and the Environment Today and Tomorrow*, Final Report. US Government Printing Office, Washington, DC, pp. 1-1 to 12-7

Sullivan G 1967 Current research trends in mined-land conservation and utilization. *Min. Eng.* 19: 63–67

Vandale A E 1967 Subsidence—A real or imaginary problem? *Min. Eng.* 19: 107–16

H. L. Cannon
[US Geological Survey, Reston, Virginia, USA]

Molybdenum Resources

Molybdenum metal was first identified in 1781 after work by Bergman and Hjelm. Most molybdenum is now produced from the mineral molybdenite (MoS_2), the largest deposits of which occur in the USA, Canada and Chile. Molybdenum is important industrially as an alloying component in special steels, and its compounds are used as dyestuffs, catalysts for organic reactions and in lubricants.

1. Geochemistry and Mineralogy

Molybdenum has an atomic number of 42, an average atomic weight of 95.94 and belongs to group VIB of the periodic table. Average crustal abundance is 1.5 ppm, varying from 0.4 ppm in basic rocks to 2.3 ppm in acidic rocks. Molybdenum has seven stable isotopes with masses 92, 94, 95, 96, 97, 98 and 100 in proportions that vary from 9.04% (^{94}Mo) to 23.78% (^{98}Mo) (Oyarzun 1978). The atomic radius of Mo is 1.39 Å, and the ionic radius of Mo^{4+} is 0.68 Å and of Mo^{6+} is 0.65Å.

Chemically, molybdenum exhibits a variety of oxidation states, but in nature only the tetravalent (chalcophile) and hexavalent (lithophile) species are important. In deep-seated environments tetravalent molybdenum is characteristic, whereas hexavalent molybdenum is important in surficial processes.

Mo^{4+}, W^{4+} and Re^{4+} exhibit similar ionic radii, and Mo^{6+} and W^{6+} have the same radii. The ability for substitution determines chemical behavior to a large degree in the magmatic environment.

The only molybdenum-bearing mineral of economic importance in the magmatic environment is molybdenite, MoS_2, although other sulfide minerals have been reported in which molybdenum is combined with copper or iron (castaingite and femolite, respectively). Molybdenite occurs in the hexagonal crystal system, $6/m\ 2/m\ 2/m$. Crystals are tabular, commonly foliated, massive, or in scales. The mineral consists of linked MoS_6 octahedra that form sheets parallel to $\{0001\}$ and are loosely bonded to adjacent layers, resulting in basal cleavage similar to graphite. It has a Mohs hardness of 1–1.5, a density 4.62–4.73 g cm^{-3}, and is metallic gray in color with a greasy feel. A greenish gray streak on glazed porcelain distinguishes it from graphite. Molybdenum is found also in other mineral species as sulfides, oxides, or various hydrated complexes.

In surficial environments, molybdenum is mobile owing to the solubility of its oxygenated hexavalent complexes. Anion complexes such as MoO_4^{2-} form by oxidation of MoS_2. In acid environments, molybdenum is found in different complexes, and below a pH of 6 is represented by HMo_4^-.

2. Types of Deposits

Molybdenum deposits are of six genetic types: (a) stockwork (disseminated and/or breccia pipe) deposits, in which molybdenite is the most important mineral in the ore; (b) porphyry copper–molybdenum deposits in which molybdenum is usually in concentrations less than or equal to copper content;

(c) contact-metamorphic and tactite ores adjacent to intrusive granitic rocks; (d) quartz–molybdenum-bearing fissure veins: (e) pegmatites and aplite dikes; and (f) bedded deposits in sedimentary rocks.

2.1 Stockwork Deposits

In general, stockwork deposits contain 0.15–0.49% MoS_2 distributed throughout or near contacts of siliceous intrusives and adjacent host rocks. With few exceptions, stockworks only produce molybdenum. Simple, composite or multiple intrusions, dikes and breccia pipes localize these ores (Clark 1972). Compositionally, igneous phases vary from granodiorite to granite, rhyolite and quartz-rich porphyries.

Distribution of molybdenite occurs in (a) veinlets associated with quartz and lesser amounts of other sulfides, oxides and gangue; (b) fissure veins; (c) fine fractures containing molybdenite "paint"; (d) breccia cement; and, more rarely, (e) disseminated grains. Intrusion and mineralization commonly occur in the epizone of the crust. Oxidation products are often fixed at the surface as secondary minerals in low-pH environments and supergene enrichment is absent.

2.2 Porphyry Copper–Molybdenum Deposits

Collectively, porphyry copper deposits produce nearly 40% of world production and one third of US production. While there are many similarities between these deposits and stockworks, some of the major differences are: (a) molybdenum and on occasion rhenium, gold and silver are economic by-products from porphyry copper deposits but not stockworks; (b) temporal and spatial distribution between the two types of deposits in wide orogenic zones, such as the western USA; and (c) the lack of supergene enrichment in stockwork deposits. Ore values vary from 0.016 to 0.060% MoS_2.

2.3 Contact Metamorphic Deposits

These deposits are far less common than the two classes described above. In these ores molybdenum is often associated with copper or tungsten and tin, the elements being located at the contact of the siliceous intrusive and favorable host rock, usually limestone. Deposition is by replacement and tends to be erratic in distribution and concentration. Molybdenite values may be as high as a few percent but average about 1.0% MoS_2 in economic deposits.

2.4 Fissure-Vein Deposits

In these deposits molybdenite occurs with quartz in fracture fillings in or adjacent to siliceous intrusives in orogenic belts. Pyrite is commonly present with small amounts of base-metal sulfides, whereas gangue minerals include fluorite, calcite and rhodochrosite. Ore shoots within the veins tend to be lenticular with molybdenite values varying from 1–20% at Questa, New Mexico. Average tenor may be as low as 1–2% MoS_2.

2.5 Pegmatite and Aplite Deposits

In these deposits molybdenite occurs sporadically in crystalline aggregates with quartz, feldspar, mica and other nonmetallic pegmatite minerals. The pegmatites occur in batholithic-sized plutons. Other metals may also be present, including beryllium, iron and zinc. On rare occasions, molybdenite is found in small quantities in quartzo-feldspathic aplite dikes that cut granitic country rock. Molybdenite values occur erratically in vugs and pockets.

2.6 Bedded Deposits

Molybdenum occurs in several sedimentary rocks, being concentrated in coals, shales and phosphorites in subeconomic amounts (King et al. 1973). Molybdenum also occurs with some uranium deposits of this type. By-product molybdenum has been recovered from uranium ores in Mesozoic sedimentary rocks during beneficiation in New Mexico. Molybdenum content varies from a few ppm to 0.5%.

3. Distribution of Deposits

Over 95% of the world's recoverable reserves are located in stockwork and porphyry deposits associated with Phanerozoic orogenic belts, of which the Cenozoic era is the most important. Approximately 87% of the world's resources occur in the Cordilleran and Andean Circum-Pacific belts with some 50% in the USA sector, 20% in the Chilean sector and 14% in the Canadian sector (Frutos 1978). At present, there are some 44 deposits of all types in which molybdenum is recovered as a primary or by-product.

The USA possesses the largest productive mines. Of the stockwork type, Climax and Henderson are located in Colorado, whereas Questa is located in New Mexico. These three deposits are located in the Rocky Mountain tectonic province and are of mid-Tertiary age. Further west, 17 porphyry Cu–Mo deposits of late Cretaceous to early Tertiary age in the Basin and Range province yield molybdenum as a by-product. There is also by-product recovery at the Pine Creek contact deposit in the Sierra Nevada.

In the Mexican Cordillera, by-product molybdenum is recovered at La Caridad (and formerly at Cananea) and also in the Cumobabi stockwork-breccia deposit, all of which are in the state of Sonora and early Tertiary in age.

In the Canadian Cordillera there has been considerable production from three stockworks at Boss Mountain, Molybdenum and also at Endako, all in British Columbia. The Canadian by-product molybdenum is derived from porphyry-type deposits whose age varies from late Jurassic to early Tertiary.

In the Appalachian belt of the USA there have been several small producers in the past, mostly from fissure veins. Several stockwork and porphyry deposits have been identified as having been formed in middle-to-late Paleozoic time, but none is a significant producer at present. However, in the adjacent province of Quebec in Canada, the Gaspé porphyry copper deposit has recovered molybdenum since 1969.

The principal production in South America comes from Peru and Chile. In Peru the main producers are the Toquepala and Cuajone porphyry deposits. In Chile, similar deposits are found in Chuquicamata. El Teniente, Andina and El Abra. All these deposits were formed during the Cenozoic era.

Located in the Alpine–Caucasian Mesozoic–Cenozoic orogenic belt are the Sar Chesmeh (Iran) and Saindak (Pakistan) porphyry deposits, the Machkatitza (Yugoslavia) and the Tyrnay-Auz contact deposit in the Greater Caucasas in the USSR.

In the Perisiberian Paleozoic fold belt there are important porphyry Cu–Mo ores at Kounrad and at other localities in Kazakhstan. Further south in the Tien-Shan system (including the Almylk district) there are similar deposits. To the north, in Zayson, fold-system deposits are richer in copper. In the easternmost part of the belt, in the Mongol-Amur fold system and East Transbaykal is another important zone of molybdenum resources. The intrusions in this region are late Jurassic in age and are associated with stockwork and skarn deposits. Finally, in the Ural tectonic belt, there are stockwork and skarn deposits.

In Europe, the Caledonian orogenic belt includes the Knaben deposit in southern Norway, an old producer and now recognized as a Paleozoic porphyry Cu–Mo system.

Younger porphyry deposits are found in the southwestern Pacific region; for example, the Bougainville (Panguna) deposit, where the mineralization epoch is Pliocene, as is the Cerro Colorado deposit in Panama.

4. Recovery and Production

While 200 deposits containing molybdenum have been discovered, only 136 deposits are sufficiently well explored to provide reliable data (Sutulov 1978). These indicate 14.8 Mt of molybdenum of which 9.76 Mt are recoverable. By deposit, stockworks contain 5.95 Mt; porphyry Cu–Mo deposits 3.50 Mt; contact ores 0.23 Mt; and pegmatite–aplite deposits 0.08 Mt.

In 1981, free-world production amounted to 93.92 kt of contained molybdenum (Distler 1981), of which stockwork and porphyry deposits contributed approximately 63% and 36%, respectively. At present rates of production and in the foreseeable future, there does not appear to be a shortage of recoverable molybdenum, particularly as US production fell from 140 kt in 1981 to 65 kt in 1987 (US Bureau of Mines 1988).

Bibliography

Billhorn W W 1984 Molybdenum. *Eng. Min. J.* 185: 52–53

Clark K F 1972 Stockwork molybdenum deposits in the western Cordillera of North America. *Econ. Geol.* 67: 731–58

Distler W F 1981 Molybdenum. *Eng. Min. J.* 182: 108–11

Frutos J 1978 World survey of molybdenum deposits. In: Sutulov A (ed.) 1978 *International Molybdenum Encyclopedia 1778–1978*, Vol. 1, *Resources and Production.* Miller Freeman, San Francisco, California, pp. 375–401

King R U, Shawe D R. MacKevett E M Jr 1973 *Molybdenum*, US Geological Survey Professional Paper 820. US Government Printing Office, Washington, DC, pp. 425–35

Oyarzun J 1978 Geochemistry (of molybdenum). In: Sutulov A (ed.) 1978 *International Molybdenum Encyclopedia 1778–1978*, Vol. 1, *Resources and Production.* Miller Freeman, San Francisco, California, pp. 154–72

Sutulov A 1978 History of molybdenum. In: Sutulov A (ed.) 1978 *International Molybdenum Encyclopedia 1778–1978*, Vol. 2, *Processing and Metallurgy.* Miller Freeman, San Francisco, California, pp. 232–41

US Bureau of Mines 1988 *Mineral Commodity Summaries 1988*. USBM, Washington, DC

K. F. Clark
[University of Texas at El Paso, El Paso, Texas, USA]

N

Nickel Resources

Nickel is a critical element in the manufacture of steels and specialty alloys. It is extracted principally from the hard-rock mining of magmatic sulfide deposits and from the stripmining of lateritic soils that develop during tropical weathering. Although not currently mined, large amounts of nickel also occur in deep-sea manganese nodules. The world's estimated resources of nickel are adequate to meet foreseeable future demands.

1 Properties and Use

Nickel is a silver-white metal that has a high ductility, good thermal conductivity, toughness and strength that are similar to iron, and a high resistance to atmospheric corrosion. Its corrosion resistance and strength over large temperature ranges cause it to be widely used in metal alloys. Most is utilized in the production of stainless steels, but it is also combined with copper, molybdenum, aluminum, chromium, niobium and titanium to make specialty alloys which are used in the chemical, nuclear and aerospace industries. In addition, nickel's chemical properties make its use desirable in some batteries, catalysts, dyes and pigments.

2. Crustal Distribution and Deposit Types

Nickel's average crustal concentration of 2wt% makes it the fifth most abundant element in the earth. The distribution of nickel, however, is not homogeneous as it increases from less than 0.01% in the earth's crust, to about 0.2% in the mantle, to an estimated 4–5% in the earth's core. Additional variations are noted among the various rock types found in the crust, with 0.1–0.3% Ni in igneous ultramafic rocks, about 0.016% in mafic igneous rocks, and less than 0.006% Ni in silicic igneous rocks. Concentrations of nickel that are currently being mined are greater than one hundred times average crustal abundance (i.e., 1–4% Ni).

Nickel occurs in four principal deposit types, of which two are currently of economic importance.

2.1 Magmatic Sulfide Deposits

These deposits form when mafic and ultramafic magmas become saturated in sulfur, thus allowing an immiscible sulfide liquid to precipitate. The sulfides form droplets which fall to and are concentrated at the bottom of magma chambers. Nickel strongly prefers to bond with sulfur and is concentrated in this sulfide liquid at levels 100–1000 times that in the host magma. Although the sulfides may have up to 25% Ni, dilution by silicates generally causes deposits to contain less than 5% Ni. Other elements such as copper, cobalt and the platinum group metals also concentrate in these sulfides and may be recovered. The mineralogy of these deposits is usually mostly pyrrhotite ($Fe_{1-x}S$), pentlandite ($(Fe, Ni)_9S_8$), and chalcopyrite ($CuFeS_2$).

Most economic magmatic sulfide deposits fall into one of three subgroups. Deposits in large stratiform complexes have traditionally been the most important, largely because of the immense deposits hosted in the Sudbury complex in Canada. However, study of Sudbury indicates that a number of unique events, including the impact of a meteor, were involved in its formation and probably cause it to be unique. Examples of other less economically attractive nickel deposits in stratiform complexes are in the Stillwater complex (USA) and the Bushveld complex (South Africa).

A second subgroup are found in Archean and Proterozoic greenstone belts and occur in MgO-rich shallow intrusives and extrusives known as komatiites. Although komatiites have only been identified as a rock type since 1969, important deposits have been found within them in the Precambrian shield areas of Australia (Kambalda district), Canada (Abitibi and Thompson districts) and Zimbabwe. Favorable areas for finding additional deposits exist in other Precambrian terrains where komatiites have not yet been recognized.

The last important subgroup of deposits are in mafic intrusions or flows emplaced during continental rifting. Examples are the important deposits at Noril'sk (USSR) and the large, but subeconomic, Duluth complex (USA). Unlike deposits of the other two subgroups, which occur only in rocks older than about 1700 million years, these deposits are found in rock as young as 200 million years. The reason for restriction of some deposit types to old rocks is not clear, but may be due to reduction in the mantle's sulfur content and/or decreases in the geothermal gradient during the earth's evolution. Sulfur isotopes show that external, nonmagmatic sulfur was important in forming the younger deposits at Duluth and Noril'sk. Rift-related deposits are not common but many favorable areas exist and further exploration, particularly in areas where country-rock sulfur could be assimilated, may uncover other deposits of this type.

2.2 Laterite Deposits

Nickel-bearing lateritic soils that form from the deep tropical weathering of ultramafic rocks represent the second economically important class of nickel deposits. During chemical weathering, the 0.2–0.3% nickel held in the olivine crystal structure is released and is moved by ground water until increases in pH cause it to be fixed, either in hydrous iron oxides (mostly goethite, Fe(OH)), or in solid solution with a variety of secondary silicates composed mostly of serpentine and clay-group minerals. Profiles of ideal laterites have a layer of iron-rich or oxide ore that contains 0.5–1.5% Ni above a thinner zone of silicate-rich or garnierite ore that usually has 1.5–3% Ni and which rests on unweathered bedrock. However, many deposits are made up mostly of oxide ore and lack significant amounts of the richer silicates. Laterite deposits are typically 2–25 m thick and may contain minor amounts of cobalt and chromium as unweathered grains of chromite.

The formation of laterites by deep chemical weathering virtually restricts their occurrence to tropical climates. Thus, most deposits occur between 30°N and 30°S latitude. Important deposits occur in New Caledonia, the Philippines, Indonesia, Cuba and the Dominican Republic. Fossil laterites are rare, but economic deposits occur in the Ural Mountains (USSR) and in Greece.

2.3 Hydrothermal Deposits

Nickel, copper and cobalt occur either as sulfides or as arsenide in vein deposits. Although some deposits of this type were important when nickel was first being mined, most are no longer of economic interest.

2.4 Manganese Nodules

The slow precipitation of metals from seawater in the deep ocean results in 2–10 cm nodules that may contain about 25% Mn, 1.5% Ni, 1.5% Cu and 0.2% Co (see *Ocean Resources and Mining*). Locally, extensive blankets of residual deposits occur, the most economically interesting of these being in the north central Pacific between 0° and 20°N latitude. Although large amounts of nickel are present in some deposits, technological and legal problems have prevented their development.

3. *Production of Metals*

Sulfide ores occur mostly as steeply dipping deposits which require extraction by difficult and costly underground "hard-rock" mining techniques. However, once mined, sulfide ores are readily upgraded by flotation and magnetic separation. Most ore is then oxidized by roasting and smelted to separate an iron-rich slag from copper–iron–nickel matte, which may then be electrolytically refined.

In contrast, the surficial character of laterite deposits make them relatively easy to mine by stripping. There is, however, no effective means to concentrate and upgrade the ore, with the result that expensive and complicated pyrometallurgical and hydrometallurgical processes are often required to refine the mineralogically and chemically diverse oxide and silicate ores. Some ores may be smelted to make nickel matte or ferronickel, while nickel in other ores is concentrated by leaching in ammonia or sulfuric acid solutions.

World primary nickel production is about 700.000 t. It is expected to grow at a rate of 2.5% per year, and be between 1.1 and 1.6 Mt by the end of the century. Canada is still the largest producer, but the production from other countries has become increasingly important and has also caused a shift in production from sulfide ores which produced 95% of the nickel in 1950 to laterites which currently contribute about 35%.

4. *Ore Availability*

The amount of nickel available for production under current economic conditions is estimated to be about 50 Mt. Land-based resources, which include both currently economic material and all other material in the earth's crust which may be usable, are estimated at about 168 Mt of nickel. These resources are adequate to supply projected demand for the next 50 years. Additionally, large amounts of nickel, estimated at about 760 Mt, are in deep-sea nodules. Thus, compared with many other metals, nickel has a large resource base which should be adequate for foreseeable future demand.

In the future, the increased mining of laterite ores will continue to shift production away from sulfide deposits. Sulfide ores, however, will still be important and significant discoveries are possible, particularly for the komatiitic and Noril'sk type deposits. The short-term adequacy of land-based resources will slow development of the ocean-floor nodules but, pending resolution of legal and technological problems, they will eventually become a major source of nickel.

Bibliography

Boldt, J R, Queneau P 1967 *The Winning of Nickel: Its Geology, Mining and Extractive Metallurgy.* Van Nostrand, Princeton, New Jersey

Cornwall H R 1973 Nickel. In: Brobst D A, Pratt W P (eds.) 1973 *United States Mineral Resources*, US Geological Survey Professional Paper No. 820. US Government Printing Office, Washington, DC, pp. 437–42

Evans D I, Shoemaker R S, Veltman H (eds.) 1979 *International Laterite Symposium.* Society of Mining Engineers (AIME), New York.

Golightly J P 1981 Nickeliferous laterite deposits. *Economic Geology 75th Anniversary Volume.* Economic Geology, Lancaster, Pennsylvania, pp. 710–35

Peters H D 1975 *Nickel.* Glückauf, Essen

Sibley S F 1984 Nickel, *Minerals Yearbook 1983*, Vol. 1. US Bureau of Mines, Washington, DC, pp. 631–45

Valdrett A J 1981 Nickel sulfide deposits: Classification, composition and genesis, *Economic Geology 75th Anniversary Volume.* Economic Geology, Lancaster, pennsylvania, pp. 628–85.

M. P. Foose

[US Geological Survey, Reston, Virginia, USA]

Niobium and Tantalum Resources

Mineral deposits of niobium (columbium) and tantalum tend to show a strong geological relationship that parallels the close chemical ties of these two elements. These relationships are described as part of a discussion of the minerals principally responsible for supply of niobium and tantalum. Chief sources, nature and adequacy of supply are given.

Niobium and tantalum have strong geochemical coherence. Rocks and minerals containing one of these elements commonly also have at least a small content of the other. The niobium to tantalum ratio theoretically can be infinitely variable, as in the two important isomorphic mineral series, pyrochlore–microlite and columbite–tantalite. Ores of niobium and tantalum are found chiefly in alkalic complexes (principally carbonatites), pegmatites, placers and residual deposits. Niobium and tantalum are almost always found in nature as oxides in association with other minerals, but not in elemental form, nor as sulfides. Niobium is the more plentiful; overall crustal abundance has been estimated as 20 ppm for niobium and 2 ppm for tantalum.

Pyrochlore, $(Na, Ca, Ce)_2 (Nb, Ti, Ta)_2 (O, OH, F)_7$, and pandaite, its barium analogue, have become the main sources of niobium. Both minerals are quite low in tantalum, the niobium oxide to tantalum oxide ratio being 200:1 or greater. These minerals are commonly found in interior parts of alkalic rock complexes, frequently in association with minerals of such other elements as titanium, thorium, uranium and the rare earths. In Brazil, where extensive deposits averaging over 2% Nb_2O_5 are being mined, the occurrences are in eluvial deposits resulting from in situ weathering of syenite–carbonatite rocks, leaving an enriched concentration of magnetite, apatite and pyrochlore. In Canada, where a deposit with an average Nb_2O_5 content of about 0.7% is being worked, the occurrences are in complex ring structures of carbonatite and alkalic rocks in the Precambrian Shield.

Columbite–tantalite, $(Fe, Mn) (Nb, Ta)_2O_6$, was the principal mineral source of both niobium and tantalum prior to development of pyrochlore deposits in Brazil and Canada. In the columbite–tantalite isomorphous series, substitution is mutual for the pairs niobium–tantalum and iron–manganese. The proper term for the mineral is columbite when niobium predominates over tantalum; when the reverse is true, the proper term is tantalite. Columbite and tantalite are normally found in pegmatites and in alkali-rich, high-silica granites. Columbite–tantalite deposits are known to exist in all continents, but most deposits with relatively high niobium or tantalum content are small and erratically distributed. In many cases, economic mineral concentrations have been produced by weathering of pegmatites and formation of residual or placer deposits. Most columbite and tantalite has been obtained as a by-product of mining for other commodities, mainly tin (cassiterite), as exemplified by recovery of columbite from the Jos Plateau in Nigeria.

Tantalum supply is highly dependent upon tin mining and smelting, tantalite being coproduced with cassiterite, and tantalum- (and niobium-) bearing slags being obtained from smelting certain tin ores. Notable instances of mining principally for tantalum, in Australia and Canada, are at mines where average Ta_2O_5 grades are about 0.06% and 0.15%, respectively. Elsewhere, average grades are less. A relatively small part of tantalum supply is obtained as microlite, the tantalum-rich member of the pyrochlore–microlite series. The tantalum oxide to niobium oxide ratio is around 10:1 in microlite, which is found not in carbonatites but in pegmatites, mainly in Mozambique and Brazil (where it is referred to as djalmaite). Elements other than tin frequently associated with tantalum deposits are lithium and beryllium; minerals such as spodumene, lepidolite, beryl, pollucite and mica may be coproduced with tantalite. In recent years about as much tantalum has been recovered from tin slags as has been obtained from tantalum concentrates. The most important tantalum-bearing tin slags have been produced in southeast Asia. Slags from Thailand have been the highest in grade (about 12% Ta_2O_5 and 8% Nb_2O_5), and have supplied the greatest quantity of tantalum.

Niobium and tantalum ores are beneficiated into concentrates by various means. Flotation and magnetic separation are the primary methods used with pyrochlore ores, whereas columbite and tantalite concentrates are produced mainly by methods characteristically used to treat ores of heavy minerals (wet and dry gravity techniques, electrostatic and electromagnetic separation). Typical mineral concentrates have about a 60% pentoxide content, which is in the form of Nb_2O_5 plus Ta_2O_5 when recovering pyrochlore concentrate, and is Nb_2O_5 when recovering columbite and tantalite concentrates.

Reserves of niobium and tantalum worldwide, exclusive of those of the USSR and China for which data are speculative, are assessed on a metal content basis currently at about 7.6 million tonnes for niobium and 48 000 tonnes for tantalum (USBM 1988). With the same exclusion, additional resources are estimated both for niobium and tantalum to be over three times as large as reserves. Reserves of niobium are large enough relative to projected demand that presently defined ore supplies are expected to last well into the twenty-first century. Reserves and resources of niobium are located mainly in Brazil; Canada ranks as the second most important source. Tantalum reserves are distributed amongst Australia, Brazil, Canada, Malaysia, Thailand and various African nations (principally Nigeria and Zaire). Reserve and resource estimation for tantalum is complicated by price volatility and uncertainty as to the potential of major tin-producing areas.

Bibliography

Bering D, Eschnauer H 1978 Tantal-vorstoff und -vorkommen [Tantalum raw materials and occurrences], *Erzmetall* 31: 194–98. [Condensed English version: *Met. Bull. Mon.* April 1978: 43, 45, 47–48]

Krausz U, Schmidt H, Kippenberger C, Eggert P, Priem J, Wetting E 1982 *Investigation of Supply and Demand of Mineral Raw Materials*, XVI, *Niobium* (in German). Bundesanstalt für Geowissenschaften und Rohstoffe and Deutsches Institut für Wirtschaftsforschung, Hannover and Berlin

Krausz U, Schmidt H. Kippenberger C, Eggert P, Kamphausen D, Priem J, Wettig E 1982 *Investigation of Supply and Demand of Mineral Raw Materials*, XVII. *Tantalum* (in German). Bundesanstalt für Geowissenschaften und Rohstoffe and Deutsches Institut für Wirtschaftsforschung, Hannover and Berlin.

Manker E A 1977 The outlook for columbium in steel alloying. *Proc. Met. Bull 1st Int. Ferro-Alloys Conf.* Metal Bulletin Ltd, London, pp. 54–57

Manker E A, Perrault G 1981 Geology and mineralogy of niobium deposits. *Int. Symp. Niobium 81*, San Francisco, California, Nov. 9–11

Minerals Yearbook (annual) Columbium and tantalum. US Bureau of Mines, Washington, DC

National Materials Advisory Board 1982 *Tantalum and Columbium Supply and Demand Outlook*, NMAB-391, National Academy of Sciences, Washington, DC

Parker R L, Adams J W 1973 Niobium (columbium) and tantalum. In: Brobst D A, Pratt W P (eds.) 1973 *United States Mineral Resources*, US Geological Survey Professional Paper 820. US Government Printing Office, Washington, DC, pp. 443–54

Parker R L, Fleischer M 1968 *Geochemistry of Niobium and Tantalum*, US Geological Survey Professional Paper 612. US Government Printing Office, Washington, DC, pp. 1–40

Taylor R G 1979 *Geology of Tin Deposits*. Elsevier, New York, pp. 381–88, 425–31, 504–10

US Bureau of Mines 1988 *Mineral Commodity Summaries 1988*. USBM, Washington, DC

T. S. Jones
[US Bureau of Mines, Washington, DC, USA]

O

Ocean Resources and Mining

The mineral resources of the sea are conventionally divided into two broad categories: deposits associated with the continental margins and the coastal zone, and deposits associated with the deep sea. This convention is based on both geological and economic factors. Some deposits occur in only one of these zones. For example, placer deposits occur only in the coastal zone and along continental margins, and metalliferous sediments are found only in deep-sea environments. Other types of deposits, such as sand and gravel, occur throughout the marine environment, but it is not economically feasible to extract these resources from the deep sea. In fact, serious economic interest in exploiting deep-sea mineral deposits has only existed since about 1960. This period has witnessed the development and widespread acceptance of the theory of seafloor spreading or plate tectonics. The ideas embodied in this theory have greatly enhanced the understanding of the origin, relative age and distribution of many types of marine deposits. With a few notable exceptions, the marine mined deposits discussed below are best described as potential resources which play little or no role in today's mineral market. The realization of this potential, and hence the future of ocean mining, will be determined by the relative economic accessibility of marine minerals compared to rival sources on land, as well as by other legal and environmental considerations.

1. Continental Margin Resources

The continental margin includes the coastal zone, the continental shelf, the continental slope and, if present, the continental rise or continental borderland. It is equal in extent to about 20% of the sea area or 50% of the land area of the world. The continental margin is the transition zone between the relatively thick granitic crust of the continents and the relatively thin basaltic crust of the ocean basins. Both the continental shelf and continental rise are typically covered with sediment accumulations several kilometers thick.

1.1 Oil and Gas

Oil and gas are now the most valuable resources extracted from the sea because of their critical role in meeting the world's energy needs. The physicochemical mechanism by which organic matter is preserved, buried, converted to oil or gas and concentrated in a deposit is not completely understood, although many key limiting factors are known. Oil and gas are generated from the remains of organisms, mostly plants, which accumulate in marine sediments. Most organic matter is destroyed by fairly rapid oxidation following deposition—a process which precludes the formation of crude oil. Preservation of the organic matter is enhanced if it settles into an oxygen-poor environment, or if the sedimentation rate is high enough to bury it before it becomes completely oxidized. The organic-rich sediments must then be buried to a depth where they experience a temperature range suitable for oil generation. If burial is too shallow, the temperature will be insufficient for conversion to petroleum, whereas if burial is too deep, the higher temperatures will crack the oil to form gas or, at a later stage, carbon. Finally, any crude oil formed must encounter some kind of geological trap which restricts seepage and allows accumulation of economic quantities. Analysis of known offshore oil fields suggests that these various conditions are most common along continental margins containing relatively undisturbed sequences of Mesozoic and Tertiary sediments a few kilometers thick.

Extensive offshore drilling for petroleum hydrocarbons has been conducted since 1950, with major discoveries occurring, for example, in the Persian Gulf, the North Sea and the Gulf of Mexico. Industry has developed semisubmersible systems capable of drilling offshore exploration wells in more than 2000 m of water, although economic considerations generally limit production wells to water depths of 200–300 m. Offshore hydrocarbon deposits presently account for nearly one third of the world's estimated oil and gas resources and generate about US$80 billion in annual revenues. This amount is greater than the combined value of all marine biological resources, and is roughly equivalent to the estimated value of all other potential seabed resources combined.

1.2 Aggregates

After hydrocarbons, the most economically important resources now being recovered from the seafloor are aggregates of sand, gravel and shell debris (limestone). These deposits are composed of material which has been eroded from the adjacent land mass or transported shoreward from outer shelf areas and then reworked and concentrated by wave and current action. Annual offshore aggregate production accounts for some US$140 million, or about half of the worlds offshore nonfuel mineral production. Offshore mining operations are generally located close to the market because the transportation costs associated with bulk minerals are usually the limiting economic factor. Onshore mining of modern beaches would certainly reduce recovery costs, but would also often conflict with other uses of the coastal zone and could lead to accelerated coastal erosion.

Sands and gravels are mined extensively in Japan, the USA and the UK for use as building materials. Limestone for use in agriculture and cement manufacture is currently recovered from offshore shell beds of the Bahama Banks and Iceland, and in the Gulf of Mexico. Historically, tropical island communities have recovered calcium carbonate from modern corals, although this practice has led to the destruction of coral reefs in Hawaii and Fiji. The principal method of recovering aggregate minerals from shallow-water deposits is by dredging. Suction dredges employing both forward and trailing suction pipes have been used extensively in the recovery of sand and gravel. The leading edge of the suction pipe is usually equipped with a cutter head or high-pressure water jet to break up bottom material before it is sucked into the pipe.

1.3 Placer Deposits

Beach and stream deposits are composed primarily of detritus from eroding continental rock which has an average specific gravity of about 2.7. The mechanical reworking of these sediments by waves and currents can result in localized concentrations of the denser fraction of these mineral grains known as placer deposits. Placer deposits that are currently forming usually occur in the nearshore zone since the relatively high density of the placer minerals limits the distance they are transported to within a few kilometers of the source area. However, changes in sea level associated with the formation and melting of glaciers during the Pleistocene epoch resulted in significant lateral migrations in the position of the shoreline. Thus placer deposits are found not only on modern beaches, but also on ancient raised beaches on emerging coastlines and on submerged beaches and drowned stream channels extending out on the continental shelf.

Several economically important minerals have been recovered from placer deposits, including diamonds from beach sands of southwest Africa, alluvial gold and platinum from Alaska, magnetite sands beneath Ariake Bay off southern Japan, and ilmenite, rutile, zircon and monazite from beach sands in Queensland and New South Wales, Australia. Mining of cassiterite deposits in the "tin belt" of Southeast Asia accounts for 14% of total world tin production and generates revenues of about US$185 million for Indonesia, Thailand and Malaysia. The deposits in this region cover an extensive area, and despite strict government controls on offshore mining, many local fishing boats are also rigged to conduct illegal mining operations.

Because of their accessibility, high-value placer deposits have often been recovered with minimal personnel and the crudest of equipment. For example, individual prospectors, equipped with little more than gold pans and shovels, were responsible for the "gold rush" periods in California and Alaska. On a commercial scale, suction dredges, bucket dredges and grab dredges have all been used to mine tin from water depths of about 50 m off Indonesia and Thailand. The choice of a specific dredge design is largely governed by the location of the deposit. For example, dredges which operate off exposed coastlines must have sophisticated mooring systems and hydraulic buffer mechanisms to counteract the effect of sea swell which causes the digging point to move in relation to the seabed.

1.4 Sulfur

Most of the world's sulfur supply is derived from natural gas. However, the demand for sulfur for use in various industrial processes and in the manufacture of fertilizer is expected to continue well beyond the point where natural gas supplies are exhausted. Since about 1960 the Frasch process has been used to recover elemental sulfur from the cap rock of salt domes in the Gulf of Mexico. This involves pumping superheated water and air into the cap rock, thus causing the sulfur to melt. The sulfur–water solution is then pumped back to the surface. Commercial amounts of sulfur have been found in about 5% of the salt domes drilled in the Gulf of Mexico, although present offshore production is limited to a single operation off the coast of Louisiana. The future of offshore sulfur production may well be influenced by our ability to economically recover significant amounts of waste sulfur from pollution-control equipment.

1.5 Phosphorites

The element phosphorus is an essential nutrient for plant growth and thus phosphates are key ingredients in the manufacture of fertilizers for agriculture. Various types of marine phosphorite deposits are ubiquitous both temporally and spatially in the geologic record. Continental margin phosphorites are most important among these, inasmuch as they provide about 75% of the world's phosphate. Many details of the genesis of the continental margin deposits are not yet known, although it is widely accepted that they formed in areas of enhanced biological activity associated with ocean zones of upwelling. Other categories of marine phosphorites include: condensed sequences which form at depths of 100–1000 m on major plateaux and rises (e.g., the Chatham Rise off New Zealand); insular deposits found on some Pacific islands; seamount deposits in the Pacific; and epicontinental sea deposits, for which there are no known examples forming in modern marine environments. Both condensed sequence deposits and seamount deposits apparently form through the replacement of carbonate by phosphate on the surfaces of calcareous sediments during long periods of nondeposition, while insular deposits form through the chemical reaction between avian guano and the underlying limestone bedrock.

Most of the world's phosphate supply is recovered from subaerially exposed continental margin deposits using conventional land-based mining techniques such as dredges, scrapers, cutters and hydraulic slurry mining. Adaptations of these techniques can be used in the extraction of offshore phosphate deposits if market conditions become favorable for marine mining. Whether or not such conditions will occur is debatable. Fluctuating phosphate prices, the high cost and long lead-time required for offshore mining operations, and the existence of extensive on-land phosphorite resources will likely forestall any significant offshore phosphate mining until at least the end of the century. Beyond that, other factors may become important, especially changing patterns of land use and environmental considerations in developed countries now involved in the large-scale export of both fertilizers and agricultural products. Offshore mining of low-grade phosphate ore may also become a viable option for developing nations unable to purchase processed fertilizers on the world market, yet faced with rapid population growth and severe food shortages. These problems are exacerbated in many developing nations because of low agricultural production rates caused by highly leached tropical soils. In such areas the direct application of low-grade phosphate ores would actually promote soil conservation and eliminate the need to build costly chemical treatment plants for refining the ore.

1.6 Seawater

Seawater itself is considered a valuable marine resource in many parts of the world. The recovery of fresh water from seawater is documented at least as far back as Roman times and still continues in many arid regions, notably the Persian Gulf. This is usually accomplished by a multistage distillation process, although solar evaporation, freezing processes and electrodialysis techniques have also been used. Sea salt contains numerous chemical elements, but few are present in sufficient concentrations for economic recovery. Only bromine, magnesium and sodium chloride are now recovered on a commercial basis, along with certain potassium and calcium compounds as minor by-products. Seawater is the major source of the world's bromine supply. Processing plants in coastal areas of the USA and the UK recover bromine and magnesium by direct precipitation from seawater. Sodium chloride and related salts are recovered by an evaporation process.

There is a long and often colorful history of attempts to recover other elements dissolved in seawater. Notable among these was that of Fritz Haber, the famous chemist, who tried to extract gold from seawater to help pay Germany's debt following World War I. This and subsequent attempts to recover rare or precious metals from seawater have failed due to the enormous technological problems involved. For example, the world's oceans contain some 10 million tonnes of gold, at a value of US$500 per ounce; however, approximately 700 tonnes of seawater would have to be processed to recover just one cent's worth of gold.

1.7 Other Continental Margin Resources

In some areas it has been profitable to extend land-based mining operations into the bedrock underlying the continental margin. This has occurred with bauxite, iron and barite deposits in Australia, Finland and Alaska, respectively, and with coal mining in Japan, the UK and Newfoundland.

2. Deep-Sea Resources

The deep sea includes the major ocean basins and the rises, ridges and trenches which form their boundaries. The most-important resources found in the deep sea are ferromanganese nodules and metalliferous sediments.

2.1 Ferromanganese Nodules and Crusts

The greatest interest in deep-sea mining has centered around the recovery of ferromanganese nodules. Extensive ferromanganese nodule fields were discovered in all of the major ocean basins during the HMS *Challenger* expedition (1872–1876). The nodules are brownish-black concretions, 1–20 cm in diameter, and are primarily composed of hydrated oxides of iron and manganese. Typical nodules contain 5–15% iron and 10–30% manganese. However, the principal economic interest lies in extracting the smaller amounts (1–3%) of much more valuable metals contained in the nodules; mainly copper, nickel and cobalt. In contrast to discrete nodules, ferromanganese deposits also occur as black, semicontinuous crusts, often found on the flanks of seamounts. For example, recent research cruises have discovered that significant portions of certain seamounts and ridges near Hawaii and Johnson Island are covered with cobalt-rich crusts ranging in thickness from 2 to 4 cm. Preliminary analyses suggest the crusts form under different chemical conditions and thus exhibit a different mineralogy than that of deep-sea nodules. The crusts apparently derive their constituent trace elements directly from seawater and, compared to nodules, typically contain greater amounts of cobalt and lesser amounts of copper.

Ferromanganese nodules form in situ and commonly grow around a nucleus such as a shark's tooth, a shell fragment or a small rock. The size, shape, mineralogy and geochemistry of the nodules vary considerably both within and between the major ocean basins. These differences have caused considerable scientific debate concerning the mechanism of nodule formation and, in particular, the origin of the various trace elements which are enriched in the nodules. Different theories have been proposed suggesting that the trace elements are derived principally from continental runoff, submarine volcanic activity or diagenetic recycling of elements in the sediment column. Recent evidence indicates that nodules may acquire trace elements from any one or more of these sources, depending on local conditions.

One puzzling aspect of nodule distribution is that they are far more abundant at the sediment surface than at depth within the sediment column. Since the sedimentation rate generally exceeds the nodule growth rate, there has been much speculation concerning how the nodules manage to avoid being buried. Many believe that this is the result of the nodules being rolled along the sediment surface by bottom currents, or else being constantly nudged to the surface by bottom organisms. It has also been noted that pulsations in bottom water flow during the geological past have led to the erosion of large amounts of bottom sediments. Newly formed nodules would remain at the surface during these periods, and previously buried nodules could be exposed. In the Pacific Ocean, at least, there appears to be a close correlation between the location of these erosion surfaces and the distribution of nodule fields.

The richest nodules (i.e., those containing the greatest amounts of copper, nickel and cobalt) occur in a rectangular belt in the north Pacific Ocean between 6–20 °N and 110–180 °W. It is estimated that the Pacific Ocean alone contains some 1.5×10^9 kt of nodules—enough to supply critical metals for several hundred years if the nodules can be recovered economically. To date, five international consortia have invested more than US$650 million (in constant 1982 dollars) in the development of technologies for the exploration and evaluation of different nodule fields, and in the design of nodule recovery systems. Two methods of nodule recovery are under development, a continuous line bucket (CLB) system and a hydraulic system. The CLB system consists of dredge buckets attached to a long cable which is positioned in a loop arrangement between a surface ship and the seafloor. The surface ship will drag the loop along the bottom while simultaneously rotating the cable to recover the nodules captured in the buckets. The hydraulic system involves using either air or water to pump a mixture of seawater, nodules and associated sediments up a long flexible pipe extending to the seafloor from a surface ship. To be profitable these systems must be capable of recovering 5–15 kt of nodules per day from water depths of 5–7 km, often under hostile surface-sea conditions. Both systems have been successfully tested—a remarkable engineering achievement considering the myriad technological problems which must be solved. However, because of declining metals prices and legal uncertainties about the right to mine in international waters, neither system has been deployed on a commercial basis, and industrial investment in nodule recovery programs has declined sharply since the peak year of 1979.

2.2 Metalliferous Sediments

Metalliferous sediments enriched in iron, manganese and several other elements are forming at several locations in the deep sea. These deposits have been discovered along the East Pacific Rise and the mid-Atlantic Ridge, in the Gulf of Aden, the Gulf of California and the Red Sea, and around some volcanic islands in the Mediterranean Sea. Many of the metals associated with these sediments are derived from the hydrothermal leaching of the basaltic oceanic crust by circulating seawater under elevated temperatures. These hydrothermal solutions may also include contributions from deep-seated magmatic sources, since some of the elements enriched in the sediments are among those concentrated in the late stages of magmatic crystallization.

The metalliferous sediments believed to have the greatest economic potential are found in the Atlantis II Deep of the Median Valley of the Red Sea. The in situ value of these deposits has been estimated at over US$2.3 billion based on 1972 prices. The Red Sea deposits occur as sulfides, oxides and silicates with the sulfide fraction containing about 15% iron, 11% zinc, 2.2% copper and enriched concentrations of several other transition series metals such as silver, nickel, cobalt and vanadium. These deposits have attracted considerable scientific attention inasmuch as they represent an orebody which is still in the process of formation. Thus information gathered here should enhance our understanding of the formation and location of similar, but much older, deposits found on land. The Atlantis II Deep deposits form when normal Red Sea water percolates through and dissolves subsurface evaporite deposits, thereby attaining a high salinity. This brine then comes into contact with recent volcanic rocks, where it is heated to temperatures in excess of 200 °C. Chemical leaching of the volcanic rocks by the hot brine leads to the enrichment of several metals in the brine. Finally, convective circulation causes the brine to discharge onto the Red Sea floor and precipitate the dissolved metal load in the form of sulfides, oxides and silicates.

Prior to the recent decline in metal prices, the governments of Sudan and Saudi Arabia were negotiating with a German mining company to exploit the Red Sea deposits. The proposed recovery method was based on suction pumping, in which a vibration-cutter head with water jets attached to the end of the suction pipe would be used to mobilize and dilute the mud.

3. Seafloor Spreading, Ocean-Basin Evolution and Mineral Resources

The most significant development in the earth sciences in the present century has been the formulation and refinement of the theory of seafloor spreading or plate tectonics. According to this theory the earth's crust consists of several large plates which are moving relative to one another at a rate of a few centimeters per year. This movement is produced when new igneous rock rises by convection from the earth's interior and is added to the surface, primarily along the midocean ridge system which encircles the world. As the volcanic material cools and solidifies it forms a new portion of the seafloor. This process is continuous, and thus subsequent addition of still younger material causes the seafloor to spread apart along either side of the midocean ridge system. Since the surface of the earth is not continuously expanding, the addition of new seafloor must be compensated by the destruction of an equivalent area of crust elsewhere on the earth's surface. This occurs in zones of crustal weakness, usually oceanic trenches, where crustal plates generated at different spreading centers collide and subduction occurs (i.e., one plate is thrust under another).

The theory of seafloor spreading provides a general conceptual framework for understanding the various stages of the evolution of new ocean basins. Certain of these stages, in turn, are especially relevant to mineral-forming processes on the seafloor. The origin, relative age, and distribution of many seemingly unrelated marine mineral deposits can now be viewed as part of a single worldwide process. The initial stage of ocean-basin evolution is characterized by tensional faulting within a continental mass which produces a long, narrow graben valley. As the valley widens its floor subsides and becomes flooded with seawater, forming a narrow, shallow sea. Eventually the valley floor of continental crust is torn apart and replaced by basaltic oceanic crust as seafloor spreading begins. During this early stage a combination of volcanic activity, hydrothermal leaching of basalts and the discharge of resulting solutions in the narrow rift valley produces a complex sequence of metalliferous deposits. The best-known example of this early stage of ocean-basin evolution is the Red Sea, where new seafloor and associated hydrothermal deposits are being formed by the extrusion of fresh basalts in the central rift zone. Continued seafloor spreading causes the ocean basins to widen and deepen as the older seafloor moves away from the spreading center, cools and subsides. During this mature stage the ocean basins become wide enough so that well-developed surface and bottom current systems are established within the basins. The present day Atlantic, Pacific and Indian Ocean basins have all reached this point. The establishment of these current systems is an essential prerequisite for the formation of both manganese nodules and phosphorite deposits. Manganese nodule formation requires both the highly oxidizing conditions associated with free bottom water circulation and the relatively low sedimentation rates characteristic of wide ocean basins. The richest and most extensive nodule fields occur in the Pacific Ocean basin, far from the influence and diluting effect of continental runoff. Phosphorite deposits, on the other hand, form primarily as a result of strong upwelling currents which stimulate biological productivity and recirculating nutrients associated with continental runoff back into the surface waters. This type of circulation pattern occurs only along the eastern margins of mature ocean basins, and hence phosphorite deposits are observed to be forming at present off Peru, Chile and southwest Africa.

Portions of some continents occur well within rather than at the edges of the moving crustal plates. This is true, for example, of the eastern margins of North and South America and western margins of Europe and Africa. These areas have not been subducted during recent geological time and, because of this, sediments have been accumulating along these continental margins for at least 200 million years. These extremely thick and relatively undisturbed sediment sequences provide many of the conditions necessary for the formation and preservation of hydrocarbons.

4. Economic, Environmental and Legal Aspects of Ocean Mining

The costs of recovering marine minerals are expected to be significantly higher than those associated with land resources. Thus economic projections based on prevailing consumption rates and depletion rates of land resources suggest that most ocean minerals will not be viable economic targets for several decades, and perhaps much longer. Rapid increases in world population growth and the industrialization of developing nations might shorten this time period somewhat, but so too could it be lengthened by new land discoveries and the substitution of cheaper materials (e.g., ceramics for metals). Consequently, when viewed strictly in an economic context, the effort expended thus far on ocean-mining activities by governments, industry and academia seems disproportionate to the nearness of the goal—it appears these institutions have begun a long distance run with a full sprint instead of a measured pace. The reasons for this approach lie in recognizing that profit-and-loss considerations, while undoubtedly important, do not exclusively dictate interest in ocean mining. For example, the oil embargo of the 1970s dramatically illustrated the dependence of the USA, Japan and western European nations on foreign oil imports. These nations are even more dependent on foreign sources for various strategic minerals, and their governments may consider it prudent to subsidize ocean-mining efforts in domestic waters to reduce this dependency. Industrial investment in ocean-mining activities is also influenced by strategic economic considerations, especially when research and development costs can be distributed among several members of a consortium. Companies that initiate pioneering efforts aimed at recovering ocean resources should be in the best position to dominate the market when these resources become viable economic targets. This behavior forces competitors to at least consider similar investments, or else risk exclusion from future markets. The key to discovering mineral deposits often lies in having a thorough understanding of how they form. The formation of any mineral deposit, in turn, implies a combination of physical and chemical circumstances which, by virtue of the rarity of these deposits, represents an anomalous situation within the spectrum of geologic processes. Scientists are motivated to study such aberrations in order to gain a different perspective and an insight into these formation processes which otherwise cannot be obtained. Thus while the rationale behind scientific investigations of mineral deposits is justified on the basis of academic inquiry, the information produced is nevertheless important in an economic sense.

Concerns about the environmental impact of land mining as well as the pressures created by changing patterns of land use may ultimately serve to accelerate a shift to ocean mining. But it is important to recognize that these same considerations often will exist for ocean-mining operations. Nearshore mining will likely conflict with a variety of other possible uses of the coastal zone. Moreover, the continental shelves are the most productive areas of the world ocean. They are the habitat of economically important benthic organisms, the nursery areas of many fish species, and are also where the bulk of the world's fishing catch is obtained. Future marine miners will be required to make special provisions to minimize pollution of the marine environment and to ensure that the affected area is reinhabitable by marine species.

Future offshore mining operations will require a regulatory framework for determining rights of ownership, issuing concessions, establishing environmental impact restrictions, and other legal matters. Many nations have anticipated this situation by declaring exclusive jurisdiction over access to seabed resources that occur within 200 nautical miles of their coastlines. The areas included in these declarations are known as exclusive economic zones (EEZs). Assuming that these declarations are honored by the international community, legal access to continental margin resources will be governed by whatever nation has claimed the EEZ in which these resources occur. Two different legal systems have emerged for deep-sea resources. Most nations have accepted the legal provisions pertaining to deep-sea resources included in the Law of the Sea (LOS) Treaty recently completed under the auspices of the United Nations. However, mainly because of strong objections to these same provisions, the USA, the UK and West Germany have refused to sign the LOS Treaty. In 1984 these three nations, along with LOS signatories Belgium, France, Italy, Japan and the Netherlands signed a Provisional Understanding Regarding Deep Seabed Matters (PU) which establishes a legal system of reciprocating agreements based on their own domestic mining laws. Significant differences exist between the LOS and PU systems, but to date no nation has forced a "test case" by attempting to conduct deep-sea mining under either system.

Bibliography

Broadus J M 1987 Seabed minerals. *Science* 235: 853–60

Cronan D S 1980 *Underwater Minerals*. Academic Press, New York

Kent P E 1981 *Minerals from the Marine Environment*. Wiley, New York

Ross D A 1978 *Opportunities and Uses of the Ocean*. Springer, New York

Teleki P G, Dobson M R, Moore J R, von Stackelberg U (eds.) 1986 *Marine Minerals: Advances in Research and Resources Assessment*, NATO ASI Series C: Mathematical and Physical Sciences Vol. 194. Reidel, Boston, Massachusetts

R. M. Owen
[University of Michigan, Ann Arbor, Michigan, USA]

Open-Cut Mining

Surface mining, comprising open-cut or open-pit mining, and the related methods of strip mining, quarrying and placer mining account for approximately 60% of the world's metallic mineral production. Open-cut mining is used in mineral deposits of any size that occur close to the surface; the dominance of mineral production from open-pit mines is, however, related to large and continuous deposits in which highly mechanized and low-cost operations can be used on a large scale. In the mining of low-grade mineral materials such as disseminated copper, the resulting pit can be huge. At Bingham Canyon, Utah, an entire mountain has been mined away and changed into an excavation 4 km long, 3 km wide and nearly 1 km deep; daily ore production at Bingham Canyon is on the order of 100 kt, accompanied by the removal of approximately three times as much waste rock.

Small open-cut mines are begun with a relatively low capital investment, and the cost of mining per tonne of ore is low compared to that of small underground mines. Large open-cut mines are, on the other hand, capital-intensive, and they require several times the preproduction investment of a relatively large underground mine. Nonetheless, the production from a large open-cut mine is so much higher per worker-shift and the mining cost so much lower that orebodies can be mined at a grade that would not otherwise be economic.

In addition to higher productivity and lower operating costs, open-cut mines have advantages over underground mines with respect to simpler design, an easier and more flexible concentration of operations, projectability of geological conditions, better recovery of ore and greater safety.

1. Methods of Development and Advance

The first step in developing an open-cut mine is to expose the orebody by stripping away the overburden.

This material, comprising waste rock and soil, is moved aside or away from the site and placed in dumps so that it will not interfere with further mining operations. Overburden is removed from a series of level benches and connecting ramps, or from gently spiralling benches.

After the orebody is exposed and production has begun, it is necessary to continue the stripping of overburden and adjacent waste rock as the pit is deepened, so that a stable overall slope can be maintained. In orebodies that continue to appreciable depth, an economic limit of open-cut mining is reached when the cost of removing waste rock becomes excessive. The limiting cost depends on the value of the ore and the waste-to-ore (or stripping) ratio. A stripping ratio of 2:1 may define the limiting depth of certain open-cut mines; in mines with higher value ore and relatively lower stripping costs, the limiting ratio may be 20:1 or even higher.

In certain types of copper, uranium and gold deposits, waste rock is not removed at a total loss; ore values below the minimum acceptable grade (cutoff grade) for conventional treatment in milling plants may be recovered in part by dump leaching or heap leaching, in which solutions are percolated through the material.

An important factor in maintaining a favorable stripping ratio is the balance between the maximum recovery of ore in a steep-walled pit and the safe limit to which the walls can be steepened. In closely fractured rock, a working slope (Fig. 1) may have to be kept as low as 20° in order to maintain the stability of pit walls while mining. In especially strong (competent) rock, the working slope may be as high as 70°. The final slope, designed for the ultimate pit at the end of the mining operation, will generally be steeper than the working slope because it is not necessary to maintain benches for further access.

Bench widths in ore and waste are 12–40 m wide, depending upon the operating needs of the loading and hauling equipment. Bench faces, established to be consistent with the stability of the pit walls and the reach of the mining equipment, are at heights of 4–8 m in weak rock and 15–20 m in relatively strong rock. In strong rock with few structural flaws, bench heights of as much as 60 m can be maintained.

The design of an open-cut mine rests heavily upon the depth and economic characteristics of the orebody and upon the geotechnical characteristics of the ore, overburden and waste rock. The design must also take account of groundwater and surface-drainage requirements and the siting of dumps.

Provision for grade control is important in planning an open-pit mining operation. Large deposits often comprise smaller interbedded or irregular zones of ore and waste. In order to maintain a uniform grade of material for milling, active mining areas must be continuously sampled and designated as waste, leaching material and ore.

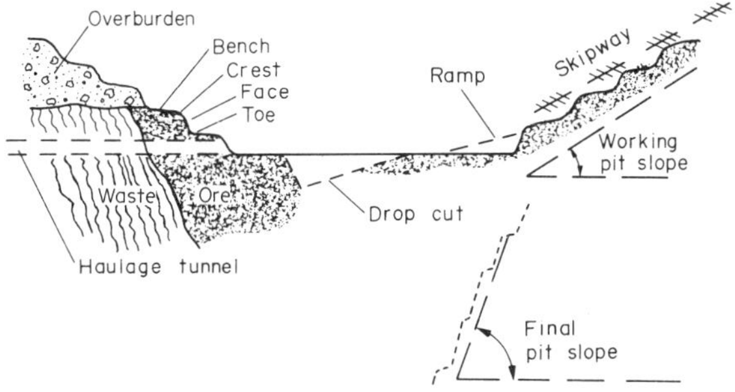

Figure 1
Open-cut mining terminology

2. *Mining Sequence and Machinery*

Ore and waste are removed from pit faces in a drill–blast–load cycle. Stripping of softer overburden and the excavation of new drop cuts into the floor of a pit may also be carried out in a continuous operation using bulldozer-rippers and tractor-scrapers.

In conventional cyclic operations, a pattern of vertical or steeply inclined blast holes is drilled into the bench by rotary or percussion drilling machines, loaded with explosives and detonated. Toe holes—near-horizontal blast holes of smaller diameter—drilled into bench faces are used in some situations. Blast holes are generally charged with ANFO (ammonium nitrate–fuel oil mixture) explosive. Nitroglycerine-type explosives are used in smaller diameter holes and in the secondary blasting of incompletely broken material. The material broken from place by blasting is removed by powershovel, front-end loader or dragline, and loaded into trucks or rail cars.

Although haulage is generally by truck or rail on ramps or spiral roadways climbing to the edge of the pit, some open-cut mines also use skipways or beltways to lift and transport the material. In some open-cut mines, haulage tunnels are used so that material can be delivered to a plant or dump site at lower elevation without first being lifted from the pit; tunnels may be entered directly by pit trains or trucks, or the broken ore can be dropped into underground crushing stations and then transferred to the haulage system in a tunnel that extends beneath the floor of the pit.

Glory-hole mining is a combination of open-cut and underground system. It is similar to open-cut mining with tunnel haulage, except that the term refers more precisely to a steep-walled and narrow pit in a vein or chimney-type orebody; a funnel-like opening or mill hole in the bottom of the pit connects to an underground haulageway.

Strip mining, in which successive strips or narrow

cuts are excavated and the waste material discharged into the mined area, is mainly applied to the production of coal from near-surface seams; flat-lying bedded deposits of other mineral materials such as bauxite, manganese ore, iron ore and phosphate rock are also mined in this way.

Strip mining is generally begun by removing the overburden and the coal or bedded ore in a long and narrow box cut. A parallel strip is then excavated and the overburden is placed into the strip previously mined. As mining continues, the overburden (spoil) from each strip is transferred to the mined area. Strip-mining equipment is commonly very large, with dragline bucket capacities and shovel dipper capacities exceeding $100\,m^3$. A large machine used in strip mining is the hydraulic bucket-wheel excavator, with which relatively soft material can be dug, conveyed and discharged into a stockpile or spoil bank in one continuous operation. Capacities of bucket-wheel excavators range from 1000 to $5000\,m^3\,h^{-1}$.

Contour mining is a type of open-cut mining used in steep hilly areas where the deposit occurs in relatively flat beds. Since only a few cuts can be made before the economic limit in stripping ratio is reached, the mine consists of one bench in a long and narrow strip following a contour line. Overburden is cast onto the adjacent slope or, in much of current practice, hauled along the bench to valley fill sites and to mined portions of the bench.

Quarrying is a term generally applied to the surface mining of dimension stone and crushed stone. The procedures are basically those of open-cut mining, except that the method of loosening the rock by cutting or blasting is dictated by the requisite size and strength of the product.

Placer mining of gravel and sand containing gold, tin, titanium and rare-earth minerals is done by dry-land methods and by dredge. Dry-land placer mining differs very little from shallow open-cut mining and strip mining, except that the gravel banks may also be loosened and removed by streams of water directed under pressure through hydraulic giant (monitor) nozzles. Dredge mining is done by floating equipment in lagoons or ponds excavated for the purpose, in rivers and offshore. Large bucket-line dredges used in mining offshore tin deposits can handle as much as $2000\,t\,h^{-1}$ of gravel, and operate in water depths of as much as 65 m. Suction dredges with submerged pumps and boosters can operate in deeper (90 m) water. Gravel and sand waste material (tailings) from the processing plant on the dredge is pumped or conveyed into the excavation behind the dredge.

Deep-ocean dredging is in the research and development stage. Mining and processing methods for the recovery of copper, nickel and cobalt from manganese nodules on the deep-ocean floor have been tested successfully in large dredging ships with air-lift systems to a depth of 3 km, and with bucket-line systems to a depth of 5 km.

3. Reclamation

A reclamation sequence as well as the mining sequence must be considered in designing open-cut mines. Mine areas and dumps are recontoured and revegetated insofar as possible during mining, as well as at the end of mining. Most governments require that land disturbed by mining must be restored as nearly as is practicable to its original condition.

In strip mining, where the depth of working is relatively shallow, the slope of spoil piles is reduced by bulldozing, soil is replaced and vegetation is planted while successive cuts are being made. Inasmuch as the last cut does not have spoil material available for filling, the cut is reduced to a gentle depression or is made into a long, narrow lake.

Open-cut mines, being deeper than strip mines, cannot be so thoroughly reclaimed. Little if any soil is available to cover the pit slopes, and the waste dumps are situated too far from the pit for practical use in backfilling. The pit area itself may not be reclaimed, but the mine dumps and mill tailings are commonly reclaimed to reduce the effects of erosion and dust. In rehabilitating dumps, the frontal slopes are often reduced and water barriers are placed at intervals. Inasmuch as the dumps do not include soil and subsoil, a blanket of seeds, nutrients, mulch and binder is sprayed onto the surface. When the seeds germinate, a protective layer of vegetation begins to form and a natural growth is permitted to develop.

See also: Mining and Processing: Environmental Problems; Underground Mining

Bibliography

Crawford J T, Hustrulid W A (eds.) 1979 *Open Pit Mine Planning and Design.* American Institute of Mining, Metallurgical and Petroleum Engineers, New York

Given I A (ed.) 1973 *SME Mining Engineering Handbook,* Vol. 2, Sect. 17. Society of Mining Engineers, American Institute of Mining, Metallurgical and Petroleum Engineers New York, pp. 1–180

Kildow J T (ed.) 1979 *Deepsea Mining.* MIT Press, Cambridge, Massachusetts

Macdonald E H 1983 *Alluvial Mining: The Geology, Technology, and Economics of Placers.* Chapman and Hall, London.

Stout K S 1980 *Mining Methods and Equipment.* McGraw–Hill, New York, pp. 119–42

Thomas L J 1977 *An Introduction to Mining.* Halsted Press, New York

W. C. Peters
[Tucson, Arizona, USA]

Ore Minerals

Ore minerals are the large and diverse group of minerals from which metals are extracted. The term ore mineral has also been used as a synonym for opaque mineral, to designate the nontransparent minerals observed in the transmitted and reflected light microscopy study of rocks. In this latter usage, minerals such as pyrite (FeS_2) and pyrrhotite ($Fe_{1-x}S$) are included as ore minerals, although they generally constitute gangue (the waste or economically unimportant minerals in an ore deposit), rather than economically important phases. Usage in the present discussion will be confined to "a mineral that carries the valuable or desired metallic constituent of any ore deposit" (US Bureau of Mines 1968). When the term is defined in this way, it is important to note that economic factors such as metal prices and the problems of mineral extraction and refining determine which minerals are ore minerals at any given time. Thus, for example, the minerals which now are sources of the rare-earth elements were not ore minerals before the rare earths had economic applications. Similarly, minerals which are now considered to be gangue could one day become ore minerals.

Table 1
Metals extracted from the mineral groups of the major ore minerals

Mineral group	Principal metals extracted
Native metals	Au, Ag, Pt, Bi, Cu
Sulfides	Cu, Pb, Zn, Hg, Co, Mo, Ni, Ag, Sb, Cd, Ga, Ge, In, Re, Tl
Oxides	Fe, Mn, Mo, Ta, Al, Ti, Sn, W, U, Nb, V, Zn
Silicates	Be, Li, Zn, Sc, U
Arsenides	As, Pt, Co, Ni
Phosphates	rare earths
Carbonates	rare earths
Tellurides	Au

1. Types of Ore Minerals

The diversity of ore minerals is demonstrated by the range of mineral groups from which metals are presently extracted (see Table 1). Generally, the precious metals occur in the native state, the base metals occur as sulfides, ferrous and ferroalloy metals and aluminum occur as oxides, light metals occur as silicates,

Table 2
Principal ore mineral sources of the major metals

Metal extracted	Principal ore mineral sources
Aluminum	bauxite, boehmite, diaspore, gibbsite
Antimony	stibnite, valentinite, senarmontite, stibiconite, bindheimite, kermesite, tetrahedrite
Arsenic	orpiment, realgar, arsenopyrite, loellingite, skutterudite, nickeline, tennantite, enargite, proustite
Beryllium	beryl, bertrandite, barylite
Cadmium	sphalerite, wurtzite
Cobalt	linnaeite, siegenite, carrollite, cobaltite, safflorite, skutterudite, glaucodot, pentlandite, co-pyrite
Copper	copper, chalcocite, covellite, bonite, chalcopyrite, enargite, digenite, cuprite, malachite
Gallium	sphalerite
Germanium	sphalerite, germanite
Gold	gold, calaverite, sylvanite
Indium	sphalerite, cassiterite
Iron	magnetite, hematite, goethite, siderite, chamosite
Lead	gaena
Lithium	spodumne, lepidolite, etalite, amblygonite
Manganese	pyrolusite, cryptomelane, psilomelane, manganite, braunite, birnessite, rhodochrosite, rhodonite
Mercury	cinnabar
Molybdenum	molybdenite, powellite, wulfenite, ferrimolybdite
Nickel	pentlandite, violarite, siegenite, linnaeite, nickeline, garnierite
Niobium	columbite–tantalite, pyrochlore, perovskite, euxenite, fergusonite
Platinum grop	ferroplatinum, sperrylit, braggite, cooperite, moncheite
Rare-earth group	monazite, bastnaesite, xenotime, apatite, euxenite, loparite, uraninite, brannerite
Rhenium	molybdenite
Scandium	thortveitite, wolframite
Silver	acanthite, silver, electrum, proustite, pyrargyrite, stephanite
Tantalum	columbite–tantalite, perovskite, loparite, euxenite, fergusonite, microlite
Thallium	sphalerite
Tin	cassiterite, stannite
Titanium	rutile, ilmenite, Ti-magnetite, ulvöspinel, leucoxene, titanite (sphene)
Tungsten	huebnerite, wolframite, ferberite, scheelite, powellite
Uranium	uraninite, pitchblende, uranothorite, coffinite, brannerite, carnotite, tyuyamunite, uranophane, thucholite
Vanadium	magnetite, ilmenite, carnotite, coulsonite, patronite, roscoelite, montroseite
Zinc	sphalerite, wurtzite, zincite, franklinite, willemite, smithsonite, hydrozincite, hemimorphite

Table 3
Properties of principal ore minerals

Ore mineral	Formula	Metal extracted	Specific gravity	Hardness VHN	Hardness Mohs	Crystal system	Appearance, remarks
Acanthite	Ag_2S	Ag	7.3	23–26	2–2½	monoclinic	metallic, silvery to dull gray
Amblygonite	$(Li, Na)Al(PO_4)(F, OH)$	Li	3.1		5½–6	triclinic	nonmetallic, white to gray
Apatite	$Ca(PO_4)_3(F, OH, Cl)$	rare earths	3.15–3.2		5	hexagonal	vitreous, pale green to brown
Argentite	Ag_2S	Ag	7.3	23–26	2–2½	isometric	always inverts to acanthite
Arsenopyrite	$FeAsS$	As	6.0–6.2	715–1354	5½–6	monoclinic	metallic, white
Barylite	$BaB_2Si_2O_7$	Be	4.0		7	orthorhombic	vitreous, colorless to white
Bastnaesite	$(Ce, La)CO_3F$	rare earths	4.8–5.2		4–4½	hexagonal	resinous yellow to red–brown
Bauxite	mixture of Al hydroxides	Al	2.0–2.55		1–3		earthy gray to tan
Bertrandite	$Be_4Si_2O_7(OH)_2$	Be	2.6		6–7	orthorhombic	vitreous, colorless to pale yellow
Beryl	$Be_3Al_2Si_6O_{18}$	Be	2.75–2.8		7½–8	hexagonal	vitreous, colorless to green–blue
Bindheimite	$Pb_2Sb_2O_6(O, OH)$	Sb	4.6–7.3		4–4½	isometric	resinous to earthy, yellow to brown
Birnessite	$(Na, Ca)Mn_7O_{14}.3H_2O$	Mn	3.0		1½	hexagonal	nonmetallic, black
Bismuth	Bi	Bi	9.8	15–18	2–2½	hexagonal	metallic, silvery with pink tint
Bismuthinite	Bi_2S_3	Bi	6.8	110–136	2	orthorhombic	metallic, silvery
Boehmite	$AlO(OH)$	Al	3.0		3	orthorhombic	earthy, white to brown
Bornite	Cu_5FeS_4	Cu	5.1	95–105	3	orthorhombic	metallic, bronze tarnishing purple
Braggite	$(Pt, Pd, Ni)S$	Pt group	~10	742–1030	5	tetragonal	metallic, minute steel-gray grains
Brannerite	$(U, Th, Ca, Ce)(Ti, Fe)_2O_6$	U, rare earths	4.2–5.4	690	4½–5½	monoclinic	vitreous to dull, black to green
Braunite	$3Mn_2O_3.MnSiO_3$	Mn	4.7–4.8	1027–1225	6–6½	tetragonal	submetallic, black to steel gray
Calaverite	$AuTe_2$	Au	9.35		2½–3	monoclinic	metallic, brassy to silvery white
Carnotite	$K_2(UO_2)_2(VO_4)_2.3H_2O$	U, V	4.1	33–37	soft	monoclinic	earthy, yellow
Carrollite	$Cu(Co, Ni)_2S_4$	Co, Cu	4.5–4.8	525–592	4½–5½	isometric	metallic, steel gray
Cassiterite	SnO_2	Sn, In	6.8–7.1	1168–1332	6–7	tetragonal	vitreous to adamantine, yellowish to brown, fluorescent
Chalcocite	Cu_2S	Cu	5.5–5.8	84–87	2½–3	monoclinic	metallic, blackish gray to black
Chalcopyrite	$CuFeS_2$	Cu	4.2	187–203	3½–4	tetragonal	metallic, yellow
Chamosite	$(Fe, Mg, Fe)_5Al(Si_3Al)O_{10}(OH, O)_8$	Fe	3.0–3.4		3	monoclinic	nonmetallic, dull green to black
Chromite	$FeCr_2O_4$	Cr	4.6	1332	5½	isometric	submetallic, black
Cinnabar	HgS	Hg	8.1	82–156	2–2½	hexagonal	nonmetallic, brilliant to dull luster, red
Cobaltite	$(Co, Fe, Ni)AsS$	Co, As	6.3	835–1131	5½	isometric	metallic, silvery gray
Coffinite	$U(SiO_4)_{1-x}(OH)_{4x}$	U	5.1	230–302	5–6	tetragonal	dull to adamantine, black
Columbite–tantalite	$(Fe, Mg, Mn)(Nb, Ta)_2O_6$	Nb, Ta	5.1–8.2	240–1021	6–6½	orthorhombic	submetallic to vitreous, black to brown
Cooperite	PtS	Pt group	9.5		4–5	tetragonal	metallic, steel gray
Copper	Cu	Cu	8.9	96–104	2½–3	isometric	metallic, copper-red but tarnishes darker
[illegible]	[illegible]	V	5.2		4½–5	isometric	metallic, bluish gray

Covellite	CuS	Cu	4.7	128–138	$1\frac{1}{2}$–2	hexagonal	nonmetallic, deep blue with red tint
Cryptomelane	KMn_8O_{16}	Mn	4.3		6–$6\frac{1}{2}$	tetragonal	submetallic, steel gray to dull black
Cuprite	Cu_2O	Cu	6.1	193–207	$3\frac{1}{2}$–4	isometric	adamantine to earthy red to black
Diaspore	AlO(OH)	Al	3.3–3.5		$6\frac{1}{2}$–7	orthorhombic	vitreous, white, greenish, pinkish, brownish
Digenite	Cu_9S_5	Cu	5.6	67–76	$2\frac{1}{2}$–3	isometric	submetallic, deep blue to black
Electrum	(Au, Ag)	Au, Ag	15–19	53–63	$2\frac{1}{2}$–3	isometric	metallic, gold yellow to pale yellow
Enargite	Cu_3AsS_4	Cu, As	4.45	283–327	3	orthorhombic	metallic, gray black
Euxenite	$(Y, Ca, Ce, U, Th)(Nb, Ta, Ti)_2O_6$	Nb, Ta, rare earths	4.3–5.9	782–813	$5\frac{1}{2}$–$6\frac{1}{2}$	orthorhombic	submetallic, black to brown
Ferberite	$FeWO_4$	W	7.5	387–418	4–$4\frac{1}{2}$	monoclinic	submetallic, black
Fergusonite	$(Y, Er, Ce, Fe)(Nb, Ta, Ti)O_4$	Ni, Ta, rare earths	5.4		$5\frac{1}{2}$–$6\frac{1}{2}$	tetragonal	vitreous to submetallic, brownish black
Ferrimolybdite	$Fe_2(MoO_4)_3.8H_2O$	Zn	4.5		1–2	orthorhombic	adamantine to earthy yellow
Ferroplatinum	(Pt, Fe)	Pt group	14–19	122–129	4–$4\frac{1}{2}$	isometric	metallic, steel gray
Franklinite	$(Zn, Mn, Fe)(Fe, Mn)_2O_4$	Zn	5.1–5.2		$5\frac{1}{2}$–$6\frac{1}{2}$	isometric	metallic to dull, black
Freibergite	$(Ag, Cu)_{12}(Sb, As)_4S_{13}$	Ag, Cu, Sb, As	4.7–5.0	252–375	3–$4\frac{1}{2}$	isometric	metallic, steel gray
Gahnite	$(Zn, Fe)Al_2O_4$	Zn	4.6		$7\frac{1}{2}$–8	isometric	vitreous, greenish to greenish black
Galena	PbS	Pb	7.6	59–65	$2\frac{1}{2}$	isometric	metallic, silvery gray, cubit habit and cleavage
Garnierite	mixture of NI silicates	Ni	2.2–2.8		2–3		vitreous to dull, greenish
Germanite	$Cu_3(Ge, Fe)(S, As)_4$	Ge	4.5		4	isometric	metallic, dark reddish gray
Gibbsite	$Al(OH)_3$	Al	2.4		$2\frac{1}{2}$–$3\frac{1}{2}$	monoclinic	vitreous to dull, white to gray
Glaucodot	(Co, Fe)AsS	Co, As	6.1	1097–1115	5	orthorhombic	metallic, grayish to reddish silver white
Goethite	α-Fe_2O_3	Fe	4.4	667	5–$5\frac{1}{2}$	orthorhombic	metallic to earthy, yellowish brown to yellow
Gold (electrum)	(Au, Ag)	Au, Ag	15–19	53–63	$2\frac{1}{2}$–3	isometric	metallic, golden yellow
Hausmannite	Mn_3O_4	Mn	4.8	536–566	$5\frac{1}{2}$	tetragonal	submetallic, brownish black
Hematite	α-Fe_2O_3	Fe	5.2	1038	5–6	hexagonal	metallic to earthy, steel gray to reddish brown
Hemimorphite	$Zn_4Si_2O_7(OH)_2.H_2O$	Zn	3.4–3.5		$4\frac{1}{2}$–5	orthorhombic	vitreous, white to colorless
Huebnerite	$MnWO_4$	W	7.2		4–$4\frac{1}{2}$	monoclinic	submetallic to resinous, yellowish brown
Hydrozincite	$Zn_5(CO_3)_2(OH)_6$	Zn	3.5–3.8		2–$2\frac{1}{2}$	monoclinic	earthy, white to pale shades of yellow
Ilmenite	$FeTiO_3$	Ti	4.7	659–703	5–6	hexagonal	metallic to dull, black
Kermesite	Sb_2S_2O	Sb	4.7	30–90	1–$1\frac{1}{2}$	monoclinic	adamantine, cherry red
Lepidolite	$K(Li, Al)_3(Si, Al)_4O_{10}(F, OH)_2$	Li	2.8–3.3		$2\frac{1}{2}$–3	monoclinic	pearly, pinkish, platelike
Leucoxene	alteration product of Ti minerals	Ti					earthy, white
Linnaeite	$(Co, Ni)_3S_4$	Co, Ni	4.8–5.0	450–613	$4\frac{1}{2}$–$5\frac{1}{2}$	isometric	metallic, steel gray
Loellingite	$FeAs_2$	As	7.4	446–560	5–$5\frac{1}{2}$	orthorhombic	metallic, silvery white
Loparite	$(Ce, Na, Ca)_2(Ti, Nb)_2O_6$	Nb, Ta, rare earths	4.0		$5\frac{1}{2}$	orthorhombic	adamantine to metallic, black, brown, yellow
Magnetite	Fe_3O_4	Fe	5.2	592	$5\frac{1}{2}$–$6\frac{1}{2}$	isometric	metallic to dull, black, magnetic
Malachite	$Cu_2(OH)_2(CO_3)$	Cu	4.0		$3\frac{1}{2}$–4	monoclinic	vitreous to dull, green to greenish black
Manganite	MnO(OH)	Mn	4.3	698–772	4	monoclinic	submetallic to dull, black to steel gray
Mercury	Hg	Hg	14.4			liquid	brilliant gray metallic liquid

Table 3—continued

Ore mineral	Formula	Metal extracted	Specific gravity	Hardness		Crystal system	Appearance, remarks
				VHN	Mohs		
Microlite	$(Na, Ca)_2Ta_2O_5(O, OH, F)$	Ta	4.3–5.7		$5–5\frac{1}{2}$	isometric	vitreous, pale yellow to brown, reddish
Molybdenite	MoS_2	Mo, Rh	4.6–4.7	8–100	$1–1\frac{1}{2}$	hexagonal	metallic, lead gray
Monazite	$(Ce, La, Th, Y, Nd)PO_4$	rare earths	4.6–5.4		$5–5\frac{1}{2}$	monoclinic	resinous to vitreous reddish to brown to yellow
Moncheite	$(Pt, Pd)(Te, Bi)_2$	Pt group	~10			hexagonal	metallic, steel gray
Montroseite	$(V, Fe)O.OH$	V	4.0	266–300		orthorhombic	submetallic, black
Nickeline (niccolite)	$NiAs$	Ni, As	7.8	363–372	$5–5\frac{1}{2}$	hexagonal	metallic, copper–red
Orpiment	As_2S_3	As	3.5	22–58	$1\frac{1}{2}–2$	monoclinic	nonmetallic, resinous, yellow
Patronite	VS_4	V			low	monoclinic	dull, greenish black to black
Pentlandite	$(Fe, Ni, Co)_9S_8$	Ni, Co	4.6–5.0	268–285	$3\frac{1}{2}–4$	isometric	metallic brassy, usually as microscopic lamellae in pyrrhotite
Perovskite	$Ca(Ti, Nb)O_3$	Nb, Ta, rare earths	4.0		$5\frac{1}{2}$	orthorhombic	adamantine to metallic black to yellow
Petalite	$LiAlSi_4O_{10}$	Li	2.4		$6–6\frac{1}{2}$	monoclinic	vitreous, white to yellow
Pitchblende	massive uraninite	U	7.5–10.0	499–598			dull, brown to black
Platinum	(Pt, Fe)	Pt-group	14–19	122–129	$4–4\frac{1}{2}$	isometric	metallic, steel gray
Polybasite	$(Ag, Cu)_{16}Sb_2S_{11}$	Ag	6.0–6.2		2–3	monoclinic	metallic, black to deep ruby red
Powellite	$Ca(Mo, W)O_4$	Mo, W	4.2		$3\frac{1}{2}–4$	tetragonal	subadamantine, yellow to brown, fluorescent
Proustite	Ag_3AsS_3	Ag, As	5.55	103–137	$2–2\frac{1}{2}$	hexagonal	submetallic, deep red
Psilomelane	mixtures of Mn oxides	Mn	3.7–4.7	208–813	5–6	monoclinic	submetallic, black to steel gray
Pyrargyrite	Ag_3SbS_3	Ag	5.85	107–144	$2\frac{1}{2}$	hexagonal	adamantine to submetallic, deep red
Pyrite	FeS_2	Co, Ni	5.0	1505–1620	$6–6\frac{1}{2}$	isometric	metallic, brassy
Pyrochlore	$(Na, Ca)_2Nb_2O_6(OH, F)$	Nb	4.3	542–665	$5–5\frac{1}{2}$	isometric	vitreous, yellowish brown to black
Pyrolusite	MnO_2	Mn	4.75	196–243	2–6	tetragonal	metallic to dull, black to dark steel gray
Realgar	AsS	As	3.5	47–60	$1\frac{1}{2}–2$	monoclinic	nonmetallic, resinous, red
Rhodochrosite	$MnCO_3$	Mn	3.45–3.6		$3\frac{1}{2}–4$	hexagonal	vitreous, pale pink to deep red
Rhodonite	$(Mn, Fe, Ca, Mg)SiO_3$	Mn	3.6–3.7		$5\frac{1}{2}–6\frac{1}{2}$	triclinic	vitreous, pink to brownish red
Roscoelite	$K(V, Al, Mg)_3(AlSi_3)O_{10}(OH)_2$	V	3.0		$2\frac{1}{2}$	monoclinic	pearly, brown to greenish brown
Rutile	TiO_2	Ti	4.2–4.3	1132–1187	$6–6\frac{1}{2}$	tetragonal	adamantine, brown to yellowish to gray
Safflorite	$(Co, Fe, Ni)As_2$	Co, As	7.2	285–464	$4\frac{1}{2}–5$	orthorhombic	metallic, tin white
Scheelite	$CaWO_4$	W	5.9–6.1	387–409	$4\frac{1}{2}–5$	tetragonal	vitreous to adamantine, white to pale brown, fluorescent
Senarmontite	Sb_2O_3	Sb	5.5		$2–2\frac{1}{2}$	isometric	resinous, colorless to white
Siderite	$FeCO_3$	Fe	3.85		4	hexagonal	vitreous, pale yellow to brownish
Siegenite	$(Ni, Co)_3S_4$	Ni, Co	4.5–4.8	503–525	$4\frac{1}{2}–5\frac{1}{2}$	isometric	metallic, steel gray
Silver	Ag	Ag	10.5	55–63	$2\frac{1}{2}–3$	isometric	metallic, silvery but tarnishes black

Skutterudite	$(Co, Fe, Ni)As_{2-3}$	As, Co	6.1–6.9	792–907	$5\frac{1}{2}$–6	isometric	metallic, tin white
Smithsonite	$ZnCO_3$	Zn	4.3–4.5		4–$4\frac{1}{2}$	hexagonal	vitreous, white, yellow, brown
Sperrylite	$PtAs_2$	Pt group	10.5	1080–1145	6–7	isometric	metallic, tin white
Sphalerite	(Zn, Fe, Cd, Ga, Ge, In, Tl)S	Zn, Cd, Ga, Ge, In, Tl	3.9–4.1	218–227	$3\frac{1}{2}$–4	isometric	nonmetallic, resinous, pale yellow to black depending on iron content
Spodumene	$LiAlSi_2O_6$	Li	3.0–3.2		$6\frac{1}{2}$–$7\frac{1}{2}$	monoclinic	vitreous to dull, colorless, gray, green
Stannite	Cu_2FeSnS_4	Sn	4.4	190–326	4	tetragonal	metallic, steel gray
Stephenite	Ag_5SbS_4	Ag	6.2	26–124	2–$2\frac{1}{2}$	orthorhombic	metallic, steel gray
Stibiconite	$SbSb_2O_6(OH)$	Sb	3.5–5.5		3–7	isometric	earthy to vitreous, white to yellow
Stibnite	Sb_2S_3	Sb	4.5–4.6	42–153	2	orthorhombic	metallic, silvery gray
Sylvanite	$AgAuTe_4$	Au, Ag	8.0–8.2	91–104	$1\frac{1}{2}$–2	monoclinic	metallic, silvery, white
Tantalite–columbite	$(Fe, Mg, Mn)(Nb, Ta)_2O_6$	Nb, Ta	5.1–8.2	240–1021	6–$6\frac{1}{2}$	orthorhombic	submetallic to vitreous, black to brown
Tennantite	$(Cu, Zn)_{12}(As, Sb)_4S_{13}$	As, Cu	4.6–4.8	297–354	3–$4\frac{1}{2}$	isometric	metallic, steel gray
Tetrahedrite	$(Cu, Zn, Ag)_{12}(Sb, As)_4S_{13}$	Ag, Cu, Sb	4.7–5.0	285–322	3–$4\frac{1}{2}$	isometric	metallic, steel gray
Thortveitite	$(Sc, Y)Si_2O_7$	Sc, rare earths	3.6		6–7	monoclinic	vitreous, grayish green
Thucholite	U-bearing hydrocarbon	U					earthy, black
Titanite (spene)	$CaTiSiO_5$	Ti	3.5		5–$5\frac{1}{2}$	monoclinic	adamantine to resinous, yellow, gray, black
Tyuyamunite	$Ca(UO_2)_2(VO_4)_2.5–8H_2O$	U, V	3.3–3.6	37–44	2	orthorhombic	resinous, greenish yellow
Ulvöspinel	Fe_2TiO_4	Ti	4.8			isometric	metallic, iron black
Uraninite	UO_2	U, rare earths	7.5–10.0	792–813	5–6	isometric	submetallic, black to brown
Uranophane	$Ca(UO_2)_2Si_2O_7.6H_2O$	U	3.8	96–115	$2\frac{1}{2}$	monoclinic	vitreous, pale yellow to brownish
Uranothorite	$(Th, U, Fe)SiO_4$	U	4.1–6.7		$4\frac{1}{2}$	tetragonal	vitreous to resinous, yellow to brown
Valentinite	Sb_2O_3	Sb	5.8		$2\frac{1}{2}$–3	orthorhombic	adamantine, white, gray, yellow
Violarite	$FeNi_2S_4$	Ni	4.5–4.8	241–373	$4\frac{1}{2}$–$5\frac{1}{2}$	isometric	metallic, light gray to slightly violet
Willemite	Zn_2SiO_4	Zn	3.9–4.2		$5\frac{1}{2}$	hexagonal	vitreous, white to green to red, fluorescent
Wolframite	$(Fe, Mn)WO_4$	W, Sc	7.0–7.5	312–342	4–$4\frac{1}{2}$	monoclinic	submetallic, grayish to brownish black
Wulfenite	$PbMoO_4$	Mo	6.5–7.0	211–333	3	tetragonal	resinous, orange, yellow, brown
Wurtzite	ZnS	Zn, Cd, Ga, Ge, In, Tl	4.0–4.1	146–269	$3\frac{1}{2}$–4	hexagonal	resinous, brownish to orange
Xenotime	YPO_4	rare earths	4.4–5.1		4–5	tetragonal	vitreous to resinous, yellow to reddish brown
Zincite	(Zn, Mn)O	Zn	5.7	190–219	4	hexagonal	subadamantine, yellow to orange-red

and rare earths occur as phosphates and carbonates. The principal ore mineral sources of metals are listed in Tables 2 and 3. Table 2 provides a brief summary of those minerals which serve as major sources of each metal. Table 3 presents the principal ore minerals and a summary of some of their important properties. These listings are confined to the most widespread, common and important minerals, and are not intended to include all of those minor minerals which may contribute locally to ores. The most complete listing of all mineral species is given by Fleischer (1983).

2. Distribution

Ore minerals only constitute ore deposits where they occur in sufficient quantity to be economically exploitable. Most of them also occur in minor quantities in ore deposits of other minerals, and as accessory minerals in common rocks. Thus magnetite, which constitutes a major ore of iron when it occurs in large quantities and in significant concentrations, is a common minor phase in many ores of copper and zinc, and is a ubiquitous accessory phase in many igneous and metamorphic rocks. Only a trace of the total amount of most metals in the earth's crust is concentrated in the ore minerals found in ore deposits; most of the metals are dispersed in trace to minor amounts in common rocks, occurring in the rock-forming minerals or as disseminated grains of ore minerals. For example, the amount of zinc in ore deposits has been estimated at 1/10 000 000 of the zinc in the earth's crust.

The degree of natural concentration of ore minerals determines whether a specific occurrence is an ore deposit. The minimum metal concentration required within an area of the earth's crust for an ore deposit to result varies from about 25wt% for iron and aluminum, to about 0.5–3wt% for copper, zinc or lead, to about 0.008wt% for gold. The economic viability of any given deposit also depends on the size of the deposit, its geographic location, its depth and shape, and the political and tax structures of the governing area. Fortunately, many ore minerals occur in association or paragenesis with one another, and permit the simultaneous extraction of several metals, not one of which may be economically extractable on its own. There are many characteristic ore mineral associations or parageneses, which are further described in Ramdohr (1980), Oelsner (1966), Stanton (1972) and Craig and Vaughan (1981).

3. Identification and Characterization

The identification and characterization of ore minerals may be carried out using a variety of methods; the techniques briefly described below are most commonly used.

3.1 Megascopic Examination

Many ore minerals, if present as grains coarser than 1 mm diameter, can be identified by their characteristic color, luster, specific gravity and hardness. In general, they tend to be metallic to submetallic in luster and have a moderate to high specific gravity.

3.2 Microscopic Examination (Reflected Light)

Most of the principal ore minerals are opaque to translucent, and are commonly studied by means of reflected light microscopy. Samples are either cut and polished directly, or cut, cast in an epoxy resin and then polished. Incident light reflected from the mineral surface may be observed visually in the microscope or may be quantitatively measured as a function of wavelength using a photometer system. Most minerals may be identified by their color, reflectance, textures, anisotropy (if present) and hardness (either quantitatively measured in terms of Vickers microhardness (VHN), or visually estimated by polishing relief).

3.3 Microscopic Examination (Transmitted Light)

Transparent to translucent ore minerals are readily studied in standard thin sections or doubly polished thin sections. Color, textures, anisotropy (if present) and refractive index are used in identification.

3.4 X-Ray Diffraction

All crystalline ore minerals yield x-ray diffraction patterns which are usually sufficiently characteristic to permit unequivocal identification. Extensive solid solution is common in many ore minerals, hence deviation of precise x-ray peak positions from their reported values is frequently observed.

3.5 Electron Microprobe Analysis

The electron microprobe enables quantitative chemical analysis to be performed on areas as small as a few micrometers across, and has become a standard tool in ore mineral identification. It has permitted the recognition of scores of new mineral species and also proven valuable in mapping the distribution of chemical elements within minerals.

3.6 Phase Diagram Study

Equilibrium phase diagrams are now available for many of the chemical systems containing the principal ore minerals. These diagrams provide valuable clues as to which ore minerals may be present, given a certain amount of basic chemical or mineralogical data. They may also provide insight into the origin or postdepositional history of ore mineral assemblages.

4. Ore Mineral Textures

The textures of ore minerals may provide information on their mode of formation, on postdepositional modification (annealing, exsolution, replacement, deformation) and on meteoric weathering. They provide important insight into the history of an ore during and after formation. Textures are also important in determining the requirements of processes of mineral extraction. The diversity of ore textures and the problems of their interpretation are described in detail by Ramdohr (1980) and Craig and Vaughan (1981).

See also: Geochemical Distribution of the Elements; Igneous Ore Deposits: Formation Processes; Metamorphic Ore Deposits: Formation Processes; Metallogenic Provinces; Sedimentary Ore Deposits: Formation Processes

Bibliography

Craig J R, Vaughan D J 1981 *Ore Microscopy and Ore Petrography.* Wiley-Interscience, New York

Fleischer M 1983 *Glossary of Mineral Species.* Mineralogical Record, Tucson, Arizona

Oelsner O W 1966 *Atlas of the Most Important Ore Mineral Parageneses Under the Microscope.* Pergamon, New York

Ramdohr P 1980 *The Ore Minerals and their Intergrowths,* 2nd edn. Pergamon, New York

Roberts W L, Rapp G R, Weber J 1974 *Encyclopedia of Minerals.* Van Nostrand Reinhold, New York

Stanton R L 1972 *Ore Petrology.* McGraw-Hill, New York

US Bureau of Mines 1968 *Dictionary of Mining, Mineral and Related Terms.* USBM, Washington, DC

J. R. Craig
[Virginia Polytechnic Institute and State University, Blacksburg, Virginia, USA]

Ores of the Future

Several possible sources of unconventional metallic and nonmetallic resource materials are available to replace the current traditional nonrenewable ores. Although most of these geologically known resources are subeconomic, research on their acquisition and utilization, as well as quantification of their location and extent, has begun for some materials. Replacement sources for current ore-grade materials are constrained mainly by the limits of human ingenuity in developing the scientific and technological skills needed to conserve *all* possible resource materials, to use ever-lower-grade conventional and unconventional materials, to discover, recover and process new types of known unconventional resource materials as soon as economic conditions become more favorable, and to continue research efforts that will lead to the identification of unconceived materials.

1. Definitions

Conventional ores are currently extractable traditional types of naturally occurring minerals (metals and nonmetals) of economic value. *Unconventional future ores* are naturally occurring or man-made accumulations of metallic and/or nonmetallic materials that are not technologically and/or economically recoverable at present but may be expected to be recoverable ultimately. Barton (1983) has broadened this category to include unconceived potential ore materials.

Geological availability indicates the presence or predicted presence of a resource material in the crust of the Earth, in either known or unknown chemical or physical form, on the basis of geological experience or from geological-occurrence models. *Economic availability* indicates extraction of a resource at a profit. Changes in economic conditions, technology, or legal and political restraints may convert subeconomic materials into economically available resources, or the reverse.

Reserves and *hypothetical* and *speculative resources,* discussed in detail elsewhere (see *Mineral Resources: Definitions, Uses, Classification and Future Availability*), are defined more generally here as follows. A *reserve* is an inventory of the usable and economically available part of known resource materials; it is also sometimes called an *ore.* The categories measured, indicated or inferred designate the degree or precision of measurement of the known inventory; that is to say, the amount of the reserve or resource is inferred, but not its existence. *Deliverable resources* are the subeconomic part of the conventional types of known resources; they are already available or discoverable as potential reserves. *Hypothetical resources* are undiscovered conventional or unconventional materials that occur as predicted extensions of known deposits in existing mining districts or in other favorable geological regions. *Speculative resources* are undiscovered materials that may exist in conventional or unconventional types of deposits in unexplored favorable geological terranes or in unrecognized new types of occurrences.

2. Conventional and Unconventional Sources of Materials

A classification of conventional and unconventional resource materials is given in Fig. 1. Conventional ore materials and sources include those currently being mined and those of closely related types that are marginally economic and deliverable resources at present.

Most of the unconventional resource materials listed in Fig. 1 are known but not currently deliverable, and although their total extent can only be inferred at present, their potential as hypothetical and speculative resources is recognized. Unconventional sources fall into the four somewhat overlapping categories discussed below.

2.1 Conservation Opportunities

Conventional resource conservation and recycling methods include scrap-metal processing, metal recovery and energy generation from solid waste, minimization of waste by improving the performance and

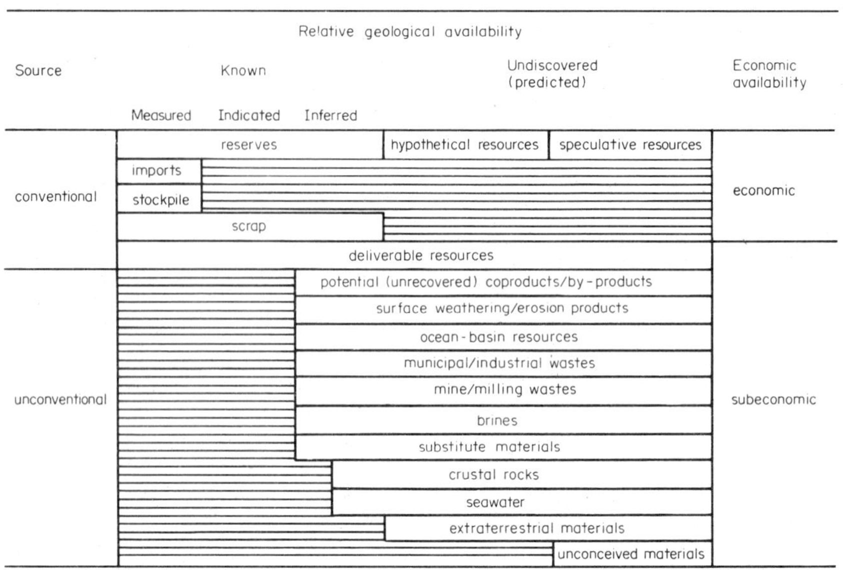

Figure 1
Classification of types of resources and their availability

corrosion resistance of metals, and more efficient shaping and finishing of materials. Only about 60% of a given reserve is recovered in conventional extraction and processing; improved conservation efforts, especially recovery techniques, should continue to make more of these materials available.

Little attention has been directed towards the recovery of the large coproduct and by-product resources presently discarded or left unmined during the extraction and processing of conventional ores. Traditional mining has been focused on one, or at best only a few, of the metals that may be present. For example, sphalerite and barite, which occur locally in cleats of coalbeds in the Illinois basin, could provide a substantial source of zinc and barium by-products. Moreover, the underclay deposits associated with Appalachian basin coalbeds could become an extensive coproduct source of aluminum, rather than being wasted in mine dumps. Table 1 gives the coproduct potential for several elements associated with the fertilizer mineral phosphate in Idaho. A substantial fraction of the US production of these metals could be realized from this single source, if recovered. Only vanadium is presently produced; recovery is, however, only about 25% of the amount present. Formations rich in organic materials, such as oil shale, are known to contain abundant metals (e.g., vanadium, nickel, zinc, selenium and molybdenum), which could be recovered as a by-product of the oil. Large amounts of rutile, a source of titanium, occur unrecovered in the mine tailings at Bingham, Utah; and several rare-earth elements are present but unrecovered in the tailings at Climax, Colorado, Municipal and industrial wastes, both solid and effluent, also are potential sources of unrecovered resource materials. For example, gold was recovered from sewage solids in Palo Alto, California; the gold occurred in industrial wastes from the extensive local electronics industry.

2.2 Technology Dependence

Technological extension of the mineral resource base by using lower-quality conventional or synthetic materials offers one of the greatest possibilities for increasing future mineral supplies, provided that sufficient energy and water are available for processing. Technological breakthroughs will continue to expand exploration concepts and targets, promote more efficient mining and metallurgical methods, provide for recovery and temporary stockpiling of coproduct or by-product materials, and develop environmentally acceptable techniques for the disposal of barren or toxic wastes. Chemical research will enlarge the area of synthetic materials capable of replacing all or part of the metal currently used in some processes.

The following are examples of unconventional sources of materials that depend on new technology. Profitable mining of deep-ocean-basin metal-bearing nodules and crusts may soon be possible, even though not all the contained metals may be economically recoverable in the early mining stages. Surface weathering and erosion have formed low concentrations of metals in soils, laterites and placers in unexplored tropical areas, both on land and along continental shelves; their exploitation requires new exploration

Table 1
Potential coproduct resources in phosphate rock of the Phosphoria formation, Idaho (after Gulbrandson 1977)

Element	Average concentration (ppm)	Potential if recovered in 1975[a] (t)	Percentage of US production in 1974
Cadmium	90	600	20
Fluorine	24000	1.6×10^5	147
Rare earths (including yttrium)	1000	6700	37
Selenium	30	200	69
Uranium	110	740	8.3
Vanadium	800	5400	120

[a] Based on production of 6.7 Mt that is projected at 13.6 Mt in year 2000

and recovery technology. The metal-bearing brines in rift zones, such as those discovered in the Red Sea and Salton Sea geothermal areas, are known to contain base and precious metals whose recovery should become routine. Deeply buried or extensive low-grade occurrences that cannot be mined by conventional methods should be amenable to in situ solution mining following fracturing by conventional geopressure or explosive blast techniques. Another possible application of solution mining is the recovery of overlooked or marginal low-grade material from water in flooded old mine workings. Finally, extension of chemical synthesis, which already can convert abundant nonmetallic substances into a wide variety of products, will provide materials whose use may prolong the life of metal products, or whose use as a partial or complete substitute may extend metal resources.

One important further technological factor will be the availability of lower-cost energy—possibly from such new sources as hot-dry-rock geothermal, solar, fusion or unconventional organic energy from the margins of ocean basins. Conventional geothermal energy from geysers probably taps only about 5% of the total energy available from this source (Heiken et al. 1981); the remaining 95% of the heat energy, which is not connected with permeable water-filled fracture systems, and which cools slowly by conduction (unless replenished by hot magma), is a challenging target for the unconventional extraction of energy by means of a man-made geothermal system. Regions of high heat flow, such as western USA, are also near terranes containing important conventional and unconventional resources.

Many unconventional sources of metallic and nonmetallic materials will undoubtedly require larger supplies of water for processing ore materials either in situ or after transportation to a site where water is available. Development of nontraditional sources of water can be obtained through improved technology for conservation and recycling of existing conventional supplies.

The most extensive, yet most difficult, unconventional materials to exploit are those dispersed in rocks of the earth's crust. The outer lithosphere contains almost limitless amounts of all the critical commodities. The relative amounts of abundant and scarcer geochemical elements listed in Table 2 are for the most part impractical to obtain under current technology, energy and environmental constraints. Figure 2 shows the typical distribution of these elements in the crust;

Table 2
Comparison of some geochemically abundant and scarce elements in average continental crustal rock[a] and in seawater[b]

Element	Crustal rock ($Mt\ km^{-3}$)	Seawater ($kt\ km^{-3}$)
Geochemically abundant		
oxygen	1280000	880000
silicon	780000	3
aluminum	220000	0.001
iron	150000	0.003
calcium	110000	410
magnesium	64000	1300
sodium	64000	11000
potassium	59000	390
titanium	17000	0.001
hydrogen	4000	110000
manganese	2000	0.002
phosphorus	3000	0.1
Geochemically scarce		
chromium	280	0.0005
nickel	210	0.007
zinc	196	0.01
copper	150	0.003
lead	35	0.00003
tungsten	4	0.0001
silver	0.198	0.0003
gold	0.011	0.00001

[a] Assumed average density of 2.8 $g\,t^{-1}$; data from Parker (1967)
[b] Data from National Research Council (1971)

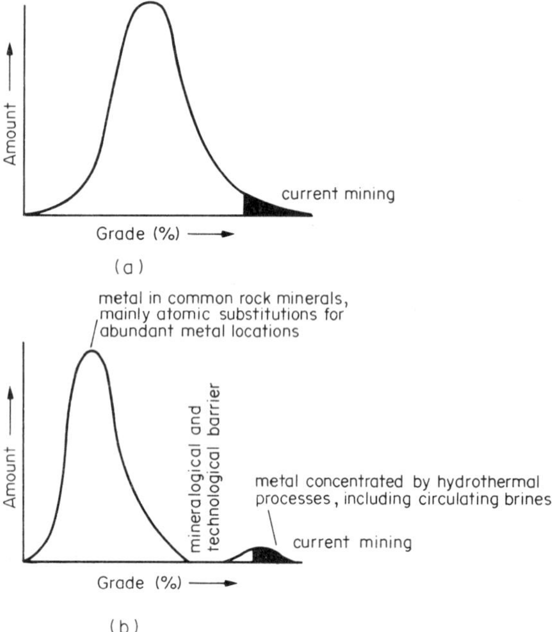

Figure 2
Distribution curves for elements in the earth's crust: (a) geochemically abundant elements amenable to current concentration techniques regardless of grade; (b) geochemically scarce elements including metals in common rock that are not amenable to current concentration techniques (after Skinner 1976. Reproduced by permission of *American Scientist*, journal of Sigma Xi, The Scientific Research Society)

the relative proportions minable today are indicated by shading. Distribution of the geochemically scarce elements (Fig. 2b) is bimodal: (a) as special and rare local concentrations in ore minerals, enriched 100 000 times the average crustal abundance; and (b) as more abundant randomly distributed atoms trapped in silicate minerals in amounts of only hundreds of ppm or less. Extraction and concentration of these randomly distributed elements in substantial amounts requires overcoming the extremely difficult mineralogical and technological barriers facing conventional methods. Once these barriers are overcome, increased knowledge of the geological environments favorable for the local enrichment of potential crustal ore materials will also be required. Indeed, current technology permits local exploitation of some of the most abundant elements where they are known to be sufficiently concentrated in chemical sediments.

Like the crust, seawater also contains significant amounts of resource materials (Table 2). These are supplied by continental erosion and by both continental and submarine volcanism. Concentration of these metals and nonmetals by evaporation is commonly used to produce many of the most abundant salts; the geochemically scarcer materials require higher technological and energy input for recovery.

2.3 Substitute Materials

Substitutions of these plentiful for less abundant resources, and of renewable for nonrenewable resources, have met with only limited success. Perhaps even greater opportunities lie in functional substitution, in which a desired function is performed in a completely different way. Examples of functional substitution are the use of adhesives in place of nails and rivets, nuclear energy instead of fossil fuels to generate electricity, and several alternatives are now known and acceptable for nearly all the major uses of mercury. Further substitutions can be expected from the utilization of abundant nonmetal–silicate mineral resources. The National Research Council (1969) reported early successes in producing nonmetallic metal-substitutes (e.g., borosilicate glasses, ceramics capable of withstanding high thermal shock, glass-ceramic materials having flexural strengths of 1.03–1.37 GN m^{-2}, ceramic armorplate, and low-density glass stronger than most metals). Silicon carbide components are now used for rocket-engine nozzles and nose cones. Silicate fibers and mine tailings have future applications in insulation and as fillers. Silicate construction materials, such as structural clays, asbestos and cement, can be expected to play an increasing resource-substitution role. Some of the applied-research opportunities leading to an expansion of the areas of functional substitution are listed by Chynoweth and Schmitt (1975).

2.4 Unconceived Materials

Though impossible to list, many of the theoretical and applied-research methods by which we may expect future resource breakthroughs to be conceived and become realities have already been developed. The basic tool for locating adequate supplies of new materials is more detailed geological, geochemical and geophysical information about the composition and structure of the Earth and its satellitic and planetary neighbors. Systematic application of the predictive principles implicit in the geochemical cycle (Barton 1983) will lead to the identification of deposits in the earth's crust that are significantly different in mineralogy, grade/tonnage and geological setting from those that are currently economic. Seismic and other geophysical technologies can probe the crust for geofractures, invisible at the surface, that may be associated with metallogenetic belts or alignments of mineral deposits. Satellite remote-sensing techniques, already useful for the discovery of organic fuels, are also proving useful in complex metallized terranes (Woodall 1984). The developing science of metallogenesis, aided by the expanding plate-tectonic concept of continent and ocean-basin formation, may provide important keys to success in recognizing and understanding the complex geological terranes that will ultimately yield the resources and resource substitutes for future years.

Bibliography

Bailly P A 1976 The problems of converting resources to reserves. *Min. Eng. (N.Y)*28: 27–37

Barton P B Jr 1983 Unconventional mineral deposits: A challenge to geochemistry. In: Shanks W C 1983 *Cameron Volume on Unconventional Mineral Deposits.* American Institute of Mining, Metallurgical and Petroleum Engineers, New York, pp. 3–14

Brooks D B, Andrews P W 1974 Mineral resources, economic growth, and world population. *Science* 185: 13–19

Chynoweth A G, Schmitt R W 1975 Needs, opportunities and priorities. *National Material Policy.* National Academy of Sciences, Washington, DC, pp. 132–46

Erickson R L 1973 Crustal abundance of elements and mineral reserves and resources. In: Brobst D A, Pratt W P (eds.) 1973 *United States Mineral Resources*, US Geological Survey Professional Paper 820. US Government Printing Office, Washington, DC, pp. 21–25

Gulbrandsen R A 1977 Byproduct resources in phosphate ore of southeastern Idaho. *Geol. Soc. Am. Abstr. Programs* 9(6): 728

Heiken G, Murphy H, Nunz G, Potter R, Grigsby C 1981 Hot dry rock geothermal energy. *Am. Sci.* 69: 400–7

Krauskopf K B 1967 Source rocks for metal-bearing fluids. In: Barnes H L (ed.) 1967 *Geochemistry of Hydrothermal Ore Deposits.* Holt, Rinehart and Winston, New York, pp. 1–33

National Academy of Sciences 1975 *National Mineral Policy.* National Academy of Sciences, Washington, DC

National Research Council 1969 *Nonmetallic Materials.* National Academy of Sciences, Washington, DC

National Research Council 1971 *Marine Chemistry.* National Academy of Sciences, Washington, DC, pp. 4–5

Parker R L 1967 Composition of the earth's crust. In: Fleischer M (ed.) 1967 *Data of Geochemistry*, 6th edn., US Geological Survey Professional Paper 440-D. US Government Printing Office, Washington, DC, pp. D1–D19

Pilger A 1976 The importance of lineaments in the tectonic evolution of the Earth's crust and in the occurrence of ore deposits in middle Europe. *Proc. 1st Int. Conf. New Basement Tectonics*, Utah Geological Association Publication No. 5. Utah Geological Association, Salt Lake City, Utah, pp. 555–64

Skinner B V 1976 A second iron age ahead? *Am. Sci.* 64: 258–69

US Geological Survey Professional Paper 940. US Government Printing Office, Washington, DC

Watson K 1985 Remote sensing—A geophysical perspective. *Geophysics* 50: 2595–610

Woodall R 1984 Success in mineral exploration: Confidence in science and ore deposit models. *Geosci. Can.* 11: 127–32

E. W. Tooker
[US Geological Survey, Menlo Park, California, USA]

P

Petroleum: Oil and Gas Fields

The buoyancy of gas and of oil in water led explorationists more than a century ago to a perception of the anticlinal trapping of oil and gas. In its simplest form, the anticlinal theory postulates that when a reservoir bends upwards, more or less into the shape of an inverted bowl, gas already present in pores, vugs or fractures will rise to the top, followed by oil. A century or so of exploration, not surprisingly, has brought considerable sophistication to concepts of the trapping of oil and gas. The stratigraphic parameter, in particular, has moved definitely into prominence, especially under the stimulus of seismic stratigraphy. The anticline itself, however, whether simple or whether considered together with stratigraphy, continues to influence strongly the critical decisions made in contemporary exploration.

1. Anticlinal Fields

1.1 Ghawar

The Saudi Arabian oil field of Ghawar, the world's largest, is an anticlinal structure. Core drilling first revealed the shallow geometry of this whole anticline. Full drilling then revealed the structure at the level of the prolific "Arab D" reservoir. Consequently, the anticlinal structure of Ghawar may be relatively or virtually unfaulted, a remarkable condition in an anticline that strikes slightly east of north for at least 250 km, that covers an area of some 2300 km^2 and that originally contained a reported 70 billion barrels or more of producible reserves. Dips on the flanks of this long anticline measure as much as 12°, a phenomenon that may be due in part to compaction and draping, and in part to the slow repeated rise of a crystalline Precambrian ridge.

1.2 Samotlor and Urengoy

Two fields in the USSR merit comparison with Ghawar for the structural simplicity of their anticlines and for the magnitude of their reserves. Samotlor represents oil and Urengoy represents gas. Both lie in the large west Siberian basin. At Samotlor, the structure is sufficiently periclinal to be classed as a dome, with a radius of some 25 km and with a drop from the center to the oil–water contact of only 100 m in 20–30 km (Fig. 1). The oil–water line does not cross its nearest contour around the dome, possibly because porosity and permeability are sufficient to effect a hydrodynamic equilibrium. All reservoirs at Samotlor consist of Lower Cretaceous sandstones with total initial reserves of more than 15 billion barrels of oil, and with a cumulative production of nearly 8 billion barrels by the end of 1981. The gas field of Urengoy, one of the world's largest, occurs in an anticline whose geometry is notably similar to that of Ghawar. Like Ghawar, the length of Urengoy exceeds 200 km, and like Ghawar, its strike is nearly north (Fig. 2). Dips on the flanks of its anticline increase with depth, but even at a depth of 3000 m they do not exceed 300 m in 15 km. If faults are present, their throw is unlikely to exceed 50 m. Production comes mainly from Middle Cretaceous sandstones and dolomites. Initial reserves amount to an imposing 7.1×10^{12} m^3 of gas.

1.3 Painter Reservoir

In the thrust belt ("overthrust belt") of the western USA, several oil and gas fields demonstrate the ability to withstand the stress of deformations. The Painter Reservoir field, between thrust faults above and below, exhibits a tightly overturned and thrust limb, yet is one that does not affect the essentially horizontal gas–oil and oil–water contacts in the sandstone reservoir (Fig. 3). The thrust belt in which Painter Reservoir lies stretches from Canada to Mexico.

1.4 Groningen

In western Europe, the Groningen gas field, which is a large area with major reserves, belongs to the class of essentially undeformed anticlines whose structural geometry is moderately broken by faults. Studies of regional geology point to Groningen as a culmination on a large area that rose possibly as far back as the early Paleozoic. Groningen received gas and held it under a seal of Permian (Zechstein) evaporites. Groningen's Lower Permian reservoir, in fluviatile and eolian facies of sandstone, is broken by faults that rise from underlying Carboniferous strata (Fig. 4). Diagenesis of lipids and allied organic matter in Carboniferous vegetal remains presumably furnished gas that rose through faults to fill the Lower Permian reservoir. Spread over an area of over 800 km^2, the initial reserves of natural gas at Groningen amount to more than 1.4×10^{12} m^3, and possibly to more than 1.7×10^{12} m^3.

1.5 Cantarell Complex

In southern Mexico, the Cantarell Complex in shelving Atlantic waters provides an example of strongly broken structure in competent beds and of abundant reserves of crude oil. Unconformities are present in the geological column (Jurassic, Cretaceous, Tertiary) but the trapping of oil is largely by updip closure against faults. Figure 5 shows the presence of both normal and reverse faults. Strike-slip faulting also may be present. Brecciation of lowest Tertiary (Paleocene) reservoirs

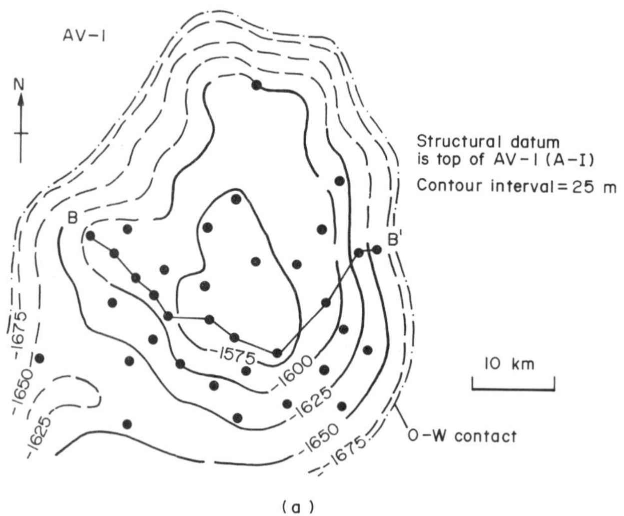

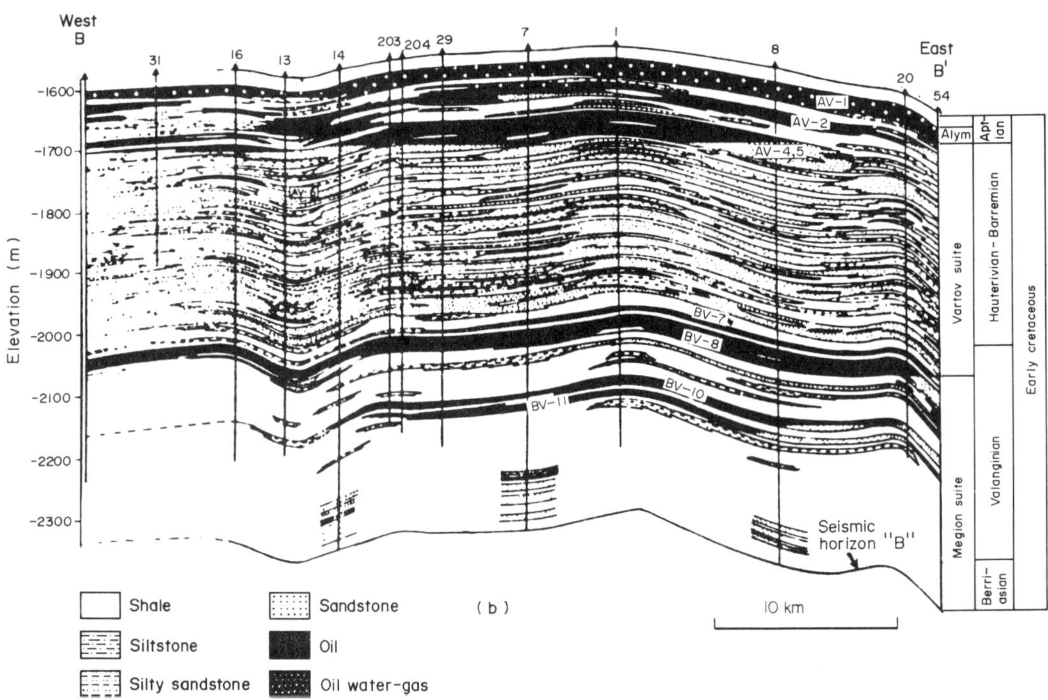

Figure 1
Samotlor oil field, west Siberian basin. Part (a) shows a contour map: contours on top produce sand; oil–water (O–W) contact is approximately 100 km vertically from center of the "dome," at a distance of 20–30 km. Part (b) is a lithostratigraphic cross section of the field. (After Meyerhoff 1982. © Petroleum Exploration Society of Australia, Melbourne. Reproduced with permission)

permits the accumulation of prolific quantities of oil, but important oil is present also throughout the drilled column. Initial producible reserves of some 8 billion barrels of oil accumulated in the Cantarell Complex. These form part of the remarkable contribution that the offshore area of southern Mexico has made to the record surge in Mexico's total reserves since the late 1960s.

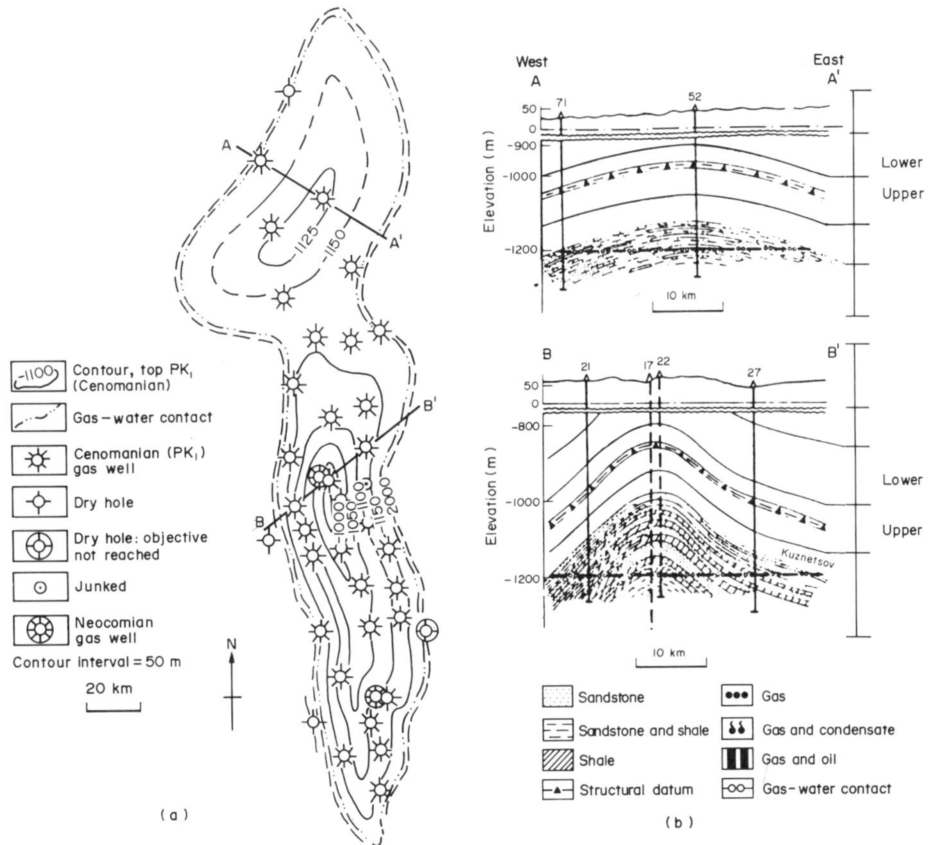

Figure 2
Urengoy gas field, west Siberian basin. Part (a) shows a contour map: contours on top produce horizon; gas–water contact is indicated. Part (b) illustrates lithostratigraphic cross sections of the field. (After Meyerhoff 1982. © Petroleum Exploration Society of Australia, Melbourne. Reproduced with permission)

1.6 West Bay

Many anticlines raised or cored by an upwelling of salt, and sometimes by a rise of serpentine, basalt or shale, are important producers of oil and gas. Among typical occurrences connected with salt are those in the region of the Gulf of Mexico, in the North Sea, in the Middle East and in the USSR. As the exigencies of economics force many countries to drill deeper for oil and gas, they may find that salt has played a part in the development of structures previously regarded as normally flexed or draped. Representative of structures where the rise of salt is in a partially early and only moderately kinetic phase is the West Bay field in Louisiana (Fig. 6). The acute apex of the salt breaks open an extremely complex pattern of faults that help to control localized reservoirs of oil and gas.

2. *Anticlinal–Stratigraphic Fields*

2.1 Bohai

In oil and gas fields where anticlinal structure and stratigraphic modification combine to form the trap for accumulation, the importance of each parameter varies from field to field. In the Bohai basin of coastal China the most prolific of its producing fields contain oil trapped in "buried hills," whose crests and/or flanks are sealed by a profound unconformity. The "central buried hill" of Fig. 7, held up by Proterozoic to Ordovician carbonates, lies under a capping of Eocene strata. This unconformity thus represents a hiatus of some 400 million years. During that long period, paleokarstic porosity was free to develop in the ancient carbonates under the unconformity, and free to

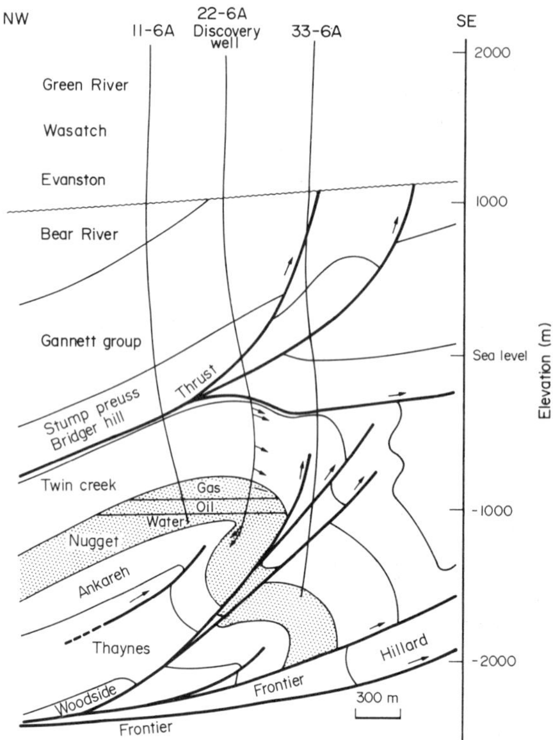

Figure 3
Painter Reservoir field, northwestern USA: horizontal gas–oil and oil–water contacts in crestal sector of overturned and thrust reservoir. (After Lamb 1980. © American Association of Petroleum Geologists, Tulsa, Oklahoma. Reproduced with permission)

receive oil that drained downward and inward from Tertiary strata. As is typical in coastal China, source beds in the Tertiary column formed in a continental paleoenvironment. Those that supplied the buried hills of the Bohai basin formed mainly from Oligocene lacustrine muds. These source beds also supplied oil to penecontemporaneous Tertiary sandstone reservoirs, either above buried hills or on their flanks. Buried hills in the Bohai basin already have produced more than a billion barrels of oil.

2.2 *Prudhoe Bay*

The stratigraphic parameter has played an important part in the trapping of oil and gas at Prudhoe Bay in Alaska. The purely structural factor consists of a broad anticlinal nose with gentle plunge to the west and strong faulting to the north. This open structural condition becomes closed, however, by the eastward onlapping of Lower Cretaceous black shales across progressively eroded Jurassic, Permian, Pennsylvanian and Mississippian formations (Fig. 8).

Reservoirs include Mississippian and Pennsylvanian limestones, but Permian and Triassic sands and conglomerates in deltaic facies yield the main production at Prudhoe Bay. The Lower Cretaceous (Barremian) black shale constitutes a critical element. Not only does this formation provide stratigraphic closure for the trapping of oil and gas but also, by consensus, source material for the hydrocarbons in underlying Paleozoic and Mesozoic reservoirs. The resulting accumulation of oil and gas occupies some 2000 km^2, and its initial reserves amounted to some 10 billion barrels of oil together with probably more than 0.5×10^{12} m^3 of natural gas above its oil. Prudhoe Bay

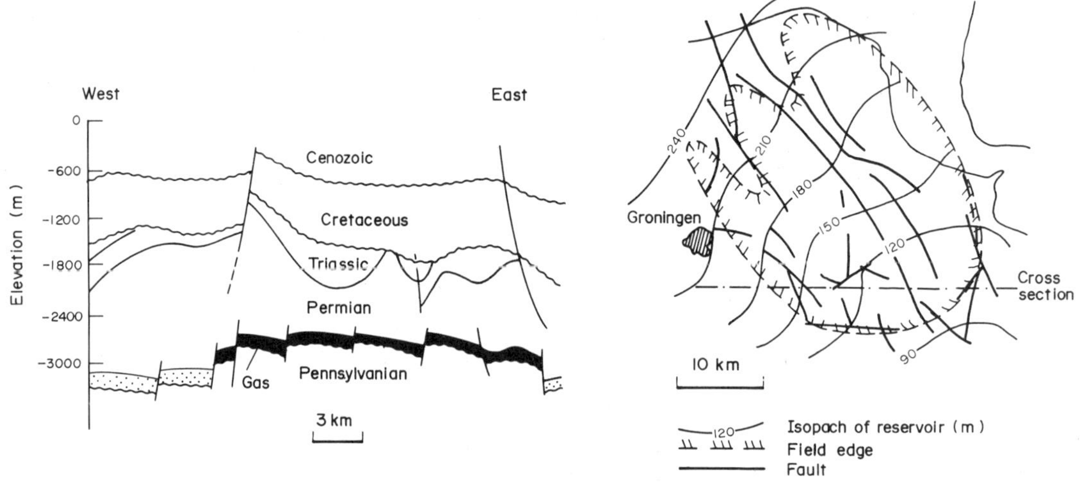

Figure 4
Groningen gas field, the Netherlands: block faulting of reservoir across broad anticline. (After Perrodon 1983. © Societé Nationale Elf-Aquitane, Pau, France. Reproduced with permission)

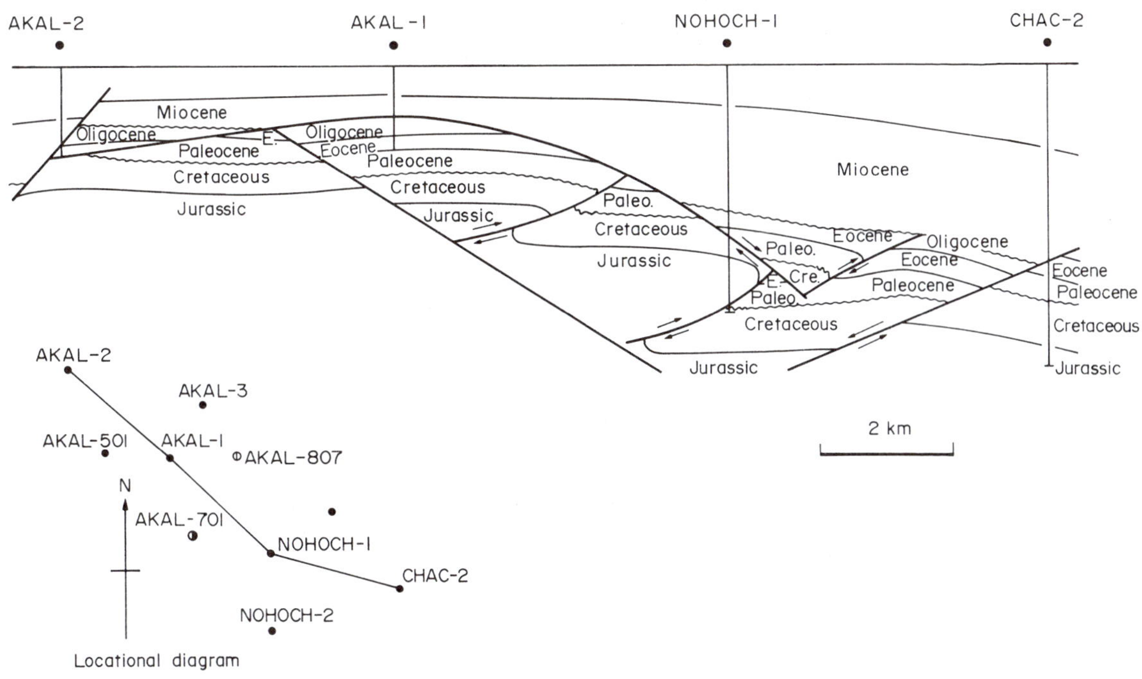

Figure 5
Cantarell complex (marine), southern Mexico: competent carbonate reservoirs broken by normal, reverse and strike-slip faults. (After Acevedo 1980. © American Association of Petroleum Geologists, Tulsa, Oklahoma. Reproduced with permission)

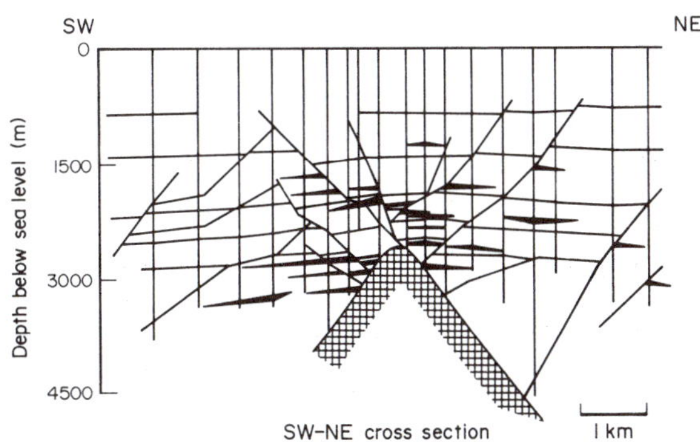

Figure 6
West Bay field, Louisiana: faults radiate from apex and flanks of diapiric salt to trap multiple reservoirs

is easily the largest field in the USA, and owing to its combined reserves of oil and gas, the largest field in North America.

2.3 Bolivar Coastal

In the largest accumulation in the New World, namely the Bolivar Coastal field of Lake Maracaibo in western Venezuela, trapping of oil occurs under complex structural and stratigraphic conditions (Fig. 9). The clear stratigraphic marker of this major accumulation occurs as an Eocene–Miocene unconformity. Above this unconformity, the production of oil in monoclinal Miocene sands may result from local stratigraphic trapping. Below this same unconformity, structure becomes prominent in the form of many high-angle normal faults that cut petroliferous Eocene sandstones. Anticlines add to the structural factor of Eocene production. Filling of the Eocene column up to the underside of the unconformity, however, may represent an expression of an important if partial stratigraphic control. Change in elevation of the oil–water contact from fault block to fault block may also indicate partial stratigraphic control. Production from the Eocene column is prolific, and it forms the main contribution to an anticipated cumulative total of possibly more than 20 billion barrels of oil from the Bolivar Coastal field.

2.4 Brent

The Brent field, situated in the Viking graben of the central North Sea (UK sector), represents one of many within this graben in which a stratigraphic parameter of unconformity is present. The structural parameter

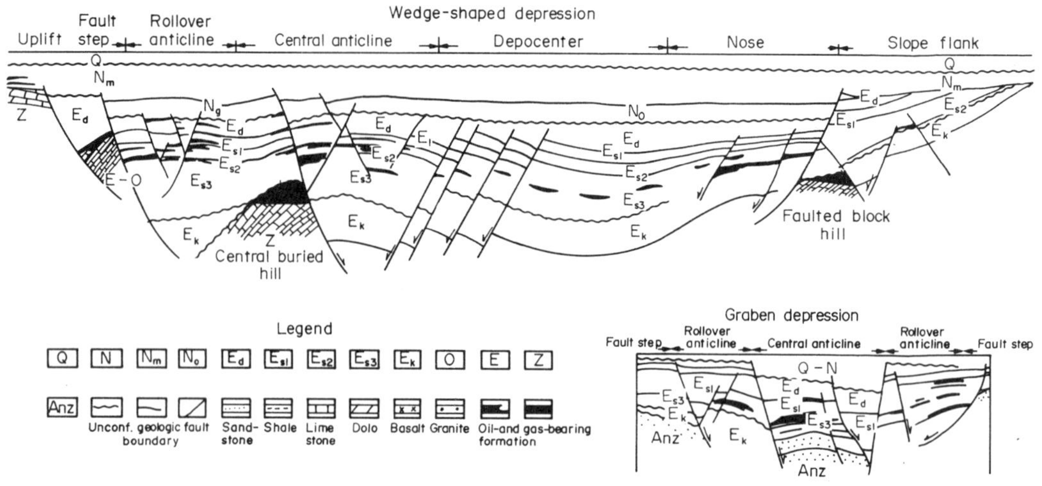

Figure 7
Bohai fields, Bohai basin, coastal China: major accumulation in carbonates of "buried hills," and in sandstones against faults of local horsts and grabens. Q, Quaternary; N, Neogene; N_m, Neogene (Miocene); N_o, Neogene (Oligocene); E_d, Eogene (Dongying); E_{s1}, Eogene (Shahejie-1); E_{s2}, Eogene (Shahejie-2); E_{s3}, Eogene (Shahejie-3); E_k, Eogene (Kongdian); O, Oligocene; E, Eogene; Z, Lower Paleozoic; Anz, mainly Proterozoic (After Perrodon 1983. © Societé Nationale Elf-Aquitane, Pau, France. Reproduced with permission)

of Brent and of its neighbors consists of block faulting. Figure 10 shows a paleotopography on the unconformity, and also a truncation of Jurassic and possibly latest Triassic reservoir sands against an incipient paleoscarp side. Late Jurassic and Cretaceous shales provide cover and source, both on the crest and on the flanks of the unconformity. The two main reservoirs at Brent may themselves contain partial source material for gas caps because of their vegetal material, if only in their minor partings of coal. Approximately 2 billion barrels of recoverable oil were present initially in the Brent field.

2.5 Niger Delta

In the petroliferous province of the Niger delta (west Africa) there are more than 140 oil fields. By 1986, at least two fields had produced more than 400 million barrels each; at least one field had produced more than 200 million barrels; at least seven fields had produced more than 100 million barrels each; and cumulative production from this petroliferous province as a whole had exceeded 11 billion barrels. In Fig. 11, four models illustrate the role of faulting in fields of the present Niger delta, especially the role of growth faulting. Geometry of many of the faults is listric (concave upward), and the faults themselves increase considerably in number when the crest of an underlying diapir collapses. Multiple sand reservoirs occur, many more than are shown in the models, and many are relatively thin. Minor stratigraphic trapping is probably important in a regime of Cenozoic sedimentation such as this one. Known examples include sands interpreted as simple bar sands or point-bar sands, or sands interpreted as turbidites in the paleoprodelta. Still another type of stratigraphic trap consists of clayey paleogullies that truncate reservoirs and trap their oil.

3. Stratigraphic Fields

3.1 East Texas

The oil or gas field in which the stratigraphic parameter is total occurs only rarely. Cases often occur, on the other hand, in which the stratigraphic parameter is so important that it is justifiable to classify the field as stratigraphic. The classic example in this category is the east Texas oil field, renowned in the literature for its discovery at a random location (made by C. M. ("Dad") Joiner) in 1930. The east Texas field strikes slightly east of north for some 65 km, with an average width of some 8 km. Except for plunging local anticlines north and south, the trapping of oil has taken place by eastern truncation of the western-dipping upper Cretaceous Woodbine Sand reservoir below an unconformity (Fig. 12). The east Texas oil field had produced some 5 billion barrels of oil by the late 1980s.

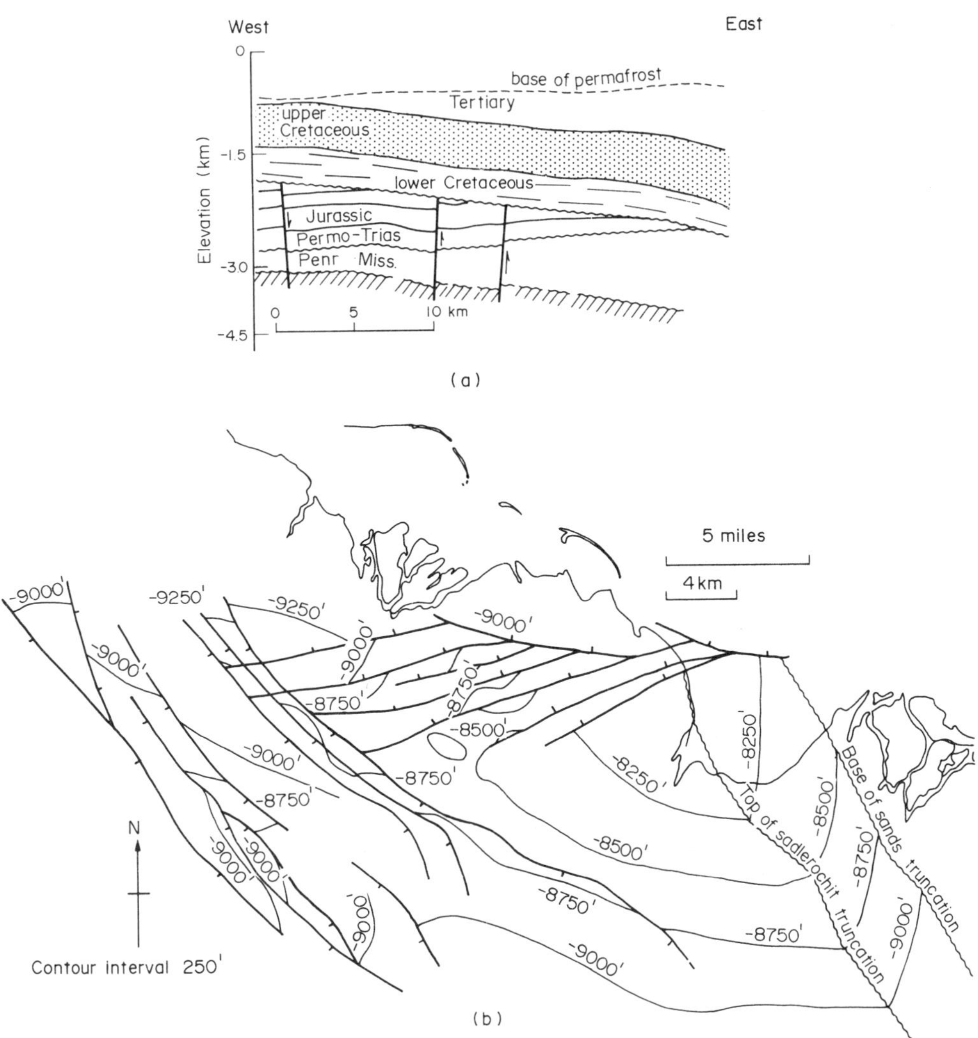

Figure 8
Prudhoe Bay field, Alaska: part (a) illustrates the eastward overlap of lower cretaceous across progressively older truncated reservoirs (after Perrodon 1983. © Societé Nationale Elf–Aquitane, Pau, France. Reproduced with permission); part (b) shows that the faulted anticline of reservoirs plunges gently westward (after Morgridge and Smith 1972. © American Association of Petroleum Geologists and Society of Exploration Geophysicists. Reproduced with permission)

3.2 Hugoton

Accumulation of gas in the Hugoton field of southwestern Kansas may represent the closest approach to a completely stratigraphic trap in the whole of North America. The contours in Fig. 13, drawn on top of one of the producing reservoirs, give no indication of significant structure. An "ideal" cross section of the Hugoton field (Fig. 13) indicates that Permian carbonates form the reservoirs and that closure for the trapping of gas occurs by irregular shaling-out of the carbonates updip. These carbonates consist of either drusy dolomites or oolitic limestones, and the character of their porosity may be diagenetic rather than original. The pattern of shaling-out contributes to hydrodynamic constraint, and to restriction of the accumulation by water, both undip and downdip. A length of more than 70 km, however, compensates adequately for this restraint. Hugoton, discovered in

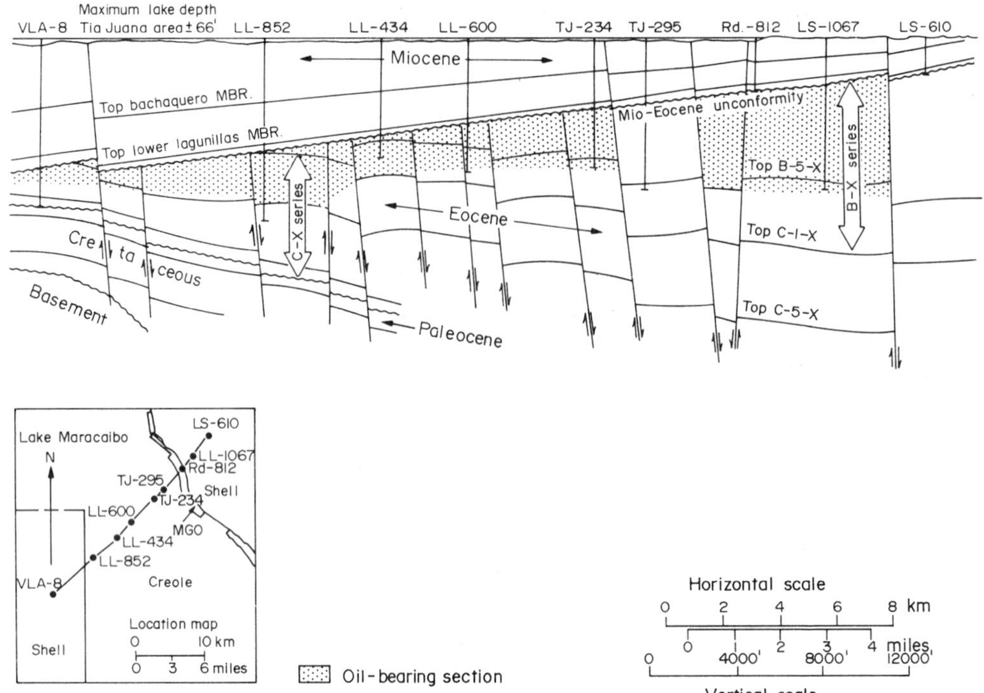

Figure 9
Bolivar Coastal field, Lake Maracaibo, western Venezuela: block faulting of main Eocene reservoirs under Miocene unconformity and Miocene overlapping reservoirs. (After Borger and Lenert 1959. © Fifth World Petroleum Congress, New York. Reproduced with permission)

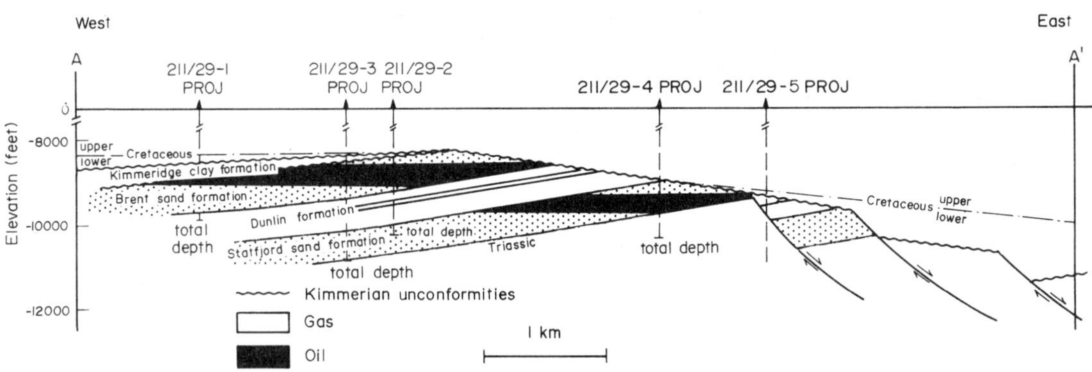

Figure 10
Brent field, North Sea: fault block: truncation of reservoirs at underside of incipient paleoscarp on late Mesozoic unconformity. (After Bowen 1975. © Wiley, New York. Reproduced with permission)

1927 or earlier, reportedly contained initial producible reserves of some 1.5×10^{12} m^3 of gas when fully delineated.

4. Conclusions

Ghawar, Burgan in Kuwait, and Samotlor are spectacular examples of the trapping power of major simple anticlines. These three together have trapped nearly 150 billion barrels of oil. The anticline of Urengoy has trapped 7×10^{12} m^3 of natural gas. Hundreds of lesser structures add their total of production to this same class of simple anticlines. Among major examples of anticlines which are less simple, strongly fractured or intensely folded are the Kangan and Gach Saran anticlines in Iran, Groningen and the Cantarell com-

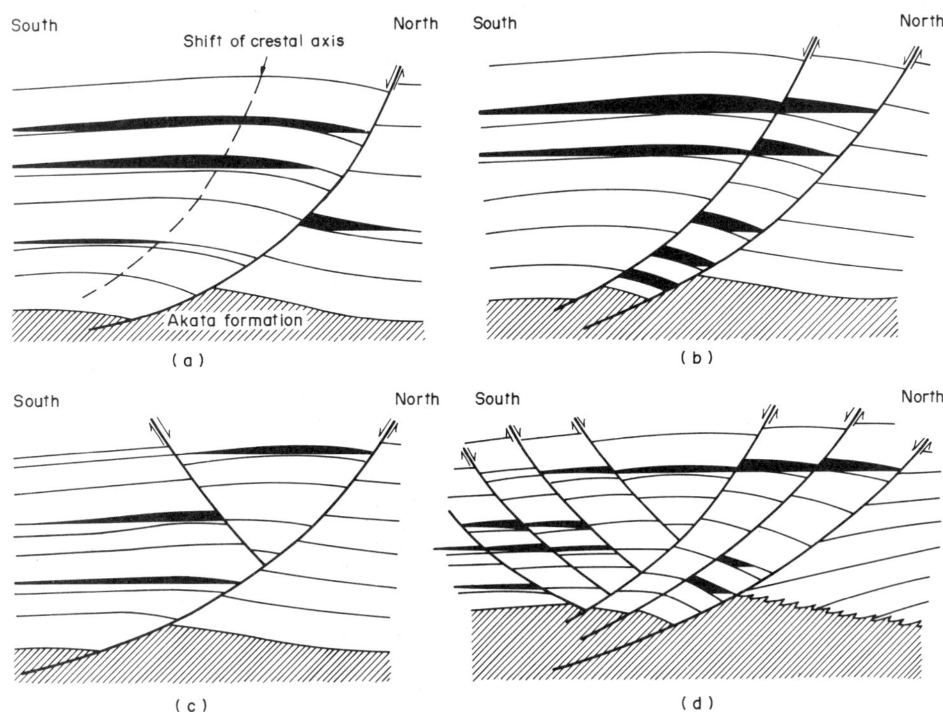

Figure 11
Niger delta fields, west Africa: models of Cenozoic reservoirs closed by listric and antithetic faults: (a) simple rollover structure; (b) structure with multiple growth faults; (c) structure with antithetic fault; and (d) collapsed chest structure. (After Whiteman 1982. © Graham & Trotman, London. Reproduced with permission)

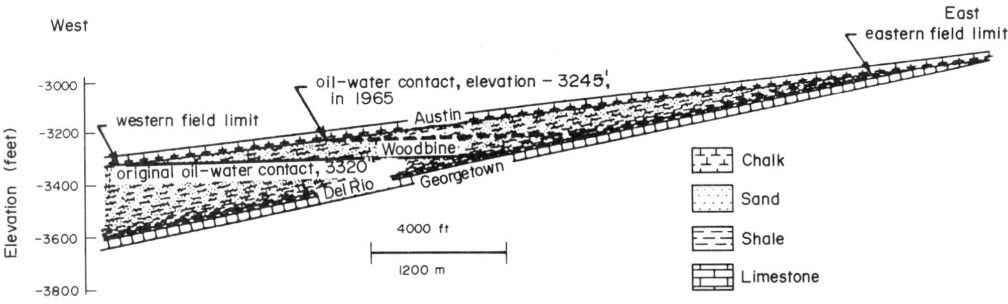

Figure 12
East Texas field: truncation of reservoir (Woodbine Sand) below onlap of Austin Chalk; western plunging anticlinal nose. (After Landes 1970. © Wiley, New York. Reproduced with permission)

plex. In all, the anticlinal concept has guided explorationists to considerable success for more than a hundred years.

The stratigraphic parameter becomes as important as the anticlinal parameter in fields such as Hassi R'Mel in Algeria and Prudhoe Bay, as well as in areas such as Bohai and the Niger delta. The more truly stratigraphic fields, as in east Texas and Hugoton, demonstrate that major accumulations can occur without the trapping power of the anticline. The anticline, either simple, relatively simple, relatively complicated or strengthened by a stratigraphic para-

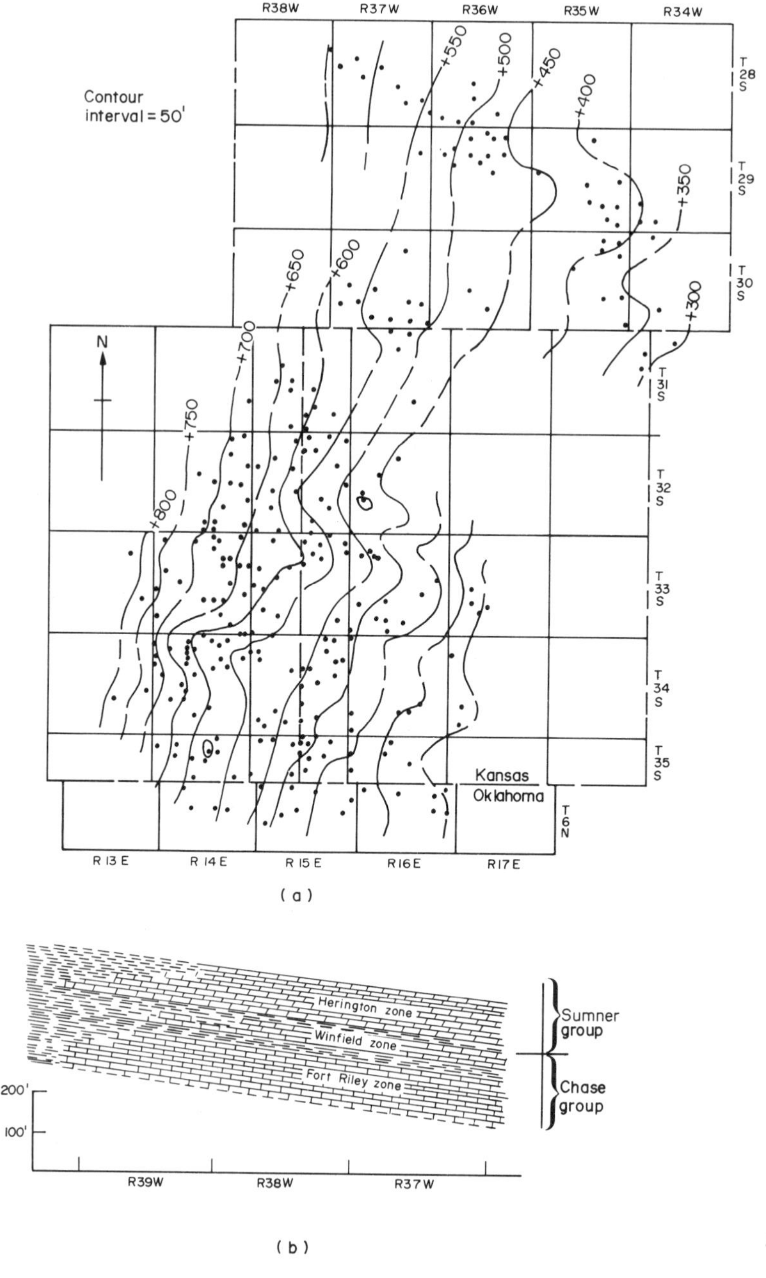

Figure 13
(a) Hugoton field, Kansas, Oklahoma: stratigraphic trap on monocline by updip shaling out of carbonates. (b) Ideal cross section of Hugoton field showing clastic increase westward (After Landes 1951. © Wiley, New York. Reproduced with permission)

meter, appears likely to continue strong as a concept in the search for additional hydrocarbons. Neither land nor the well-drilled continental shelves, however, seem likely as places to repeat the huge successes of Ghawar, Burgan and Samotlor; more likely targets may consist of the continental slopes, if drilling intensifies on them in the decades to come.

For the more immediate future, explorationists may increase their emphasis on the stratigraphic contribution. The resolving power of modern reflection seismography often brings into remarkably clear view the paleotopography of unconformities and the presence of traps on their undersides. For many countries, the intensive drilling of prospects with indicated stratigraphic parameters has become imperative. The results might not reverse the ongoing slide of these countries into exploratory maturity, but possibly they may slow its pace.

See also: Hydrocarbons: Origin, Migration and Accumulation; Petroleum: World Resources

Bibliography

Acevedo J S 1980 Giant fields of the southern zone, Mexico. In: Halbouty M T (ed.) Giant oil and gas fields of the decade 1968–1978. *Mem. Am. Assoc. Pet. Geol.* 30: 339–85

Borger H D, Lenert E F 1959 The geology and development of the Bolivar Coastal field at Maracaibo, Venezuela. *Proc. 5th World Petroleum Congr.*, Sect. 1. Fifth World Petroleum Congress, New York, pp. 481–98

Bowen M J 1975 The Brent oil-field. In: Woodland A W (ed.) 1975 *Petroleum and the Continental Shelf of North-West Europe*, Vol. 1, *Geology*. Wiley, New York, pp. 353–61

Lamb C F 1980 Painter Reservoir field—Giant in the Wyoming thrust belt. In: Halbouty M T (ed.) Giant oil and gas fields of the decade 1968–1978. *Mem. Assoc. Pet. Geol.* 30: 281–88

Landes K K 1970 *Petroleum Geology of the United States*. Wiley, New York

Meyerhoff A A 1982 *Petroleum Basins of the Union of Socialist Soviet Republics and the People's Republic of China and the Politics of Petroleum*. Petroleum Exploration Society of Australia, Melbourne

Morgridge D L, Smith W B Jr 1972 Geology and discovery of Prudhoe Bay field, eastern Arctic Slope, Alaska. In: King R E (ed.) 1972 *Stratigraphic Oil and Fields—Classification, Exploration Methods and Case Histories*. American Association of Petroleum Geologists and Society of Exploration Geophysicists, Tulsa, Oklahoma, pp. 489–501

Perrodon A 1983 Dynamics of oil and gas accumulations. *Bull. Cent. Rech. Explor. Prod. Elf-Aquitaine, Mem.* 5

Staff of Kuwait Oil Company Ltd 1953 Kuwait. In: Illing V C (ed.) 1953 *The Science of Petroleum*, Pt. 1, Vol. 6. Oxford University Press, Oxford, pp. 99–100

Stauble A J, Milius G 1970 Geology of Groningen Gas Field, Netherlands. In: Halbouty M T (ed.) 1970 *Geology of Giant Petroleum Fields*. American Association of Petroleum Geologists, Tulsa, Oklahoma, pp. 359–69

Thralls W H, Hasson R C 1956 Geology and oil resources of eastern Saudi Arabia. In: Guzman E J (ed.) 1956 *Mexico, Int. Geological Congr. XX: Symposium Sobre Yacimientos de Petróleo y Gas*, pp. 11–32

Whiteman A 1982 *Nigeria: Its Petroleum Geology, Resources, and Potential*. Graham and Trotman, London

M. Kamen-Kaye
[Cambridge, Massachusetts, USA]

Petroleum: World Resources

1. Petroleum and its Phases

The two roots of the word petroleum, *petra* and *oleum*, translate as "rock oil" which was a term used in the western world before 1850 to denote natural seepages of flammable liquid hydrocarbons. The term rock oil continued to be used informally into the early years of the twentieth century, but yielded gradually to the word "petroleum." For technical purposes, however, the modern concept of petroleum includes not only the familiar liquid phase of "crude oil" (and its "heavy oil") but also two phases that are lighter and heavier. The lighter phase is natural gas and the heavier phase is almost solid bitumen (with a subphase of solid asphalt). Therefore, a study of global reserves of petroleum, although concerned with crude oil, must also address itself to reserves of natural gas and bitumen. Reserves of natural gas have risen spectacularly since the 1950s whereas reserves of bitumen have changed little over the same period; but as world production of crude oil and natural gas decline, the phase of bitumen may find more place in a global economy of petroleum.

The gradation of bitumen to crude oil, and crude oil to natural gas, is obviously a gradation of heavy complexes of hydrocarbons to light complexes. For oil, specific gravities of the heaviest phases can be as great or greater than that of water. Specific gravities of the lighter phases of crude oil can be as little as 0.74 or less. The decimal fractions of specific gravity are cumbersome to use and consequently a scale of numbers issued by the American Petroleum Institute (API) has superseded them almost completely. Whereas specific gravities decrease as crude oil becomes lighter, corresponding API degrees increase in value. The correspondence between API degrees and specific gravities are shown in Table 1.

Below 10 degrees on the API scale, petroleum, whether viscous or virtually solid, sinks in water. The viscous phase or the virtually solid phase of petroleum are commonly referred to as tar or bitumen.

2. World Reserves of Crude Oil

For purposes of discussing reserves, the consensus of industry and government is followed which defines "conventional crude oil" as consisting of oil lighter than 10 degrees API. Available reserves of crude oil for the world as a whole can thus be placed at approximately 500 billion (10^9) barrels, distributed as shown in Table 2.

All of the data given in Table 2 are less than those generally quoted, but in the author's opinion, many estimates of available reserves are optimistic. The predominance of the Middle East, however, is clear: the data in Table 2 show 300 billion barrels for the Middle East and 250 billion barrels for the rest of the world.

Table 1
Correspondence between API degrees and specific gravities

API degrees	Specific gravity
60	0.74
50	0.78
40	0.83
30	0.88
20	0.93
10	1.00

Table 2
Distribution of available reserves of crude oil

Region	Barrels $\times 10^9$
Middle East	300
Latin America	80
Communist bloc	50
Africa	40
North America	30
Western Europe	10
Asia-Pacific	10

2.1 Middle East

Of the 300 billion barrels attributed to the Middle East, Saudi Arabia alone supplies an almost legendary 100 billion barrels or more. Almost as legendary is a possible remaining 50 billion barrels from a single oil field: Ghawar, the world's largest oil field, is a structure in Saudi Arabia that strikes north–south for at least 250 km, and produces oil from carbonate reservoirs at depths no greater than 2100 m. Kuwait, whose southern border is common with Saudi Arabia and which is a country covering an area that is minuscule compared to the area of Saudi Arabia, contains the world's second largest oil field. This is Burgan, a domoid anticline with an area of approximately 570 km^2 and with reserves of some 40 billion barrels. Depths to the producing reservoirs are remarkably shallow, less than 1500 m over most of this structure. These two individual oil fields in the Persian Gulf geosyncline (synclinorium) thus can supply approximately 100 billion barrels of reserves to the world's total of 500 billion barrels or more. The crude oil in such shallow horizons also gives them a tremendous economic advantage.

Elsewhere in the Middle East, Iran and Iraq each may have contained original proved reserves as great as 45 billion barrels. Since the 1930s, however, Iran has produced more than 30 billion barrels of crude oil. The corresponding figure for Iraq is more than 17 billion barrels. Iran and Iraq, for the present, draw on reserves in a relatively mature phase. The amount on which each can draw (roughly equal for both countries) could be some 25 billion barrels.

The remaining large contributor to crude-oil resources in the Middle East is Abu Dhabi, the chief sheikdom and capital of the United Arab Emirates, towards the southern end of the Persian Gulf. Abu Dhabi has produced approximately 8 billion barrels of crude oil since the mid-1960s. Some 20 billion barrels possibly remain for continued production.

2.2 Latin America

Mexico and Venezuela are the two major contributors to reserves in Latin America. Both countries are mature producers of crude oil, but the leadership in reserves has changed hands in recent years. In Venezuela, reserves of crude oil had remained in the range 15–20 billion barrels for several decades. In Mexico, the discovery of prolific reefal and reefoid carbonates in southern areas and their marine waters in the 1970s raised reserves of crude oil to an astonishing 50 billion barrels. A cumulative production of some 13 billion barrels for Mexico is attributable in large part to discoveries made before the 1970s, and therefore has made small inroads on reserves available at the end of 1987. In Venezuela, there were no extremely large additions to early discoveries. Exploitation, however, proceeded intensely, and by 1987 had resulted in a cumulative production of some 40 billion barrels, an amount exceeded by no other country in the non-communist world except Saudi Arabia (52 billion). The net result of Venezuela's unusually large cumulative production has been to draw down available reserves of crude oil to a figure of 18 billion barrels or less. Venezuela itself, however, claims more than 30 billion.

2.3 Communist Bloc

The USSR is the only member of the communist bloc of nations whose production and reserves have attained global prominence. After nearly a century of development of crude oil and a cumulative production of some 70 billion barrels, the USSR may have reached a relatively mature stage in its total of available reserves. To take one indication of maturity, the giant Samotlor oil field was in its twenty-second year of production in 1987, and had produced some 8 billion barrels of oil from an original reserve of 17 billion barrels. Large petroliferous areas of the USSR remain to be explored, but any serious challenge to the prolific production of the Middle East may never materialize. The available reserves of the USSR may amount to some 40 billion barrels although some estimates reach 70 billion barrels.

China is the only other country in the communist bloc whose reserves merit consideration in a global perspective. Its giant oil field, Daqing, is notable not only for its production of more than 5 billion barrels of oil up to 1987 but also for its remarkable combination of continental source beds and continental reservoir beds. At least two other oil fields or oil-field areas in China have produced more than 1 billion barrels apiece. Remaining reserves, not firmly reported, may not be large. The figure of 10 billion barrels may be too high and commonly quoted figures higher than 10 billion do not appear reasonable. Large areas of northwestern China remain to be explored intensively. These northwestern areas, already productive in part, may yield sufficient oil over the next decade or two to maintain available reserves at about the present levels.

2.4 Africa

The available reserves of crude oil in Africa are low compared to the huge area of the continent. The major contributors to reserves, namely Libya, Nigeria and Algeria, have produced prolifically, but they are now in a mature phase. Reserves for Africa as a whole may have decreased to 10 billion barrels or less by the end of 1987, but most contemporary estimates show much higher figures.

2.5 North America

The two contributors to this large area, the USA and Canada, have reached a mature phase with regard to petroleum. Total discovery of crude oil in the USA amounted to some 160 billion barrels by 1987, but considerably more than a century of production has reduced available reserves to little more than 20 billion barrels. In Canada, significant production of crude oil is less than 40 years old, and available reserves amount to about 6 billion barrels.

2.6 Western Europe

Production of crude oil in western Europe remained unimportant until the end of the 1960s. Groningen in the Netherlands, a major discovery of gas with some 2×10^{12} m^3 of reserves, had indicated that large quantities of hydrocarbons existed on the southern rim of the North Sea basin. By the end of the 1960s, exploration had moved out into the North Sea itself, and by the late 1970s, initial reserves of crude oil under the waters of the North Sea had reached a total of $\sim$10 billion barrels. Since that time, cumulative production in the UK sector of the North Sea has passed 6 billion barrels, to leave a reserve of $\sim$5 billion barrels. Cumulative production in the Norwegian sector has amounted to 2 billion barrels, to leave a reserve of $\sim$9 billion barrels.

2.7 Asia-Pacific

Two areas contribute to the standing of the Asia-Pacific region. The larger contribution comes from the mature petroliferous province of Indonesia–Malaysia; the smaller contribution from the more recent province of the "Bombay High" in marine waters off the western coast of India. The total amount of crude oil discovered in Indonesia–Malaysia is approximately 20 billion barrels. A cumulative production of nearly 14 billion barrels thus leaves some 6 billion barrels for available reserves. Production from the marine waters of the Bombay High, from the late 1970s to the late 1980s, amounted to about 2 billion barrels. The balance for available reserves may amount to 3 billion barrels.

3. World Reserves of Natural Gas

Since the late 1950s, discoveries of several deposits of gas, each larger than 6×10^{12} m^3, have recharged the global supply to a considerable degree. Deliverability of these gigantic accumulations may constitute a problem in some cases. In other cases, a single great field such as Urengoy has enabled the USSR to pipe natural gas as far as central and western Europe. In earlier years, discoveries like Hugoton in the USA, Groningen in the Netherlands, Hassi R'Mel in Algeria and others, had brought proved global reserves to a level of some 20×10^{12} m^3 by 1960. Discoveries made since that time may have raised that figure to as high as 85×10^{12} m^3 for the closing years of the 1980s. Data for the major geographical regions are given in Table 3.

3.1 Middle East

Deeper drilling since the 1970s has revealed gigantic accumulations of gas in carbonates of the Permian (latest Paleozoic) Khuff Formation. In Iran, the result has been the discovery of possibly 11.3×10^{12} m^3 of gas at or around Kangan near the eastern shore of the Persian Gulf. Confirmation of this discovery would mean that this one district holds some four times as much natural gas as that now available in the USA. Other reserves of natural gas from the Khuff Formation are as follows: approximately 2.8×10^{12} m^3 in Qatar; 2.8×10^{12} m^3 in Saudi Arabia; and 2.27×10^{12} m^3 in Abu Dhabi. These reserves, relatively little produced into the late 1980s, remain largely available.

3.2 Communist Bloc

Major reserves of natural gas in the communist bloc consist essentially of those within the USSR. The published figure of approximately 42.5×10^{12} m^3 may represent original reserves rather than reserves still available in the late 1980s. Even reserves which are still available, however, probably are considerably larger than those of the Middle East. At least for the near future, the USSR will continue to lead the world in reserves. One field alone has made a considerable contribution over the years. This is Urengoy, a sharply

Table 3
Distribution of global reserves of natural gas

Region	Reserves 10^{12} m^3
Communist bloc	42.48
Middle East	19.82
North America	7.64
Africa	4.81
Western Europe	3.96
Asia-Pacific	3.40
Latin America	3.11

folded anticline some 200 km long, with proved ultimate reserves of some 7.1×10^{12} m^3.

To China, at least one report has assigned 0.85×10^{12} m^3 of gas, but (this figure) may need downward revision.

3.3 Africa

Comparatively few details are available concerning the figure of 4.8×10^{12} m^3 assigned to Africa. Data such as 10^{12} m^3 for Nigeria and 0.85×10^{12} m^3 for Libya, may include a considerable amount of gas cap; that is, gas above oil in a common reservoir. In such cases the production of oil draws down much associated gas, which is thus lost to reserves. In Algeria, the big single element of reserves is the Hassi R'Mel gas field which was mentioned in Sect. 3. Hassi R'Mel, by far the largest gas field in the whole of Africa, has been producing natural gas steadily since the 1960s. Its original reserves of some 10^{12} m^3 have probably decreased considerably. Gas fields discovered later than Hassi R'Mel may have added a figure of the order of 10^{10} m^3 to the reserves of Algeria, but a reported figure of 2.8×10^{12} m^3 for this country may be an overestimation.

3.4 North America

Hugoton, discovered in the heartland of the USA in 1918, was one of the most prominent gas fields of the world, with initial reserves of some 1.4×10^{12} m^3. There has been no successor of this stature in the USA, and reported reserves of some 5×10^{12} m^3 for the nation as a whole represent the sum of contributions from hundreds of small and medium fields. The rate of discovery of gas in the USA was of the order of 0.2×10^{12} m^3 per year in the late 1980s. Consumption during that same period, however, was rising toward 0.6×10^{12} m^3 per year. The drain on reserves is clearly a matter of concern. In Canada, where the population is relatively small, domestic consumption of natural gas is correspondingly small. Important resources, consequently, are available in Canada for significant delivery to the USA if the published figures of 2.6×10^{12} m^3 for "proved reserves" are correct.

3.5 Latin America

Individual fields with natural-gas reserves of the order of 10^{11} m^3 seem not to exist in Latin America. It is assumed, therefore, that fields no more than moderately large are the joint contributors to the published national "proved reserves" as follows: Argentina, 0.6×10^{12} m^3 Venezuela, 10^{12} m^3 and Mexico, 2×10^{12} m^3.

3.6 Western Europe

In the 1950s, Groningen in the Netherlands stood out as a unique phenomenon. Its initial reserves of 1.4–1.7×10^{12} m^3 of natural gas moved western Europe onto the global scene. Groningen continues to rank high as a discovery, and is also prominent owing to its remaining reserves, despite considerable production and distribution since the 1960s. The Netherlands is possibly capable of providing some 10^{12} m^3 of gas in the future. By the 1970s, however, Groningen was no longer alone. Both the UK on the western side of the North Sea basin and Norway on the eastern side had gained prominence in gas. Fields for the most part an order of magnitude less than Groningen in size had brought proved reserves of approximately 10^{12} m^3 to Norway and 0.6×10^{12} m^3 to the UK. Heavy domestic consumption in the UK has lowered its reserves significantly, but a recent resumption of exploration and discovery has slowed the decline. Norway, with a low domestic consumption, and with good prospects of relatively large future discoveries of natural gas, has suffered little decline of reserves.

3.7 Asia-Pacific

A steady accession of discoveries has occurred across this widespread area since the 1960s. None of these has been of phenomenal size, but many have occurred in nearshore waters as marine technology has developed. Available reports show several countries with proved reserves above 0.3×10^{12} m^3 of natural gas. Indonesia and Malaysia lead, each with approximately 0.85×10^{12} m^3 of natural gas. Pakistan, India and Australia follow, each with approximately 0.42×10^{12} m^3. The figure for Bangladesh may be around 0.3×10^{12} m^3.

4. World Reserves of Bitumen

Except for the great spread of sands in the Athabasca region of northwestern Canada, where only bitumen (tar) occurs, a distinction between bitumen and heavy oil does not appear in reports of other regions. The remarkable "Petroliferous Belt of the Orinoco" ("La Faja Petrolífera del Orinoco") in Venezuela contains many billion barrels of bitumen in the subsurface. Its potential producible reserves, however, lie much more in its heavy oils that, although viscous, might rise to the surface by the stimulation of steam at high pressure. This is the well-known in situ method of recovery. If in situ methods can succeed in the Orinoco region, a claim of one or two hundred billion barrels of reserves is within reason. Whether this method is practical may not be known for several decades. In either case, it is unlikely that bitumen in any significant amount can be moved from subsurface reservoirs to surface either in this region or elsewhere. In the USSR, available reports credit Melekess with some 100 billion barrels of bitumen, with no mention of heavy oil. Unless the bitumen of Melekess is minable at surface or below an overburden of less than 60 m its contribution to reserves is problematic.

For bitumen *sensu stricto*, only the tremendous tar sands of Athabasca in northwestern Canada can make an indisputable contribution to global reserves. Two mining plants now at work in open pit in the

Athabasca region could recover a cumulative 1.5 billion barrels of bitumen in the 1990s. The economic limit of overburden for the near future amounts to about 40 m, and within that limit it appears that a total reserve of some 35 billion barrels of bitumen exists. In the remainder of the Athabasca prospect, the volume of bitumen in place below 180 m amounts to several hundred billion barrels. Recovery of this deeper material by in situ methods seems remote at present.

5. *Outlook for World Reserves of Petroleum*

5.1 Crude Oil

Insofar as this is the age of the internal combustion engine, we must assume that crude oil will continue to play a pivotal role in the economic, social and political development of the world. With global consumption of crude oil moving toward a total of 60 million barrels a day, available reserves would decrease by more than 20 billion barrels per year. By simple arithmetic, within 25 years present reserves of crude oil would be consumed and this highlights the contrived character of a "glut of oil" in 1987.

For the western world, the economics of discovery are nothing less than hostile. Despite high technological capability, nations such as the USA, Canada, Mexico, Venezuela and the UK eventually may be forced into politically unpopular measures to sustain intensive exploration in the hope of stemming declines in their reserves. The USSR eventually may face a similarly uncomfortable necessity, possibly exacerbated if that nation's technology continues to lag.

The future of crude oil, as the general press has repeatedly stressed, lies critically with the countries of the Middle East, especially if their reserves should amount to 500 billion barrels instead of the 300 billion barrels assumed in this article. The Middle East, at either figure, is obviously less pressed than other regions, but within a few decades it will become necessary to mount a search for new reserves in this area. Whatever the ultimate result of that search might prove to be, it would play an important part in carrying the world relatively far into the twenty-first century.

5.2 Heavy Oil

This phase of petroleum, transitional between oil of natural or pumped flow and semisolid bitumen, lies in the range 10–15 API degrees. The possible importance of heavy oil in the future of petroleum stems in large measure from two main sources. First is the heavy oil of the "Petroliferous Belt of the Orinoco" in Venezuela. Claims of several hundred billion barrels of hydrocarbons in place for this belt appear reasonable. Less clear is the ratio of essentially immovable bitumen to heavy oil. Before billions of barrels of heavy oil can be forced to surface, the in situ method must be proved not only petrophysically but also economically feasible. The first billion barrels of heavy oil from the Orinoco region may not appear until after the first decade of the twenty-first century. In northwestern Canada, apparently large amounts of heavy oil exist and are already under limited production by moderate in situ methods.

5.3 Natural Gas

The shape of the future for natural gas is not unlike that for crude oil in terms of global economics, social development and politics. The unbalanced pattern of distribution follows quite closely that of crude oil. Within this pattern the only notable difference consists of an exchange of places at the head. The USSR held first place in reserves of natural gas during the late 1980s by a considerable margin. If countries in the Middle East should follow an aggressive program of deep drilling in the widespread Khuff Formation, however, the Middle East eventually could overtake the USSR. Natural-gas reserves of the order of 10^{12} m^3 in the Middle East could extend global use for a significant period.

5.4 Bitumen

As already stated, to discuss bitumen is to discuss the minable tar sands of Athabasca in northwestern Canada. In this case a sharp distinction exists within the phase of bitumen itself. The minable bitumen, with approximate reserves of 30 billion barrels, is technologically readily available for conversion to crude oil and for addition to global reserves, especially to those of North America as a whole. The several hundred billion barrels of bitumen in place below 180 m at Athabasca unfortunately are likely to resist in situ attempts to dislodge them even more than the recalcitrant billions of heavy oil in the Orinoco region of Venezuela. At Athabasca, in fact, the prolific volumes of bitumen below the reach of mining may prove so obdurate as never to become a relevant factor in the world's future reserves of petroleum.

6. *Conclusion*

Unless economic or political factors change in a way that encourages intensive and incessant programs of exploration and development, the age of petroleum in the western world could enter difficult times. The Middle East and the USSR could conceivably sustain the rest of the world and themselves on a spartan basis well into the twenty-first century. That these two regions would consent to such an arrangement, however, is highly doubtful. The viability of the western world in petroleum must rest upon its own efforts, redoubled especially in the technological sphere. The western world should be investing sums an order of magnitude greater than those of today on research into special methods of recovery, irrespective of the fact that these methods (oil shale included) are at present largely uneconomical. As reserves decline, the uneconomical processes of today could become the mainstay of tomorrow.

See also: Hydrocarbons: Origin, Migration and Accumulation; Petroleum: Oil and Gas Fields

Bibliography

Alayeto M B E, Louder L W 1974 The geology and exploration potential of the heavy oil sands of Venezuela (the Orinoco Petroleum Belt). In: Oil sands of the future. *Mem. Can. Soc. Pet. Geol.* 3: 1–18

Anon 1986 Worldwide report. *Oil Gas J.* 84(51/52): 33–73

Atwater G I 1980 Petroleum. In: Goetz P W (ed.) 1980 *The New Encyclopaedia Britannica, Micropaedia*, Vol. 14. Benton, Chicago, Illinois, pp. 164–75

Govier G W 1974 Alberta's oil sands in the energy supply picture. In: Oil sands of the future. *Mem. Can. Soc. Pet. Geol.* 3: 35–49

Grunau H R 1984 Natural gas in major basins worldwide attributed to source rock type, thermal history and bacterial origin. *Proc. 11th World Petroleum Congr.*, Vol. 2. Wiley, Chichester, pp. 229–49

Halbouty M T 1979 World ultimate reserves of crude oil. *Proc. 10th World Petroleum Congr.*, Vol. 2, *Exploration Supply and Demand.* Heyden, London pp. 291–301

Halbouty M T 1984 Reserves of natural gas outside the communist bloc countries. *Proc. 11th World Petroleum Congr.*, Vol. 2. Wiley, Chichester, pp. 281–92

Janisch A 1981 Oil sands and heavy oil: Can they ease the energy shortage? In: Meyer R F, Steele C T (eds.), Olson J C (asst. ed.) 1981 *The Future of Heavy Crude and Tar Sands.* McGraw-Hill, New York, pp. 33–41

Jardin D 1974 Cretaceous oil sands of western Canada. In: Oil sands of the future. *Mem. Can. Soc. Pet. Geol.* 3: 50–67

Masters C D, Root D H, Dietzman W D 1984 Distribution and quantitative assessment of world crude oil reserves and resources. *Proc. 11th World Petroleum Congr.*, Vol. 2. Wiley, Chichester pp. 229–49

Meyer R F, Steele C T (eds.), Olson J C (asst. ed.) 1981 *The Future of Heavy Crude and Tar Sands.* McGraw-Hill, New York

Meyerhoff A A 1979 Proved and ultimate reserves of natural gas and natural gas liquids in the world. In: *Proc. 10th World Petroleum Congr.*, Vol. 2, *Exploration Supply and Demand.* Heyden, London, pp. 303–11

Meyerhoff A A 1982 *Petroleum Basins of the Union of Socialist Soviet Republics and the People's Republic of China, and the Politics of Petroleum.* Petroleum Exploration Society of Australia, Melbourne

Mossop G D, Kramers J W, Flach P D, Rotenfusser B A 1981 Geology of Alberta's oil sands and heavy oil deposits. In: Meyer R F, Steele C T (eds.), Olson J C (asst. ed.) 1981 *The Future of Heavy Crude and Tar Sands.* McGraw-Hill, New York, pp. 197–207

Root D H, Atanasi E D, Turner R M 1987 *Statistics of Petroleum Exploration in the Non-Communist World Outside the United States and Canada*, US Geological Survey Circular 981. USGS, Washington, DC

Schmerling L 1982 Petroleum. In: Parker S P (ed.) 1982 *McGraw-Hill Encyclopedia of Science and Technology*, Vol. 10. McGraw-Hill, New York, pp. 75–77

Walters E J 1974 Review of the world's major oil sand deposits. In: Oil sands of the future. *Mem. Can. Soc. Pet. Geol.* 3: 240–62

Zarrug A Y, Bois C 1984 Potential oil reserves in the Middle East and North Africa. *Proc. 11th World Petroleum Congr.*, Vol. 2. Wiley, Chichester, pp. 261–75

M. Kamen-Kaye
[Cambridge, Massachusetts, USA]

Pigments: Industrial Minerals

Mineral pigments are insoluble substances used to impart coloring, opaqueness or protecting capabilities to, or in combination with, a wide variety of metals and nonmetallic materials. Pigments are widely used in products such as paints and enamels, rubber systems, plastics, concrete and clay building materials, wood finishes, drugs, papers, leather finishes, inks, animal feeds, metal primers and electronic components.

The use of mineral pigments for coloring purposes, particularly iron oxide, goes back to prehistoric times. Until the late nineteenth century, only natural minerals were used for pigments. In the early part of the twentieth century, chemical methods were developed for producing what are now known as synthetic pigments, and iron oxides were the predominant materials produced. Synthetic production offers greater uniformity and more colors than can be achieved with natural minerals. However, some natural iron oxides continue to be used because of lower cost or uniqueness of color.

Pigments are generally classified by mineral or color. White is the predominant color in both total volume and product value. Titanium dioxide (TiO_2) is the principal white pigment mineral although there are many others such as zinc oxide (ZnO), silica (SiO_2), kaolin ($Al_2O_3.2SiO_2.2H_2O$), white lead (lead carbonate, sulfate or silicate), antimony trioxide (Sb_2O_3), whiting ($CaCO_3$), gypsum ($CaSO_4$), magnesium silicate ($Mg_3(SiO_3)_4$) and carbonate ($MgCO_3$), mica ($K_2O.3Al_2O_3.6SiO_2.2H_2O$), barium sulfate ($BaSO_4$) and other aluminum silicates.

Nonwhite pigments include an unlimited variety of materials, producing an infinite spectrum of colors. The principal natural coloring pigment, iron oxide, produces a variety of red, yellow, orange and brown colors. Synthetic iron oxides are produced usually from iron salts, which are generally obtained from waste products and alkalis. For example, large quantities of ferrous sulfates are obtained from steel-industry pickle liquors, titania production or scrap iron. Synthetic iron oxides offer a broader range of colors, extending to blacks. The largest US consumers of iron-oxide pigments are the coatings (37%) and construction-materials (22%) industries. The latter includes concrete, cement, bricks, roofing tiles and mortar.

Other colors are produced by chromium oxide (green), cadmium sulfide (yellow, orange, red), ferriferrocyanide (blue), mercuric sulfide (red), cobalt oxide (blue), nickel antimony titanates (yellow), manganese phosphate (violet), lead chromate (yellow), lead chromate–molybdate–sulfate (orange, red) and others.

Pigments used in protective (anticorrosive) coatings, where color is less important, include red lead (Pb_3O_4), lead silicochromate ($PbSiO_3.3PbO$ and $PbCrO_4.PbO$ on SiO_2 cores), zinc chromate ($Zn_5CrO_{12}H_8$), strontium chromate ($SrCrO_4$), calcium–strontium–zinc molybdates ($CaSrZnMoO_4$), calcium plumbate (Ca_2PbO_4) and specular hematite (Fe_2O_3).

Pigments of a specialized type which display luminescence (phosphors) have a variety of uses. These pigments are able to convert absorbed energy to photons of emitted energy in excess of thermal radiation, usually in the visible or near-visible range. Minerals used for luminescence in areas such as paints, plastics, inks, paper coatings, textile markings and crayons include zinc sulfide and zinc cadmium sulfide, with activators such as copper, silver or manganese. Phosphorescent pigments which provide an afterglow in products such as paints, plastics and tapes are made from such minerals as zinc and cadmium sulfides activated with copper, and calcium and strontium sulfides activated with bismuth or copper. Luminescent pigments can be designed to be highly temperature sensitive, and have achieved notable usage in sophisticated infrared scanning devices to map surface temperatures.

See also: Pigments: Titania, Iron Oxides and Other Minerals

Bibliography

Hancock K R 1975 Mineral pigments. In: Lefond S J (ed.) 1975 *Industrial Minerals and Rocks*, 4th edn. American Institute of Mining, Metallurgical and Petroleum Engineers, New York, pp. 335–57

M. A. Schwartz
[US Bureau of Mines, Washington, DC, USA]

Pigments: Titania, Iron Oxides and Other Minerals

The two most widely used pigment minerals are titania and iron oxide, both of which exist in a number of forms. This article describes the nature and characteristics of these and other pigment minerals which can be synthesized or exist in nature.

1. Titania

Titania (titanium dioxide, TiO_2) is the most widely used white pigment and is available in two crystal forms, both characterized by high refractive indices, 2.55 for anatase and 2.76 for rutile. Since 1957, with the introduction of the chloride process for recovering titania, high-grade rutile has become the predominantly used material.

The production of titania pigment in the USA in 1986 amounted to about 832 kt, valued at about US \$12 billion. This accounts for 33% of the world's capacity, followed by the Federal Republic of Germany (13%), the UK (9%), Japan (8%) and France (7%).

From a geological viewpoint, titanium is the ninth most abundant element of the lithosphere, comprising approximately 0.62% of the earth's crust. Until the early 1940s, the two principal titanium minerals, ilmenite ($FeTiO_3$) and rutile (TiO_2), came from sand deposits. Today, ilmenite production from rock deposits exceeds that from sand deposits. However, rutile is still produced only from sand deposits, and is only found in economic quantities in Australia. In the USA, the principal titanium ore reserves are in New York, Virginia, Florida, New Jersey and Georgia.

Present technology uses ilmenite for making titania pigment by the older sulfate process, either directly or through a high-titania slag as an intermediate form. Rutile is used with the newer chloride process and is preferred because less waste is produced. However, ilmenite can be blended with the rutile and used for chloride processing. A third mineral of commercial interest, leucoxene, can be included in the blending. It has no specific composition but is an alteration product of ilmenite from which a portion of the iron has been leached.

In the sulfate process, ilmenite (43–65% TiO_2) or titania slag (70–72% TiO_2) is converted to water-soluble sulfates, and the TiO_2 is recovered by hydrolysis. Impurities such as chromium, vanadium, niobium, manganese and phosphorus can adversely impair pigment properties, so care is required in selecting the ilmenite ores. In the chloride process, rutile (95% TiO_2) or high-TiO_2 ilmenite and rutile combinations (60–95% TiO_2) are converted to volatile chlorides and the TiO_2 is recovered by vapor-phase oxidation of titanium tetrachloride. The chloride process is continuous, is inherently simpler and has less waste material to dispose of as it uses feed materials with higher TiO_2 and lower iron contents than the sulfate process. The latter uses lower-cost raw materials but has a greater quantity of waste materials (spent acid and iron sulfate) which have not been economically recoverable. Efforts are being made to utilize lower-grade feed materials with the chloride process.

Commercial uses for titania pigments are in paints (49%), paper products (22%), plastics and synthetic products (12%), rubber products (3%), ceramics and glass (3%), fluxes (2%) and other applications (9%). Future needs for titania pigments are expected to increase continuously as the standard of living in developing countries rises.

2. Other White Mineral Pigments

White lead, $2PbCO_3.Pb(OH)_2$, the world's oldest white pigment material, varies in lead carbonate content from 62 to 80%. During the nineteenth century, it was the most important pigment used in oil-base paints. Even today it is used with other white pigments to improve adhesion, flexibility, resistance to water and general durability. It is produced by reacting lead or litharge (PbO) with acetic acid to produce soluble basic lead acetate, which is then reacted with CO_2 to form basic carbonate white-lead.

Basic sulfate white-lead is similar to the carbonate but is only used together with other pigments, zinc oxide in particular. Basic silicate white-lead pigments include several types, ranging from a core of silica covered with monobasic lead silicate and sulfate, to a coreless material composed wholly of complex lead silicates.

Zinc pigments comprise two compounds, zinc oxide and sulfide, which are used in several forms and modifications. Commercial use dates back to the early nineteenth century. In paints, zinc oxide aids in mixing and grinding, controlling consistency and sealing, and improves drying or hardening of the film (to reduce discoloration and chalking), self-cleaning and mildew resistance. Zinc sulfide, although more expensive, has a higher refractive index and better hiding power. It was widely used in combination with barium sulfate (lithopone) until the titania pigments took over the market.

Other white mineral pigments that also serve as extenders or fillers include whitings ($CaCO_3$), gypsum ($CaSO_4$), magnesium silicate ($H_2.Mg_3(SiO_3)_4$), magnesium carbonate ($MgCO_3$), kaolin ($Al_2O_3.2SiO_2.2H_2O$), mica ($K_2O.3Al_2O_3.6SiO_2.2H_2O$) and silica ($SiO_2$).

Table 1
Percentage iron-oxide consumption, by end use (1983)

End use	Natural	Synthetic
Coatings (industrial finishes, paints, varnishes, lacquers)	25	38
Construction materials (cements, mortar, preformed concrete, roofing granules)	25	23
Ferrites and other magnetic and electronic applications	4	14
Colorants for plastics, rubber, paper, textiles, glass, ceramics	16	14
Industrial chemicals (such as catalysts)	2	5
Animal feed and fertilizers	12	1
Foundry sands	14	
Other (including cosmetics and jeweller's rouge)	2	3

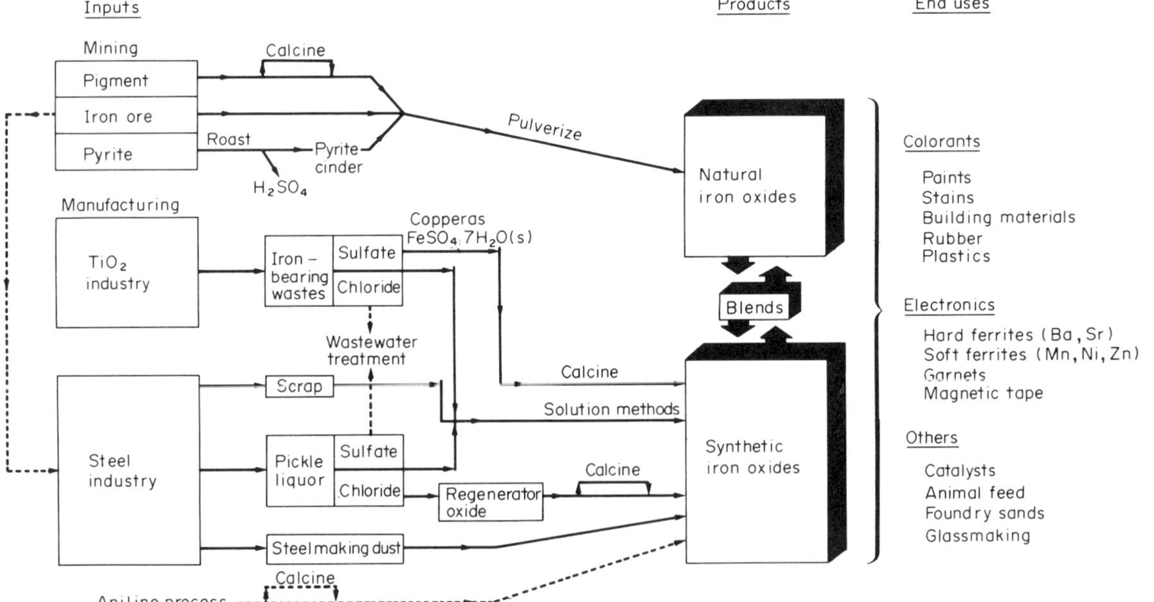

Figure 1
Relationship between iron oxide producers and other materials industries

3. Iron Oxide

Iron-oxide pigments are produced both from the ore and synthetically. In 1986 shipments by domestic producers of finished natural iron-oxide pigments in the USA amounted to about 42.9 kt, valued at about US$12 million, and output of synthetic iron oxide amounted to about 77.8 kt, valued at about US$119 million. Worldwide principal producers of natural iron-ore pigments in order of decreasing production include India, Spain, Cyprus, Germany, France, Austria and Brazil. Uses for iron-oxide pigments are listed in Table 1.

Four natural iron-oxide minerals are generally used for pigments. Hematite, the stable form of ferric oxide, Fe_2O_3, is the basis compound for red pigments. It has a rhombohedral crystal structure and is antiferromagnetic. Goethite ($2Fe_2O_3.2H_2O$), the second most common iron-bearing mineral, is associated with limonite ($2Fe_2O_3.3H_2O$), these being the basis compounds for natural yellow pigments (ochers, siennas and umbers). Goethite has an orthorhombic crystal structure and is antiferromagnetic, Magnetite, Fe_3O_4, is abundant worldwide and best known for its magnetic behavior. It has an inverse spinel structure and is the basis compound for black pigments.

Synthetic iron-oxide pigments offer the advantage of precise control of particle size and shape and chemical purity. Iron salts and alkalis are used to prepare the many shades of red, yellow, brown and black. The two principal salts of commercial importance are ferrous sulfate and ferrous chloride, usually obtained as waste materials from steel-mill pickle liquors and ilmenite processing plants, as well as from neutralizing scrap iron with sulfuric acid. The three methods used to manufacture synthetic iron oxides are (a) thermal decomposition of iron salts, (b) precipitation of iron salts accompanied by oxidation (Penniman–Zoph process), and (c) nitrobenzene reduction in the presence of iron (Aniline process). As shown in Fig. 1, the titania and steel industries are important sources for the iron-containing materials from which iron-oxide pigments are produced.

See also: Pigments: Industrial Minerals

Bibliography

Jolly J L W, Collins C T 1980 *Iron Oxide Pigments, 2: Natural Iron Oxide Pigments—Location, Production, and Geological Description*, US Bureau of Mines Information Circular 8813, USBM, Washington, DC

Jones T S 1978 *Iron Oxide Pigments, 1: Fine-Particle Iron Oxides for Pigment, Electronic, and Chemical Use*, US Bureau of Mines Information Circular 8771. USBM, Washington, DC

Lynd L E 1980 Titanium, *Mineral Facts and Problems*, US Bureau of Mines Bulletin 671. USBM, Washington, DC

Lynd L E, Hough R A 1983 Titanium. *Minerals Yearbook 1983*. US Bureau of Mines, Washington, DC

Patton T C (ed.) 1973 *Pigment Handbook*. Wiley, New York

Pettifer 1981 Iron oxide minerals: The other uses. *Ind. Miner.* 161: 53–63

Spinrad W I Jr 1983 Iron oxide pigments, *Minerals Yearbook 1983*. US Bureau of Mines, Washington, DC

Thomas A W 1980 *Colors From the Earth*. Van Nostrand Reinhold, New York

Watson I 1979 Iron oxide pigments: Colorful competition between natural and synthetic. *Ind. Miner.* 143: 43–51

M. A. Schwartz
[US Bureau of Mines, Washington, DC, USA]

Plate Tectonic Settings for Metallic Ore Resources

As a rule, metals are widely dispersed in the earth's crust or, if abundant, are chemically bound in the structure of refractory rock-forming minerals. Economically exploitable concentrations (ore deposits) are deposits in which metals are sufficiently concentrated, physically accessible and extractable by a reasonable expenditure of energy. They are rare and unequally distributed in the earth's crust. The processes whereby metals are concentrated are as varied as the geochemical behavior of metals. Their inhomogeneous distribution can be explained in a descriptive sense by the concept of metallogenic provinces (see *Metallogenic Provinces*) or, in a more fundamental sense, by combining various models of ore genesis with current theories of crustal evolution and dynamics. The following discussion describes how the current model of the earth's crust, plate tectonics, correlates with the origin and distribution of several types of metallic mineral deposits. This correlation is useful for studies of mineral genesis and in exploration for metallic ore deposits.

1. Structure of the Earth and Definition of Plates

The earth is divided into three zones—core, mantle and crust (see Fig. 1). These zones are distinguished by their physical and chemical properties, being, in fact, the products of petrochemical differentiation processes during the early history of the earth and of the same processes which continue, to a lesser extent, today. From a resource viewpoint, the crust is the principal accessible reservoir for metals, but the mantle, in particular the upper mantle, dominates crustal dynamics in the plate tectonic model.

The crust and mantle are distinguishable chemically, mineralogically and physically. Of the physical properties, the most significant are density and the elastic constants, E and μ, as these are the independent variables which determine seismic velocities. Petrochemically, the mantle is enriched in Fe, Mg and Ca (i.e., mafic) over the Si, Al and alkali-rich (i.e., sialic) crust. This static model, however, cannot characterize the earth as a dynamic body.

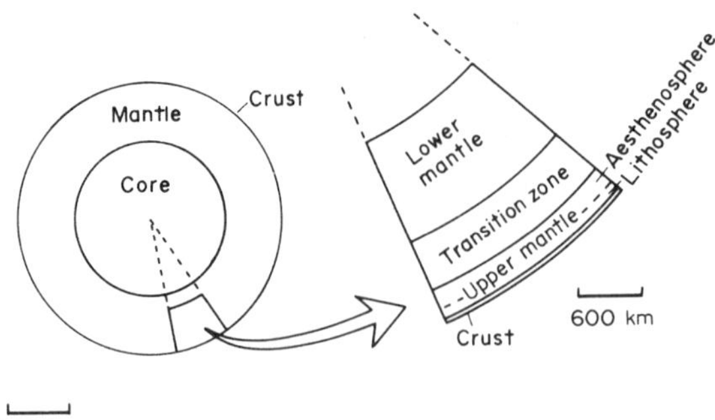

Figure 1
Internal structure of the earth with detail of mantle and crust including tectonic units of lithosphere and aesthenosphere

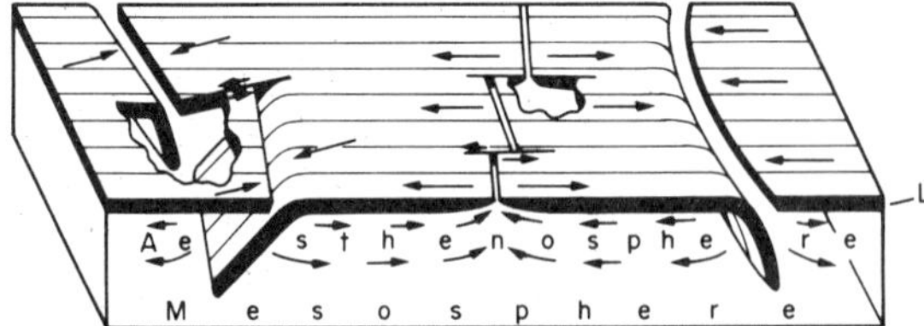

Figure 2
Schematic block diagrams of the earth's upper mantle and crust. Arrows on lithosphere (L) indicate relative movement of adjoining crustal blocks, arrows in aesthenosphere indicate compensating flow. Subduction zones with arc-to-arc transform on left, spreading center with ridge-to-ridge transforms in center, and simple subduction zone on right (after Isacks et al. 1968)

The mantle, in fact, consists of three concentric shells (see Fig. 1). The outermost, the upper mantle, is characterized by both horizontal and vertical zoning of various ultramafic rocks, such as peridotites, eclogites and dunites. At the boundary with the transition zone, a major phase transition (presumably pyroxene combining with olivine to produce garnet) takes place and is, in turn, followed at depth by a series of phase transitions and perhaps compositional changes through the transition zone to the lower mantle. The lower mantle is significantly denser owing to either a higher Fe/(Fe+Mg) ratio, a higher-order coordination of Fe and Mg in the silicate phases or a combination of these two. Within the upper mantle, at a depth of approximately 100 km and extending to depths of several hundred kilometers, is the low-velocity zone. This layer corresponds to a small percentage of partial melting in hydrated mantle material. The mantle and crust above the low-velocity zone acts as a rigid unit, called the lithosphere. The zone of partial melting is a layer of plastic material, the aesthenosphere.

Figure 2 shows the relationship between the aesthenosphere and the lithosphere, and demonstrates the concept that convective flow in the aesthenosphere results in predominantly lateral movement of the lithosphere. Given the constraints of a constant radius for the earth and the dynamics of convective flow in the aesthenosphere, the lithosphere becomes divided into discrete plates, moving relative to one another. The boundaries between these plates are of three types, and the characterization of these plate boundaries (see Sects. 2–4) is of significance to ore genesis. Finally, it is important to distinguish between environments where specific mineral deposits are formed and environments where, by accidents of geological history, they are now found. Deposits are described in terms of their environments of genesis and their present distribution noted.

2. *Constructive Plate Boundaries*

Where upwelling aesthenospheric convection brings hot mantle close to the surface, the lithosphere is subjected to tension, resulting in splitting the lithosphere into two plates which move away from one another (see Fig. 2). The combination of the upwelling of hot mantle, the introduction of H_2O and the pressure release associated with crustal tension results in partial melting of the mantle convection plume. Intrusion of mafic, mantle differentiates, primarily in the form of sheeted swarms of diabase dikes, and basaltic volcanism follow. The surface expression of these tensional boundaries are rift valleys, similar to the Rio Grande and East African Rifts, when continental plates undergo extension, and midocean ridges and associated submarine volcanoes and volcanic islands when oceanic crust is involved (see Fig. 3 for location of modern constructive boundaries, which are also called divergent zones or spreading centers).

The result of mantle-derived intrusion and volcanism along divergent plate margins is that these boundaries constitute the loci of creation of new oceanic crust. This new crust, which is compositionally a basalt, is emplaced at the spreading center and then, sequentially, it is rifted anew and younger ocean crust introduced. In combination with continuing aesthenospheric convection, this newly generated oceanic crust moves, on the rigid lithospheric plate, away from the zone of tension. Numerous experiments have confirmed that (a) the ocean floor is in motion; (b) the oceanic crust becomes progressively older as one moves away from zones of divergence; and (c) new, mantle-derived crust is being emplaced at these plate boundaries. Significant to ore genesis is the fact that these spreading centers, as appreciable crustal heat anomalies, drive large-scale aqueous convection cells.

Three fundamental classes of metallic mineral deposits are formed in spreading centers (see Table 1). The first is base-metal sulfide deposits which are generated by the high crustal heat flow over zones of

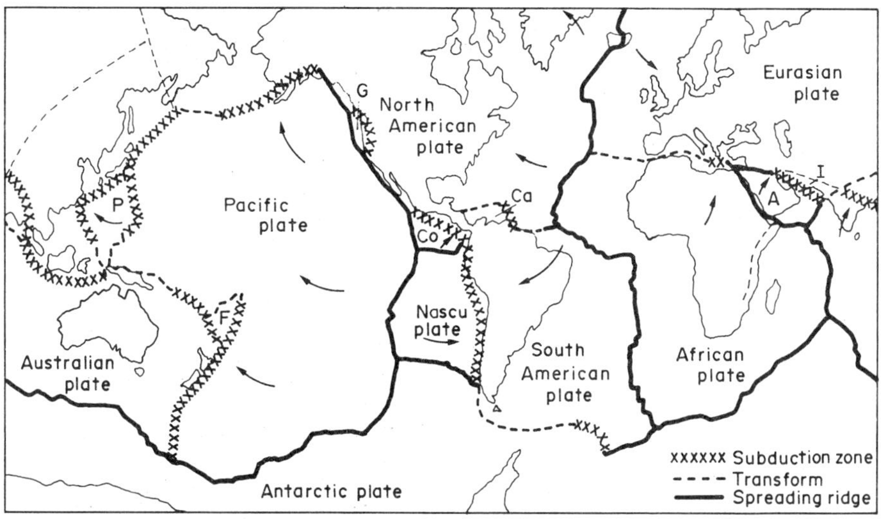

Figure 3
Major plates and plate boundaries on the earth: P, Philippine plate; F, Fiji plate; G, Gordo plate; Co, Cocos plate; Ca, Caribbean plate; A, Arabian plate; I, Iranian plate. Significant detail is missing in the Caribbean, Mediterranean Sea and southwestern Pacific owing to the scale of the map. Arrows indicate directions of modern plate movement

Table 1
Ore deposits associated with divergent zones

Type of deposit	Host rock	Example
1. Massive sulfide deposits:		
Cu–Fe–Zn	Basalts of ophiolite suite	E. Pacific Rise at 21°N Troodos Massif, Cyprus
Cu–Pb–Zn	Continentally derived sediments in aulocogens	Red Sea brines Sullivan, BC
2. Magmatic deposits:		
Podiform chromite	Ophiolitic peridotites	Moa, Cuba
Cu–Ni–Pt sulfides	Ophiolitic peridotites	Acoje, Luzon
3. Disseminated W–U–MO–Sn	Granites of alkaline affinities	Nigerian Sn granites

mantle upwelling. This heat drives large-scale seawater convection through the ocean crust with resulting hydrothermal alteration, metal scavenging by hot Cl-rich brines, and ultimately the precipitation of metal sulfides on the contact of the metal-rich brines with seawater at or near the seawater/sediment interface. In many cases, these deposits, while formed at spreading centers, are found at destructive plate boundaries, where they are emplaced tectonically. These deposits are unique in the geological record as they constitute the only metallic mineral deposits that have been directly observed in the process of formation.

Metal sulfide deposits are forming on the sea floor on the East Pacific Rise at 21°N. At this locality, geologists have observed vents and mounds of Fe, Cu and Zn sulfides associated with metalliferous gels, metal sulfates, such as barite, and numerous hydrated silicate phases, both crystalline and amorphous. Significant Ag values are reported. These deposits are analogues to the important class of Cu–Zn deposits known as Cyprus-type, typified by the pyritic massive sulfide deposits in submarine basalt sequences in the Troodos Massif, Cyprus. These deposits are in tectonically uplifted slabs of oceanic crust, known as ophiolites. The *tectonic* emplacement occurs in convergent zones (see Sect. 3), but the deposits itself is formed by exhalative processes on the ocean floor during the formation of new oceanic crust.

In areas of divergence, where the initial rifting occurs beneath continental terrains, similar hydrothermal mobilization of metals occurs. Modern examples include the brines forming in depressions on the floor of the Red Sea and in subsurface brines over the landward extension of the East Pacific Rise in the Salton Sea area of California. These brines range in temperature from 45 °C to 220 °C in the deep geothermal brines of the Salton Sea. They are saline (up to 225000 ppm) and base-metal-rich (e.g., 81 ppm Fe, 82 ppm Mn, 0.26 ppm Cu, 3.4 ppm Zn and 0.63 ppm Pb in the Atlantis Pool of the Red Sea; 2090 ppm Fe,

1560 ppm Mn, 8 ppm Cu, 84 ppm Pb, 790 ppm Zn in Bore Hole No. 1 IID in the Salton Sea area). In all likelihood, these brines are richer in Cu and Pb than their oceanic equivalents owing to the availability of these more lithophile metals in the underlying continental crust, continentally derived basin fill or even in marine evaporite sequences associated with early rifting. Sawkins (1984) calls these Sullivan-type deposits after the large, rich, Ag, Pb and Zn deposit of Sullivan, British Columbia. Similar deposits include those of Rammelsberg and Megen, West Germany; Ducktown, Tennessee; and Mt Isa, Queensland. These are all found in thick sequences of continentally derived sediments which accumulated in rifted basins associated with early stages of continental breakup. In most cases these basins do not evolve into ocean basins and are often referred to as "failed arms" or aulocogens.

The ophiolite suite, i.e., rocks that are recognized as fragments of ocean crust and upper mantle emplaced into a tectonic mélange associated with constructive plate margins, hosts the second class of deposit associated with rifting, i.e., Cr and Cu–Ni–Pt deposits associated with ultramafic rocks ($SiO_2 < 45$ wt%) in mountain (orogenic) belts. The Cr deposits consist of chromite in pod-shaped serpentinite and peridotite masses, as typified by the chromite ores of Moa, Cuba. These peridotites are not strictly ophiolites, in that one or more component rock types of the ophiolite suite is missing. Recent models of orogenic peridotite genesis view them as incomplete fragments of an oceanic crust/mantle slab. However, there is the possibility that these podiform peridotites are generated from the mantle in a continental setting during mountain building. At Acoje, on Luzon, near the base of a chromite-bearing peridotite mass, nickel and platinum sulfides with Cu are present in economic concentrations. Presumably these peridotite-hosted deposits represent magmatic segregations, either as a crystal fraction or as an immiscible liquid sulfide phase.

The final class of deposit associated with divergent plate boundaries are disseminated, vein and contact metasomatic deposits of selected higher-order transition metals, especially W, Mo, Sn, Nb, Ta, the rare earths and U, with such "large ion lithophile" elements as lithium and small, high-charge cations like beryllium and boron. These elements are associated with a suite of granitoid rocks known as alkaline rocks, in which K_2O and Na_2O concentrations together exceed the Al_2O_3. These are derived from highly evolved, well-differentiated magmas, commonly, though not exclusively, associated with continental rifting. Tin and Nb in alkali granites of the Jos Plateau of Nigeria and the metal-rich carbonatite (intrusive, $CaCO_3$-rich magma) of Palabora in South Africa are examples of this class of deposit.

3. *Destructive Plate Boundaries*

As the plates move away from the spreading centers, they must eventually "collide" with plates moving away from other spreading centers (see Fig. 2). At these boundaries, one slab of lithosphere is drawn beneath another, into the aesthenosphere, producing a convergent plate boundary or subduction zone. As the crust is drawn into the aesthenosphere, two processes occur. The first is the heating of the subducted slab, accompanied by the addition of H_2O, resulting in partial fusion of the oceanic crust. Figure 4 is a schematic representation of this setting, indicating that magmas generated along the subducted plate produce upward migrating diapirs of intermediate to felsic composition (i.e., $SiO_2 > 52$wt%). On reaching the surface behind the subduction zone, these magma systems produce volcanic islands, known as island arcs. The petrochemical, isotopic, seismologic and geophysical evidence for this lithosphere configuration is overwhelming. The second process involves the tectonic emplacement of faulted blocks of the subducted slab on overriding plate. This results in the

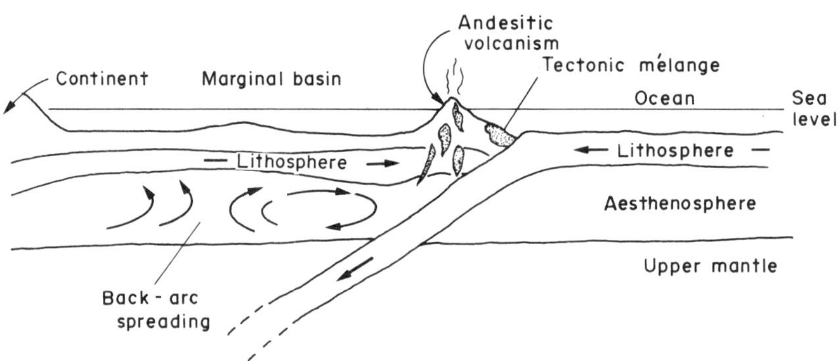

Figure 4
Schematic cross section of typical interocean island arc, subduction zone

uplift, or obduction, of portions of oceanic crust and upper mantle into extremely complex terrains known as mélanges. It is in these mélanges that many of the ore deposits whose genesis was described in the previous section are found. As a complicating factor, it is not uncommon for "back-arc spreading" to occur behind the island arc, producing marginal basins of distinctly oceanic character. These are often mantled by sediments owing to their proximity to either continental or arc sources of sediment. Finally, the "arrival" of a continental mass on either plate can alter the history of subduction and the chemistry of intrusives, volcanics, and the ore deposits.

Two unique classes of ore deposits are formed in these environments. Table 2 lists the principal ore deposits found in convergent zones.

Several types of massive sulfide deposits are known from island arc environments. Massive Cu and Cu–Zn deposits in association with submarine, mafic volcanics, especially basalts, and thick graywackes are known from island arcs in Japan, Norway, and Cuba. These deposits, known as the Besshi-type, are formed from volcanically generated brines during arc-related volcanism in a manner similar to the Cyprus-type deposits of divergent plate margins. They are distinct from a larger group of massive sulfide deposits associated with submarine felsic volcanism known as Kuroko deposits after the Kuroko district of Japan. These are vertically zoned, Cu–Zn–Pb and Ag deposits found in association with dacitic to rhyolitic domes and often associated with barite and gypsum. This class of deposits includes such districts as Noranda, Quebec; Timmons, Ontario; Bathurst, New Brunswick; Captains Flat, New South Wales and Huelva, Spain. These deposits are formed by submarine brines. In the arc environment, these brines may migrate tens of kilometers to produce massive sulfide accumulations which are not directly associated with volcanic centers or exhalative vents.

Table 2
Ore deposits associated with convergent zones

Type of deposit	Host rock	Example
1. Massive sulfide:		
Cu–Zn	Mafic volcanics and graywackes	Besshi, Japan
Pb–Zn–Cu	Felsic-intermediate volcanics	Kuroko, Japan
2. Disseminated deposits:		
Porphyry coppers	Granite intrusives	Panguna, Bougainville
Porphyry Mo	Granite intrusives	Endako, BC
Porphyry U–W–Sn–Mo	Granite intrusives	Hercynian granites, France

The second type of deposit associated with convergent boundaries are large, low grade, disseminated ores, commonly referred to as porphyry-type deposits. These deposits, in the 100 million to perhaps billion tonne range, consist of host rocks which show extensive hydrothermal alteration and significant postmagmatic mobilization of metals. The host rock is usually a granite body which is viewed as the thermal drive for migration of metalliferous fluids and as the source of the metals. The intruded country rock is frequently altered and mineralized. The most common metal carried in economic amounts in island arc-related porphyry deposits is Cu. Porphyry copper deposits are associated with active subduction in the Andes, on New Guinea and Bougainville, in the Philippines and the Caribbean. These ore-bearing granites are clearly produced by melting on the subducted crustal plate. Molybdenum porphyries are associated with older, subduction-related intrusive activity in British Columbia. Disseminated and vein deposits of Sn, U, Mo and W—similar to the association noted in rift environments—are probably produced during waning subduction where significant continental crustal involvement is present. Examples would be the Hercynian (250–350 million years old) granites of France and the Sn granites of Bolivia. Continental involvement in subduction makes this ore genesis significantly more complicated.

4. *Conservative Plate Boundaries*

The third class of plate tectonic boundary (see Fig. 2) are strike-slip or transform faults where segments of plates move laterally past one another with no new crust being added (as at constructive plate boundaries) or crust being consumed (as at destructive plate boundaries). No ore deposits are known to be generated along or are uniquely associated with laterally displaced plate boundaries.

5. *Intraplate Environments*

Within lithospheric plates, numerous types of ore deposits are found. Some of these mineral deposits may have been associated with plate tectonic regimes early in earth history that are difficult to correlate with modern interplate boundaries because of the complexity of the intervening geological events or because of changes in tectonic mechanisms during the evolution of the earth. Thus, for example, Au deposits in Precambrian greenstone belts have certain arc affinities, but greenstone belts remain enigmatic in terms of their origin. The trailing edge of continents, as they move away from spreading centers, are ideal settings for placer deposits of Au, ilmenite and rutile, or monazite, but it is doubtful whether the dynamic process of plate tectonics is necessary to produce such deposits.

Other intraplate environments of great economic significance for metal production, such as the environments that host stratabound Pb and Zn in carbonate rocks or Precambrian ultramafic lopoliths, are not controlled by plate tectonic regimes.

6. Application to Models of Ore Genesis and Mineral Exploration

The correlation of certain classes of deposits with specific tectonic settings has had a significant influence on mineral exploration. The impact has mostly been felt in reviving interest in syngenetic ore deposits, i.e., those mineral concentrations that are formed at the same time as their host rocks. Many of these are volcanogenic and/or exhalative, and the effect has been to shift the emphasis from postmagmatic, intrusive-related models of ore genesis to studies of brine migration and geochemistry and of submarine, anoxic basin dynamics. New tectonic regions, such as island arcs, have become active areas for exploration. In return, ore deposits are increasingly seen as potential indicators of paleoenvironments and tectonic setting.

See also: Metallogenic Provinces; Ore Minerals

Bibliography

Dewey J F, Bird J M 1970 Mountain belts and the new global tectonics. *J. Geophys. Res.* 75: 2625–47

Evans M 1980 *An Introduction to Ore Geology*. Elsevier, New York

Hekinian R, Feurier M, Bischoff J L, Picot P, Shanks W C 1980 Sulfide deposits from the East Pacific Rise near 21°N. *Science* 207: 1433–44

Institute of Mining and Metallurgy 1977 *Volcanic Processes in Ore Genesis*, Special Paper No. 7. Geological Society of London, London

Isacks B, Oliver J, Sykes L R 1968 Seismology and the new global tectonics *J. Geophys. Res.* 73: 5855–99

Ishihara S, Takenouchi S (eds.) 1980 *Granitic Magmatism and Related Mineralization*, Mining Geology Special Issue No. 8. Society of Mining Geologists of Japan, Tokyo

Mitchell A H, Bell J D 1973 Island-arc evolution and related mineral deposits. *J. Geol.* 81: 381–405

Mitchell A H G, Garson M S 1981 *Mineral Deposits and Global Tectonic Settings*. Academic Press, London

Rona P A 1973 Plate tectonics and mineral resources. *Sci. Am.* 229(1): 86–95

Sawkins F J 1984 *Metal Deposits in Relation to Plate Tectonics*. Springer, Berlin

Strong D F (ed.) 1976 *Metallogeny and Plate Tectonics*, Special Paper No. 14. Geological Society of Canada, Toronto

Walker W 1976 *Metallogeny and Global Tectonics*, Benchmark Papers in Geology No. 29. Dowden, Hutchinson, and Ross, Stroudsburg, Pennsylvania

Wylie P J 1971 *The Dynamic Earth*. Wiley, New York

P. G. Feiss
[University of North Carolina, Chapel Hill, North Carolina, USA]

Platinum Group Metals Resources

The platinum group elements (PGEs) are ruthenium, rhodium, palladium, osmium, iridium and platinum. Their earliest use is thought to have been about 700 BC when a single forged grain of platinum was used as a hieroglyphic character by an Egyptian artisan to decorate an etui. Natural alloys of PGEs frequently occur as inclusions in ancient gold jewelry from Egyptian (12th Dynasty onwards), Roman and Byzantine times (Ogden 1976). These PGE alloys had been concentrated in gold-bearing placers and were inadvertently incorporated into the goldwork. The Indians of Ecuador and Colombia, prior to the exploration of the Americas by Europeans, made crude articles of platinum and gold–platinum alloys. The name platinum is attributed to Don Antonio de Ulluoa y Gracia de la Torre, who mentioned the occurrence of "Platino del Pinto" in describing his travels in South America in 1748. (Platino is the Spanish diminutive for silver and Pinto is the name of the platinum-bearing river.) The identification and characterization of the six PGEs spanned about 100 years: platinum in 1750 (Wood, Brownrigg and Watson), palladium and rhodium in 1802–1804 (Wollaston), iridium and osmium in 1805 (Tennant), and ruthenium in 1844 (Klauss) (McDonald and Hunt 1982).

1. Production

Colombia, where mining of platinum-bearing placers began in the eighteenth century, was the only source of PGEs until the discovery in 1824 of rich platinum placers in the Ural mountains made Russia the leading producer until the 1920s. Around this time, by-product recovery of PGEs from the Ni–Cu deposits of the Sudbury district, Canada, and commercial PGE production from the Merensky Reef, South Africa, began. By 1936 Canada had become a leading producer, and South Africa had become the largest by 1956–1957. In the 1940s, the decline in supply of PGEs from USSR placers was offset by recovery from new Ni–Cu mines in northwestern Siberia and the Kola Peninsula. Today the USSR and South Africa are the leading producers of PGEs, with Canada third (Table 1). It should be noted, however, that although production of PGEs from South Africa and the USSR is nearly the same in total, the ores from South Africa have a Pt:Pd ratio of 2.5, compared with 0.4 for ores from the principal producers in the USSR.

Total world production of PGEs up until the end of 1981 was estimated to be 3842 t; post-1972 production accounted for 48% of this total.

2. Reserves and Resources

Crustal abundances are not known precisely but are estimated to be 1–10 ppb Pd, 5 ppb Pt, 1 ppb Ir, Os, Rh and Ru. The Merensky Reef of the Bushveld

Table 1
World PGE production, reserves and resources

Country	Production (t)						Reserves (t)						Resources (t)					
	1978[a]	% of total	1981[b]	% of total	1987[c]	% of total	1980[a]	% of total	1981[b]	% of total	1987[c]	% of total	1980[a]	% of total	1981[b]	% of total	1987[c]	% of total
USSR	94.9	48.1	104.2	49.0	119.8	48.1	6621	17.0	5910	19.0	5910	10.5	6621	9.8	6221	16.7	6220	9.3
South Africa	89.0	45.1	93.3	43.9	115.1	46.2	30170	82.2	24571[d]	79.0	49800	88.5	44477	70.1	30170	80.8	59100	88.8
Canada	10.8	5.5	12.4	5.8	9.3	3.7	280	0.8	249	0.8	249	0.4	218	0.3	280	0.8	280	0.4
Japan[e]	1.1	0.6	1.1	0.5														
Colombia	0.4	0.2	0.5	0.2									124	0.2				
Australia	0.4	0.2	0.3	0.2														
USA	0.3	0.2	0.2	0.1	1.5	0.0	31	0.1	31	0.1	249	0.4	9300	14.7	498	1.3	778	1.2
Zimbabwe			0.3	0.1									3100	4.9				
Yugoslavia	0.2	0.1	0.2	0.1														
Others	0.1	0.1	0.1	0.1					342	1.1					155	0.4		
Total	197.2	100.0	212.6	100.0	249.0	100.0	36702	100.0	31103	100.0	56300	100.0	36702	100.0	31103	100.0	66560	100.0

[a] Jolly (1980) [b] Loebenstein (1983) [c] Loebenstein (1988) [d] Including the UG2 horizon [e] Refinery recovery from imported ores

Complex in southern Africa accounted for an estimated 88% of the world's reserves in 1988 (Table 1). This estimate excludes two other PGE-bearing horizons in the Bushveld Complex, the UG2 and Platreef, which represent the bulk of the resources. The third-place ranking of the USA is mostly (91%) owing to estimates of the PGE content of deposits in Montana and Minnesota. The more promising Montana deposit, in the Stillwater Complex, has the same Pt grade as the Merensky Reef, but has four times the Pd grade and twice its 1 m thickness.

3. Mineralogy

The low crustal abundance of PGEs and the usual occurrence of their minerals as small (< 1 mm) inclusions hindered knowledge until the development of in situ microscopic analytical tools such as the electron microprobe. There are 75 platinum group minerals (PGMs) recognized as species, and 28 unnamed PGMs which probably represent new species (Cabri 1981a). Considerably more Pd minerals are known; proportions of total species for the PGEs are Pd 48%, Pt 21%, Ru and Ir 9% each, Rh 8% and Os 5%.

The PGEs form alloys and a variety of natural compounds with elements of groups IV (Sn, Pb), V (As, Sb, Bi) and VI (S, Se, Te). Oxides, hydroxides and chlorides are not well known as minerals. PGEs also occur as a solid solution in common minerals such as alloys, tellurides, selenides, arsenides, sulfarsenides, sulfides, oxides and silicates. However, there are few reliable quantitative solid-solution data for most minerals because of the difficulty of making in situ analyses and the reliance on bulk analyses of mineral separates. The best-documented data are for pentlandite, cobaltite, gersdorffite, nickeline, maucherite, melonite, Au and Au–Cu alloys.

Different types of mineral deposits have characteristic PGM suites, but sperrylite ($PtAs_2$, see Fig. 1) is found in all deposits. Other relatively common PGMs, in various geological environments, are: braggite ((Pt, Pd)S), cooperite (PtS), Pt–Fe alloy and laurite (RuS_2) in Merensky-type deposits; laurite, braggite, cooperite and an unnamed Pt–Rh–Cu sulfide in layered chromitite deposits; Pt–Fe and Pt–Ir alloys in Alaska-type deposits and derived placers; iridosmine (Os, Ir), osmiridium (Ir, Os) and rutheniridosmine (Ru, Os, Ir) in Alpine-type deposits and derived placers; and michenerite (PdBiTe) and moncheite ($PtTe_2$) in magmatic Ni–Cu sulfide deposits. The Ni–Cu sulfide deposits of the Noril'sk–Talnakh areas, USSR, probably contain the largest suite of PGM species, including several Pb- and Sn-bearing PGMs uncommon in other deposits.

4. Geology

PGEs occur in a wide variety of geological environments, but Naldrett and Duke (1980) estimate that more than 99% of primary PGE production comes from sulfide ores of magmatic origin. Cabri (1981b) has proposed the tentative classification of PGE deposits shown in Table 2. A more comprehensive classification was proposed by Cabri and Naldrett (1984) in which PGE deposits are initially divided into those of the Sulfide Association and the Oxide-Silicate Association. The Merensky Reef, Noril'sk and Sudbury types fall in the former association, whereas the Alaskan and Alpine types belong to the latter. Currently, the most important is the Merensky type, with the layered chromitites representing the largest resource. Merensky-type deposits are an integral part of large basaltic bodies that usually intruded into continental rocks. Layering is prominent, with PGEs restricted to layers 10–200 cm thick containing minor Fe–Ni–Cu sulfides. Layered chromitites represent a different horizon in the same petrotectonic setting as the Merensky type. Alaskan-type deposits are moderate-sized, concentrically zoned ultramafic complexes generally intruded in active orogenic zones. The Alpine-type deposits represent dismembered portions of ophiolite sequences emplaced in the solid state during tectonism and are often associated with podiform chromitites. The PGEs are also recovered as by-products from

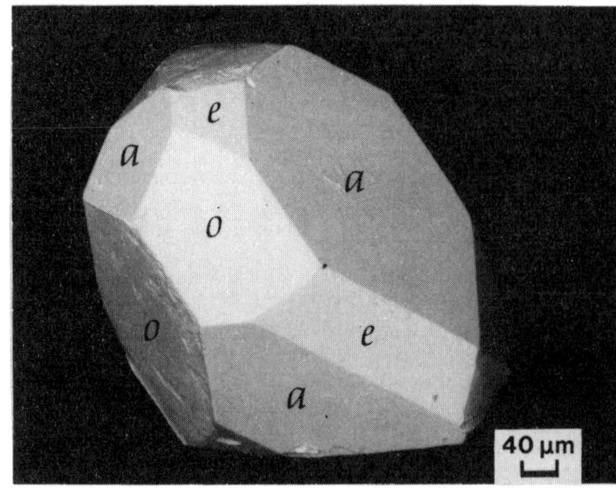

Figure 1
Scanning electron micrograph of a sperrylite crystal: a, cube face; o, octahedron face; e, pyritohedron face

Table 2
Primary classification of PGE deposits (after Cabri 1981b)

PGE-dominant deposits	PGE coproduct and by-product deposits
Merensky type	Alpine type and derived placers
Layered chromitites	Ni–Cu and Ni magmatic sulfide
Alaskan type and derived placers	Miscellaneous deposits

numerous different Ni–Cu and Ni sulfide deposits. The most important are the Noril'sk deposits which occur in the intrusive equivalent of flood basalts associated with intracontinental rifting. The important Sudbury area deposits occur in an elliptical structure associated with noritic rocks.

5. Mineral Beneficiation and Extractive Metallurgy

The PGEs are usually recovered from ores by physical concentration such as gravity separation or flotation, and upgraded using additional physical, hydrometallurgical or pyrometallurgical techniques. Gravity concentration is usual for alluvial deposits, with some magnetic separation where Pt–Fe alloys are abundant. Gravity concentration is still practised at Rustenberg platinum mines (Merensky-type), South Africa, to recover the coarser-sized PGMs prior to concentration of PGEs by conventional processing. Various gravity and upgrading treatments are required to recover PGEs from the low-grade (6–7 ppb) Witwatersrand Au–U deposits.

In deposits with major or minor Ni and Cu, the PGEs are recovered from residual products mainly by dissolution and reprecipitation as insoluble salts, complexes or metals.

Recoveries of PGEs, which are very dependent on grade and mineralogy, tend to be poor and variable. Merensky-type deposits grade from 3–20 ppm PGE and estimated losses are of the order of 25%. The greatest loss is thought to occur during milling (crushing, grinding, flotation) (Newman 1973). Average grades for Sudbury deposits are estimated to be about 1 ppm PGE, and recoveries about 70–75%. The lowest recovery ($\sim 50\%$) is thought to occur from the Witwatersrand Au–U conglomerates.

6. Uses and Strategic Implications

Industrial usage of PGEs surpassed that for jewelry about 30–40 years ago and has continued to increase because PGEs have special physical and chemical properties: they are excellent catalysts, are refractory and are chemically inert over a wide temperature range. Important industrial applications include exhaust catalysts, manufacture of a wide range of chemicals, (e.g., nitric acid for fertilizer production), pharmaceuticals, laboratory and process equipment, petroleum refining, hydrocracking, isomerization, the glass and ceramic industries, electrical and electronic devices, and dental and medical uses.

Present confinement of major PGE production to southern Africa and the USSR is of concern to industrial nations. The USA is the largest consumer of PGEs, but obtains only about 10% of its needs from secondary (recycled) metal. However, strategic implications aside, South African producers can quickly adjust production to suit demand. This is in marked contrast to the principal producers in the USSR and Canada, where PGE production depends on Ni (and Cu) demand.

Bibliography

Cabri L J (ed.) 1981a *Platinum-Group Elements: Mineralogy, Geology, Recovery*, CIM Special Vol. 23. Canadian Institute of Mining and Metallurgy, Montreal

Cabri L J 1981b Nature and distribution of platinum-group element deposits. *Episodes* 1981(2): 31–35

Cabri L J, Naldrett A J 1984 The nature of the distribution and concentration of platinum-group elements in various geological environments. *Proc. 27th Int. Geological Congr.* 10: 17–46

Canadian Mineralogist (May 1979): issue devoted to Nickel Sulfide and Platinum-Group Element Deposits

Economic Geology (November 1976): issue devoted to Platinum-Group Elements

Economic Geology (September–October 1982): issue devoted to Platinum-Group Elements

Jolly J H 1980 *Platinum-Group Metals.* US Bureau of Mines Bulletin 671. USBM, Washington, DC, pp. 1–24

Loebenstein J R 1983 Platinum-group metals. *Mineral Commodity Profiles.* US Bureau of Mines, Washington, DC, pp. 1–20

Loebenstein J R 1988 Platinum-group metals. *Mineral Commodity Summaries 1988.* US Bureau of Mines, Washington, DC, pp. 118–19

McDonald D, Hunt L B 1982 *A History of Platinum and its Allied Metals.* Europa Publications, London

Naldrett A J, Duke J M 1980 Platinum metals in magmatic sulfide ores. *Science* 208: 1417–24

Newman S C 1973 Platinum. *Trans. Inst. Min. Metall. A* 82: 52–68

Ogden J M 1976 The so-called 'Platinum' inclusions in Egyptian goldwork. *J. Egypt. Archaeol.* 62: 138–44

L. J. Cabri

[Canada Center for Mineral and Energy Technology, Ottawa, Ontario, Canada]

Portland Cement Raw Materials

Portland cement is manufactured by firing crushed mixtures of a wide range of raw materials in kilns to form a clinker. The clinker is cooled and then finely ground with about 5wt% gypsum ($CaSO_4.2H_2O$) or anhydrite ($CaSO_4$), and usually with a small amount of organic air-entraining agent to make it resistant to frost action. The most usual combination of materials fed into cement kilns is 14wt% SiO_2, 3wt% Al_2O_3, 3wt% Fe_2O_3, 75wt% $CaCO_3$, 4wt% $MgCO_3$ and 0.6wt% alkali (sodium and potassium oxides). Of these constituents, the first four are essential; the latter two, $MgCO_3$ and the alkalis, are tolerated as inevitable impurities.

The raw materials used are mostly natural rock but several industrial by-products may be used when available. By far the most common raw material used for the manufacture of portland cement is limestone and most cement plants are located at their sources of limestone. The required constituents may be almost wholly within the limestone, as in the case of cement rock, or may be transported from some distance to the cement plant.

1. Portland Cement Production

Portland cement, because it is basic to most major construction, is produced in most countries of the world. World production for 1983 was 925 642 kt. The largest production, by the USSR, was 128 000 kt, followed by the People's Republic of China, 108 250 kt, Japan, 80 650 kt and the USA, 64 725 kt (Davis and Johnson 1983). Production of portland cement is primarily related to the amount of heavy construction work, such as highways, dams and buildings, that is being carried out. Portland cement is produced in 117 countries and most areas underlain by limestones are geologically capable of producing the raw materials.

There are five basic types of portland cement and all consist of similar compounds. The typical constituents from which the five primary types of portland cement are made are summarized in Table 1, along with the critical physical characteristics of each type. The oxides of Table 1 are fired in kilns at temperatures of 1425–1650 °C, the CO_2 is driven off and the oxides are combined to form a clinker which consists essentially of the compounds listed in Table 2.

Normally, portland cements also contain 4–6wt% gypsum or anhydrite, which is added after firing as an essential constituent to regulate the setting time of concretes and mortars. Organic air-entraining agents, to the extent of a percent or so, are also usually present in portland cement.

A typical oxide composition of type I portland cement with gypsum added is 63.8wt% CaO, 20.7wt% SiO_2, 6.2wt% Al_2O_3, 2.5wt% MgO, 2.4wt% Fe_2O_3, 2.2wt% SO_3, 1.3wt% alkali and 1.0wt% H_2O.

2. Raw Materials

The raw materials which are used in portland-cement manufacture must be inexpensive and abundant. Table 3 summarizes most materials which have been used in cement manufacture. The cement-manufacturing plant is generally located at the source of $CaCO_3$, its largest constituent. Other necessary but minor constituents, such as iron oxides or calcium sulfate, may be transported to the plant from considerable distances.

Barton (1967) has given two useful summaries pertaining to raw materials used for portland cement in the USA. These statistics are probably not very different on a worldwide scale and on this basis his tabulations, in modified form, are produced in Tables 4 and 5. Barton's statistics remain fairly accurate for the USA although certain other minor constituents such as fluorspar (CaF_2), pumicite (frothy volcanic glass), $CaCl_2$, staurolite ($HFeAl_5Si_2O_3$), fly ash (silica-rich coal ash) and diatomite (microscopic silica shells) have been used, as available and as needed. Most constituents used in portland cement are impure, with the most common impurities also being common constituents in portland cement. Thus most raw materials are sources of more than one required oxide.

In using materials for portland-cement manufacture, care must be taken that nonessential constituents are not excessive. Examples of the most common of such constituents are the alkalis (sodium and potassium oxides) and magnesia. The alkalis do not improve the quality of portland cement and quantities larger than 0.6% total by weight Na_2O equivalent (percentage of $Na_2O + 0.658$ times the percentage of K_2O) tend to react with some forms of silica (opal, chalcedony, cristobalite, tridymite and finely divided quartz) to create hydrous alkali silicates which cause expansion of concrete leading to deterioration. Portland cements with higher alkali contents (2wt% or more) are manufactured, but care must be taken to use nonreactive aggregate or to neutralize the expansive reactions in some way. Excessive magnesia in portland

Table 1
Typical analysis of constituents (wt%) for the five basic types of portland cement

	Constituents (wt%)						Critical physical characteristic
	SiO_2	Al_2O_3	Fe_2O_3	$CaCO_3$	$MgCO_3$	Alkalis	
Type I	14.1	4.2	1.6	75.8	3.5	0.9	standard
Type II	14.7	3.2	2.6	74.7	4.2	0.6	moderate heat
Type III	13.7	3.6	2.0	77.3	3.0	0.4	high early strength
Type IV	16.3	3.3	3.0	72.5	4.2	0.7	low heat
Type V	12.4	1.6	1.0	76.0	3.5	0.5	sulfate resistant

Table 2
Analysis of compounds (average wt%) in type I portland cement clinker (after Ames and Cutcliffe 1983)

Compound	Oxide composition	wt%
Tricalcium silicate (alite)	$(CaO)_3SiO_2$	45
Dicalcium silicate (belite)	$(CaO)_2SiO_2$	27
Tricalcium aluminate	$(CaO)_3Al_2O_3$	11
Tetracalcium aluminoferrite	$(CaO)_4(Al_2O_3)(Fe_2O_3)$	8

Table 3
Sources of raw materials used in the manufacture of portland cement clinker (after Ames and Cutcliffe 1983)

Sources	Raw materials
Calcium carbonate	limestones
	lithified limestones, chalk, marble, marl, oyster shells (reef), coquina shells, aragonite sand, slag
Silica	sand, sandstone or quartzite
	clay and claystone
	shale
	loess
	slag
	mill fines
	fly ash
Alumina	shale
	clay and mud
	loess
	slag
	fly ash
	bauxite
	alumina process waste (red mud)
	aluminum dross
	staurolite
	mill fines (granite)
	pumice or other volcanic material
Iron	iron ores
	blast furnace flue dusts
	pyrite cinder
	mill scale
	fly ash

cements may also create expansive reactions and therefore the amount of MgO in the final product should generally be less than about 4.0wt%.

Any other constituent not essential to cementitious reactions is also considered to be deleterious. For example, sulfur trioxide should not be present in quantities more than 3.0–3.5wt%. Another example is strontium (as $SrSO_4$) which may be excessive in some limestones.

2.1 Calcium Carbonate

The most common source of calcium carbonate is limestone, with probably over 95% of the world's CaO sources being variations of limestone or its metamorphic equivalent, marble. Calcium-carbonate-rich rocks occur in nearly all countries.

Table 4
Percentage of raw materials used in cement manufacture in the USA in 1965 (after Barton 1967)

Raw material	Percent
Cement rock	16.5
Limestone and shells	67.9
Marl	0.5
Sand and sandstone	1.5
Iron materials	0.6
Clay, shale and schist	9.4
Blast-furnace slag	0.8
Gypsum	2.7
Miscellaneous items	0.1

Table 5
Raw material mixes used at the 181 cement plants in the USA in 1965 (after Barton 1967)

Ingredients	Percentage of plants using mix
Cement rock or limestone only	18.8
Cement rock and limestone	7.2
Marl and limestone	0.6
Shells, marl and limestone	0.6
Limestone and/or cement rock plus clay or bauxite	29.8
Limestone or cement rock plus shale	25.4
Shells and clay	5.5
Limestone plus shale and clay	3.9
Cement rock or limestone plus slag	3.9
Limestone plus clay and slag	1.7
Limestone plus shale and slag	1.7
Shells, marl and clay	0.6
Shells, clay and shale	0.6

Limestones result from cementation or lithification of accumulations of calcium carbonate as calcite or aragonite in the form of shells, oolites or lime muds with most limestones originating from varying combinations of these materials. Chalks are poorly cemented accumulations of microscopic calcium carbonate shells; marls are poorly cemented impure, often clay-rich, lime muds, usually rich in shells, and may be deposited either in fresh or marine water; and calcitic marbles are the metamorphosed and recrystallized versions of these materials.

Ideally, the easiest material to use as a raw material for portland cement is cement rock. This is an impure limestone with a composition such that the removal of carbon dioxide leaves an oxide composition which approximates the ideal composition of common type I or type II portland cements without the gypsum or anhydrite. Rock of this type occurs in the Lehigh

Valley of Pennsylvania where it has long been used for portland-cement manufacture. Table 6 gives three typical partial compositions of cement rock from the Lehigh Valley, along with the composition of the "average" limestone for reference.

Rock raw materials in nature are seldom compositionally homogeneous, even where cement rock is used with no additives, other than calcium sulfate and air-entraining agents. As a result, careful quarrying and blending of rock layers from within the quarry are usually needed for quality control.

Loosely consolidated materials such as oyster shells, coquina or coral-shell accumulations or oolitic aragonite (sands consisting of spherical aragonite grains), all occurring in shallow marine conditions, are also sources of calcium carbonate in some marine coastal areas.

Slags which result from processing iron ore to iron may be rich enough in calcium oxide to be a source of the CaO constituent. Certain slags have compositions which are relatively close to that of portland cement and the manufacture of portland cement from them may be a relatively easy process. A typical slag from an air-cooled blast furnace consists of 32–42wt% SiO_2, 7–16wt% Al_2O_3, 32–45wt% CaO and 5–15wt% MgO (modified from McCarl et al. 1983). A slag of this type which is low in MgO (less than 6 or 7wt%) can be combined with high-calcium limestone or marble to approximate an ideal portland-cement composition.

2.2 Silica

Relatively pure silica in the form of quartz (SiO_2) occurs naturally as sand or sandstone, or their metamorphosed or recrystallized equivalent, quartzite. In addition, silica is the major constituent of clay, claystone, silt, shale, granites and volcanic glasses as well as loess (wind-blown dust accumulations).

In each of these latter cases alumina is the other major constituent and therefore these sedimentary materials are also sources of alumina. Somewhat arbitrarily, those sources of silica in which the silica is over four times the weight percentage of alumina will be summarized in the chemical analyses in this section. Lower-ratio materials will be summarized in Sect. 2.3. The reason for this separation is that the average portland cement contains about four times as much silica as alumina. If the percentage of alumina in portland-cement feed is approximately correct, and silica must be raised, material in which the weight percentage of silica is over four times greater than that of alumina would have to be added so as not to raise the alumina excessively. Typical compositions of siliceous materials which may be used are given in Table 7.

Commercial waste products such as slag, mill fines and fly ash are also commonly sources of silica. Mill fines from the manufacture of granite products have the composition of granite (Table 7). According to Barton, typical fly ash consists of 44–51wt% SiO_2, 13–26wt% Al_2O_3, 7–15wt% Fe_2O_3, 1.6–12wt% CaO; 0.9–2.7wt% MgO and 0.2–16wt% C, and has an ignition loss of 9–17% (Clausen 1960).

Siliceous limestones, often rich in quartz sand and/or cherts, may be blended with limestones lower in silica, especially where both silica-rich and silica-poor limestone are in close proximity.

2.3 Alumina

Most sources of alumina are also sources of silica and several have been discussed in Sect. 2.2. Some typical

Table 6
Three typical partial compositions of cement rock from the Ordovician Jacksonburg formation, Lehigh Valley, Pennsylvania, together with the composition of "average" limestone for comparison

	Composition (wt%)			
Constituent	Copley[a]	Whitehall[b]	Lehigh[c]	Average Limestone[d]
$CaCO_3$	71.8	79.3	74.6	76.03
$MgCO_3$	4.3	4.0	4.8	16.58
SiO_2	14.8	13.2	14.0	5.19
Al_2O_3	6.6	2.1	·3.9	0.81
Fe_2O_3	1.6	1.1	1.3	0.54

[a] Copley Cement Company; Deasy et al. 1967 [b] Whitehall Cement Company; Deasy et al. 1967 [c] Lehigh Portland Cement Company; Deasy et al. 1967 [d] Mason 1966

Table 7
Typical compositions (wt%) of common siliceous materials which may be used for portland-cement manufacture

Constituent	Granite[a]	Rhyolite[b]	Silica sand[c]	Sandstone[d]
SiO_2	70.18	72.35	99.70	78.33
TiO_2	0.39	0.25	0.02	0.25
Al_2O_3	14.47	13.98	0.08	4.77
Fe_2O_3	1.57	0.60	0.02	1.07
FeO	1.78	1.78		0.30
MnO	0.12			
MgO	0.88	0.30	0.01	1.16
CaO	1.99	1.30	0.01	5.50
Na_2O	3.48	5.04	0.01	0.45
K_2O	4.11	3.92		1.31
H_2^+	0.84	0.05	0.10	1.63
H_2^-		0.45		
P_2O_5	0.19	trace		0.08
CO_2				5.03
BaO				0.05
S				0.07
Total	100.00	99.33	99.95	100.00

[a] Average of 546 analyses (Rankama and Sahama 1950 p. 166) [b] Rhyolite obsidian, Newberry volcano, Oregon (Williams 1942) [c] Commercial silica sand, Silverado sandstone, Orange, California (Murphy 1975) [d] Average sandstone (Mason 1966)

compositions of alumina-rich raw materials which may be used in portland-cement manufacture are summarized in Table 8.

2.4 Iron Oxide

Typical sources of iron oxides are the iron ores and various industrial by-products. Iron-ore minerals are limonite (approximately 85.5wt% Fe_2O_3 and 14.5wt% H_2O), magnetite (69.0wt% Fe_2O_3 and 31.0wt% FeO) and hematite (Fe_2O_3). The actual ores are highly variable in composition with varying amounts of SiO_2, Al_2O_3, CaO, MgO, P_2O_5 and MnO as the major impurities. Total impurities in iron-rich materials usable for cement manufacture are probably less than 25wt%.

Typical industrial by-products which are sources of iron are:

(a) pyrite cinder from the manufacture of sulfuric acid from pyrite, 67.5–86.7wt% Fe_2O_3 with the major impurities of 6.3–13wt% SiO_2 and 2.1–2.8wt% Al_2O_3 (Clausen, 1960);

(b) flue dust from steel manufacture, 80–90wt% Fe_2O_3; and

(c) mill scale from the surface of iron in iron manufacturing, 93–98wt% Fe_2O_3.

Red mud, a very fine, wet waste product resulting from the manufacture of aluminum, may also be used. It contains hydroxide of iron, along with some aluminum hydroxide and silica; the exact composition depends on the actual process employed and the nature of the raw materials.

Table 8
Average compositions (wt%) of common sources of alumina for portland cement

Oxide	Shale[a]	Phyllite[b]	Kaolinite[c]	Bauxite[d]
SiO_2	58.38	45.20	46.77	3–16
Al_2O_3	15.47	37.02	37.79	55–70
Fe_2O_3	6.07	0.27	0.45	4–25
FeO		0.06	0.11	
TiO_2		1.26	0.02	2–35
CaO	3.12	0.52	0.13	
MgO	2.45	0.47	0.24	
Na_2O	1.31	0.36	0.05	
K_2O		0.49	1.49	
MnO			trace	
SO_3				
H_2O^-	5.02	1.55	0.61	10–20
H_2O^+		13.27		
CO_2	2.64	trace	12.18	
Total	97.71	94.69	100.21	

[a] Composite of 78 shales (Rankama and Sahama 1950 p. 222) [b] Composite of 11 phyllites (metamorphosed shales), central Norway (Rankama and Sahama 1950 p. 222) [c] Kaolinite, St. Austall, UK (Kerr et al. 1950) [d] Bauxite, range of compositions, France (Shaffer 1983)

As such small quantities of iron are usually needed, and the iron content of additives is so high, the impurities in the sources of iron are not usually a significant factor in portland-cement quality.

2.5 Calcium Sulfate

Calcium sulfate for portland-cement manufacture is obtained from gypsum and anhydrite rock. The latter are rarely pure as mined because they contain varying amounts of clay, calcite, dolomite, salt and silica. Typical gypsum or anhydrite rock as mined contains 85–98wt% $CaSO_4.2H_2O$ or $CaSO_4$, respectively. Both are mined and available in the quantities required for portland cement in most countries of the world. Controlled calcining of gypsum forms plaster of paris, $CaSO_4.\frac{1}{2}H_2O$, which may then be ground for use in wallboards or plasters. Anhydrite is not as valuable as gypsum because it cannot be calcined to form plaster of paris. Portland-cement manufacturers can use somewhat less-pure calcium sulfate, and anhydrite is not a problem. Hence, they can use some less-expensive, lower grades of gypsum or anhydrites. The largest production of gypsum is from the USA, Canada, France, Iran and the USSR (Anon 1983).

3. Raw Materials for Special Portland Cements

Portland cement may be mixed with some other materials to enhance or alter the characteristics of portland-cement concretes for special purposes.

A "portland granulated blast-furnace slag" consists of portland cement plus 25–65wt% very finely pulverized blast-furnace slag. Granulated blast-furnace slag results from rapid quenching of hot (1371 °C) slag in steam or water. Slag portland cements are resistant to attack by sulfate in water and air, are not alkali reactive with silica-rich aggregates and have low permeability.

Portland–pozzolan cements contain finely divided reactive silica and are often used when portland cements are very high in alkali and therefore tend to be reactive with silica in mineral aggregates. The 15–40wt% of finely ground silica-rich materials may rapidly react chemically with the calcium and alkali hydroxides in portland cement, using up the alkali and minimizing later deleterious reactions which could occur by reaction of alkalis with silica on aging. Some typical siliceous pozzolanic materials are fly ash, clay, diatomaceous earth, opaline chert and shale, tuff, and volcanic ash or pumicite.

See also: Calcium Aluminate Cements; Portland Cements, Blended Cements and Mortars

Bibliography

Ames J A, Cutcliffe W E 1983 Construction materials. Cement and cement raw materials. In: Lefond S J (ed.) 1983 *Industrial Minerals and Rocks*, 5th edn. American Institute

of Mining, Metallurgical and Petroleum Engineers, New York, pp. 133–59
Anon 1983 Gypsum-optimism within a broad market. *Ind. Miner.* 193: 19–59
Barton W R 1967 Raw materials for manufacture of cement. In: Faber J H, Capp J D, Spencer J D (eds.) 1967 *Fly Ash Utilization, Proc. Edison Electric Institute–National Coal Association–Bureau of Mines Symp., Pittsburgh, Pennsylvania, March 14–16, 1967.* US Bureau of Mines Information Circular 8348. USBM, Washington, DC, pp. 46–51
Clausen C F 1960 Cement materials. In: Gilson J L (ed.) 1960 *Industrial Minerals and Rocks*, 3rd edn. American Institute of Mining, Metallurgical and Petroleum Engineers, New York, pp. 203–31
Davis L L, Johnson W 1983 Cement. *Minerals Yearbook 1983.* US Bureau of Mines, Washington, DC
Deasy G F, Griess P R, Bolazik R F, Burtnett J W 1967 *The Atlas of Pennsylvania Mineral Resources*, Mineral Resources Report 50. Pennsylvania Geological Survey, Harrisburg, Pennsylvania
Kerr P F, Hamilton P K, Pill R J, Wheeler G V, Lewis D R, Burkhardt W, Reno D, Taylor G L, Mielenz R C, King M E, Schieltz NC 1950 *Analytical Data on Reference Clay Materials*, Report No. 7. American Petroleum Institute, Project 49, Columbia University Preliminary
McCarl H N, Eggleston H K, Barton W R 1983 Aggregates—Slag. In: Lefond S J (ed.) 1983 *Industrial Minerals and Rocks*, 5th edn. American Institute of Mining, Metallurgical and Petroleum Engineers, New York, pp. 111–31
Mason B 1966 *Principles of Geochemistry*, 3rd edn. Wiley, New York, p. 153
Murphy T D 1975 Silica and silicon. In: Lefond S J (ed.) 1975 *Industrial Minerals and Rocks*, 4th edn. American Institute of Mining, Metallurgical and Petroleum Engineers, New York, p. 153
Patterson S H, Murray H H 1983 Clays. In: Lefond S J (ed.) 1983 *Industrial Minerals and Rocks*, 5th edn. American Institute of Mining, Metallurgical and Petroleum Engineers, New York, pp. 585–651
Rankama K, Sahama T G 1950 *Geochemistry*, University of Chicago Press, Chicago, Illinois
Shaffer J W 1983 Bauxitic raw materials. In: Lefond S J (ed.) 1983 *Industrial Minerals and Rocks*, 5th edn. American Institution of Mining, Metallurgical and Petroleum Engineers, New York, pp. 503–27
Williams H 1942 *The Geology of Crater Lake National Park, Oregon.* Carnegie Institute, Washington, DC, No. 10, p. 149

J. R. Dunn
[Dunn Geoscience Corporation, Albany, New York, USA]

Portland Cements, Blended Cements and Mortars

Portland and blended cements are vital to civil engineering and construction. The world's annual production of these elements is about 1 billion tonnes (10^{12} kg) and they are the binding ingredient for about 10 billion tonnes of concrete. Portland cements are water-reactive inorganic powders formed by intergrinding portland cement clinker and ~5% of gypsum. Blended cements are blends of portland cements with other finely divided inorganic materials such as fly ash and granulated blast-furnace slags. When mixed with 1 to 2 times their volume of water, cements give pastes which harden to stone-like products. Their essential ingredients are anhydrous calcium silicates.

1. Cement Industry Nomenclature

Because the compositions of their raw materials and products are usually expressed in terms of oxides, cement chemists have adopted a shorthand notation using a single letter to represent a unit of an oxide. The commonly used symbols are:

$A=Al_2O_3$	$f=FeO$	$N=Na_2O$
$C=CaO$	$H=H_2O$	$S=SiO_2$
$\bar{C}=CO_2$	$K=K_2O$	$\bar{S}=SO_3$
$F=Fe_2O_3$	$M=MgO$	$T=TiO_2$

Using this notation, tricalcium silicate, Ca_3SiO_5, is written C_3S, and calcium hydroxide, $Ca(OH)_2$, is written CH. Phases which cannot be written solely in oxide notation may be written in a mixed notation. For example, the compound with the formula $Ca_{11}Al_{14}O_{32}\cdot CaF_2$ is written $C_{11}A_7\cdot CaF_2$.

2. Mechanisms of Cementing Action and the Microstructure of Hardened Cement Pastes

The setting and hardening of cement–water pastes results from the formation of hydrated solid phases of greater volume than the anhydrous cement. The hydrated phases grow into the available water-filled space to give a rigid porous structure. For both portland and blended cements, the main hydration products at normal ambient temperatures are a poorly crystallized calcium silicate hydrate (usually designated as C–S–H, although its composition is approximately $C_3S_2H_3$) and calcium hydroxide. While these products generally adapt well to the available space, some hydration products grow as large crystals which displace other particles to accommodate their growth. Examples are brucite, $Mg(OH)_2$, formed by hydration of MgO, and ettringite, $C_3A\cdot 3C\bar{S}\cdot H_{32}$.

The ability of a cement paste to develop strength depends largely upon the formation of a low-porosity solid; the exact composition and morphology of the hydration products appear to be less important. Intrinsic factors affecting the engineering performance of a cement are the particle size distribution, the compositions and quantities of the phases present and the distribution of phases between particles of different sizes and shapes. However, the ability to predict the performance of cements in concrete reliably from the characteristics of the cements and the conditions of use has not been attained.

3. Chemical and Mineralogical Compositions of Ingredients of Portland and Blended Cements

3.1 Portland Cement Clinker

Portland cement clinker is a synthetic product composed essentially (see Table 1) of impure tricalcium silicate (C_3S) and β-dicalcium silicate (β-C_2S) together with lesser quantities of impure tricalcium aluminate (C_3A) and a ferrite solid solution (fss) approximating to the composition of tetracalcium aluminoferrite (C_4AF). Small quantities of other phases may also be present. The phases in the clinker are similar to those which would be expected from the portland cement area of the C–A–S phase diagram (Fig. 1), allowance

Table 1
Some phases commonly found (in impure form) in portland cement clinkers

Chemical name of pure phase	Common name for impure phase	Chemical composition	Industry designation	Frequent impurities	Number of crystalline forms known in clinker
Tricalcium silicate	alite	Ca_3SiO_5	C_3S	A, M	3
β-Dicalcium silicate	belite	Ca_2SiO_4	β-C_2S	A, M, K	3
Tricalcium aluminate	C_3A	$Ca_3Al_2O_6$	C_3A	M, N, S	2
Tetracalcium aluminoferrite	C_4AF[a]	$Ca_4Al_2Fe_2O_{10}$	C_4AF	M, N	2
Magnesium oxide	periclase	MgO	M	f	1
Calcium oxide	free lime	CaO	C		1
Calcium sulfate	anhydrite	$CaSO_4$	$C\bar{S}$		1
Sodium potassium sulfate	aphthitalite	$(Na,K)_2SO_4$	$(N,K)\bar{S}$		

[a] Frequently referred to as brown millerite or the ferrite solid solution (fss) phase; compositions between C_2F and C_4A_2F have been identified in portland cement clinkers

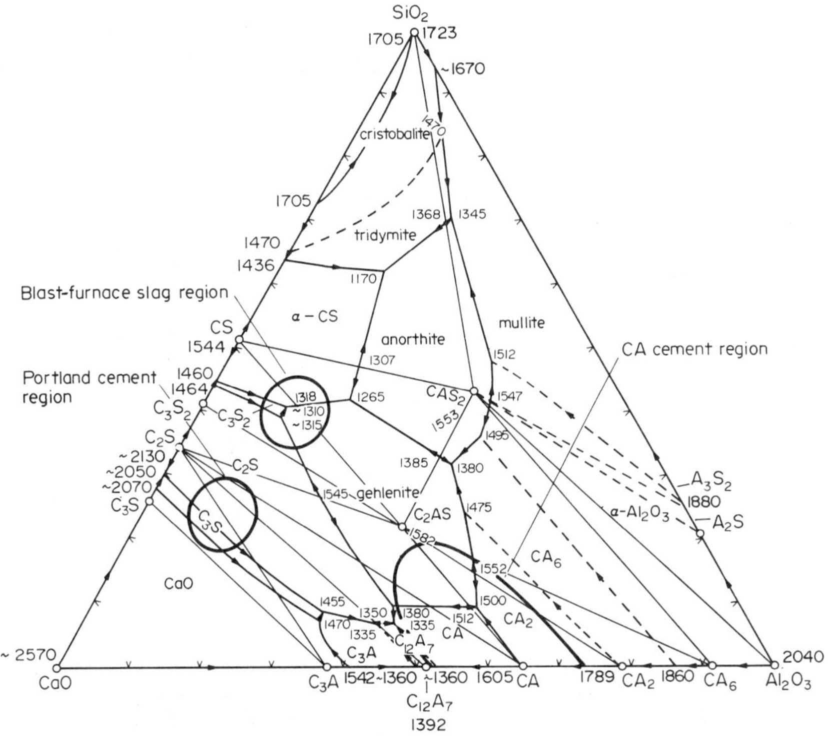

Figure 1
C–A–S phase diagram showing cement regions (after Taylor 1964; courtesy of Academic Press)

being made for the effects of F and other oxides, of which M, N, K and S are the most common. Other factors which may affect the nature of the clinker and the phases present are nonattainment of high-temperature equilibrium and the rate of cooling.

The approximate compound composition of a clinker can be calculated using the Bogue potential compound composition formulae, which, for the case where $A/F \geqslant 0.64$, are:

$$\%C_3S = 4.071(\%C) - 7.600(\%S) - 5.718(\%A) - 1.430(\%F) - 2.852(\%\bar{S})$$

$$\%C_2S = 2.867(\%S) - 0.7544(\%C_3S)$$

$$\%C_3A = 2.650(\%A) - 1.692$$

$$\%C_4AF = 3.043(\%F)$$

Similar formulae have been deduced for the less common case where $A/F < 0.64$.

3.2 Gypsum

Gypsum ($CaSO_4 . 2H_2O$ or $C\bar{S}H_2$) is interground with most cements containing portland cement clinker to control an otherwise too rapid reaction of the tricalcium aluminate with water. Because gypsum affects strength development and drying shrinkage, the quantity added should be close to the optimum for the proposed application of the cement. Natural gypsum is usually preferred, but less hydrated calcium sulfates and certain by-product gypsums have been used.

3.3 Granulated Blast-Furnace Slag

Slags used with portland cements are obtained by rapid quenching (granulation) of the molten slag from iron blast furnaces. They are predominantly glassy silicates or aluminosilicates with calcium and magnesium as the main cations present. Ignoring other oxides than C, A and S, their ranges of compositions are indicated in Fig. 1. Some crystallization of the slag usually takes place during cooling. As a rough guide, the more basic the slag and the higher its glass content, the more suitable it is for use in cement manufacture.

3.4 Pozzolans

Pozzolans are finely divided aluminosiliceous materials which, though having no cementing action with water alone, combine with calcium hydroxide in the presence of water to form cementing compounds. They are used in quantities of ~ 10–40% in portland–pozzolan cements. Naturally occurring pozzolans include volcanic glasses and certain clay-like rocks which require heat treatment to develop their full pozzolanic activity. Artificial pozzolans include fly ash, a by-product from the combustion of pulverized coal. Fly ash consists mostly of spherical particles, 1–50 μm in diameter, of a silicate glass with smaller quantities of refractory phases such as mullite, hematite, magnetite and wustite. Crystals of these phases may be included in the glassy particles. The fly-ash glasses are usually less basic than the glassy material in granulated blast-furnace slags.

4. Chemistry of Portland Cement Hydration

When the cement is mixed with water, the water quickly dissolves any alkali sulfates present. Also, as a result of reactions with C_3S and other cement compounds, the water becomes saturated or supersaturated with respect to calcium hydroxide and gypsum. The reactions then stop almost completely for a period of several hours referred to as the induction or dormant period; the causes are not fully understood, but they may be the formation of a low-permeability coating on the cement grains or slow processes of nucleation of calcium hydroxide, a calcium silicate hydrate or both. At the end of the induction period, the reactions accelerate until, typically, about a quarter of the cement has reacted. The reactions then gradually slow down due to diffusion control by the formation of an increasingly thick layer of hydration products around the cement grains.

Some major hydration products are listed in Table 2 and the reactions which result in their formation are given in Table 3. The complexity of the system of simultaneous and consecutive reactions which take place during cement hydration is indicated by Fig. 2. The main products are C–S–H and calcium hydroxide. The C–S–H does not have a well-defined stoichiometry, though its C/S ratio is usually in the range 1.5 ± 0.2. It appears to have a clay-like silicate-layer structure and its specific surface area, as deduced from water vapor adsorption data, is $\sim 1.5 \times 10^5\ m^2\ kg^{-1}$;

Table 2
Some hydrated phases commonly found in portland cement pastes

Name of pure phase	Common name	Chemical composition	Cement industry designation
Calcium hydroxide	hydrated lime	$Ca(OH)_2$	CH
Calcium sulfate dihydrate	gypsum	$CaSO_4 . 2H_2O$	$C\bar{S}H_2$
Calcium silicate hydrate gel	C–S–H gel	$Ca_xSi_yO_{x+2y} . xH_2O$	$C_xS_yH_x$
Calcium aluminate sulfate 32-hydrate	ettringite	$Ca_3Al_2O_6 . 3CaSO_4 . 32H_2O$	$C_3A . 3C\bar{S} . H_{32}$
Calcium aluminate sulfate 12-hydrate	monosulfate	$Ca_3Al_2O_6 . CaSO_4 . 12H_2O$	$C_3A . C\bar{S}H_{12}$

Table 3
Some reactions of portland cement compounds with water

$2C_3S+6H$	$\longrightarrow$	$2C_xSH_x+(6-2x)CH$ $(1.3<x<1.7)$
$2C_3S+4H$	$\longrightarrow$	$2C_xSH_x+(4-2x)CH$
C_3A+6H	$\longrightarrow$	C_3AH_6
$C_3A+3C\bar{S}H_2+26H$	$\longrightarrow$	$C_3A.3C\bar{S}.H_{32}$
$C_3A+C\bar{S}H_2+10H$	$\longrightarrow$	$C_3A.C\bar{S}.H_{12}$
$C_4AF+2CH+6C\bar{S}H_2+5OH$	$\longrightarrow$	$2C_3(A,F).3C\bar{S}.H_{32}$
$2C_3A+C_3A.3C\bar{S}.H_{32}+4H$	$\longrightarrow$	$3C_3A.\bar{S}.H_{12}$

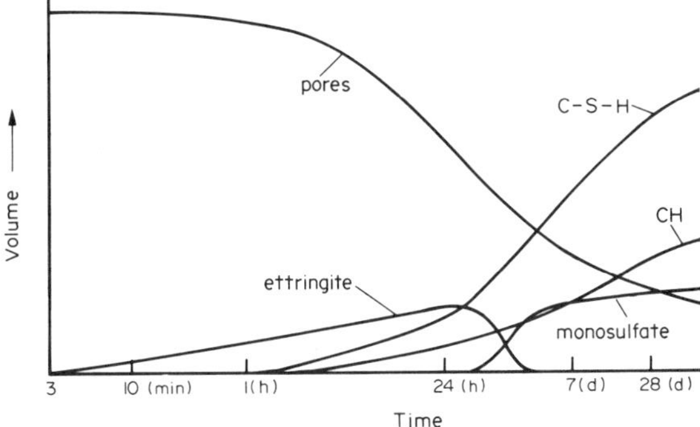

Figure 2
Schematic representation of the course of reactions of a portland cement with water; exact volumes depend on cement composition and water–cement ratio

the area determined by nitrogen adsorption is much lower. The highly hydrated hexacalcium aluminate trisulfate, ettringite, $C_3A.3\bar{S}.H_{32}$, which is formed at early ages, is transformed into tetracalcium aluminate monosulfate hydrate, $C_3A.C\bar{S}.H_{12}$, after all the gypsum has reacted and the sulfate concentration in the aqueous phase falls.

The ferrite solid solution reacts more slowly with water than do the other cement compounds. It can form analogues of ettringite and the tetracalcium aluminate monosulfate hydrate in which at least part of the Al is replaced by Fe. It may also form $C_4(A,F)H_{13}$. The aqueous phase in a cement paste has a pH of 12.5 or higher, depending on the concentrations of sodium and potassium ions dissolved in it. The reaction rates depend upon the exact natures of the cement compounds and their states of subdivision, as well as the temperature and the concentrations of any soluble materials added to the mixing water to retard or accelerate setting. The calcium silicate hydrates formed in portland cement and related systems at elevated temperatures (e.g., $\geqslant 100\,°C$) may be significantly different in composition and degree of crystallinity from those formed at lower temperatures (see Fig. 3).

5. *Chemistry of Blended Cement Hydration*

The hydration of the portland cement component of blended cements is similar to that of the portland cement by itself, though at a given fineness the rate of strength development is likely to be somewhat slower because of the slower reactions of the other components. The high pH and the sulfate content of the aqueous phase resulting from the portland cement hydration reactions activate the slags or pozzolans in the cements. Once activated, slags appear to react independently to form additional quantities of calcium silicate hydrates. On the other hand, pozzolans can form calcium silicate hydrates only by reaction with the calcium hydroxide liberated from the portland cements. The C–S–H from blended cements is similar to that from portland cements.

6. *Performance Requirements*

If a cement is to perform satisfactorily in general concrete construction, most or all of the following performance requirements should be met by mortars and concretes which contain it:

(a) adequate rate of strength development,

(b) ease of mixing and placing,

(c) dimensional stability (soundness)—no detrimental volume changes,

(d) low creep and low drying shrinkage,

(e) durability—capacity to withstand normal environments above and below ground level,

(f) nonreactivity with aggregates,

(g) noncorrosivity to reinforcing steel,

(h) low heat evolution,

(i) uniform appearance, and

(j) capacity for control of setting and rheological properties by means of admixtures.

Portland and blended cements can be manufactured to meet essential requirements for most applications. The rate of strength development at early ages can be increased by finer grinding of the cement or increasing the contents of C_3S and C_3A. The ease of mixing and placing usually offers no problems, provided that the particle size distribution is broad and the gypsum additions are sufficient to prevent flash setting due to rapid hydration of C_3A. Dimensional stability is usually acceptable unless a large quantity of ettringite forms after setting or if, in products which are to be autoclave-cured, the cement contains more than a small amount ($\sim 2\%$) of periclase (MgO). Creep and drying shrinkage are related phenomena which are at a minimum when the optimum quantity of gypsum is contained in the cement.

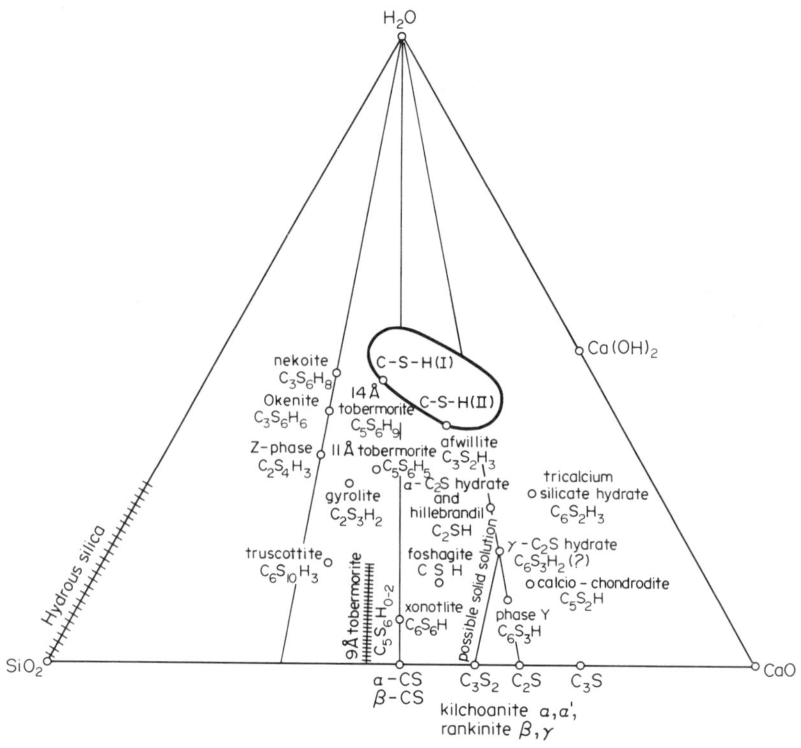

Figure 3
Phase diagram showing some calcium silicate hydrates found in hydrated portland cements and related systems at ambient and elevated temperatures (after Taylor 1964; courtesy of Academic Press)

The chemical durability of hardened cement paste is usually good in natural environments, except where sulfate groundwaters and soils are encountered. Pastes of cements with higher C_3A contents are more prone to sulfate attack as the result of the formation of large ettringite crystals within the structure. For cements which are to be used with certain siliceous sands and aggregates, a low-alkali portland cement or a portland–pozzolan cement is generally desirable if the possibility of disruption due to alkali–aggregate reactions is to be avoided. Hardened cement pastes form a good corrosion-protective cover for reinforcing steel because the high pH of the paste passivates the steel unless chloride is present or neutralization of the paste by atmospheric carbon dioxide has reached the level of the steel.

The heat evolution accompanying the hydration of a cement depends primarily on its composition. Cements which are high in C_3A and C_3S evolve more heat than those which are low in these constituents. The appearance of a hardened cement paste is affected by the composition of the cement and the structure of the hardened material. Differences in porosity due to different water–cement ratios and different degrees of compaction can lead to differences in appearance; also, soluble salts, such as sodium sulfate, in the pore water may be deposited on the surface due to evaporation under certain environmental conditions, giving a white incrustation (efflorescence).

The effects of intentionally added soluble materials in the mixing water on the setting and rheological properties of cement pastes are technically important. Certain organic compounds, such as lignosulfonates, can retard the setting and improve the flow of cement pastes. The effects are due to adsorption on the surfaces of the cement grains and hydration products. Acceleration of setting and hardening reactions is sometimes achieved by addition of calcium chloride to the mixing water, though this may lead to accelerated corrosion of steel reinforcement.

Frequently, a specific performance objective can be met in different ways with different cements. For example, high early strength could be achieved either by using a finely ground portland cement or by curing the product at an elevated temperature. As another example, a concrete which is likely to be exposed to sulfate groundwaters could be made either with a portland cement selected for its low C_3A content or a portland–pozzolan cement known to be sulfate-resistant. Because of the large number of composition and particle size variables which can be controlled in cement manufacture, the number of performance options which can be exercised is very large. In practice, the number is limited by the cost of manufacturing and storing many types of cement.

7. *Standards for Portland and Blended Cements*

In the USA the most widely used standards for cements are those established by the American Society for Testing and Materials (ASTM). The ASTM standard specification for portland cement, ASTM C 150, covers five main types with differing performance characteristics and also recognizes some optional subcategories. The chemical and physical requirements for the five types are given in Table 4.

The ASTM standard specification for blended hydraulic cements. ASTM C 595, is newer than the portland cement specification and rests on a less complete base of knowledge. Needs for conservation of energy and materials have increased interest in blended cements, and improvements in the standards are being brought about to facilitate substitutions between blended cements and portland cements. The categories of blended cement recognized in ASTM C 595 are indicated in Table 5. Cements manufactured in different countries tend to be similar, but the standards may differ substantially in detail. An important factor in the selection of cements is uniformity. This is covered in ASTM C 917 (established in 1982).

8. *Methods of Characterization of Cements*

For specification purposes, portland cements are characterized in terms of their oxide compositions using wet chemistry or instrumental methods such as atomic absorption, x-ray fluorescence or direct-coupled plasma spectrometry. Assuming that the only phases present are tricalcium silicate, dicalcium silicate, tricalcium aluminate, tetracalcium aluminoferrite, gypsum, periclase and free lime, the phase composition is calculated from the elemental analysis using a

Table 4
Types of portland cement defined in ASTM C150

Type	Characteristics	Limits on clinker phases	Typical potential compound composition (%)			
			C_3S	C_2S	C_3A	C_4AF
I	general purpose		53	23	10	8
II	moderate heat evolution, moderate sulfate resistance	$<8\%\ C_3A$	50	24	7	11
III	high early strength		57	18	10	8
IV	low heat evolution	$\geqslant 40\%\ C_2S$, $\leqslant 35\%\ C_3S$, $\leqslant 7\%\ C_3A$	30	44	6	12
V	sulfate resistant	$<5\%\ C_3A$	44	30	4	14

Table 5
Types of blended hydraulic cement defined in ASTM C595

Limits on % of		Type	Name	Use
Pozzolan	Slag			
<15	a	I(PM)	pozzolan-modified portland cement	general purpose
15–40	a	IP	portland–pozzolan cement	general purpose
15–40	a	P	portland–pozzolan cement	general purpose where strength at early ages is not important
	0–25	I(SM)	slag-modified portland cement	general purpose
	25–65	IS	portland blast-furnace slag cement	general purpose
	≥60	S	slag cement	for use with portland cement in concrete or with lime in masonry mortars

[a] May contain slag because the cement may be made by blending the pozzolan with either a portland or a portland blast-furnace slag cement; in practice, slag is seldom present

modified Bogue formula. The results reflect errors in the assumptions made.

In addition to composition, cements are characterized in terms of fineness as indicated by specific surface area, particularly as determined by the Blaine air permeability method. The characterization of blended cements is less definitive than that of portland cements because there is less knowledge about the phases present in blast furnace slags and pozzolans. Chemical analyses of blended cements are carried out only to establish that the oxide compositions are being maintained within a defined range. It is recognized that, for purposes of quality assurance and prediction of performance, as well as for research, it would be desirable to be able to characterize cements in terms of phase composition and the distribution of the phases by particle size and shape. This is not yet practicable.

For research purposes, techniques used for determining the phases present in both portland and blended cements include x-ray diffraction, thermal analysis, petrography and scanning electron microscopy complemented by energy-dispersive x-ray analysis. These techniques can be applied to whole cements or cements which have been separated into size fractions. Progress is being assisted by the development of techniques for selective dissolution of cement phases. For example, the calcium silicates in portland cements can be preferentially dissolved by solutions of salicylic or maleic acids in methanol to leave the aluminate phases.

9. Uses of Portland and Blended Cements

Uses of portland and blended cements (Table 6) can be divided into factory and field applications. In factory applications, the operations can be better controlled and elevated-temperature curing with steam is frequently used to accelerate strength development and shorten the production cycle. In field use, by far the greatest quantity of cement is used in ready-mixed concrete delivered to the worksite in a mixer truck.

Table 6
Current uses of cements in the USA

	Usage (%) (estimate)
Cast-in-place concrete	
foundations	36
slabs	14
highways	13
columns	7
masonry mortar	4
canal linings	2.5
tunnel linings	2
other	8
Precast concrete	
concrete block	4
pipe	2
panels	2
beams	2
fiber-reinforced products	2
other precast	1.5

10. Manufacture of Portland and Blended Cements

Manufacture of portland cement clinkers begins with the grinding of a carefully proportioned mixture of previously crushed raw materials. Common raw materials, and the main oxides they contribute, are limestone (CaO), oyster shells (CaO), marl (CaO, SiO_2), cement rock (CaO, SiO_2, Al_2O_3), sand (SiO_2), iron ore (Fe_2O_3) and mill scale (Fe_2O_3). The raw meal is usually fed through a preheater into a rotary kiln in which it is gradually heated to ~1500 °C by a slightly oxidizing countercurrent flow of the combustion gases from a pulverized coal, oil or natural gas flame. During passage through the kiln, which rotates at 1–2 rev min^{-1}, the minerals in the raw meal are progressively dehydrated and calcined (decarbonated) and undergo phase changes, including melting, and chemical reaction before being discharged into a clinker cooler. At the highest temperature, the liquid phase makes up ~25% of the volume of the material; it is predominantly a calcium aluminate and ferrite melt and the solid phases are the calcium silicates. The viscosity of the melt depends on its composition: MgO and alkali metal oxides have large effects. The distribution of the minerals (compounds) in the clinker is controlled by the raw materials, their fineness, the homogeneity of the raw meal and the operating conditions of the kiln and cooler. The clinker is normally in the form of roughly spherical nodules ~5–50 mm in diameter. It is stored in stockpiles until needed to make the finished cement by grinding in a ball mill with ~5% of gypsum. Cement is stored in silos from which it may be shipped in bulk by truck, rail or barge. In 1986, the average selling price of portland cements in the USA was in the range US$55.48 per tonne.

The manufacture of portland cement clinker is energy-intensive and accounts for ~0.8% of the nation's energy consumption. About 80% of the energy is used in the pyroprocessing, with lesser quantities for crushing and grinding. The theoretical energy requirement for the formation of portland cement clinker from typical ground raw materials is ~1750 kJ kg^{-1}; the most efficient of present day plants use about twice this.

Blended cements are formed by intergrinding portland cement clinkers with granulated blast-furnace slag or fly ash and a small quantity of gypsum. They may also be formed by blending of the separately ground ingredients.

11. Research Needs

Cement research needs can be divided into two categories—those of cement manufacture and cement use. Improved knowledge of the basic properties of cements and of the mechanisms and kinetics of the hydration reactions are required to help make concrete a more predictable material. In the high-temperature reactions in which portland cement clinker is formed, there is need for greater knowledge about effects of the raw materials and their processing on the formation and characteristics of the clinker, and the relationship between the clinker characteristics and the properties of the finished cement. With growing interest in the use of blast-furnace slags, fly ash, and other waste and by-product materials in cement manufacture, there is a need for improved methods for characterizing these materials and identifying the characteristics which lead to superior performance in cement and concrete.

12. Application of Computers in Cement Manufacture and Use

The use of computers in cement science and technology is growing rapidly. Computers are being applied to control of most aspects of cement manufacturing operations and useful mathematical models of the operations have been developed. As the capabilities of computers have grown, the models have become more comprehensive and realistic. Manufacturing processes modelled include raw-material proportioning and blending, grinding, and processes in kilns, including clinker formation. The models have been used for improving the manufacturing process in terms of energy use, product uniformity, reduced grinding costs and lengthened kiln-lining life.

A difficulty in developing models relating the steps in cement manufacture to cement performance lies in the difficulty of determining the particle size distribution of the cement and the distribution of phases between particles of different sizes. There are similar problems in the characterization of raw meal and in the determination of the composition and structure of clinker. In application to the finished cement, computers are being used to control analytical instrumentation and process the data, and attempts are being made to develop mathematical models which can predict the performance of cements in concrete.

13. Portland and Blended Cement Mortars

Mortars are mixtures of cements with sand, or other fine aggregate, and water. The sand–cement ratio is usually ~ 3. Mortars are used extensively in masonry construction as the bedding materials for the masonry units and as a rendering applied to vertical surfaces for architectural purposes or to increase water tightness. The rheological properties of masonry mortars are important in their use, a "buttery" consistency usually being desired by the mason. Beneficial effects are frequently obtained by including in the mixture hydrated lime, flow modifiers, such as methyl-cellulose, and air-entraining agents. Important properties of masonry mortars are water retentivity of the plastic material when brought into contact with the masonry units and relatively low shrinkage after hardening. Quality assurance testing of cements for strength and dimensional stability is usually carried out on standard mortars made with a standard quartz sand.

See also: Calcium Aluminate Cements

Bibliography

American Society for Testing and Materials 1981 *Annual Book of Standards*, Pt. 13, *Cement; Lime; Ceilings and Walls*. American Society for Testing and Materials, Philadelphia, Pennsylvania

Duda W H 1977 *Cement Data Book*, 2nd edn. Macdonald and Evans, London.

Lea F M 1970 *The Chemistry of Cement and Concrete*, 3rd edn. Chemical Publishing Company, New York

McKee H J 1973 *Introduction to Early American Masonry Stone, Brick, Mortar and Plaster*. National Trust for Historic Preservation, Washington, DC.

National Materials Advisory Board 1980 *The Status of Cement and Concrete R & D in the United States*. NMAB-361. NMAB, Washington, DC

Proceedings of the International Congresses (Symposia) on the Chemistry of Cements: Paris 1980, Moscow 1974, Tokyo 1968, Washington 1960, London 1952, Stockholm 1938

Ramachandran V S, Feldman R F, Beaudoin J J 1981 *Concrete Science Treatise on Current Research*. Heyden, Philadelphia, Pennsylvania

Taylor H F W (ed.) 1964 *The Chemistry of Cements*. Academic Press, London.

US Bureau of Mines 1980 *Mineral Facts and Problems*, US Bureau of Mines Bulletin 671. USBM, Washington, DC

World Cement Standards. Cembureau, Paris, France

G. J. Frohnsdorff
[National Bureau of Standards, Washington, DC, USA]

Prices of Industrial Minerals: History

Commodities classified as industrial minerals cover such a wide range of materials and values that it is difficult to make general observations on their price history. The more common materials, such as crushed stone, sand and gravel, have relatively low unit values that have varied from about US\$1 to US\$5 t^{-1} over the past 35 years. The more exotic industrial minerals, such as long-fiber asbestos, may be valued up to US\$1800 t^{-1}, and industrial diamonds at US\$3–5 per carat (US\$15–25 million t^{-1}). Most of the industrial mineral commodities have been valued from US\$10 to US\$100 t^{-1} over the past 35 years.

The value of an industrial mineral depends largely on how it is used, with more precise requirements of chemical purity, crystalline perfection, or other physical properties such as hardness generally associated with higher prices. Less exacting specifications would be associated with lower priced materials.

The values shown in Table 1 give a 30-year history of prices (1950–1980) for most of the important industrial minerals. In cases of significant variations due to grading or specifications, there are some ranges presented. This can be noted for asbestos, graphite, mica, nitrogen compounds and talc. Clays are subdivided into four commodities according to significant differences in their usage; attapulgite, used as an absorbent in cat litter and floor sweep; ball clay, used in ceramics; bentonite, used as a bonding or suspension agent; and kaolin, with a wide variety of uses from ceramics and fillers to paper coating. Kaolin used as a coating material for high-grade paper normally sells for prices up to ten times those shown for the average grades based on tonnage consumed. Other clays show similar historical price variations.

One general observation on industrial minerals prices is that increased economic activity normally creates increased demand for industrial minerals that is not always met with new supplies. In 1960, many of the commodities shown in Table 1 had higher unit values than in 1970; 1960 was a year of economic rebound from the 1958 recession, and industrial mineral producers realized higher prices than normal owing to supply shortages in many areas. The long-term price trend in real (constant purchasing power) dollars was not in line with changes in the value of the dollar until relatively recently. Thus, the prices show an upward trend through the mid-1970s that was not in keeping with the rate of inflation. The real prices of industrial minerals thereby created values that made them bargains as compared with other commodities. Since 1975, the prices of most industrial minerals have kept ahead of the rate of inflation, but even in 1980 they were still relatively low when compared with other commodity price increases during the 1950–1980 period.

Most of the prices shown in Table 1 are FOB (free-on-board) or point-of-production prices. Since transportation costs can be significant for bulk materials, the delivered prices are often two to three times greater than the prices shown. This is especially true for long-distance transportation from point of production to point of use.

Individual commodities show significant price variations owing to swings in demand and shortages of supply over the 30-year period. Potash has seen significant supply variations over this period. When supplies were plentiful, such as after completion of several new large Canadian mines, the prices were significantly depressed. At other times, when demand outpaced supply, the prices escalated.

The highest grades of asbestos held relatively constant in nominal price over an extended period (1960–1980), indicating a falling demand and relatively constant availability of the longer fiber material. Abrasive materials such as corundum, diamond, emery, garnet and pumice have been subject to significant competition from synthetic materials. In the case of corundum, a natural mineral composed of aluminum oxide, the availability of alumina produced from bauxite has virtually displaced most of the demand for the natural material.

Prices for bauxite, used as an abrasive, are dominated by its major use as an ore for aluminum. Ilmenite and rutile prices vary in relation to the demand for paint and pigments. Phosphate rock, potash and nitrogen compounds relate directly to the demand for chemical fertilizers. Gypsum, cement, crushed stone, and sand and gravel are tied to the level of construction activity.

Several of the listed commodities in Table 1 relate to glass manufacturing, specifically borates and soda ash, but are also used in laundry detergents and chemicals. Sulfur prices relate to the market for sulfuric acid, and also to fertilizer demand owing to the significant use of sulfuric acid in phosphate processing.

Barite (or barytes) is especially important in oil-well drilling muds, and demand is likely to grow as long as drilling activity is increasing. Ball clay, feldspar, kyanite, wollastonite and zircon follow the demand for various ceramic and refractory materials. Lime has recently become more closely associated with the basic oxygen process of steelmaking, and its demand is likely to demand more directly on the demand for steel and chemicals, its major use over the 30-year period. Salt is closely tied to the manufacture of chloralkali chemicals.

Diatomite is largely consumed as a filter aid and specialized carrier or filler. Fluorspar is closely associated with aluminum, steel and chemicals. Graphite is used in electrical applications and high-temperature ceramics, as well as a lubricant. Mica is used in paint and various electronic components. Perlite and vermiculite are closely tied to insulation uses. Talc is used as a filler and in cosmetics, such as face powder and make-up. The other materials listed in Table 1 are basic chemical sources. Zeolites and a few other industrial minerals are not listed owing to their wide price fluctuations.

In all cases, the demand for industrial minerals depends on their ultimate end-use (derived demand). They have limited usefulness except in other products or after further processing. Supply is largely controlled by the availability of deposits of suitable ores and the timing of mine development and processing capacity by the producers. Prices vary according to growth in both supply and demand.

Over the past thirty years, the markets for industrial minerals have tended to show more stable and orderly growth than markets for metals and other commodities. Prices have increased somewhat below the rate of

Table 1
Industrial mineral prices[a]

Commodity	Price (US$ t^{-1}) 1950	1960	1970	1980
Asbestos				
average grade	56.00	72.00	85.00	356.00
lowest grade	30.00	40.00	60.00	120.00
highest grade	960.00	1800.00	1800.00	1800.00
Barite or barytes	28.20	19.85	16.50	31.40
	6.50	16.65	9.85	14.75
	27.05	52.40	82.95	186.30
	0.46	0.47	0.37	0.58
	13.80	19.75	18.00	51.00
	18.10	24.75	35.30	49.60
	13.55	14.85	18.75	27.55
	9.70	13.05	16.30	27.60
	18.75	18.75	26.50	38.60
	160.00	240.00	23.40	43.00
	3.23	4.06	3.67	4.95
	44.00	54.60	60.25	172.40
	13.25	19.20	22.00	83.00
	6.90	10.50	14.65	35.25
	38.85	46.30	57.35	100.35
	94.00	103.00	113.30	143.30
		308.65	369.30	626.15
		330.70	405.65	535.75
		20.95	39.70	153.25
	3.05	4.00	4.10	8.45
	17.00	22.65	20.70	54.10
	3.80	6.00	5.28	13.30
	55.00	61.00	80.00	141.00
	12.30	14.75	16.00	51.40
	1.00	27.60	23.40	44.00
	6.00	13.20	1.75	6.00
	2.00	32.00	45.00	120.00
	8.00	88.00	55.00	125.00
	7.05	10.55	12.90	28.75
	6.00	7.20	5.80	24.25
	2.00	28.00	22.00	86.00
	4.10	2.35	1.70	4.65
	7.00	105.00	204.00	485.00
	.10	29.50	34.00	57.00
	.08	1.12	1.30	3.10
	.65	38.00	23.10	79.60
	.20	45.20	20.00	68.90
	.35	1.55	1.90	3.30
	.55	67.70	24.25	60.00
	.75	24.60	22.75	87.85
	00	10.00	5.00	10.00
	00	40.00	250.00	280.00
Vermiculite	11.25	17.22	25.14	79.01
Wollastonite	22.00	43.00	38.50	86.00
Zircon	52.35	52.00	60.65	182.00

[a] FOB (free-on-board) mine or processing plant prices in the USA, unless otherwise indicated [b] Price given in US$ kg^{-1} [c] Price given in US$ per carat [d] CIF (cost, insurance and freight) prices

change of prices for all other goods and services, but the price increases have been more or less in line with other minerals in the long-run. While prices are unlikely to follow the trend of fuels and energy materials, they are likely to continue to increase at a more rapid pace in the future, as more countries develop their domestic economic capacity and their demand for commodities becomes increasingly competitive with that of the industralized nations.

See also: Prices of Metals: History

Bibliography

Anon 1981 Materials prices. *Eng. News-Rec.* 207 (19): 66–67

Lefond S J 1981 Another disappointing year for industrial minerals. *Min. Eng.* (*N.Y.*) 33: 562–92

Lefond S J (ed.) 1983 *Industrial Minerals and Rocks*, 5th edn. American Institute of Mining, Metallurgical and Petroleum Engineers, New York, p. 1446

Levine S 1971 The year ahead, 1972. *Rock Prod.* 74 (12): 53–61

McCarl H N, Maclean J A, Wright D T 1980 *Non-Metallic Minerals*, Mahon's Industry Guides for Accountants and Auditors, Guide 16. Warren, Gorham and Lamont, Boston, Massachusetts

Minerals in the US Economy: Ten-Year Supply–Demand Profiles for Nonfuel Mineral Commodities (1966–1975) and (1968–1977). US Bureau of Mines, Washington, DC

US Bureau of Mines 1974 *Minerals Yearbook 1972*, Vol. 1. USBM, Washington, DC (1974 ASI Microfiche: 5604–1)

US Bureau of Mines 1976 *Commodity Data Summaries.* USBM, Washington, DC

US Bureau of Mines 1977 *Mineral Commodity Profiles.* USBM, Washington DC (1977 ASI Microfiche: 5606–2)

US Bureau of Mines 1978 *Minerals Yearbook 1976*, Vol. 1. USBM, Washington, DC (1978 ASI Microfiche: 5602–20)

US Bureau of Mines 1978–1981 *Mineral Commodity Summaries*, Annual Editions 1978–1981. USBM, Washington, DC

US Bureau of Mines 1980 *Mineral Facts and Problems*, US Bureau of Mines Bulletin 671. USBM, Washington, DC (1981 ASI Microfiche: 5608–62)

US Bureau of Mines 1981 *Minerals Yearbook 1978–79*, Vol. 1. USBM, Washington, DC (1981 ASI Microfiche: 5604–41)

H. N. McCarl
[University of Alabama, Birmingham, Alabama, USA]

Prices of Metals: History

The behavior of metal prices often seems erratic and hard to predict. In the short run, prices fluctuate with random shocks in supply and demand, making it difficult to detect any regular behavior patterns. However, when examining the history of prices over a century, systematic tendencies can be seen and useful generalizations about price trends, both for individual metals and for the metals taken as a group, can be made.

Long-run secular price trends are discussed in Sect. 1 below, where it is pointed out that many metals exhibit U-shaped price paths, with metal price declines owing to technological improvements, new discoveries and economies of scale, followed by price stability. The stable price phase may eventually be succeeded by one of price increases, as the depletion of high-grade orebodies raises mining costs. Different metals, however, can be in different phases of this cycle. For example, tin prices have risen during most of this century whereas aluminum prices have consistently fallen. Therefore, when an aggregate price index for the metals is constructed, it appears to be flat.

Section 2 discusses the cyclical patterns that are frequently superimposed on the long-run metal price trends. Metal prices often exhibit cyclical tendencies in addition to short-run random fluctuations. These cycles, which can be ten years or more in length, are related to stock depletions and to industry investment behavior. The presence of long-run cycles distinguishes metal price behavior from the behavior of fuel prices, which are not characterized by cyclical patterns.

The final section summarizes the principal conclusions that can be drawn from an examination of metal price history, and contrasts this historic behavior to the price patterns of the fuels and nonmetallic, nonfuel minerals.

1. Secular Trends

The prices discussed here (and those shown in the figures) are for US markets. However, because most metals are internationally traded, US prices cannot systematically deviate from world prices for long periods of time. Therefore, US prices are probably not far from world-average prices and the general conclusions drawn from an examination of their behavior would not change if some other price series were studied.

Current-dollar price paths for many metals are broadly similar to one another. In general, prices reached a minimum around the turn of the century. There were sharp price increases around World War I, dramatic declines during the depression, and increases again after World War II. The 1950s and 1960s, which were periods of relative price stability, were followed by the commodity boom in the early 1970s, when the prices of many metals reached unprecedented highs. This rise was possibly owing to fears of impending resource scarcity. After a subsequent decline in prices, there was a second major boom in the 1980–1981 period. It is likely that this second surge was associated with speculative frenzy in the silver market. Just as the mid-1970s boom was followed by collapse, so the boom of the early 1980s has given way to depressed prices in the mid-1980s.

Current dollar prices are highly correlated with the economy-wide rate of inflation. To detect patterns that are unique to the metals, therefore, it is necessary to correct for more general price fluctuations. That is, any trends that are purely inflationary or recessionary must be removed. This is achieved by dividing each price series by the US Wholesale Price Index (1967 = 1). The result is a set of constant dollar prices—prices expressed in terms of a hypothetical dollar of constant value. These constant dollar prices are discussed throughout the remainder of this article. Other deflating indices, such as the Consumer Price Index or the Wholesale Commodity Price Index, could also have been used. However, aggregate price indices tend to move together, and the conclusions drawn here would not be affected by the choice of deflator.

Each metal has a unique price path. However, certain general tendencies emerge from the diverse behavior patterns. The most important factors that affect metal prices are growth of metal demand, changes in mining and metallurgical techniques, discovery of new deposits and the depletion of high-grade orebodies. The first three factors predominate in the early phases of a metal's life cycle, whereas the fourth may be more important later on. Thus, it can be expected that real metal prices will initially fall but will eventually rise.

In the initial phase of a metal's commercial history, many factors lead to falling costs. Metal mining and processing is characterized by substantial economies of scale. For example, one mine can supply 10% of current consumption of some metals. When the market is new, mines and smelters tend to be too small to be fully efficient. The growth in demand as additional metal uses are discovered, therefore, allows mining companies to exploit cost-lowering economies of scale.

Learning-curve effects can also be important. As companies gain experience in mining, smelting and refining, they discover new techniques that lower costs. Therefore, costs may fall with cumulative metal production. In addition, there may be improvements in extractive and metallurgical technology that are not specific to an industry, but may be applicable across broad groups of industries. For example, the discovery of froth flotation lowered processing costs for many sulfide minerals.

Technical change can take place on the production or on the consumption side of a market. If it occurs on the production side, it may lower costs and therefore price. If it occurs on the consumption side, it may increase metal demand and allow mining companies to exploit economies of scale. In either case, the change can exert downward pressure on price.

The discovery of new deposits may increase metal supply and cause prices to fall. Even when new discoveries do not lower prices immediately, the expansion of known reserves may postpone the onset of higher costs owing to the depletion of high-grade ore deposits.

There are therefore many factors that can lead to falling real metal prices. Empirical studies such as that by Barnett and Morse (1963) have verified that the trend in the majority of mineral-commodity prices was negative during the first half of this century. However, the period of falling prices is often followed by one of stable prices. The stable-price period can be very long, as for lead and zinc, or fairly short-lived. In this stable period, production growth rates are steady, but not dramatic, and the average grade of ores mined falls only slowly.

A final phase of rising prices can be expected for most metals as high-grade deposits are depleted. Newly discovered deposits will tend to have lower metal contents, be more deeply buried, be of lower grade, or be in thinner veins. Often in the later phases of a metal's commercial history, the rate of introduction of cost-lowering techniques is insufficient to offset the rise in costs owing to depletion. This has long been the case, for example, for tin and mercury. The consequent rise in price triggers substitution away from the metal and therefore consumption growth is small or negative in this final phase.

There is a general inverse relationship between metal price and ore grade. In the mining and milling of many ores, it is the tonnage of rock processed that determines cost. Mining and milling costs are therefore proportional to the reciprocal or ore grade. In contrast, refining and smelting costs are virtually independent of the initial metal content of the ore. Therefore, the cost of producing the refined metal increases hyperbolically as ore grade falls (Page and Creasey 1975). Present information on tonnage–grade distribution curves is very limited, but it is not necessarily the case that tonnage increases as grade falls (Skinner 1976). There is therefore no reason to be optimistic about the future reversal of the upward trend in real prices that can already be seen for many metals.

The copper industry illustrates the historic counterbalancing influences of improvements in technology and deterioration in ore quality in determining production cost. The average grade of copper ores mined in the USA declined from about 2% in the early 1920s to 0.6% in 1983 (Jolly 1985). In spite of this decline in grade, real copper price fell until about 1940. The fall in price was made possible by technological developments in the early part of the century, particularly the advent of large earth-moving equipment, which made possible the strip mining of extremely low-grade orebodies, and the discovery of froth flotation, which made concentration of low-grade sulfide ores very economical. However, by 1940 the switch to the new technology had reached its natural limits and since that time, with no fundamentally new technological developments in the industry, the decline in grade has become the dominant factor in determining cost.

Petersen and Maxwell (1979) document the history of the relative impact of ore grade and technology on the cost of other metals (silver, tin, lead, zinc and iron) and claim that these factors have been much more important than the discovery of new deposits in the historical determination of metal prices.

Figure 1 shows deflated prices of all of the major metals with the exception of gold. Gold is not included because its price was linked to the dollar for most of the period. Its deflated price is thus proportional to the reciprocal of the deflating index. Superimposed on the price paths are fitted linear and quadratic trends. The fitted linear trends underestimate relative prices of all the major metals in the last few years of the period, because prices of all commodities shown have passed the minimum points on their fitted U-shaped curves and have begun to increase. Evidence of the statistical significance of the trend reversals (from negative to positive) can be found in Slade (1982a).

An examination of the fitted quadratic price trends reveals three basic price paths: falling, stable and rising prices. The first class is best illustrated by aluminum, a modern metal whose ores are very abundant. Growth rates for aluminum consumption have been high as new uses have been found, and technological advances combined with economies of scale have lowered prices over most of its commercial history. However, even the price of aluminum has begun to rise in recent years.

The second class is best illustrated by lead and zinc. The rate of growth of consumption of these metals has not been as high as that of aluminum, and technological advances and grade declines, which have both been modest, have almost exactly offset each other in determining lead and zinc production costs.

Finally, the third class is best illustrated by tin, an ancient metal with slowly increasing or declining consumption rates. Tin ores were not suitable for froth flotation, as were the ores of many sulfide minerals, and the decline in the grade of tin ores mined has been both steady and sizable. Copper and silver also fell into the third category in recent years. These categories, however, should not be thought of as three distinct price paths, but perhaps should be considered to be different phases in the life cycles of the respective mineral commodities.

2. *Cyclical Behavior*

Metal markets are well known for being unstable. The instability of these markets has been studied by various groups, including industry members, consumers and governments throughout the world. For example, the US Domestic Policy Review of Nonfuel Minerals (1979) found that both prices and production of nonfuel minerals in the USA are much more volatile than prices and production in the economy as a whole. Each of these groups is concerned with the nature and causes of severe price fluctuations and the gluts and shortages that characterize these markets. An understanding of mineral-commodity price oscillations is important to the developed nations, because they depend on imports of many minerals as basic inputs to industrial production. It is even more important to developing nations, because they may depend on sales of mineral commodities as their principal source of foreign exchange.

There are many reasons why metal prices may be volatile. In the short run, price elasticities of supply and demand are generally low. Short-run supply elasticities are low because mineral industry capacities cannot be varied quickly or at low cost. Demand elasticities are low because metals are typically used as intermediate inputs, and their cost often constitutes a small fraction of the price of final goods. Therefore, moderate price changes may have little effect on metal production or consumption in the short run. Prices are also unstable because income elasticities of demand for these commodities are generally high. Income elasticities are high because metals are consumed almost entirely for industrial purposes. Therefore, changes in economic activity may lead to sizable fluctuations in metal demand. Finally, prices are volatile because mineral-industry investment projects are typically large, and because there are often long time lags between the initiation and completion of these projects, perhaps as long as ten years. Such long lags may cause very long-run price cycles, as discussed below.

One explanation for the observed fluctuations in metal prices relates to the nature of investment in these industries. Mineral-industry investment projects typically involve large sums of money. For example, more than one billion dollars is often required to bring an integrated mine-through-refinery operation online. In addition, a single investment project may account for a large fraction of productive capacity in the industry. Therefore, when a new mine is opened, the additional output may depress price.

It is often noted that investment projects are decided upon when prices are rising, but are completed when prices are falling. If new projects that were initiated during periods of high prices contribute to metal price declines (by expanding output), price cycles will be roughly twice the investment period in length. That is, the peak-to-trough period will be equal to the investment period. The length of the investment period will then be an important determinant of metal price behavior.

This period consists of several stages: exploration and deposit definition, execution of an engineering feasibility study, procurement of project financing, and, finally, construction of the facility and any needed infrastructure. The total time required can be fifteen years or more; however, several of these steps can be shortened. Many companies own well-explored deposits that can be developed when economic conditions are favorable, and financing is facilitated if internal funds are available. Therefore, particularly in the USA, mines can be brought online in as little as five years. Thus, price cycles of ten years or more in length

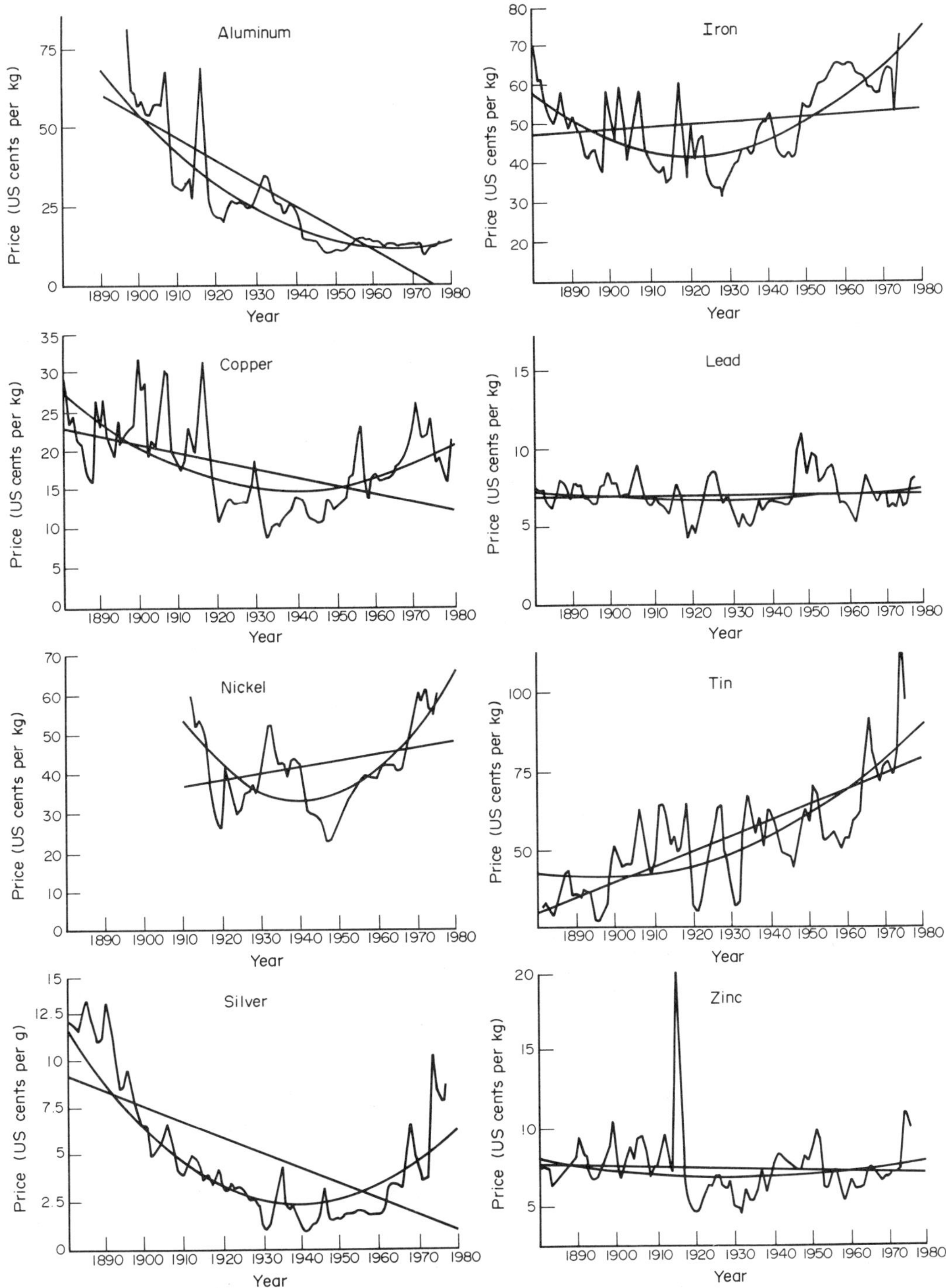

Figure 1
History of deflated prices and fitted linear and quadratic trends for the major metals

might be expected. This prediction is borne out in Slade (1982b).

There are thus many causes of metal price instability. Even in industries where list prices have been remarkably constant there is often considerable variation between list and transaction prices, and transaction prices fluctuate much more than list prices. For example, the molybdenum list price did not change over a five-year period (1969–1973); however, there was substantial discounting of the list price and, therefore, the transaction price was not stable.

3. *Summary and Conclusions*

The earlier analysis of the long-run relative price movements of the metals revealed that the price paths of many of these commodities are U-shaped. Similar long-run patterns are found in the prices of many fuels (Slade 1982a). In contrast, no such systematic behavior is found in the price history of nonfuel, nonmetallic mineral commodities.

Analysis of intermediate-run relative price movements of the metals revealed that cyclical behavior is superimposed on the long-run price trends of most metals. In contrast to the metals, no systematic cyclical behavior seems to occur in fuel or nonfuel, nonmetallic mineral-commodity prices, perhaps because productive capacity in these industries can be increased in smaller increments and with shorter time lags.

See also: Economic Theory of Mineral Resources and Markets; Prices of Industrial Minerals: History

Bibliography

Barnett H J, Morse C 1963 *Scarcity and Growth: The Economics of Natural Resource Availability.* Johns Hopkins, Baltimore, Maryland

Jolly J L W 1985 Copper. *Mineral Facts and Problems,* US Bureau of Mines Bulletin 675. USBM, Washington, DC, pp. 197–221

Manthy R S 1978 *Natural-Resource Commodities: A Century of Statistics, Prices, Output, Consumption, Foreign Trade, and Employment in the United States. 1870—1973.* Johns Hopkins, Baltimore, Maryland

Page N J, Creasey S C 1975 Ore grade, metal production, and energy. *J. Res. U.S. Geol. Surv.* 3: 9–13.

Petersen U, Maxwell R S 1979 Historical mineral production and price trends. *Min. Eng. (N.Y.)* January: 25–34

Skinner B J 1976 A second iron age ahead? *Am. Sci.* 64: 258–69

Slade M E 1982a Trends in nonrenewable natural-resource commodity prices: An analysis of the time domain. *J. Environ. Econ. Manage.* 9: 122–37

Slade M E 1982b Cycles in nonrenewable natural-resource commodity prices: An analysis of the frequency domain. *J. Environ. Econ. Manage.* 9: 138–48

US Domestic Policy Review of Nonfuel Minerals 1979 *Background Papers: Draft for Public Review and Comment of the Report on Nonfuel Minerals Policy Review.* US Department of the Interior, Washington, DC

M. E. Slade
[University of British Columbia, Vancouver, British Columbia, Canada]

R

Rare-Earth Metals Resources

The rare-earth elements (as defined by the International Union of Pure and Applied Chemistry) include the group IIIA elements (Sc, Y) and the lanthanide elements (La, Ce, Pr, Nd, Pm, Sm, Eu, Gd, Tb, Dy, Ho, Er, Tm, Yb and Lu).

Introductory textbooks view the rare-earth elements as being so chemically similar to one another that collectively they can be considered as one element. To a certain degree this is correct—many uses are based on this close similarity—but vast differences in their behavior and properties do exist. For example, the melting points of the end members of the lanthanide series differ by a factor of almost two (La 918 °C, Lu 1663 °C).

1. Abundances

The term *rare* implies that these elements are scarce, but in fact they are quite abundant and exist in many workable deposits throughout the world. The 16 naturally occurring rare earths fall into the 50th percentile of elemental abundances. Cerium, the most abundant, ranks 28th and thulium, the least abundant, ranks 63rd. The light lanthanides (lanthanum through europium) are more abundant than the heavy lanthanides (gadolinium through lutetium). Furthermore, the elements of even atomic number are more abundant than their neighbors of odd atomic number.

2. Minerals and Ores

The three major minerals of some 160 known to contain rare earths are bastnasite, monazite and xenotime. The first two are sources of the light lanthanides and account for about 95% of the rare earths currently utilized. Xenotime and also uranium tailings are a source of the heavy lanthanides and yttrium.

Bastnasite, a fluorocarbonate, accounts for $\sim$43% of the rare earths utilized in the free world; most of it is mined at Mountain Pass, California. The People's Republic of China has vast bastnasite deposits which are greater than all the other known bastnasite deposits.

About 52% of the rare earths consumed in the free world is derived from monazite, a rare-earth phosphate containing 5–10% ThO_2. The five major free-world countries producing monazite are Australia (32% of total production in 1986), Brazil (18%), Malaysia (18%), the People's Republic of China (12%) and India (12%). Other countries mining monazite are the USA, the USSR, Thailand, Sri Lanka and the Republic of South Africa.

Xenotime is the major ore source for the heavy lanthanides and yttrium. It is a phosphate containing $\sim$3% each of U_3O_8 and ThO_2. The major producers of xenotime are Malaysia, the People's Republic of China and Australia.

The other rare-earth commercial ores are apatite (the USSR) and Longnan clay (Jiangxi Province, the People's Republic of China), and uranium tailings (Canada). These three ores plus xenotime account for $\sim$5% of the rare earths mined in 1986.

3. Extractive Chemistry and Metallurgy

Rare-earth oxide (REO) is concentrated from bastnasite by a hot froth flotation technique and then treated with HCl to dissolve the calcite. The separation of individual elements is carried out by oxidizing Ce^{3+} to Ce^{4+} and dissolving the remaining trivalent rare earths, which are then separated by liquid–liquid solvent extraction.

Both monazite and xenotime are usually cracked by heating the ore in an autoclave at 150 °C with a 70% NaOH solution. After cooling, the addition of H_2O removes the soluble Na_3PO_4 and the Th is left behind as $Th(OH)_4$ by carefully dissolving the REO in HCl. The REO is either solid (e.g., as a mixed rare-earth chloride) or separated into the individual rare earths by a solvent extraction process.

The anhydrous mixed rare-earth chloride is reduced electrolytically to give a product known as mischmetal. Most of the individual rare-earth metals are prepared by reducing the fluoride with Ca metal. However, Sm, Eu, Tm and Yb are prepared by heating the oxide with La and subliming the pure metal.

4. Uses and Markets

On a volume basis, $\sim$95% of rare-earth usage is in the mixed form. The individual elements account for the remaining 5% of volume but over 50% of the monetary value. The major uses of the mixed rare earths in 1987 include those listed below.

(a) Approximately 29% of the rare earths are used in glass/ceramics areas as decolorizing agents, antibrowing agents, glass-polishing compounds, glass and ceramic coloring agents, additives to structural ceramics such as stabilized zirconia and Si_3N_4, and in optical lenses and glasses.

(b) About 36% are used as catalysts, primarily in the cracking of oil to form gasoline and in automobile exhaust catalysts.

(c) About 31% are used in metallurgy as an alloying agent to desulfurize steels, as a nodularizing agent in ductile iron, as lighter flints and as alloying agents to improve the properties of superalloys and magnesium, aluminum and titanium alloys.

(d) The remaining 4% of the rare earths are used in the electronic/magnetic field as phosphors (color television and fluorescent lighting), $Nd_2Fe_{14}B$ and $SmCo_5$–Sm_2Co_{17} permanent magnets, garnet bubble storage devices, oxygen sensors, microwave ferrites and garnets, lasers, and hydrogen storage materials.

Bibliography

Gschneidner K A Jr (ed.) 1981a *Industrial Applications of Rare Earth Elements*, ACS Symposium Series No. 164. American Chemical Society, Washington, DC

Gschneidner K A Jr 1981b Rare earth speciality inorganic chemicals. In: Thompson R, Matterson K (eds.) 1981 *Speciality Inorganic Chemicals*. Royal Society of Chemistry, London, pp. 403–43

Gschneider K A Jr, Eyring L (eds.) 1978–1988 *Handbook on the Physics and Chemistry of Rare Earths*, Vols. 1–11. North-Holland, Amsterdam

K. A. Gschneidner Jr
[Iowa State University, Ames, Iowa, USA]

Recycling of Demolition Wastes

Owing to the large volume and generally low specific value of demolition materials, the degree to which they are recycled depends not only on where and in what condition and quantity they are found, but also on the materials themselves. In this article, the origin of demolition materials is first discussed because of its strong effect on the potential for recycling. The approximate flows of different materials from different sources are given, and then the manner in which the principal materials are recycled is reviewed. The present data come from US and European studies, but the proportions of the materials involved should be broadly applicable to the demolition of similar structures elsewhere.

The origins of demolition materials can be classified into two groups: pavements and structures. Pavements include highways, sidewalks, parking lots, malls, airport runways, quays and stockyards, where a hard, strong surface has been laid over the ground to withstand the passage of motor vehicles, people or animals. Structures include buildings, bridges, tunnels, ramps and walls. The distinction is significant because pavements are usually demolished only for repair, and not for a complete change of form and use, as is often the case for structures. As energy, materials and landfill space become relatively more expensive, there is an increasing tendency for pavement materials to be recycled in situ and be incorporated into the new pavement. The proportion of pavement materials being recycled in this way is, however, still small; it is almost certainly under 10% of available wastes.

In situ recycling is more usual with asphalt pavements, which can be lifted by travelling machines, crushed, screened, mixed with new binders, and pressed or rolled back into place. Concrete, whether from pavements or structures, can also be crushed and screened, and that fraction falling within certain size limits can be incorporated into new concrete in place of aggregate (Nixon 1976, Frondistou-Yannas and Ng 1977). The strength of concrete made in this way is typically reduced to 80% of that made with crushed stone, but is normally adequate for use in pavements. However, machinery capable of crushing large pieces of concrete is not portable enough to be made part of pavement-resurfacing equipment. The presence of steel reinforcing mesh in pavement concrete, or heavier reinforcing bars in most structural concrete, is a complicating factor that only a few plants are equipped to handle, and most demolition concrete is used as coarse-fill or base material. At present in the USA and the European Economic Community (EEC), the demand for this material is such that it can be dumped on site free of charge (Wilson et al. 1979, Environmental Resources Ltd. 1980). Crushed and graded concrete can be sold. Contractors wishing to dispose of mixed demolition wastes, particularly mixtures including wood and other degradable or combustible material, must often transport them considerable distances (typically 50 km in the neighborhood of large cities) and pay a tipping charge. There is thus an increasing incentive to recycle as large a proportion of demolition materials as possible.

Demolition materials from structures can be classified into those from residential and those from nonresidential buildings. In North America, residences are usually constructed either mostly of wood or mostly of masonry. The resulting demolition flow has slightly more wood than bricks by mass for the USA (see Fig. 1). It is expected that the proportions for Canada would be similar. In the EEC, the proportion of wood in demolition materials is almost negligible compared with brick (Environmental Resources Ltd. 1980). The ready availability of wood in the USA results in very little recycling of wood as lumber, fuel or chips for building board, despite the considerable proportion of wood in residential demolition materials. Companies formed to process demolition wood in the USA have ceased operations. In contrast, Mexican demolition firms operating near the US border have removed all lumber into Mexico (Wilson and Malarkey 1974).

In most countries, a high proportion of the unbroken bricks in demolition wastes is recycled. It is still profitable in the USA for contractors to employ people to remove mortar from bricks by hand, as some machines that were developed to clean bricks were found to be uneconomic (Wiesman and Wilson 1976).

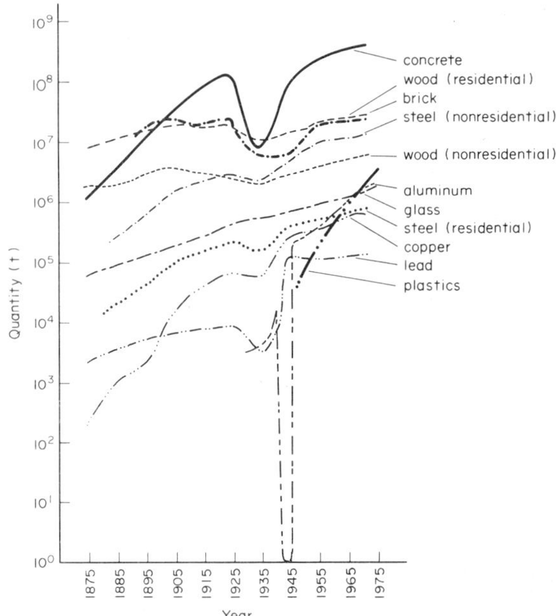

Figure 1
Quantities of materials used in building construction in the USA

Table 1
Estimated quantities of materials from the demolition of buildings in the USA (mean values for the 1970s)

Material	Quantity ($kt\,year^{-1}$)	Percentage share
Steel and iron	477.4	1.57
Aluminum	3.0	0.01
Copper	15.2	0.05
Lead	18.2	0.06
Glass	100.3	0.33
Brick and clay	4564	15.01
Concrete	19255	66.33
Wood	5972	19.64
Plastics	0.9	<0.01
Total	30405	100

Most of the "heavy" (i.e., thick-section) metals in demolition wastes, both ferrous and nonferrous, are recycled. This category includes copper pipe but not insulated copper wiring, although machinery has been developed to strip insulation from wiring. However, while there is always a market for recycled copper, brass, lead and aluminum which will repay the contractor's work in separating these metals from the overall waste stream, scrap prices can fall low enough in times of recession to make the recycling of steel and all but the heaviest nonferrous sections uneconomic. At these times the larger contractors often stockpile steel beams, for example, until market conditions improve.

Table 2
Estimated quantities of materials from the demolition of buildings in the EEC (mean values for the 1970s)

Material	Quantity ($Mt\,year^{-1}$)	Percentage share
Masonry	40.85	57
Concrete	26.52	37
Wood	1.43	2
Steel	0.22	0.3
Other	2.65	3.7
Total	71.67	100.0

Other demolition materials such as glass, plaster and plastics are recycled only to a very small extent.

The quantities of materials which have been used in the USA in building construction (excluding pavement, bridges and tunnels) are shown in Fig. 1. The average age of demolished buildings has been approximately 60–80 years for the older cities and 40–45 years for the newer cities of the west and south. The resulting quantities of materials produced in the 1970s from building demolition are estimated for the USA in Table 1 and for the countries of the EEC in Table 2. Because of the frequent immediate reuse as base or fill, there are no good data for the quantities of pavement (especially highway) demolition materials.

See also: Recycling of Metals: Technology

Bibliography

Davidson T A, Wilson D G 1982 US building-demolition wastes: Quantities and potential for resource recovery. *Conserv. Recycling* 5: 113–32

Environmental Resources Ltd. 1980 *Demolition Waste.* Construction Press, New York

Frondistou-Yannas S A, Ng H T S 1977 *Use of Concrete Demolition Wastes as Aggregates in Areas that have Suffered Destruction: A Feasibility Study,* NSF Report AEN 75-23221. National Science Foundation, Washington, DC

Nixon P J 1976 The use of materials from demolition in construction. *Resour. Policy* 2: 276–83

Saylak D, Gallaway B M, Epps J A 1976 Recycling old asphalt–concrete pavements. *Proc. 5th Mineral Waste Utilization Symp.* Illinois Institute of Technology Research Institute, Chicago, pp. 16–26

Wiesman R M, Wilson D G 1976 *An Investigation of the Potential for Resource Recovery from Demolition Wastes,* Report AEN 75-14197. National Science Foundation, Washington, DC

Wilson D G, Davidson T A, Ng H T S 1979 *Demolition Wastes: Data Collection and Separation Studies,* NSF Report RA-790407. National Science Foundation, Washington, DC

Wilson D G, Malarkey J M 1974 *The Resource Potential of Demolition Debris*, NSF Report 74-SP-0889. National Science Foundation, Washington, DC

D. G. Wilson
[Massachusetts Institute of Technology, Cambridge, Massachusetts, USA]

Recycling of Metals: Technology

Collection and processing of scrap are the first two steps in recycling. Scrap can be classified into three categories: (a) home, revert or run around; (b) prompt industrial; and (c) postconsumer or obsolete. Purchased scrap consists of the last two categories. Since home scrap is recycled within the plant where it is generated and requires no additional processing prior to melting, only purchased scrap will be discussed. Prompt industrial scrap is usually new, clean material generated during an industrial operation (e.g., trimmings which result from the stamping of automobile fenders) or reject products (e.g., defective cans). Prompt industrial scrap is usually sold by the generator and may be purchased by either scrap processors or final consumers. Obsolete scrap can come from a variety of sources (e.g., discarded automobiles, outdated electronic gear, old buildings, railroad cars and ships).

Since scrap comes from a variety of sources in many different forms, it must be processed to facilitate efficient use. The primary roles of the scrap processor are to collect, accumulate, sort, grade, prepare, market and distribute scrap. Sorting is done following identification of the scrap. Preparation may involve any of a number of activities, such as shearing, baling or briquetting.

1. Ferrous Scrap

Ferrous scrap supplied a slowly increasing proportion of the iron consumed in the USA between 1976 and 1981 (Table 1). Almost all of this scrap was used to produce steel and cast iron. Small amounts have been sold for copper cementation (about 450 kt per year) and for production of ferroalloys.

Ferrous scrap is usually identified and sorted according to source or prior use. As ferrous scrap is low in value compared with nonferrous scrap, stainless steel and specialty alloy scrap, only simple identification techniques are economic. When mixed scrap is received, it is sold as such and commands an appropriately lower price.

Processing of ferrous scrap depends on the requirements of the ultimate consumer and the form in which it is received by the scrap processor. Oxyacetylene torches are used to cut up obsolete railroad cars and ships, structural steel from buildings, and other very large objects. Shears reduce large pieces of scrap (e.g., structural members) to a dimension suitable for furnace charging, usually 1–1.5 m long. Automobiles pressed into logs also are sheared. Two types of shears are used: guillotine and alligator shears. The guillotine shear is the larger of the two, consisting of a movable head which is hydraulically forced past a stationary anvil. The scrap to be cut rests on and protrudes beyond the anvil. Alligator shears function like giant scissors, except only the upper part is movable.

Table 1
US consumption of ferrous scrap

Year	Consumption (kt)	Percentage of total ferrous consumption
1977	41040	35
1979	44970	37
1981	39640	36
1983	32230	39
1985	37270	39
1987	40290	42

Loose, light-gauge scrap is frequently baled or briquetted. Examples of this type of material include sheet remaining after a stamping or punching operation, and turnings (metal removed in a machining operation). In addition, entire automobiles and appliances are sometimes baled. Balers consist of hydraulic cylinders that compress the scrap by pushing it against fixed walls of a cube. Steel briquetters have a pair of rotating drums with a series of cavities around the periphery of each drum into which the scrap is compressed. The scrap (e.g., turnings) assumes the shape and dimensions of the cavities.

Shredders are used mostly for preparing light scrap (e.g., automobiles and large home appliances). They consist of a series of steel hammers mounted on a rotating drum. The hammers pass by a series of fixed anvils. Feeding the car, appliance or other scrap between the hammers causes it to be quickly ripped into fist-sized chunks. Magnetic separation removes the ferrous metal from nonferrous metals and nonmetallic materials such as glass, rubber, plastics and dirt. The resulting ferrous product has the advantages of increased purity and improved consistency, handling, transportation and melting characteristics. In 1980 there were an estimated 200 automobile shredders operating in the USA with a capacity of more than 13 Mt per year.

The ferrous fraction of municipal solid waste presents a unique case. This material may be sold directly to steel mills for blending with other scrap materials, to a detinner, or to a scrap processor. The cans must be cleaned of paper and food waste by the seller or scrap processor prior to detinning. The detinner will leach (dissolve) the tin away from the steel using an

$NaOH-NaNO_3$ solution. This process also removes any aluminum present. The resultant ferrous scrap has considerably increased value, and the tin can be recovered from solution by electrolysis.

Lower purity scrap usually is restricted to applications with less stringent chemical and mechanical property specifications. These include the manufacture of reinforcing bars and manhole covers. High-purity scrap can be used to make any grade of steel or cast iron.

2. Copper

Copper-bearing scrap materials are among the oldest used by man, and indications are that these materials were recycled by the earliest civilizations. Extensive copper recycling has continued up to the present time, when copper-bearing scrap accounted for almost 50% of all copper raw materials in 1979 and 1980 (Table 2). Depending on the purity of the scrap and the melting technique employed, using scrap to produce copper requires only 3–40% of the energy needed to produce pure copper from ore.

Copper-bearing scrap is divided into a number of grades. No. 1 copper scrap must be unalloyed, as it is used directly to make wirebars. No. 2 copper scrap is refined electrolytically and therefore may contain some impurities, though a copper content of at least 94–96% is normally required. Light copper also passes through an electrolytic refinery; its copper content must exceed 92%. Refinery-grade brass must contain at least 61.3% copper, while copper-bearing material generally contains less than 60% copper. These last two grades normally are melted in a blast furnace and blown in a converter prior to electrolytic refining.

Copper-bearing scrap can be used by tube mills, refiners, brass mills, ingot makers and specialty companies. The end-use determines the preparation of the scrap by the processor. As tube mills do not refine electrolytically, metallic contamination of scrap to be used by tube mills is unacceptable. However, small amounts of nonmetallic contaminants may be acceptable, depending on the user's ability to fire refine. Refiners can produce wirebar directly from No. 1 copper scrap, but No. 2 and light copper scrap must first be cast into anodes and refined electrolytically before it can be used for this purpose. Brass mills usually use No. 1 copper and brass scrap in their melts in combination with pure copper.

The composition of brass scrap must be known and kept within narrow limits so as to keep the product made from the scrap within composition specifications. Brass and bronze ingot makers supply alloys of known, fixed composition to foundries. Hence, they can and do melt a wide variety of copper-bearing scrap. Specialty uses for copper-bearing scrap include chemicals, electroplating, castings (both copper and gray-iron), copper–aluminum alloying and steelmaking.

Table 2
US consumption of copper scrap

Year	Consumption (kt)	Percentage of total copper consumption
1977	1085	40
1979	1553	47
1981	1418	45
1983	1084	41
1985	1139	41
1987	1212	41

Prompt industrial copper-bearing scrap, which accounts for 60–70% of recycled copper, is generated by a variety of industrial operations. For example, scrap turnings from brass screw machines amount to up to 65% of the brass feed. Other sources include stamping plants, wire-drawing rejects, die shavings and short rod wire lengths. Postconsumer scrap includes metal in surplus and obsolete items from communication, rail, electronics, power utility and defense operations; metal in disused automobiles and ships; and metal in spent shell casings.

Identification may be made using the material's color, spectroscopy, spot testing with acids, chemical analysis or, in some cases, magnetism. Highly trained and experienced sorters are required for the first three methods. Together with the shape of scrap and its past use, the color of a clean surface may be all that is needed to determine the alloy type. Spectroscopy offers only a semiquantitative analysis of an alloy, but this information is often sufficient for sorting scrap into alloy groups. Spot testing involves observing the reaction, or lack thereof, between various reagents and the metal.

A variety of techniques is used to prepare copper-bearing scrap materials by the scrap processor. Traditional processes include shearing to the size required by the customer, briquetting or crushing of thin and lightweight materials (e.g., turnings and borings), magnetic separation to remove iron, cleaning and degreasing.

One new technique is chopping of copper wire scrap. Large, stiff pieces of wire are presheared prior to chopping. Bales of wire must be broken down and are usually cut up with an alligator shear. The major steps then consist of chopping in three stages, with magnetic separation to remove ferrous contaminants following the first stage and sizing following the second stage. The third stage handles unliberated wire from the air separation process which follows sizing. Both the coarse and intermediate size fractions are processed by air separation to produce four fractions: (a) clean metal, (b) liberated metal with some insulation contamination, (c) unliberated wire, and (d) clean insulation.

Shredded automobiles also yield copper-bearing scrap. Following magnetic recovery of the ferrous metals, the nonferrous rejects can be processed by a number of techniques. The nonmetallics can be removed by air separation and water elutriation, and/or heavy media separation. The copper-bearing metals can then be hand picked after the aluminum and zinc are floated in a heavy medium.

Cryogenics (i.e., cooling scrap to temperatures as low as −185 °C) can be used to recover copper from items such as small motors, transformers, generators, and their components. When exposed to these very low temperatures, certain materials, including steel and plastics, become brittle, whereas copper remains malleable. Therefore, crushing of an item immediately after cooling will allow relatively large pieces of copper to be separated from the other, finer materials by screening.

3. *Aluminum*

The increase in aluminum recycling has been dramatic. In 1960, recovery of secondary aluminum was about 300 kt but by 1976 that figure had increased to over 1 Mt, accounting for 21% of total aluminum production (Table 3). Recycling during the period 1976–1980 fluctuated slightly, ranging from 20–22% of total aluminum consumption. However, in 1981 the rate rose to 34% as production decreased and secondary recovery increased.

Prompt industrial aluminum scrap is produced by a variety of industrial operations, such as foil rolling, canmaking, airplane manufacturing and wire drawing. It is usually kept sorted and moves in bulk to the scrap processor or recycler. Prompt industrial scrap is, therefore, easy to handle.

Major suppliers and potential sources of obsolete scrap include industry, government and municipal solid waste. Products from which aluminum scrap is obtained are automobiles, household appliances and furniture, used cans and containers, trucks, aircraft, boats and buildings.

Scrap sorted by alloy class brings the highest prices, so the first step in scrap processing is identification.

Table 3
US consumption of aluminum scrap

Year	Consumption (kt)	Percentage of total aluminum consumption
1977	1456	26
1979	1612	26
1981	1789	28
1983	1773	35
1985	1762	33
1987	1986	37

The first step in the identification process is to consider the nature (e.g., sheet, plate, forging, casting) and source of the scrap. If more information is needed, a spectroscope may be used. In the case of castings, chemical spot tests are used to determine whether the alloy is based on aluminum, magnesium or zinc.

Once the scrap has been sorted, one or more processing techniques may be employed to prepare it for shipment. Borings and shredded material may be shipped loose for bulk handling and unloading. Other types of scrap, such as clippings and sheets, usually require baling or briquetting. Large pieces of aluminum are often sheared to a size specified by the customer.

In addition to the technology mentioned above, there are five other processes for recovering aluminum: wire chopping, shredding, aluminum can recycling, sweating and dross processing. Depending on the type of wire metal, wire chopping recoveries are usually in the range 55–65%. The technology is similar to that described for copper. However, air classification of the coarse and intermediate aluminum wire size fractions gives three products, not four fractions. These products are clean metal, plastic and middlings (a mixture of some metal with insulation). The middlings are sent to a tertiary chopper or granulator to separate the metal from the insulation. The chopped middlings are then recycled to join the product of the second-stage chopper before sizing. Wire chopping produces a well-accepted product containing about 99.5% aluminum.

Most shredded aluminum scrap is obtained from three sources; shredded automobiles, scrap cans and aluminum shredded for handling in the scrapyard. For every tonne of shredded steel scrap produced from disused automobiles, about 13.5 kg of aluminum is recovered. With the potential recovery in 1979 being about 135 kt of aluminum from the 10 Mt of shredded automobile scrap, actual recovery was approximately 100 kt. Given projections for increased use of aluminum in cars, these numbers should increase.

Following automobile shredding, the material undergoes magnetic separation. A number of techniques exist to recover the aluminum from the nonmagnetic shredded material. Initially, the nonmetallic materials can be recovered by air classification, water elutriation, eddy current, or a friction sliding board to produce a heavy fraction. Alternatively, a sink–float technique, in which the density of a heavy medium is controlled between that of the metallic and nonmetallic materials, can be used. Finally, a second sink–float tank, using a medium of higher density, can be used to float the aluminum and zinc from the other metals. Final separation is made by hand picking or in a sweat furnace.

Scrap aluminum cans come from both prompt industrial and postconsumer sources. Collections of cans from postconsumer sources exceeded 165 000 t in 1979, about 30% of the aluminum used for cans that year. There are three levels of aluminum can recycling:

(a) minimal, with a scale and sorting table; (b) intermediate, with scales, magnetic sorters and balers or flatteners; and (c) full-scale, with shredders, material handling and truck or rail car loading equipment in addition to (b). New separation processes (e.g., the eddy current linear-induction motor) are the key to recovering aluminum from municipal solid waste.

Sweating is the process of separating materials based on differences in their melting points. In this process the scrap is heated to a temperature between the melting points of the higher and lower melting point materials. Sweating should be used only to recover aluminum from contaminated scrap. This scrap should: (a) be free of contamination with other low-melting metals (e. g., zinc and magnesium, which should have been sweated off first); (b) be at least 6 mm thick so that the particles will melt into droplets large enough to flow in the furnace; and (c) contain at least 30% (preferably 50–60%) aluminum. Ingots produced from the sweated aluminum may not contain more than 1% iron, 1% magnesium and 1–3% zinc depending upon requirements.

There are three kinds of sweating furnaces: (a) direct-fired sloping hearth, (b) direct-fired rotary, and (c) indirectly fired rotary. The direct-fired sloping hearth is the original style of sweating furnace. It consists of a refractory brick lined box with a sloping floor and a tap hole at its lowest point. The aluminum in the scrap charge is melted and then runs down the sloping hearth to the tap hole and into an ingot mold or collection chamber. The scrap on the hearth must be manually raked and rabbled to free any trapped aluminum which cannot flow off the hearth. Manual raking is unnecessary if a rotary hearth furnace is employed. These may be heated directly, with a burner firing into a kiln from the discharge end, or indirectly with a ceramic-lined, stainless steel tube rotating in a temperature-controlled furnace. Both styles of rotary furnaces are inclined, with weep holes at their lower (discharge) end. The aluminum runs down the hearth through the weep holes and is collected, while the remaining solid material passes over the weep holes and is discharged.

Dross recovery is the separation of metallic aluminum from the nonmetallic residues produced whenever aluminum of any type is melted. The dross is a mixture of aluminum oxide, dirt, spent fluxes and entrained aluminum. The aluminum content, which depends on the care that was taken to prevent oxidation in and after removal from the furnace, can range from zero up to 70%. When the aluminum content is sufficiently high, usually in excess of 40–45%, the dross may be sent directly to a rotary furnace where it is heated along with a flux consisting of equal proportions of NaCl and KCl, and up to 5% cryolite. However, the majority of drosses are milled and screened at about 20 mesh (0.84 mm) with the oversize processed in the rotary furnace and the undersize discarded or sold to the steel industry for use as hot-topping compound.

Wrought aluminum scrap of sufficient purity can be sold directly to a primary smelter where it is used to make new alloy. Economics are the primary incentive for the dramatic increase in this practice since the mid-1960s. For example, construction of an aluminum reduction plant with a capacity of 73 kt per year would cost about 250 million US dollars, whereas construction of the same size plant producing can stock from can scrap would cost about 16–20 million US dollars (1983). In addition, recycling aluminum scrap requires only about 5% of the energy needed to produce new metal.

Most secondary aluminum smelters produce recycled secondary ingot from both wrought and cast scrap (i.e., an ingot with a known composition). This material is used primarily for aluminum castings. In some cases notch bar or shot is produced for use as a deoxidizer or alloy addition. Scrap (e.g., from wire chopping) can also be used in these applications. Other possible uses for aluminum from wire chopping include production of powder, pharmaceuticals, and paints and coatings.

4. Zinc

The USA is the world's largest consumer of zinc. From 1976 to 1981 the fraction of domestic consumption met from scrap increased from 24 to 29% (Table 4). One reason for the relatively low level of zinc recycling is that about 45% of zinc consumption is used dissipatively (e.g., for galvanizing).

Zinc is recovered primarily as alloy scrap, which may contain zinc, copper, aluminum or magnesium as the major constituent. Zinc is also recovered from skimmings, drosses and fumes. The major sources and uses of zinc scrap materials are presented in Table 5. Galvanizing provides the largest single source of scrap zinc, in the form of skimmings drosses, fumes and spent fluxes. These zinc-containing wastes are sold directly to recyclers. Zinc die-casters also produce a dross, as well as rejected castings. The availability of fragmentized die-cast scrap has grown markedly since

Table 4
US consumption of zinc scrap

Year	Consumption (kt)	Percentage of total zinc consumption
1977	330	24
1979	370	27
1981	305	29
1983	349	28
1985	347	27
1987	383	28

Table 5
Major sources and uses of zinc scrap materials

Type of scrap	Sources	Uses
Fragmentized die-cast	nonmagnetic portion of crushed automobiles and applicances	distillation
New industrial die-cast	rejected castings from die-cast operations	die-cast alloy distillation
Old die-cast	old, oxidized and contaminated accumulators, mostly from automobile wrecking	distillation
Die-cast skimmings	ashes, drosses from melting die-cast scrap	distillation chemicals
Aircraft forming dies	forming dies from aircraft and automobile industry	die-cast alloy distillation
Zinc skimmings	hot-dip galvanizing kettles	distillation chemicals (fluxes), agricultural compounds
Sal-ammoniac skimmings	hot-dip galvanizing kettles	chemicals (fluxes)
Zinc dross (top)	top of continuous galvanizing kettles	distillation
Zinc dross (bottom)	bottom of continuous and hot-dip galvanizing kettles	distillation
Brass ingot makers fume	baghouse collectors	chemicals (agricultural compounds)
New zinc, painted, oxidized, corroded	printing plates, battery casings, addressograph plates, punchings from rolled zinc	remelt zinc brass ingot makers galvanizing
Old zinc, painted, oxidized, corroded	roofing, boat anodes, jar tops, spent plating anodes	remelt zinc distillation
Prime die casters dross	die-casters kettle (high metallic skim)	die-cast alloy distillation
Remelt die-cast (blocks)	sweating iron-containing and fragmented die-cast	distillation

the early 1970s as a result of the tremendous expansion of automobile and appliance shredding. However, owing to the smaller amount of zinc used in the automobile during the last few years as a result of increased use of plastics, future zinc recovery from automobile shredding is expected to diminish considerably.

The major role of the scrap processor is in the recovery of zinc from shredded automobiles and appliances. If this scrap zinc is recovered by mechanical means (e.g., flotation and/or screening) rather than by sweating, a number of problems are created for the consumer. If the zinc content of this material is too low, operation of the recycling furnaces is inefficient, resulting in higher production costs. Excessive aluminum and finely divided zinc in the scrap increases the amount of skimmings and therefore the loss of zinc. The recovery of zinc is decreased by the presence of glass, rubber, rocks and trash. The degree of purity required is determined by agreement between the scrap processor and the consumer.

5. Magnesium

Consumption of magnesium-based scrap for 1976–1981 is shown in Table 6. In 1980, consumption of magnesium-based scrap increased significantly both in absolute terms, and as a percentage of total magnesium consumption.

Table 6
US consumption of magnesium scrap[a]

Year	Consumption (t)	Percentage of total magnesium consumption
1977	7818	7
1979	8893	7
1981	7644	6
1983	7424	7
1985	6140	5
1987	4703	4

[a] Not including magnesium in aluminum-based alloy scrap or magnesium used to make aluminum-based alloys

Magnesium, scrap or primary, has both non-structural and structural applications. Nonstructural uses are as alloying agents, as a reducing agent for the production of other metals, as an addition to nodular cast iron, as anodes for the cathodic protection of steel structures and as a scavenger for sulfur in blast-furnace iron destined for primary steel production. Except when used as an alloying agent, the composition of magnesium scrap does not need to be closely controlled. However, when magnesium scrap is used for structural purposes, composition of the scrap is critical. For example, high-grade casting alloys may contain no more than 0.05% copper and 0.05% nickel, with some alloys limited to 0.002% nickel. Therefore, scrap processors must take great care to remove these metals or their alloys from the magnesium scrap. Removal is usually done manually.

Because magnesium and aluminum are very similar in appearance, sorting by visual identification of these two metals, or their alloys, is extremely difficult. A reliable method of distinguishing between magnesium and aluminum is to apply a 5% silver nitrate solution to a scratched surface of the metal in question. The presence of magnesium is indicated by the surface turning black. Aluminum is not affected.

Magnesium-bearing wastes, such as skimmings, generally have not been recycled owing to the low prices offered by magnesium smelters for them.

6. *Lead*

More than half the lead consumed in the USA is recycled from scrap (Table 7). In 1980, 86% of lead scrap was obsolete and 83% of the obsolete scrap was obtained from recycled batteries.

Lead scrap is of three types: whole battery scrap (both prompt and depleted batteries); prompt industrial scrap (such as drosses, skimmings and lugs); and other scrap (such as cable sheathings, bearings, babbits and solders). Of these types, only batteries require significant processing to recover the lead.

Batteries are processed to drain the contained sulfuric acid and to liberate the lead. Four techniques are used to accomplish this, of which two, sawing and guillotining, are similar. Both remove the top of the battery, including the posts and connectors of the battery section, from the rest of the case and the plates. The remainder of the battery is hit against a striker bar to allow plates, separators and acid to fall into a conveyor. Acid drains off the conveyor into a sump where it is neutralized with lime or ammonia. The battery top is crushed and the metallic posts and connectors are separated from the crushed casing material by air classification. A more recent concept in processing scrap batteries is to break the whole battery and separate the battery plates from the casing, followed by heavy media and hydrometallurgical techniques to separate the grid metal from the paste. An alternative method involves breaking the battery to drain the acid, then charging the entire battery to the smelting furnace.

Although about 60–65% of the lead used in batteries is recycled, the same is not true of lead used in other products. For example, only about 10% of the lead used in solder, 25% of the lead consumed by the cable industry and 30% of the lead in bearings is recycled. The reasons for this lower recycling rate are diverse. Most solder applications are dissipative (i.e., the solder is used in very small quantities bonded to other metals). Cable sheathing, while easy to recover, has an average life of 40 years. Recovery of lead in bearings frequently requires disassembly of a large piece of machinery, which is uneconomic.

Virtually all lead scrap is sold to a secondary smelter and/or refiner. The resulting product finds its principal use in the production of new batteries. However, the resulting lead also may be used to produce other alloys for type metals, solders, shot and bearing metals.

7. *Nickel Alloy and Stainless Steel Scraps*

Although treated together because of the importance of nickel, nickel alloys and stainless steels are fundamentally different. Scrap nickel alloys can be used in stainless steel production. However, because of the high iron content, stainless steel scrap cannot be used as a feed material for making nickel alloys. Table 8 presents data on the rate of nickel recycling.

Nickel alloys are usually classified as follows.

(a) Superalloys possess high strength at high temperatures. A typical application of these alloys, such as the Inconels, Hastelloys and Waspaloys, is in gas turbine engines. Nickel content can vary from 20 to 80%. Most of the prompt industrial scrap is generated by the aerospace industry in the form of turnings, rejected turbine components, and casting stubs and sprues. A major source of obsolete scrap is jet aircraft engines.

(b) High-nickel alloys and heating alloys are used in corrosion- or oxidation-resistant applications. Nickel contents vary from 20 to 100%. As usual,

Table 7
US consumption of lead scrap

Year	Consumption (kt)	Percentage of total lead consumption
1977	758	53
1979	801	59
1981	641	55
1983	504	44
1985	616	54
1987	710	58

Table 8
US consumption of nickel scrap

Year	Consumption (kt)	Percentage of total nickel consumption
1977	45.8	22
1979	52.1	23
1981	41.4	27
1983	45.2	25
1985	48.7	25
1987	42.3	23

prompt industrial scrap results from metal forming and fabrication. Obsolete scrap is recovered from discarded industrial equipment.

(c) Pure nickel scrap is recovered as depleted anodes from electroplating. This represents only a small portion of the nickel available for recycling.

(d) Other nickel-bearing wastes include Monels, small amounts of special alloys, grinding swarfs, sludges and spent catalysts. Their importance is not sufficiently significant to warrant further treatment.

Stainless steel alloys are classified into three broad categories: 200, 300 and 400 series. As the 400 series contains only chromium it will not be treated. The 300 series contains about 18% chromium and about 8% nickel. Other elements such as niobium and molybdenum may also be present. Applications of 300 series stainless steels are found in kitchen utensils, hospital equipment, chemical plants and the aerospace industry. In the 200 series, 5–10% manganese is added to the alloy to promote the formation of austenite. Because nickel serves the same function, the nickel content can be reduced to under 4%.

The users of superalloys and stainless steels have rigid specifications for these alloys, and so reliable segregation and identification of this type of alloy scrap is essential to its recycling and composition. The source or form of the scrap is often a good indication of its identity. In addition, four relatively simple procedures (magnetic testing, spark testing, chemical spot testing and x-ray fluorescence) can be used for initial identification. If more reliable information or a guaranteed composition is required, a sample is taken for chemical analysis.

The magnet test is conducted with a small permanent magnet. Unless it is used to separate two alloys of known composition, it can provide only an approximate or initial classification of alloys. Spark testing can be used to differentiate among iron- and nickel-based alloys. However, the reliability of this procedure depends strongly on the skill of the operator. Spot tests include susceptibility to acid attack and reaction of test reagents to form specific colors.

Following identification of the scrap, it is processed for sale. Processing steps may include cleaning and/or sizing. Turnings are usually crushed in a ring-type or hammer mill crusher and then sampled for chemical analysis. Superalloy scrap is degreased with trichloroethylene, screened to remove fines and processed through a magnetic separator. The screened fines are sold to a nickel refinery, while the prepared turnings are sold to a superalloy manufacturer or a stainless steel mill or are exported, depending on their composition. Stainless steel scrap may be crushed and, if series 200 or 300, passed through a magnetic separator, but degreasing is rarely economic and screening is never performed.

Superalloy solids are cut with a plasma torch, and the surface is cleaned by sand blasting. With the exception of cleaning the surface by a combination of steel and grit blasting, solid, stainless steel scrap is handled in an identical manner to solid, plain, carbon-steel scrap.

Stainless steel dusts and grinding swarfs can be pelletized or briquetted prior to charging back to the steelmaking furnace. Alternatively, service companies convert this material to alloy ingots of known composition suitable for charging to the steelmaking furnace.

8. Titanium

Titanium alloys are classified as alpha, beta and alpha–beta. Alpha alloys include unalloyed metal or low-alloyed titanium grades. Typical alpha alloys contain aluminum and tin or palladium. Of the beta alloys, the alloy Ti–13V–11Cr–3Al was widely used during the 1960s, but has since been displaced by Ti–8V–4Mo–6Cr–4Zn–3Al. The alpha–beta alloys are the most important, with a production volume comparable to that of the alpha and beta alloys combined. Typical alpha–beta alloys are Ti–6Al–4V, Ti–4Al–3Mo–1V and Ti–8Mn.

Physically, titanium scrap is classified as chips (light) and solids (heavy). This material is recycled to the smelter in the ratio 1:2. The processing method employed by scrap processors is determined by the physical properties of the scrap.

Table 9 presents data on the tonnage and degree of recycling of titanium from 1976 to 1981. Note that 5% of total ingot weight becomes home scrap in the form of croppings, collars and turnings. About 86% of total ingot weight ends up as prompt industrial scrap. Because of restrictions on contamination, only about 35% of total ingot weight is recycled to the titanium-melting furnace. The remainder is used for alloying in the steel and aluminum industry, exported, or is lost as grinding and other wastes. A small amount of obsolete scrap, mostly from aircraft maintenance, is also recycled.

Titanium scrap must meet very strict standards for purity, especially for aerospace applications, because

Table 9
US consumption of titanium scrap[a]

Year	Consumption (t)	Percentage of total titanium consumption[b]
1977	9880	40
1979	12690	37
1981	13420	32
1983	9500	39
1985	13350	41
1987	16360	48

[a] Excludes titanium scrap used as an alloying addition [b] Total consumption includes only scrap and sponge; pigment is excluded

there is very little refining action during melting. Contaminants that must be handled by processing include surface oxides resulting from exposure to high temperatures during use, hydrocarbons from metal-working lubricants and pieces of cutting tool bits (usually tungsten carbide).

Chips are crushed in a hammermill to improve their handling and their packing density for shipping and subsequent briquetting. The crushed scrap is then solvent-vapor degreased in trichloroethylene. This also removes tramp impurities and any foreign metal pieces entrained in the grease. Next, magnetic separation is used to extract any ferrous or ferromagnetic impurities that are present. A representative sample is then taken for chemical analysis. Care should be taken in handling and storing turnings as they are easily flammable, particularly if they are light and oily.

Solids are subjected to minimal processing by the scrap processor. Following magnetic separation each piece is analyzed by x-ray spectroscopy, segregated by grade and shipped to a melter. The melter cuts the heavy scrap with a plasma torch, descales the titanium by abrasion, further cleans the surface with caustic soda and pickles the surface with nitric acid containing some hydrofluoric acid.

9. Precious Metals

This section covers gold, silver and platinum-group metals recycling. With the exception of periods, such as early 1980, when the price of precious metals was exceptionally high, most secondary precious metal bearing materials have been obtained from industry. Once the secondary precious metals bearing scrap is put into a concentrated or segregated form, it is sold to an appropriate refinery where it is converted into pure metal or alloys identical to those obtained from primary sources. Consumption of gold, silver and platinum-group metals scrap is presented in Tables 10, 11 and 12, respectively.

Precious metals are obtained from a variety of secondary sources. Film users, such as hospitals and printers, generate scrap film and hyposolutions from which silver is recovered. Electrical and electronic components may contain gold, silver and/or platinum-group metals (e.g., as contacts or solders). Catalytic converters from automobiles contain platinum and palladium, although converters on newer models also contain rhodium. Precious-metals scrap also is generated during plating and jewelry manufacture, and by scientific and dental laboratories. Aircraft and aerospace equipment contains precious metals in spark-plugs and brazing alloys.

Table 10
US consumption of gold scrap

Year	Consumption (kg)	Percentage of total gold consumption
1977	32340	21
1979	52100	35
1981	50070	49
1983	55480	58
1985	49820	52
1987	63260	63

Table 11
US consumption of silver scrap

Year	Consumption (t)	Percentage of total silver consumption
1977	3098	65
1979	2378	49
1981	2607	72
1983	2102	58
1985	2254	61
1987	2126	58

Table 12
US consumption of platinum-group metals scrap

Year	Consumption (kg)	Percentage of total platinum consumption
1977	6077	12
1979	9620	11
1981	12182	12
1983	9429	16
1985	8043	11
1987	5177	8

Precious metals are sometimes found in conjunction with base metals, most frequently in an assembled part (e.g., electrical equipment and automobile catalytic converters). The function of the scrap processor is to remove as much of the base metals and nonmetals as possible.

Complicated devices may be chopped or shredded to produce a fairly uniform material. This material may be subjected to magnetic separation, screening, air classification, and/or incineration to produce precious metal concentrates. When appropriate, hand dismantling may be practised.

The material is sampled and analyzed prior to shipment. Preparation for sampling would be by grinding to fine powder or by melting and either taking a dip sample or pouring into an ingot from which saw cuttings or drill turnings may be obtained.

Automobile catalytic converters are removed from junked cars and then sold for further processing. The processor opens the stainless steel containers and removes the platinum–palladium covered alumina balls or honeycomb that constitute the catalyst. The stainless steel is sold to a stainless steel producer and the catalyst is sold to a precious-metals refiner specializing in platinum-group metals.

See also: Recycling of Demolition Wastes

Bibliography

Alter H 1982 Toward a national policy for secondary material symposium on resource recovery and environmental issues of industrial solid wastes. *Resources and Conservation.* Elsevier, Amsterdam

Arthur D Little Inc. 1979 *Proposed Industrial Recovered Materials Utilization Targets for the Metals and Metals Products Industry.* US Department of Energy, Washington, DC

Battelle Columbus Laboratories 1972 *Identification of Opportunities for Increased Recycling of Ferrous Solid Waste.* National Technical Information Service, Springfield, Virginia

Battelle Columbus Laboratories 1977 *A Study to Identify Opportunities for Increased Solid Waste Utilization: I. General; II. Metals.* National Technical Information Service, Springfield, Virginia

Carrillo F V, Hibpshman M H, Rosenkranz R D 1974 *Recovery of Secondary Copper and Zinc in the United States.* US Bureau of Mines, Washington, DC

Fine P, Rasher H W, Wakesberg S (eds.) 1973 *Operations in the Nonferrous Scrap Metal Industry Today.* National Association of Recycling Industries, New York

Kusik C L, Kenehan C B 1978 *Energy Use Patterns For Metal Recycling.* US Bureau of Mines, Washington, DC

Rheimers G W, Rholl S A, Dahlby A B 1976 Density separations of nonferrous scrap metals with magnetic fluids. *Proc. 5th Mineral Waste Utilization Symp.* US Bureau of Mines, Washington, DC

Siebert D L 1970 *Impact of Technology on the Commercial Secondary Aluminum Industry.* US Bureau of Mines, Washington, DC

Spendlove M J 1961 *Methods for Producing Secondary Copper.* US Bureau of Mines, Washington, DC

US Bureau of Mines 1980 *Minerals Yearbook 1980.* USBM, Washington, DC

US Bureau of Mines 1988 *Mineral Commodity Summaries 1988.* USBM, Washington, DC

US National Association of Recycling Industries 1975 *Recycled Lead in the United States: A Study of Current Market Trends and Projections.* USNARI, New York

US National Association of Recycling Industries 1977 *Recycling of Nickel Alloys and Stainless Steel Scrap.* USNARI, New York

US National Association of Recycling Industries 1977 *Wire and Cable Chopping and Recovery of Metals from Shredder Operations.* USNARI, New York

US National Association of Recycling Industries 1978 *Recycling of Zinc.* USNARI, New York

US National Association of Recycling Industries 1979 *Recycling Precious Metals.* USNARI, New York

US National Association of Recycling Industries 1980 *Recycling Copper and Brass.* USNARI, New York

US National Association of Recycling Industries 1981 *Recycling Aluminum.* USNARI, New York

US National Association of Recycling Industries 1982 *Recycled Metals in the 1980s.* USNARI, New York

R. S. Kaplan
[US Bureau of Mines, Washington, DC, USA]

H. Ness
[US National Association of Recycling Industries, New York, USA]

Remote Sensing in the Search for Mineral Deposits

Remote sensing refers to the detection and analysis of phenomena from a distance through the use of aerial and satellite-borne sensor systems. These sensors include photographic cameras, electronic sidelooking radars, electrooptical scanning radiometers and line or area array cameras. Image data provided by the sensors are subjected to both visual and computer-assisted analysis techniques to yield:

(a) information needed for geological reconnaissance;

(b) maps of topographic features and structures associated with mineral deposits;

(c) maps of lineaments indicating regional stress fields;

(d) reflectance patterns unique to altered rocks associated with mineral deposits; and

(e) basic control for detailed geological mapping.

Today, the demands for energy and mineral resources require intensified exploration on a global scale and remote sensing provides a means to accomplish this task.

1. Remote-Sensing Systems and their Applications for Mineral Exploration

Geologists have used remote-sensing techniques since the 1940s to aid the search for mineral deposits (Colwell 1983). Until 1970 most studies were conducted with black-and-white stereoscopic aerial photographs at scales of 1:40 000 to 1:60 000 by means of visual photointerpretation and conventional mapping techniques. In these studies the geologist would produce maps of lineaments and attempt to correlate the patterns of linear features with field data on mineralized zones, core samples, geophysical surveys and other relevant information. This procedure is used extensively today, but the primary sources of data have become color infrared and/or color photographs obtained from high-altitude aircraft and image data recorded from satellites orbiting the earth.

High-altitude aerial photographs at scales of 1:80 000 to 1:30 000 have ground resolutions of 3–5 m for low-contrast features and permit the reconstruction of three-dimensional terrain models with the aid of a stereoscope or plotting instrument. They may be easily rectified to conform to a map base and *x*, *y*, *z* terrain coordinates can be determined to within a few meters. Satellite photographs, on the other hand, provide the synoptic view desired by geologists but are of small scale and inferior resolution. Early photographs obtained with hand-held cameras on the Gemini and Apollo missions of the 1960s indicated the potential of space photographs for geological studies (Lowman 1968, Rowan 1975) and subsequently led to the use of high-resolution and/or multispectral film cameras on the Skylab (US), Salyut (USSR) and Soyuz (USSR) missions of the 1970s.

In the future, the role of film cameras in space will diminish in favor of electrooptical and electronic sensors; however, both the USA and the European Space Agency plan to employ specially constructed photogrammetric cameras (the large format camera and the metric camera, respectively) on the Shuttle–Spacelab missions of the future to obtain stereoscopic photographs with ground resolutions of 10–20 m. These photographs will meet exploration geologists' requirements for high-resolution coverage from which three-dimensional terrain information can be extracted. A major problem with photographic camera systems, however, is the limited film supply, which makes global coverage impractical.

The above problems are reduced by the use of electrooptical and/or electronic sensor systems. In 1972, the first Earth Resources Technology Satellite (ERTS-1) was launched by the United States National Aeronautics and Space Administration (NASA). Two electrooptical sensor systems were employed: the return beam vidicon (RBV) camera system and the multispectral scanner (MSS). Of these, the MSS has provided a wealth of multidate global coverage. The ERTS program was tremendously successful. In 1975 the name was changed to Landsat. Landsats 2 and 3 were placed in orbit in 1975 and 1978, respectively, and provided additional global coverage which is used by mineral exploration companies throughout the world (Carter et al. 1980). Landsat image data are available in both film and computer-compatible tape (CCT) formats from the US Geological Survey's EROS Data Center in Sioux Falls, South Dakota, and from other centers around the world.

The worldwide coverage of Landsat, and the ability to process the digital data by computer to produce images of enhanced resolution and contrast, generated a new era of geological and mineral exploration by means of remote-sensing techniques (Siegal and Gillespie 1980). In addition, because the image data are in digital formats, they can be merged with other digital data from aeromagnetic, geochemical, seismic and topographic surveys to create an information base from which three-dimensional graphics can be created to aid in the delineation of mineral deposits. These techniques are currently being investigated by major exploration companies who anticipate receiving data of improved spatial and spectral resolution from satellite programs of the 1980s such as the Landsat 4 and 5 programs of the USA, the SPOT program of France and the Marine Observation Satellite (MOS) program of Japan. The Landsat 4 and 5 programs, for example, feature a sensor known as the thematic mapper (TM) that records data of 30 m resolution in the visible and infrared portions of the electromagnetic spectrum.

2. Principles of Remote Sensing for Mineral Exploration

The basis for mineral exploration by remote-sensing techniques may be considered in terms of the ability (Smith 1977, Sabins 1978) to:

(a) detect and map lineaments associated with mineralized zones; and

(b) detect and map unusual terrain reflectance patterns associated with altered rocks.

Mineralized zones are commonly associated with linear features which are the surface expression of structural displacements in the earth's crust. The term lineament is normally employed to represent the alignment of morphological features such as stream valleys, escarpments or mountain valleys. Tonal contrasts owing to differences in vegetation, soil moisture or rock type may also form lineaments. Many lineaments are the surface expression of fault and fracture zones.

Mineral prospecting by means of remote sensing involves the detection and mapping of lineaments, which may then be correlated with the location of known mines, mineral deposits or other survey data. In general, the detection of lineaments is a function of the spatial resolution and scale of the image data, and the area covered by a single image.

More specifically, 23×23 cm format high-altitude aerial photographs with area coverage ranging from about 340 to 900 km^2 are excellent for the detection and mapping of lineaments < 10 km in length. While such photographs provide excellent resolution of fine detail, the synoptic view desired by the geologist searching for regional patterns of linear features is lacking. Landsat MSS and TM images, although of coarse resolution (79 m) and small scale, cover an area of $> 34\,000$ km^2 per scene and enable the geologist to map linear features from 10 to more than 200 km in length which may be related to regional structural regimes or to crustal or plate tectonic elements (Fig. 1). Mosaics and maps prepared from Landsat images are particularly useful for regional mapping (Fig. 2).

Mapping of the lineament patterns is the first step in analysis. After maps have been prepared, the lineaments can be digitized and processed by computer to determine length, orientations, density per areal unit

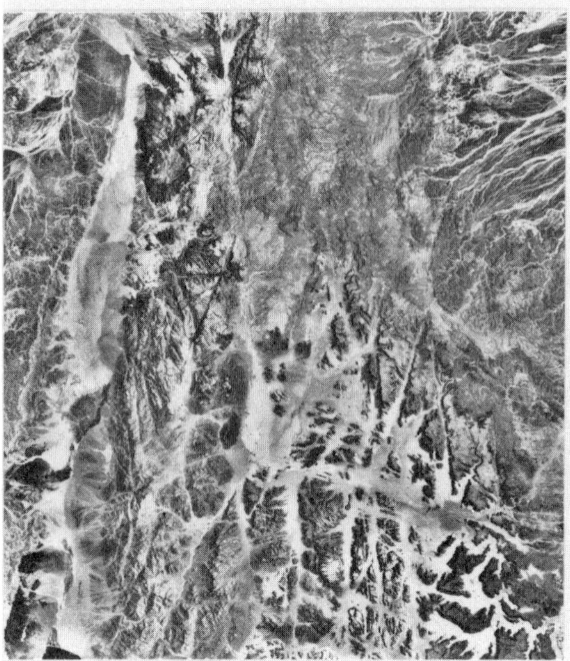

Figure 1
Digitally enhanced black-and-white image showing the geological structure of a 185 × 185 km area in Jordan recorded in the near infrared (0.8–1.1 μm) range by the Landsat MSS (courtesy of P Chavez, US Geological Survey)

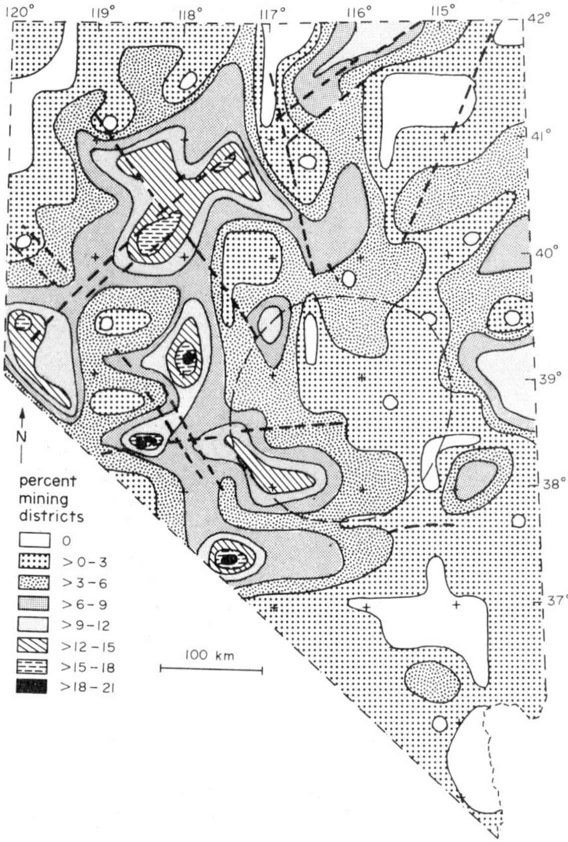

Figure 2
Contour map showing the areal distribution of metal mining districts in Nevada in relation to the major lineaments (dashed lines) and a large circular feature derived from analysis of Landsat-1 images and synthesis of geophysical evidence (contour interval is 3%) (after Rowan 1975)

and number of intersections per areal unit. Both two- and three-dimensional maps can be prepared quickly with appropriate computer graphics routines and, if required, other survey data can be registered to the plots to facilitate the location of mineralized zones.

Landsat data also provide an excellent means of constructing reconnaissance maps or for targeting areas requiring further detailed studies by aerial or ground surveys. When satellite, aircraft and ground data are available for the same area, it is feasible to use multistage sampling techniques to aid in the location and mapping of mineral deposits.

The above techniques are particularly well suited to arid or semiarid areas enjoying cloudfree weather for a good portion of the year and which are relatively void of vegetative cover. If the area of interest is rarely cloud-free and/or has a dense vegetative cover, an alternative remote-sensing technique must be considered. Airborne side-looking radar (SLR) and synthetic aperture radar (SAR) systems operating in the microwave portion of the spectrum, for example, are considered all-weather sensors and have been extensively used to generate image data of tropical countries. Pulses of energy are directed towards the ground from the antenna of the aircraft and, upon reflection from the terrain, are returned to the antenna and recorded on film or tape. To some extent, SLR–SAR sensors will penetrate vegetative cover to create an image of the underlying topography (Fig. 3). Lineaments are most likely to be detected when data records are produced from pulses striking the terrain at right angles to the trends of the topographic features (Fig. 4). Satellite-borne radar systems of considerable interest to geologists include the Seasat synthetic aperture radar and the Shuttle imaging radars (SIR-A, -B) (Elachi 1980, Taranik and Settle 1981).

The second major approach to identifying mineralized zones or ore-bearing rocks is based on the detection of unusual spectral reflectance patterns in the visible and near or middle infrared regions which are unique to specific rocks (particularly altered rocks) associated with mineral deposits. All objects reflect radiation and the reflectance patterns can be used as a

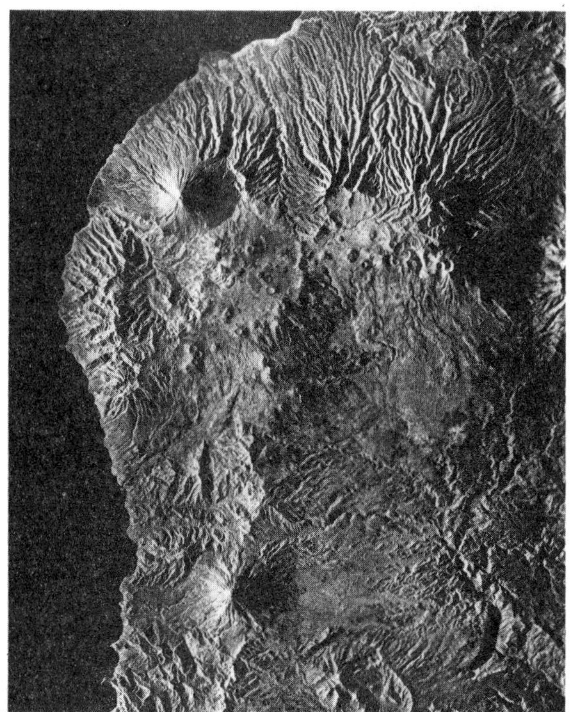

Figure 3
SLR image mosaic of the volcanic structures of Flores Island, Indian Ocean, recorded through total cloud cover from an altitude of 12 000 m by Aero Service; area shown in approximately 37 × 46 km (Courtesy of Goodyear Aerospace Corporation)

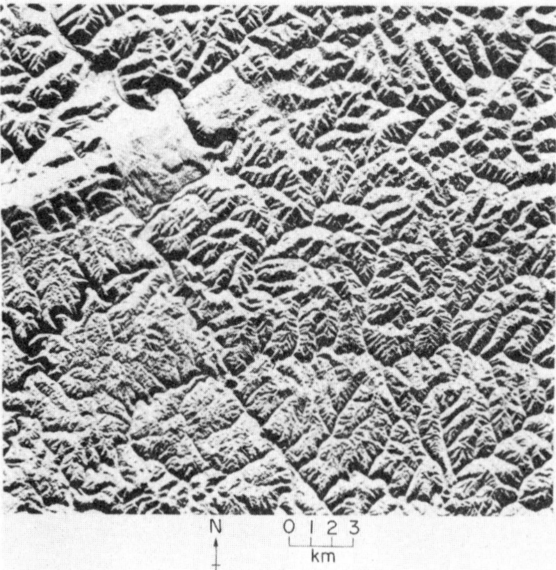

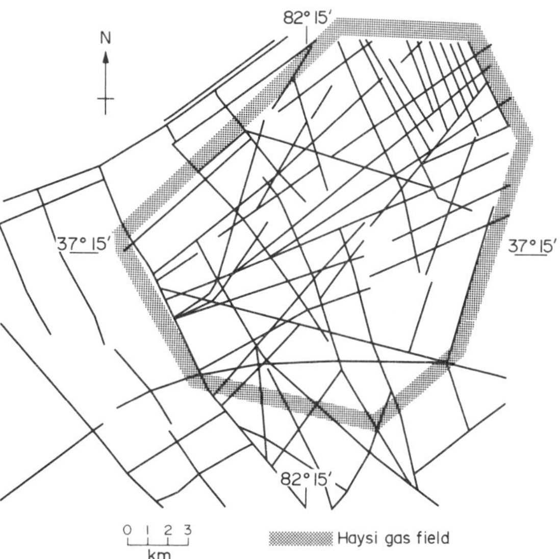

Figure 4
Side-looking radar (SLR) image of the Haysi, Virginia–Kentucky, gas field recorded by Aero Service (top); and lineaments compiled from SLR image by G. Owens, Columbia Gas System, Columbus, Ohio (bottom); same scale for both figures (after Goetz and Rowan 1981)

means of identification of the source object. Field-acquired reflectance spectra for some rocks and natural features are shown in Fig. 5. As viewed from space by a camera or sensor system, these reflectance spectra may appear somewhat different owing to the transmission properties of the atmosphere and the sensitivity characteristics of the film or detectors.

To date, most studies of spectral reflectance have concentrated on the visible and near and short wavelength infrared (SWIR) zones of the electromagnetic spectrum, extending over the wavelength interval from 0.4 to ~3.0 μm. Goetz and Rowan (1981) have discussed in some detail the significance of the reflectance properties of rocks and minerals. Referring to Fig. 5, it is evident that peak reflectance occurs in the infrared between 1.2 and 1.8 μm. Important zones are the high reflectance at ~1.6 μm, which is indicative of altered rocks containing clay, and the region between 2 and 2.5 μm, in which carbonates and layered silicates such as clays and micas exhibit sharp spectral absorption bands. Reflectance and emittance at longer wavelengths may also permit diagnosis of minerals. For example, the 3–5 μm region contains spectral bands which may aid in the discrimination of nitrates and sulfates, and emittance patterns in the midinfrared region beyond 8 μm may aid in distinguishing between silicate and nonsilicate rocks.

Recognition of the uses of spectra for the identification of rocks and minerals has led to the development of satellite sensor systems and experiments which will

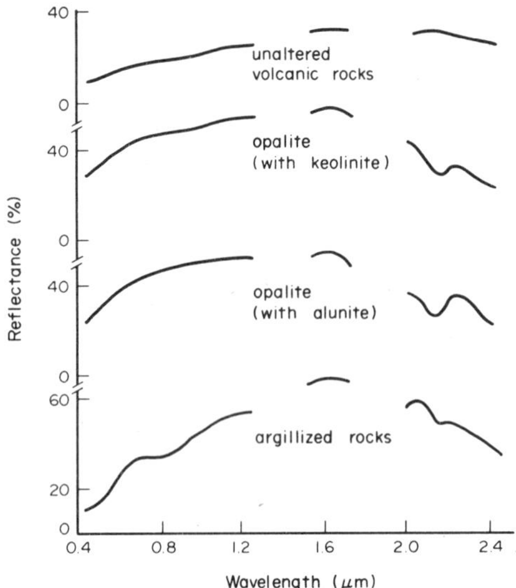

Figure 5
Representative spectral reflectance curves of altered and unaltered rocks in the Cuprite, Nevada, mining district (after Goetz and Rowan 1981)

record terrain reflectance patterns. For example, an instrument known as the Shuttle multispectral infrared radiometer (SMIRR), designed by the Jet Propulsion Laboratory and onboard the second Shuttle mission in 1982, was used to record the spectra of terrain features over the wavelength interval of 0.5–2.35 μm for a total linear distance of 80 000 km. These spectra are being correlated with known geological structures, field spectra and Landsat images in an effort to determine which spectral bands will aid in the discrimination of mineral deposits from space.

3. Data Analysis Techniques

Image data are recorded in both film and digital formats. Although traditional visual interpretation techniques with standard film images are preferred for most geological interpretation tasks, the demand today is for enhanced image products prepared from digital data according to user specifications (Condit and Chavez 1979). Digital image data are normally recorded as discrete picture elements (pixels) arranged in regular rows (lines) and columns and stored on magnetic tape (Fig. 6). Each pixel has a digital number

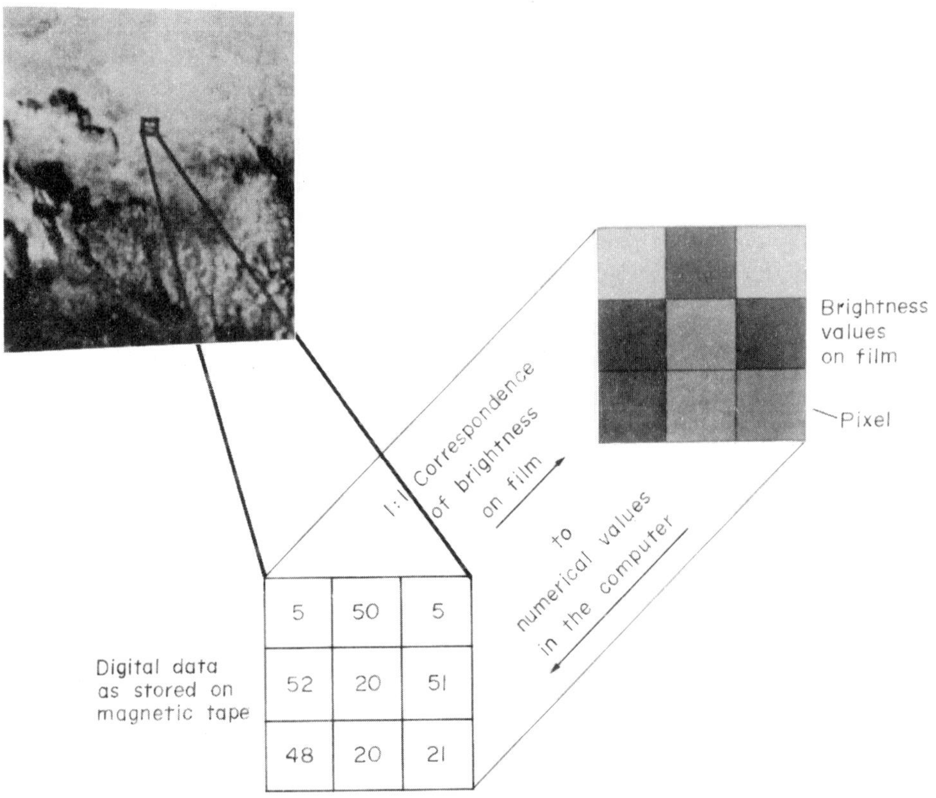

Figure 6
Relationships between pixel brightness (gray levels) on hard-copy print and values of DN (digital numbers) on magnetic tape for a digital image (Condit and Chavez 1979)

(DN) which represents the reflectance from the corresponding terrain parcel. Because the image is created from an array of DNs, it is possible to use a computer to perform radiometric and geometric corrections on the data set. These include the possibility of formatting the pixels in a defined coordinate system, thus facilitating registration of the image data with existing map bases. Once corrections have been implemented, algorithms are available (a) to enhance the contrast; (b) to conduct spatial filtering in order to improve the resolution of edges and linear features; (c) to create pseudocolor images; (d) to produce ratio images which enhance reflectance differences between altered and unaltered rocks; (e) to undertake thematic classification which may isolate specific rock types or vegetative indicators of rocks or minerals; and (f) to merge the image data with other exploration data obtained from seismic, magnetic, gravity and geochemical surveys. In summary, digital processing permits full exploitation of the spectral and spatial resolution of the image data.

4. Future of Remote Sensing

The availability of high-resolution image data from satellite programs of the 1980s will further stimulate the use of remote-sensing techniques for mineral exploration. Subject areas which will receive particular attention include: (a) advantages of global high-resolution stereoscopic coverage; (b) digital-processing techniques for exploiting image data in the short and middle infrared wavelengths; (c) assessments of imaging radars; and (d) applications of merged data sets. Correspondingly, it is expected that the demand for geologists trained in remote-sensing and image-processing techniques will continue to grow.

See also: Mineral Deposits: Examination and Evaluation

Bibliography

Carter W D, Rowan L C, Huntington J F (eds.) 1980 *Remote Sensing and Mineral Exploration.* Pergamon, Oxford

Colwell R N (ed.) 1983 *Manual of Remote Sensing.* American Society of Photogrammetry, Falls Church, Virginia

Condit C D, Chavez P S Jr 1979 *Basic Concepts in Computerized Digital Image Processing for Geologists,* US Geological Survey Bulletin 1462. US Government Printing Office, Washington, DC

Elachi C 1980 Spaceborne imaging radar: Geologic and oceanographic applications. *Science* 209: 1073–82

Goetz A F H, Rowan L C 1981 Geologic remote sensing. *Science* 211: 781–90

Lowman P D Jr 1968 *Space Panorama.* Reinhard A Müller, Feldmeilen/Zurich

Rowan L C 1975 Applications of satellites to geologic exploration. *Am. Sci.* 63: 393–403

Sabins F F Jr 1978 *Remote Sensing Principles and Interpretation.* Freeman, San Francisco, California

Siegal B S, Gillespie A R 1980 *Remote Sensing in Geology.* Wiley, New York

Smith W L (ed.) 1977 *Remote Sensing Applications for Mineral Exploration.* Dowden, Hutchinson and Ross, New York

Taranik J, Settle M 1981 Space shuttle: A new era in terrestrial remote sensing. *Science* 214: 619–26

R. A. Welch
[University of Georgia, Athens, Georgia, USA]

Resource Appraisal

Resource appraisal refers to the evaluation of mineral materials in the ground, both discovered and undiscovered. Attention is centered on materials in such form, concentration and location that they might be extractable under foreseeable economic and technological conditions.

Among the users of such appraisals are mineral exploration planners, economic analysts, land-use planners and policy makers. Each group looks for aspects that are most pertinent to its own field and time frame of interest. There is, however, no such thing as an all-purpose resource appraisal. Any type of appraisal has its conceptual limitations.

1. Terminology

Discussions of resources have long been bedevilled by a vague and inconsistent use of terms. For more than a century, efforts have been made to establish definitions of the chief concepts, and of their subdivisions, that would be widely accepted and applied. Unfortunately this has resulted in a multitude of definitions, of which none has won general acceptance, largely because of differences in purpose and emphasis. Even the commonly used expressions *resource base*, *resources* and *reserves* convey different things to different people. Their most usual meanings are described below.

Resource base refers to the total amount of a mineral material presumed to exist in the earth's crust, regardless of technological feasibility and costs of extraction. Limits may be specified, such as geographic area or depth.

Resources are the part of the resource base that is considered extractable under current and foreseeable technological and economic conditions. (In this context, the foreseeable future usually does not extend further than 25 years or so; beyond that, the effects of changes in technology and economics become totally conjectural.)

In practical terms, resources may be thought of as mineral concentrations—the known "deposits" as well as the merely speculative—that represent realistic opportunities for supply during the next few decades.

Reserves are restricted to those resources that, at a given time, have been fairly well measured and are expected to be profitably minable in the immediate future, given current technology. The term reserves is applied to tonnages of economically minable rock (ore reserves) as well as to quantities of any particular mineral commodity contained in such rock (e.g., copper reserves).

2. Resource Base—Some Indications of Size

The concept of resource base provides a useful backdrop to resource appraisal. It affords a perspective that helps in envisioning the immense mineral supply possibilities in the remote future through exploration and technology, as well as some of nature's hurdles that may have to be cleared.

Column 1 of Table 1 shows that a considerable number of useful metals collectively make up <0.77% of the continental crust. Nonetheless, on a human scale the absolute amounts are still vast. For most of these metals, the elemental abundance as a fraction of the total mass of the earth's crust (taken as 24×10^{18} t) is in the range 10^8–10^{11} times the amount of current annual world production.

Table 1
Crustal composition and minable grades

Element	Composition of continental crust[a] (wt%)	Lowest grade economically minable in 1975, as multiple of average crustal grade[b]
Oxygen	45.20	
Silicon	27.20	
Calcium, magnesium, sodium, potassium, hydrogen	11.97	
Aluminum	8.00	2
Iron	5.80	3
Titanium	0,86	16
Phosphorus	0.10	70
Manganese	0.10	190
Subtotal	99.23	
Copper	most of balance	56
Cobalt		80
Nickel		100
Lithium		240
Carbon		310
Uranium		350
Zinc		370
Molybdenum		770
Gold		1000
Silver		1330
Tin		2000
Chromium		2100
Lead		3300
Tungsten		4000
Mercury		11200

[a] Skinner (1976, p. 260) [b] Cooke (1976, p. 678)

Conveniently the constituent elements are unevenly distributed in the crust. For most metals, however, concentrations have to be very high to be of interest to miners (see Table 1, column 2). Society is willing to pay the cost of extraction only for those deposits that are large, of preferred mineralogical compositions and in accessible locations. Such deposits are geological rarities.

3. Possible Link Between Ore Concentrations and Crustal Abundance

Many geologists believe that the size and number of a metal's largest ore deposits are related to its crustal abundance (Brobst 1979). It has even been suggested (McKelvey 1973) that a rough proportionality exists between the reserves and many well-explored nonfuel commodities and their crustal abundance, the tonnages of reserves being equal to 10^9–10^{10} times the crustal abundance (expressed in percent). However, this remains difficult to explain, given the many changing economic factors that affect the size of reserves over time.

4. Grade Distributions in the Earth's Crust

The analytical sampling of the crust carried out to date is insufficient for a decisive test of any of the hypotheses that have been developed concerning the relation between the tonnage of rock and its content of any individual metallic element. This content or "grade" is usually expressed as weight percent.

The total low-to-high spectrum of grades at which a metal occurs in the earth's crust may be divided into small equal-sized ranges. A plot of the total amounts of metal in each of these ranges would, in the case of geochemically abundant metals such as aluminum, iron and titanium (see Table 1), follow a pattern that many geologists envisage as approximating a symmetrical bell-shaped curve. In the case of the scarcer metals, it has long been assumed that a closer approximation would be given by a log normal distribution, that is, by an asymmetrical bell-shaped curve with a peak at low grade and a long drawn-out tail towards the high grades. However, Skinner (1976) has challenged this assumption by proposing that, for the relatively scarce metals, the grade distribution might have two distinct peaks.

Skinner built his argument on the fact that metallic elements generally occur combined with other elements in crystalline compounds. The production of metals involves their liberation from such compounds by physical or chemical processes. Separation is easiest in the case of sulfides, oxides, hydroxides and carbonates. Silicate minerals, which make up the bulk of the earth's crust, are more difficult to break down. Furthermore, in common silicate rock, the scarce elements are present chiefly as randomly distributed atoms trapped by isomorphous substitution within minerals of the geochemically abundant elements; for example, a zinc atom may replace a magnesium atom. To release such zinc atoms would involve a complicated and energy-hungry process, well beyond the realm of current practicality.

These basic differences in mineralogical associations led Skinner to suggest that the amounts of scarce metals occurring at different grades would show bimodal distributions as illustrated in Fig. 1. The large peak represents the distribution of a scarce element in common silicates, the small peak its distribution in nonsilicate minerals. Skinner suggested that a point would be reached—for some metals, soon—where a scarce metal was no longer available in adequate quantities in the kinds of deposits from which the metal could be recovered relatively easily. If Skinner is right, mankind may eventually (although the time scale is by no means clear) be forced to rely on low-grade silicate minerals for supplies of certain metals. These minerals may be physically inexhaustible in human terms, but they would yield their metals only at high costs, unless progress is made to the use of cheaper, cleaner, and more plentiful forms of energy and acceptable ways of processing immense volumes of rock are developed.

The question of whether progressively larger quantities of the scarce elements exist at lower and lower grades or whether gaps occur is of vital importance to long-range metal supply. However, data available for studies on this subject (see Sects. 6, 7) come chiefly from the high end of the grade spectrum, which for economic reasons has attracted the greatest interest. This represents not only a very small but also a biased sample of the crust. Radical extrapolation from such samples toward lower grades in the crust is difficult to justify.

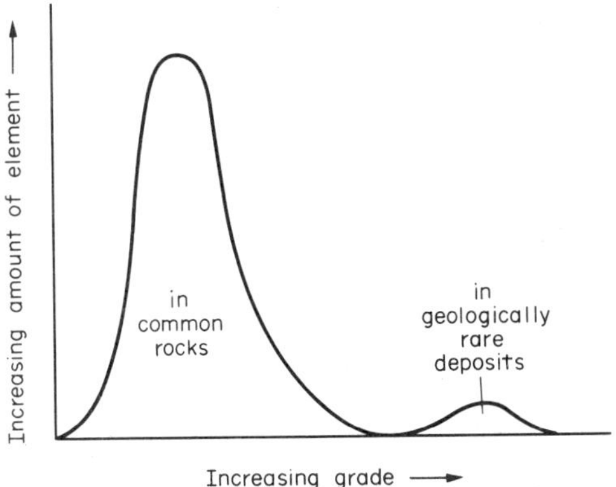

Figure 1
Possible distribution, by grade, of geochemically scarce metals (Skinner 1976)

5. Deposit Size Versus Deposit Grade

The smaller peak in Fig. 1 includes the amount of a geochemically scarce metal contained in all nonsilicate deposits collectively. Focusing on individual deposits within this group, studies have been carried out to determine the relationship between deposit tonnage (total mineralized material) and average grade. This has been done by deposit "type," a designation encompassing all deposits found in the same general geological environment and characterized by similar mineral assemblages, such as porphyry copper deposits.

The results of such studies by the US Geological Survey in the 1970s have been mixed. Whereas it is commonly believed among mineral scientists and engineers that, for a given deposit type, larger deposits may be expected to occur at lower average grades, such an inverse relationship between size and grade has actually been found only in some deposit types. For most, grades and tonnages of the deposits covered by the studies were statistically independent (DeYoung and Singer 1981).

For any deposit type in which deposit tonnage and grade are in fact independent, the total low-grade tonnage of mineralized rock is likely to contain less metal than the total high-grade tonnage. If more metal exists at lower grade, which may well be the case, it would have to be in other deposit types, some of which may have remained unrecognized to date.

Singer and DeYoung (1980) observed that typically more than half of the total known metal in any one deposit type is contained in the largest 10% of the deposits. This means that there can be no great expectations of large numbers of small undiscovered deposits, as they could contribute relatively little to total metal supply.

The implication of the above findings is that the high-grade end of the curve in Fig. 1 may well have several peaks for some metals, each peak reflecting the metal content of a deposit type. Thus, although such analyses do not allow projection very far below observed grades, they have provided some grounds for questioning the comfortable assumption of continuity in grade–tonnage relations from higher to lower grade.

6. Other Tonnage–Grade Relationships

The studies referred to in Sect. 5 sought a measure of statistical predictability of average deposit grade, given the size of a deposit, or vice versa, and found none for many deposit types. Other researchers, with a different purpose in mind, have developed mathematical equations linking tonnage and grade in a general way, pertaining alike to parts of deposits, entire deposits and groups of deposits within deposit types. For example, Lasky (1950), starting with the highest-grade material and successively adding tonnages of mineralized rock of lower and lower grade, observed that as the cumulative tonnage of mineralized rock grew at a constant *geometrical* rate, the average grade of this cumulative tonnage decreased at an *arithmetical* rate. This relationship, which has since become as Lasky's law, also applies only to certain high-grade sections within the smaller peak of Fig. 1. It cannot be validly extrapolated far enough toward lower grades to shed light on Skinner's idea of a grade gap.

DeYoung (1981) has warned that the mathematical and geological limitations, assumptions and complexities underlying mathematically expressed relationships of this type are not widely appreciated. Moreover, he has noted that cumulative tonnage in Lasky's law has, by several later authors, been taken to refer to metal instead of to mineralized rock, a misinterpretation that may well have contributed to the common notion that the lower the grade, the more there is in any deposit type. DeYoung stressed, however, that this notion is implied neither by Lasky's law nor by the mathematical expressions of tonnage–grade relationships formulated by later researchers.

7. Approaches to Resource Appraisal

Resource appraisals may be divided into two main types:

(a) Primarily geoscientific appraisals, whose chief purpose is to provide long-range land-use planners, exploration planners and mineral supply analysts with the best possible geoscientific judgements on the likely distribution and character of undiscovered mineral resources by region. Mostly applied to regions in which there has been little or no mineral development, such appraisals may be only qualitative. Economic constraints imposed tend to be loose and implicit.

(b) Geoscientific–economic appraisals, whose aim is to provide policy analysts with estimates of the magnitudes (as now perceived) of the sources of short-range and longer-range mineral supply, so that appropriate efforts may be directed toward exploration and development and toward technology to aid in recovery. Sources of supply that can be quantified reliably are confined mostly to those from which supplies may come only within the next 15–20 years. This type of appraisal tends to be conducted on the scale of entire countries. Economic considerations are stated explicitly.

Harris (1984) is the only single source providing a complete state-of-the-art coverage of all aspects of mineral resource appraisal. The quarterly *Materials and Society* dedicated a special issue (1984) to the history, state-of-the-art and applications of resource appraisal; this issue provides a convenient overview of the subject. McLaren and Skinner (1987) brought together up-to-date resource assessments in the context of relevance to future world development.

7.1 Primarily Geoscientific Appraisals

Worldwide observations have shown that mineral deposits typically occur in particular geological environments and have characteristic relationships to their host rocks. This is explained by the assumption that all are the end result of the same series of earth processes, although we are far from knowing all the processes that have taken place and from understanding their effects. Thus, recognition of favorable geological environments provides the key: evaluating a region's potential for undiscovered mineral resources involves determining the presence or absence of geological features believed favorable for the existence of particular deposit types.

The result of such evaluations may be shown on maps indicating regions favorable for the occurrence of deposits of specified metals or other mineral commodities. Refinements may be introduced to show degrees of favorability of occurrence, such as high, moderate, low or nil, and a distinction may be made between certain deposit types. It is usually implied that the favorability refers to occurrence of deposits discoverable with current exploration tools, but the questions of how much may actually be discovered, and when, are not generally addressed in these resource appraisals.

Regional resource appraisals in such form are increasingly in demand for the classification of land according to the future uses that may be made of it. Singer and Ovenshine (1979) have described the variety of approaches taken by the US Geological Survey in assessing metal resources in Alaska, where large regions had not even been mapped on a reconnaissance level. Tracts were delineated—144 for the whole of Alaska—in which the occurrence of various types of mineral deposits was considered conceivable. For each tract, a tabulation was provided giving succinct descriptions of: deposit types; contained metals; geological settings; geological, geochemical and geophysical indications of favorability; extent and adequacy of exploration and geological knowledge; and, for most of the large-tonnage deposit types, estimates of the number of deposits and indications of tonnages and grades extrapolated from models based on well-explored prototype mineral districts elsewhere in the world. Where the level of available geoscientific information warranted it, experienced geologists were asked to make subjective judgements, in terms of probabilities, of the number of deposits by type likely to occur in each tract.

Shawe (1981) prepared a compilation of mineral resource appraisal procedures that may be applied to wilderness areas and to quadrangles of the Conterminous United States Mineral Assessment Program (CUSMAP), patterned after the Alaskan Mineral Resource Assessment Program. His report also contains examples of how results may be portrayed. Appraisals of this sort call for integrated efforts of team members such as field geologists, geophysicists, geochemists and geostatisticians, who may be aided by specialists in, for example, isotope geology, remote sensing and paleontology. Thus, a wide range of geoscientific thinking is brought to bear on the description of the probabilities of existence and distribution of many types of mineral deposits and on the estimation of amounts. In such regional appraisals, estimates may be made of tonnage and grade ranges and of the chemistry and mineralogy of deposits, but explicit economic considerations tend to be avoided.

Among all mineral commodities, uranium is one for which the potential for as yet undiscovered resources has been subjected to exceptional scrutiny on an international scale. A report prepared jointly by the Organization for Economic Cooperation and Development (OECD) Nuclear Energy Agency and the International Atomic Energy Agency (1978) reviewed the possible extent and location of "speculative" resources of six different deposit type categories for 185 countries. Speculative resources were defined as resources that, mostly on the basis of indirect indications and geological extrapolation only, were thought to exist in deposits discoverable with existing techniques. They were restricted further to resources that, if discovered in 1978, would have been considered exploitable at costs of less than US$130 per kg of uranium. Areas were identified that were believed to be favorable for the existence of such uranium resources, and the potential for occurrences was ranked as low, moderate, moderate to high, high, and high to very high, or very high. For the purpose of obtaining an aggregate measure of this potential, tonnage ranges were then applied to these rankings. The report took care to emphasize that judgements of speculative resources should be viewed only as a measure of the then current state of geological knowledge, with all its inherent uncertainties. The tonnage ranges given in this 1978 report were critically reassessed and updated in 1983 (OECD Nuclear Energy Agency and the International Atomic Energy Agency 1983 p. 25).

7.2 *Geoscientific–Economic Appraisals*

In multidisciplinary resource appraisals, two aspects have generally been considered separately, namely (a) the degree of assurance about an estimated quantity actually being present in the earth's crust, and (b) the degree of economic attractiveness. Given these two dimensions, the simplest subdivision of resources is into four main categories as shown in Fig. 2. This should be viewed as a dynamic picture, the arrows portraying the directions of possible movement between the various categories.

The clean separation of discovered from undiscovered resources along the horizontal axis of Fig. 2 is a simplification. Inferred extensions of known orebodies and nearby mineralization surmised on the basis of analogous geology are examples of material that lies in a transition zone between the definitely discovered and the entirely speculative.

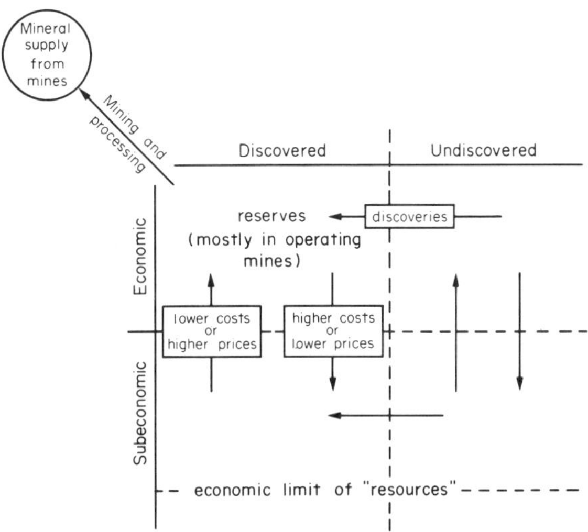

Figure 2
The essentials of a resource appraisal

The level of assurance about the physical existence of certain tonnages of given grades (regardless of economic considerations) is determined by the degree of confidence in geological continuity between observation points and in assumed geological analogies. The assurance level may cover a continuous spectrum from very high to nil, but in practice it is generally expressed in discrete gradations described by terms such as proved, measured, probable, indicated, possible, inferred, speculative and many others. The use of a series of such terms expressing diminishing confidence levels is an improvement on the simplistic division discovered–undiscovered, but the dividing lines between them are difficult to define satisfactorily, so that what is "proved" to one person may be "probable" to another.

The distinction between economic and subeconomic, along the vertical axis of Fig. 2, is also vague. This distinction can be made properly only on the basis of mining feasibility studies that consider all the details of mining methods, costs and revenues. Subdivisions in the economic dimension are therefore highly judgemental as well, largely because of uncertainty about future costs and prices. The explicit criterion used for subdivision in the economic dimension is generally one of the following: (a) price, (b) cost, (c) cost–price ratio or (d) probability of becoming economically minable within a given time period. For some metals, economic subdivision is complicated by association with other metals in the same deposits, which may enhance the economics of exploitation.

Detailed knowledge about reserves, the discovered-and-economic category in Fig. 2, together with past production, is invaluable as a basis for extrapolation in resource appraisal. However, the cost of upgrading knowledge about the exact tonnage of new discoveries to the "reasonable assurance" level of reserves is high. So also is the cost of economic analysis. Therefore, the necessary drilling and analysis to establish reserves are normally undertaken only to the extent needed for production planning. As to methodologies for computing reserves, which cannot be discussed here, the reader may consult Popoff (1966) for conventional methods and Clark (1979) for geostatistical methods.

Especially imprecise is the outer limit of resources (bottom of Fig. 2) determined by foreseeable technology and economics, twins that respond closely to one another. Among technological improvements that would extend the economic limit of resources are techniques for locating deposits at greater depths and methods for better extraction of ores and recovery of the mineral commodities from them. These would represent changes in the technology of raw material *production*. The economic limit of the resources of a commodity can also be affected by the technology of material *use*, for example, by lowering the demand for it, and thus its price, through the use of another material in its place.

Therefore, the economic limit of resources reflects an analyst's perceptions about the future evolution of technology and other factors influencing prices. It is worth noting that this dependence on perceptions has generally made resource estimates grow over time; while some deposits were being exhausted, the growth in possibilities for the future envisaged by analysts has tended to more than offset the losses.

A well-known practical framework using the same two dimensions as in Fig. 2 is shown in Fig. 3, the classification used by the US Bureau of Mines and the US Geological Survey for reporting resources of individual mineral commodities. Variations on this principle have been devised in many countries, maintaining the same two dimensions but applying slightly different criteria for separating the categories, in efforts to make the divisions more practical or more informative.

	Identified resources			Undiscovered resources	
	Demonstrated		Inferred	Probability range (or)	
	Measured	Indicated		Hypothetical	Speculative
Economic	reserves		inferred reserves		
Marginally economic	marginal reserves		inferred marginal reserves		
Subeconomic	demonstrated subeconomic resources		inferred subeconomic resources		

Figure 3
A scheme for classifying resources of a mineral commodity (US Geological Survey 1980)

The proliferation of resource classification schemes has worsened the terminological confusion that has always plagued resource appraisals. Difficulties in interpretation of descriptive terms such as those shown in Fig. 3 are compounded if some important details and assumptions remain unstated (Zwartendyk 1976). For example, in the case of undiscovered or subeconomic resources, appraisals usually describe the mineral content of rock in place, but, in the case of reserves, they tend to report only the mineral content of the ore that would be recovered in the mining process.

On an international basis, a common classification scheme with a common set of clearly defined terms has been applied only in appraising uranium resources, specifically those whose existence is assumed with some confidence. To date, quantified estimates have been provided biennially at two or more levels of assurance about existence, and in two or more cost of categories (OECD Nuclear Energy Agency and the International Atomic Energy Agency 1983). Within the USA, the potential for undiscovered resources has been studied more thoroughly for uranium than for any other mineral commodity. Within the US Department of Energy, assessments of geological favorability have been subjected to economic analysis to describe "uranium potential" by forward cost (US Department of Energy 1980).

Given the lack of international consistency and comparability of resource appraisals in general, the United Nations has made an endeavor to standardize mineral resource classification by proposing a common set of definitions and terminology that would be suitable for reporting to the UN about mineral resources (Schanz 1980). To sidestep a choice from the multilingual profusion of existing terms, a letter–number system was designed to label the resource categories (Fig. 4). The term *reserves* (corresponding to the recoverable part of R-1-E) was deliberately avoided entirely to minimize possible confusion, in several languages, between resources and reserves.

8. Geostatistics

Geostatistics refers to the application of statistical methods to problems in mining and geology. Applied to the mining problem of reliably estimating ore in known deposits, geostatistical methods have proved their worth, for instance in estimation of grade distributions between control points through kriging, a sophisticated interpolation technique (Clark 1979). A basic assumption in such methods is that the difference in value between two positions in a deposit depends only on the distance between them and their relative orientation.

Apart from applications in ore estimation, geostatistics has been used also to assess links between regional geological variables and mineral concentrations. However, as statistical analysis of geological

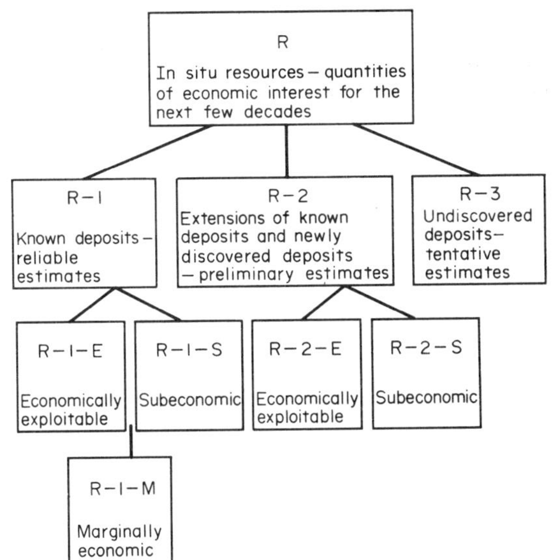

Figure 4
The United Nations scheme for classifying resources

variables distributed in three dimensions has evolved only since the early 1960s, it is too soon for estimates of undiscovered resources obtained through this technique to have been adequately tested by actual production. Many geologists believe that, on the basis of only indirect geological evidence, credible quantitative estimates cannot be made of totally speculative resources well away from known deposits. However, others are convinced that it is valid to evaluate undiscovered resources geostatistically to produce quantitative values with probability limits, provided that sufficient objective data are available for such an approach. Brinck has even estimated resources ignoring geology and using the statistical (log normal) distribution of element concentration in the earth's crust (Harris 1984).

Given our imperfect knowledge about the associations between mineral deposits and certain other geological features presumed to have been caused by the same earth processes, geostatisticians believe that the most appropriate way to view the observed associations is in terms of probabilities. For regional resource appraisals, the geostatistical approach has consisted essentially of observing statistically nonrandom features in the geology of mineral deposits and of inferring the probability of occurrence of undiscovered deposits on the basis of similar observed geological features.

Deposit models developed in this way employ human experience with economic deposits and observations of the anomalies of metal concentration as a basis for inferring the existence of undiscovered economic and subeconomic deposits. In contrast, models based on the geochemical distribution of elements (referred to in Sect. 2) draw their inferences from all or large parts of the earth's crust. Harris (1984) has described both types of models in detail. Applications of geostatistics are no longer unusual; such techniques were employed in most of the regional mineral resource appraisal methods analyzed by Singer and Mosier (1981) in a review of more than 100 publications describing a great variety of such methods.

A useful twofold division was followed by Harris and Agterberg (1981) in a comprehensive perspective of approaches to appraisal methods:

(a) those that estimate or describe one or more of the components by a mathematical relationship; and

(b) those that rely on subjective judgements of a geologist but require them to be expressed quantitatively, usually as probabilities or percentile values.

The subjective (judgemental) approach has been used more commonly, typically for large regions, as it takes less time and can usually integrate more miscellaneous information than can be captured formally and in uniform quality for a mathematical model. However, subjective appraisals done through unconstrained judgements may lack credibility because they are not reproducible. To lessen this weakness, much experimentation has taken place to construct explicit appraisal systems in which a geologist is asked questions in a step-by-step logical sequence. This provides a record of the individual judgements made and of how these were synthesized by subsequent computations. It has also shown the limitations of the mind's capabilities when faced with the complexities of evidence, geoscience and inference, and has pointed to the deficiencies in geoscience when it comes to explaining the relations between earth processes and mineral concentration.

Harris and Agterberg (1981) concluded that the raising of the reliability of resource appraisals will require significant advances not only in the accuracy with which we can relate geology to mineral concentrations but also in geological and resource data required to achieve that accuracy. Current resource data primarily describe discovered economic resources, an unrepresentative sample of what occurs in nature. Anything too small or low-grade to be of economic interest tends to go unreported, but knowledge about it is vital for the development of geoscientific theory.

9. Existence Versus Availability of Resources

Resource appraisals presented in the form of aggregated resource categories are snapshots without a time dimension. They represent some analysts' current perceptions of the relative magnitudes of various resource categories. As such, they illustrate or imply possible targets for exploration and for applying technology.

Appraisals that confine themselves to estimating the likelihood of existence of resources—as distinct from their availability—are a necessary foundation for judgements pertaining to future supply, but they are not sufficient for that purpose. When it comes to finding answers about actual mineral supply flows, which have a time dimension, rates of discovery and production must also be considered. Estimates of resources in the ground should not be (but often are) mistaken for amounts that will be available at acceptable prices when and where needed.

Much useful information gathered on individual deposits or deposit types for a resource appraisal is not visible in the aggregated totals displayed in classification schemes such as those in Figs. 3 and 4. To progress towards judgements about rates of future production, analysts cannot use these aggregated totals; they must go back and continue the analysis by individual deposits and deposit types, because production rates are constrained by deposit characteristics and place of occurrence.

For discovered resources, estimates can be made, with an indication of the degree of reliability, of the timing and future rates of production on a deposit-by-deposit basis. Consequently, for discovered resources, existence and the likely rate of future availability can be combined in a portrayal such as Fig. 5 (Zwartendyk 1981). This shows future supply from reliable sources (category I), from likely sources that can be reasonably counted on (category II) and from additional sources (category III) for which, at this stage, the timing and rate of supply cannot be estimated.

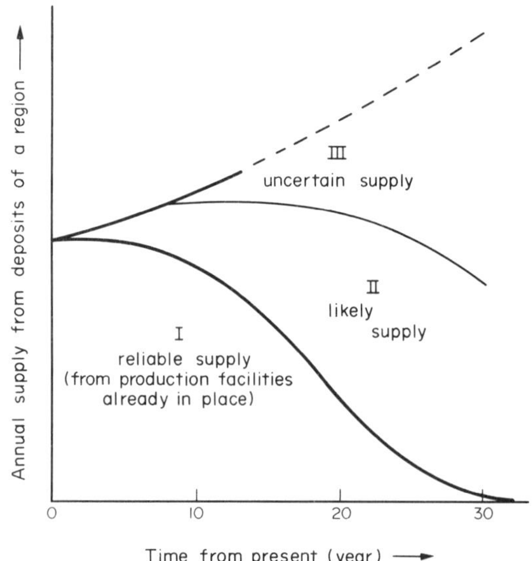

Figure 5
Mine supply of a mineral commodity

For undiscovered resources, size, quality characteristics and exact location of individual deposits are matters of pure speculation. Even more so are the timing and rate of future production from them. Long-term supply from undiscovered resources can be forecast only on a generalized statistical basis, extrapolating from past experience in geologically comparable settings. In any case, because the entire process of exploration–discovery–deposit delineation–economic assessment–mining development usually takes at least 10 years, undiscovered resources are unlikely to contribute much to supply in the immediate future. For some 10–15 years, most of the mineral supply will have to come from already identified deposits and their likely extensions.

Resource appraisal is no longer an exercise based solely on geological judgements. The determination of the amounts of minerals likely to be available for those living in the distant future has always been shrouded in uncertainty and this has not changed of late. However, with regard to possible supply sources for the coming decades alone, evaluation techniques have evolved markedly, receiving growing attention in the 1970s from geoscientists, engineers and economists. The resulting flow of ideas has contributed to setting resource appraisal on its way from a tentative art to a respectable science.

See also: Mineral Deposits: Examination and Evaluation

Bibliography

Brobst D A 1979 Fundamental concepts for the analysis of resource availability. In: Smith V K (ed.) 1979 *Scarcity and Growth Reconsidered.* Johns Hopkins University Press, Baltimore, Maryland, pp. 106–42

Clark I 1979 *Practical Geostatistics.* Applied Science, London

Cook E 1976 Limits of exploitation of nonrenewable resources. *Science* 191: 677–82

DeYoung J H 1981 The Lasky cumulative tonnage-ore relationship—A re-examination. *Econ. Geol.* 76: 1067–80

DeYoung J H, Singer D A 1981 Physical factors that could restrict mineral supply. *Econ. Geol.* 75th Anniversary Vol: 939–54

Harris D P 1984 *Mineral Resources Appraisal. Mineral Endowment, Resources, and Potential Supply: Concepts, Methods and Cases.* Oxford University Press, New York

Harris D P, Agterberg F P 1981 The appraisal of mineral resources. *Econ. Geol.* 75h Anniversary Vol.: 897–938

Lasky S G 1950 How tonnage and grade relations help predict ore reserves. *Eng. Min. J.* 151(4): 81–85

McKelvey V E 1973 Mineral resource estimates and public policy. In: Brobst D A and Pratt W P (eds.) *United States Mineral Resources*, US Geological Survey Professional Paper 820. US Government Printing Office, Washington, DC, pp. 9–19

McLaren D J, Skinner B J (eds.) 1987 *Resources and World Development: Energy and Minerals, Water and Land*, Dahlem Konferenzen. Wiley, Chichester

Materials and Society (Vol. 8 1983): issue devoted to Resources Assessment: History, State of the Art and Applications
OECD Nuclear Energy Agency and International Atomic Energy Agency 1978 *World Uranium Potential.* Organization for Economic Cooperation and Development, Paris
OECD Nuclear Energy Agency and International Atomic Energy Agency 1983 *Uranium—Resources, Production and Demand.* Organization for Economic Cooperation and Development, Paris
Popoff C C 1966 *Computing Reserves of Mineral Deposits: Principles and Conventional Methods,* US Bureau of Mines Information Circular 8283. USBM, Washington, DC
Schanz J J 1980 The United Nations' endeavour to standardize mineral resource classification. *Nat. Resour. Forum* 4: 307–13
Shawe D R 1981 *US Geological Survey Workshop on Nonfuel Mineral-Resource Appraisal of Wilderness and CUSMAP Areas,* US Geological Survey Circular 845. USGS, Washington, DC
Singer D A, DeYoung J H 1980 What can grade-tonnage relations really tell us? In: Guillemin C, Lagny P (eds.) 1980 *Ressources minérales.* Bureau de recherches géologiques et minières (BRGM), Orléans, pp. 91–101
Singer D A, Mosier D L 1981 A review of regional mineral resource assessment methods. *Econ. Geol.* 76: 1006–15
Singer D A, Ovenshine A T 1979 Assessing metallic resources in Alaska. *Am. Sci.* 67: 582–89
Skinner B J 1976 A second iron age ahead? *Am. Sci.* 64: 258–69
US Department of Energy 1980 *An Assessment Report on Uranium in the United States of America.* GJO–111(80). US DOE, Grand Junction, Colorado
US Geological Survey 1980 *Principles of a Resource/Reserve Classification for Minerals,* US Geological Survey Circular 831. USGS, Washington, DC
Zwartendyk J 1976 Problems in interpretation of data on mineral resources, production and consumption. *Nat. Resour. Forum* 1: 7–15
Zwartendyk J 1981 Economic issues in mineral resource adequacy and in the long-term supply of minerals. *Econ. Geol.* 76: 995–1005

J. Zwartendyk
[Energy, Mines and Resources Canada, Ottawa, Ontario, Canada]

S

Sedimentary Ore Deposits: Formation Processes

Sedimentary rocks are formed by the breakdown and weathering of preexisting rocks, followed by transport of the sediment and material in suspension and solution to the sites of deposition. Transport of colloids is of minor importance. Deposition of detrital material occurs where the gradient or water velocity of a stream is decreased and the suspended material is dropped. Precipitation of material in solution is generally by supersaturation, which may result when the stream reaches the sea at a different temperature or Eh–pH condition; bacteria may also cause deposition. In some waters, especially lagoons or continental lakes, evaporation occurs resulting in the supersaturation of dissolved material, which is precipitated as salt, gypsum or anhydrite, and evaporite deposits are formed. The oceanic occurrences of manganese nodules that form on the ocean bottom by accretion may be sedimentary deposits. In addition, many sulfide deposits once thought to be of hydrothermal origin, especially massive sulfide deposits and stratabound deposits, actually formed as ocean-bottom deposits from the hydrothermal fluids associated with submarine volcanic extrusions (Sillitoe 1973).

Deposits of sedimentary origin make up at least 97% of the value of all mineral deposits, including immense sedimentary deposits of iron ore, coal and the stratabound continental and submarine sulfides. The nonmetallic sedimentary deposits, which include sand, gravel, evaporites (salt, gypsum, potash) and others, have an economic value close to triple that of all massive, fissure vein and porphyry sulfide deposits.

1. Deposits Formed by Solution Transport

1.1 Precipitates

Bicarbonated waters are very effective solvents of limestone, iron, manganese and phosphorus. The ionic state of the metals, however, is important. Ferrous sulfate ($FeSO_4$), for example, is fairly soluble but ferric Fe^{3+} iron is relatively insoluble.

Garrels and Christ (1965) have shown that the deposition of iron in seawater containing calcium carbonate is controlled by Eh–pH conditions (Fig. 1). When slightly acid river waters transport iron to seawater with a pH of 7.8 and an oxidizing Eh, iron precipitates out as siderite ($FeCO_3$) and hematite (Fe_2O_3). This also occurs in a weak acid environment although the alkaline seawater is more effective in promoting precipitation. In near-shore acid muds, iron concentrates as pyrite (FeS_2) (Fig. 1).

Colloidal iron is deposited almost immediately upon contact with oppositely charged electrolytes of seawater. Extensive beds of marine iron ores of the Clinton type may thus be formed. In these deposits, some iron replaces oolites, or coats or replaces shell fragments. These characteristics are seen in the Clinton oolitic iron deposits in Alabama; similar deposits are found in the Lorraine and Luxembourg deposits of Europe, and in the USSR.

1.2 Banded Iron Formations

The largest world iron deposits are iron oxides of the Lake Superior type referred to as banded iron formations (BIF). They are almost all Precambrian in age and formed during a single epoch about 2000 to 2800 million years ago. The tonnage of iron deposited during this time of sedimentation is enormous, estimated at 10^{14} to 10^{15} t.

Iron formations have been defined by James (1954) as "a chemical sediment, typically thin-bedded or laminated, containing 15% or more iron of sedimentary origin and commonly but not necessarily containing layers of chert."

In the Lake Superior region of the USA, the iron formations are up to 300 m thick but they have been strongly folded, crumpled and faulted in most of the ranges, and the rocks have been metamorphosed into slates, quartzites, schists, cherts and jaspers (Fig. 2). The rhythmic banding of iron oxide and silica may be the result of several factors, such as seasonal variations, lack of oxygen and increased carbon dioxide in the early stages of the earth's atmosphere. Under tropical conditions, in areas of low relief, the soil and streams are acidic, resulting in the leaching and transport of silica from the rock (Fig. 2). During wet seasons, the abundance of rain water dilutes the ground water, raising the pH, and allowing the transport of iron oxide. These periodic bands are quite thin, sometimes averaging only about 1 mm in thickness. Later, siliceous and carbonate minerals are leached by weathering, resulting in an increase in iron concentration. Minable BIFs should be more than 55% Fe. Lower-grade magnetite ores can be magnetically separated and pelletized into higher-grade material.

Large BIF deposits are worldwide in occurrence, being found in the USA, Canada, Brazil, Venezuela, the USSR, Australia, India, Africa and China.

1.3 Manganese Deposits

Manganese may be deposited as a minor constituent in iron ores, as a sedimentary deposit relatively free from iron, and in the ocean as nodules.

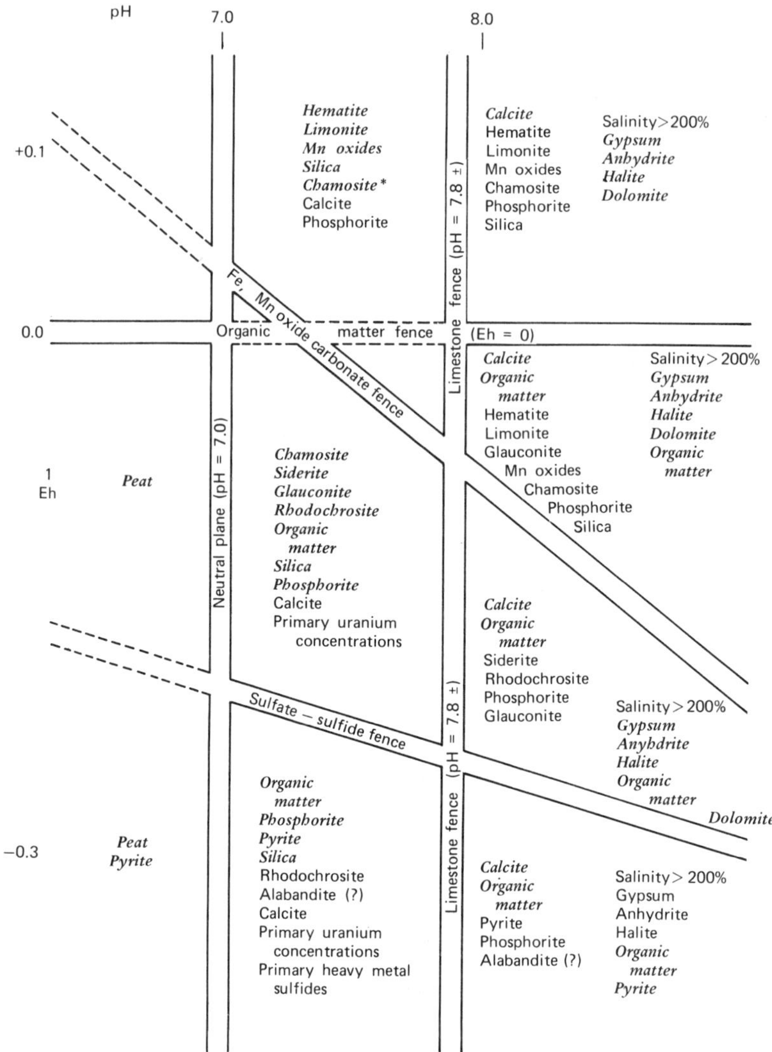

Figure 1
Fence diagrams showing relations of chemical sediments to Eh and pH controls for normal seawater conditions (adapted from Krumbein W C, Garrels R M 1952 *J. Geol.* 60: 1–33)

Many iron and manganese oxides which occur together also form at the same pH and Eh, but other manganese minerals which occur separately from iron form at a lower pH or higher Eh than iron oxides (Fig. 1). Neutral pH conditions in a slightly reducing environment result in precipitation of both as carbonates. Stratiform manganese deposits are not abundant, and are attributed to low-temperature hydrothermal waters related to centers of volcanicity. They lack the large amount of iron that would be present if the metals were derived from the weathering of continental rocks.

Large deposits of this kind formed in the USSR, South Africa, Brazil, Gabon and India. The chief suppliers of manganese to the USA are Gabon (23%), South Africa (20%) and Brazil (18%). There are small but relatively pure deposits of sedimentary manganese carbonate in Newfoundland, Arkansas, Minnesota, South Dakota, California, Nevada, the Appalachian states, Wales and Belgium. The beds seldom exceed a meter or so in thickness or contain more than 20% manganese. The smaller occurrences supply protores, which upon weathering may yield economic deposits of secondary oxides. Manganese is a strategic element

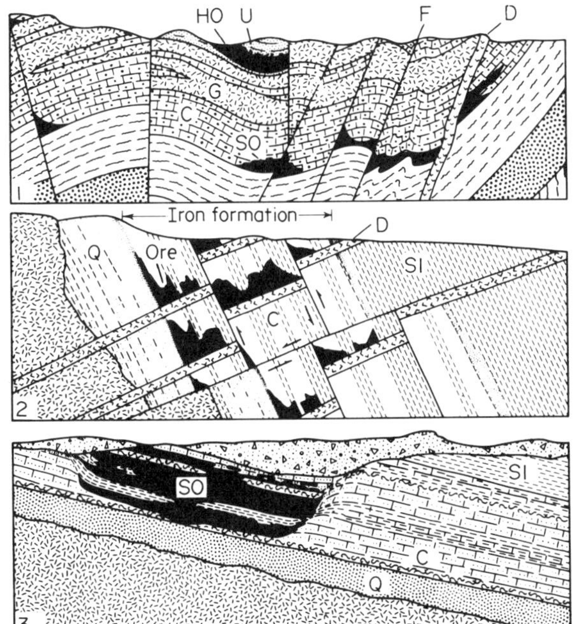

Figure 2
Diagrammatic sections in Lake Superior iron districts showing the occurrence of ore: (1) Marquette range, Michigan; (2) Gogebic range, Michigan and Wisconsin; (3) Mesabi range, Minnesota. HO, hard crystalline ore; SO, soft ore; F, fault; D, dike; G, gabbro sill; C, siliceous iron carbonate; Si, slate; Q, quartzite (after Lovering 1943)

as each tonne of steel requires about 12 kg of manganese for its production.

The largest known manganese deposits of the world are in the USSR at Chiaturi, Georgia, and Nikopol, Ukraine. The ore occurs in oolites and nodules in Tertiary beds of sand and clay from 1 to 4 m thick. Similar deposits occur in the Urals, Siberia, Milos in Greece and India.

A future potential source of manganese is manganese nodules that are widespread but concentrated in places at the bottom of the Atlantic, Pacific and Indian Oceans at long distances from shore. The nodules contain ~20% Mn, ~15% Fe, fractions of more than a few percent of Cu, Ni, Zn and Co, with lesser amounts of Mo, Pb and Cr. The nodules vary in diameter between less than 1 mm to more than a few centimeters. They are concretionary and show growth rings in cross section. Their rate of growth is extremely slow; a 20 mm diameter nodule may have been growing for more than ten million years. As the oceans are not saturated in any metal, the probable origin and growth of the nodules and the source of the metals may be solutions from submarine volcanic activity.

1.4 Phosphorus Deposits

Phosphorus is dissolved from some rocks, enters the soil from which it is extracted by plants eaten by animals and disposed of via their excreta and bones. These in turn undergo resolution, reach the sea and are deposited ultimately as phosphate.

Economic deposits of phosphate are formed under marine conditions as phosphorite. The beds range in age from Cambrian to Pleistocene and extend with remarkable uniformity over thousands of square miles. One formation, the Permian Phosphoria formation, covers portions of the states of Utah, Idaho, Wyoming and Colorado.

A study of the Phosphoria formation led Krauskopf (1967) to suggest a process of accumulation: the sinking of organic material from shallow marine zones into deeper zones where sunlight did not penetrate, followed by the dissolving of phosphorus in organic material to phosphate. When upwelling occurs, especially where the continental shelf meets the continental slope, the phosphate-rich water results in increased organic activity and phosphate accumulates in concentrations of more than 50%.

Direct inorganic precipitation of apatite may occur, according to Krauskopf (1967), by the upwelling of cold water to shallow areas where greater organic activity exists. Although calcium carbonate ($CaCO_3$) precipitates readily under these conditions, phosphorite only precipitates in a pH range that is very narrow and much lower than that for $CaCO_3$. These critical conditions are unusual, which explains why such deposits are uncommon.

2. *Bacteriogenic Processes in Forming Sedimentary Deposits*

2.1 Sulfur

The sulfur of sedimentary rocks is derived (a) from sulfates in rock, (b) from partial oxidation of hydrogen sulfide (H_2S), (c) from volcanic emanations, and (d) by anaerobic bacterial reduction of sulfates in solution.

Sulfur is deposited in bodies of water where reducing conditions and anaerobic sulfur-fixing bacteria, *Desulfovibrio desulfuricans* and *Clostridium nigrificans*, exist. These bacteria reduce sulfate in solution to hydrogen sulfide, which is partially oxidized exothermally to native sulfur. Sedimentary sulfur deposits are often associated with gypsum ($CaSO_4.2H_2O$) and anhydrite ($CaSO_4$), which are the source of the sulfate.

The first economic sulfur deposits were those found on Sicily but these no longer have any significant production. They lie in isolated basins up to 7.5 km long and almost 1 km across. The deposits are in a cellular limestone, interstratified with bituminous shale and gypsum, overlain by beds of marl, clay, limestone and sandstone, and folded and faulted into complex structures. The sulfur is disseminated in cells in the limestone and also occurs in discrete layers up to a few centimeters in thickness. Sulfur content ranges from 12 to 50% and averages 26%. Sedimentary sulfur deposits also occur near Knibyser, Sukeierto and Chekur in the USSR, and in other parts of the world.

In marine basins, sulfate-reducing bacteria reduced sulfate in solution to hydrogen sulfide using bitumen (probably original organic material) as their energy source. The bacteria assimilated $(SO_4)^{2-}$ and hydrocarbons into their systems and released H_2S and CO_2. The H_2S oxidized to native sulfur; CO_2 reacted with Ca from the $CaSO_4$ and produced relatively pure cellular limestone.

Evidence that this process occurred is provided by the isotopic composition of the sulfur isotopes ^{34}S and ^{32}S, and the stable carbon isotopes. The isotopic equilibrium reaction rates indicate that bacterial, or organic, reactions are more rapid for the lighter isotopes of each element. ^{12}C is therefore prevalent in the limestone and ^{32}S is prevalent in the sulfur, in contrast to the source gypsum and hydrocarbons.

Sulfur in the cap rock of inland and offshore salt domes and sulfur recovered from sour crude oils primarily from the Gulf Coast area provides the massive amounts of sulfur now consumed. Only those domes that have associated hydrocarbons are producers of sulfur. This would be expected because organic matter is required as an energy source for the bacteria. Anaerobic sulfate-reducing bacteria acquire their oxygen from sulfate solutions. The sulfate is reduced to H_2S, which in the limited oxygen bearing caprock is oxidized to native sulfur. At the same time, the hydrocarbon energy source, such as CH_4, is oxidized to CO_2, which reacts with the Ca^{2+} ions released from $CaSO_4.2H_2O$ (gypsum) to form $CaCO_3$. Stoichiometrically, about one quarter of the caprock becomes native sulfur and about three quarters become white massive calcite.

2.2 Sulfide Deposits

It has long been observed that pyrite forms syngenetically in black shales, but the formation of other sulfides by similar processes has been slower in acceptance. Yet there are classic base metal sulfide deposits that appear to have had a sedimentary and bacterial origin.

The Nairn pyrite deposits in Australia are prime examples of massive pyrite deposits that are undoubtedly of sedimentary origin. No intrusive bodies are near. Graphite is common, probably derived from organic marine hydrocarbons that provided an energy source for the sulfate-reducing anaerobes and were later reduced to graphite by metamorphism.

Many other deposits may have bacteriogenic origins: the Mount Isa Cu–Pb–Zn deposits of Queensland, Australia; phosphate deposits formed in nearshore shallow marine basins; possibly the stratabound deposits of the Zambian copperbelt; the Kupferschiefer and Mansfeld deposits of Europe; and red beds and some stratabound uranium deposits.

Pyrite readily forms by bacteriogenic (or inorganic) H_2S reacting with indigenous hematite in sediments as follows:

$$9H_2S + 2Fe_2O_3 + 3O_2 \rightarrow \underset{\text{(pyrite)}}{4FeS_2} + SO_2^{4-} + 2H^+ + 8H_2O$$

3. Deposits Formed by Evaporation

These deposits are the result of evaporation from enclosed lagoons, bays and inland evaporite basins; salt is the most common product (Mattox 1968). Some examples are: the Great Salt Lake, Death Valley, playas of the Great Basin of Utah and Nevada, and Searles and Owens dry lakes. Such evaporite deposits occur worldwide and provide a large variety of evaporite minerals. Gypsum ($CaSO_4.2H_2O$) and anhydrite ($CaSO_4$) are derived entirely from evaporite deposits. Either form may change into the other after deposition, but gypsum commonly changes to anhydrite by loss of water of hydration through deep burial. Gypsum is an economic mineral and the source of plaster of Paris ($CaSO_4.\frac{1}{2}H_2O$); anhydrite is of little value.

As some evaporite formations exceed scores of meters in thickness, there must be unusual conditions of repetitive influx of seawater to the evaporite site and sufficient time between influxes to allow evaporation and precipitation. Furthermore, gradual subsidence of the evaporite site must occur to allow the thicker deposits to form.

Brines are collected from the Great Salt Lake primarily for magnesium. Brines are pumped from depth in playas, especially in Utah, Nevada and California and produce potash, magnesium and especially lithium after evaporation.

4. Sedimentary Deposits Formed by Sabkha Process

4.1 Sabkha

This is an Arabic word for barren, uninhabitable, evaporite flats bordering landlocked seas (Renfro 1974). A coastal sabkha forms at the margins of a large body of water where the ground water table is very near the surface and the nearby land has a flat surface but slopes gently towards the sea. Evaporation of the water results in a hydraulic gradient toward the sabkha. The sabkha becomes an interstitial mud flat covered with a leather-like mat of sediment-entrapping, blue-green algae that consists of a fetid, organic ooze containing sulfur-reducing anaerobes. The bacteria-derived H_2S precipitates any metals carried into the sabkha. A thin sulfide deposit extending laterally over large distances then gradually builds up.

4.2 Zambian Copper Belt

The Zambian copper belt, formerly known as the Rhodesian copper belt, was discovered in 1927. It is a stratigraphically controlled mineralized belt 225 km long and 50 km wide extending from east central Zambia northward and slightly northwest into Zaire. 15% of the world's copper is derived from this belt as well as a significant production of cobalt.

These disseminated copper deposits are confined to certain stratigraphic (*Series de Mines*) horizons throughout the entire length of the belt. The mineralization occurs in cross-bedded, feldspathic sandstones and slates containing specks, coarse blebs and veinlets of chalcocite (Cu_2S), bornite (Cu_5FeS) and pyrite (FeS_2).

The origin of this mineralized belt is controversial but is largely thought to have a sedimentary, probably diagenetic origin. The *Series de Mines* lies unconformably above the basement rocks and granites from which the ore-bearing fluids were once thought to have been derived, thus the older granitic rocks could not be the source of the ore-bearing fluids. These deposits, in fact, could have a sabkha origin.

5. Sedimentary Deposits Formed by Redox Processes

5.1 Copper

One of the larger sedimentary copper deposits is the famed Kuperschiefer of Germany, which lies in the Permian Kuperschiefer basin with an area of 58 000 km^2. The formation is a thin, black cupriferous shale, about 1 m thick. It was deposited as a shallow, marine, organic mud, full of land plants washed in from adjacent coasts. Copper is the main sulfide metal contained in this formation but Fe, Pb and Zn are also present as sulfides. Other metals are Ag, Ni, Co, Mρ and V.

There is no totally adequate explanation for the origin of this type of deposit. Some geologists suggest a syngenetic origin (i.e., the metals were deposited at the same time as the enclosing sediments). The metal-bearing solutions, soluble metal sulfates derived by oxidation of distant lodes, from as far as the Hartz mountains, were converted to sulfides in the bottom muds of the Kuperschiefer sea by hydrogen sulfide derived from sulfate-reducing bacteria. Others suggest a metalliferous source from submarine hot springs from which complex ions may have been transported. The White Pine shale deposits in Michigan; Mt Isa, Queensland, Australia; and even the Zambian copper belt, Africa, may have similar origins.

5.2 Red Beds, Copper Deposits

These deposits exist predominantly in the southwestern USA but are also found in Nova Scotia, New Brunswick, in the USSR, southwest of the Ural mountains, in Siberia at Ukokan, and in Bolivia at Corocoro. Many of the deposits in the USA occur in sandstone-type uranium deposits; the Silver Reef mine north of St George, Utah, is a red-beds deposit containing silver. Although these deposits did not form syngenetically, they are found in sedimentary formations.

An understanding of the origin of these low-temperature copper deposits is provided by Rose (1976), who determined the Cu–O–H–S Eh–pH diagram. He demonstrated that the solubility of copper at 25 °C is 6.3 ppm at pH 5.67, and only 0.06 ppm at pH 6.17 and decreases even more at higher pH. With chlorine added to the system, there is a tenfold increase in the solubility of copper as a chloride complex ion at the expense of Cu^{2+} and Cu. Cu_2S is associated with uranium whether the deposit is a red-beds copper deposit or a sandstone-type uranium deposit.

5.3 Uranium

Even though uranium is concentrated in hydrothermal vein deposits, such as those at Yellowknife, Northwest Territories, Canada, and Shinkolobwe, Zaire, sedimentary deposits of uranium are economically far more important. The uranium in these deposits is epigenetic, however (i.e., formed later than the enclosing sediments), as indicated by radioactive dating of the ores.

Uranium is extremely soluble in the hexavalent form $(UO_2)^{2+}$. In this form, it can be leached from associated volcanic ash in the sediments or from granites surrounding the sedimentary basins. It is concentrated in permeable sandstone lenses associated with organic matter consisting of limbs, twigs and trash cellulose material. The organic material reduces U^{6+} to the tetravalent relatively insoluble (UO_2) ion as one or more of scores of uranium minerals, pitchblende or uraninite being most common.

The process of deposition, or reduction from U^{6+} to U^{4+}, may be more complex than suggested. Inorganic processes or environmental conditions may be important, especially where organic debris is lacking. The economic grade of uranium in these deposits is low, cutoff grade being about 0.1% U_3O_8, but a high price of uranium makes high-volume deposits worth mining.

See also: Igneous Ore Deposits: Formation Processes; Metamorphic Ore Deposits: Formation Processes

Bibliography

Garrells R M, Christ C L 1965 *Solutions, Minerals, and Equilibria*, Harper Geoscience Sciences. Harper and Row, Scranton, Pennsylvania

Guilbert J M, Park C F Jr 1986 *The Geology of Oil Deposits*. Freeman, New York

James H L 1954 Sedimentary facies of iron-formation. *Econ. Geol.* 9: 235–93

Krauskopf K B 1967 *Introduction to Geochemistry*. McGraw-Hill, New York

Lovering T S 1943 *Minerals in World Affairs*. Prentice-Hall, Englewood Cliffs, New Jersey

Mattox R B (ed.) 1968 *Saline Deposits*, Geological Society of America Special Paper 88. Geological Society of America, Boulder, Colorado
Renfro S T 1974 Genesis of evaporite-associates stratiform metalliferous deposits. *Econ. Geol.* 69: 33–45
Rose A W 1976 The effect of cuprous chloride complexes in the origin of Red-bed copper and related deposits. *Econ. Geol.* 71: 1036–48
Sillitoe F H 1973 The environments of formation of volcanogenic massive sulfide deposits. *Econ. Geol.* 68: 1321–36
White W W, Wright J C 1954 The White Pine copper deposit. *Econ. Geol.* 49: 675–716

M. L. Jensen
[University of Utah, Salt Lake City, Utah, USA]

Silicon Dioxide Hazards

Most of the work on SiC fiber-reinforced composites has been done on titanium (6Al, 4V)–SiC/W fiber composites. Below 350 °C, SiC-coated boron fibers are superior to SiC fibers as reinforcement for the titanium matrix. Above 350 °C, titanium–SiC fiber composites have favorable properties in comparison wih other composite materials. At high temperatures, however, an interfacial reaction between titanium and the SiC fiber occurs. In the reaction zone, Ti_3Si, Ti_5Si_3 and TiC_{1-x} are formed.

Workers involved with mineral extraction and the processing or use of siliceous materials, such as building stone, have been exposed to the risk of inhalation of dust containing free crystalline silica (FCS); the widespread use of power tools in recent times for drilling, grinding and polishing has increased this risk. Through the centuries a high incidence of lung disease has occurred in workers who have had prolonged exposures to FCS dust. Such disease has been known by many names which indicate the type of dust exposure: miners' phthisis, grinders' asthma, grit consumption, potters' rot, rock tuberculosis, stonemasons' disease and so on. The understanding of the role of silica in such diseases has developed since Kussmaul first reported the presence of silica in the lungs of workers with dust disease in 1866. The term silicosis was first used by Visconti in 1870. Worldwide interest in silicosis in the mining industry arose at the beginning of the twentieth century, and in 1915 Collis postulated crystalline silica to be the cause of lung diseases, including a predisposition to tuberculosis. Since 1929, "accelerated" and "acute" variations have been differentiated from "chronic" silicosis, and the biological effects of synthetic and amorphous silicas have also received much attention.

1. Forms of Silicon Dioxide

Silicon dioxide exists in a variety of crystalline and amorphous (noncrystalline) forms and also as a mixed crystalline and amorphous material. As indicated, silica is widespread in nature; it is also produced synthetically in large amounts, especially as high-purity quartz and in several amorphous forms. Table 1 lists the crystalline forms of silicon dioxide, which are identified by use of microscopy, infrared spectroscopy and x-ray diffraction.

Also classified with crystalline silicas are cryptocrystalline and microcrystalline materials, such as flint, agate, tripoli, rottenstone and novaculite, which contain quartz-like crystal. All of the crystalline silicas, except stishovite, can produce silicosis (fibrosis in the lung) when excessively inhaled as FCS dust.

Amorphous silica is distinguished from crystalline forms by x-ray diffraction. The absence of a diffraction pattern in this analytical procedure indicates a predominantly random structure with little or no regularity (crystallinity). (Crystalline content is readily measured above 1%, but with more difficulty below 0.1%. Regions of crystallinity below 10 nm may go undetected.)

Specific solution rate in selected solvents (see Table 2) has also been used to characterize silica surfaces (Baumann 1981). Crystalline silicas may have "perturbed" surface layers that dissolve faster than the underlying mass. Amorphous silicas have a variety of

Table 1
Types of crystalline silica

Type	Silicon coordination number	Crystalline form	Specific gravity	Comments
Quartz	4	hexagonal	2.63–2.66	insoluble in water
Tridymite	4	rhombic or hexagonal	2.26	reacts with HF very slightly soluble in strong alkali
Cristobalite	4	cubic (fcc) or tetrahedral	2.32	causes fibrosis
Coesite (rare)	4	tetrahedral	> 2.66	less soluble than quartz causes fibrosis
Stishovite (rare)	6	octahedral (rutile)		not toxic to macrophages no fibrosis

Table 2
Specific solution rates of silica

Form	SiO_2 dissolved in 0.1 M NaOH at 23 °C (μg m^{-2} per day)
Crystalline silica particulate	
coesite	~ 100
quartz	~ 300–400
Amorphous silica particulate	
vitreous silica	~ 3000
diatomaceous earth (natural)	2000–4000
silica gel	5000–11000

silanol (SiOH) group structures and have various compositions and degrees of randomness.

Vitreous or fused silica is prepared by melting quartz to eliminate crystallinity and then cooling it in a manner to obtain an amorphous solid structure (some traces of crystallinity may remain). Diatomaceous earth (diatomite or kieselguhr) is found in natural deposits where the silica skeletons of diatoms (microscopic plants) have been laid down on an ancient sea or lake bed. This material is typically 88% SiO_2 with a specific gravity range of 1.9–2.35 and high silanol content; natural material may have small levels of crystalline silica content.

Two types of commercial synthetic amorphous silica are prepared by acidification of sodium silicate solutions under precisely controlled conditions. Silica gel is prepared by adding a silicate solution to an acid to form a hydrosol of silicic acid, $Si(OH)_4$, which gels to form a silica structure filled with salts and water. After dewatering and washing, the amorphous material is dried below 200 °C. (Aerogels are silica gels from which water has been displaced with alcohol before special drying.) Precipitated amorphous silica (PAS) is prepared by adding acid to a stirred silicate solution to precipitate silica, which is filtered, washed and dried below 200 °C. Higher-purity amorphous silicas have been prepared using ion-exchange resins to prepare silicic acid hydrosols which are spray dried.

Drying at low temperature is necessary to preserve an amorphous structure, since amorphous silica is converted to crystalline silica when thermal processing allows it:

$$\text{amorphous silica} \xrightarrow{>870\,^\circ\text{C}} \text{tridymite} \xrightarrow{>1570\,^\circ\text{C}} \text{cristobalite} \quad (1)$$

The calcining of diatomaceous earth can generate 50% cristobalite. (The presence of certain impurities reduces the temperature and increases the rate at which a crystalline silica can form.)

A third type of commercial amorphous material, fumed silica, can be prepared by the high-temperature, vapor-phase hydrolysis of a purified chlorosilane in a hydrogen/oxygen flame, followed by rapid cooling. The resulting "smoke" is a high-purity amorphous silica particulate.

Both amorphous and crystalline silicas are based on SiO_2, but they can differ considerably in chemical impurities, in silanol group content, in surface nature, in particle size, size distribution and agglomeration and in biological effects on lung tissue. Iler (1981) has reviewed the characterization of synthetic amorphous silica. Table 3 indicates analyses of typical commercial amorphous silicas.

As a general rule, amorphous silicas appear to be more benign in interaction with the lung than are the crystalline silicas, although this is not always the case. (Certain so-called "amorphous" silicas have been reported to be fibrogenic; most of these have been shown to have significant crystallinity. However, some truly amorphous silicas have been strongly fibrogenic in the lungs of test animals.) Biological effects should therefore be determined for the various kinds of amorphous silica materials. The amorphous silicas have greater solubility than the crystalline silicas. After inhalation and deposition in the lung, amorphous silicas will tend to be cleared from the lung by body processes after exposure has ceased. In contrast, crystalline silica particulate is not readily cleared and can have a very long residence time in the lung.

2. *Silicosis*

Silicosis is the progressive fibrosis of the lung which results from the excessive inhalation of FCS particulate < 5 μm but mostly < 3 μm in diameter. Epidemiological evidence indicates a more rapid progression of the disease for heavier exposures to higher-purity FCS dust of submicrometer diameter. The following order has been indicated for the rate of induction of lung fibrosis by respirable silica particulate: tridymite ~ cristobalite > quartz ~ coesite > vitreous (or fused) silica. Table 4 indicates the characteristics of exposure that result in the various types of silicosis.

Table 3
Analysis of synthetic amorphous silicas

	Fumed silica	Silica gel	PAS
Crystallinity by x-ray diffraction	nil	nil	nil
SiO_2 content (%)	~ 97	~ 90	~ 87
Ignition loss (–SiOH dehydration and H_2O loss) (%)	~ 2	~ 9	~ 11
Some impurity contents (ppm)[a]			
S	200	500	3150
Ca	90	360	2000
Al	26	161	1734
Fe	47	104	890
Ti	61	215	445

[a] Al and Fe have been reported to reduce the fibrogenic potency of a silica

Table 4
Classification of silicosis

Type/description	Years to a positive x ray	Exposure
Simple		
FCS particulate reaches critical concentration at some lung locations to produce discrete whorled collagen nodules; also multicentric nodules (mostly < 5 mm diameter). Little lung tissue involved	see below[a]	excessive low-level inhalation of FCS respirable particulate
Chronic		
Steady nodular size growth and proliferation. Conglomeration of nodules and obliteration of lung fine structure by fibrosis. Hyalination. Pleura can also be involved	20	prolonged excessive inhalation of dust with < 30% FCS
Accelerated (rare)		
Initiation of many collagen nodules, accompanied by progressive massive fibrosis	4–8	higher exposures; higher-purity FCS (sandblasting)
Acute or Diffuse (rare)		
Fast initiation of many small collagen nodules. Diffuse fibrosis develops rapidly throughout both lungs. Fluid, granules and macrophages in alveoli. Lung walls thickened. Lung consolidation but volume maintained	1–3	repeated very heavy FCS exposure in confined area with inadequate protection (tunnelling with pure quartz vein involved)

[a] The development of isolated small silicotic nodules involving a relatively small area of the lung is neither disabling nor symptomatic. Disease may be arrested at this point, but it may proceed to become chronic silicosis

Wright (1978) has reviewed the anatomy and physiology of the lung and its reaction to inhaled solid particulate (pneumoconiosis, including silicosis). Ziskind et al. (1976) have reviewed the diagnosis of silicosis. A principal clinical symptom of established chronic silicosis is the shortness of breath upon physical effort. X rays of the lungs can show the development and progression of fibrosis. Pulmonary function tests for airway obstruction, capacity loss and blood–lung gas exchange can be used to detect progressive lung disfunction even earlier than x-ray diagnosis.

The FCS content of the normal lung is 0.2 g; silicotic lungs may contain 15–20 g of respirable-size FCS particulate. (Crystalline silica is not readily cleared by the lung and has a long residence time.) About 14.7% of the ash of normal lung tissue is silica, but it is in the range of 29–48% for silicotic lungs.

An inhaled mixture of FCS and certain other dusts, such as iron oxide or coal, will produce a pneumoconiosis different from the classical silicosis. The fibrosis present in such disease, however, is believed to be due to the effect of FCS. Silicotuberculosis patients experience an increased degree of fibrosis as compared with silicosis patients for the same FCS exposure. Chronic FCS exposure adversely affects lung-tissue defenses against the tubercle bacillus and like infections. Mycobacterial lung infections in conjunction with silicosis are decreasing for the chronic form, but increasing for the accelerated and acute forms of silicosis.

3. Mechanism of Silicosis Development

Excessive levels of inhaled respirable FCS particulate, reaching the terminal bronchioles and alveoli deep in the lung, can cause silicosis. The alveolar macrophages, lung scavenger cells of 10–50 μm size, engulf deposited particulate, isolate it from lung tissue and transport it for lung clearance. For most particulates this system works satisfactorily in protecting the lungs, but not for crystalline silica, which is cytotoxic to the macrophages.

The normal life span of macrophages is weeks; after ingestion of FCS, it can be hours or days. The macrophages rupture and die, spilling out their cell contents, including the collected silica particulate. Other macrophage cells take up the task of controlling these same particles, and these in turn rupture and die. It is believed that the dying macrophage cells release a chemical factor which, as it reaches a critical concentration, stimulates the production of collagen by the underlying lung or lymphoid tissue. In any case, collagen is irreversibly laid down in fibrotic nodules of scar tissue which enclose the particulate-silica sites.

The rate of stimulation of collagen production appears to be the key to the rate of development and spread of fibrotic tissue, giving the silicosis results outlined in Table 4. The same basic mechanism is believed to be operating in all forms of silicosis, from the simple to the acute. As fibrotic nodules mature, hyalination produces a hard glassy surface; the pleura (the sacs surrounding the lungs) can be thickened and bridged to lung tissue by fibrosis. Fibrosis also bridges the established nodules and destroys the fine structure of the lung, conglomerating and immobilizing it until there is insufficient functioning lung tissue to sustain normal physical activity. The result is a developing pulmonary incapacity and finally death.

The cytotoxity to macrophages of FCS particulate is believed to result from the interaction between cell membranes and the special silanol surface geometry of the crystalline silicas with four-coordinated silicon atoms. The surface silanol groups of the FCS denature macrophage membranes, rupturing the macrophage cell walls. Stishovite, an octahedral crystalline silica (see Table 1), is not cytotoxic to macrophages, nor fibrogenic (Ziskind et al. 1976). When cytotoxic FCS particulate is coated with a very thin layer of amorphous silica, it is no longer cytotoxic to alveolar macrophage cells of rabbits, but it becomes cytotoxic again upon solution removal of the amorphous coating (Baumann 1981). In animal tests some amorphous silicas have shown fibrogenicity in animal lungs that has not been shown to occur in exposed humans (see Groth et al. 1981).

4. Silicon Dioxide Hazard Control

Silicosis, the fundamental hazard in the use of crystalline silica, can be controlled by limiting the inhalation exposure of workers to respirable silica particulate. The current Occupational Safety and Health Administration standard (1987) in the USA allows exposure to respirable quartz particulate based on the following 8 h time weighted average (TWA) as determined by particle count (millions of particles per cubic foot of air, mppcf) or particulate weight per cubic meter of air:

$$\text{8 h TWA exposure} = \frac{250}{\%\text{quartz} + 5}\ \text{(mppcf)}$$

$$\text{or}\ \frac{10}{\%\ \text{quartz} + 2}\ (\text{mg m}^{-3}) \qquad (2)$$

Cristobalite and tridymite are controlled at half the level for quartz. Amorphous silica and diatomaceous earth (natural) are controlled at an 8 h TWA of 20 mppcf (80 mg m^{-3} (% $SiO_2)^{-1}$).

The Threshold Limit Value (TLV) Committee of the American Conference of Government Industrial Hygienists (ACGIH) (1984) has recommended several silica limits, in contrast to its inert nuisance particulate recommendation (see Table 5). These silica exposures are considered safe for almost all exposed workers for 8 h per day, 40 h per week for a working lifetime.

Even though many improvements in safety have occurred and the total control of silicosis as an industrial disease is possible, in practice conditions are still inadequate in some cases and the disease persists; for example, a National Institute of Occupational Safety and Health study of silica flour mills in the USA found as late as 1979 that control of FCS dust was satisfactory in only a small minority.

Table 5
1984–85 ACGIH threshold limit values (or "intended changes") for silica

Form	Respirable dust (< 5 μm) (mppcf)	Respirable dust (< 5 μm) ($mg\ m^{-3}$)	Total dust ($mg\ m^{-3}$)
Cristobalite or tridymite	1.5	0.05	0.15
Quartz or fused silica	3	0.1	0.3
Tripoli	3	0.1	0.3
Diatomaceous earth, natural uncalcined (< 1% quartz)		1.5 (5)	(10)
Amorphous silica	(20)	3(5)	6(10)
Nuisance particulate (inert, < 1% quartz)	(30)	5	10

Bibliography

American Conference of Governmental Industrial Hygienists 1984 *Threshold Limit Values for Chemical Substances and Physical Agents in the Work Environment with Intended Changes for 1984–85.* American Conference of Governmental Industrial Hygienists, Cincinatti, Ohio

Banks D E, Morring K L, Boehlecke B A 1981 Silicosis in the 1980s. *Am. Ind. Hyg. Assoc. J.* 42: 77–79

Baumann H 1981 Characterization of silicon dioxide surfaces by successive determination of the solution rate. In: Dunnom 1981, pp. 30–47

Brody A R, DeNee P B 1981 Biological activity of inorganic particles in the lung. *CRC Crit. Rev. Environ. Control* 11(3): 277–99

Dunnom D D (ed.) 1981 *Health Effects of Synthetic Silica Particulates: A Symposium,* ASTM STP 732. American Society for Testing and Materials, Philadelphia, Pennsylvania

Groth D H, Moorman W J, Lynch D W, Stettler L E, Wagner W D, Hornung R W 1981 Chronic effects of inhaled amorphous silica in animals. In: Dunnom 1981, pp. 118–43

Iler R K 1979 *The Chemistry of Silica: Solubility, Polymerization, Colloid and Surface Properties, and Biochemistry.* Wiley, New York, pp. 769–83

Iler R K 1981 The surface chemistry of amorphous synthetic silica—Interaction with organic molecules in an aqueous medium. In: Dunnom 1981, pp. 3–29

National Institute of Occupational Safety and Health 1974 *Occupational Exposure to Crystalline Silica: Criteria for a Recommended Standard,* NIOSH 75–120. US Department of Health, Education and Welfare, Washington, DC

Stokinger M E 1981 The halogens and the nonmetals boron and silicon. In: Clayton G D, Clayton F E (eds.) 1981 *Patty's Industrial Hygiene and Toxicology,* 3rd edn., Vol. 2B: *Toxicology.* Wiley, New York, pp. 3011–21

Wright G W 1978 The pulmonary effects of inhaled inorganic dust. In: Clayton G D, Clayton F E (eds.) 1978 *Patty's Industrial Hygiene and Toxicology,* 3rd edn., Vol. 1: *General Principles.* Wiley, New York, pp. 165–202

Ziskind M, Jones R N, Weill H 1976 Silicosis. *Am. Rev. Respir. Dis.* 113: 643–65

J. M. Nielsen
[General Electric Corporate Research and Development, Schenectady, New York, USA]

Silver Resources

The uses of silver are varied and demand for the metal is high, especially for image processing and electronics applications. Ore deposits of silver are rare, as unique geochemical processes are required for their formation. Solution, underground and open-pit mining methods are used to recover the ores, and milling and smelting are required to obtain the pure metal. North America has about 42% of the world's silver reserves and about 30% of the world's resources. India has 30–60% of the world's primary reserves. The reserve base of the USSR is over one third that of North America and is expected to influence future world stocks.

1. History

Silver has been regarded as a metal of value by most civilizations. The early Egyptians measured its value as two fifths the weight of gold, and its use by the Chinese and the Persians dates from around 2500 BC. Rich mines of Laurium, Greece, helped the Greeks build a powerful navy, which defeated the fleet of Xerxes I of Persia in 480 BC. Silver acted as a monetary base for the Roman Empire, and its dearth during the period of that empire's decline may have had a significant effect on the subsequent collapse. Later discoveries of rich silver deposits in central Europe allowed its important economic role to continue. Silver from the mines of the New World was also of great economic significance, two and a half billion ounces eventually came from the "richest hill on earth" at Potosi, Bolivia. More recently, discoveries such as the Comstock lode in 1859 helped open up the West in the USA and placed the USA first in world silver production for 29 years. In this period, Canada and Mexico also became major world producers of the metal and benefited from the great wealth of their rich silver mines.

2. Geochemistry and Mineralogy

The pure metal is soft with a Mohs hardness of 3.0 and a Brinell hardness of 25 $kg\,cm^{-2}$; it is ductile and has a density of 10.53 $g\,cm^{-3}$, a melting point of 961 °C and a boiling point of 2180 °C. Its atomic number is 47, two stable isotopes being known; these have mass numbers 107 and 109 and relative abundances of 51.4 and 48.6%, respectively. The most common oxidation state of silver is +1, and the ionic radius is 0.113 nm. It is a chalcophile element, readily combining with sulfur or related ions rather than uniting in silicate minerals. Crystals of the native metal are isometric, with cubes and octahedrons as common faces. Silver is the most reactive of the noble metals, dissolving in oxidizing acids such as nitric and sulfuric.

Crustal abundance of silver is estimated at 70 ppb with varying quantities present in the various types of igneous rocks (Fig. 1). Rocks of intermediate composition (andesites and diorites) contain 70 ppb and both felsic and ultramafic rocks contain less than 50 ppb. Silver is most abundant in basalt (100 ppb). Major ore bodies of silver, or silver-bearing ores, are associated with igneous rocks of intermediate composition.

Major mineral sources of silver are native silver, acanthite and argentite (both Ag_2S), antimony sulfosalts (miargyrite ($AgSbS_2$), pyrargyrite (Ag_3SbS_3) and stephanite (Ag_5SbS_4)) and cerargyrite (AgCl). Currently most silver is recovered as a by-product of the mining of other metals. Copper, lead and zinc mining provides 35%, 14% and 13%, respectively, of

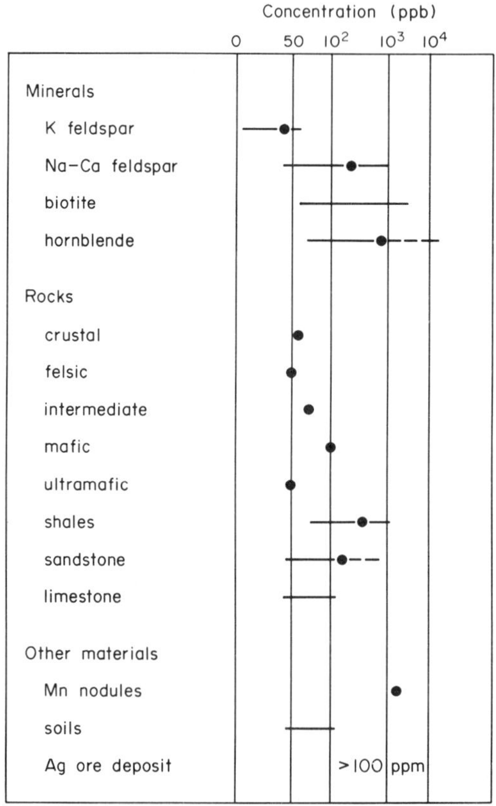

Figure 1
Silver content in minerals, rocks and other materials: ●, average; ——, range; – – –, estimated range

silver production, with smaller quantities obtained from gold, cobalt, manganese, nickel and uranium mining.

3. *Ore Deposits*

Heyl et al. (1973) use combined metal and geological association as a classification basis for silver ore deposits, permitting a ready evaluation of silver resources (Table 1). Epithermal (i.e., low-temperature, shallow-seated) deposits have been major producers. Long-term producers, containing large tonnages of silver ore, are in the mesothermal group (i.e., moderate temperature, moderate depth deposits). Major silver production also comes from low-grade ores of silver in by-product silver deposits such as the great Mississippi Valley type deposits of Missouri, the large porphyry copper deposits of the western Cordilleras, and massive sulfide deposits as at Timmins, Ontario. Although the former two are of very low grade, the large quantities of ore mined make these a major source of silver.

4. *Mining*

Terrain, orebody size, shape, depth and grade, reserves, character of the enclosing rocks, access and other economic factors unique to the deposit determine the mining method employed. At the Sunshine mine in Idaho, USA, silver ores are recovered by underground mining through a shaft, development drifts along the veins, and a cut-and-fill stoping method. Ore is transported to the main shaft and hoisted to the surface for milling.

Open-pit mining methods are used at the rich Kidd Creek mine in Timmins, Ontario. On a much larger scale, in by-product silver deposits such as the porphyry copper deposits in Arizona and Utah, electric scoop shovels of up to 15 m^3 capacity and haulage trucks up to 250 t capacity are utilized. Such large-scale mining affords significant by-product silver production from the very low grade deposits.

5. *Processing*

Numerous treatment or processing methods have been used to extract silver from its ores. These include amalgamation, cupellation, the Patio process, the Augustin process and the Patera process. Cyanidation was introduced in about 1900 and resulted in improved recovery of silver. The development of the flotation process revolutionized milling and most silver-containing minerals are now recovered in this way, although cyanidation is still used in many places to enhance silver recovery from low-grade ores.

Silver is separated from lead by smelting according to the Parkes process. Silver in copper ores is recovered from the anode slime produced during electrolytic refining of the impure copper, and is smelted to produce silver bullion. Silver in bullion is generally expressed by its fineness, or as parts per thousand: fine silver is pure silver described as 1000 parts fine, sterling silver is 925 fine with 75 parts copper, and commercial silver bullion is 999–999.9 fine.

6. *Production and Uses*

Mine production of silver in the USA in the period 1960–1987 is shown in Fig. 2 together with average price per ounce. The rapid price rise in 1979 resulted in a consumption decrease of almost 30% in one year.

Scrap silver from jewelry, works of art, coins, batteries, electronics, industrial film and other sources makes a significant impact on silver availability. India's silver, estimated at 2–4 billion ounces in coins, decorations, works of art and jewelry, could strongly affect the amount of silver in world markets as prices edge upward.

Table 1
Classification of silver deposits according to Heyl et al. (1973)

Silver as major component	By-product silver deposits
Epithermal	porphyry copper deposits
veins, lodes and pipes	copper–zinc–lead replacement deposits and vein clusters
disseminated and breccia deposits	massive sulfide deposits
silver–manganese deposits	lead–zinc replacements
silver–lead–zinc replacement deposits	Mississippi Valley and Alpine-type lead–zinc and fluorspar deposits
silver–copper–barite deposits	copper deposits in sandstone and shale
Mesothermal	native copper deposits
silver–lead–zinc–copper deposits	gold deposits in veins, conglomerate and placers
cobalt–silver, cobalt–uraninite–silver and cobalt–silver zeolite deposits	nickel and magnetite deposits
sandstone silver deposits	
seafloor muds and hot-spring deposits	

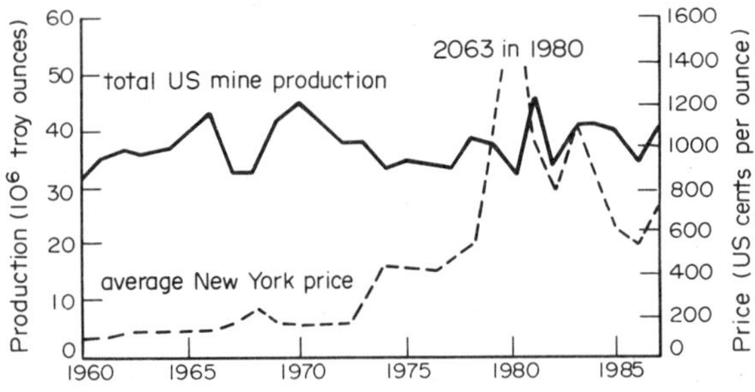

Figure 2
US production and average price of silver 1960–1987

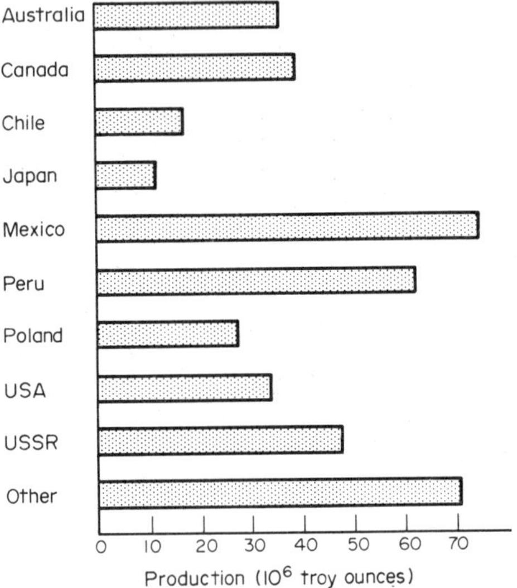

Figure 3
World mine production of silver (1986)

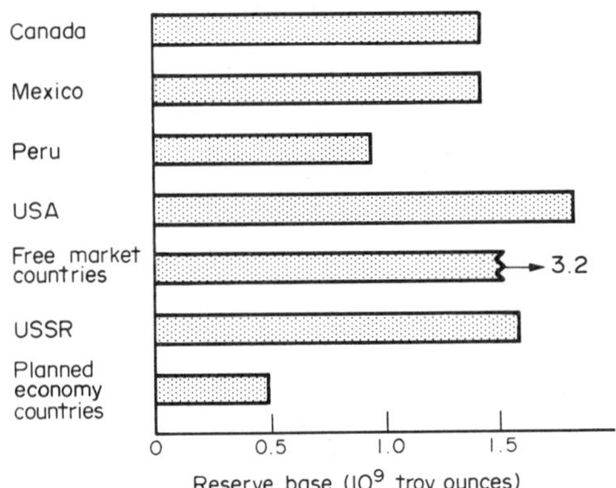

Figure 4
Reserve base of silver (1986)

Silver uses for 1987 in the USA included 45% for photographic materials; 26% for electrical and electronics; 13% for sterling ware, electrical and electronics materials, and jewelry; 6% for brazing alloys and solders; and 10% for other uses. The price of the metal often determines its use: with high prices substitutes are sought (e.g., the large consumption of silver by the photographic industry may decrease as new image-preservation systems of a less costly nature are developed). The unique reflectivity, electrical and heat conductivity, and resistance to oxidation at high temperatures make silver a much needed metal in the electronics industry.

7. *Distribution of Resources*

By-product and primary silver reserves in the USA may be as much as 1.8 billion ounces, and in North America, 3.2 billion ounces. The world total reserves are estimated at 10.8 billion ounces. Silver resources, based on much more tenuous data, suggest 4.2 billion ounces for the USA, 8.6 billion ounces for North America and 16.5 billion ounces for the free world. The USSR has total estimated resources of 7 billion ounces (Figs. 3, 4). Long-term primary resource development is expected to continue in the USA, Mexico, Canada and South America, as well as in the USSR.

Bibliography

Boyle R W 1968 *The Geochemistry of Silver and its Deposits*, Geological Survey of Canada Bulletin 160. Supply Service of Canada, Ottawa

Freuh A J, Vincent E A 1978 Silver. In: Wedepohl K H (ed.) 1978 *Handbook of Geochemistry*, Vol. 2. Springer, Berlin, Chap. 47

Heyl A V, Hall W E, Weissenborn A E, Stager H K, Puffett W P, Reed B L 1973 *Silver*, US Geological Survey Professional Paper 820. US Government Printing Office, Washington, DC

Jensen M L, Bateman A M 1979 *Economic Mineral Deposits*, 3rd edn. Wiley, New York

Lovering T S 1943 *Minerals in World Affairs*. Prentice–Hall, New York

Reese R G 1986 Silver. In: *Minerals Yearbook 1986*, Vol. 1. US Bureau of Mines, Washington, DC, pp. 837–56

US Bureau of Mines 1988 Silver. In: *Mineral Commodity Summaries 1988*. USBM, Washington, DC

P. D. Proctor
[Brigham Young University, Provo, Utah, USA]

Slag

The production of metal by melting requires metal ore, fluxing material (usually limestone to assist the melting process) and coke. The yield consists of molten metal and molten slag which normally floats on the metal because of its lower specific gravity. The molten slag is removed from the production area for disposal. The term slag as used in this article refers to the by-product material from the production of iron and steel.

The earliest use of slag as a construction material dates back to Roman times, but it is only since the 1920s that slag has gained wider use as a construction aggregate. Prior to about 1920 slag was a waste product that represented a waste disposal problem. Only after World War II did blast-furnace slag begin to be used increasingly as a construction aggregate. In 1985 the US construction industry used nearly 17.1 Mt of iron and steel slag with over 70% coming from the production of iron. An additional 1.8 Mt were used outside of the construction industry.

The benefits of using slag are threefold. First, the use avoids environmental concerns associated with large-scale disposal problems; second, it eliminates the costs to the metal producers that are associated with the disposal of large volumes of waste material; and third, it supplies an additional source of construction raw materials.

1. Distribution

The bulk of slag production in the USA is located around the Great Lakes in Detroit, Chicago, Gary, Cleveland and Buffalo. Other ironmaking and steelmaking centers include Pittsburgh, Philadelphia, Baltimore, Birmingham and Bethlehem. Besides these major steel production areas, there are several locations where steel is produced at "mini mills," e.g., Florida, Louisiana, New Jersey, Oklahoma and Washington State.

Imported steel has had a major impact on the production and availability of slag in the USA. In 1980 there were 56 plants producing iron in blast furnaces. By 1985 there were 38 domestic blast furnaces. During the same period, the number of open-hearth steel furnaces decreased from 11 to 5; basic oxygen steel furnaces increased in number from 16 to 22 and the number of electric steelmaking furnaces grew from 19 to 40.

2. Production

Slag that is used by the construction industry comes mainly either from the production of iron (where it is known as blast-furnace slag) or from the production of steel (where it is referred to as steel slag). In both cases the furnace yields two products: molten metal and slag. Slag comes from the furnace at a temperature of around 1500 °C and resembles molten lava. Blast-furnace slag supplies about three times as much material as steel slag.

2.1 Blast-Furnace Slag

The blast furnace is loaded with iron ore, flux stone (limestone and/or dolomite) and coke for fuel. The flux stone serves two purposes: initially it reduces the temperature necessary to melt the ore, then, as the iron begins to melt, the slag (molten limestone) serves as a scavenger to collect (in solution) the impurities from the ore and from the coke. The slag consists essentially of silica and alumina impurities from the iron ore, and lime and magnesia from the flux stone. Since slag is lighter than iron, it floats on the molten metal. The molten slag is removed from the blast furnace and undergoes one of three cooling processes, each of which yields a different product: air-cooled, granulated or expanded slag.

(*a*) *Air-cooled blast-furnace slag.* The great majority of blast-furnace (iron) slag is dumped into slag pits and allowed to air-cool. In 1985 air-cooled slag accounted for about 90% of all blast-furnace slag in the USA. The resulting product is a vesicular, fine-grained, mainly crystalline material. After cooling, it undergoes crushing and screening similar to other crushed-stone materials that are used for construction aggregates. It is used for road-base material, coarse aggregates in asphalt and portland cement concrete, railroad ballast and roofing aggregate.

(*b*) *Granulated blast-furnace slag.* Granulated slag is produced by quenching the molten material in water. The rapid cooling results in a material which is glassy compared with the more crystalline nature of the air-cooled variety. The granulation process yields a material that requires little or no crushing. It has cementitious properties which result in it being used as a stabilized base material beneath pavements, around bridge abutments and in building foundations. Ground granulated slag is also used as a partial replacement for cement in concrete construction. The manufacture of cement clinker from granulated slag requires less energy than the corresponding process using conventional cement-making raw materials, and the resulting cements have a better resistance to corrosion by seawater and by sulfate exposure.

(*c*) *Expanded blast-furnace slag.* Expanded slag is the material that results from treating the slag with smaller quantities of water than are required for granulation. Two common methods of expanding involve pouring the slag either into a stream of water or onto a water-cooled rotating paddle wheel. The first method yields an angular product and the latter method a more spherical product (referred to as pelletized). Both products are more vesicular than air-cooled slag because expanding techniques increase the air content in the particles. The main use of expanded slag is as a lightweight concrete aggregate. In addition to a weight saving, expanded slag also offers excellent insulation and fire-resistance properties.

2.2 Steel-Furnace Slag

Steel is produced in one of three types of furnaces: open-hearth furnaces, basic-oxygen furnaces (BOF) and electric-arc furnaces. All three use scrap steel as a partial source of metal, and fluxing material. The open-hearth furnace uses limestone for flux and molten iron from the blast furnace supplements the scrap steel as the source of metal. The BOF process is much faster than the open-hearth process (45 min vs. 6 h to produce 300 t of steel). Scrap steel and molten iron supply the metal, but this process differs from open-hearth melting in that oxygen is "blown" into the furnace to speed the melting process, and lime (not limestone) is usually added as a flux. Electric furnaces use scrap steel and limestone in the production of steel. Their advantages are a smaller capital investment and reduced air pollution.

Steel slag differs significantly from blast-furnace slag. It contains more iron, lime and other oxides than blast-furnace slag, as shown in Table 1. Unlike blast-furnace slag, which can undergo three cooling processes, steel slag undergoes air-cooling only, which results in a material that is darker colored and denser than blast-furnace slag mainly because it contains more iron. Steel slag does not have as many uses as blast-furnace slag because of its different weight and chemistry. Its main uses are as a highway base material, railroad ballast and some fill applications.

3. Properties

Properties vary between the three types of blast-furnace slag and also between blast-furnace slag and steel-furnace slag. The chemical and physical properties govern, to a large extent, the end uses for which slag may be a suitable raw material. The alteration of properties may prove to be a goal for future research efforts.

3.1 Chemical Properties

The chemical characteristics of slag are controlled by the requirements of the metal production. Since the metal product is of primary importance, the technology required to produce the iron and steel dictates the composition of the slag. The use of slag as a construction material came as a result of marketing efforts, not to any change in the chemical characteristics of the by-product.

Table 1 presents the average compositions for blast-furnace slag, BOF steel slag and electric-arc steel slag. Steel slag often contains free or unreacted lime. The presence of free lime, coupled with the higher iron content, creates two problems. It interferes with desired chemical reactions when used with portland cement, and hydration of the free lime causes expansion. Therefore steel slag is not used as a concrete aggregate and the expansion problem means it is unsuitable as a structural fill material (beneath buildings).

Table 1
Chemical composition (in %) of iron and steel slag

	Blast furnace slag	Basic-oxygen furnace slag	Electric-arc furnace slag
SiO_2	36.0	14.1	16.9
CaO	39.0	44.4	41.2
MgO	12.0	7.4	10.2
Fe_2O_3	0.5		
FeO		22.4	18.1
(Fe)		(20.7)	(13.6)
Al_2O_3	10.0	1.9	8.1
MnO	0.4	5.7	3.6
S	1.4	0.2	0.2
P_2O_5		0.8	0.6
P		0.3	0.4

3.2 Physical Properties

The specific gravity of blast-furnace slag ranges from 2.0 to 2.5 and depends mainly on the amount of void space (vesicules) in the particles, and the chemical composition. The unit weight of blast-furnace slag depends in part on the cooling method. Air-cooled blast-furnace slag has a loose unit weight of 1120–1360 $kg\,m^{-3}$ influenced by the gradation of the particles. Granulated slag ranges from 830 to 950 $kg\,m^{-3}$. The density of expanded slag is 560–880 $kg\,m^{-3}$ for coarse particle sizes and 800 to 1000 $kg\,m^{-3}$ for fine particle sizes.

Blast-furnace slag is very durable as a construction aggregate material. It resists weathering, shows very little degradation due to freezing and thawing, and has a low coefficient of thermal expansion (about $1\times10^{-5}\,°C^{-1}$ from ambient temperature to melting temperature). Its resistance to polishing under traffic conditions makes it desirable as a high-friction aggregate for asphalt pavements.

The specific gravity of steel slag is about 30–50% greater than that of blast-furnace slag because of its higher iron content. The unit weight of steel slag ranges from 1660 to 2000 $kg\,m^{-3}$. Its surface texture and weight make it suitable as a road-base material and railroad ballast. Some is also used for asphalt paving as it has the properties necessary for a high-friction aggregate.

3.3 Mineralogical Properties

The mineral content of both iron and steel slags depends on the chemical composition of the slag and its cooling history. Crystallization proceeds more slowly in slags with a high silica content than in those which contain more lime and magnesia. These properties are analogous to those of acidic and basic igneous systems.

Little or no crystallization takes place in expanded or granulated blast-furnace slag because of the rapid cooling. The following minerals have been identified in air-cooled blast-furnace slags:

Akermanite $Ca_2(MgFeAl)_2O_7$
Pseudowollastonite $CaOSiO_2$
Periclase MgO
Anorthite $CaAl_2Si_2O_8$
Forsterite Mg_2SiO_4
Merwinite $Ca_3Mg(SiO_4)_2$
Gehlenite $Ca_2Al_2SiO_7$
Calcium orthosilicate Ca_2SiO_4
Olivine $(MgFe)_2SiO_4$
Monticellite $CaMgSiO_4$
Pyroxene XO (Si_2O_6)
Oldhamite CaS
Manganous sulfide MnS
Ferrous sulfide FeS

Although air-cooling is slower than water-quenching, it is still relatively fast and the resulting crystal size rarely exceeds 10 mm. The presence of glass, even in air-cooled slag, testifies that crystallization is incomplete.

Calcium orthosilicate is undesirable because its low-temperature phase is unstable. When the lime content of the slag is more than 10% greater than the silica content, calcium orthosilicate may form. Partial substitution of dolomite for limestone in the flux diminishes the likelihood that calcium orthosilicate will form in the cooled slag.

Steel slags undergo air-cooling like most blast-furnace slag. However, steel slag is more fluid, and cooling produces a higher glass content than is found in blast-furnace slags. The mineral content of steel slags has not been as widely published as that of blast-furnace slag.

4. Uses

Slag is used outside the construction market as a raw material for the manufacture of glass (amber quality) and for glass wool, soil conditioner, roofing material and sewage treatment. The use of slag in the construction industry is described in detail by Lewis (1986).

While iron and steel slags are the most commonly used, slag from the processing of other metals can be used in some instances. The production of copper, nickel, zinc and ferroalloys all yield slag. Some of this material has been used for the same range of construction materials as iron and steel slags. Production data for other slags are not reported because the volumes produced are much less than those of iron and steel slags.

In 1985 the estimated world production of blast-furnace slag was 113.4 Mt and 51.3 Mt of steel slag was produced. In several countries, such as Canada, slag is still thought of as a waste product rather than a resource. As a result, the data for international slag production and consumption are incomplete.

Uses of slag in other countries are generally not as diverse as in the USA. The manufacture of cement is apparently the major use of slag in France, the FRG, Italy, Japan and South Africa. However, the uses within the UK parallel the US consumption patterns.

See also: Construction Materials: Crushed Stone; Construction Materials: Fill and Soil; Construction Materials: Granules; Construction Materials: Lightweight Aggregate

Bibliography

Eggleston H K 1968 *Slag—The All-Purpose Construction Aggregate.* National Slag Association, Washington, DC

Hogan W T 1984 *Steel in the United States: Restructuring to Compete.* Lexington Books, Lexington, Massachusetts

Lewis D W 1986, Slag utilization. In: Bever M B (ed.) 1986 *Encyclopedia of Materials Science and Engineering.* Pergamon, Oxford, pp. 4460–61

McCarl H N, Eggleston H K, Barton W R 1983 Construction materials, aggregates—slag. In: *Industrial Minerals and Rocks*, 5th edn. Society of Mining Engineers, Littleton, Colorado, pp. 111–31

Mickelson D P 1985 Slag—Iron and steel. In: *Minerals Yearbook.* US Bureau of Mines, Washington, DC

National Slag Association 1982 *Processed Blast Furnace Slag—The All-Purpose Construction Aggregate.* National Slag Association, Washington, DC

Seagall R T 1970 Slag—Michigan's All-Purpose Construction Aggregate. In: *Proc. 6th Forum on the Geology of Industrial Minerals*, Misc. Pub. #1. Michigan Geological Survey, pp. 117–26

H. L. Bourne
[Independent Consulting Geologist,
Northville, Michigan, USA]

Soil Additives: Industrial Minerals

Inorganic soil additives are divided into three categories: primary, secondary and trace nutrients or micronutrients (Bryant and Kincheloe 1983). Although all three types must be present for plants to grow, some plant nutrients are more important than others. Primary additives consist of nitrogen, phosphorus and potassium, and are highly important. Secondary soil additives or plant nutrients are less significant in the amount added. They are commonly added to the primary nutrient carrier, and consist of calcium, magnesium and sulfur. Trace additives or micronutrients are those essential to plants, but are only used in minute quantities; they consist of boron, iron, manganese, copper, zinc, molybdenum and chlorine.

1. Primary Nutrients

1.1 Nitrogen

Nitrogen is perhaps the most important fertilizer constituent. It accounts for about half of the total nutrients applied as fertilizers in the USA. Almost all nitrogen fertilizers are derived from ammonia and natural gas remains the preferred feedstock. Although nitrogen in the air is unlimited, the quantity, location, availability and price of the resources that provide hydrogen, such as natural gas, petroleum or coal, are the limiting factors in ammonia production (Davis 1985).

In the past, virtually all nitrogen applied to the soil was in the form of animal manures. However, manure is not an important commercial source of fertilizers today, and the major organic sources of nitrogen are now urea and urea–formaldehyde; both are synthetic products.

1.2 Phosphorus

In 1840, Justus von Liebig demonstrated that the fertilizer value of bones was significantly increased by treatment with sulfuric acid. When this process was repeated using apatite, a phosphate-bearing mineral, a product known as ordinary superphosphate (20% P_2O_5) was produced. Three products currently dominate the world market in phosphates; they are normal superphosphate ($Ca(H_2PO_4)_2 . H_2O + CaSO_4 . 2H_2O$); concentrated superphosphate ($Ca(H_2PO_4)_2 . H_2O$) and complex fertilizers, including monoammonium phosphate ($(NH_4)_2HPO_4$); and nitric phosphates.

Over 80% of the world's reserves of phosphorus occur in the USA, northwest Africa and the USSR. Morocco leads with 59% of the known reserves, followed by the USA with 16% (Stowasser 1985). China has produced sizable tonnages in recent years, but primarily for domestic use. (see *Soil Additives: Phosphate, Potash and Sulfur*).

1.3 Potassium

Growing plants absorb potassium as the K^+ ion. Potassium is unlike the other major mineral additives, in that its exact metabolic role in plants is not known, but it is indispensable as a nutrient and it is required in large quantities by most plants. Europe, the USA and the USSR have been the major consumers of potassium fertilizers.

Muriate of potash (KCl) is the dominant source of fertilizer potassium throughout the world. Other important, but much less significant, sources are sulfate of potash (K_2SO_4) and sulfate of potash–magnesia or langbeinite ($K_2SO_4 . 2MgSO_4$).

Economic potash deposits are essentially restricted to thick accumulations of marine chloride evaporites, which yield high-grade, low-tonnage ore bodies. Canada and the USSR have most of the world's known reserves amounting to nearly 80% (Searls 1985). Eastern Europe, namely the USSR and the GDR, is the largest producing area (see *Soil Additives: Phosphate, Potash and Sulfur*).

Potassium and phosphorus are applied in a very similar manner. Most is spread with the seed at planting time. Muriate of potash is usually mixed with phosphorus carriers, and the two nutrients are applied simultaneously, commonly with nitrogen.

2. Secondary Nutrients

Only a small part of the total production of secondary nutrients is used in agriculture, in contrast to primary nutrients. However, secondary nutrients are important to plant growth, and are normally added to a mix of the primary constituents.

2.1 Calcium

Calcium is significant not only as a plant nutrient, but also because of the very favorable influence it has on maintaining the desired physical and biological conditions of the soil. Calcium is the dominant exchangeable base found in soil, normally accounting for 50–80% of the total base exchange.

Calcium is absorbed by plants as the Ca^{2+} ion. Inside the plant it tends to accumulate in the older tissue. Much of the calcium in plants is permanently fixed in the cell walls as a salt which gives cell walls their strength. Calcium is also important in the utilization of nitrates, as they make their way along the pathways toward protein formation.

Most soils, even acid ones, contain sufficient calcium for plant growth and calcium-deficient plants are seldom observed. However, many soils are too acid for optimum plant growth, and here, the soil pH can be raised to a desirable level by addition of calcium. Certain crops, such as peanuts and tomatoes, require large amounts of soluble calcium. Gypsum ($CaSO_4 . 2H_2O$) is commonly used where these crops are grown, and is usually applied directly rather than as a component of mixed fertilizer. However, calcite ($CaCO_3$) is the most important calcium carrier.

2.2 Magnesium

Magnesium, like calcium, is required both as a nutrient and as a soil component. The desired level of magnesium in soils is usually about one tenth that of calcium. Magnesium is absorbed by a plant as the Mg^{2+} ion, and its role inside the plant is of major significance because it is the only mineral constituent of the chlorophyll molecule. Magnesium is also important in phosphorus metabolism.

Magnesium sources which are soluble or available for use by plants are hydrated magnesium sulfate and langbeinite. Other magnesium sources which are only slightly water soluble, but will become soluble over a period of 2–4 years, are dolomite ($CaMg(CO_3)_2$), brucite ($MgO . H_2O$), magnesite ($MgCO_3$), basic slag, periclase (MgO) and magnesium–ammonium phosphates.

2.3 Sulfur

Sulfur is an essential plant nutrient. Normal superphosphates, which are produced by treatment of phosphate rock with sulfuric acid, contain about 12% sulfur in the form of gypsum ($CaSO_4 . 2H_2O$). Some plants require almost as much sulfur as they do phosphorus. Sulfur enters the plant as the $(SO_4)^{2-}$ ion and inside the plant it becomes a necessary component of protein.

The usual carriers of sulfur in today's fertilizers are normal superphosphate, potassium phosphate, langbeinite and gypsum. Ammonium thiosulfate ($(NH_4)_2S_2O_3$) is perhaps the most widely used sulfur carrier in the fluid fertilizer industry, and contains about 26% sulfur in the sulfate form. It is completely soluble in liquid nitrogen–phosphorus–potassium mixture, and is important in the fluid-fertilizer industry as these mixtures are normally sulfur-free and not compatible with most sulfur carriers (see *Soil Additives: Phosphate, Potash and Sulfur*).

3. Trace Nutrients and Fertilizers

Trace nutrients or micronutrients are essential to plants, but only in minute quantities. They almost always occur in soils in very small quantities, and include boron, copper, iron, manganese and zinc. Chloride is an essential nutrient in plants, but there are no known areas of soil deficiencies. The leading trace nutrient is zinc, followed by manganese. Most micronutrient carriers are incorporated in multinutrient mixed fertilizers. They may be mixed in chemically manufactured fertilizers, in blends or in fluids.

Boron is added to fertilizers as water soluble borax, modified borax ($Na_2B_4O_7 . 10H_2O$), colemanite ($Ca_2B_6O_{11} . 5H_2O$) and frits (synthetic borosilicate glasses). The use of colemanite and frits has increased because of their lower solubility and lower possibility of toxicity.

Although most soils contain sufficient iron, it is not always in a form available to the plants. Therefore, fertilizers contain specific iron carriers, such as partially soluble sulfates and chelated iron. Iron may also be added as finely ground hematite (Fe_3O_4), limonite ($Fe_2O_3 . nH_2O$) and pyrite (FeS_2).

Manganese is a trace element supplied in fertilizers in the form of soluble sulfates, variable insoluble oxides, and even the chelated form. Usually, the less soluble oxides are least effective. The chief minerals which supply manganese for fertilizers are pyrolucite (MnO_2), psilomelane (mixed manganese oxides) and by-product manganese oxide (MnO).

High organic soils are also often copper deficient. Copper is usually added to the fertilizer mixture as a sulfate, oxide or even as chelated copper. Chalcocite (Cu_2S), chalcopyrite ($CuFeS_2$) and cuprite (Cu_2O) are probably the most important mineral sources.

Zinc is usually applied with a fertilizer mixture as a sulfate, sulfide, oxide, or in the chelated form. Most of these carriers have been prepared from sphalerite (ZnS).

Excesses of molybdenum can be as detrimental as inadequate amounts, and so it is used in many fertilizers in small quantities. The usual carriers are sodium molybdate, ammonium molybdate, and sometimes molybdenum oxide. The chief source of these molybdenum compounds is the mineral molybdenite (MoS_2).

Bibliography

Bryant J P, Kincheloe S 1983 Fertilizer minerals In: Lefond S J (ed.) 1983 *Industrial Minerals and Rocks*, 5th edn., Vol. 1. American Institute of Mining, Metallurgical and Petroleum Engineers, New York, pp. 233–41

Davis C L 1985 Nitrogen (ammonia). *Mineral Facts and Problems*, US Bureau of Mines Bulletin. US Bureau of Mines, Washington, DC, pp. 532–62

Searls J P 1985 Potash. *Mineral Facts and Problems*, US Bureau of Mines Bulletin. US Bureau of Mines, Washington, DC, pp. 617–33

Stowasser W F 1985 Phosphate rock. *Mineral Facts and Problems*, US Bureau of Mines Bulletin. US Bureau of Mines, Washington, DC, pp. 579–94

G. S. Austin

[New Mexico Bureau of Mines and Mineral Resources, Socorro, New Mexico, USA]

Soil Additives: Phosphate, Potash and Sulfur

Phosphate rock, potash and sulfur are mined principally to provide nutrients for soils. More than 90% of the world's production of phosphate rock and potash, and about half of the sulfur produced, are used as soil additives. This article concentrates on the sources and mining of these minerals and their applications as soil additives; other soil additives such as nitrogen-bearing fertilizers and trace additives such as calcium, magnesium, boron, iron, manganese, copper, zinc, molybdenum and chlorine are discussed elsewhere (see *Soil Additives: Industrial Minerals*).

1. Phosphate Rock

Phosphorus occurs in igneous and sedimentary rocks, and in fresh and salt water. The average content of phosphorus in these rocks, expressed as P_2O_5, is less than 1%. However, there are 200 minerals containing more than 1% P_2O_5; the most important are of the apatite family. Fluorapatite ($Ca_5(PO_4)_3F$) contains about 40% P_2O_5 (Emigh 1983).

Phosphate rock is the term used for phosphorus-bearing materials found in sedimentary deposits, and apatite for phosphorus-bearing minerals found in igneous deposits. Phosphorite is a deposit of phosphate, directly or indirectly of sedimentary origin, and is of economic interest.

1.1 Sources

The USA, the USSR and northwest Africa supply the majority of the world's supply of phosphate rock. The principal producing areas of the world are Florida in the USA, the Kola peninsula and the Kara-Tau district in the USSR and Morocco respectively. In the USA, there are three major phosphate-rock producing areas: Florida and North Carolina; the western states of Idaho, Montana, Utah and Wyoming; and Tennessee.

The USA produces 27% of the world's output (Russell 1987), enough to satisfy its own demands and at the same time provide competition in the world market for the closely knit north African exporters. While Florida's production is currently down and 36 Mt of phosphate rock was mined in 1986 (about 70% of the total US production), newer producing areas in south Florida are now initiating production. Undoubtedly a diversification of sources of phosphate will occur in the future, but the USA (and Florida in particular) will continue to produce significant amounts for some time to come (Harben 1981). Morocco leads the world in phosphate exports and is planning to have a rock capacity of nearly 55 Mt by the year 2000, with about 60% exported and 40% for downstream processing (Russell 1987).

1.2 Mining

In central Florida, where near-surface sedimentary deposits are mined, exploration drilling patterns may vary from five to ten holes per km^2. Initially, core samples are obtained for processing in the laboratory to simulate beneficiation plant processing. The data obtained characterize the overburden, depth of matrix, grade of ore and concentrate, volume of matrix mined per km^2, and tonnage of product per km^2. The phosphate-bearing matrix ranges from about 0.3 to 15 m in thickness, averaging 5 m. The overburden covering the matrix is quartz sand and clay, and averages 6 m in thickness. Before mining, the land is drained to eliminate swampy conditions, and vegetation is removed. Electric-power drag lines strip the overburden and mine the matrix. Drag lines work in cuts that may be 45–75 m wide and range from a few hundred meters to more than 1000 m long. The overburden is stacked on natural ground level adjacent to the cut. The drag line deposits the matrix in previously prepared sluice pits. Hydraulic guns slurry the matrix and it is pumped to a washing plant. Florida "hardrock" or phosphorus-containing material that would have to be broken mechanically, is not currently mined. However, several companies are recovering phosphate-bearing tailings from dry hardrock washing plants by scraping the dried surface layers. Mining techniques in North Carolina and North Africa are similar to those used in Florida.

All phosphate ore produced in the western states of the USA is strip mined except for that obtained from an underground mine in Montana. Most of the strip mines in southeast Idaho use scrapers and bulldozers to remove the overburden, and mine economically recoverable ore. The ore from the underground mine in Montana is broken in stopes, removed through chutes into several adits, and hoisted to the surface. In southeast Idaho the main bed of ore is selectively mined for consumption in wet-process phosphoric-acid plants; some furnace shale used in electric and other furnaces is beneficiated to increase its grade and, after calcining, is treated in wet-process acid plants. In Utah, phosphate rock is quarried after an overlying limestone cap is drilled, blasted and removed. The thickness of the ore is about 6 m and the total overburden ranges in thickness from 15 to 30 m.

Phosphate ore in Tennessee is stripped and mined with large drag lines. The loose clay overburden ranges from 2 to 6 m thick. Ore thickness averages 2 m, but may be as much as 8 m. If the ratio of overburden to ore exceeds 3:1, the deposit is not mined.

In parts of Indonesia and Morocco, underground mining employs room-and-pillar methods and the relatively soft ore does not require drilling and blasting; however, large quantities of timber posts are used to prevent roof slabs from falling. In the USSR the open-stope method of underground mining is used.

The bulk of the world's phosphate rock (about 80%) is mined using open-pit methods, this being the least expensive way to move large volumes of material.

1.3 Uses

Approximately 95% of the world's phosphate rock production is consumed in fertilizers. In the USA over 90% is used for agricultural purposes, and the remainder is used for producing elemental phosphorus by melting in electric furnaces with a balance charge of coke and silica (Stowasser 1985). The phosphorus obtained is volatized, condensed and collected under water. Principal outlets for furnace acid are in the production of food-grade dicalcium phosphate and sodium tripolyphosphate, a detergent builder.

Essentially all phosphate rock produced in Florida, North Carolina, and part of the production of the western states is used to produce wet-process phosphoric acid. Phosphate rock is digested with sulfuric acid to produce phosphoric acid and waste impure gypsum. When phosphate rock is mixed with sulfuric acid in controlled but lesser quantities, the phosphorus content is converted to water-soluble normal or ordinary superphosphate. If phosphate rock is acidulated with phosphoric acid, high analysis (46% P_2O_5) triple superphosphate is produced.

2. Potash

2.1 Sources

Potash is the generic term for a variety of potassium-bearing minerals, ores and refined products, and its importance is primarily due to the potassium requirement of growing plants (Adams 1983). The term potash originally referred to potassium carbonate produced from the leaching of wood ashes. In the 1840s, potash was discovered to be essential for plant growth, and during the latter part of the 1800s, nearly all potash came from German deposits. Most of the potash produced since that time has been derived from evaporite deposits; approximately 95% has been used in fertilizers. The potassium content of ore minerals or products is customarily expressed in percent K_2O, a unit of measure which is neither a natural nor a synthetic compound of potassium.

Potassium is one of the most abundant elements in the earth's crust and is present in silicate minerals in igneous, sedimentary and metamorphic rocks. Potassium is also a major constituent of many surface and subsurface brines. Economic potash deposits are, however, essentially restricted to widespread thick accumulations of marine chloride evaporite deposits. These deposits yield high-grade large-tonnage orebodies, many of which are suitable for low-cost mining and beneficiation. The products of these deposits are ideal for use in the fertilizer industry because of the relative solubility of potassium chloride and sulfate evaporite minerals.

The marine evaporite deposits are precipitated in assemblages or sequences of assemblages as determined by the composition of the brines and the solubility relationships of the salt systems. Near the end of the evaporation cycle the brines tend to precipitate sulfate, chloride and supersaline mineral assemblages, including potassium and magnesium. Subsequent fluxing or movement within the accumulating mush of brine and crystals may change the composition of the deposits.

Evaporite deposits are composed of a sequence of tabular beds with a characteristic proportion of minerals. Beds may be a few centimeters to more than a meter in thickness and may be present over hundreds of square kilometers. The principal potash deposits of a district may be associated with the areas of the greatest halite accumulation, peripheral to these areas or in nearby subbasins, depending upon the tectonic history of the area.

The deposits near Carlsbad, New Mexico, have dominated the US production market since the 1930s, however in 1962 Canada replaced New Mexico as the principal source of potash used in the USA (Siemers and Austin 1979). At present, reserves are concentrated in Canada and in the USSR, each of which contain about 40% of the world's known reserves.

2.2 Mining

Underground potash deposits have frequently been discovered as the result of exploration for oil. Commercial potash beds are often regular, tabular bodies with a wide lateral extent, which makes underground potash mining relatively straightforward.

Potash ore can be detected in drill holes by the radioactivity of the potassium isotope, which produces γ-rays. For pure minerals and low concentrations of potassium the amount of γ-ray emission correlates well with the potassium content of the ore. Mixtures of potassium minerals and high concentrations of potassium require the use of correction factors. A neutron log of the same drill hole is often used to distinguish between sylvite (KCl) or langbeinite ($K_2SO_4.2MgSO_4$) (the principal ore minerals) and the other potash minerals.

The room-and-pillar method of mining has been used extensively in underground mines in the USA and Canada, and it is widely used throughout the USSR and Europe. Timbering is not often used, but roof bolting is practised at various locations within the mine. The first pass through the mine collects about 70% of the ore (Singleton 1978). With the room-and-pillar configuration, miners can re-enter an area and "rob" pillars and remove as much as 90% of the total available ore. The normal method is by undercutting, drilling, and blasting with ANFO (ammonium nitrate–fuel oil mixture) or with the use of continuous miners, either boring or rotating drum types. Ore in underground mines is normally loaded on shuttle cars which move the ore to belts for haulage to the shaft.

Solution mining is used in some operations, such as one in Utah and several in the Canadian province of Saskatchewan. Potash is also removed by longwall mining (an adaptation from coal mining) in Europe and also in the USSR.

2.3 Processing

Most marketable potash products are concentrated by (a) flotation, (b) selective dissolution of the gangue minerals, or (c) precipitation of the potash mineral from brine. The finely ground contaminants are largely clays and quartz, and they are removed in a desliming circuit. In several operations near Carlsbad, New Mexico, the slurry of sylvite and halite is passed through a flotation cell where reagents attach the sylvite grains to bubbles in the froth, causing the sylvite to float away from the halite. The sylvite is drawn off near the top of the cells, dried and stored.

Solution mining is applicable only to relatively soluble potassium chloride minerals and is profitable only for relatively thick but also deep high-grade sylvinite ores.

2.4 Uses

Potash is used mainly in agriculture. It cannot be reclaimed or recycled, and there is no known substitute. The principal potash products are muriate of potash (KCl), potassium sulfate (K_2SO_4) and potassium–magnesium sulfate ($K_2SO_4.2MgSO_4$). They are sold on the basis of chemical composition and particle size. Most products are sold on the basis of a guaranteed minimum potash content in standard coarse and granular grades. Muriate of potash, the highest grade potash, accounts for about 95% of US consumption. The low chlorine content and lower solubility of potassium sulfate products make them desirable for certain crops and soil conditions. Products produced by solution and precipitation are often pure, and hence they are preferred for certain applications, including the chemical industry.

3. Sulfur

3.1 Sources and Mining

Sulfur is generally classified as elemental or nonelemental for statistical reporting. Elemental sulfur is sulfur produced in the pure uncombined form, and includes production from native deposits as well as that recovered in elemental forms from oil and natural gas or other sulfur-bearing materials. Nonelemental sulfur is that which is utilized commercially in combinations with other elements, such as sulfides and sulfates.

Sulfur may be available in fertilizers with phosphates (as superphosphates with 12% sulfur) or in the sulfate form in langbeinite, gypsum and other sulfate-bearing minerals.

Native sulfur is produced from deposits of three general types: (a) the caprock over salt domes, (b) stratiform deposits, and (c) volcanic deposits (Barker 1983). Salt domes and other similar piercement structures are known to occur in various parts of the world, but economic sulfur deposits have been discovered only in the salt domes of the Gulf Coast region of the USA and Mexico. In these deposits, and within an envelope of barren caprock and barren anhydrite over salt, is a sulfur-bearing limestone. The normal melting point of sulfur is 119 °C. Thus superheated water injected under pressure into the sulfur-bearing formations will melt the sulfur, which is found in the cavernous limestone. The molten sulfur is removed by bleeder wells piped to the surface and into evaporating ponds where the sulfur solidifies. This technique, known as the Frasch process, was first developed in the 1940s.

Elemental sulfur also occurs in subsurface deposits similar to salt dome deposits in that the sulfur is found with secondary limestone; it is associated with gypsum- and anyhydrite-bearing formations laid down in a shallow basins. The sulfur in these deposits is generally believed to be derived from hydrocarbon-reducing sulfates and assisted by bacterial action. The same process used in the salt dome deposits (the Frasch process) is also used in these operations with the result that native sulfur is formed from melting the ore at depth and solidifying it on the surface.

Although sulfur of volcanic origin is widely distributed and is known in most volcanic regions of the world, its main production comes from the mountain ranges that border the Pacific Ocean, particularly in South and Central America, Japan, the Philippines, Taiwan and New Zealand. Mining of volcanic deposits is limited primarily to Japan, Turkey, Morocco, Italy and the Andes countries of South America. The volcanic sulfur deposits include those originating from a variety of volcanic activity; they may replace or impregnate previously existing material precipitated as elemental sulfur in hot volcanic lakes formed by sublimination from gaseous or fumarolic activity near volcanic centers or as sulfur flows originating from previously formed deposits that have been remelted. Mining methods used include tunnelling, room-and-pillar, open pit, cut-and-fill, or various types of stoping. Processing of ore to separate sulfur from volcanic ash is by various combinations of melting, distillation, conglomeration, flotation and solvent extraction, or combinations of these.

Elemental sulfur may also be recovered from sour natural gas or organic sulfur compounds contained in crude oil. Refineries in general are equipped to recover sulfur only from the lighter refinery products, and much of the sulfur in crude oil remains in the refinery emission gases or the residual fuel oils to be ultimately admitted into the atmosphere when these oils are burned. In general, the process of recovering sulfur from sour natural gas or crude oil consists of first separating the H_2S from other gases and then converting it to elemental sulfur.

3.2 Uses

Sulfur is increasingly important not only to farmers because of more widespread soil deficiencies, but also to the phosphate industry. It is apparent that the use of fertilizers is increasing and therefore the utilization of sulfur as a fertilizer mineral will also increase. It is believed that elemental sulfur from either salt domes, bedded deposits or the volcanic rocks of the world will become increasingly important in supplying sulfur to soils.

Bibliography

Adams S S (Revised by Hite R J) 1983 Potash. In: Lefond 1983, pp. 1049–84

Barker J M 1983 Sulfur. In: Lefond 1983, pp. 1235–74

Bryant J P, Kincheloe S 1983 Fertilizer minerals. In: Lefond 1983, pp. 233–41

Emigh G D 1983 Phosphate rock. In: Lefond 1983, pp. 1017–47

Harben P 1981 Four Corners phosphate mine: Progress and potential. *Ind. Miner.* 164: 61–67

Harris G T, Harre E A 1979 *World Fertilizer Situation and Outlook 1978–85*. International Development Center and Tennessee Valley Authority, Muscle Shoals, Alabama

Lefond S J (ed.) 1983 *Industrial Minerals and Rocks*, 5th edn. American Institute of Mining, Metallurgical and Petroleum Engineers, New York

Russell A 1987 Phosphate rock—Trends in processing and production. *Ind. Miner.* 240: 25–59

Siemers W T, Austin G S 1979 Industrial rocks and minerals of New Mexico. *Annual Report 1977–1978*. New Mexico Bureau of Mines and Mineral Resources, Socorro, New Mexico, pp. 43–60

Singleton R H 1978 *Potash*, US Bureau of Mines Mineral Commodity Profiles No. 11. US Bureau of Mines, Washington, DC

Stowasser W F 1985 Phosphate rock. *Mineral Facts and Problems*, US Bureau of Mines Bulletin. US Bureau of Mines, Washington, DC, pp. 579–94

G. S. Austin
[New Mexico Bureau of Mines and Mineral Resources, Socorro, New Mexico, USA]

Statistical and Computer Models in Mining and Exploration

Many uses of computers and statistics have developed in exploration and mining geology during the last thirty years, and particularly since the late 1970s when microcomputers have become generally available. Because words have never been as well-matched to exploration and mining geology as graphic displays of data, much of the current emphasis is on computer graphics, database development and manipulation, and man–machine interaction. Many new devices, including touch sensing, terminals and microcomputers make communication easier, quicker and more accurate than in the early days of computers.

1. Exploration

1.1 Mineral-Deposit Models

Models of mineral deposits have been employed for many years, although in the past they were given names such as hypotheses or case histories. More recently, models have been systematically developed so that they can be compared with one another and so that deposits can be assigned to them. This work has required data files stored on computers and statistical analyses implemented on computers. One collection of models contains 85 descriptive deposit models and 60 grade–tonnage models (Cox and Singer 1986). Grade–tonnage models investigate the relationship between grade and tonnage of ore; the various mathematical functions that have been proposed are statistically analyzed.

1.2 Exploration Planning

Many mining companies and some governmental agencies use computers for exploration planning. The most widely used techniques are those of operations research, including queueing, linear programming, networking and the Project Evaluation and Research Technique–Critical Path Method (PERT-CPM) (Wignall and De Geoffroy 1987).

1.3 Resource Evaluation

Shortages of mineral commodities including petroleum have placed emphasis on resource evaluation. Because of the large amounts of data and the multivariate statistical analyses required, computers have been essential for these investigations.

Classification and evaluation of gridded areas have been of major concern in this work; later concepts have included subjective probability and probabilistic interpretations of inferences made by geologists applying scientific methodology consistently in regional evaluation (Harris 1984). Thus, the discipline imposed by the inflexibility of the computer has helped to emphasize the importance of systematic geological observation; the computer has made possible a consistent application of scientific methodology.

Resource evaluation from a governmental perspective has been extensively investigated, particularly in recent years in the USA and Canada, reflecting their large areas and well-developed science and technology (Cargill and Green 1986). A typical approach, tested in Ontario and Quebec, Canada, provided a geomathematical evaluation of copper and zinc potential. First, a database was established for geological and geophysical parameters measured in small cells. Through multivariate statistics, these measurements were compared to metal content for cells containing ore bodies and predicted metal endowment in other cells. The study demonstrated that meaningful data could be obtained from geological maps, that these data were incomplete and that effective statistical methods could be devised (Agterberg 1988).

The idea of unit regional value is that "the resources produced in some specific region may be measured in terms of the amount of resource produced and its value; if these measures are cumulated over the period of production and prorated over the area of a region, . . . they yield the unit regional weight and unit regional value of resources produced in the region" (Griffiths 1978). If these measures are related to one another and compared for relatively developed and undeveloped parts of the earth, predictions can be made about the kind and amount of mineral resources that can be developed in various regions.

The discovery-process model estimates undiscovered resources in a partially explored region based upon characteristics of the discovery process (Drew et al. 1980). Another model is a classification technique named characteristic analysis (Botbol et al. 1978). In explored areas, deposit modelling has been used successfully for resource estimation (Sinding-Larsen and Vokes 1978). All of these models have been successful for various geological situations; further work will define their generality more clearly.

1.4 Sampling Designs for Exploration

Probabilistic models for detecting ore deposits have been devised for exploration either on grids or on equally spaced traverse lines; data may be collected periodically or continuously. The designs are used most often either for drilling campaigns or for geophysical or geochemical surveys.

The models are well developed for two-dimensional situations, particularly if the ore deposits or other targets can be represented by geometrical forms, such as ellipses, susceptible to mathematical analysis. Three-dimensional models are less developed than the two-dimensional ones and apply to a narrower range of physical situations. One model explores for three-dimensional folded targets (Malmqvist and Malmqvist 1979). Another model uses the area of influence of mineralization and Bayesian statistics to select targets for univariate or multivariate data and multiple drill holes (Singer and Kouda 1988).

Statistical models are available for circles, semicircles, spheres and hemispheres. These models allow structural data associated with the host rocks and with the ore deposits themselves to be incorporated into the search models.

Dynamic program models have been devised to cope with sequential exploration, which is generally done using Bayesian statistics (Wignall and De Geoffroy 1987).

1.5 Exploration Geochemistry

Large-scale exploration geochemical surveys require the collection of thousands or tens of thousands of samples, which may be analyzed for 50 or more chemical elements or other variables. Therefore, for data handling, statistical analysis and preservation of information, computers are essential.

Sampling designs and methods of statistical analysis implemented by computers include those detailed in the previous sections. To determine *thresholds*, which are cutting values separating chemical-element concentrations (or other variables) believed to be related to mineralization from those believed to be unrelated, probability graphs are widely used (Sinclair 1976).

Besides their original variables, which are generally multivariate, exploration geochemists define additional variables, including ratios, sums and differences, as well as variables derived by multivariate statistical techniques including factor, cluster, regression, discriminant and cluster analysis (Howarth 1983).

Regional geochemical surveys require a systematic collection of samples, analytical work and presentation of results. Regional surveys have been completed for several small countries and extensive areas of large countries. In one survey for England and Wales, statistics were used for quality control and geological analysis; computers did the statistical analyses and processed the data for plotting (Webb et al. 1978).

1.6 Expert Systems and Artificial Intelligence

Expert systems are computer programs which store information for dealing with facts, inferences and judgements. They carry out consultations, produce maps, give advice, answer questions and explain reasoning. Thus they store expertise and judgemental knowledge for the purpose of guiding users through evaluations or diagnoses. Expert-systems software tools are becoming widely available, so a wide range of projects are now possible at reasonable cost. Expert systems have promise for storing expert knowledge, putting it into usable form and distributing it.

In the 1960s, artificial-intelligence scientists tried without success to develop general-purpose programs. Then, in the 1970s, these scientists made a "conceptual breakthrough" by recognizing the following principle: "To make a program intelligent, provide it with lots of high-quality, specific knowledge about some problem area" (Waterman 1986, p. 4). The resulting programs were named *expert systems*, developed through *knowledge engineering*, involving collaborative work among *knowledge engineers*, who design and build the expert systems, and *domain experts*, who have subject-matter expertise (Waterman 1986).

The first expert system in geology was PROSPECTOR; in the language of computing and information sciences it is an expert system, developed through concepts of artificial intelligence (Campbell et al. 1982).

PROSPECTOR was developed by a team of geologists and computer scientists (knowledge engineers). The geologists devised models of ore deposits (Sect. 1.1) and, with prompting from the knowledge engineers, ascribed probabilities and linkages in reasoning to them. The knowledge engineers developed inference nets to relate the models one to another and to the exploration process.

Describing PROSPECTOR, Duda et al. (1979) write that:

> The Prospector system is intended to emulate the reasoning process of an experienced exploration geologist in assessing a given prospect site or region for its likelihood of containing an ore deposit of the type represented by the model he or she designed. The performance of Prospector depends on the number of models it contains, the types of deposits modelled, and the quality and completeness of each model .... Besides the running program, there appear to be several other benefits to this type of expert system approach. The model design process challenges the model designer to articulate, organize, and quantify his expertise. Without exception, the economic geologists who have designed Prospector models have reported that the experience aided and sharpened their own thinking on the subject matter of the model.

Expert-system applications are well suited to resource evaluation (Sect. 1.3). Many of the databases needed to make decisions are already in place, and the public, government and exploration industry need clear, rational and unbiased explanations of decisions that a good expert system gives. For resource evaluation and planning, the formulation of a rational, definable mineral policy is the focus of much activity. Many methods exist for the quantitative analysis of resources. Economic modelling, environmental modelling, statistical estimation and environmental impact statements are all examples. The integration of these studies into an easily explained and understood policy is difficult and cumbersome. Expert-systems technology is clarifying the interactions of these different methods; the technology can be tested, improved and updated as problems are found or policy changes (Koch et al. 1988).

The understanding of the object relationship, implicit in geological cross sections and maps, requires expert-system programming methods for practical implementation. Expert systems are being developed for the following applications in computer vision, remote sensing and geographic information systems: interpretation of aerial photographs; hyperspectral classification in remote-sensed imagery; image segmentation; intelligent front-ends for data integration (different sensors, ancillary data from maps and images, digitization, etc.); and analysis of data from digital terrain models (Koch et al. 1988).

The newer machines in the mineral-resource industry are controlled by electronic sensors. These measure temperature, pressure, voltage, frequency, weight, velocity and other values. Computers interpret the sensor values to regulate machines and processes. The same computers can warn operators when sensor values go beyond acceptable limits. In modern systems, numerous sensors quickly deliver large quantities of information. When things go awry, many operators have trouble interpreting the copious readings. Sensor expert advisors can be expected to give the quality of advice that an experienced operator gives on his better days. This is advantageous because advice will be available when the experienced operator is not available. In the hypothetical example of a hydrothermal production well, an expert system would be set up to continuously monitor the electronic sensors. At the first signs of trouble, the proposed sensor advisor would comment about the developments and offer advice. In contrast, at present, experienced operators may get late night calls when emergencies are highly developed, if they can be found at all. Although this example is hypothetical, expert sensor advisors are being used in proprietary situations today (Koch et al. 1988).

1.7 Remote Sensing

Remote sensing, in the usual sense, refers to data collection by satellite imagery or aerial photography. Because the data collected are voluminous, statistical methods are needed for sampling and analysis, and computers are needed to perform the analyses and to manage and store the data. Integration of multivariate statistical analysis, image processing and expert systems is necessary (Fabbri and Kasvand 1988). As with other methods, analysis of remote-sensing data has been done with microcomputers (Papacharalampos 1988).

1.8 Geological Mapping

In geological mapping, notes may be taken on forms suitable for keypunching for data entry into computers. Alternatively, notes can be taken directly into devices that provide machine-processible data. The disadvantage of these procedures is that they routinize geological mapping, so that data that do not fit the form may be ignored. The advantages are that data are collected systematically and language problems are minimized, as when field work is done by teams of geologists with different languages.

Once the data are in a computer, they can be readily organized and statistically analyzed. By entering the data into a geographic information system (GIS), computer-generated maps can be produced to relate data collected by different geologists or on different traverses. Expert-system techniques can collate variates or rock names, and sorting and indexing become easy.

1.9 Databases

Databases for resources were first developed extensively during the 1970s. Typically, geologists and engineers wrote the specifications for the bases, which were completed by computer and information scientists.

One worldwide database for economic and subeconomic ore deposits consists of more than 10 000 data bits, recording grades, tonnages and geometric data (locations, strikes, dips, overburden thicknesses, etc.) (Wignall and de Geoffroy 1987). Another database for North America contains 4015 deposits; data include name, location (latitude, longitude), major and minor commodities, type of deposit, geological environment and age (Guild 1981).

The US Geological Survey maintains the Computerized Resources Databank (CRIB), which contains some 50 000 records locating and describing mineral deposits. The US Geological Survey also has an exploration geochemical database originally compiled by the US Department of Energy for about 650 000 sites in two thirds of the USA; samples were analyzed for as many as 60 chemical elements and other variables. These are only examples of some of the many computerized databases that are now available (Peters 1987).

The current tendency is to combine databases and geographic information, including topography, drainage and culture, in geographic information systems. Development of these systems has proven to be more difficult than anticipated because of statistical and computational problems.

In a somewhat different category are bibliographic databases, which are required to accumulate the often scattered geological data pertinent for exploration, particularly for areas in developing countries where pertinent old data may be found in scientific journals of other countries, reports of colonial geological surveys, and so on. Identifying and searching these databases requires the participation of specialist librarians in cooperation with geologists who can define the parameters of the search in terms of areas and/or commodities.

Databases require combination by geographic regions, mineralization type, or in other ways. One method is analysis of variance, followed by multiple comparison tests if required to establish subgroups (Wignall and De Geoffroy 1987).

2. Mining Geology

2.1 Geostatistics

Geostatistics is a specialized branch of statistics developed by Matheron (1963) and his students to analyze *regionalized variables* that are linked to one another by distance and direction (or in time). Geostatistics may be considered an extension of the weighting schemes such as inverse distance, long used in ore-reserve calculations, and of trend-surface analysis, developed in the USA and South Africa in the 1950s and 1960s (Clark 1979, David 1977, Journel and Huijbregts 1978).

2.2 Other Uses of Statistics in Ore-Reserve Estimation

While geostatistics developed, so did classical and other statistics for ore-reserve estimation, as computers became more and more available (Krige 1978). The statistical valuation of diamondiferous deposits, a difficult subject because of the extremely low grade of these deposits and their high variability, has become practicable with powerful computers to process data.

2.3 Economic Evaluation of Mineral Deposits

Current economic evaluations of mineral deposits are made with computerized models, to allow for testing various scenarios corresponding to changing conditions over the life of a mine, including the preproduction, production and shut-down stages. Among the variables considered are grade and tonnage of ore, dilution of ore by waste rock, removal of overburden, plant construction cost, cost of production (mining, milling, overheads, etc.), recovery of metals in the concentrator, preproduction development period, metal prices, revenue from sales, interest rates and income tax considerations (depreciation, depletion, etc.). By purposefully selecting variable values from statistical distributions or in other ways, decision makers devise suboptimal plans (Gentry and O'Neil 1984, Peters 1987, Stermole and Stermole 1987).

3. Outlook

Today, computers and statistics are most widely used in exploration and mining geology to develop and process models that embrace the reductionist views most prevalent in twentieth-century science. While the model makers may believe that they apply "objective" rather than "subjective" methods, many fail to recognize that the methods, the commodities for which to search, the areas to be searched, the effort to be expended and other decisions are made purely or in a large part on subjective grounds. For the complexity of mineral exploration, systems models of such workers as Capra (1982) may be more appropriate.

In the future, continued rapid growth in applying statistics and computers in exploration and mining geology is anticipated. The role of Matheron's geostatistics and classical statistics will become more clearly defined. Many additional methods of data analysis, particularly those of the exploratory data-analysis school of Tukey (1977), will be employed widely and be familiar to all practitioners. Enough information will be at hand from completed exploration studies in which statistics and computers played a part to allow appraisal of the models and their refinement.

Bibliography

Agterberg F P 1988 Application of recent developments of regression analysis in regional mineral resource evaluation. In: Chung et al. 1988, pp. 1–28

Botbol J M, Sinding-Larsen R, McCammon R B, Gott G B 1978 A regionalized multivariate approach to target selection in geochemical exploration. *Econ. Geol.* 73: 534–46

Campbell A N, Hollister V F, Duda R O, Hart P E 1982 Recognition of a hidden mineral deposit by an artificial intelligence program. *Science* 217: 927–29

Capra F 1982 *The Turning Point*. Bantam, New York

Cargill S M, Green S B (eds.) 1986 *Prospects for Mineral Resource Assessments on Public Lands*, US Geological Survey Circular 980. USGS, Washington, DC

Chung C F, Fabbri A G, Sinding-Larsen R (eds.) 1988 *Quantitative Analysis of Mineral and Energy Resources*. Reidel, Dordrecht

Clark I 1979 *Practical Geostatistics*. Applied Science, London

Cox D P, Singer D A 1986 *Mineral Deposit Models*. US Geological Survey Bulletin 1693. US Government Printing Office, Washington, DC

David M 1977 *Geostatistical Ore Reserve Estimation*. Elsevier, Amsterdam

Davis J C 1986 *Statistics and Data Analysis in Geology*. Wiley, New York

De Geoffroy J G, Wignall T K 1985 *Designing Optimal Strategies for Mineral Exploration*. Plenum, New York

Drew L J, Schuenemeyer J H, Root D H 1980 *Resource Appraisal and Discovery Rate Forecasting in Partially Explored Regions*. US Geological Survey Professional Paper 1138. US Government Printing Office, Washington, DC

Duda R, Gaschnig J, Hart P 1979 Model design in the PROSPECTOR consultant system for mineral exploration. In: Michie D (ed.) 1979 *Expert Systems in the Microelectronic Age*. Edinburgh University Press, Edinburgh

Fabbri A G, Kasvand T 1988 Automated integration of mineral resource data by image processing and artificial intelligence. In: Chung et al. 1988, pp. 215–36

Gentry D W, O'Neil T J 1984 *Mine Investment Analysis*. American Institute of Mining, Metallurgy and Petroleum Engineers, New York

Griffiths J C 1978 Mineral resource assessment using the unit regional value concept. *J. Math. Geol.* 10: 441–72
Guild P W 1981 *Preliminary Metallogenic Map of North America.* US Geological Survey Circulars 858-A,B. USGS, Washington, DC
Harris D P 1984 *Mineral Resources Appraisal.* Clarendon, Oxford
Howarth R J (ed.) 1983 *Statistics and Data Analysis in Geochemical Prospecting.* Elsevier, Amsterdam
Journel A G, Huijbregts C 1978 *Mining Geostatistics.* Academic Press, New York
Koch G S Jr, Campbell A, Fabbri A, Lanyon W, Pereira H G, Shulman M J, Sinding-Larsen K S, Stokke P R, Toister A, Watney W L 1988 Workshop on mineral and energy resource expert system development. In: Chung et al. 1988, pp. 703–13
Krige D G 1978 Lognormal-de Wijsian geostatistics for ore evaluation. *South African Institute of Mining and Metallurgy Monograph Series, Geostatistics 1.* SAIMM, Marshaltown, Transvaal
Malmqvist K, Malmqvist L 1979 A feasibility study of exploration for deep-seated sulphide ore bodies. In: O'Neil T J (ed.) 1979 *Proc. 16th Symp. Application of Computers and Operations Research in the Mineral Industry.* University of Arizona, Tucson, Arizona
Matheron G 1963 Principles of geostatistics. *Econ.Geol.* 58: 1246–66
Papacharalampos D 1988 GEMS: A microcomputer-based expert system for digital image analysis. In: Chung et al. 1988, pp. 529–42
Peters W C 1987 *Exploration and Mining Geology.* Wiley, New York
Ramani R V (ed.) 1986 *Proc. 19th Symp. Application of Computers and Operations in the Mineral Industry.* Society of Mining Engineers, Littleton, Colorado
Sinclair A J 1976 *Application of Probability Graphs in Mineral Exploration,* Special Volume No. 4. Association of Exploration Geochemists, Rexdale, Ontario
Sinding-Larsen R, Vokes F M 1978 The use of deposit modelling in the assessment of potential resources as exemplified by Caledonian stratabound sulfide deposits. *J. Math. Geol.* 10: 565–80
Singer D A, Kouda R 1988 Integrating spatial and frequency information in the search for Kuroko deposits of the Hokuroku district, Japan. *Econ. Geol.* 83: 18–29
Stermole F J, Stermole J M 1987 *Economic Evaluation and Investment Decision Methods.* Investment Evaluations Corporation, Golden, Colorado
Tukey J W 1977 *Exploratory Data Analysis.* Addison-Wesley, Reading, Massachusetts
Waterman D A 1986 *A Guide to Expert Systems.* Addison-Wesley, Reading, Massachusetts
Webb J S, Thornton I, Thompson M, Howarth R J, Lowenstein P L 1978 *The Wolfson Geochemical Atlas of England and Wales.* Clarendon, Oxford
Wignall T K, De Geoffroy J 1987 *Statistical Models for Optimizing Mineral Exploration.* Plenum, New York

For information in addition to the bibliography, see the proceedings of the series of symposia on the Application of Computers and Operations Research in the Mining Industry (APCOM) that started in 1961 and have been held nearly annually since. The proceedings of the 19th symposium (Ramani 1986) lists the publishers and availability of earlier volumes in the series.

G. S. Koch Jr.
[University of Georgia, Athens, Georgia, USA]

Stockpiling

Webster's Unabridged Dictionary had no listing for "stockpiling" in its 1922 edition, but in 1951 it defined a stockpile as "a reserve supply of an essential raw material—processed food, or the like, accumulated within a country for use during a war-induced shortage." By 1973, "war-induced shortage" had evolved to "a shortage caused by emergency conditions (as war)."

Private firms, especially those involved in mineral-intensive industries, maintain stocks of raw materials both as normal business inventories and as a hedge against shortages, price escalations or other contingencies. Governments also stockpile and the USA maintains an elaborate national defense stockpile for use in an "emergency with respect to the national defense" as well as a strategic petroleum reserve. Stockpiling in the USA is detailed elsewhere (see *Stockpiling in the USA*).

Broadly speaking, stockpiling is only one, albeit important, component of national materials policy. This article deals primarily with the stockpiling of nonfuel strategic minerals in six industrialized countries that are mainly raw-materials importers: France, the FRG, Japan, Sweden, Switzerland and the UK. Other countries have also considered various approaches to the problem. Spain, for example, planned in 1979 to develop a US$350 million industry-held, government-financed stockpile of critical nonfuel minerals, but has not yet acted because of budget limitations. Other smaller countries have elected to rely on world markets or on industrial suppliers in their larger neighbors, particularly within the European Economic Community (EEC). There does not, however, seem to be much synchronizing of stockpile-type programs even within the Community. There are sharing provisions for oil under the International Energy Agency accords, but it is unclear whether and to what extent any one country's stocks of nonfuel minerals might be available to an ally under emergency conditions.

Most of the Communist nonmarket economies also have significant stockpiles. Some are incidental to normal and supplemental inventories at state enterprises: some are stored at production sites, some held pending favorable export marketing opportunities, and some maintained as mobilization or emergency reserves. Soviet mineral policy, for example, continues to pursue the goal of maximum self-sufficiency even at uneconomic costs. Although there is an extensive literature on the mineral industries of China and the

USSR, there is insufficient publicly available information on the types, purposes and holdings of stockpiles in the major Communist countries to warrant coverage in this article.

1. Overview

The USA maintains by far the world's largest strategic stockpile (if the state holdings of Communist countries are disregarded), largely because it inherited one at the end of World War II, but other countries have outstripped the USA in other minerals vulnerabilities risk management techniques. France and the UK have developed relationships with their former colonies as material suppliers through the Lomé Conventions and their related Stabex and Sysmin programs. Japan likewise has a network of supplier relationships tied up through investments, long-term contracts, aid programs, price stabilization schemes and the like. Access to resources in the postcolonial era has generally played a much larger role in the foreign policies of these countries than has been the case for the USA. Like Switzerland and the Netherlands, for example, the FRG has dozens of bilateral investment treaties with developing countries providing for national treatment and other standards, while the USA is still struggling to initiate a few such treaties. However, the FRG's plan for a cooperative government–industry stockpile has been shelved for the time being because of budget limitations and disagreements over control.

On the other hand, France and the UK maintain modest governmental (or quasigovernmental) stockpiles of critical nonfuel minerals, and Japan, Sweden and Switzerland have what might be considered "private" stockpiling policies. Japan operates through associations of private firms which maintain inventory levels under the guidance of the Ministry of International Trade and Industry (MITI) and receive government-subsidized financing in which stocks are pledged to the Metal Mining Agency. These serve both as a hedge against shortage and a cushion to maintain processing capacity during slack times. The Swedish government maintains an elaborate national economic defense mechanism, including extensive stocks of food, supplies, machines, minerals and fuels. In addition, its income-tax treatment of business inventories (partly designed to avoid taxing inflationary increases in value) allows accelerated tax write-offs for reduced inventory valuation, encouraging higher levels of stockholding for private firms.

Switzerland has both mandatory stocks financed by the government and voluntary stockpiles which are facilitated by government credit arrangements, of which half can be retained by the holder and half are subject to emergency allocation. Their combined value is about two fifths that of the current US national defense stockpile—in a country that has only 3% of the US population!

Both the Organization for Economic Cooperation and Development and the EEC have sought to coordinate members' efforts in this field. The former is in the process of gathering and comparing data, but its eventual report is unlikely to be made public. The Community has also performed useful analyses but is blocked because some countries, notably France, consider policies in this field a purely national responsibility. France, however, did play a key role in the negotiations of the first and second Lomé Conventions between the Community and former colonies and dependencies of its member countries.

2. France

Faced with a dearth of domestically produced nonfuel minerals, France's mineral-intensive industries (which contributed 5%) of its gross domestic product in 1980) rely primarily upon foreign producers for an adequate supply (55%) of their principal manufacturing inputs. In the light of the recently acquired independence of many of the mineral-rich countries (including its own former African colonies), the influence of the Organization of Petroleum Exporting Countries (OPEC) on energy supplies and prices, and several contemporary commodity shortages, the government of France undertook a major analysis of that country's vulnerability to sudden supply disruptions of raw materials. The strategic analysis, geological surveys and domestic mineral inventories were published in 1978 (Bureau des Recherches Géologiques et Minières 1978).

Consistent with the general recommendations put forth in that report, industrial policymaking as outlined in the government's seventh five-year plan (from 1976 to 1980) was revised to pay more attention to the economic benefits attainable through better management of imported raw materials. This "calculable risk" study of mineral supply shortages illustrates the combination of foreign-policy considerations and economic and political forces that drive international minerals markets.

To this end, France has pursued several options, including: (a) improving conservation and recycling through technology developed in government and business research programs; (b) increasing domestic production through incentives to mining companies for long-term financial and commercial exploration; (c) promoting economic, political and scientific (transfer) agreements with France's major suppliers of raw materials, as demonstrated by the Lomé Convention; and (d) to a lesser extent, stockpiling. All these measures serve to reduce dependence, ensure availability (judged more important than price) and offset foreign-exchange costs.

Of the major Western European countries, France is one of the few to devote substantial financial resources to the creation and maintenance of a strategic stockpile. Hampered by budget constraints in the last few years, however, expansion plans devised in 1975 have not lived up to expectations, so that what was originally to be a US$681–851 million materials stockpile has reached only US$340 million. Controlled by the GIRM (a centralized purchasing agent of the Ministry of Industry), stockpiled minerals are purchased at favorable market prices and sold to domestic buyers when market supplies dry up.

Significantly, the French assign degrees of vulnerability. The "very great" category includes silver, platinum, diamonds, phosphates, zirconium, titanium, cobalt and vanadium. The "great" class includes antimony, copper, manganese, molybdenum and tungsten. "Medium" includes bauxite, chromium and tin, while France considers itself merely "weak" in iron, nickel, lead, zinc and potash.

Determined not to establish a special budget for the stockpile or a specialized purchasing agent, and opposed to the financing of commodity stockpiles through bank loans, the French government addressed its stockpile shortfalls with the creation of the CFMP (Caisse Française de Matières Premières) under the jurisdiction of the Ministry of Industry for issuing and administering bonds. The intention was to use funds raised in this manner to increase holdings in copper, lead, tungsten and chromium—none, curiously, in the "very great" vulnerability category. So far CFMP has issued 500 million French francs in 14% ten-year bonds, which were fully subscribed. Long-term goals, revised in 1980, envisioned spending up to three billion French francs for the stockpile, but overall progress remains inconclusive. Only 169 million were earmarked for purchases in the 1980 budget, a figure which has undoubtedly been further reduced under the current government. Therefore, the same tensions between planners and budget officials are evident as have characterized the US nonfuel minerals policy in the past three decades.

3. *Federal Republic of Germany*

The FRG consumes over 10% of the world's annual mineral production but produces less than 1%; thus its dependence level is somewhat above that of the UK and France. Moreover, because of common EEC trade policy, FRG refined-metal exports are shipped primarily to other EEC countries, thus distorting the dependency of Western European nations on imported raw materials.

The economic policy of the FRG is to rely entirely on market forces in order to obtain the socially optimum supply and demand equilibrium. Eschewing large-scale, statist-type industrial policymaking, the German government prefers to treat different sectors of industry as separate branches, which merit special attention only if they affect the country's security or international competitiveness (either directly or through forward linkages). While focusing upon industries of the future such as aerospace, electronic data processing and nuclear energy, the FRG also follows a national tradition of high standards in science and technology, including metallurgical technology.

Although worldwide leaders in minerals processing and refining, German firms have not been able to transfer this dominance into overseas mining operations, where they have been overshadowed by American, British, French and Japanese firms. At the crux of German concern over raw materials are long-term imbalances in the development of supply and demand (owing to insufficient investment for exploration and development) and the growing appetite of the FRG's mineral-intensive industry.

The German government can more easily deal with the problem of overseas investment. Indeed, it has been a leader, along with the Netherlands and Switzerland, in negotiating bilateral investment treaties with developing countries, and well ahead of the USA in this respect. However, a second supply problem that deeply troubles Germans is the politically motivated supply shortfalls that might be caused by "other OPECs" and general third-world intervention in resource markets, for these factors are essentially out of their hands.

Despite their declared position of nonintervention in commodity markets, concern over such nonmarket forces has stirred considerable debate in the FRG over whether to stockpile strategic minerals. The 1979 proposals for both governmental and private-sector stockpiling efforts have been shelved for budgetary reasons, but may be reconsidered as the economy recovers. The government and industry never agreed on who should have operating control, but the essence of the scheme proposed was for voluntary participation of companies in a consortium with the government. Each company would receive incentives in the form of interest-free loan and tax credits to add 3–12 months to its normal inventory of materials such as chromium, cobalt, manganese and vanadium and store it until an emergency, when the extra stocks would be split 50–50 between the companies and the government. Costs would have run to about US$100 million per year for five years.

4. *Japan*

No other country has such an absolute dependence upon the importation of raw materials as Japan. In contrast to the early periods of Japan's modern history, when minerals such as base metals, gold and even coal were major export earners, the burgeoning demands of Japanese "miracle" industries, coupled with depletion of the scattered home reserves, have made it necessary to import over 90% of the nation's crude nonfuel minerals requirements.

The Japanese acceptance of state involvement in industrial affairs, shared by government and corporate executives alike, has ensured that formulation and periodic reevaluation of raw materials management policies, objectives and philosophies are prime elements of industrial policy. Furthermore, cooperation between the two principal actors in the procurement of overseas raw materials permits Japan to exercise a considerable degree of monopsonist power when negotiating with foreign producer nations. Indeed, the strength and diversity of equity investments, loans and grants, long-term production contracts, price-stabilization agreements and development assistance projects that Japan has established with a variety of developed and developing country mineral suppliers have led its industrial policymakers to feel relatively secure in the area of raw materials despite their overall dependence.

In essence, Japanese industry has accepted the high costs of capital investments in offshore exploration and mining facilities, and of stabilized prices through long-term contracts, in order to obtain reliable supplies to fuel its sustained economic growth.

Although it has invested management efforts and funds in offshore suppliers of strategic minerals, the Japanese government has also participated in three nonferrous-metal stockpiling associations: (a) the Metallic Minerals Stockpiling Association, for various forms of refined copper, zinc and lead; (b) the Light Metal Stockpiling Association, for primary aluminum; and (c) the Special Metals Stockpiling Association, for nickel, chromium, tungsten, cobalt and molybdenum (including ore and intermediates). Strategically, these stockpiling associations provide the Japanese with a cushion to fall back on in their negotiations with raw-material suppliers. However, the more explicit objectives of the stockpiles are less strategic in nature.

First of all, by protecting the economies of minerals-supplying countries when demand from Japanese industries falls below projected levels, stockpiling of surplus contracted materials enables Japan to ensure continued flows, avoid international friction and preserve the integrity of long-term contracts. Conversely, in the event of a severe supply shortage, timely releases of stockpiled material to the consuming industries would permit production to continue with reduced or no loss of export earnings. Finally, the stockpiles have provided relief to the domestic smelting and refining industry, insofar as excess production is purchased from them, stocked during surplus periods and returned to them when they must meet increased demand.

Japan has made an explicit decision to use its stockpiles to ensure price and supply stability, maintain assured gross domestic product (GDP) and sustain relationships with key suppliers, and it is willing to act to limit vulnerability to disruptions in overseas supply. The Japanese have also considered more direct government stockpiling of certain critical materials beyond those covered by the associations, utilizing facilities of MITI and the Metal Mining Agency of Japan; but at the time of writing, details on what has been done are not available. This is one of the areas in which Japan provides a textbook example of business, government and labor cooperation to meet a major national problem.

5. Sweden

Given the traditional neutrality of Sweden, as demonstrated in World War II, economic defense is considered a part of the country's national security posture. In what are called "storage villages," the government stores, maintains, rotates and upgrades supplies of food, fertilizers, clothing, medical supplies, chemicals, and solid and liquid fuels as well as metals, machines and tools. Sweden spends 300–500 million kronor (about US$ 65 million) annually on these activities, not counting petroleum. Total capital expenditures to date run to several billion kronor.

The National Board of Economic Defence maintains stockpiles of a large number of mineral raw materials to be used in the event of war or a blockade. In addition, there are critical peacetime stockpiles of chromium ore, manganese ore, cobalt and vanadium, to be used if supply lines are disrupted in peacetime. A recent government study proposed that molybdenum, tungsten, nickel, titanium, niobium, tantalum and borates be added to the peacetime stockpiles, and that critical peacetime stockpiles of manganese, chromium and cobalt be expanded. Peacetime stockpiles equivalent to three-months' imports were recommended.

Strategic stockpiling in Sweden was initiated in the 1930s to strengthen national security in the event of war. After World War II, the Korean crisis led to a further buildup of such stockpiles. Present strategic stocks, administered by the National Board of Economic Defense, are calculated to enable the country to cope with a wartime cutoff for about one year.

Minerals stocked under this system (strategic stockpiles in a traditional sense) include chromium, cobalt, nickel, manganese, vanadium, titanium, molybdenum and tungsten (ore, alloys and metal) as well as tin, zinc, aluminum and a number of industrial minerals.

In addition, Sweden is in the process of building up stocks for needs that may arise in a peacetime crisis. The purpose is to maintain, as far as possible, normal levels of production, employment and exports. The new program includes certain alloy metals such as chromium, cobalt, manganese and vanadium. The stocks will correspond to normal use of chromium and cobalt for two months and one month for manganese and vanadium.

Recent Royal Commissions on defense and minerals policies have suggested that the peacetime program be developed further. For instance, stocks of chromium, vanadium, manganese, cobalt, molybdenum, nickel and titanium should be increased to three-months' normal use. It has also been suggested that private industry might be induced to carry a larger share of stockpiling for peacetime purposes.

In addition to these economic defense activities, Sweden has a de facto private stockpiling system in the way in which materials inventories are treated under Swedish tax laws. Inventories are valued at purchase (FIFO) or replacement value, whichever is lower, and can be written down by as much as 60% annually after adjustment for obsolete or unmarketable items under various procedural rules. Taxable income is also reduced by the amount of this write-off in inventory valuation; given Sweden's high tax rates, some corporations use the tax savings to purchase larger stocks than they would otherwise hold. Others prefer to pass the tax savings on to shareholders through higher dividends.

6. *Switzerland*

Switzerland, with another long and fiercely maintained tradition of neutrality and a dearth of most natural resources, stockpiles imported foodstuffs to last one year and fuel for at least six months. In addition, the Swiss have established a system of voluntary stockpiling programs administered by the Federal Office for Economic Defense Measures. Importers of strategic raw materials and industrial items are encouraged to hold stocks corresponding approximately to a normal twelve-month supply. As an incentive to maintain these voluntary levels, under the government stockpiling program importers are offered substantially lower interest rates, tax rebates and preferential treatment in times of war or other emergencies. Items of primary interest to Switzerland are iron and steel, nonferrous metals, chemical raw materials, plastics and textiles. Reportedly, defense stocks of these commodities are high.

Expenditure for national defense in Switzerland currently amounts to 2.2% of its gross national product, of which military items represent 1.9% and civil defense takes 0.3%. This must be regarded as low, even if allowance is made for a certain cost reduction stemming from unpaid voluntary services under the militia system. Procurement outlays have increased tremendously and are therefore being staggered.

The system of stockpiling strategic reserves has been maintained and strengthened throughout the years following World War II. Its basic features are as follows. For essential goods on which Switzerland has a critical import dependence, compulsory stocks are maintained by private firms, grouped into "consortia" according to economic sectors, such as food, fodder, fertilizer, seeds, fuel and detergents. The Swiss maintain safety stocks in decentralized warehouses for normal nine-months' requirements of the total population. The cost of financing and handling is covered by a special import levy, based on consumption level prices.

In addition, industry is encouraged to form voluntary stockpiles in essential raw materials and intermediates, such as iron and steel, nonferrous metals, cotton, antibiotics, pulp and fuel rods. These stocks can be financed by bank credits at favorable rates, with federal guarantees, up to 90% of purchase value. For a long time, the rule was to value them at prewar price levels (i.e., constant prices). Now this level is usually negotiated, both for federal and cantonal corporation taxes. Strategic inventories are thus treated preferentially. The capital risk rests with the individual firm. The value of these safety stocks, compulsory and voluntary inventories taken together, amounts to 10–12 billion Swiss francs (roughly US$5 billion). This is equivalent to three times the Swiss current expenditure on national defense, showing the importance the Swiss assign to their stockpiling. (For comparison, the US strategic stockpile was valued at US$10.9 billion in 1982.)

Through careful economic management, the social utility of such reserves in goods can potentially be much higher than might be revealed by current consumption and import patterns. If an international crisis creates critical shortages through cutoffs, boycotts and the like, the monetary reserves of the banking system alone cannot hedge against material distortions in supply. The total safety stocks amount to approximately a quarter of official monetary reserves. The real needs in an emergency may have to be reduced drastically and rapidly to survival levels.

Through planning and preparing basic responses, the Swiss hope to gain valuable time for adjustment, including ensuring their continued autonomy through indigenous production via the Swiss food plan, largely based on the success of Plan Wahlen during World War II. All these emergency preparedness measures are linked and include provisions for fuel, which is critically important to a modernized economy, especially under wartime or siege conditions.

The Swiss have pragmatically solved one of the problems that plague stockpile planning: in an emergency, who gets what? The mandatory stocks are available under special conditions for disposition on government order, but 50% of the voluntary stocks remains with the holder, while the other 50% is subject to emergency allocation by the government. This could serve as a useful example for the USA.

7. *United Kingdom*

The UK, the founder of the industrial revolution, is without a significant raw materials base and has long been a highly dependent consumer of strategic minerals, this being a major factor in its long colonial history. Confirmation of the British government's intention to set up a strategic minerals stockpile came on February 14, 1983, in the form of a House of Commons statement. It was initially clouded in secrecy, apparently to take advantage of low prevailing prices and avoid disruption of markets, and there are still only speculative press reports as to the scope of the program.

The British divide strategic materials according to criticality, which refers to the role which a specific mineral commodity plays in the country's reliance on imported raw materials. The House of Lords Select Committee Report on Stockpile Minerals (House of Lords 1982) identified four minerals of prime strategic importance: chromium, manganese, phosphate rock and platinum. It also listed eight minerals less important strategically (because of substitutability) but still supplied by sources outside the EEC: antimony, cobalt, molybdenum, nickel, niobium, tantalum, titanium and vanadium. Costs are not known, but one earlier British plan would have involved £100 million in subsidies to private-sector companies.

Focusing primarily on southern African imports, the UK stockpiling scheme is designed to mitigate their short-term disruption by targeting levels equivalent to a three-months' supply. The possibility of disruption of these imports has long been recognized but never seriously acted upon, owing to severe budget restrictions. Apparently, given the increasing ability of the USSR to intervene both in source countries and on the sea lanes, past strategies (such as development of strategic minerals within the EEC, aid to third-world countries as in the Lomé Convention's Sysmin scheme and promoting recycling and substitution) have been deemed insufficient to meet real danger.

Because until recently there was no British government-run strategic stockpile or privately operated reserve system, it is unclear why both the government and (to a lesser extent) mineral-intensive industries did not perceive the need to prepare for a war or peacetime crisis. Unlike the French, whose dependence most resembles their own, the British viewed threats to overseas mineral supplies primarily as a foreign-policy problem. As long as the large British mining houses continued exploration and development in the supplier countries, government efforts other than maintaining relatively friendly relations were not deemed necessary.

This view accorded with Prime Minister Thatcher's effort to cut back on government involvement in the national economy, but the outlook may be changing as Britain's dependence and vulnerability increase. Because the mining houses are mainly living on past investments and not looking to the future, supply interruptions of "strategic minerals" are of mounting concern in the UK.

8. Summary

The shortages and price escalations during the early 1970s, especially the oil exporters' ability to manipulate petroleum prices to unprecedented levels, gave the major impetus to stockpiling policies in the post-World-War-II period. However, the outbreak of minor crises in the Middle East, Africa, the Falklands and Central America, plus instabilities in the domestic situation of many key supplier countries, uncertainties over the future of southern Africa and a propensity on the part of the USSR to be involved in resource issues and markets for strategic and political as well as commercial reasons, will continue to make stockpiling options of interest to policymakers in the key industrial nations.

9. Acknowledgements

The author is indebted to a translation of an article *Emergency Stockpiling in Sweden* by K-G Bartoll of the National Board of Economic Defense; and also thanks Dr Carl Eugster of Basle, Switzerland, for his unpublished analysis of strategic preparedness and stockpiling in that country.

See also: Critical and Strategic Materials; Stockpiling in the USA

Bibliography

Bureau des Recherches Géologiques et Minières 1978 *Annual Report*. Bureau des Recherches Géologiques et Minières, Paris

Holt J L, Stanley T W 1983 *Nonfuel Mineral Stockpiling Policies: The Private Stockpiling Alternative*, Contemporary Issues No. 6. International Economic Studies Institute, Washington, DC

House of Lords 1982 *20th Report of the Select Committee on the European Communities, Strategic Minerals*, House of Lords Papers and Bills, Session 1981–82. HMSO, London

International Economic Studies Institute 1977 *Raw Materials and Foreign Policy*. Westview, Boulder, Colorado

Kaplan J J, Stanley T W 1977 *Dependence and Vulnerability: The United States and Imported Raw Materials*, Contemporary Issues No. 3. International Economic Studies Institute, Washington, DC

Minerals Yearbook, Vol. 2, *Area Reports International*. US Government Printing Office, Washington, DC

Stanley T W 1982 *A National Risk Management Approach to American Raw Materials Vulnerabilities*, Contemporary Issues No. 5. International Economic Studies Institute, Washington, DC

US Congress, Office of Technology Assessment 1985 *Strategic Materials: Technologies to Reduce U.S. Import Vulnerability*. US Government Printing Office, Washington, DC

US Congressional Budget Office 1983 *Strategic and Critical Nonfuel Minerals: Problems and Policy Alternatives*. US Government Printing Office, Washington, DC

US Federal Emergency Management Agency (semiannual) *Stockpile Report to the Congress*. US Government Printing Office, Washington, DC

US General Accounting Office 1982 *Foreign Governments' Stockpile Policies*, Report to the Chairman, Preparedness Subcommittee of Senate Armed Services Committee (GAO/ID83-16). US Government Printing Office, Washington, DC

T. W. Stanley
[International Economic Studies Institute, Washington, DC, USA]

Stockpiling in the USA

Stockpiling food in times of plenty for use in future times of need antedates humanoid origins. Formal guidance for food stockpiling was set forth by the highest authorities in the "2-year stockpile rule," as divined by Joseph four millenia ago: one fifth of the produce of the land set aside in each of the seven good years (Genesis Chap. 41). In later centuries stockpiles not only of food but also weapons, including the ingredients for chemical warfare, were accumulated in secure places. *Histoire d'une Forteresse*, written shortly after the Franco-Prussian War by the noted architect and military engineer Colonel E Viollet-le-Duc, pointed out the elaborate precautions taken in earlier times to have stocks of food, water and weapons on hand in major fortified areas, to which the peasantry in time of war could retreat to be armed to repel invaders.

Isaiah prophesied the beating of swords and spears into plowshares, but the reverse process often also took place. After a major war, bronze cannon were melted down and recast as heavy fences. Later such bronze objects were remelted for cannon again, thus serving useful purposes in any season of man's affairs. In World War I many materials and fuels were in short supply. A variety of controls and coordinated allied efforts were needed to channel supplies to priority programs. Supplies of materials were expanded and new industries created. For example, synthetic nitrates were produced for explosives, forming the basis of the peacetime nitrate fertilizer industry.

Again, in World War II, materials and fuels were in short supply. The Axis nations accumulated supplies of strategic materials before the war, which made up in part for their natural resource shortages. In his book *Wartime Economy of the USSR*, Voznesensky, Deputy Premier and Chief of the State Planning Commission during World War II, stated that a program for the accumulation of state and mobilization reserves—especially oil products, ferrous metals and foodstuffs—was planned and implemented before the start of World War II under direct instructions from Stalin, and that such stockpiles played a significant role in the Soviet war effort. He pointed out that in modern warfare, victory depends upon reserves of production capacity, manpower, raw materials and other commodities.

1. Strategic Considerations

Assessments of vulnerability must consider international security arrangements. Western European nations and Japan are even more dependent upon imports for many strategic materials than is the USA: US net import reliance of selected minerals and metals is shown in Table 1. Also imported were substantial quantities of antimony (from China, South Africa, Bolivia and Mexico), bismuth (from Mexico, Peru, the UK and Canada), gallium (from Switzerland, France, the FRG and the UK), iodine (from Japan and Chile), ilmenite (from Australia, Canada and South Africa), mercury (from Spain, Algeria, Japan and Turkey), rhenium (from Chile, the FRG and Italy), rutile (from Australia, Sierra Leone and South Africa) and vanadium (from South Africa, the EEC, Canada and Austria). In contrast, the Comecon bloc is largely self sufficient—the result of a long-pursued policy of autarky on the part of the USSR, which, with an area of 22 million km^2, is ideally situated for such a policy. Further, not just import dependence but also many other factors must be considered. Materials on hand in strategic stockpiles, other government stocks and industry stocks are often significant. The use of flywheels to deliver power to offset sudden increases in load is accepted practice in machinery design, and the use of surge bins is standard practice in materials handling. Expanding such analogies to national economies points to the desirability of maintaining larger than normal working inventories to deal with unexpected interferences with normal supplies which may be occasioned by wars, politicoeconomic disruptions or such natural disasters as earthquakes, landslides, hurricanes and floods.

Escalating costs and higher interest rates in recent years have tended to discourage the private sector from maintaining large inventories. Tax incentives for the maintenance of larger than normal working inventories have been in effect in Sweden for a number of years. In the Swedish system a large fraction of the acquisition costs of raw materials can be written off in computing taxable income for the year in which materials are acquired. Such incentives can cover the wide variety of materials in everyday use and permit informed judgments by numerous buyers and sellers, rather than requiring direct market intervention by government entities. Switzerland, too, is among countries providing some tax incentives to encourage holding of inventories.

Several Western European nations have recently considered stockpiling imported materials, but the costs have been a major deterrent to most. In Japan, the Ministry of International Trade and Industry (MITI) sometimes helps to finance major metal stockpiles, but these are more in the nature of assistance to domestic smelters to carry stocks in periods of economic downturn rather than strategic stockpiles per se. Stockpiles in nations with centrally planned economies are considered state secrets. As a consequence of the dislocation of oil markets in the 1970s, many nations adopted a variety of programs to increase oil stocks.

2. Early Stockpiles in the USA

The USA was early to recognize the importance of stockpiles in the form of known mineral deposits in the ground. Commodore Perry's Far East naval expedition of 1852–54 had specific instructions to look for

Table 1
US net import reliance for selected nonfuel minerals materials

Material	% imported 1987	Major foreign sources 1982–87
Niobium	100	Brazil, Canada, Thailand, Nigeria
Graphite	100	Mexico, China, Brazil, Madagascar
Manganese	100	South Africa, France, Brazil, Gabon
Mica (sheet)	100	India, Belgium, France, Japan
Strontium	100	Mexico, Spain, China
Yttrium	100	Australia
Bauxite and alumina	97	Australia, Guinea, Jamaica, Surinam
Tantalum	92	Thailand, Brazil, Australia, Malaysia
Diamonds (industrial)	89	South Africa, UK, Ireland, Belgium, Luxembourg
Platinum-group metals	88	South Africa, UK, USSR
Fluorspar	88	Mexico, South Africa, China, Italy
Cobalt	86	Zaire, Zambia, Canada, Norway
Tungsten	80	Canada, China, Bolivia, Portugal
Chromium	75	South Africa, Zimbabwe, Turkey, Yugoslavia
Nickel	74	Canada, Australia, Norway, Botswana
Tin	73	Brazil, Thailand, Indonesia, Bolivia
Potash	72	Canada, Israel, GDR, USSR
Zinc	69	Canada, Mexico, Peru, Australia
Cadmium	66	Canada, Australia, Mexico, FRG
Barite	63	China, Morocco, India, Chile
Silver	57	Canada, Mexico, UK, Peru
Asbestos	51	Canada, South Africa
Gypsum	37	Canada, Mexico, Spain
Silicon	33	Brazil, Canada, Norway, Venezuela
Iron Ore	28	Canada, Brazil, Venezuela, Liberia
Copper	25	Chile, Canada, Peru, Zaire
Aluminum	24	Canada, Japan, Venezuela, Brazil
Cement	20	Canada, Spain, Mexico
Iron and steel	19	EEC, Japan, Canada, Republic of Korea
Lead	15	Canada, Mexico, Peru, Australia, Honduras
Salt	12	Canada, Mexico, Bahamas, Chile

Source: US Bureau of Mines 1988

deposits of high-quality steam coal in the far Pacific. In later years, the Army and the Navy were granted preference purchase rights to coal in Alaskan deposits and before World War I the USA established naval petroleum reserves in the ground in California and Wyoming, and naval oil shale reserves in Colorado and Utah.

As early as 1921 the US War and Navy Departments were interested in a program for the stockpiling of strategic and critical materials. During the period 1933–36 these departments and the National Resources Board made recommendations concerning defense stockpiles, but it was not until the Naval Appropriation Act of 1938 that US$3.5 million was made available for stockpiling, followed by US$0.5 million in financial year 1939 and in 1940. With these funds the Navy purchased small stocks of tin, ferromanganese, tungsten ore, chromite, optical glass and manila fiber. In late 1938 and early 1939 the Army and Navy Munitions Board, aided by government departments, made studies leading to Public Law 76-117, June 7, 1939—the first legislation for the primary purpose of establishing stockpiles of strategic and critical materials for national defense, and also authorizing exploration and development of domestic resources by the Department of the Interior. Prewar appropriations under this Act were US$10 million in August 1939, US$12.5 million in March 1940, and US$47.5 million in June 1940. Also in August 1939 the Commodity Credit Corporation of the Department of Agriculture was authorized to exchange surplus US agricultural products for foreign strategic and critical materials.

However, by mid-1940 it became evident that the 1939 Stock Piling Act, designed primarily for acquisitions in a period of relative peace, was inadequate. Instead, stimulation of the flow of domestic and foreign materials directly into munitions and essential industrial production was required. Accordingly, on June 25, 1940, the Reconstruction Finance Corporation (RFC, which dated from 1932) was given broad powers and vastly greater financial authority. The

RFC then proceeded to charter the Metals Reserve Company, the Defense Plant Corporation and the Defense Supplies Corporation to acquire, store, produce and sell strategic and critical materials. Stockpiles were totally inadequate at the start of US active participation in World War II. Some of the larger holdings were: 290 000 t of chromite, 815 752 carats of industrial diamonds, 523 900 t of manganese ore, 818 500 kg of quartz crystals, 50 842 t of tin, 6574 t of tungsten ore and 80 466 t of zinc concentrates.

During World War II, shortages of materials required major priorities and allocations programs, conservation and use limitation directives, domestic and foreign supply expansion programs and allocations of ships to priority cargoes. Mindful of the problems encountered in the war, the Army and Navy Munitions Board in March 1944 adopted the following definition: "Strategic and critical materials are those materials required for essential uses in a war emergency, the procurement of which, in adequate quantities, quality, and time is sufficiently uncertain for any reason to require prior provision for the supply thereof." Also during World War II, extensive cooperative measures with allied nations were undertaken to expand supplies in strategically accessible nations and to move them to high-priority production programs.

3. Post-World-War-II Planning

After the end of World War II, increased attention was paid to defense planning and industrial mobilization readiness. In 1946 the US Congress reviewed and strengthened the Stock Piling Act and materials were acquired through direct purchases, transfers of war surpluses, procurement through use of foreign counterpart funds generated by US foreign aid programs and barter of US surplus agricultural commodities. The legislative intent of the 1946 Stock Piling Act was much broader than just acquiring strategic and critical materials for a national stockpile. A second major purpose of the Act was to encourage the conservation and development of sources of these materials within the USA.

In 1947 the National Security Act established the National Security Council (NSC), the Central Intelligence Agency (CIA) and the National Security Resources Board (NSRB). The NSRB was charged specifically "to advise the President concerning the coordination of military, industrial, and civilian mobilization, including . . . programs for the effective use in time of war of the Nation's natural and industrial resources . . . ; policies for unifying, in time of war, the activities of Federal agencies and departments engaged in or concerned with . . . materials; the relationship between potential supplies of, and potential requirements for, . . . resources . . . ; (and) policies for establishing adequate reserves of strategic and critical material, and for the conservation of these reserves . . . " The importance of appropriate planning with US allies led in 1949 to the formation of the Joint United States–Canada Industrial Mobilization Committee to coordinate plans for industrial mobilization "in order that the most effective use may be made of the productive facilities of the two countries."

4. Korean War Activities

The start of the Korean War on June 25, 1950, found US strategic and critical materials stockpiles valued at only US$1.6 billion, compared with stockpile goals valued at US$4.1 billion. It soon became apparent that the government needed a much broader range of authorities to deal with the emergency. Consequently, Title I of the Defense Production (DP) Act of 1950 provided specific authority for priorities and allocations and Title III provided broad authority for expanding supplies of materials, making specific provisions for:

(a) loans or guarantees of loans for exploration, development and mining of strategic and critical metals and minerals;

(b) purchases of or commitments to purchase metals, minerals and other materials, for government use or resale;

(c) the encouragement of exploration, development and mining of strategic and critical minerals and metals; and

(d) the development of substitutes for strategic and critical materials.

Coordination with allied nations took place through the International Materials Conference, which, through a number of committees, stimulated needed production and assisted in implementing allocation programs.

The Defense Production Act of 1950, as amended, has long been the basis for defense mobilization efforts and will remain in effect through to September 30, 1989. The creation or enlargement of domestic productive capacity, if it can be done at reasonable cost, should in most cases provide greater flexibility than stockpiles. In fact, 1 t of domestic productive capacity is considered equal to 3 t of stockpiled material under current stockpile planning. DP Act supply expansion programs as of June 30, 1956, reached US$8.4 billion in gross transactions with probable ultimate net cost of US$0.9 billion. Major materials programs in US$ billion included:

aluminum	1.5	tin	0.2
titanium	0.9	molybdenum	0.2
rubber	0.9	magnesium	0.1
nickel	0.9	cobalt	0.1
copper	0.7	niobium–tantalum	0.1
manganese	0.4	other materials	0.4
tungsten	0.4	(more than 50)	

In a few years these programs doubled US aluminum production, increased US copper mine capacity by a quarter, initiated US nickel mining, created the titanium industry, quadrupled US tungsten mining and greatly expanded the world niobium–tantalum mining and processing industries as well as expanding supplies of many other materials for production needs and stockpiles. The latest major nonfuel use of the DP Act was a US$83 million loan in 1967 in the Vietnam War period to facilitate the Duval Sierrita copper mine development.

Also in the Korean War period, accelerated amortization ("rapid writeoffs" of capital expenditure), authorized by Sect. 168 of the Internal Revenue Code, was a powerful device to encourage capital investment in areas of national concern. This incentive alone was sufficient to add nearly one quarter to US steel capacity. In the period from October 1950 through December 1956, industrial expansion valued at US$37 billion was certified, of which 61%, or US$22 billion, was eligible for rapid amortization. The total amounts in US$ billion then certified in materials-related industries were as follows:

mining	2.2
primary metal industries	5.7
chemical and allied products	2.9
products of petroleum and coal	3.0
pipeline transportation	1.4

Use of the US tax laws to encourage domestic production of strategic minerals is not limited to periods of war or national emergency. As a general rule, minerals considered more strategic are allowed higher percentage depletion rates. Exploration for minerals is also specifically encouraged by the tax laws.

5. *Post-Korean-War Activities*

In 1954 the Congress created a "supplemental stockpile" of strategic materials to be acquired by barter of surplus agricultural commodities. Several billion dollars of materials were acquired by this route and were later merged into the total government stockpiles.

Under the 1946 Stock Piling Act, administration policy in the period from 1946 to 1958 based stockpile planning on a 5-year emergency, from 1958 to 1973 on a 3-year emergency, from 1973 to 1976 on a 1-year emergency and from 1976 to the present again on a 3-year emergency, following a major interagency review coordinated by the then Federal Preparedness Agency (FPA), successor to the NSRB. The FPA was subsequently in 1979 merged into the newly created Federal Emergency Management Agency (FEMA). FEMA currently oversees stockpile management, while the General Services Administration (GSA) buys, sells, rotates and stores stockpile materials.

In 1979 the new Strategic and Critical Materials Stock Piling Revision Act of 1979, Public Law 96-41, reaffirmed the need for stockpiling, conservation and development of domestic sources. The Stock Piling Act states " . . . that the natural resources of the United States in certain strategic and critical materials are deficient or insufficiently developed to supply the military, industrial, and essential civilian needs of the United States for national defense . . . " and that the purpose of the Act is " . . . to provide for the acquisition and retention of stocks of certain strategic and critical materials and to encourage the conservation and development of sources of such materials within the United States and thereby to decrease and to preclude, when possible, a dangerous and costly dependence by the United States upon foreign sources for supplies of such materials in times of national emergency." The Stock Piling Act further provides that "the purpose of the stockpile is to serve the interest of national defense only and is not to be used for economic or budgetary purposes" and that "the quantities of the materials stockpiled should be sufficient to sustain the United States for a period of not less than 3 years in the event of a national emergency."

The Act provided definitions as follows: "For the purposes of this Act: (1) the term 'strategic and critical materials' means materials that (A) would be needed to supply the military, industrial, and essential civilian needs of the United States during a national emergency, and (B) are not found or produced in the United States in sufficient quantities to meet such need. (2) The term 'national emergency' means a general declaration of emergency with respect to the national defense made by the President or by the Congress." Section 8 of the new Act also reaffirmed the injunction to develop domestic resources, assigning that responsibility directly to the President (a trend in more modern legislation), who then by Executive Order 12155 of September 10, 1979, reassigned responsibility for minerals to the Secretary of the Interior and for agricultural materials to the Secretary of Agriculture. Additional reenforcing policy statements dealing with materials and the national security are provided by the National Materials and Minerals Policy, Research and Development Act of 1980 (94 Stat. 2305; 30 USC 1601 ff.), the Mining and Minerals Policy Act of 1970 (84 Stat. 1876; 30 USC 21a) and many other more narrowly focused laws.

6. *Planning in the 1980s*

On April 5, 1982, the President transmitted to the Congress his "National Materials and Minerals Program Plan and Report to the Congress." His letter of transmittal stated "This national minerals policy recognizes: the critical role of minerals to our economy, national defense, and standard of living; the vast, unknown and untapped mineral wealth of America and the need to keep the public's land open to appropriate mineral exploration and development; the critical role of government in alerting the Nation to minerals issues and in ensuring that national decision-makers take into account the impact of their decisions on minerals policy; and the need for long-term, high potential pay-off research activity of wide generic application to improve and augment domestically available materials," and "This policy is responsive to America's need for measures to diminish minerals vulnerability by allowing private enterprise to preserve and expand our minerals and materials economy."

The report stated "It is the policy of this Administration to decrease America's minerals vulnerability by taking positive action that will promote our national security, help ensure a healthy and vigorous economy, create American jobs, and protect America's natural resources and environment." It enumerated several major areas of action including :

(a) increased availability of lands for exploration, development and multiple use;

(b) reform of excessively burdensome or unnecessary regulations and statutes;

(c) increased stockpiling, including use of exchanges and barter and ensuring the quality of the stockpile;

(d) encouraging private sector materials research and development, stimulating coordination between industry and government in technical innovation and focusing government-financed research and development on long-term, high-risk technology with the best chance for wide generic application to materials problems; and

(e) improved minerals and materials information collection, analysis and dissemination.

On July 22, 1982, the President issued National Security Decision Directive 47 which states ". . . It is the policy of the United States to have an emergency mobilization preparedness capability that will ensure that government at all levels, in partnership with the private sector and the American people, can respond decisively and effectively to any major national emergency with defense of the United States as the first priority."

The section on industrial mobilization of NSDD 47 states "It is the policy of the United States to have a capability to mobilize industry in order to achieve timely and sufficient production of military and essential civilian material needed to prosecute successfully a major military conflict, to lend credibility to national strategic policy, and to respond to national security emergencies." The industrial mobilization program will:

> "improve the capability of United States industry to meet current and mobilization requirements by identifying production and supply deficiencies and initiating actions to overcome them;
>
> "increase the capability of industry and infrastructure systems, including transportation and energy, to meet national security needs through use of improved guidance on resource-claimancy, continued use of import and export controls, and appropriate use of Defense Production Act authorities in cases where the free market cannot be reasonably expected to provide the required national security capability in a timely manner;
>
> "provide for assessment of the impact on the industrial base resulting from existing and proposed agreements for coproduction of defense material, related offset arrangements with our allies, and other reciprocal trade agreements; and
>
> "ensure the availability of strategic and critical materials by primary reliance on the National Defense Stockpile; the President may authorize the use of DPA Title III in those instances where the free market cannot be reasonably expected to provide the required national security capability in a timely manner."

7. *Stockpile Status in the USA 1987*

In 1983, the US President ordered a major review of strategic stockpiling led by the National Security Council (NSC) and involving a dozen major federal agencies. The NSC analyses included assessment of US military, industrial and essential civilian needs as well as foreign demand, trade patterns and shipping losses in a three-year conventional global military conflict—commonly considered to require the largest supplies of materials. After two years of interagency study and review, the US President, on July 8 1985, proposed significant modernization, concluding that significantly lower goals were appropriate but that an additional supplemental reserve of materials already owned would be held to provide assurance against shortages during a period of military conflict. The proposed new goals were valued at about US$0.7 billion and the supplemental reserve at US$6.0 billion, compared to previously existing goals valued at US$16.6 billion and total inventories valued at US$10 billion. Table 2 provides details. However, the US Congress resisted such changes and the US President's proposals were still under consideration as of mid-1987. The US President also emphasized the need to determine requirements for new high-technology materials for future weapons systems. These could be items already on the stockpile list or new candidates for some form of government program not necessarily limited to stockpiling. Candidate items could include rhenium, scandium, gallium, lanthanum, neodymium, strontium and high-quality optical glass (ultralow-expansion fused silica quartz and substitutes).

As listed in Table 2, 94 materials were singled out by the US Government as "basic stockpile materials," largely on the basis of import vulnerability, but it would be fallacious to conclude that only these are strategic. For example, neither steel nor iron ore, coke, coking coal and limestone are on the stockpile list, yet steel is considered a highly strategic material in every

Table 2
FEMA's stockpile goals (desired inventory mix) and inventories as of September 30 1986, and the US President's proposed stockpile modernization as of July 8 1985

	Units	Goal	Inventory	New goal	Supplemental reserve
Aluminum metal group	ST Al	7150000	4278912		
(alumina)		0	0		
(aluminum)	ST	700000	2080		
(bauxite, metal grade, Jamaica type)	LDT	21000000	12457740		4278912
(bauxite, metal grade, Surinam type)	LDT	6100000	5299597		metal eq.
Aluminum oxide, abrasive grain group	ST abrasive grain	638000	259124		208139
(aluminum oxide, abrasive grain)	ST	0	50904		
(aluminum oxide, fused, crude)	ST	0	249867		
(bauxite, abrasive grade)	LCT	1000000	0		
Antimony	ST	36000	37420	4585	
Asbestos, amosite	ST	17000	34011		
Asbestos, chrysotile	ST	3000	10705		
Bauxite, refractory	LCT	1400000	274229		274926
Beryllium metal group	ST Be metal	1220	1089		437
(beryl ore 11% BeO)	ST	18000	17656		
(beryllium copper master alloy)	ST	7900	7387		
(beryllium metal)	ST	400	290		
Bismuth	lb	2200000	2081298		
Cadmium	lb metal	11700000	6328809		
Chromium, chemical and metallurgical group	St Cr metal	1353000	1303379	199300	594123
(chromite, chemical grade ore)	SDT	675000	242414		
(chromite, metallurgical grade ore)	SDT	3200000	2130938		
(chromium, ferro, high carbon)	ST	185000	524352		
(chromium, ferro, low carbon)	ST	75000	318942		
(chromium, ferro, silicon)	ST	90000	58357		
(chromium, metal)	ST	20000	3763		
Chromite, refractory grade ore	SDT	771000	391414		180000
Cobalt	lb Co	85400000	53109681	22570000	6000000
Copper	ST	1000000	29048		29048
Cordage fibers, abaca	lb	155000000	0		
Cordage fibers, sisal	lb	60000000	0		
Diamond, industrial group	carat	29700000	34563531		
(diamond dies, small)	piece	60000	25473		
(diamond, industrial, crushing bort)	carat	22000000	22001344		
(diamond, industrial, stones)	carat	7700000	11049555		7950000
Fluorspar, acid grade	SDT	1400000	895983		
Fluorspar, metallurgical grade	SDT	1700000	411738		
Germanium	kg	30000	0	146049	
Graphite, natural—Ceylon, amorphous lump	ST	6300	5497	5086	415
Graphite, natural—Malagasy, crystalline	ST	20000	17838	13996	
Graphite, natural—other than Ceylon and Malagasy	ST	2800	2803	2237	
Iodine	lb	5800000	7372112		5500000
Jewel bearings	piece	120000000	74205118		
Lead	ST	1100000	601025		300000
Manganese dioxide, battery grade group	SDT	87000	208165		
(manganese, battery grade, natural ore)	SDT	62000	205154		
(manganese, battery grade, synthetic dioxide)	SDT	25000	3011		
Manganese, chemical and metallurgical group	ST Mn metal	1500000	1933338		869667
(manganese ore, chemical grade)	SDT	170000	171806		
(manganese ore, metallurgical grade)	SDT	2700000	3166496		
(manganese, ferro, high carbon)	ST	439000	690324		
(manganese, ferro, low carbon)	ST	0	0		

Table 2—continued

	Units	Goal	Inventory	New goal	Supplemental reserve
(manganese, ferro, medium carbon)	ST	0	29057		
(manganese, ferro, silicon)	ST	0	23574		
(manganese, metal, electrolytic)	ST	0	14172		
Mercury	flasks	10500	169226		
Mica, muscovite, block, stained and better	lb	6200000	5212361	246400	200000
Mica, muscovite film, 1st and 2nd qualities	lb	90000	1178755	18700	
Mica, muscovite splittings	lb	12630000	14652181	14391100	
Mica, phlogopite block	lb	210000	130745	85000	
Mica, phlogopite splittings	lb	930000	1518951	482600	
Molybdenum group		0	0		
(molybdenum disulphide)		0	0		
(molybdenum, ferro)		0	0		
Morphine sulphate and related analgesics	AMA lb	130000	71303		
(crude)	AMA lb	0	31795		
(refined)	AMA lb	130000	39508		
Natural insulation fibers	lb	1500000	0		
Nickel	ST Ni+Co	200000	37222		5000
Niobium group	lb Nb metal	4850000	2713469		2532419
(niobium carbide powder)	lb Nb metal	100000	21372		
(niobium concentrates)	lb Nb metal	5600000	2019218		
(niobium ferro)	lb Nb metal	0	930911		
(niobium metal)	lb Nb metal	0	44851		
Platinum group metals, iridium	Troy oz	98000	29590		
Platinum group metals, palladium	Troy oz	3000000	1264602		
Platinum group metals, platinum	Troy oz	1310000	452641		
Pyrethrum	lb	500000	0		
Quartz crystals	lb	600000	1848532	26500	1800000
Quinidine	Av. oz	10100000	2473109		
Quinine	Av. oz	4500000	3246164		
Ricinoleic/sebacic acid products	lb	22000000	12524242		
Rubber	t	864000	127446		127455
Rutile	SDT	106000	39185		
Sapphire and ruby	carat	0	16305502		
Silicon carbide, crude	ST	29000	80550		
Silver (fine)	Troy oz	0	130005707		87500000
Talc, steatite block and lump	ST	28	1081		
Tantalum group	lb Ta metal	7160000	2642073	1900700	1023320
(tantalum carbide powder)	lb Ta metal	0	28688		
(tantalum metal)	lb Ta metal	0	201133		
(tantalum minerals)	lb Ta metal	8400000	2837943		
Thorium nitrate	lb	600000	7121812		
Tin	t	42700	180889		150000
Titanium sponge	ST	195000	36831	3900	21100
Tungsten group	lb W metal	50666000	74215353		52215245
(tungsten carbide powder)	lb W metal	2000000	2032942		
(tungsten, ferro)	lb W metal	0	2025361		
(tungsten metal powder)	lb W metal	1600000	1898831		
(tungsten ores and concentrates)	lb W metal	55450000	80013111		
Vanadium group	ST V metal	8700	721		722
(vanadium, ferro)	ST V metal	1000	0		
(vanadium pentoxide)	ST V metal	7700	721		
Vegetable tannin extract, chestnut	LT	5000	12746		
Vegetable tannin extract, quebracho	LT	28000	126649		
Vegetable tannin extract, wattle	LT	15000	15001		
Zinc	ST	1425000	378316		85000

Source: Federal Emergency Management Agency Key: AMA lb, anhydrous morphine alkaloid (pounds); Av. oz, avoirdupois ounce; LCT, long calcined ton; LDT, long dry ton; long ton; LT, long ton; SDT, short dry ton; ST, short ton; 1 flask = 76 pounds

modern economy. Many materials have been recognized as stockpile candidates ever since World War I, while newer ones have been added to the list only after research and development have proved their utility, such as titanium. Of the materials listed, 81 are of mineral origin and 13 of agricultural origin. Despite the vast synthetic rubber industry, natural rubber is still preferred for tires for aircraft and off-the-road vehicles. Quebracho is used not only in tanning leather but also in oil well drilling muds. Abaca is preferred for some hawsers. Materials have been removed from the stockpile list because substitutes or alternatives were found. For example, hog bristles for paint brushes have been replaced by tapered nylon and Egyptian extra-long-staple cotton by domestic sea island cotton or synthetic fibers.

Serious consideration must be given to the forms in which materials are stockpiled, particularly in view of the need for "surge capability" in the defense industrial mobilization base. Upgraded forms (e.g., aluminum metal instead of bauxite, ferroalloys or pure metals instead of ores) would increase stockpile costs but would also stockpile energy, transportation, production facilities, manpower, and time—all likely bottlenecks in time of mobilization.

Other government stockpiles (with holdings as of mid-1987) are the US Department of Energy's strategic petroleum reserve (515 million barrels), the US Bureau of Mines' helium stockpile (one billion cubic meters) and the US Treasury Department's gold stocks (262 million troy ounces).

8. Related Programs

In addition to strategic stockpiles, various price support programs, especially for agricultural commodities, have been significant at various times in the last several decades. These price support programs sometimes result in the acquisition by the US Government of substantial quantities of certain commodities. Further, from time to time suggestions have been advanced that there should be peacetime US "economic stockpiles" which would attempt to moderate the peaks and valleys of cyclic demand for materials. As of mid-1987, however, no such deliberate schemes have been put into operation other than for certain agricultural commodities. In addition, there have been a variety of international schemes proposing "buffer stocks" which would act as economic stockpiles. The USA did participate in the 5th International Tin Agreement and in the International Natural Rubber Agreement and it has also participated in international discussions involving other materials.

See also: Stockpiling

Bibliography

FEMA, FPA, OEP, OCDM, ODM, Munitions Board 1947–1987 *Stockpile Report to the Congress* (published semiannually). Federal Emergency Management Agency, Washington, DC

Morgan J D 1958–1987 Mining Annual Review: The United States. In: *Mining Journal* (*London*)

US Bureau of Mines 1985 *Mineral Facts and Problems*. USBM, Washington, DC

US Bureau of Mines *Minerals Yearbook*, annual editions 1939–1986. USBM, Washington, DC

US Congress Joint Committee on Defense Production 1951–1976 *Annual Report of the Joint Committee on Defense Production*. US Government Printing Office, Washington, DC

Stockpile item are regularly reported in various trade publications, including:

Mining Journal, Mining Magazine, Metal Bulletin, Metals Week, Engineering and Mining Journal, American Metal Market, Mining Engineering, Wall Street Journal

In recent years the National Materials Advisory Board of the National Research Council/National Academy of Sciences/ National Academy of Engineering has prepared a number of studies dealing with stockpiling and/or major strategic materials. Published by the National Academy Press, Washington, DC, these include:

NMAB-335	*Contingency Plans for Chromium Utilization*
NMAB-378	*Considerations in Choice of Form for Materials for the National Stockpile*
NMAB-381	*Assessment of Selected Materials Issues*
NMAB-385	*Analytical Techniques for Studying Substitution Among Materials*
NMAB-391	*Tantalum and Columbium Supply and Demand Outlook*
NMAB-392	*Titanium: Past, Present, and Future*
NMAB-403	*Priorities for Detailed Quality Assessments of the National Defense Stockpile*
NMAB-406	*Cobalt Conservation Through Technological Alternatives*
NMAB-424	*Quartz for the National Defense Stockpile*

J. D. Morgan
[US Bureau of Mines, Washington, DC, USA]

Structural Clay Products

The unique nature of the structural clay products industry can be explained in social and economic terms. This industry now produces a large number of individually identifiable products under the general categories of bricks, tiles, pipes and industrial specialities. The large variety of raw materials and forming methods explains the wide range of physical properties found among these products.

1. Structural Clay Products Industry

The nature of the structural clay products industry today is profoundly affected by the traditions established over a period of at least 100 000 years. Building bricks were first made when the urbanization of society began. The making of tiles and pipes for the construction of cities followed as soon as skills in the working of clays developed. Up to the nineteenth century, and in some places much later, these products were manufactured as needed near sites of building construction. An industry for the purpose of supplying structural clay products to all who required them did not exist in early periods of civilization, but the art and manual skills for making these products developed. When the industrial revolution took place in the eighteenth century, the manufacture of structural clay products in factories was a gradual and natural transition from unorganized manual skills to an industry using mechanical and electrical power. In making this transition, much of the manual arts and skills were simply transferred to machine and equipment design. Today, the scientific revolution is gradually intruding into the structural clay products industry in the form of a better understanding of raw materials and products and in the form of automation.

Partly because of its historical development and partly because of its close association with the building industry, the structural clay products industry is very conservative. It resists change in type and quality of its products and methods of manufacture as long as there is a market for them and profits are being made. Many factories have simply gone out of business and closed their doors rather than modify a product line or develop new techniques of manufacture. In fact, whole sections of the industry have disappeared over the years when a product went bad or the market was taken over by a substitute material. The building industry, which is supplied by the structural clay products industry, is characteristically conservative also. This association affects the structural clay products industry in the type of market available. Buildings, whether residential, industrial or public, are large investments for their owners; therefore, they are usually unwilling to try new materials and tend to stay with that which has proven satisfactory in the past.

This conservatism causes the structural clay products industry to be engineering oriented rather than research oriented. The industry is concerned with efficiency and process control from the mining of raw materials to the packaging of products for shipment. In order to keep the unit price low, there is an intense interest in eliminating labor costs in the manufacturing process by automation. The other major concern of this industry is the ability to use various kinds of hydrocarbon fuels and to use them inexpensively and efficiently. All of these production problems are approached from an engineering point of view, and are mostly handled on the production line.

Unlike our new-technology industries that have their origins in the research laboratory, the structural clay products industry is not research oriented. The industry as a whole does not put much of its profits into basic and applied research because it has no expectation of the development of new products, and it sees little need for the improvement of the products produced currently. In addition, the industry has little confidence in the outcome of research and not enough patience to allow a research program to be productive.

The large variation in the sizes of the companies producing structural clay products in the USA is also not conductive to proper research endeavor. The efforts that small companies can make is often too small to be effective, and large companies are unwilling to make a considerable investment that would assist their competitors as well as themselves. Cooperative efforts have been tried on a national scale in the highly diverse brick section of the industry, but they have not been successful.

Several well-developed countries meet the research needs of the structural clay products industry by government sponsorship of the pertinent programs. Such an effort would be beneficial in the USA because there are problems in this industry that should be addressed. Some of these problems are related to a better characterization of raw materials, a better understanding of clay plasticity, an identification of the phases in the final products that contribute to strength and durability, a precise description of pore structure, the adsorption of molecular species on internal surfaces and the phenomenon of secondary moisture expansion.

The structural clay products industry could be classified as energy intensive as are most branches of the classical ceramic and metallurgical industries. A large portion of the total cost of the product is in the form of energy use and fuels consumed. Electric power is used to run the heavy machinery required for mineral dressing and forming the products. In addition, electricity operates the automation and air circulating fans connected with the dryers and kilns. A large part of the total energy consumed is in the form of heat from the burning of natural gas, oil or coal that is required to dry the products from the plastic state and to fire them to maturity. Temperatures between 980 and 1260 °C are obtained in the kilns in order to convert the raw materials into durable products with specified physical properties.

As in the case of the producers of most materials the structural clay products industry has little control of the end use of its products. Once bricks and structural tiles are placed in the walls of buildings, it is an extremely complex task to trace the causes of successes or failures. When failures do occur, it is difficult to determine whether the clay products, the design of the building, the customer use or the quality of construction were at fault. The best defense this industry has against complaints is to make products good enough to withstand the most serious stress situations that can be anticipated.

2. *Size and Economics of the Industry*

The size of the structural clay products industry in the USA is relatively small, but it is an important supplier to the building construction industry. The Statistical Abstract of the USA published by the US Bureau of the Census lists 47 000 workers employed in this industry in 1976. The same source gives the dollar value of the products shipped in 1976 as US$595 000 000 and the value added by manufacturing as US$930 000 000.

The structural clay products industry is economically unstable in that the volume of business fluctuates widely with the economic climate of the country. During periods of recession, many plants have to be shut down to prevent intolerable inventories, but in periods of economic growth these same factories may be back ordered up to a year and a half. These trends follow similar fluctuations in the building construction industry. Figure 1 shows these relationships and trends over a period of 26 years. The quantity of facing bricks shipped was selected as an indicator of the activity in the whole structural clay products industry because facing bricks constitute a little more than half of the dollar value of all structural clay products used. New housing units started include private, corporate and government programs, and it was selected as a measure of the total building construction of the period.

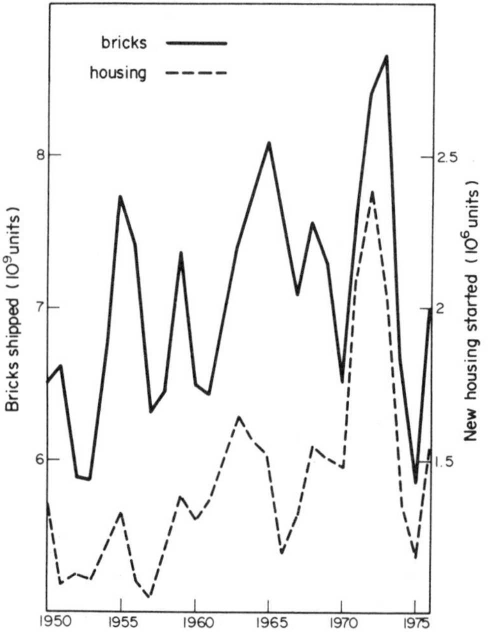

Figure 1
Fluctuations in the business activity of the structural clay products industry and the building construction industry

Although the average output of facing bricks has been relatively constant over the past 30 years, some other types of structural clay products have had different histories. The production of structural facing and partition tiles has decreased steadily over this period. This decline was probably due to the substitution of lower-cost materials of a different kind. Sewer and drainpipe production reached a peak in 1964 and has been in a decline since that year; however, it still remains the third largest section of the structural clay products industry with an annual dollar value of shipments in 1976 of US$123 000 000. The second largest portion of the whole industry in dollar value of shipments is the production of floor and wall tiles. In 1976, this value stood at US$186 000 000. The actual production of these tiles has remained reasonably constant over the past 20 years.

Perhaps the large initial capital investment for new factories to produce such low-cost products discourages growth in the structural clay products industry. The cost of a new, single-line production factory is from 6 to 10 million dollars. Brick plants must produce 80 000–100 000 standard brick equivalents per day for each production line in order to make a satisfactory profit.

3. *Classification of Types of Products*

The various identifiable types of structural clay products are outlined in Table 1. The largest sections of this classification are bricks, tiles and pipes. Under each of these sections there are many varieties that have their own unique properties appropriate to the intended use. The forming methods used are a part of this classification because the appearance and physical properties are different for the various methods.

4. *Forming Methods*

The forming of structural clay products by extrusion induces inherently unique physical properties. The most common extrusion machinery uses an auger in a vacuum chamber to force the plastic, clay-containing raw materials through a shaping die. Because of the flaky morphology of the clay minerals and the frictional resistance of the clay going through the die under considerable pressure, laminations are induced into the extruded column. These laminations are caused by the extensive orientation of the clay particles under the pressure of the auger flights, and slip planes which develop between the center of the extruded column and the inner die surfaces. Such preferred orientation and concentration of clay particles in the vicinity of the laminations give a directional character of the crystalline microstructure and the pore structure of the products. These directional characteristics are often reflected in directional strength and absorption. The laminations can be planes of weakness and planes of open pores. In the

Table 1
Types of structural clay products

I. Bricks
  A Facing (C62–75a, C126–76, C216–79)[a]
    1 Extruded
      a Solid, smooth, textured or glazed
      b Cored, smooth, textured or glazed
      c Panel, smooth, textured or glazed (C652–77)
    2 Sand molded, natural, colored or glazed
    3 Dry pressed, solid
  B Paving, solid—extruded
    1 Sidewalk, smooth or textured (C902–79a)
    2 Street or highway, smooth or textured
  C Industrial floor, solid, smooth or textured—extruded (C410–78)
  D Chimney–extruded
  E Industrial chimney lining—extruded
  F Sewer and manhole—extruded (C32–79)

II. Tiles
  A Structural for walls—extruded
    1 Non-load-bearing, smooth or textured (C56–76)
    2 Load-bearing, smooth or textured (C34–75)
    3 Facing (C212–75)
      a Smooth
      b Textured
      c Glazed (C126–76)
      d Acoustical
      e Lightweight
      f Through-the-wall
    4 Screen, smooth or textured (C530–75)
  B Floor
    1 Large-cored, smooth or textured—extruded (C57–78)
    2 Quarry—plastic pressed
    3 Mosaic, unglazed and glazed—dry pressed
  C Wall, unglazed and glazed—dry pressed
  D Roofing—extruded and plastic pressed
  E Flue linings—extruded (C315–78c)

III. Pipes—extruded
  A Sewer and drain, plain or glazed (C700–78a)
    1 Bell and spigot end
    2 Plain end
  B Perforated subdrainage (C700–78a)
    1 Bell and spigot end
    2 Plain end
  C Chemical resistant, plain or glazed
  D Conduit, plain or glazed
  E Drain tile
    1 Plain (C4–75)
    2 Perforated (C498–75)

IV. Liner plates, curved or flat, plain or glazed—extruded (C479–77)

V. Filter blocks for trickling filters, plain or glazed—extruded (C159–78)

VI. Chemical resistant tower packings (C515–76)

[a] Designations in parentheses refer to the appropriate specifications of the American Society for Testing and Materials

best industrial practice, such laminations are completely healed by maintaining adequate vacuum on the auger chamber, optimum particle-size distribution of the coarse filler grains and the proper water content in the plastic mix. Some large sewer pipes are extruded by a ram instead of an auger; therefore, laminations are less well developed and often completely unnoticeable in these products.

The sand molding of plastic clay is generally referred to as the soft-mud process. The clay material is made up in a softer plastic condition than for extrusion; then, bricks are formed by ramming the soft clay into sand-coated, wooden molds. Although this is a very ancient method for making face bricks, it is a highly automated process in our modern factories. This method has the advantage of a random distribution of crystalline phases and pores, and there is no tendency toward laminations. Bricks made in this way are intentionally less perfect in size and shape than those made by extrusion.

Clay materials for plastic pressing are prepared to the state of maximum plasticity or maximum plastic strength, and the tempered clay is vacuum extruded and cut into regular bats in preparation for pressing. In relation to the other forming methods, the plastic state is intermediate between that for extrusion on the drier side and sand molding on the wetter side. The forming of the ware is done by pressing the plastic clay bat between upper and lower molds made from plaster of paris. This method produces some distortion of the pore structure, but the product is relatively free of laminations and is of uniform strength.

Thin floor and wall tiles are dry pressed to shape in large hydraulic presses under a pressure of about 34.47 MPa (5000 psi). Some small amount of water containing a binder is added to the clay batch; so the forming operation should more accurately be termed damp pressing. This method of forming produces a different type of pore structure from that produced by plastic methods. Laminations are possible, but they are usually avoided by the proper selection of clays and compacting pressure. The major advantage of dry pressing is the more uniform size and shape of the product, since all drying shrinkage has been eliminated.

The porosity of the final product is more dependent on the raw materials used and the heat treatment than on the forming method. Firing of clay products shifts the pore-size distribution towards larger sizes and reduces the total pore volume. This change in pore structure is accompanied by shrinkage.

5. *Raw Materials*

The wide variation in the types of clay raw materials used for the manufacture of structural clay products multiplies the classification by a very large number if the esthetics of the products are considered. Both clay and argillaceous shale deposits, located near the factories, are used as nature provided them with little or no beneficiation. The exception is in the floor and wall tile industry. Here, the clay bodies are prepared in exact compositions from several beneficiated raw materials. This industry uses mixtures of specific clays, feldspar, flint (pulverized quartz), pyrophyllite and talc.

Regardless of whether natural or beneficiated raw materials are used, there is a general range of mineral compositions that must be adhered to for satisfactory production. There are three types of minerals that must be present in proper amounts for all clay products. The clay content can range from 35% to 55%. Inert fillers, such as quartz, will generally make up from 25% to 45% of the mixture. Finally, reactive fillers or fluxing materials are required in the range from 25% to 55%. Natural raw materials must contain these elements in acceptable proportions in order to be useful to the structural clay products industry. For this reason many plants elect to blend two or more naturally occurring raw materials to achieve the desired physical properties and visible appearance.

A list of the mineralogical–geological types of natural raw materials used in the USA indicates the variations that can occur in each product line. This can be illustrated by reference to fired color:

(a) crude, residual, kaolinitic clays—white;

(b) kaolinitic–illitic fireclays—buff;

(c) illitic–kaolinitic shales—red;

(d) illitic clays—red;

(e) illitic shales—red;

(f) illitic–chloritic shales—red;

(g) argillaceous–calcareous clays—pink to buff; and

(h) argillaceous–calcareous shales—pink to buff.

The clay minerals in the above list are kaolinite and illite. Kaolinite is a refractory clay that requires other reactive fillers or fluxes to produce satisfactory products; however, illite acts as its own flux on firing to 1050 °C and higher temperatures. A third clay mineral, montmorillonite, is sometimes found in small quantities in clay deposits, but it is not an essential ingredient, and it does present some special problems in production.

6. *Physical Properties*

As one might expect from the variations in forming methods and raw materials, the physical properties of structural clay products vary widely but still remain within certain broad limits. Table 2 gives the ranges of values for strength and water absorption for five general types. The ranges of values mean that in some locations these products can be found with average values in these ranges. Any specific product will have a much narrower spread of values around its mean. The compressive strength of some bricks will be as high as 137.9 MPa (20 000 psi), and the modulus of rupture is usually around 6.9 MPa (1000 psi). The absorption of water by building bricks is preferred to be in the 5% to 8% range. Structural tiles have compressive strengths below that of bricks made from the same material because of their difference in geometry. Sewer pipes can have crushing strengths up to 68.9 kN m^{-1} (10 000 lb ft^{-1}), but drainpipes are usually weaker and have higher absorptions because of their different end use.

7. *Specifications*

The generally accepted standard specifications for the physical properties of bricks and structural tiles are too broad and relatively ineffective. For these reasons, many organizations rely on the past performances of specific products as a guide to reliability, and this practice is subject to many dangers, including undetected changes in the products and purchasing frauds. The main reason for the broadness of current specifications is that an attempt was made to write one set of specifications for all products regardless of their types of raw materials or methods of manufacture. An important secondary explanation for the broadness is

Table 2
Physical properties of structural clay products

Product	Compressive strength (lb in.$^{-2}$)	Compressive strength (MPa)	Modulus of rupture (lb in.$^{-2}$)	Modulus of rupture (MPa)	Absorption (%)
Bricks, facing	1250–20000	3.62–137.90	500–2000	3.45–13.79	2–25
Bricks, paving	2500–20000	17.24–137.90	1500–2500	10.34–17.24	1–6
Tiles, structural	500–10000	3.45– 68.95			2–28
	Crushing strength (lb ft^{-1})	Crushing strength (kN m^{-1})			
Pipes, sewer	1200–10000	17.5–145.8			1–8
Pipes, drain tile	680–5000	9.9– 72.9			5–16

that the appropriate relationships between physical properties and actual service conditions are not known. This ignorance reflects the lack of research in the field of structural clay products. At it is, the specifications are not much protection for the customer and only a vague guide for manufacturers.

See also: Traditional Ceramics: An Overview

Bibliography

American Society for Testing and Materials 1980 *Annual Book of ASTM Standards.* American Society for Testing and Materials, Philadelphia, Pennsylvania

Bloor E C 1959 Plasticity in theory and practice *Trans. Brit. Ceram. Soc.* 58: 429–53

Brownell W E 1976 *Structural Clay Products.* Springer, New York

Emrich E W 1973 History and development of ceramic wall tile bodies in the United States. *Am. Ceram. Soc. Bull.* 52(9): 687–88

Grim R E 1968 *Clay Mineralogy*, 2nd edn. McGraw-Hill, New York

Grimshaw R W 1971 *The Chemistry and Physics of Clays*, 4th edn. Wiley–Interscience, New York

National Clay Pipe Institute 1972 *Clay Pipe Engineering Manual.* National Clay Pipe Institute, Washington, DC

van Olphen H 1963 *An Introduction to Clay Colloid Chemistry.* Interscience, New York

W. E. Brownell
[New York State College of Ceramics,
New York, USA]

T

Thorium Resources

Thorium is a relatively scarce element which is available in ample supply to meet present demand. Thorium abundance in the earth's crust is estimated as 6–12 ppm, with a figure of 7.2 ppm for igneous rocks. Thorium is considered to be three times as abundant as uranium, with which it is closely associated in igneous rocks. It concentrates in the residual magmatic fluid and is found in pegmatites and disseminated in granites and igneous rocks. Granites have an average concentration of 25 ppm but some have up to 64 ppm. In the sedimentary cycle, thorium remains in residual deposits or in fluvial and beach placers as resistant minerals such as monazite in association with Nb, Ti and Zr. In 1987, US ThO_2 consumption was reported as 57×10^3 kg, which was satisfied by by-product ThO_2 from processing monazite for rare earths.

Thorium reserves and resources are very uncertain because the small demand for thorium and its recovery as a by-product have inhibited prospecting and drilling. US resources are estimated as 42×10^6 kg as by-product, 97×10^6 kg in ores of content >0.1% and 500×10^6 kg in total. Global identified resources total 1855×10^6 kg. Monazite beach sand reserves are estimated to contain 400×10^6 kg in India and 118 $\times 10^6$ kg in Brazil.

Future demand for thorium is expected to remain level through 1990 and then decline because of environmental concerns and the development of non-radioactive substitutes. Thorium has advantages in thermal reactors, but it is not possible at present to forecast this type of utilization.

Bibliography

Kirk W 1980 Thorium. *Minerals Yearbook 1980*, Vol. 1. US Bureau of Mines, Washington, DC, pp. 821–26

Hedrick J B 1988 Thorium. *Mineral Commodity Summaries 1988*. US Bureau of Mines, Washington, DC, pp. 166–7

Rodriguez P, Sundaram C V 1981 Nuclear and materials aspects of the thorium fuel cycle. *J. Nucl. Mater.* 100: 227–49

Searl M F, Platt J 1975 Views on uranium and thorium resources. *Ann. Nucl. Energy (UK)* 2: 751–62

Staatz M H, Olson J C 1973 Thorium. In: Brobst D A, Pratt W P (eds.) 1973 *United States Mineral Resources*, US Geological Survey Professional Paper 820. US Government Printing Office, Washington, DC, pp. 468–76

D. T. Peterson
[Iowa State University, Ames, Iowa, USA]

Tin Resources

Tin is a soft, malleable, faintly bluish-white, heavy metal (specific gravity 7.29). It is nontoxic, easily melted (232 °C), has a low coefficient of friction, extreme fluidity in the molten state and excellent wetting characteristics. Tin is rarely used in the pure, elemental form, but because of its ability to alloy with other metals, it is one of the most widely used metals and touches many facets of our daily lives. Its most important use is as tin plate, which accounts for about 21% of the tin used in the US and about 40% of world use. The US produces about 25% of its consumption needs from recycled tin; virtually all of the remainder is supplied by imports of primary tin.

1. History

Few metals have played as important a role in the development of civilization as tin. Ancient coppersmiths discovered that adding tin to copper yielded an alloy (bronze) that was easier to cast and more workable than copper and produced superior weapons and tools hard enough to retain a cutting edge. The earliest dated use of tin in bronze is about 3000 BC (Muhly 1973), and pure tin ingots were produced between 1500 and 1100 BC (Muhly 1978).

Because ancient tin-producing and tin-consuming cultures were geographically separated, tin greatly influenced patterns of early trade routes. Tin became a standard metal of commerce during the Bronze Age, which began between 3500 and 3000 BC and ended about 1000 BC when bronze was replaced by iron. Early sources of tin for the Near East cultures remain an enigma, but by the late Bronze Age (1600–1200 BC) and certainly by 700 BC the bronze industry of Greece utilized tin that came from Cornwall in southwestern England. The metal was designated by the Greek word *kassiteros*; cassiterite is the main ore mineral of Cornish tin.

Early tin at Cornwall was in large part produced from alluvial deposits (placers); later, and during the Roman occupation, tin was produced from lode mines. The East Indies and Malay states became producers of alluvial tin in the eighth century, and Malaysia, Indonesia and Thailand are today's major sources of tin.

2. Uses and Supply

The USA is the world's largest tin consumer but produces virtually no primary tin. In 1987 it consumed about 36 900 t of primary tin, or 22% of world production. Principal uses of tin in the US are: electrical, 20%;

cans and containers, 17%; construction, 16%; transportation, 14%; and other, 33%.

Worldwide, the largest use of tin is in metal containers for food, beverages and other commodities. Substitution of aluminum, plastics and tin-free steel by the packaging industry, as well as new manufacturing processes for two-piece beverage cans, is reducing the US demand for tin plate. The US Bureau of Mines (Harris 1978) forecasts that by the year 2000 the use of tin in cans and containers will decrease by about 25%. However, this decrease in demand for tin may be matched by an increase in the demand for tin for other purposes—particularly chemicals. Organotin chemicals are effective pesticides, fungicides and wood preservatives, and have low toxicity. The use of tin catalysts in polyurethane production, organotin chemicals in poly(vinyl chloride) (PVC), and molten tin used in processes to recycle nuclear fuel are examples of the growing new chemical uses of tin.

Compared with other metals, tin consumption has grown slowly but continually throughout this century. Although some materials can be used as substitutes for tin, their use often reduces efficiency; some substitutes such as aluminum are largely imported and may themselves become critical materials. The US is the world's largest producer of recycled tin, which provides about 25% of its consumption.

US production of tin is negligible. In 1987, the only domestic production of primary tin was from small placer operations in Alaska. The principal sources of US primary tin supplies for the period 1983–86 were: Brazil, 24%; Thailand, 17%; Indonesia, 13%; and Bolivia, 12%.

3. Geochemistry

Tin has an atomic number of 50, an atomic weight of 118.7, and ten isotopes between mass numbers 112 and 124. It has two common valences, +2 and +4, with ionic radii of 0.93 Å and 0.71 Å, respectively. Tin has two forms: the more common white tin of tetragonal crystal symmetry, and the gray tin of cubic symmetry. The allotropic change occurs at 13.2 °C. Alloying tin with any other metal prevents this change from white to gray tin.

Tin is sparsely distributed in the earth's crust; its abundance is in the range of 2–3 ppm. The metal is preferentially concentrated by magmatic differentiation processes and shows a worldwide affinity for felsic granitic rocks or their extrusive equivalents, whose average tin content is about 3.5 ppm. Intensely altered granites (greisens) containing 0.3–0.7% tin are presently being mined in Australia.

4. Geology

On a continental scale, most tin deposits occur in elongate foldbelts of thick platform sedimentary and volcanic rocks that were deposited on a cratonic crust. Associated granitoids are generally nonfoliated and postdate major folding. Tin deposits, in general, are temporally and spatially associated with late products of felsic magnetic activity.

The tin dioxide mineral cassiterite (SnO_2), containing 78.6% tin, is the only significant commercial tin ore mineral. Bolivian mines produce tin-bearing sulfosalt and sulfide minerals, but the mineralogical complexity of these ores creates difficulty in beneficiation and less than 50% of the tin from these deposits is recovered.

Tin is produced from both lode and placer deposits. Worldwide, placer deposits furnish about 70% of commercial production. Economic placer grades depend on a variety of factors (e.g., cost of labor, transportation, availability of water, mining methods); as a general rule, placers containing 0.01–0.02% tin can be mined profitably by gravel pump operations in Southeast Asia. The grade of tin lode ore mined today depends on the value of by-products recovered and ranges from 0.3 to 1.2% tin. Bolivian ores now average about 0.7% tin, as compared to earlier operations when mill-head grades of more than 5% were common.

The following summaries include the most common deposit types—in terms of production, conditions of formation, petrology, environment and mineral association.

4.1 Lode Deposits

(*a*) *Pegmatite deposits.* Most tin-bearing pegmatites occur in Precambrian shield terrains, and cassiterite is recovered as a coproduct along with columbite, tantalite, beryl, spodumene and wolframite. The productive pegmatites occur in areas of deep tropical weathering (central and southern Africa, Brazil) which contributes to lower mining costs. Worldwide, tin production from pegmatites is minor.

(*b*) *Subvolcanic deposits.* Traditionally, this category (also called "telescoped" or xenothermal deposits) is typified by the rich tin–silver veins of southern Bolivia. The deposits are associated with felsic Tertiary volcanic and/or hypabyssal and intrusive centers. The veins contain, in addition to cassiterite, abundant sulfosalt and sulfide minerals containing Ag, As, Bi, Pb and Zn. The presence of rare tin minerals makes recovery difficult. Individual Bolivian mines have produced more than 500 000 t of tin; associated porphyry tin deposits, although exceedingly large (in excess of 100 million tonnes), contain only 0.1–0.3% tin and are not now economic producers.

(*c*) *Pneumatalytic-hydrothermal deposits.* These deposits include lodes of the Cornwall type, greisen veins, massive greissenized granite (Reed 1982), and quartz–cassiterite–sulfide veins. They represent the major part of the world's lode tin deposits. Lodes generally occur within or near the apical parts (cupolas or cusps) of biotite or biotite–muscovite granite and typically are mineralogically complex fissure fillings in diverse types of country rock or greisen veins. Most economically viable greisen deposits contain 5–50 million tonnes of ore at grades ranging between 0.1 to 0.4% tin (Menzie et al. 1985).

(*d*) *Replacement* (*including skarn*) *deposits.* Replacement and skarn deposits are characteristically associated with igneous contacts. Carbonate replacement deposits consist chiefly of cassiterite and sulfide minerals and contain 1–25 million tonnes of ore at grades ranging between 0.5 and 1% tin. Tin minerals in skarn deposits are erratically distributed. In addition to cassiterite, nonrecoverable tin is present in other tin-bearing silicate minerals. Other common minerals in tin skarns include magnetite, pyroxene, garnet, tourmaline, fluorite and various sulfide minerals. In the Lost River Area, Alaska, tin is a potential by-product of beryllium and fluorite production from replacement veins and skarn deposits. At Renison Bell (Tasmania), one of the world's largest operating underground tin mines (5000–6000 t of tin per year), the main deposits are quartz–cassiterite–sulfide replacement lodes in carbonate strata.

(*e*) *Rhyolite-hosted deposits* (*Mexican type*). Deposits of this type consist of discontinuous fracture fillings of cassiterite and wood tin (a colloform variety of cassiterite), along with specular hematite and chalcedony, in Tertiary alkali-feldspar rhyolite, and as disseminated cassiterite in volcanic breccia. Grades are generally erratic and quite low. Most deposits contain between 230 and 3900 t of ore at grades between 0.14 and 1.04% tin. Production of tin from these deposits has been limited to small placer operations that have produced a few tens of tonnes of tin each. Many rhyolite-hosted tin deposits lie in the mid-Tertiary volcanic province of the Sierra Madre Occidental located in central and northern Mexico. A few deposits are present in the western USA, such as those in New Mexico's Black Range (Reed and Tooker 1979).

(*f*) *Massive sulfide deposits.* Fine-grained disseminated cassiterite also occurs in stratabound bodies of mixed sulfide minerals mined for Pb, Zn and Cu. Although, worldwide, massive sulfide deposits are minor tin producers, for many years the largest tin-producing mines in North America were at Sullivan (British Columbia) and Kidd Creek (Ontario), where tin is a by-product.

4.2 Placer Deposits

Because cassiterite is both heavy and relatively stable in the surficial environment, it is commonly concentrated in placers—deposits that form over or near bedrock source areas where weathering and erosional processes remove lighter rock materials and gravity assists in concentrating heavy minerals. Tin placers locally contain recoverable amounts of other heavy minerals such as columbite–tantalite, wolframite, ilmenite, monazite and xenotime. Alluvial placers are the richest. They occupy both modern and ancient stream beds, and in Southeast Asia many alluvial placers now lie beneath seawater and are mined by seagoing dredges. Exceedingly low-grade placer deposits with a grade of less than 0.01% tin can be mined economically by dredging.

5. Reserves

The USSR and the People's Republic of China, for which values are estimated, together with Malaysia, Indonesia, Thailand, Brazil and Bolivia, produce about 83% of the global supply and also contain 83% of the world's reserves (Carlin 1988). Reserves are sufficient to meet world requirements well into the twenty-first century. Southeast Asia, Bolivia and Brazil, which supply most of the US imports, contain about 67% of the world's tin reserves. US primary tin production is negligible, and domestic reserves would provide about one-year supply. In 1987, the strategic stockpile contained sufficient tin for about four-year supply; thus the US must continue to depend on other nations for its primary tin requirements.

Bibliography

Carlin J F 1983 Tin. *Minerals Yearbook 1983*, Vol. 1, *Metals and Minerals.* US Bureau of Mines, Washington, DC

Carlin J F Jr 1988 Tin. *Mineral Commodity Summaries 1988.* US Bureau of Mines, Washington, DC, pp. 168–69

Harris K L 1978 Tin. *Mineral Commodity Profiles 16.* US Bureau of Mines, Washington, DC

McKerrell H 1978 The use of tin–bronze in Britain and the comparative relationship with the Near East. *The Search for Ancient Tin.* US Government Printing Office, Washington, DC, pp. 7–24

Menzie W D, Reed B L, Singer D A 1985 Models of grades and tonnages of some lode tin deposits. In: Hutchinson C (ed.) 1985 *Proc. Int. Symp. Geology of Tin Deposits*, Nanning, People's Republic of China, 1984

Muhly J D 1973 Tin trade routes of the bronze age. *Am. Sci.* 61: 404–13

Muhly J D 1978 New evidence for sources of trade in bronze age tin. *The Search for Ancient Tin.* US Government Printing Office, Washington, DC, pp. 43–48

Reed B L 1982 Tin greisen model. *Characteristics of Mineral Deposit Occurrences*, US Geological Survey Open-File Report 82–795. US Geological Survey, Washington, DC, pp. 55–61

Reed B L, Tooker E W 1979 *Preliminary Map of Tin Occurrence Areas in the Coterminous United States*, US Geological Survey Open-File Report No. 79–576L. US Geological Survey, Arlington, Virginia

B. L. Reed

[US Geological Survey, Anchorage, Alaska, USA]

Titanium Resources

Titanium is the ninth most abundant element in the earth's crust. In spite of its abundance, deposits of economic grade and volume are relatively scarce. Titanium combines with a number of other elements to form about 46 different minerals, but it is most commonly found in combination with iron, forming the mineral ilmenite ($FeTiO_3$), or as rutile, anatase or brookite (TiO_2). Less common minerals are perovskite ($CaTiO_3$) and sphene ($CaTiSiO_5$).

Titanium production comes from two different types of deposit. Hard-rock (primary) deposits furnish large tonnages of ilmenite of lower TiO_2 content (35–40%). Heavy-mineral (specific gravity >2.9) beach sand deposits provide ores of higher TiO_2 content, including altered ilmenite (60–75% TiO_2), leucoxene (76–90% TiO_2) and rutile (95% TiO_2). Ilmenite with 50–60% TiO_2 may be further processed to produce materials containing 80–90% TiO_2.

Titanium ores are produced from hard rock deposits in Canada, the USSR, Norway, Finland and the USA. Ores with higher TiO_2 are produced from heavy-mineral beach sand deposits in Australia, South Africa, the USA, India and Sri Lanka. Sierra Leone produces rutile from alluvial deposits. Future titanium ores may be produced from anatase deposits in Brazil or from perovskite deposits in the USA.

Most of the titanium ores are used to manufacture white TiO_2 pigment. Small amounts are processed into titanium metal. Rutile is also used as a welding rod flux.

1. Mineralogy of Titanium

The most abundant form of titanium, the mineral ilmenite, has the theoretical composition 49% TiO_2 and 51% FeO, but usually occurs with some of the iron converted to the ferric state. Only in oxygen-free environments, such as on the Moon, is ilmenite found in unaltered form. In the earth's crust, ilmenite alters to mixtures of TiO_2, FeO and Fe_2O_3, with the TiO_2 content increasing from 50 to 75% as the mineral oxidizes and iron is leached out by groundwater (Fig. 1).

Altered ilmenite is usually amorphous, but the leucoxene stage begins to show the definite crystalline structure of rutile. The name "pseudorutile" has been proposed for ilmenite alteration products containing 75–92% TiO_2. The final alteration products show either rutile or anatase structure and contain >92% TiO_2. The minerals which form as a result of this alteration are called "secondary minerals" as opposed to primary or unaltered rutile or anatase. Primary and secondary rutile crystallizes in the tetragonal system, distinguishing it from the other two forms taken by titanium dioxide: anatase (orthorhombic) and brookite (monoclinic). Rutile is the high-temperature form of TiO_2 and anatase the low-temperature form.

2. Deposits

2.1 Primary Ilmenite Deposits

Primary ilmenite is most commonly associated with subsilicic rocks (gabbros, diorites and anorthosites) made up largely of feldspars with amphiboles and pyroxenes. Most of the ilmenite was presumably of magmatic origin or formed as high-temperature vein deposits. It is often found as dikes cutting anorthosite.

Large deposits of ilmenite occur in Canada. The best known of these are St. Urbain, Ivry, Lac Tio and St. Charles in Quebec. Other deposits are found in Ontario and Newfoundland. The Lac Tio deposit near Allard Lake, which averages 32–36% TiO_2, is the most important deposit in Canada at the present time. The ilmenite from this deposit is slagged to form a high-TiO_2 product (70–75% TiO_2) and a pig-iron coproduct.

The major US deposit is near Tahawus in the Adirondack Mountains of New York State. In this deposit, large bodies of ilmenite–magnetite are found in anorthosite and gabbroic anorthosite. The ilmenite from this deposit is upgraded to a final product containing 34–38% TiO_2. Other ilmenite deposits are found at Chugwater Creek in Wyoming and in the stillwater complex in Montana.

Norway is a major hard-rock ilmenite producer. Two large deposits are found in the great intrusive region in Egersund, where mining for ilmenite began in the eighteenth century. The Storgangen deposit near Hauge was deposited in a fracture of anorthosite forming a dike. A large deposit at Tellnes was formed as an ilmenite-rich norite intrusion into anorthosite. The ilmenite is upgraded from 18% TiO_2 to 45% by flotation and acid leaching.

Finland has an ilmenite–magnetite deposit at Otanmaki. The ore occurs as lenses which are usually located along contacts between anorthosite and amphibolite.

The USSR has large deposits of ilmenite in the Ilmen Mountains (part of the Ural Mountains) and the Kola Peninsula in the Ukraine. The ilmenite occurs in close association with titaniferous magnetite. Large reserves containing up to 16% TiO_2 are reported in these deposits.

2.2 Primary Rutile Deposits

Most of the rutile in commerce is a coproduct from beach sand mining for ilmenite, zircon and other valuable heavy minerals. Australia and South Africa are the major rutile producers. Rutile is also produced from reworked residual deposits in Sierra Leone.

Primary rutile in rock deposits occurs in alkalic anorthositic complexes, but none of these is considered economic at present.

2.3 Heavy-Mineral Beach Sand Deposits

Weathering-resistant heavy minerals provide most of the high-grade ilmenite, leucoxene and rutile used by

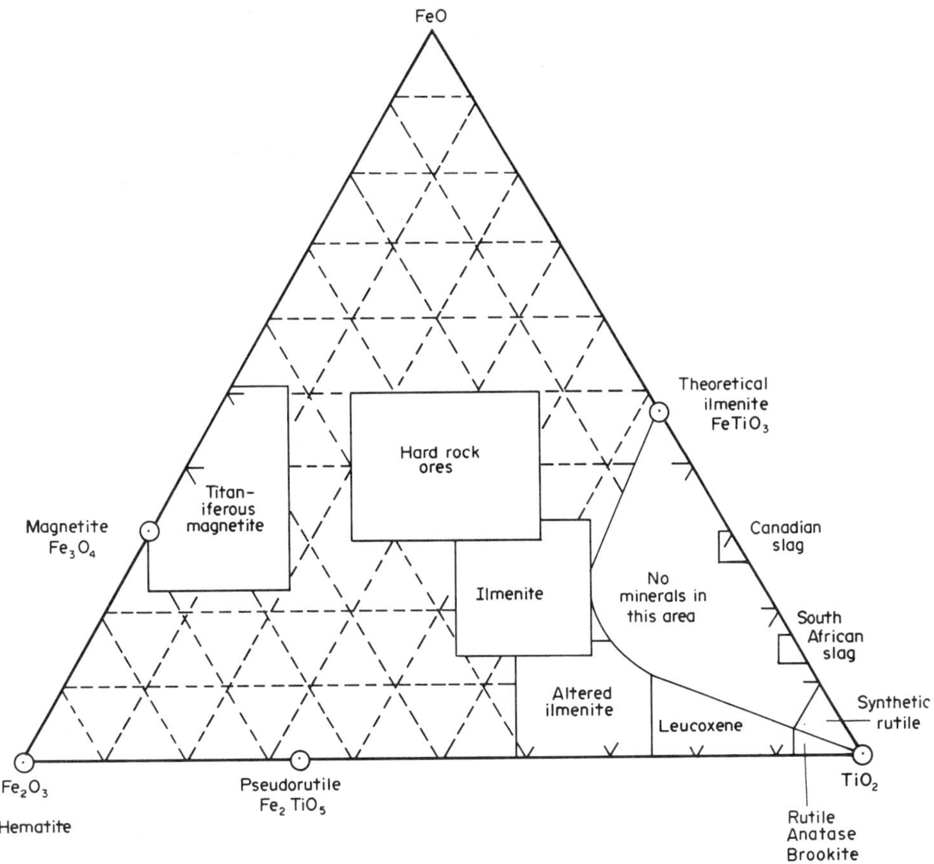

Figure 1
Ternary diagram showing approximate composition ranges of commercially important titanium dioxide minerals and ores

industry. The major heavy-mineral producing countries are Australia, South Africa, the USA, India, Sri Lanka, Malaysia and Thailand. All these deposits are of Pleistocene to Recent age.

The Australian deposits occur on both the east and west coasts of the continent. The major east coast deposits are located between Newcastle and Rockhampton. The ilmenite from these deposits is too low in TiO_2 content to be used in pigment manufacture, but rutile and zircon are produced as coproducts. The major west coast deposits are located near Bunbury to the south and at Eneabba 225 km north of Perth. These deposits furnish ilmenite, leucoxene, rutile, zircon and monazite.

South Africa's heavy-mineral deposits occur in a high dune at Richards Bay ~160 km north of Durban. The TiO_2 content of the ilmenite is too low for pigment manufacture, so it is slagged to make a product containing 85% TiO_2 with pig iron as a coproduct. Rutile is concentrated to 95% TiO_2 and exported with zircon as a coproduct.

Major US heavy-mineral deposits are located in north central Florida and southeast New Jersey. The Florida mines produce ilmenite (64% TiO_2), leucoxene–rutile mixture (82% TiO_2) and rutile (95% TiO_2), with zircon, staurolite and monazite as coproducts. The New Jersey deposits produced ilmenite containing 60% TiO_2 during the 1970s and early 1980s, but have since shut down. There are also known but undeveloped heavy-mineral deposits in Georgia, Tennessee and the Carolinas.

India produces ilmenite and rutile from deposits near Quilon in Kerala State. Large deposits have also been reported in the Chatrapur area.

Sri Lanka, Malaysia and Thailand produce small amounts of heavy minerals, in the latter two as byproducts from tin mining.

2.4 Other Titanium Ore Sources

There are several titanium minerals in large deposits which may replace the conventional ores used today. Of these the one with the most potential is anatase from Brazil. Anatase-rich segregations in several carbonatites have been found in the states of Minas Gerais and Goiás. Some development has started on those located near the town of Araxá.

Large reserves of the mineral perovskite ($CaTiO_3$) occur in alkalic rocks in Colorado, but there are no pigment processes that use this mineral economically at the present time.

There are extremely large reserves of titaniferous magnetite throughout the world. Technology to use these may be developed sometime in the future.

3. Mining and Processing

Mining and processing methods vary with the type of deposit. Most of the hard rock deposits are mined by open-pit methods. The ore is blasted and then picked up with power shovels or front-end loaders and loaded into trucks for hauling to a crusher. After the ore is reduced to a size where the impurities are liberated, the titanium minerals are concentrated using magnetic separators or froth flotation.

Heavy-mineral beach sands are mined with dredges except in areas where lack of water requires dry mining methods. In these cases scrapers, bulldozers and draglines are used. The heavy minerals are separated from quartz using gravity concentrators (spirals, cones or sluices). The heavy-mineral concentrates are then fractionated using magnetic and high-tension separators. Ilmenite and altered ilmenite are magnetic, whereas rutile is nonmagnetic.

Low-grade ilmenite sands may be chemically treated to produce synthetic rutile. This is done in Australia, Japan and the USA. High-TiO_2 materials (70–85% TiO_2) are also produced from low-grade ilmenite by slagging. Hard rock ilmenite is slagged in Canada and ilmenite sand is slagged in South Africa.

4. Uses of Titanium Ores

The primary use for titanium ores is the manufacture of white TiO_2 pigment. Until 1950 most TiO_2 pigment was made using the sulfate process. In this process ground ilmenite is reacted with sulfuric acid to produce titanyl sulfate and ferrous sulfate (copperas). The titanyl sulfate is then converted to the anatase form of TiO_2. Rutile cannot be used with this process because it is insoluble in sulfuric acid. Beginning in the 1950s, a process using elemental chloride was developed to manufacture TiO_2. This process can use ilmenite, rutile (both primary and man-made) or slag. These are reacted with elemental chlorine to form $TiCl_4$ and $FeCl_3$. The $TiCl_4$ is oxidized to TiO_2 in the form of rutile and the chlorine from this step is recycled back to the head feed and used repeatedly. Chloride pigment is reported to have qualities that are superior to those of sulfate pigment.

Small amounts of titanium ores are converted into titanium metal used in aircraft manufacture, desalination tubing, condensers for heat exchangers and elsewhere.

Some rutile is used in manufacturing welding rod fluxes. A small amount of ilmenite is used in well-drilling mud and as a sandblast abrasive.

5. Future Prospects

The US Bureau of Mines has identified 10^9 t of TiO_2 occurring as ilmenite and 280×10^6 t as rutile and anatase. However, only a small fraction of these vast reserves is considered economic to mine at present. The high-grade ore sources obtained from beach sands will be mined out sometime in the early part of the twenty-first century. Development of new deposits to replace these will be influenced by the needs of the various consuming industries. Environmental regulations will also be a key factor in the development of new titanium deposits. New processes to beneficiate low-grade ores into high-grade products may be required to replace the beach sand ores. Development of the Brazilian anatase deposits may offset loss of some of the high-grade ores. Tar sand waste, phosphate sand tailings or sand and gravel mine wastes may also provide high-TiO_2 ores for the future.

Bibliography

Baker G 1962 *Detrital Heavy Minerals in Natural Accumulates.* Monograph Series No. 1. Australasian Institute of Mining and Metallurgy, Melbourne

Baxter J L 1976 *Heavy Mineral Deposits of Western Australia.* Mineral Resources Bulletin No. 10. Geological Survey of Western Australia, Perth

Garnar T E Jr 1981 Heavy minerals industry of North America. In: Coope B M (ed.) 1981 *Proc. 4th Int. Industrial Minerals Congress.* Metal Bulletin, London, pp. 29–42

Gillson J L 1949 Titanium. In: Dolbear S H et al. (eds.) 1949 *Industrial Minerals and Rocks (Nonmetallics other than Fuels)*, 2nd edn. American Institute of Mining, Metallurgical, and Petroleum Engineers, New York, Chap. 49, pp. 1042–73

Klemic H, Marsh S P, Cooper M 1973 Titanium. In: Brobst D A, Pratt W P (eds.) 1973 *United States Mineral Resources*, US Geological Survey Professional Paper No. 820. US Government Printing Office, Washington, DC, pp. 653–65

Lynd L E 1975 Titanium. In: Lefond S J (eds.) 1975 *Industrial Minerals and Rocks (Nonmetallics other than Fuels).* American Institute of Mining. Metallurgical, and Petroleum Engineers, New York, pp. 851–80

Rose E R 1969 *Geology of Titanium and Titaniferous Deposits of Canada*, Economic Geology Report No. 25. Geological Survey of Canada, Ottawa

T. E. Garnar Jr.
[Titanium Consultants, Keystone Heights, Florida, USA]

Traditional Ceramics: An Overview

The American Ceramic Society defines ceramics as inorganic, nonmetallic materials which either in their formation or in use are subjected to heat. This includes the nitrides, carbides and borides. By this definition, therefore, glass, enamels and advanced ceramics are included under the general term of ceramics. The European concept of ceramics does not include glass or enamels but sets them separately in their own industrial classes. However, even the European concept of ceramic materials includes "advanced" or "new" ceramics.

The term *ceramics* comes from the Greek word *keramos*, which means burned clay or pottery. Clay-base products today are still the most important economic sector of production, use and sales throughout the world of all ceramic products.

The term *traditional ceramics* as used here refers to the more familiar sectors of the ceramics industry, such as the articles used in the home, in industry and in the construction trade. This in itself does not preclude the use of the latest technologies in the traditional ceramic industries, since the economic incentive for more efficient processes and cheaper products is driven by refined technologies. There is a tendency to confuse the application of the word "traditional" in ceramics with low technology. This is absolutely not the case. The application of sophisticated tooling, process controls and characterization instrumentation provides for an efficient traditional ceramics industry that is economically viable and useful to international trade.

To understand and appreciate the economic impact of traditional ceramics, it must be realized that in all its forms traditional ceramics is a substantial part of the ~US$50 billion total ceramic business in the USA. Of this business, 50% is clearly visible as products such as dinnerware, glass, sanitary ware and electrical porcelain. The remainder is "invisible," in the form of ceramics produced for in-house use by manufacturers of products such as radio and television sets and computers.

In contrast to the heavy impact of traditional ceramics, advanced ceramics have a world market of US$4.5 billion; that is, <10% of the total US ceramics market. The US share of the advanced ceramics market is currently US$1.5 billion. The future looks very bright, however, with a projected world market in advanced ceramics of US$12 billion by the early 1990s (Sanders 1984).

1. History

Ceramics is probably the oldest of all technologies, and the development of the ceramics industries chronologically follows the development of civilization. The oldest man-made ceramic products were clay pots. These were produced as far back as Neolithic times. These skills developed rather uniformly around the world. The great museums of China, Egypt, Europe and the USA all contain examples of prehistoric pottery of indigenous man dating back to ~6000–3000 BC.

As man learned to use the combination of ceramic materials and fire, he developed a very useful product. As civilization developed, ceramics became more essential as building materials and for proper sanitation. Thus the various fields of traditional ceramics emerged in response to the growing needs of a more sophisticated civilization.

2. Classification

A very useful technical classification of clay-based ceramic materials has been provided by Hennicke (1967). This groups ceramics by physical characteristics rather than by industrial distinctions (Fig. 1).

In the class of traditional ceramics used in this article, sectors of the ceramics industry that are non-clay-based are also considered. Hennicke classes these as products derived from special ceramic materials and again distinguishes them by physical characteristics (Fig. 2).

3. Common Forming Methods

The most commonly used forming methods throughout the various areas of traditional ceramics are slip casting, jiggering, extrusion, pressing (with variable moisture content), isostatic pressing and fusion casting. The properties of the product depend very much upon the technique of processing.

3.1 Slip Casting

There are two basic techniques used in the slip-casting process: the indirect and direct methods. The indirect method requires that the minerals to be used in the composition are blunged, then screened and passed through a ferrofilter to remove grit and iron contamination. The mix is then filter-pressed at pressures between 0.4 and 1.0 MPa. The resulting filter cakes retain ~24% of water. In some factories they are stored for a period of time and allowed to "age." Subsequently they are reblunged in a tank containing enough water and deflocculants to produce a fluid "slip." The normal slip has a specific gravity of ~1.70–1.83 and its viscosity is in the range 200–800 MPa s. The viscosity is controlled with a deflocculant such as sodium silicate. The slip is then poured into a plaster mold, where the excess water is drawn off to form a rigid casting at the surface of the mold. The excess slip is then drained off after the proper amount of time to permit the requisite thickness of the cast.

In the direct method, the materials are placed directly in a blunger and mixed with the electrolytes to the proper specific gravity and flowability. In some instances the ball clays are washed and screened before

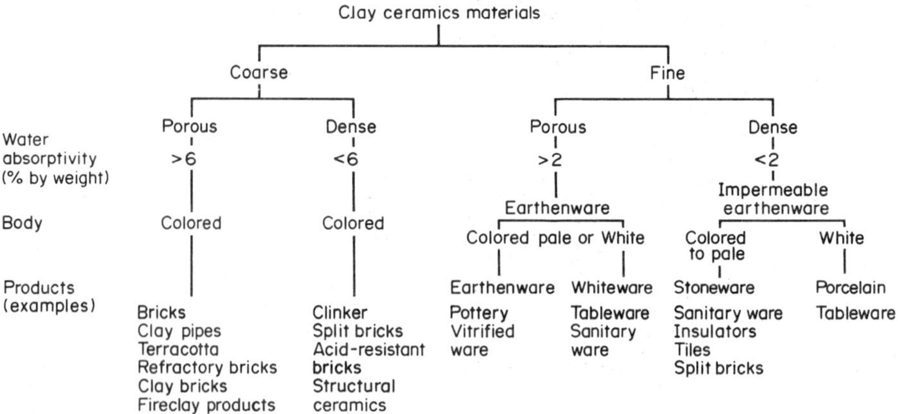

Figure 1
Classification of clay-based ceramics (Hennicke 1967)

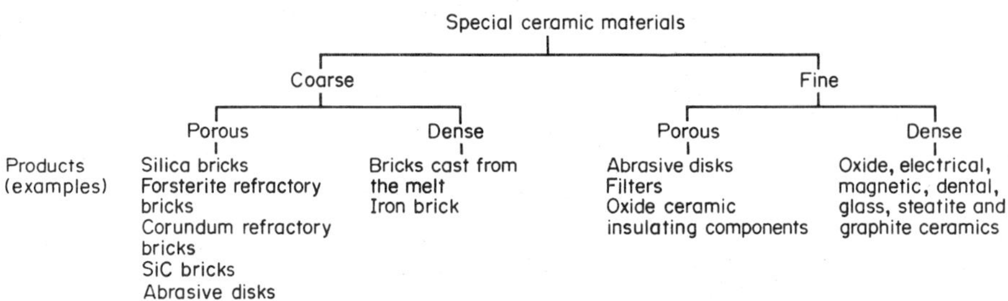

Figure 2
Classification of special ceramics (Hennicke 1967)

the addition of the other material. However, in using this technique there is always the probability of a variation in the salts contained in the raw materials, so more care must be exercised in the proper deflocculation and retirement of the salts. Normally the interfering ions are the sulfates and chlorides. Minor levels can be tolerated, but an excess (>300 ppm) must be chemically eliminated with barium carbonate.

3.2 Jiggering

The jiggering technique depends upon the proper preparation of the plastic clay mass. The clays are thoroughly blunged and screened to remove grit and foreign materials, followed by magnetic separation to remove iron-bearing impurities. Screens of ~80–100 μm are commonly used for this purpose. This screened and washed slurry is then filter-pressed, most commonly at 1.1 MPa. The filter presses are manually unloaded and the filter cake is stored for use. The filter cake is then extruded, cut into the proper sizes, placed on a rotating plastics head and brought into contact with a profiled steel tool. Normally the formed piece in the "enclosing mold" is passed through a hot air dryer and is ready then for firing.

3.3 Extrusion

Here the clay mass is fed into an extruding mill, which forces the clay through a die at the end of a pug mill to give the desired shape to the piece. This is then dried ready for firing.

3.4 Pressing

Perhaps the most used forming technique for traditional ceramics is the pressing process. The prime example is for the very common whiteware compositions of clay, feldspar and flint. This technique can use a very wide range of moisture contents from 0 to 15%. The amount of water depends upon the desired pressing quality and the fineness of grain of the constituent materials. The most common process for preparation is to take air-floated materials that are dry, to mix them in a muller for 5–10 min and then to add the proper amount of water for the desired consistency. Another technique is to use a rotating cone mixer.

When mixing is complete, the powder is fed to a granulator and subsequently to a screen so that a controlled size of material is available for the pressing machines. Depending upon the intricacy of the part, the press can be either hydraulic or mechanical.

3.5 Isostatic Pressing

Preparation of the powder for isostatic pressing is largely accomplished by the use of spray dryers. This means the preparation of a slip that normally has a specific gravity of 1.6–1.75 and a controlled viscosity. This slip is fed into a spray dryer of the nozzle or rotating disk type and the powder is collected at the bottom of the dryer. Properly done, this produces a very uniform powder that is dense, spherical and quite free-flowing. Normally it retains $<0.1\%$ moisture. The powder is then put into proper tooling and placed in an isostatic press, where the pressure is delivered to the formed part uniformly from all directions. This is particularly advantageous when making large parts. An example is large refractory shapes.

4. Drying

After the forming process, the parts must be dried before being placed in kilns for firing. This is because the formed piece must dry from the interior to the exterior. In general terms, this means that at the beginning of the drying cycle the relative humidity in the dryer must be high, until the interior of the formed piece is at a high enough temperature to expel the moisture from this region first. The relative humidity is then gradually reduced until all the moisture is removed. The piece is then ready for introduction into the kiln.

5. Firing

Firing can be accomplished by a kiln utilizing electricity, gas, coal, wood or oil as the energy source. In recent years, because of the increased cost of oil, countries such as Brazil have developed techniques for the conversion of wood and agricultural products into gases for firing kilns. In Brazil, 21% of energy consumption is accounted for by the ceramics industry, including the manufacture of cement, glass and construction products. Therefore, the development of alternative fuel sources for firing of ceramic products is a very essential feature of the economy. Hydroelectric power is also being developed in those countries where this is a viable choice.

The firing of ceramic ware has been the subject of many publications. The basic principles are simple, yet often violated. The stages of firing can be regarded as

(a) the water smoking period, which includes complete drying of the ware as well as the evolution of all binders used for green strength;

(b) the loss of chemically combined water; and

(c) a consolidation phase which can arise either from the melting of some of the composition or, in the case of pure oxide systems, the growth of crystalline phases.

Care should always be exercised when heating the ceramic piece if a crystalline inversion is to be expected. Normally there is a volume change during such an inversion and cracking will arise if proper rates of temperature change are not used during this period. Again, in general terms the shrinkages that occur from the green ware to the final fired state can be anything from 5 to 20% and allowance must be made for a general consolidation of the piece without developing stresses high enough to fracture the ware. Although experience has been the most instructive method of developing firing techniques, the principles of physics and chemistry can certainly be applied if diffusivity, thermal expansion and strength data over the temperature range and the viscoelastic properties are known. Firing techniques in traditional ceramics industries range from the simplistic to the technologically advanced.

A wide spectrum of firing techniques is used around the world. At one end of the spectrum would be the Xian district of China, where roofing tiles and bricks are still fired in caves dug from the loess formations. Here advantage has been taken of a geological formation which serves as the refractory for the kiln. The ware is placed inside the cave, the entrance is sealed and channels are dug from the roof to serve as chimneys. Coal is burned at the base of the entrance. This is in contrast to the very sophisticated automated temperature- and atmosphere-controlled tunnel kilns that are used in the manufacture of a variety of ceramic ware in the more highly developed nations.

6. Conclusion

The important point to be kept in mind, no matter what area of traditional ceramics or advanced ceramics is being considered, is that in any potential use there is a preferred microstructure of the ceramic composition that will best suit the desired application. The interrelation between the mineral selected, the forming method and the firing technique is extremely important in the development of the desired microstructure. A most essential consideration for the ceramist is the proper characterization of each of these steps to arrive at such a microstructure.

As stated previously, advanced ceramics has its roots in the traditional ceramic materials and processes. It is an extension of existing knowledge based on unique compositions and advanced ceramic understanding. However, it is important to recognize that the traditional ceramics industries have refined their materials and processes to the point where each branch has become highly sophisticated. Table 1 illustrates the trend of change in the established ceramics

Table 1
Typical established ceramic industry sectors compared with new sectors[a]

Established and evolving	New and rapidly changing
Refractories	Structural heat-engine ceramics
Whitewares	Wear-resistant specialty ceramics
Electrical ceramics	Electronic ceramics
Flat and container glass	Optical communication glass
Cement and concrete	Composites involving cement
Mineral resources	Synthetic powders and fibers
Enamelled metal	Ceramic-fiber-reinforced metals
Co–WC cutting tools	Ceramic cutting tools
Ceramic nuclear fuels	Nuclear waste disposal ceramics

[a] Wachtman and McLaren (1984)

sectors and their evolution into new and rapidly changing sectors. It can be seen from this table that the established ceramics sectors have the greatest opportunity to develop the new technology.

Traditional ceramics has its roots in antiquity and advanced materials have their roots in traditional ceramics. It becomes increasingly more difficult to know where the boundary between traditional and advanced ceramics lies, for the processes, materials and considerations are essentially the same in both cases. Mankind is becoming increasingly dependent upon material resources. Ceramics plays a key role in materials selection opportunities.

See also: Glass: An Overview; Structural Clay Products

Bibliography

Hennicke H W 1967 *Ber. Dtsch. Keram. Ges.* 44: 209–11
Kingery W D 1960 *Introduction to Ceramics.* Wiley, New York
Kingery W D (ed.) 1963 *Ceramic Fabrication Processes.* MIT Press, Cambridge, Massachusetts
Lawrence W G, West R R 1982 *Ceramic Science for the Potter.* Chilton, Pennsylvania
Onoda G Y (ed.) 1978 *Ceramic Processing Before Firing.* Wiley, New York
Sanders H J 1984 High-tech ceramics. *Chem. Eng. News* July 9: 26
Wachtman J B, McLaren M G 1984 *New Ceramics,* Advances in Materials Series. United Nations Industrial Development Organization, Vienna

M. G. McLaren
[Rutgers University, New Brunswick, New Jersey, USA]

Tungsten Resources

Tungsten has many applications; it has desirable refractory, hardening, alloying and catalytic properties. Uses of its chemical compounds are well established. Tungsten is uniquely suited for use in filaments, electron tubes, high-speed steel alloys and tungsten carbide wear-resistant parts; it has potential applications in the fabrication of components for space vehicles and nuclear reactors. Tungsten end uses are metalworking, mining and construction machinery, lighting applications, electrical equipment and transportation.

1. Geochemical Abundance

Tungsten's average crustal abundance has been estimated at 1.0–1.3 ppm. Granitic igneous rocks contain the most tungsten (average 1.5 ppm), mafic rocks have less (0.5–1.0 ppm), and ultramafic rocks have an even lower content (0.1–0.8 ppm). Sedimentary rocks are estimated to have a tungsten content of 1.0–2.0 ppm; clastic rocks contain more tungsten than do carbonate rocks. Data for metamorphic rocks are fragmentary.

2. Ore Minerals

Tungsten ore minerals can be categorized into two groups: (a) the wolframite series (ferberite, wolframite, hubnerite); and (b) the scheelite series (scheelite and powellite). These ore minerals form primary ores with a variety of coproducts or may be obtained themselves as a coproduct.

3. Production and Consumption

About 50% of the world's estimated tungsten resources are located in the People's Republic of China. Other countries with significant resource potential are Australia, Austria, Bolivia, Brazil, Burma, Canada, Malaysia, North Korea, Peru, Portugal, the Republic of Korea, Thailand, Turkey, the USA and the USSR.

Estimated 1987 world tungsten concentrate production remained near the same level as that of 1986. Annual world production and consumption have been estimated at, respectively, 41.4 and 42.5×10^6 kg of tungsten content. Reliable world concentrate production is dominated by about 24 mines with a production capacity of 300 t per day. Numerous small mines are active only during periods of high prices and price support. Tungsten concentrates have been produced in many countries including the USA, Canada, Tasmania, South Korea, Brazil, Portugal, Bolivia, Sweden and Austria.

4. Ore Deposit Models

Tungsten deposits have a close temporal and spatial relationship with intrusive granitic rocks and their related porphyries. Probably 95% of all tungsten concentrate production has been from three types of deposits: (a) contact metamorphic (Pine Creek roof pendant); (b) quartz veins ± scheelite ± wolframite series (People's Republic of China; western USA); and

(c) stockworks and related deposits (Climax mine, Colorado). Many of these deposits are formed in subduction, or collision related, plate tectonic environments.

There has been sporadic production from pegmatites (Pershing County, Nevada), breccia zones (Yellow Pine mine, Idaho), hot spring deposits (Golconda, Nevada), epithermal manganese veins (western USA; Bolivia), placer deposits (People's Republic of China; Atolia, California), and supergene replacement deposits (rare). Large resources are potentially recoverable as a by-product of alkaline brine production at Searles Lake, California, where it has accumulated from the leaching of tungsten deposits in the upper drainage basin.

Stratiform tungsten deposits have been reinterpreted recently as volcanic–exhalative (northeast Brazil, Argentina, Europe, South Korea); similar deposits elsewhere are of disputed origin. These occur in metallographic provinces of Precambrian to early Paleozoic age.

Tungsten skarn deposits (TSD) and quartz–wolframite deposits (QWVD) are two significant classes of tungsten deposits. Associated deposits include tin and tungsten skarns, and zinc skarns.

Tungsten skarn deposits are characterized by scheelite in calc–silicate contact metasomatic rocks. These deposits occur in contact and roof pendants of granite batholiths and thermal aureoles of apical zones of granitic stocks that have intruded carbonate rocks. Ore controls are carbonate rocks in thermal aureoles of intrusions. Their geochemical signature is W, Mo, Zn, Cu, Sn, Bi, Be and As. The granitic rocks which occur in orogenic belts are of synorogenic or late-orogenic age, and although they are mainly Mesozoic, they may be of any age. Tungsten skarn deposit mineralogy is scheelite ± molybdenite ± pyrrhotite ± sphalerite ± chalcopyrite ± bornite ± arsenopyrite ± magnetite ± traces of wolframite, fluorite, cassiterite and native bismuth. Alteration mineralogy consists of a broad suite of calc–silicate minerals. The tonnages of TSD range from 0.05 to 22 million tonnes and average 1.1 million tonnes. Tenor ranges from 0.34 to 1.4% WO_3 and averages 0.67% WO_3.

Quartz–wolframite deposits consist of wolframite, molybdenite and minor base-metal sulfides in quartz veins associated with granitic stocks which have intruded sandstone, shale and their metamorphosed equivalents. The age of these veins is Paleozoic to late Tertiary. These deposits fill tensional fractures in epizonal granitic plutons and their wallrocks. The associated granitic rocks were derived from the remelting of continental crust. Tin–tungsten veins and pegmatites may also be associated. Albitization, K-feldspar (± REE), greisenization and chloritization are alteration zones. Wolframite persists in soils and stream sediments and is sought in stream-sediment surveys. The QWVD geochemical signature is W, Mo, Sn, Bi, As, Cu, Pb, Zn, Be and F. The tonnages of QWVD range from 0.045 to 7.0 million tonnes and average 0.56 million tonnes. Tenor ranges from 0.6 to 1.4% WO_3 and averages 0.9% WO_3.

5. *Exploration*

Common exploration techniques are the "lamping" of panned concentrates and bedrock with shortwave ultraviolet (uv) light (scheelite fluoresces), and the geochemical analysis of heavy-mineral stream and soil samples. Multielement bedrock lithogeochemistry and geochemical pathfinders (F, As, P, Mo, Cu, Bi, Sb, Sn) have also proved useful in some regions. Biogeochemical techniques using woody plant parts have been used as a guide to tungsten deposits with limited success. Normal waters have a low tungsten content (ppb); tungsten mobility in surface environments is intermediate to low at normal pH values. Until fairly recently, low-ppm chemical analysis techniques have been unreliable. Remote-sensing techniques may indicate alteration zones or granitic intrusives.

6. *Mining*

Tungsten mines are comparatively small and rarely produce more than 2000 t per day of ore, with most producing approximately 300 t of ore per day. A typical mine consists of an integrated mining and milling operation producing concentrates of 60–75% WO_3.

Most tungsten mines are underground; open-pit mining is generally shortlived and related to startup of production. Mining methods are governed by the shape and size of the ore body, and the competence of the ore, country rock and alteration zones. Mining methods are generally: open-, shrink- and sublevel stoping, cut and fill, and square set. Except for structural complications and alteration zones, mining problems are minimal. Mineralization is commonly erratic and variable.

Benefication is by crushing and grinding to liberation size, followed by concentration. Impurities are removed by leaching, roasting, magnetic separation or, rarely, by high-tension separation. Flow sheets are based on gravity (supplemented by flotation) because the principle ore minerals (scheelite and wolframite) are heavy. Overgrinding results in sliming because of the brittle nature of wolframite and scheelite; as a result, concentration is attempted at the coarsest possible level. Low labor and high energy costs permit hand sorting and panning in developing countries. Concentrates are sold on international markets with 60–70% WO_3; prices are quoted on a short ton unit basis.

Few mine safety, health or environmental problems are associated with tungsten mining.

7. Outlook and Demand

The demand for tungsten is closely related to the general economic growth. Most tungsten is used in cutting tools; other applications for unalloying tungsten are electrical devices, and heat- and wear-resistant nonferrous alloys, consumption of which parallels electrical machinery and durable goods demand.

Coatings (aluminum oxide, titanium oxide, titanium nitride) have improved the cutting and wear resistance of cemented tungsten carbide tool inserts. Coating use is expected to increase. The extended wear of inserts decreases the rate of replacement, and thus the growth of tungsten consumption. In addition there will be a slow increase in the substitution of cemented tungsten carbide-based products of titanium carbide-based cutting tools, by ceramic cutting tools and wear parts, and by polycrystalline diamond. Limited scrap is recycled.

See also: Tin Resources

Bibliography

Barabanov V F 1971 Geochemistry of tungsten. *Int. Geol. Rev.* 13: 332–44

Barton P B 1986 Commodity/geochemical index. In: Cox D P, Singer D A (eds.) 1986 *Mineral Deposit Models*, US Geological Survey Bulletin 1693, Appendix C. US Government Printing Office, Washington, DC, pp. 303–17

Cox D P 1986 Descriptive model of W skarn deposits. In: Cox D P, Singer D A (eds.) 1986 *Mineral Deposit Models*, US Geological Survey Bulletin 1693. US Government Printing Office, Washington, DC, pp. 55–57

Cox D P, Bagby W C 1986 Descriptive model of W veins. In: Cox D P, Singer D A (eds.) 1986 *Mineral Deposit Models*, US Geological Survey Bulletin 1693. US Government Printing Office, Washington, DC, p. 64

Hobbs S W, Elliot J E 1973 Tungsten. In: Brobst D A, Pratt W P (eds.) 1973 *US Mineral Resources*, US Geological Survey Professional Paper 820. US Government Printing Office, Washington, DC, pp. 667–78

Jones G M, Menzie W D 1986 Grade and tonnage model of W veins. In: Cox D P, Singer D A (eds.) 1986 *Mineral Deposit Models*, US Geological Survey Bulletin 1693. US Government Printing Office, Washington, DC, pp. 65–66

Krauskopf K B 1978 Tungsten (Wolfram). In: Wedepohl K H (ed.) 1978 *Handbook of Geochemistry*, Vol. II/5. Springer, Berlin, pp. 74-B–74-O

Larsen L P, Lowrie R L, Leland G R 1971 *Availability of Tungsten at Various Prices from Resources in the United States*, US Bureau of Mines Information Circular 8500. USBM, Washington, DC

Little H W 1959 *Tungsten Deposits of Canada*, Economic Geology Series 17. Geological Survey of Canada, Ottawa

Maucher A 1976 The strata-bound cinnabar–stibnite–scheelite deposits. In: Wolf K H (ed.) 1976 *Handbook of Stratabound and Stratiform Ore Deposits*, Vol. 7, *Regional Studies and Specific Deposits*. Elsevier, New York, pp. 477–503

National Materials Advisory Board 1973 *Trends in Usage of Tungsten*, NMAB-309. National Technical Information Service, US Department of Commerce, Springfield, Virginia

Reid J C 1986 Computer-aided lithogeochemistry: An application for prospecting for Precambrian Stratabound scheelite, Northeastern Brazil. In: Ramani R V (ed.) 1986 *Application of Computers and Operation Research in the Mineral Industry*. Society of Mining Engineers, Littleton, Colorado, Chap. 48, pp. 531–41

Singer D A 1986 Summary statistics of grade-tonnage models. In: Cox D P, Singer D A (eds.) 1986 *Mineral Deposit Models*, US Geological Survey Bulletin 1693, pp. 293–302

Shunso I 1981 The granitoid series and mineralization. *Econ. Geol.* 75th Anniversary Volume, pp. 458–84

Smith G R 1987 Tungsten. *Mineral Commodity Summaries 1987*, US Bureau of Mines, Washington, DC, pp. 172–73

Yih S W H, Wang C T 1977 *Tungsten: Sources, Metallurgy, Properties, and Applications*. Plenum, New York

J. C. Reid
[North Carolina Geological Survey, Raleigh, North Carolina, USA]

U

Underground Mining

While open-cut mining is the predominant method of producing the world's minerals, most deposits are so situated that underground methods must eventually be used. In steeply dipping vein and bedded deposits, very little production can be obtained from an open cut; the cost of removing larger quantities of waste rock from the walls of the deposit in order to maintain a stable working slope with increased depth is economically prohibitive. Even in disseminated and massive deposits, where open-cut mining can be done on a large scale and at a low cost per tonne of material, the ratio of waste to ore increases with depth, until a condition is reached where mining can continue only if there is a change to underground methods.

Large mines with individual capacities of 5000–75 000 t per day of ore account for a much larger percentage of underground mineral production than do the more numerous medium-sized mines with capacities of 500–1000 t per day, and small mines with capacities of less than 100 t per day. Developments in mining technology indicate that underground mines with a productive capacity of as much as 100 000 t per day can be expected in the near future, and that the maximum depth of conventional mining, now of the order of 2–3 km, can be expected to reach depths as great as 5 km. In situ mining, a new technique based on the leaching of ore minerals from deep orebodies, and the extraction of the pregnant solutions from boreholes, may extend future economic depths of mining to 8–9 km.

Underground mines are categorized by the means of access to the orebody and the type of support required for mine workings. Some mines are entered by tunnels, others by deep shafts. Some workings are in rock strong enough to require little or no artificial support, other workings require support by rock bolts, timber, waste rock or mill tailings. Some orebodies and wall-rock are so weak that a caving method can be used, in which the deposit is allowed to collapse in a specific pattern as it is mined.

1. Development Methods

Access to a mineral deposit (Fig. 1) may be obtained by a horizontal tunnel or *adit*, driven to the orebody from a portal in a favorable topographic site, so that an acceptable tonnage of ore can be mined above the adit level. Access may be obtained by a vertical shaft sunk in the orebody, or, preferably, in solid rock near the orebody. Access may also be obtained by an incline (decline), by an inclined shaft or by a gently inclined access spiral in which low-profile rubber-tired trucks can haul material directly from the mine working face to the surface. In deep orebodies, a vertical shaft is generally most economical, and a mine will commonly have several shafts to accommodate production, personnel and ventilation.

Production shafts—rectangular or circular openings 6–10 m in diameter—are commonly sunk by conventional methods, incorporating cycles of drilling, blasting and removal of the broken rock. Ground-support systems in shafts use timber, steel and concrete. Bored shafts, sunk by mechanical drilling machines, are generally 1.5–5 m in diameter; bored shafts are most commonly used for escapeways or ventilation, and have not as yet been drilled to depths beyond 1 km.

Figure 1 shows development workings in a vein deposit. Level (horizontal) workings in the mine consist of drifts along the orebody, and cross-cuts driven to intersect the orebody. Drifts may be within the ore zone or may be driven for greater stability in solid rock on the lower or footwall side of a dipping ore zone.

Levels at successive depths within the mine are connected by raises—steeply inclined or vertical workings driven upward at appropriate intervals for access to specific blocks of ore. A *winze*—an inclined or vertical working driven downward from a level—is more costly, because the broken rock must be hoisted rather than collected at a lower level. Extraction of the ore is begun when a complete network of levels, sublevels and raises for access, haulage and ventilation has been developed and equipped for mining.

2. Mining Sequence

Mining operations in development workings or headings and ore-extraction *stopes* have generally been done in a conventional cycle of drilling, blasting, mucking (removal of broken rock) and the installation of timber or of roof bolts for ground support. Continuous mining operations, first used in coal mining, have been adopted in noncoal mines where the nature of the rock permits development workings to be bored with large rotary drilling machines, and mining to be done by rotary or similar machinery. The resulting opening is generally smoother and stronger than would be obtained in a conventional cyclic operation, thus providing less resistance to the flow of ventilation air and lessening the need for ground support. The practice of raise boring is well-established; shaft boring is less common but is finding increased use, especially for smaller-diameter escapeways and ventilation shafts. Tunnel-boring machines and mechanical rock-cutting equipment have been used in the driving of level

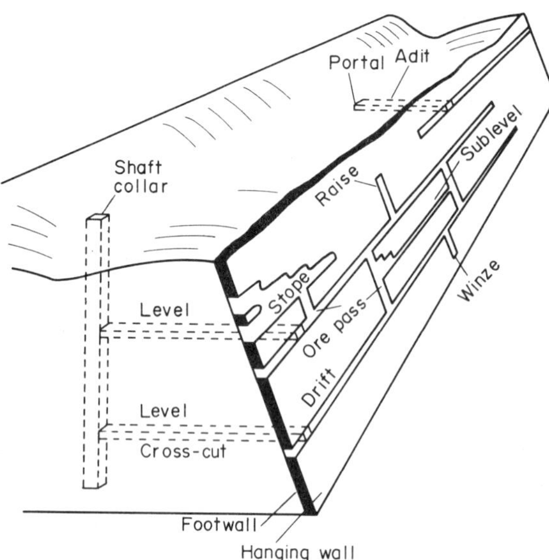

Figure 1
Underground mining terminology

workings, but they do not perform as well in hard rock and ore as they do in coal and in softer materials such as potash and salt.

Waste rock and ore broken from development headings and from mining faces in stopes may be collected and transferred to various types of haulage units by air- or electric-operated mechanical loaders (mucking machines), by cable-guided scraper systems or by mobile conveyors. Haulage to a transfer point or to the mine portal has commonly been done by electric-, diesel- or compressed-air-powered locomotives with trains of ore cars. In many modern mines, where ground support is provided by roof bolts rather than timbers, there is room for trackless haulage, a versatile practice in which electric- or diesel-powered trucks and shuttle cars on rubber tires can be loaded and driven where needed.

The use of diesel engines with exhaust-conditioning devices in underground mining has permitted the widespread use of LHD (load–haul–dump) units. These highly maneuverable and powerful rubber-tired machines can be driven to a heading or mining face, loaded with broken ore and rock, and driven for several hundred meters on levels or inclines to locations for transfer of the material to larger haulage units.

Ore and waste are commonly removed from the mine by hoisting in shafts, but material can also be removed in a conveyor-belt system or in an extension of the trackless underground haulage system through large-diameter adits, inclines or access spirals.

In hoisting arrangements, ore and waste are loaded into skips from underground storage pockets, lifted to the headframe at the surface, and dumped into ore bins or waste-disposal systems.

3. Production Methods and Patterns

Table 1 lists principal methods of underground mining relative to orebodies in strong (competent) rock that will support openings (stopes or near-horizontal rooms), weaker rock requiring at least temporary artificial support, and weak (incompetent) rock that can be allowed to cave during mining. The strength, attitude and dimensions of the ore, as well as the strength of the wallrock, are factors in the choice of a mining method. Since very few ore deposits and wallrocks are uniform, most mines use more than one method.

3.1 Open Stoping

Open stoping is used in situations with strong wallrock. In overhand open stoping, the ore is mined upward from a level. In breast stoping the ore is mined horizontally. In underhand stoping (a less-common practice) the ore is mined downward, with the broken material discharged into an ore pass and collected at a lower level. Even in the strongest ore, the width of an open stope in a dipping orebody is limited to approximately 30 m of unsupported span. In practice, occasional pillars of low-grade ore may be left, roof bolts may be used in fractured areas to maintain stability, and *stulls* (short timbers) may be wedged between the stope walls for stability and access.

3.2 Sublevel Stoping

Sublevel stoping in steeply dipping orebodies with strong ore and strong wallrock begins with the partitioning of a large block of ore into a series of thinner blocks or slices by driving sublevels. The ore is drilled and blasted by miners in the sublevels so that each

Table 1
Underground mining methods

With naturally supported openings:
open stoping
sublevel stoping
longhole stoping
room-and-pillar mining
shrinkage stoping
With artificially supported openings:
stull stoping
cut-and-fill stoping
square-set stoping
longwall mining
top slicing
Caving methods:
block caving
sublevel caving

sublevel retreats an echelon slightly ahead of the next-higher level; this pattern allows the broken ore to fall directly to the bottom of the stope. The back in each sublevel is safer and more accessible than the high and remote back in a simple open stope. Crown pillars or floor pillars of unmined ore are left at the top of the completed stope to support the next major level above. Rib pillars are commonly left for stability near the raises at the ends of the stope. A series of funnel-like draw points or mill holes in a lower or sill pillar is used to collect the broken ore for loading into the haulage system.

3.3 Longhole Stoping

Longhole stoping is a variety of sublevel stoping in which larger blocks of ore can be taken by drilling and massive blasting from locations in more widely spaced sublevels. Carefully planned rings of blast holes are drilled upward and downward from each sublevel, with holes as much as 30 m in length, to provide a series of fan-like patterns for breaking large tonnages of ore.

3.4 Room-and-Pillar Mining

Room-and-pillar mining, in which a regular pattern of rooms (stopes) alternates with undisturbed pillars, is applicable to flat-lying or gently dipping deposits. It is an especially low-cost method, in which fast-moving trackless equipment can be used to advantage. Thin-bedded or tabular deposits can be mined in one operation in the full-face method; thicker deposits require benching and mining in a second stage after the first advance. An important design criterion is the maximum safe width of the rooms, generally of the order of 5 m in softer rock and 20 m in harder rock with appropriate roof bolting. Fifty to seventy percent of the ore is extracted during mining, but extraction can sometimes be increased to 90% or more by "robbing" pillars and allowing the roof to cave.

3.5 Shrinkage Stoping

Shrinkage stoping is an overhand method, in which a major portion of the broken ore accumulates in the stope and helps support the walls during mining. It is suited to steeply dipping veins in wallrock that would slough if left open, but which is still strong enough not to collapse and dilute the ore. When ore is broken it has a swell factor of 1.2–1.7; the excess volume is shrunk from time to time by drawing enough broken ore to allow access to the unbroken ore in the back. Miners work at the top of the broken ore. When all of the ore in the stope has been broken by drilling and blasting, it is drawn. The empty stope may be left open or it may be filled with mill tailings or waste rock. Extraction is high, amounting to 75–90%, but dilution from waste rock is a problem because there is no opportunity for leaving waste pillars or low-grade pillars.

3.6 Stull Stoping

Stull stoping is similar to open stoping, except that it is a method requiring artificial support for the walls. Support is provided by stulls wedged at intervals between the hanging wall and footwall of the stope. Wallrock must be fairly competent; even so, the maximum width that can be mined in this way is about 5 m.

3.7 Cut-and-Fill Stoping

Cut-and-fill stoping (Fig. 2) is practised in steeply dipping orebodies with weak walls. Fill material (sand or mill tailings) is built up as overhand stoping progresses. The compacted fill material provides a suitable floor for trackless mining machinery in wide stopes. Vein widths as great as 30 m have been mined in this way; the limit in minable width is a function of the ability of the ore to stand across a stope back. Resuing is a less-mechanized type of cut-and-fill stoping that is used in narrow veins; in this method, ore and waste are broken separately and the waste rock is left in the stope as fill.

3.8 Square-Set Stoping

Square-set stoping is a labor-intensive method in which a timber set (a skeletal box of interlocking timber) of the order of 1.5–3 m in diameter is installed in the opening left by the drilling and blasting of each small block of ore. The stopes eventually become a network of interlocked timber sets. Square-set stoping is a long-established method of mining relatively weak orebodies in slabby wallrock, and it is still in use where the ore is rich enough to permit a high mining cost. In many mines, square setting has been superseded by larger scale and lower grade ore caving methods, or by a mechanized cut-and-fill method. When stopes are completed by square setting, they are commonly filled with mill tailings or waste rock to provide longer-term stability while mining continues in other stopes.

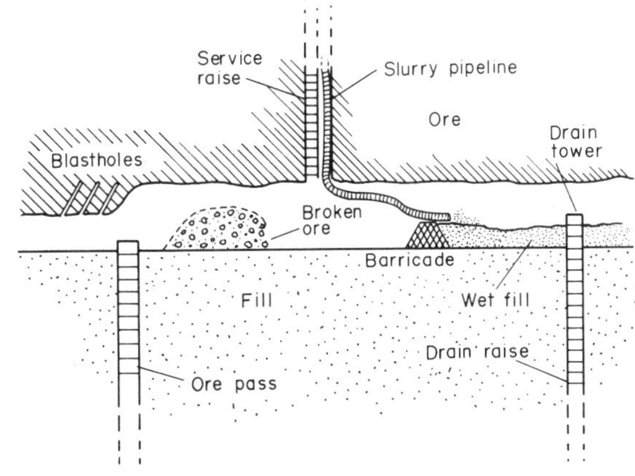

Figure 2
Cut-and-fill stoping (longitudinal geometry)

Pillars remaining between filled square-set stopes are commonly recovered by cut-and-fill or top-slicing methods.

3.9 Longwall Mining

Longwall mining is principally a coal-mining method, but it is also used in gently dipping bedded ore deposits with relatively weak overlying rock and a relatively strong or competent floor. Roof support is only temporary, and the roof is allowed to cave behind the active area. Roof-support units, sometimes with adjustable hydraulic legs, are moved ahead as a mining face or wall 100–200 m long is advanced. At great depths, where rock pressures are too high to permit room-and-pillar mining with safety, longwall mining is done in a carefully designed pattern in which pressures are high in advance of the face, low at the face, and again high in the caved and abandoned area.

3.10 Top Slicing

Top slicing is a method used in thick ore deposits with weak ore and weak wallrock; it is also used in the recovery of ore pillars between filled stopes. Top slicing is actually a caving method of mining, but caving does not proceed until each successive horizontal cut (or slice) is taken; the overlying waste material is then allowed to subside onto a flexible mat of timber. Mining continues downward slice-by-slice, beneath an accumulation of waste and timber matting. Like square-set mining, top slicing is labor intensive and requires a supply of inexpensive timber.

3.11 Block Caving

Block caving is a method with a low cost per tonne, particularly suited to large bodies of weak ore in relatively weak wallrock. Closely fractured and brittle ore is needed, and the overlying material must be capable of subsiding as the ore is removed. A block-caving mine is developed by driving workings on a haulage level, on a grizzly level at a higher elevation, and on an uppermost undercutting level. Caving of the block is induced by drilling and blasting on the undercut level. In some mines, the breaking of the block must also be induced by additional cutoff raises and slots around the edges. Once the ore begins to cave, the block continues to break by tension, and the caving progresses for as much as 1000 m to the surface, where a depression forms. Continued caving and crushing in the ore block provides broken ore for drawpoints on the undercut level. Finger raises feed broken ore from the undercut level to the grizzly level, and transfer raises feed broken ore to the haulage level. Careful monitoring and control of the entire operation is essential, so that no part of the system will hang up or will "pipe" through to waste rock prematurely. While the mining cost in block caving is low, the capital cost is the highest among all underground mining systems. Production rates are characteristically high, reaching 50–75 kt per day.

3.12 Sublevel Caving

Sublevel caving is similar to sublevel open stoping and longhole stoping, but the walls are allowed to collapse and follow the mining of the ore in successive downward increments. Sublevel caving is best suited to the mining of well-fractured ore that is not free-caving but will nonetheless break into relatively small fragments. As with block caving, production rates are high, recovery is high, and there is a substantial amount of dilution by wallrock.

4. In Situ Leaching and Borehole Mining

In situ leaching is a mining method in which hydrometallurgical extraction takes place with the orebody. It is applied to copper and uranium deposits that are uniformly permeable in their natural state, or deposits that can be made permeable by some method such as explosive shattering, block caving or hydraulic fracturing. The permeable orebody must in addition be isolated, either naturally or by artificial barriers, from permeable zones in the wallrock, so that the leaching solutions can be controlled.

In situ leaching grew out of a much older practice of recovering copper by means of heap leaching or dump leaching at the surface. Cement copper, the product obtained by reaction of shredded scrap iron with copper-bearing acid water from leaching operations or from mine drainage, has been obtained for many years.

Copper mining in situ makes use of sulfuric acid as a leaching agent, with assistance from certain bacteria that oxidize copper and iron sulfides in acid solutions. Uranium mining in situ is done with acid solutions or with ammonium carbonate solutions. Uranium concentrate is recovered from pregnant solutions by ion exchange. Copper may also be recovered in this way, with further recovery by electrowinning.

Preparation of orebodies for in situ leaching operations is by conventional mine development workings or by boreholes. Boreholes are arranged in a pattern of injection, recovery and monitor wells. In situ mining by borehole methods is not a common practice, but it has attracted special attention as an environmentally safe method of recovering minerals from deep-seated, low-grade or small deposits that would not otherwise be economic for conventional mining.

Borehole hydraulic mining or slurry mining is an erosional technique that breaks ore into small fragments without decomposing them. The work is done by water jets generated in a mining tool within the borehole; the slurry of eroded ore is pumped to the surface. Borehole hydraulic mining is not yet practised on a commercial basis, but it has been tested and found to be an attractive possibility for extracting ore from near-surface deposits that are friable or easily fragmented.

An increasing proportion of the world's minerals will be produced by large and deep underground mines. Conventional underground mining technology is advancing in productivity and in mine capacity. In situ leaching of ore deposits, still in its infancy, can be expected to provide access to depths and orebodies beyond the reach of current mining practice.

See also: Ocean Resources and Mining; Open-Cut Mining

Bibliography

Dravo Corporation 1974 *Analysis of Large Scale Non-coal Underground Mining Methods.* US Department of Commerce, National Technical Information Service, Springfield, Virginia

Given I A (ed.) 1973 *SME Mining Engineering Handbook*; Vol. 1. Society of Mining Engineers of AIME, New York, pp. 1–262

Hustrulid W A (ed.) 1982 *Underground Mining Methods Handbook.* American Institute of Mining, Metallurgical and Petroleum Engineers, New York

Schlitt W J (ed.) 1982 *Interfacing Technologies in Solution Mining.* American Institute of Mining, Metallurgical and Petroleum Engineers, New York

Stout K S 1980 *Mining Methods and Equipment.* McGraw-Hill, New York, pp. 143–96

Thomas L J 1977 *An Introduction to Mining: Exploration, Feasibility, Extraction, Rock Mechanics.* Halsted Press, New York

W. C. Peters
[Tucson, Arizona, USA]

Uranium Resources

Uranium is a metal unique not only for its physical and chemical properties but also, more importantly, for its nuclear properties. Its natural fissionable and fertile isotopes provide material for explosive devices, as well as for peaceful energy production and medicinal uses. Almost every natural material contains a trace of uranium and anomalous concentrations are widespread, but ores are restricted to specific geological settings. The mining, milling, isotope enrichment and use of uranium can cause health problems that have evoked political and social controversy. In the mid-1970s during the energy crisis, very high projections for the demand of uranium were made, but several nuclear accidents and the ensuing economic recession resulting in slower economic growth and demand for electricity led to projections for one-third less uranium requirements in 1986.

1. Definition and Uses

Uranium is a silvery white metal that consists of the three semistable radioactive isotopes ^{238}U, ^{235}U and ^{234}U. It is an important energy source because fission of ^{235}U releases large amounts of energy. This readily fissionable nuclide constitutes only about 0.7% of natural uranium. Most of the remaining 99.3% is ^{238}U and about 0.005% is ^{234}U. The isotope ^{238}U is not readily fissionable, but it is fertile material that under neutron bombardment converts to fissionable ^{239}Pu.

Before 1942, uranium was used chiefly for coloring glass and ceramic glazes. An ample supply for these purposes was obtained by recovering uranium from ores mined for radium and vanadium. In 1942, controlled nuclear fission was demonstrated, making possible two new and vastly more significant uses for uranium: as an explosive for weapons and as a source of heat from reactors to drive steam generators.

Among the electrical-power reactor designs, those that yield less fissionable material than consumed are called converter reactors, whereas those that yield more are called breeder reactors. Only 1–2% of the potential energy in uranium can be utilized in a converter reactor, whereas a breeder reactor can utilize 60–80%. The ^{235}U in natural uranium must be enriched from about 0.7% to about 3.5% to be fuel in a converter reactor; enrichment leaves a tails product or a depleted uranium material with about 0.20% ^{235}U. Because of the density of depleted uranium metal and its pyrophoric properties on impact, it is used chiefly for armor-piercing projectiles and for counterweights and ballast weights. In 1979, demand for depleted uranium almost equalled that of enriched uranium in the USA (Kirk 1980).

Highly enriched material (20–93.5% ^{235}U) is used in high-temperature thorium-powered reactors that require the creation of ^{233}U, the third fissionable isotope, from ^{232}Th.

Spent fuel from nuclear reactors may be reprocessed for future use. One reprocessing product is ^{239}Pu, which can be reused (especially in breeder reactors) but perhaps more significantly is required for nuclear explosive devices. Because of this use and worldwide concern over proliferation of nuclear-warhead capability, reprocessing is controlled by the International Atomic Energy Agency.

2. Mineralogy

Uranium occurs only in minerals that contain oxygen. Uranium in black ores formed by reduction is tetravalent, being most commonly found as the minerals uraninite (pitchblende) and coffinite, which are generally associated with pyrite. Tetravalent uranium also substitutes for thorium in minerals such as monazite and uranothorite, and for calcium in carbonate fluorapatite. It also occurs in organic compounds in many coal-bearing rocks and marine black shales.

Under oxidizing conditions, hexavalent uranium forms oxide, vanadate, arsenate, silicate, sulfate and carbonate minerals, most of which are hydrous and many of which are bright yellow or green. The yellow vanadates, carnotite and tyuyamunite, are most common. Copper, gold, silver, nickel, chromium, molybdenum and selenium minerals accompany some uranium ores.

3. Geochemical Cycle

Uranium is a ubiquitous and mobile element. Anomalous concentrations of uranium are widespread in continental crustal rocks but rare in oceanic crust. Economically minable ores, containing 250–5000 times the mean crustal concentration of 2 ppm uranium are not abundant. Deposits occur in nearly every major rock type, but major economic resources are few in kind. Many different igneous, metamorphic and sedimentary processes have mobilized and concentrated uranium during fairly restricted intervals of geological time.

Precambrian rocks either directly host uranium deposits or indirectly were sources of uranium for deposits in younger rocks (Nash et al. 1981). The earliest part of the Precambrian era is the Archean (3.8–2.5 billion years ago), whose rocks contain no known uranium deposits because the crust then was thin and unstable, and conditions were unfavorable for either uranium concentration or preservation. In the early Proterozoic (2.5–1.6 billion years ago), the crust became thicker and more stable, and under the anoxic atmosphere, placer concentrations of tetravalent uranium minerals formed in distinctive quartz–pebble conglomerates of the first sedimentary basins on the new stable crust. A combination of sedimentary and magmatic processes during this time probably created uranium-rich regions, generally referred to as uranium provinces.

About 2.2 billion years ago the earth's atmosphere became oxygenated, and uranium minerals could be oxidized, releasing uranium to aqueous solutions for transport to new sites of deposition, mainly by reduction. Within the stable platforms, broad shallow basins accumulated sandstone, limestone and shale sequences that host uranium deposits. During the early to middle Proterozoic period, carbonaceous shale appears to have accumulated uranium by adsorption and reduction from uraniferous oxidized ground and surface waters. These uraniferous shales probably became the protoliths for unconformity-related vein and other types of deposits formed after metamorphism. Uranium in other settings was recycled by magmatic, hydrothermal and erosional processes from middle Proterozoic time onward to form other types of uranium deposits.

The next major change to affect uranium concentration was the development of land plants in Devonian time (0.4 billion years ago). Land plants provided organic and bacterial reductants for uranium in continental sandstone formations of Carboniferous, Permian, Jurassic, Cretaceous and Tertiary ages.

Uranium continues to be remobilized and in the past million years minable deposits have formed in the zone of evaporation close to the surface.

4. Types of Deposits and their Resource Importance

Uranium deposits have been named after some distinguishing feature, such as host rock, mineralizing process or structural and stratigraphic setting. No acceptable taxonomic classification of deposits is available. Important deposits consist of quartz–pebble conglomerate, unconformity-related vein, disseminated, contact, classical vein, sandstone, sedimentary breccia, volcanogenic and calcrete deposits (Nash et al. 1981). In addition, uraniferous phosphorite yields significant by-product uranium, and marine black shale contains large low-grade (0.006–0.03% U) tonnages not economic for uranium alone.

Quartz–pebble conglomerate uranium deposits are restricted to basal early Proterozoic fluvial and deltaic beds lying unconformably on Archean granitic and metamorphic basement rocks of Precambrian shield areas. These well-developed, extensive deposits at Witwatersrand, South Africa, and Blind River–Elliot Lake, Ontario, Canada, represent about 20% of world resources. Less-developed deposits are known in other shield areas, such as in Brazil and India. The Witwatersrand deposits contain 0.03–0.06% U and yield uranium mainly as a by-product of gold mining, although some uranium is a coproduct or main product. The Blind River–Elliot Lake deposits contain 0.10–0.14% U, little gold and some by-product thorium.

Nearly all identified unconformity-related vein uranium deposits are in early Proterozoic marginal marine metasedimentary rocks that are unconformably overlain by middle Proterozoic unmetamorphosed fluvial quartz sandstone, chiefly in the Pine Creek geosyncline, Northern Territory, Australia, and Athabasca Basin, Saskatchewan, Canada. These ores, recognized and explored only since 1968, account for nearly 15% of world uranium resources. The deposits are typically high grade (0.35–10.0% U) and contain 20 000–200 000 t U. The ores are mainly in repeatedly faulted and chloritized zones in graphite–mica schist, biotite–garnet schist and dolomitic marble sequences. Most deposits contain only uranium as uraninite and minor coffinite associated with hematite; a few deposits contain significant amounts of nickel, cobalt and gold.

Disseminated uranium deposits occur in granitic and pegmatitic rocks formed in deep, hot metamorphic environments. The prime example is the Rössing deposit, Namibia, which contains fine uraninite, betafite and secondary uranium minerals disseminated in late Proterozoic or Cambrian alaskitic–pegmatitic granite. This deposit contains about 125 000 t U in rock averaging about 0.03% U. Disseminated deposits account for more than 5% of world resources.

Contact uranium deposits occur in metamorphic contact zones around intrusive granite, as typified by the Midnite deposit, Washington, USA, and Mary Kathleen deposit, Queensland, Australia. At the

Midnite mine, uraninite and coffinite occur as disseminated grains, replacements of rock minerals and fracture fillings in graphitic and pyritic metapelite of middle Proterozoic metasedimentary rocks intruded by upper Cretaceous granite. At the Mary Kathleen deposit, uraninite is disseminated in rare-earth–thorium-bearing allanite found in garnet breccia at the contact of early Proterozoic metasedimentary rock with intrusive granite. Contact deposits contain 5000–15 000 t U in rock averaging about 0.20% U; they contribute less than 1% to world resources.

Classical hydrothermal uranium vein deposits occur in narrow near-vertical fault and fracture zones extending to as deep as 1000 m chiefly in Proterozoic metamorphic (gneiss, schist) and granitic rocks. Uraninite and coffinite richly fill cavities and are disseminated in adjacent rock; associated minerals can be simple, such as hematite or iron sulfide, or complex mixtures of silver, cobalt, nickel, bismuth, selenium, arsenic and other minerals. Important deposits include Schwartzwalder, Colorado; Jachymov, Czechoslovakia; Shinkolobwe, Zaire; and Massif Central, France. Classical veins contain 5000–15 000 t U in rock averaging as much as 1% U and constitute nearly 10% of world resources.

Sandstone uranium deposits are widespread in continental and marginal-marine rocks of Carboniferous age and younger. Important examples include Carboniferous deltaic sandstones, Niger; Permian and Cretaceous sandstones, Argentina; Permian siltstone, France; and Triassic, Jurassic and Tertiary sandstones, USA. Sandstone deposits constitute about 32% of world resources and 95% of US resources (US Department of Energy 1980). The host rocks are chiefly fluvial or lacustrine, medium- to coarse-grained, arkosic or tuffaceous, quartzose sandstone. Uraninite and/or coffinite either fill interstices or replace organic matter (plant remains, humate cement) and rock grains unevenly to form tabular and crescent-shaped bodies. Individual orebodies are small to medium size and average 0.05–0.25% U, but collectively they can form deposits that total more than 100 000 t of contained uranium. One percent or more vanadium or copper occurs in some deposits, and molybdenum, selenium and other metals are common locally and may be recovered.

Sedimentary breccia deposits are of two varieties; one consists of stratabound and remobilized ores in a Proterozoic continental breccia environment (Roberts 1983), and the second consists of rich ore in collapse-breccia pipes in upper Phanerozoic sandstone and shale formations (Wenrich 1985). Both are polymetallic, commonly having associated copper, iron and precious metals. The first is represented by a single gigantic deposit, Olympic Dam, South Australia, with a relatively low uranium content of 0.05% U but containing 1 000 000 t U. Alone it represents nearly 10% of the world resources. The breccia-pipe deposits are known only in northern Arizona, USA, and they average about 0.55% U in bodies containing 500–2000 t U.

Volcanogenic uranium deposits occur mainly in Tertiary tuff, rhyolite, mafic flow rocks and volcanoclastic sediments. Those in lacustrine sediments are transitional into sandstone uranium deposits. Important examples are Peña Blanca district, Chihuahua, Mexico; Marysvale district, Utah, USA; and Maureen deposit, Queensland, Australia. Uranium minerals occur as disseminated grains, impregnation of host rock and fracture fillings, and commonly are associated with fluorine, molybdenum, beryllium, lithium, mercury and rare-earth elements. The deposits average 0.05–0.10% U and contain 1000–10 000 t U; they represent <1% of world resources, although they are the chief resource in Mexico.

Surficial uranium deposits represent <2% of world resources and occur in upper Pleistocene to Holocene surficial sediments rich in organic matter and surficial clastic sediments cemented by carbonate or sulfate minerals (calcrete, gypcrete). Calcrete uranium deposits contain hexavalent uranium minerals, chiefly carnotite, that fill voids and fractures. The largest deposit, Yeelirrie, Western Australia, contains about 42 000 t U in rock averaging about 0.13% U; calcrete and gypcrete deposits also occur in Namibia, Somalia and Mauritania. Deposits in organic-rich sediments have formed during the past 10 000 years or so and thus emit very little radioactivity. This, coupled with an absence of distinct uranium minerals, obscured their detection until the early 1980s. They may represent a large worldwide low-cost resource.

5. *Production, Resources and Demand*

Production of uranium from the WOCA (World Outside Centrally Planned Economic Areas) countries totalled about 777 000 t U in the period 1948–1984 with about 39% from the USA, 22% from Canada, 15% from South Africa, 13% from other African countries, 6% from Europe and 3% from Australia (Organization for Economic Cooperation and Development, Nuclear Energy Agency and the International Atomic Energy Agency 1986). In 1985, production was about 35 000 t U, which was about 2 000 t below reactor requirements, and production is expected to rise to only 36 500 t in 1986.

WOCA resources, defined as reasonably assured resources (RAR) at a cost of < US$80 per kg U, were estimated at 1.6 million tonnes U in 1986 (Organization for Economic Cooperation and Development, Nuclear Energy Agency and the International Atomic Energy Agency 1986). About 90% of these reserves are in Australia, Brazil, Canada, Namibia, Niger, South Africa and the USA. Resources at a cost between US$80 and US$130 per kg U are estimated to be 0.65 million tonnes U. Estimated additional resources (undiscovered) in the low-cost and high-cost categories total 1.33 million tonnes U.

Demand projections for uranium for electrical power since the 1970s have steadily declined because of political decisions against nuclear power and a real drop in the rate of growth of electricity demand. Reactor requirements in 1995 are estimated to be about 55 000 t U. Projected reactor requirements, ameliorated by large-scale introduction of breeder reactors in 2010 and using the low-uranium-consumption scenario, place the estimated demand at 39 000–121 000 t U by 2025. Projected long-term production capabilities should be able to meet demands until 2000, but new discoveries of uranium ores will be needed in the early 2000 s.

Bibliography

Kirk W S 1980 Depleted uranium. *Mineral Facts and Problems*, US Bureau of Mines Bulletin 671. USBM, Washington, DC, p. 903

Nash J T, Granger H C, Adams S S 1981 Geology and concepts of genesis of important types of uranium deposits. In: Sims P K (ed.) 1981 *Economic Geology Seventy-Fifth Anniversary Volume 1905–1980*. University of Texas, El Paso, Texas, pp. 63–116

Organization for Economic Cooperation and Development, Nuclear Energy Agency and the International Atomic Energy Agency (OECD, NEA, IAEA) 1986 *Uranium—Resources, Production, and Demand*. OECD Publications, Paris

Roberts D E 1983 The Olympic Dam copper–uranium–gold deposit, Roxby Downs, South Australia. *Econ. Geol.* 78: 799–822

US Department of Energy 1980 *An Assessment Report on Uranium in the United States of America*, US DOE GJO-111(80). US DOE, Washington, DC

Wenrich K J 1985 Mineralization of breccia pipes in northern Arizona. *Econ. Geol.* 80: 1722–35

W. I. Finch
[US Geological Survey, Denver, Colorado, USA]

V

Vanadium Resources

Vanadium has a crustal abundance of 100–150 ppm. In nature, 3- and 4-valent vanadium are rather insoluble, and 5-valent vanadium is rather soluble. Silicic igneous rocks contain about 25 ppm vanadium and mafic rocks about 200 ppm, mostly in iron and ferromagnesian minerals. Magmatic magnetite deposits associated with mafic rocks commonly contain a few thousand ppm vanadium. Much vanadium in magmas is trivalent (insoluble) and not available for hydrothermal transport and the formation of hydrothermal ore deposits.

During the normal weathering of rocks, vanadium remains insoluble and goes into the resulting clay minerals, which ultimately form shales; most shales contain about 100 ppm vanadium, whereas sandstones and limestones contain much less. If weathering is strongly oxidizing, pentavalent vanadium forms. This species can dissolve in natural water, from which it can be precipitated and concentrated by: (a) reaction with aluminum and iron minerals in bauxite and sedimentary iron deposits; (b) reaction with oxidized copper, lead and zinc minerals, forming base-metal vanadate deposits; and (c) a reducing agent, such as fossilized organic material, which could form vanadium deposits in sandstone or accumulations in carbonaceous shales and phosphate deposits. Vanadium can accumulate in crude oils, tars and asphaltites when these carbonaceous shales are converted to liquid hydrocarbons.

Recognizable vanadium ore minerals are sparse, consisting of a few vanadates and oxides (e.g., vanadinite, carnotite, montroseite), a silicate (roscoelite) and a sulfide (patronite). Much vanadium is recovered from deposits that contain no recognized vanadium mineral.

Magmatic titaniferous magnetite deposits currently supply about 80% of the world's vanadium. In 1987, world production came from South Africa 15.5 kt (52.5%), USSR 9.5 kt (32.2%) and China 4.5 kt (15.3%). Ore grades in these deposits vary widely: 16–60% Fe, 1.5–38% TiO_2 and 0.1–1.7% V_2O_5 are typical ranges. Similar deposits occur in many countries. Most deposits are large, and the vanadium resources represent an ample supply at the present rate of world use.

Sources of vanadium in the USA in 1987 included recovery from petroleum residues, 2.2 kt or about 30% of domestic production, and significant amounts also recovered from ferrophosphorus and iron slags and utility ash. Other sources include phosphate rock and uraniferous sandstones and siltstones. Significant amounts are also found in bauxite and carboniferous materials such as petroleum and coal. Resources in these and similar occurrences range from small to large.

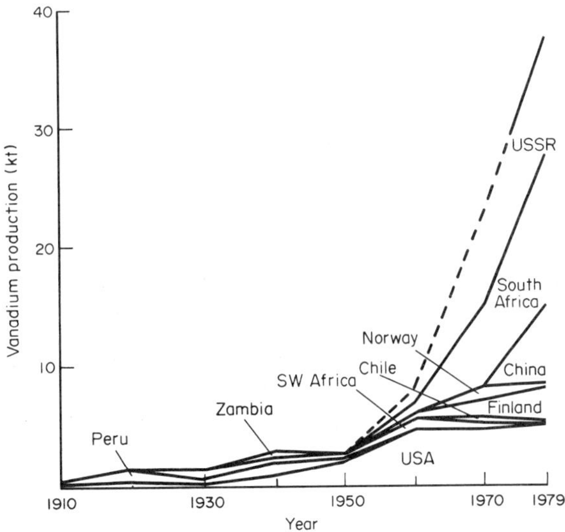

Figure 1
Vanadium production in the period 1910 to 1980: estimated values are shown by the dotted line

Vanadium-bearing carbonaceous shales, including some oil shales, contain ~1% V_2O_5. Although these deposits have not yielded vanadium commercially, some deposits are very large and contain very large vanadium resources. Large deposits are known in North and South America, Europe, Asia and Australia.

Bibliography

Fischer R P 1973 *Vanadium*, US Geological Survey Professional Paper 820. US Government Printing Office, Washington, DC, pp. 679–88

Fischer R P 1975 *Geology and Resources of Vanadium Deposits*, US Geological Survey Professional Paper 926. US Government Printing Office, Washington, DC, pp. A1–14, B1–10

Polyakov A Y 1959 Principles of the metallurgy of vanadium (in Russian). *Tr. Gos. Nauchno-Tekh. Izdatel. Lit Chernoy Tavetna. Met.* 1–137. English summary in *US Bur. Mines Miner. Trade Notes* 53(5): 48–50 (1961)

Rose E R 1973 *Geology of Vanadium and Vanadiferous Occurrences of Canada*, Economic Geology Report (Geological Survey of Canada) No. 27. Geological Survey of Canada, Ottawa

US Bureau of Mines 1988 *Mineral Commodity Summaries 1988*. USBM, Washington, DC

R. P. Fischer
[US Geological Survey, Grand Junction, Colorado, USA]

Well-Drilling Materials: Clay and Nonclay Minerals

In modern drilling operations, clay and nonclay minerals are used as additives in drilling fluids. Clays provide the colloidal base of nearly all water-based drilling fluids, and are also used to a lesser extent in oil-based drilling fluids. Nonclays such as barite, iron oxide, iron carbonate and galena are used to increase the density of drilling fluids when subsurface pressures are high. At present, the majority of drilling fluids consist of water with barite and bentonite added.

1. Clays

During the early periods in which oil wells were drilled in the Texas Gulf Coast area, sufficient clay was usually present in the formation to form a good colloidal base in the circulating fluid. The need to add a suspending agent was minimal.

Experimentation with heavy minerals to increase mud density led to the need for an efficient suspending medium. The need was further complicated when oil exploration expanded in areas in which unconsolidated sands were encountered.

The first established practice of intentionally adding clay to a circulating fluid occurred in California. Bentonite was successfully used to prevent severe caving while drilling a well near Kettleman, California. This initial success led to widespread interest in the use of clays to overcome hole problems.

1.1 Bentonites

These minerals are used primarily to increase the suspending power of water-based muds and to impart to them the property of laying down a filter cake on the wall of the bore-hole of permeable formations. The suspending power is needed not only to assist the passage of cuttings out of the hole, but also to allow weighting agents to be stably suspended.

Both natural and activated bentonites are used. Natural bentonites consist predominantly of sodium montmorillonites, which swell rapidly in fresh water. Activated bentonites are made from naturally occurring calcium montmorillonites by reaction with sodium carbonate.

Bentonites are fine-grained minerals consisting essentially of montomorillonite, a name proposed in 1847 for a highly plastic clay discovered in Montmorillon, France. Montmorillonite is by far the best-known member of the smectite group of clay minerals. Since bonding between layers of smectites is weak, they can easily be cleaved. Partly because of weak bonding and partly because of high repulsive potentials on the surface of the layers (arising from isomorphous substitutions), water can enter between the layers. Thus, smectites have a lattice that can expand, which greatly increases their colloidal activity. This unique characteristic gives bentonites their remarkable efficiency as a suspending agent for drilling fluids.

Bentonites were used in small volumes for various purposes until 1928, when their use as a suspending medium for drilling fluids was discovered. Before this period, production of bentonites had varied from 70 to 100 t year^{-1}. The drilling fluid market created an instant demand for bentonites, and by 1980 the annual market had grown to approximately 2 178 000 t (1 452 000 t in the USA).

The size of the modern bentonite market was made possible by large deposits of sodium bentonite in the northwestern part of the USA. The Mowry Shale of Cretaceous age in Wyoming and South Dakota provides a large portion of the finest grades of naturally occurring bentonite in the world. Various other beds in the Mowry Shale and horizons in the younger formations have been mined in the Bighorn Mountains in Wyoming and at a few locations in Montana. Bentonites produced in the northwest of the USA are commonly classed as Wyoming bentonite. They are established as the standard by which all other bentonites are judged.

The only known sodium bentonite deposits of commercial importance outside the USA are in India. Sodium bentonite is mined in the Kutch and Bhavnager districts of Gujarat State, with production estimated to be in excess of 200 000 t year^{-1}. The total production does not go into drilling fluids, but this market is expected to increase as producers refine processing techniques to meet drilling fluid specifications.

Deposits of calcium bentonite are relatively widespread throughout the world. In most cases, the calcium ion is exchanged for sodium before this material is used in drilling fluids. When cost and logistics are favorable, this base-exchanged bentonite is extensively used. At one time, it supplied almost the entire North Sea market.

Calcium bentonite is extensively mined in the USA, even though adequate supplies of sodium bentonite are available. Part of the calcium bentonite is used in drilling fluids, but amounts are virtually impossible to estimate. Some of the world's highest-quality calcium bentonite occurs in the southern USA. Deposits are commercially mined in Mississippi, Alabama, Georgia and Texas. The beds in Alabama and Mississippi are of Cretaceous age, beds in Texas are Tertiary and those in Georgia are Miocene. A relatively small amount of this calcium bentonite finds its way to the drilling fluids market, because of a prevailing preference for naturally occurring sodium bentonite.

1.2 Organophilic Clays

Oil-based muds are made from mineral oil—usually a distillate in the diesel boiling range—to which are added solids and liquids to confer suitable properties. Barites are used to increase the density of oil-based muds, and care must be exercised to avoid settling in such cases. Various organic materials may be added to create suspending properties in oil-based muds, but the development of clay compositions capable of forming gels in oil, similar to those formed by bentonite in water, was a major contribution to the technology of oil-based muds.

The exchangeable cations of bentonite can be replaced by certain amine compounds, resulting in organophilic materials which will suspend inert solids in oil. Organophilic bentonites are very important to oil-based drilling fluids, but the size of the market is limited; the total world production of organophilic bentonites for use in drilling fluids is estimated to be 7000–9000 t $year^{-1}$.

1.3 Attapulgite

Attapulgite is principally used to impart suspending properties to water-based drilling fluids which contain high concentrations of soluble ions. In the presence of electrolytes, bentonite will not maintain a high viscosity in water. Under these conditions, attapulgite is often selected as a suspending agent. It is almost unaffected by dissolved ions.

Attapulgite particles are completely different in structure and shape from bentonite particles. They are essentially bundles of laths, which separate into individual particles when mixed vigorously with water. There are few atomic substitutions in the structure, and therefore the surface charge on individual attapulgite particles is low. Also, their specific surface area is low. Consequently, the rheological properties of attapulgite suspensions are dependent on mechanical interference between the long laths, rather than on electrostatic interparticle forces.

Drilling into salt deposits poses particular problems. The soluble salts not only contaminate the drilling fluids, but the salt beds wash out owing to the erosive action of the drilling fluid. The problems of contamination and washout can be eliminated by using a salt-saturated drilling fluid.

Use of attapulgite in drilling fluids peaked in the 1950s, and has gradually declined, to the level of about 85 000–95 000 t $year^{-1}$ in the early 1980s. The primary source of attapulgite is from mines in the USA located near the border separating Georgia and Florida. Clays, which have been named palygorskite, found in the Palygorsk Range of the Ural Mountains are structurally similar to attapulgite.

1.4 Sepiolite

A clay-like variety of sepiolite is used in drilling fluids. It occurs in fibrous and elongated lath-like particles, and closely resembles attapulgite. Sepiolite contains less substituted aluminum than attapulgite, and the unit cell is somewhat larger. Many of the physical properties of the two minerals are very similar.

Sepiolite is used primarily as a suspending agent for salt-water drilling fluids. Sepiolite has been used in Europe for several years, but it has only recently attracted interest in the USA. This is partially due to availability of adequate supplies of attapulgite in the USA, and the lack of commercial quantities of sepiolite.

A significant deposit of sepiolite exists in the Amargosa Desert in Nye County, Nevada. The deposit is currently being mined, and is the sole source of sepiolite used in drilling fluids in the USA.

Laboratory studies in the mid-1970s indicated that sepiolite is more thermally stable than attapulgite, and this led to an interest in the use of sepiolite in drilling fluids for geothermal drilling. This is a relatively limited market, however, and sepiolite is not expected to attain general acceptance in areas where attapulgite is readily available and high temperature is not a factor.

The total use of sepiolite in the early 1980s in drilling fluids in the USA is estimated to be 3000–5000 t $year^{-1}$. Sepiolite is sold to meet the same specifications as attapulgite.

1.5 Asbestos

Asbestos is sometimes used as a viscosifier in both fresh-water and salt-water muds. When drilling conditions are such that only requirement for the mud is to improve cuttings recovery at the surface, the addition of small quantities of chrysotile asbestos can have the desired effect. The amount added to drilling fluids in 1980 was estimated to be 4500–6000 t. Use of asbestos for this purpose is not expected to grow in the future, because of the possible health hazards associated with its use (see *Asbestos Hazards*).

2. Nonclays

2.1 Water

Water is the most important single substance in drilling fluids. With the exception of a few wells drilled with dry air, water is the major component by volume of the drilling fluid at some time in the course of drilling a well. Even when the use of water-based mud is discontinued in favor of an oil-based mud or a foam, water continues to be significant to the properties of the drilling fluid. The unusual characteristics of water affect each step in the drilling operation, and the availability and chemical content of the make-up water must be considered when planning to drill. Reactions between water and clay surfaces and the effect of electrolytes dissolved in the water on clay–water interaction are of prime consideration when studying drilling fluid properties.

Water is usually available at a relatively low cost. At some locations, however, the quality is such that the composition of the mud must be altered, or the water must be treated before use.

Daily consumption may be as high as 60vol% of the drilling fluid system. Total water used for drilling oil wells worldwide was probably over 30 000 000 t in 1980.

2.2 Barite

Barite, the naturally occurring form of barium sulfate, is added to drilling fluids to increase density. From the standpoint of cost, it is by far the leading mineral used in drilling fluids, and it is second only to water from the standpoint of volume.

The first recorded attempt to increase the density of a drilling fluid was in laboratory tests performed in 1922, in which heavy-metal compounds such as iron oxide and lead sulfide were used. Late in this same year, barite was used to make a heavy mud. This led to the development of weighted mud systems, which were successfully used in Louisiana and Texas.

Theoretically, any substance that is more dense than water can be used to increase drilling fluid density. From a practical standpoint, however, materials added to increase density must not adversely affect other drilling fluid properties. The various finely ground solids listed in Table 1 have been used to increase the density of drilling fluids.

Barite is well suited for use in drilling fluids. It is chemically inert, relatively nonabrasive and not harmful to humans or the environment. Barite is available in large quantities and is easy to mine and process. For these and other reasons, it is preferred over all the other minerals in Table 1.

The specific gravity of pure barium sulfate is 4.5, but most commercial barite has a lower specific gravity because of impurities such as quartz, chert, calcite and anhydrite. However, some deposits of barite contain iron compounds which may actually increase the average specific gravity of the ore.

The total world usage of barite in 1980 in drilling fluids was estimated to be 4 000 000 t. The demand is expected to increase, and there is no known practical substitute. However, barite occurs throughout most of the world, and no shortage is predicted within the forseeable future.

2.3 Iron Oxides

Natural iron oxides were among the first materials used to increase the density of drilling fluids. Their use was discontinued in the early 1940s because barite was available at a lower cost. However, interest in iron compounds has been revived in recent years. Considerable amounts of iron oxides in the forms of hematite and ilmenite are being used as weighting agents for drilling fluids. The most common practice is the blending of up to 20% iron compounds with barite. This serves the twofold function of increasing the specific gravity of the finished product and extending the life of barite reserves. The use of iron oxides will probably increase, subject to the availability of barite and continued growth in the need for oil.

2.4 Siderite

Siderite gained considerable attention during the mid-1970s as a weighting additive for certain drilling fluid systems. Though siderite cannot compete with barite because of lower specific gravity and being less plentiful than barite, its acid solubility is significant to some drilling operations. Completion fluids weighted with siderite are less likely to cause permanent formation plugging, as any residual siderite can be removed with acids.

2.5 Calcium Carbonate

Calcium carbonate is useful as a drilling fluid additive, primarily because it is soluble in acids, and secondarily because it is readily available throughout most of the world. Limestone and oyster shells, the main sources of calcium carbonate used in drilling fluids, are relatively inexpensive.

Calcium carbonate can be used to increase the density of any drilling fluid, if extremely high weights are not required. The density of a drilling fluid weighted with calcium carbonate is limited to approximately 1.4 $g\,cm^{-3}$ because of the low specific gravity of calcium carbonate.

Calcium carbonate is very compatible with the oil-based drilling fluids used to drill sensitive formations. Any calcium carbonate deposited in the formation can be removed with acid.

Worldwide annual usage of calcium carbonate in 1980 in drilling fluids was estimated to be 30 000 t. Growth in demand is expected to depend on growth of the drilling industry.

2.6 Galena

Interest in galena (lead sulfide) as a weighting agent for drilling fluids dates back to 1922. Limited availability

Table 1
Materials used for adjusting the density of drilling fluids

Material	Formula	Specific gravity
Barite	$BaSO_4$	4.2–4.5
Calcite	$CaCO_3$	2.6–2.8
Celestite	$SrSO_4$	3.7–3.9
Dolomite	$CaCO_3.MgCO_3$	2.8–2.9
Galena	PbS	7.4–7.7
Hematite	Fe_2O_3	4.9–5.3
Ilmenite	$FeO.TiO_2$	4.5–5.1
Magnetite	Fe_3O_4	5.0–5.2
Siderite	$FeCO_3$	3.7–3.9

and high cost have prevented this material from becoming an accepted component of drilling fluids. Galena's high specific gravity (7.4–7.7) makes it useful as an emergency weighting material for controlling unusual subsurface pressures. Drilling fluids with a density approaching 4 g cm^{-3} can be made with galena.

Stocks of galena are sometimes maintained in areas of known abnormal pressures for use in case of emergency. Actual use of galena in drilling fluids is not expected to reach a significant volume.

2.7 Lignite

Lignite is a naturally occurring substance which has become an important drilling fluids constituent. Depending on geographic origin, it is also known by such names as leonardite, brown coal and slack. Originally introduced in 1947 as a viscosity control agent for water-based drilling fluids, lignite now functions effectively as a filtration–reduction additive, and as an oil emulsifier. It has also gained acceptance as a stabilizing agent against the effects of high temperatures on drilling fluids. Additionally, lignite serves as a raw material in certain speciality products used to treat both water-based and oil-based muds.

Not all lignites are useful as drilling fluid additives, and extensive testing is required before deposits are mined for this purpose. There are no empirical specifications for drilling fluid lignite, but effective material will generally have a low pH and a high oxygen content.

Lignite deposits occur in many parts of the world. The principal source of drilling-fluid-grade lignite in the USA is an extensive deposit in North Dakota. Deposits have also been mined in New Mexico and Texas, but neither offer the same quality or quantity as the North Dakota mines.

Lignite has been the subject of considerable research. It can be successfully sulfonated and sulfomethylated. These two modifications increase its immunity to dissolved ions, making it useful in drilling fluids mixed in seawater. Lignite is frequently added to the drilling fluid in the form of metal salt (e.g., as a sodium, potassium, zinc or alkyl ammonium salt). This practice has significant advantages and accounts for a major portion of the market.

Total use of lignite in all forms in 1980 in the drilling fluid industry is estimated to be 25 000–27 000 t. Its relatively low cost, thermal stability and unique molecular structure should assure a stable demand for lignite.

2.8 Calcium Sulfate

Calcium sulfate in the form of anhydrite, plaster of paris or gypsum serves as a source of calcium ions for drilling fluids. The practice of adding calcium sulfate to water-based drilling fluids was introduced in western Canada in the early 1950s. Intentionally treating the drilling fluids with calcium sulfate in the form of gypsum minimized the adverse effects of drilling into natural deposits of anhydrite. These gypsum-treated muds were difficult to control and were not widely used until after 1955. At this time, effective conditioners for gypsum-bearing drilling fluids were introduced. This increased the versatility of calcium-treated drilling fluids and marked the beginning of widespread interest in their usage.

Use of calcium sulfate in drilling fluids has declined in recent years. The gypsum-treated fluids of the 1950s have mostly been replaced by other systems as a result of continuing technical developments.

Current use of calcium sulfate is mostly limited to drilling fluids containing undesirable soluble carbonate ions. Gypsum is the source of calcium ions used to precipitate carbonate ions.

Use of calcium sulfate in 1980 in drilling fluids in North America was approximately 3000 t. Demand is not expected to increase significantly in the foreseeable future.

2.9 Sand

Hydraulic fracturing techniques are commonly used during completion of oil wells to increase the recovery of hydrocarbons from the subterranean formations. These techniques involve injecting a fracturing fluid down a well and into the formation, under sufficient pressure to initiate and propagate fractures into the formation. Proppant materials are generally entrained in the fracturing fluid and are deposited in the fractures to maintain the openings after pressure is released.

The most common proppant consists of screened sands of 10–20 or 20–40 mesh (2.000–0.841 or 0.841–0.420 mm). Worldwide usage of sand in 1980 as a fracture proppant was approximately 1 125 000 t.

The use of hydraulic fracturing as a method for increasing hydrocarbon yields from oil wells has increased significantly in recent years. This trend will continue as it becomes increasingly important to maximize the amount of oil or gas that can be produced from each well that is drilled.

See also: Well-Drilling Materials: Industrial Minerals

Bibliography

Abbot G A 1963 *Leonardite: A Material of Industrial Promise*, US Bureau of Mines Information Circular 8164. US Bureau of Mines, Washington, DC, pp. 80–86

American Petroleum Institute 1969 *Principles of Drilling Mud Control*, 12th edn. Petroleum Extension Service, University of Texas, Austin, Texas

Bradley W F 1940 The structure of attapulgite. *Mineralogist* 25: 405–10

Browning W C, Perricone A C 1963 Clay chemistry and drilling fluids. *First University of Texas Conf. Drilling and Rock Mechanics*, SPE Paper 540. Society of Petroleum Engineers of AIME, Dallas, Texas

Grim R E 1962 *Applied Clay Mineralogy*. McGraw-Hill, New York
Haden W L Jr, Schwint I A 1967 Attapulgite, its properties and applications. *Ind. Eng. Chem.* 59(9): 58–69
Jordan J W 1949 Organophilic bentonites I. Swelling in organic liquids. *J. Phys. Chem.* 53: 294–306
Jordan J W, Hook B J, Finlayson C M 1950 Organophilic bentonite II. Organic liquid gels. *J. Phys. Chem.* 54: 1196–208
Magcobar Group, Dresser Industries 1977 *Drilling Fluid Engineering Manual*. Dresser Industries, Houston, Texas
Sloan J P, Brooks J P, Dear S F 1975 A new non-damaging acid-soluble weighting material. *J. Pet. Technol.* 27: 15–20

F. Turner
[Dresser Industries, Houston, Texas, USA]

Well-Drilling Materials: Industrial Minerals

Industrial minerals used for oil well drilling are mostly added to the circulating fluid used to clean the bore hole during drilling operations.

The circulating fluid, usually called mud, is essential to successful rotary drilling. During drilling, the fluid is moved by reciprocating pumps and flows in a continuous circuit from a surface pit to the well bottom and back to the surface pit. It travels down the inside of the drill pipe, exits through openings in the bit, and surges back to the surface in the annular space between the drill pipe and the well bore.

The term drilling fluid encompasses all of the compositions used to aid the production and removal of cuttings from a bore hole in the earth. Use of drilling fluids is applicable to all forms of earth excavation, but this practice is usually more closely associated with drilling bore holes, and specifically with oil exploration.

Modern oil wells are drilled with rotating bits practically without exception, and rotary drilling techniques have indirectly changed the course of civilization because of their importance to the petroleum industry. Many people consider rotary-drilling technology unique to exploration for oil, when in fact the practice is several thousand years old. The exact date that a rotating bit was first used to drill a hole is not known, but there is evidence that the Egyptians used this method during construction of the pyramids. Bore holes to depths of 6 m dating from 3000 BC have been found in hard rock near the pyramids, and two drill cores from this site are in the Egyptian collection at the Metropolitan Museum of Art in New York. However, rotary-drilling techniques have developed into a science almost completely as a result of the exploration for oil.

From a practical standpoint, the modern petroleum industry was born on August 27, 1859, when Edwin L Drake drilled near a stream called Oil Creek, near Titusville, Pennsylvania, and discovered oil at a depth of 21 m. However, Drake did not use the rotary-drilling techniques in use today. Fluid rotary-drilling systems were used to obtain a dependable source of water in the Dakota territory during the mid-nineteenth century. The first recorded case of oil being produced from a well drilled with rotary water flush equipment was in 1894 at Corsicana, Texas. However, the well was originally planned as a source of water, and the oil was encountered unexpectedly.

One of the first documented accounts of a clay–water mixture being used in a well drilled solely to find oil was during the drilling of the famous Spindletop well near Beaumont, Texas. The popular story contends that some cattle wading in the shallow pit which held the water supply produced a muddy fluid. When this muddy water was used as the circulating fluid, it sealed off permeable sands and stopped loss of fluid to the formation. This marked the beginning of active interest in the use of rotary drilling for oil exploration and the practice spread rapidly throughout the coastal plains bordering the Gulf of Mexico and the state of California.

If drilling fluid is defined as a material used to aid tools in the creation of bore holes, this practice also far antedates the petroleum industry. Wells were drilled in China about 1100 BC, some hundreds of meters deep, near the border of Tibet. Water was poured into these wells to soften the rock and to aid in the removal of cuttings. Water is still the principal constituent of drilling fluids in use today.

Water was the first circulating fluid used to aid rotary drilling, but it was soon observed that a mud-laden fluid would better restrain unconsolidated sands and seal formation pores. The practicability of adding mud to drill holes was demonstrated in 1913, when it was discovered that use of a mud-laden fluid during drilling sealed each gas-bearing stratum as it was encountered. First attempts to improve circulating fluids consisted of the addition of surface clays dug near the well site. These first drilling muds encountered many difficulties, and the search for a satisfactory circulating fluid became a continuing effort.

Development of the use of drilling fluid into a science began in 1921, with the first attempt to control "mud properties" through the use of additives purchased specifically for that purpose. However, the wells drilled during this period placed few rigorous demands on the circulating fluid. Consequently, muds remained relatively simple over the next 10 years. The early 1930s marked the beginning of the necessity to cope with greater depths, higher temperatures and pressures, and the invasion of contaminants from the formation.

As drilling became more complicated, the science of drilling fluids kept pace, and the circulating fluids used today for rotary drilling may well be scientifically compounded fluids. The composition of individual drilling fluids varies widely, with geological conditions usually dictating the composition of the mud used.

There are two basic drilling fluid systems: water-base and oil-base. Muds with water as the continuous liquid are called water-base muds, and those with oil as the continuous liquid phase are called oil-base muds. There are several subgroups under these two general classifications. The type selected depends on the particular drilling operation. Economics, contaminants, available material, pressures and temperatures are all significant to the formulation of a drilling fluid. For economic reasons, a water-base mud is used whenever possible.

A drilling fluid consisting of oil as the continuous liquid phase is much more expensive than one having water as the continuous liquid phase. However, the entire exploration and recovery operation must be considered, and oil-base muds can be justified under certain conditions. They are economically feasible for drilling (a) certain troublesome shales; (b) in high temperatures; or (c) when water may damage producing zones.

Initially the primary function of a drilling fluid was to bring cuttings freed by the rotating bit from the bottom of the hole to the surface. To function satisfactorily in current sophisticated drilling programs the circulating fluid must:

(a) remove cuttings from the bottom of the hole and carry them to the surface;

(b) hold cuttings and other solid particles in suspension when circulation of the fluid is interrupted;

(c) release cuttings and sand at the surface;

(d) control formation pressures;

(e) build a semipermeable coating on the wall of the bore hole;

(f) cool and lubricate the drill pipe and drill bit;

(g) minimize adverse effects on the adjacent formation;

(h) assure maximum information about the formation penetrated;

(i) partially support the weight of drill pipe; and

(j) transmit hydraulic power to the drill bit.

Not surprisingly, no single drilling fluid combines the wide variety of properties required of it in the optimum way for all drilling situations, particularly when cost is also taken into account. Consequently, a wide variety of systems has evolved, ranging from water—used only in the least demanding circumstances—to extremely complex formulations containing up to ten selected components.

Without exception, all drilling fluids must have suspending properties adequate to remove cuttings freed by the drill bit. The usual practice is to add bentonite, attapulgite or sepiolite to a water-base drilling fluid. Naturally occurring clays will not produce suspending properties in an oil-base drilling fluid. However, certain high-grade bentonites can be chemically altered to render them organophilic. These organophilic bentonites will effectively impart suspending properties to an oil-base drilling fluid.

Controlling formation pressures is one of the primary functions of drilling fluids. As a particular drilling operation encounters increasing formation pressures, the drilling-fluid density must be increased accordingly. The mineral most widely used for this purpose is barite. Other minerals used to increase drilling-fluid density include iron oxide, siderite, calcium carbonate and galena.

Lignite, calcium sulfate and sand, while not used in all drilling fluids, fulfil certain specific requirements, and are very important to the drilling-fluid industry.

See also: Well-Drilling Materials: Clay and Nonclay Minerals

Bibliography

Brantly J E 1971 *History of Oil Well Drilling.* Gulf Publishing, Houston, Texas

Gray G R, Darley H C H, Rogers W F 1980 *Composition and Properties of Oil Well Drilling Fluids,* 4th edn. Gulf Publishing, Houston, Texas

Gray G R, Young F S Jr 1973 25 years of drilling technology—A review of significant accomplishments. *J. Pet. Technol.* 25: 1347–54

F. Turner
[Dresser Industries, Houston, Texas, USA]

Whiteware: Component Minerals

Whiteware is the generic term for a class of ceramic products, usually white or nearly white in color, that are produced primarily from the traditional ceramic raw materials of clay, flint and feldspar. Compositions in this class find application as sanitary ware, wall and floor tiles, dinnerware, artware, technical porcelains and in numerous other uses. An in-depth understanding of the component minerals is essential not only for achieving the correct constitution of a whiteware, but also for establishing successful processing parameters. A general flowsheet for whiteware manufacture includes: selection of raw materials based on physical and chemical criteria; either wet or dry processing to achieve intimate mixtures of the components; forming by liquid, plastic or dry methods; drying; firing at high temperature to produce the desired properties; any one or more of numerous finishing operations on the fired ware; and optional glazing and decorating.

1. General Principles

Whitewares are usually formed from mixtures of naturally occurring minerals, although synthetic materials with equivalent properties are finding increased use. Raw material particle size distribution and particle shape are important physical factors. A major portion of the particles must be fine to provide plasticity, dry strength and reactivity during firing, and the particle size distribution should be extended to contribute good rheological characteristics and to minimize water requirements and drying shrinkage. The physical properties of the components primarily dictate the properties of the unfired body, but fired properties are also affected to a lesser extent.

The components of a commercial whiteware body should have the following attributes.

(a) Their composition should be consistent with a white fired color (i.e., low amounts of coloring elements, specifically Fe, Ti, Cr and Mn) and the desired fired properties of the product. Surfactants, defoaming agents, deflocculants, flotation agents and soluble impurities should not be present. Fluorine, an environmental pollutant, should also be absent.

(b) The materials should meet particle size specifications based on previously stated considerations, should be free flowing without excessive lumps or caking, and should have a constant water content.

(c) Delivered cost should be low, consistent with performance in the body and in line with competitive materials. Supply must be dependable with large reserves, consistent over time and capable of clean bulk deliveries, and the supplier should be dedicated to the ceramics market.

The clay component of whiteware compositions provides the property of plasticity necessary for many forming operations. When wet, the two-dimensional plate-like form of the clay minerals allows great amounts of shear strain to develop between parallel adjacent particles when a stress is applied. The same overlapping clay plates that supply the great plasticity also produce mechanical strength in the body when the water is removed (green strength), owing to the van der Waals forces between the small plates. The degree of plasticity and green strength of a body increases as the particle size of the clay component decreases. A useful means for measuring the surface area of the whiteware body, and hence a measurement of plasticity and strength, is the methylene blue adsorption (Phelps and Harris 1967). Methylene blue is a cation which absorbs quantitatively on negative clay surface sites and is thus a measure of these properties.

The nonplastic components of a body have two prime functions. First they form a refractory skeleton, which maintains the object shape and prevents slump during firing. This is an important function of silica. Secondly, they provide elements that form low-melting compounds or glasses with part of the more refractory components. This is an important function of feldspar, mica and nephelene syenite. All nonplastics are ground to a particle size of the order of < 50 μm.

Components of a body may contribute to more than one property; for example, kaolin forms a refractory skeleton and adds to the plasticity of a sanitary body. Mica is primarily a flux, but also adds to plasticity.

2. Whiteware Components

2.1 Plastic Components

The clay minerals consists of two-dimensional silicate (tetrahedral) and aluminate (octahedral) sheets weakly bonded together in the third dimension to form stacks. Since the bonds between the plates are weak, the layers separate under certain physicochemical conditions. The degree of action needed to separate the plates depends on the strength of the interplate bonding and varies from one clay mineral to another. Talc requires active grinding to develop characteristic clay properties, whereas most kaolin clays need only be dispersed in water to develop good properties.

The most important clay group, the kaolinite minerals, are divided into primary and secondary categories based on the geological conditions of deposition. Both types are derived from feldspars by the removal of alkali either by weathering or by natural acid leaching. Primary kaolins, such as those found in southwest England, northwest Spain and Brittany in France, are formed in situ from granite. Secondary kaolins, such as the deposits in Georgia, USA and in the area between Madrid and Valencia in Spain were formed from secondary quartz and feldspar, but are primary with respect to kaolin formation. The truly secondary ball clays are finer and appear in a number of beds which reveal the sequence of deposition. These clays confer plasticity to the whiteware body beyond that of a kaolin of similar particle size, apparently owing to the colloidal organic content of these clays. For example, some of the fine kaolins of Florida and South Carolina approach the ball clays in particle size but not in plasticity.

The smectites are another important class of clays, of which bentonite is perhaps the best-known member. Relatively small additions to whiteware compositions produce plasticity, dry strength and reactivity owing to the very fine particle size of these clays. Each clay plate in this class of clays consists of three layers: an octahedral sheet sandwiched between two tetrahedral sheets. Exchangeable cations are present between the layers as well as on the outside of the plate. Thus, the basal spacing is increased and water is easily absorbed between the plates, which results in interlayer swelling. This swelling can cause a doubling of the clay volume when four layers of water are absorbed. These clays are rarely free from undesirable impurities, either fluxing or coloring, although some very white bentonites and hectorites are known in the southwest USA. Smectites, particularly montmorillonite, are common minor components in the nonexpanding clays. If they appear in too great a proportion, say 1%, high viscosity results, which produces low casting slip densities, reduced body permeability and increased production losses. Pyrophyllite is used to control shrinkage on firing, and talc is used to produce cordierite for high strength and thermal shock resistance in the body.

The structure of the clay minerals is described in detail by van Olphen (1963). A short summary is provided here for the major commercial clay types. The basic structural units of clays are the two-dimensional tetrahedral sheets of silicon tetrahedra and the two-dimensional octahedral sheets of sixfold-coordinated aluminum atoms. These sheets, as shown in Fig. 1, are laminated together, either in alternating sequence to form two-layer clays or in the sequence tetrahedral–octahedral–tetrahedral to form the three-layer clay minerals. Various ions are either adsorbed onto the clay surface or reside between the clay plates, assisting in bonding. Five major types of clays are summarized in Table 1.

2.2 Quartz

Quartz provides a refractory structure, increases the viscosity of the glass component on firing and, like all nonplastic additions, improves the rheology for the wet forming process. It does this by increasing the shearing gradient across the plastic layers for a given shear across the bulk.

Naturally occurring quartz is relatively pure, and if a quartz sand is coated with hydrous iron oxides, these are easily removed by sodium dithionite bleaching at pH 3. The surface of sand may also be cleaned by autogenous attrition and subsequent washing. Impurities of fine particle size can be removed in a mechanical classifer, such as the Aitkins spiral raked classifier (Taggart 1945). Calcined and ground flint is used occasionally in whiteware bodies and contributes somewhat to plasticity. A different thermal expansion from that of quartz is exhibited owing to the presence of cristobalite. Where pegmatites are used as the source of feldspar, part or all of the quartz can also come from this source.

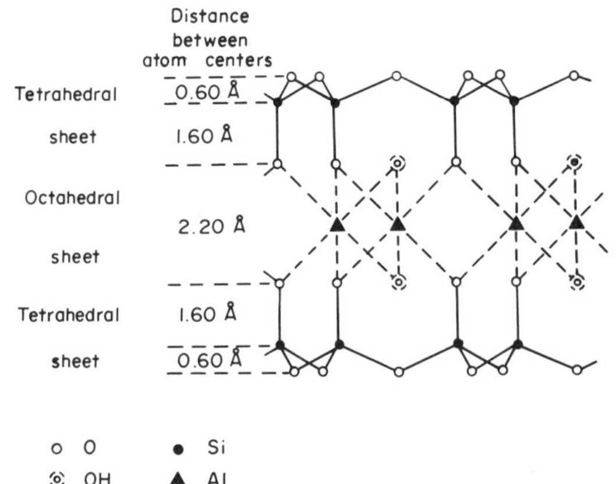

Figure 1
Layer structure of the clay minerals

Table 1
Structure of major types of clays

Clay	Number of layers	Ion substitutions
Kaolinite	two	
Pyrophyllite	three	
Talc	three	$3Mg^{2+}$ for $2Al^{3+}$ in the octahedral sheet
Montmorillonite	three	some Mg^{2+} for Al^{3+}
Hectorite	three	some Li^{+} for Mg^{2+}

2.3 Fluxes

The major fluxing elements in whitewares are potassium, sodium and calcium, plus magnesium as dolomite or talc in the bodies. A preponderance of potassium over sodium in porcelain gives a more viscous glass, and the body shows a higher firing temperature and a wider firing range. The main flux sources are orthoclase feldspar (potassium), albite (sodium), pegmatites (50:50 potassium and sodium) and nephelene syenite (sodium). The value of calcium as a flux in tiles, or with magnesium as dolomite, is to control expansion and reduce cost.

The main impurities in the feldspathic rocks are ferrous salts, which are largely removable by dry magnetic methods at a liberation size of 300 μm. Fluorides can be removed from feldspars by flotation with fatty amines using hydrofluoric acid as a suppressor for the feldspars (Pryor 1965).

3. Mineral Production Techniques

Before the advent of highly mechanized mineral-processing techniques, the raw materials supplied to the ceramics industry were essentially run-of-the-mine materials with little or no postmine processing. Many recent advances in raw material beneficiation have led to greater supplies of useful materials and to higher quality materials overall. The major production techniques for whiteware mineral components are: impurity removal by magnetic, physical and chemical techniques; fine-particle beneficiation; plasticity improvement for clays; and physical comminution by autogenous milling and sand grinding.

3.1 Impurity Removal

Impurities in ceramic clays normally consist of coloring oxides such as iron and iron–titania. These impurities react during firing to reduce the whiteness of the body, producing a light tan hue. High-intensity magnetic removal of paramagnetic impurities such as colored ferrous titania and mica from kaolin was originally developed for manufacturing high-purity kaolins for paper-coating applications. A high magnetic field (~ 2 T) is applied to a fine wire-wool matrix, and paramagnetic particles are collected as the clay is passed over the matrix (Clark 1974, Balentine et al. 1972). A superconducting solenoid magnet machine (Watson et al. 1975) has been designed and proved experimentally on mineral materials. Increases in economy and reliability are features of this new technology. A promising superconducting deflection system without matrix has been devised for coarser particles (Cohen and Good 1976). This system consists of closely situated, opposed-field superconducting electromagnets. This geometry results in the development of a strong magnetic field external to the axial cylinder of the magnets, which is useful in deflecting coarse paramagnetic particles without the benefit of a matrix.

Particle classification by size offers a convenient method for removing impurities when significant size or density differences exist between the product and the impurity. In kaolin processing, quartz and ferrous mica are common impurities which are larger in size than the clay and are easily removed by hydraulic classifiers. Hydroclones are widely used for this function since they are simple to operate and require minimal deflocculant. Hydroseparators are also used but require full deflocculation of the clay. The use of deflocculants in processing ceramic kaolins is controversial since the deflocculant serves to alter the particle size distribution. Furthermore, the subsequent flocculants that are used to facilitate filter pressing can cause problems in the ceramic casting operation. Deflocculants commonly used in kaolin processing are sodium carbonate, sodium silicate, tetrasodium pyrophosphate and sodium polyacrylate.

Superficial iron can in many cases be removed from clay by adding 0.5wt% sodium dithionite.

3.2 Fine-Particle Beneficiation

Although particle size classification has been practised for a long time in the minerals processing industry, application of flotation and classification of clay particles is more recent. Fractionation of kaolins to produce separate coarse and fine products is current commercial practice. Flotation of very fine particles is a relatively new technology developed for paper kaolins but which will certainly become standard practice for ceramic materials. Froth flotation is widely used for the purification of pegmatites and talcs (Greene et al. 1961). The result of these improved beneficiation processes is that higher-grade raw materials are being produced throughout the particle size range of whiteware bodies, from 75 μm down to the colloidal range.

However, there are practical limits to the application of fine-particle benefication. Normally ball clays and smectites are not treated because of the subsequent difficulties and cost of filtering such materials. Occasionally in Third World situations, low-grade ball clays must be upgraded by these techniques to produce satisfactory properties. The filtering process for ball clays can be totally avoided if the proper amount of water for casting slips can be obtained by preparing the slip from ball-clay slurries with addition of the other components in the dry state.

3.3 Techniques to Enhance Physical Properties

The generation of enhanced physical properties in the ceramic raw material by the physical means of altering size, shape and size distribution is an important technology, and the beneficial effect of working, or shearing, plastic clay masses before ceramic processing has been known for a long time. It is immediately apparent to a potter that working the body by "throwing" and "wedging" will achieve considerably more than simply homogenization and air removal. The shearing action applied to the whiteware body causes delamination and size reduction of the previously agglomerated clay particles. When present, smectite particles remain firmly attached to kaolinite particles during this process. Furthermore, the autogenous grinding action creates clean active surfaces on the clay. This is especially true of the octahedral surfaces of kaolinite, which are generally contaminated with silica in nature.

Before 1930, mechanical wedging machines and pugmills were introduced into the German porcelain industry. The pugmill greatly enhanced the ability of manufacturers to increase plasticity through mechanical working. About 1930, the de-airing pugmill was introduced and further improvements were achieved. Although the increases in plasticity were greatly appreciated, it was still not evident that the extra energy input was used in shearing and delaminating the clay stacks.

In research directed toward improving the flow properties of kaolin for use in paper coating it was found that very substantial increases in plasticity and modulus of rupture were attained when pugmills were used under conditions of restricted water content so that a great deal of energy was absorbed, in the order of 40 kWh per tonne of kaolin (Clark 1952). Pugmills need special construction for this purpose. Heavy internal circulation is generated and the machinery must be designed to withstand the great forces inherent in such processing. A very accurate control of water content is needed in the feed to avoid failure of the machine. De-airing is irrelevant to this process.

Autogenous comminution, whether by hand wedging or by highly mechanized pugmilling, has revolutionized the production and control of the primary kaolins with respect to their rheological and mechanical properties.

3.4 Sand Grinding

A second process of comminution relevant to ceramic clay minerals is that of sand grinding. The grinding of nonplastics to particle sizes of about 10 μm by pebble milling is well known in ceramic practice. Since the efficiency of grinding to increased fineness is improved by reducing the size of the pebbles, it is possible to achieve very high efficiencies by reducing the grinding medium to the size of sand. Media as small as 300 μm can be used if the energy is imparted by stirring or vibration (Feld et al. 1960).

A sand mill is simply a pot of sand stirred by an impeller and fed with a slurry of the material to be ground. Feed particle size should be in the range of several millimeters down to a fraction of a micrometer; the efficiency in any range of size is controlled by the sand size used as the medium.

Many exotic media have been proposed with varying degrees of success, but for ceramic purposes silica sand is the most suitable. Suitable sands that are widely used are Ottawa sand in Canada and Leighton Buzzard sand in the UK. The loss of silica to the product is minuscule since these sands are particularly resistant to self-grinding. Where silica is needed in the product, clean, more friable sands can be used.

Important aspects of design and operation for high efficiency are as follows.

(a) Isometric construction for continuous processing (Brociner 1970).

(b) A high sand–liquid ratio throughout the grinding zone, held to a constant value.

(c) Low agitation speed to maintain high efficiency.

(d) Abrasion-resistant construction.

(e) Efficient removal of medium from the product stream by screens or settlement traps.

(f) Optimum concentration of mineral in the water for most efficient grinding, consistent with the requirements in the product. This concentration must be determined experimentally. Moderate particle size reduction of a coarse kaolin may involve energies of 20–80 kWh per tonne of product. Fine comminution of talc may require 160 kWh per tonne of product.

3.5 Chemical Alterations

Chemical alterations of natural clays are an additional means of developing desired characteristics in whiteware mineral components. In addition to the previously mentioned techniques for removing impurities such as iron from clay by chemical action, it is also possible to coat reactive clays produced in comminution processes with property-altering agents. A second interesting mineral modification by chemical means is the reaction of montmorillonite with alumina to form chlorite (Bidwell et al. 1977), a reaction noted as far back as 1950 but studied only recently for the practical purpose of improving paper-coating clays. This process could be of great interest in removing the undesirable effects of montmorillonite on body permeability and on the control of slip viscosity.

4. Conclusion

In conclusion, ceramic whitewares are a diverse group of products with a wide functional range, from artware to highly technical electrical materials. A wide range of formulations permits such diverse products to be made from the traditional ceramic materials of clay, flint and feldspar. Recent developments in this technology have permitted great flexibility in formulation and processing. By understanding the characteristics of a given composition, alterations in body properties or raw material substitutions can be readily engineered. Advances in raw material processing have greatly diversified the selection of available mineral components and at the same time increased their quality.

See also: Traditional Ceramics: An Overview

Bibliography

Balentine J W, Iannicelli J, Whitley J B 1972 Improving the brightness of kaolin clays by the addition of oxalic acid and its salts, British Patent No. 1,258,964 (5 January 1972)

Bidwell J I, Adams R W, Jepson W B 1977 Treatment of clay minerals. US Patent No. 4,045,235 (30 August 1977)

Brociner R E 1970 Grinding of minerals. US Patent No. 3,624,796 (8 May 1970)

Clark N O 1952 Improvements in/or relating to the treatment of clays, precipitated chalk and the like, British Patent No. 706,429 (29 May 1952)

Clark N O 1974 Processing and disposal of mineral slimes. In: Jones M J (ed.) 1974 *Proc. 10th Int. Mineral Processing Congr.* Institute of Mining and Metallurgy, London, pp. 913–28

Cohen H E, Good J A 1976 A superconducting magnet system for a very high intensity magnetic mineral separator. *IEEE Trans. Magn.* 12: 493–97

Feld I L, McVay T N, Gilmore H L, Clemmens B H 1960 *Paper-Coating Clay from Coarse Georgia Kaolins by a New Attrition-Grinding Process*, US Bureau of Mines Report of Investigations 5697. USBM, Pittsburgh, Pennsylvania

Greene E W, Duke J B, Hunter J L 1961 Froth flotation method. US Patent No. 2,990,958 (4 July 1961)

Phelps G W, Harris D L 1967 Specific surface and dry strength by methylene blue adsorption. *Am. Ceram. Soc. Bull.* 47: 1146–50

Pryor E J 1965 *Mineral Processing*, 3rd edn. Elsevier, New York
Taggart A F 1945 *Handbook of Ore Dressing*. Wiley, New York
van Olphen H 1963 *An Introduction to Clay Colloid Chemistry for Clay Technologists, Geologists, and Soil Scientists*. Interscience, New York
Watson H P, Clark N O, Windle W 1975 A superconducting magnetic separator and its application in improving ceramic raw materials. *Proc. 11th Int. Mineral Processing Congr.* University of Cagliari, Cagliari, Sardinia, pp. 795–812

N. O. Clark
[Imperial College, London, UK]

R. L. Lehman
[Rutgers University, Piscataway, New Jersey, USA]

Z

Zinc Resources

Zinc is a bluish-white metal when freshly cast, but on exposure to air it develops a thin corrosion-resistant gray oxide, an important characteristic in its use by industry. In 1987 in the USA, construction accounted for 50% of the consumption of slab zinc; transportation, 23%; machinery, 12%; electrical, 10%; and other industries, 10%. The main uses of slab zinc in the USA were: galvanizing, 50%; diecasting alloys, 27%; brass and bronze, 13%; and other, 10%. Zinc in compounds and dusts were consumed by the chemical, agricultural, rubber and paint industries (US Bureau of Mines 1988).

1. Geochemistry, Crustal Distribution and Ore Tenors

Zinc has atomic number 30 and atomic weight 65.38. It has a specific gravity of 7.13, a melting point of 419.58 °C and a boiling point of 907 °C. The ionic radius of zinc (Zn^{2+}) in fourfold coordination is 0.71 Å, and in sixfold coordination is 0.74 Å. Zinc is predominantly chalcophilic and weakly oxyphilic. Estimates of the average crustal abundance range from 65 to 95 ppm. Concentrations in igneous rocks range from about 130 ppm in basalt and syenite, through nearly 72 ppm in rocks of intermediate composition, to 36–60 ppm in silicic rocks; sedimentary rocks such as sandstone, shale and carbonates average about 16, 95 and 20 ppm, respectively. The chief ore mineral of zinc is sphalerite (Zn, Fe)S. Most zinc ores also contain economically recoverable quantities of lead, copper, silver, gold, barium, sulfur or fluorine. Moreover, trace metals associated with and recovered from sphalerite concentrates include silver, cadmium, germanium, gallium, indium and thallium. Most zinc ores contain 2.5–20 wt % Zn, most commonly 3–10%.

2. Production

The record of world zinc production is one of exponential growth since the end of the nineteenth century, interrupted only by periods of political and economic stress. Mine production of zinc in the 1970s grew at a slightly lower rate; it increased from 5.3 Mt in 1970 to 7.1 Mt in 1987. The principal zinc producers in 1987 included: Canada, 1.48 Mt; Australia, 0.69 Mt; Peru, 0.53 Mt; Mexico, 0.29 Mt; the USA, 0.24 Mt; and the USSR and other centrally planned economies, 1.80 Mt. Other important producers were Japan, Sweden, Poland, Spain, Ireland, the People's Republic of China, North Korea and the FRG (US Bureau of Mines 1988).

3. Reserves and Resources

Major world identified economic and promising subeconomic resources of zinc in 1977 according to Briskey and Morris (1980) (here corrected to 266 Mt) include: North America, 117 Mt; South and Central America and Mexico, 14 Mt; Europe, >53 Mt; Africa, 22 Mt; Asia, >34 Mt; and Australia, 26 Mt. Jolly (1983) incorporated marginally economic and some subeconomic resources into an estimate of 290 Mt that includes: Canada, 56 Mt; the USA, 53 Mt; Australia, 39 Mt; South Africa, 14 Mt; the USSR, 13 Mt; India, 12 Mt; Peru, 12 Mt; Spain, 10 Mt; Mexico, 8 Mt; the People's Republic of China, 7 Mt; Zaire, 7 Mt; Ireland, 6 Mt; Iran, 6 Mt; Japan, 5 Mt; Poland, 4 Mt; Brazil, 3 Mt; Central America, 1 Mt; and others, 34 Mt. Wedow et al. (1973) estimated that recoverable and subeconomic, identified and undiscovered zinc resources total roughly 5085 Mt. Scrap zinc provides about 5% of total supply, but has major potential for expansion.

4. Consumption and Adequacy of Supplies

Twelve countries accounted for 73% of slab (smelted) zinc consumed in 1982, including: the USSR, 1.05 Mt; the USA, 0.80 Mt, Japan, 0.70 Mt; the Federal Republic of Germany, 0.37 Mt; France, 0.26 Mt; the People's Republic of China, 0.26 Mt; Italy, 0.20 Mt; the UK 0.18 Mt; Poland, 0.14 Mt; Belgium, 0.13 Mt; Brazil, 0.11 Mt; and India, 0.10 Mt. World demand for primary sources of zinc is expected to increase from 5.9 Mt in 1982 to between 7.8 and 14.3 Mt by 2000, with a most probable demand of about 10.3 Mt (Cammarota 1980). These projections, which anticipate a 2% annual growth of zinc consumption, result in a most probable cumulative demand of 182 Mt by the turn of the present century. Identified economic and promising subeconomic zinc resources cited above are more than adequate to meet anticipated cumulative demand; recoverable and subeconomic identified and undiscovered resources are many times larger.

5. Geological Occurrences of Principal Types of Ore Deposits

Most exploitable primary zinc ores can be divided into five principal categories based on differences in their geological occurrences: (a) volcanic-hosted submarine exhalative massive sulfide deposits; (b) sediment-hosted submarine exhalative deposits; (c) strata-bound carbonate-hosted deposits; (d) strata-bound sandstone-hosted deposits; and (e) vein-, replacement- and contact-metasomatic deposits.

Volcanic-hosted submarine exhalative massive sulfide deposits are composed almost entirely of pyrite (FeS) and pyrrhotite ($Fe_{1-x}S$), with varying amounts of chalcopyrite ($CuFeS_2$), sphalerite and galena (PbS). They typically occur as multiple stratiform lenses within rhyolitic to basaltic submarine volcanic and associated rocks, and commonly overlie or are adjacent to discordant feeder zones containing low-grade stringer and disseminated deposits in altered volcanic rocks. Most ore bodies contain between 0.1 and 10 Mt of ore, although groups of related deposits commonly aggregate more than 100 Mt. Ore bodies typically contain 4–5% Zn, 1–2% Cu and 1–1.5% Pb; however, those in exclusively basaltic terranes typically have slightly more than 2% Cu, less than 1% Zn and no significant Pb. Major deposits include the Matagami, Timmins and Mattabi districts in Canada, the Kuroko deposits in Japan and the deposits on Cyprus.

Sediment-hosted submarine exhalative deposits of zinc and lead are stratiform basinal accumulations, mainly of fine-grained pyrite and pyrrhotite, sphalerite, galena, sporadic barite and minor chalcopyrite, deposited concordantly and interbedded with euxinitic sediments including black shale, siltstone, sandstone, chert, dolomite and micritic limestone. Stockwork and disseminated deposits and hydrothermally altered rocks representing the feeder zone of these deposits are commonly present beneath or adjacent to the stratiform ore bodies. Deposits generally contain 2–200 Mt of ore, either individually or in aggregate, and have grades between 6 and 31% Zn + Pb + Cu, with Zn between 3 and 19%. Typical examples include Meggen and Rammelsberg, the Federal Republic of Germany; Mount Isa and McArthur River, Australia; and Sullivan, Tom and Howards Pass, Canada.

Strata-bound carbonate-hosted zinc–lead deposits are most commonly composed of sphalerite, galena, pyrite, barite, fluorite (CaF_2) and chalcopyrite, as well as dolomite ($CaMg(CO_3)_2$), calcite ($CaCO_3$) and quartz (SiO_2). These minerals typically fill primary and secondary voids developed in favorable beds or horizons within thick sequences of shallow-water dolomite or limestone. Ore bodies are usually ramose, to roughly rectilinear, to linear intricately interconnected networks of mineralization, ranging in size from several tens of thousands of tonnes to over 20 Mt of ore; districts containing well over 200 Mt of ore are not uncommon. Ore grades usually range from 3 to 10% Zn + Pb, but some deposits contain only zinc, with grades commonly between 3 and 6%. Important examples are those of the Mississippi Valley and adjacent regions of the USA; Pine Point and Newfoundland Zinc, Canada; and Silesia, Poland.

Sandstone-hosted lead–zinc deposits are typically strata-bound and commonly stratiform concentrations, mainly of fine- to medium-crystalline galena, with scattered smaller amounts of sphalerite, pyrite, barite, fluorite and minor chalcopyrite deposited in multiple, thin, sheet-like ore bodies. Deposits are contained within marine, terrigenous and continental arenaceous sedimentary rocks, consisting largely of quartzitic and arkosic sandstone, conglomerate, grit and siltstone. The ore minerals commonly occur as regularly distributed subspherical clots, or as evenly distributed disseminations, but typically they form more massive ore bodies where localized by sedimentary structures. Exploitable ore bodies or closely spaced clusters of ore bodies generally contain 1–80 Mt of ore, with 3.5–5% Pb and 0.5–0.75% Zn. Typical examples include Laisvall, Vassbo and Guttusjö, Sweden; Largentière, France; Bou-Sellam, Morocco; and Salmon River, Canada.

Vein-, replacement- and contact-metasomatic deposits generally are temporally and spatially related to nearby granitic plutons of granodioritic or quartz monzonitic composition; the contact-metasomatic deposits are mostly restricted to the contact aureoles of the plutons. Replacement- and contact-metasomatic ore bodies typically replace carbonate rocks near the plutons, whereas vein deposits are largely open-space fillings in fractures in any variety of host rock. Most deposits consist of multiple tabular to podiform or pipe-shaped ore shoots, or occur as extensive highly irregular and branching masses. Ores are typically composed of coarsely crystalline aggregates of pyrite, sphalerite, galena, chalcopyrite and numerous complex base- and precious-metal sulfide, sulfosalt, arsenide and antimonide minerals. Individual deposits usually contain several hundred thousand to a few million tonnes of ore, with 10–30% Pb + Zn in highly varying proportions, although many districts have produced 5–50 Mt of lower-grade ore. Important districts include Bingham, Tintic and Coeur d'Alene in the USA; Fresnillo and Taxco in Mexico; Clausthal and Freiberg in the Federal Republic of Germany; and Kamioka, Chichibu, Nakatatsu and Obori in Japan.

Bibliography

Briskey J A, Dingess P R, Smith F, Gilbert R C, Armstrong A K, Cole G P 1986 Localization and source of Mississippi Valley-type zinc deposits in Tennessee, USA, and comparisons with Lower Carboniferous rocks of Ireland. In: Andrew C J, Crowe R W A, Finlay S, Pennell W M, Pyne J F (eds.) 1986 *Geology and Genesis of Mineral Deposits in Ireland*. Irish Association for Economic Geology, Dublin

Briskey J A, Morris H T 1980 Lead and zinc resources of the United States and of the world. *US Geological Survey Research 1979*, US Geological Survey Professional Paper 1150. US Government Printing Office, Washington, DC, pp. 2–3

Cammarota V A Jr 1980 Production and uses of zinc. In: Nriagu J O (ed.) 1980 *Zinc in the Environment*. Wiley, New York, pp. 1–38

Jensen M L, Bateman A M 1979 *Economic Mineral Deposits*, 3rd edn. Wiley, New York

Jolly J H 1983 *Zinc*, US Bureau of Mines Mineral Commodity Profiles. USBM, Washington, DC

Lead and Zinc Statistics (updated monthly). International Lead and Zinc Study Group, London

Roskill Information Services 1978 *The Economics of Zinc*. Roskill Information Services, London
Stanton R L 1972 *Ore Petrology*. McGraw-Hill, New York
US Bureau of Mines 1988 *Mineral Commodity Summaries 1988*. USBM, Washington, DC
Wedow H Jr, Kiilsgaard T H, Heyl A V, Hall R B 1973 Zinc. In: Brobst D A, Pratt W P (eds.) 1973 *United States Mineral Resources*, US Geological Survey Professional Paper 820. US Government Printing Office, Washington, DC, pp. 697–711
Wolf K H (ed.) 1976 *Handbook of Strata-Bound and Stratiform Ore Deposits*. Elsevier, Amsterdam
World Metal Statistics (updated monthly). World Bureau of Metal Statistics, London

J. A. Briskey
[US Geological Survey, Menlow Park, California, USA]

H. Wedow Jr.†
[late of Knoxville, Tennessee, USA]

Zirconium and Hafnium Resources

Zirconium occurs most commonly in nature as zirconium silicate (zircon) and less commonly as zirconium oxide (baddeleyite). Although zircon occurs in a variety of rocks, the only commercial source of zircon is as a coproduct from beach sand mining for titanium ores (see *Titanium Resources*). Minerals from these deposits are called heavy minerals because they have specific gravities of 2.9 or higher. Australia, South Africa and the USA are the major world producers.

Zircon was first recognized as a component in alluvial and beach sands in 1895, but it was not produced in any quantity until 20 years later during World War I, when it was mined just south of Jacksonville Beach, Florida, and used as a high-temperature refractory. Australian beach sand mining expanded in the late 1930s and large-scale mining in the USA began in the 1940s and 1950s. Baddeleyite first became available as a commercial product in 1916, but never in as much quantity as zircon.

Most of the zircon sand produced in the world is used as foundry sands, refractories, ceramics and abrasives or is ground into flour. Only a small portion is used in the manufacture of zirconium metal or chemicals. Zircon is also the major source of hafnium.

1. Mineralogy of Zirconium

Commercial zirconium minerals are zircon ($ZrSiO_4$) and baddeleyite (ZrO_2). Small amounts of hafnium are always present in zirconium minerals. Zirconium and hafnium also occur in silicate and oxide minerals in combination with a number of different elements, but these are generally of academic interest only.

Zircon is extremely resistant to weathering; therefore it is often found in ancient beach sands and alluvial placer deposits. Although resistant to alteration from external sources, it is vulnerable to internal alteration by thorium and uranium, either substituting in the zircon lattice or in solid solution. Radioactive emanations from these elements cause zircon to alter gradually to an amorphous state (called metamict). During this process the mineral hydrates, the specific gravity is lowered and the color changes. Fresh unaltered zircon is white or colorless. As alteration progresses, the grains darken and are called hyacinth. In later stages of alteration an amorphous zirconium silicate forms, called malacon.

2. Occurrence of Zircon

Zircon forms as an accessory mineral in a variety of igneous and metamorphic rocks, especially those containing sodic feldspars. It is one of the earliest minerals to crystallize from a cooling magma and frequently incorporates inclusions (e.g., apatite, magnetite). Zircon in these rocks often crystallizes in tetragonal prisms with pyramidal terminations, but it also commonly occurs as rounded grains in igneous rocks. It is sometimes found in metamorphic rocks such as gneiss and schists.

Before the zircon particles can become part of a beach sand deposit, their host rocks must undergo a series of events that will ultimately liberate the zircon for transportation to a seacoast. These begin with exposure of the host rocks to subaerial weathering. The rocks are broken down into smaller fragments and transported downhill by rainwater and gravity. The rocks further decompose to a point where the smaller zircon grains are liberated from the enclosing feldspars and quartz. The mineral particles are transported by streams and ultimately end up along a marine shoreline. Here the action of waves, tidal currents and wind may remove lighter quartz, forming a heavy mineral deposit. A commercial heavy mineral deposit may contain 0.2–3% of zircon. Commercial beach sand deposits are found in Australia, South Africa, the USA, India, Sri Lanka, Malaysia, China, Thailand and Tanzania. All the commercial zircon products today are mined and separated from relatively young beach sand deposits found on or near active coast lines.

3. Australian Zircon Deposits

Australia is the major world producer and exporter of zircon sand. Deposits are found on both the east and west coasts. The eastern Australian heavy mineral beach sand deposits along the Pacific Ocean are both Pleistocene and Recent in age. Zircon and its coproduct rutile have been derived from the breakdown and erosion of well-sorted silica sandstones in the late Paleozoic and Mesozoic basins of eastern Australia. Zircon grains from these deposits range from 90 to 100 μm in average size, depending on the producing mine.

Mineral sand deposits of Western Australia are found along older beaches paralleling the Indian Ocean that lie at different elevations ranging from 10 to 200 μm above present sea level. The strand line deposits are part of a series of Pleistocene and Recent sediments that exist as a thin veneer throughout the Perth basin. These overlie older marine and continental deposits. Deposits in southwest Western Australia near the town of Bunbury have been in production for a number of years. Larger deposits of heavy minerals were discovered in 1970 near the town of Eneabba. These are richer in zircon than the deposits to the south. Zircon grains from Eneabba are coarser (188 μm average) than zircon from any deposit in the world that is now in production.

4. South African Deposits

Heavy minerals are produced from a high dune paralleling the Indian Ocean about 160 km north of Durban. The present ore body is estimated to be mined out around the year 2010. The zircon grains are rounded to irregular or prismatic in shape and are slightly finer than zircon from Trail Ridge, Florida, USA. They are about 108 μm in average size. Like Australian zircon, the grains have surface coatings that turn dark orange to brown on heating.

The mineral baddeleyite (ZrO_2) is recovered from the mill tailings at the Palabora copper mine in the Transvaal. The ZrO_2 (+ HfO_2) content is in the range 97–99%. Baddeleyite is the end product of zircon alteration and is sometimes found in alluvial deposits associated with ilmenite, zirkelite (a mixture of zircon and baddeleyite), apatite and perovskite. It is also found in Brazil, Sri Lanka, Italy and the USA.

5. United States Deposits

A small amount of zircon was produced from a beach sand mine located near Ponte Vedra, Florida, between 1918 and 1929. Zircon was later produced from a heavy mineral deposit located at Arlington, a suburb of Jacksonville, Florida, between 1943 and 1963. It was also produced from mines located near Folkston, Georgia, and Boulogne, Florida, between 1963 and 1979 when the deposits were mined out.

Since 1950 most of the zircon produced in the USA has come from the Trail Ridge deposit in north central Florida. The deposit was formed at the height of Florida's last submergence during Pleistocene time. Reworked sands transported from older features were redeposited, forming the ridge. Waves, currents and wind action removed part of the lighter silica sand, leaving an enrichment of heavy minerals in sand dunes on the southwestern flank of this prominent geographic feature. After deposition of the heavy-mineral-bearing sands, the ridge was covered with windblown silica sand which forms an overburden. Zircon from the Trail Ridge deposit is extremely uniform in particle size, grain morphology and chemical purity. The grains are 119 μm in average size and are free of surface coatings. Many grains have been slightly etched as a result of wind action during the formation of the deposit and later leaching by ground water.

More recently, in 1973, zircon production began from an old shoreline deposit near Green Cove Springs, Florida. Zircon from this deposit averages about 98 μm in size, making it similar to zircon produced from the Jacksonville area during World War II. This deposit was formed by a regressing sea in late Pleistocene time. It will be mined out some time near the end of this century.

Undeveloped zircon deposits are also known to exist in Tennessee, Georgia and the Carolinas.

6. Mining and Processing

Most of the heavy mineral operations in the world are similar. Sand ores are mined with dredges, either cutterhead–suction or in some cases bucketline. In areas where there is insufficient water for dredging, the ores are mined using elevating scrapers and bulldozers.

The heavy minerals are concentrated by removing the quartz and other light minerals using gravity concentrators. They are further fractionated using magnetic and high-tension separators.

Zircon produced in the USA is calcined to produce a white product. Zircon sands from other world sources are not calcined, because they turn yellowish orange when heated, owing to ferruginous coatings on the grain surfaces.

7. Markets and Uses

Zircon sands are used in a variety of markets ranging from foundry sands to zirconium metal manufacture. The major uses (1977) are:

foundry sands	35%
refractories	28%
flour	20%
abrasives	5%
other	12%

8. Future of Zircon

The US Bureau of Mines (1988) identified some 51 million tonnes of known zircon reserves. However, only a small portion of these are considered economic to mine at the present time. Zircon from deposits now being exploited will be mined out sometime in the early part of the twenty-first century. Development of new zircon deposits will be influenced by the needs of the various consuming industries and by the availability of alternative products of equal or lower cost. Under certain conditions future zircon supplies may come from tar sand wastes, phosphate sand tailings or sand and gravel mine wastes.

Bibliography

Baker G 1962 *Detrital Heavy Minerals in Natural Accumulates*, Monograph Series No. 1. Australasian Institute of Mining and Metallurgy, Melbourne

Baxter J L 1977 *Heavy Mineral Sand Deposits of Western Australia*, Mineral Resources Bulletin No. 10. Geological Survey of Western Australia, Perth

Calver J L 1957 *Mining and Mineral Resources*, Geological Bulletin No. 39. Florida Geological Survey, Tallahassee, Florida, pp. 15–31

Garnar T E 1978 Geological classification and evaluation of heavy mineral deposits. *12th Forum on the Geology of Industrial Minerals*, Information Circular No. 49. Georgia Geological Survey, Atlanta, Georgia, pp. 25–36

Garnar T E 1981 Heavy minerals industry of North America. In: Coope B M (ed.) 1981 *Proc. 4th Int. Industrial Minerals Congress*. Metal Bulletin, London, pp. 29–42

Grogan R M, Few W G, Garnar T E, Hager C R 1964 Milling at DuPont's heavy minerals mines in Florida. In: Arbiter N (ed.) 1964 *7th Int. Mineral Processing Congr.* Gordon and Breach, New York, Chap. 9, pp. 205–29

Klemic H 1975 Zirconium and hafnium minerals. In: Lefond S J (ed.) 1975 *Industrial Minerals and Rocks*, 4th edn. American Institute of Mining, Metallurgical and Petroleum Engineers, New York, pp. 1275–83

Klemic H, Gottfried D, Cooper M, Marsh S 1973 Zirconium and hafnium. *US Mineral Resources*, US Geological Survey Professional Paper 820. USGS, Arlington, Virginia, pp. 718–22

Kramer J W, Brown R A 1976 Survey of heavy minerals in surface mineable area of Athabasca oil sand deposit. *Canad. Min. Metall. Bull.* 69(776): 92–99

Lissiman J C, Oxenford R J 1973 *The Allied Mineral N.L. Heavy Mineral Deposit in Eneabba, W.A.*, Conference Volume. Australian Institute of Mining and Metallurgy, Parkville, Victoria, pp. 153–61

Lynd L E 1978 *US Dependence on Foreign Sources of Heavy Mineral Concentrates*, AIME Preprint 78H366. American Institute of Mining, Metallurgical and Petroleum Engineers, New York

Lynd L E 1980 Zirconium and hafnium. *Industrial Minerals Survey*. US Bureau of Mines, Washington, DC

Lynd L E 1981 Zirconium, *Mineral Commodity Summaries*. US Bureau of Mines, Washington, DC, pp. 182–83

Martens J H C 1928 Beach deposits of ilmenite, zircon, and rutile. *19th Annual Report of the Florida State Geological Survey*. Florida Geological Survey, Tallahasee, Florida, pp. 124–25

Palache C, Berman H, Frondel C 1944 Baddeleyite. *Dana's System of Mineralogy*, Vol. 1, 7th edn. Wiley, New York, pp. 608–10

Pirkle E C, Pirkle W A, Yoho W H 1974 The Green Cove Springs and Boulogne heavy mineral sand deposits of Florida. *Econ. Geol.* 69: 1129–37

Pirkle F L 1975 Evaluation of possible source regions of trail ridge sands. *Southeastern Geol. (Duke University)* 17(2): 93–114

US Bureau of Mines 1944–1970 *Minerals Yearbook*, annual editions, 1944–1970. USBM, Washington, DC

US Bureau of Mines 1988 *Mineral Commodity Summaries 1988*. USBM, Washington, DC

T. E. Garnar Jr.
[Titanium Consultants, Keystone Heights, Florida, USA]

LIST OF CONTRIBUTORS

Contributors are listed in alphabetical order, together with their addresses. Titles of articles which they have authored follow in alphabetical order. Where articles are co-authored, this has been indicated by an asterisk preceding the article title.

Ault, C H
Indiana Geological Survey
611 N Walnut Grove Avenue
Bloomington, IN 47405
USA
Construction Materials: Crushed Stone

Austin, G S
New Mexico Bureau of Mines
& Mineral Resources
Socorro, NM 87801
USA
Soil Additives: Industrial Minerals
Soil Additives: Phosphate, Potash and Sulfur

Bailey, E H
[Deceased; late of US Geological Survey, Menlo Park, California, USA]
Mercury Resources

Barna, D L
US Bureau of Mines
2401 E Street NW
Washington, DC 20241
USA
**Electronic and Optical Minerals*

Bourne, H L
PO Box 293
Northville, MI 48167
USA
Slag

Bradbury, J C
Illinois Department of Energy
& Natural Resources
615 East Peabody Building
Champaign, IL 61820
USA
Abrasives
Abrasives: Siliceous Minerals

Briskey, J A
US Geological Survey
345 Middlefield Road
Menlo Park, CA 94025
USA
**Zinc Resources*

Brownell, W E
NYS College of Ceramics
Alfred University
Alfred, NY 14802
USA
Structural Clay Products

Cabri, L J
Canada Center for Mineral
& Energy Technology
555 Booth Street
Ottawa, Ontario, K1A 0G1
Canada
Platinum Group Metals Resources

Cannon, H L
US Geological Survey
954 National Center
Reston, VA 22092
USA
**Mining and Processing: Environmental Problems*

Cannon, W F
US Geological Survey
954 National Center
Reston, VA 22092
USA
**Manganese Resources*

Carr, D D
Indiana Geological Survey
611 N Walnut Grove Avenue
Bloomington, IN 47405
USA
Construction Materials: Industrial Minerals

Clark, J P
Room 8-413
Massachusetts Institute of Technology
Cambridge, MA 02139
USA
**Critical and Strategic Materials*

Clark, K F
Department of Geological Sciences
University of Texas at El Paso
El Paso, TX 79968
USA
Molybdenum Resources

Clark, N O
Imperial College of Science and Technology
London SW 7
UK
**Whiteware: Component Minerals*

Coombs, G
Johns-Manville Sales Corporation
Research and Development Center
Ken-Caryl Ranch
Denver, CO 80217
USA
Fillers and Coatings: Siliceous Minerals
Filters, Sorbents and Ion Exchangers: Siliceous Minerals
Insulation Raw Materials

Cox, D P
US Geological Survey
345 Middlefield Road
Menlo Park, CA 94025
USA
Copper Resources

Craig, J R
Department of Geological Sciences
Virginia Polytechnic & State University
Blacksburg, VA 24061
USA
Ore Minerals

Damberger, H H
Illinois State Geological Survey
Champaign, IL 61820
USA
Coal: World Resources

Doremus, R H
Materials Science Department
Rensselaer Polytechnic Institute
Troy, NY 12181
USA
Glass: An Overview

Duncan, L R
Harbison-Walker Refractories International
2 Gateway Center
Pittsburg, PA 15222
USA
Magnesium Resources

Dunn, J R
Dunn Geoscience Corporation
12 Metro Park Road
Albany, NY 12205
USA
Portland Cement Raw Materials

Ellwood, B B
Department of Geology
University of Texas
Arlington, TX 76019
USA
Geophysical Exploration for Mineral Deposits

Eyde, T H
GSA Resources
Box 1127
Cortaro, AZ 85230
USA
Filters, Sorbents and Ion Exchangers: Zeolite Minerals

Feiss, P G
Department of Geology
University of North Carolina
Chapel Hill, NC 27514
USA
Metallogenic Provinces
Plate Tectonic Settings for Metallic Ore Resources

Finch, W I
US Geological Survey
Box 25046
Federal Center MS 916
Denver, CO 80225
USA
Uranium Resources

Fischer, R P
2200 Range View Court
Grand Junction, CO 81503
USA
Vanadium Resources

Foose, M P
US Geological Survey
954 National Center
Reston, VA 22092
USA
Nickel Resources

Fraser, G S
Indiana Geological Survey
611 N Walnut Grove Avenue
Bloomington, IN 47405
USA
Construction Materials: Sand and Gravel

Freas, R C
Franklin Limestone Company
612 10th Avenue N
Nashville, TN 37203
USA
Lime

Frohnsdorff, G J
National Bureau of Standards
Center for Building Technology
Building 226, Room B348
Washington, DC 20234
USA
**Calcium Aluminate Cements*
Portland Cements, Blended Cements and Mortars
Beryllium Resources

Garnar Jr, T E
Titanium Consultants Inc
PO Drawer 1200
Keystone Heights, FL 32656
USA
Titanium Resources
Zirconium and Hafnium Resources

Gschneidner Jr, K A
Energy & Mineral Resources Institute
Iowa State University
Ames, IA 50011
USA
Rare-Earth Metals Resources

Gualtieri, J L
US Geological Survey
Box 25046 MS 972
Federal Center
Denver, CO 80225
USA
Arsenic Resources

Hague, J M
655 Rountree Avenue
Platteville, WI 53818
USA
Lead Resources

Harper, D
Indiana Geological Survey
611 N Walnut Grove Avenue
Bloomington, IN 47405
USA
Coal: Mining

Harvey, R D
Illinois State Geological Survey
Natural Resources Building
615 East Peabody Drive
Champaign, IL 61820
USA
Coal: Geology

Herz, N
Department of Geology
University of Georgia
Athens, GA 30602
USA

Hodgson, A A
65 Wellesley Drive
Crowthorne
Berks
RG11 6AL
UK
Asbestos: Alternatives
Asbestos Fibers

Holtz, R D
School of Civil Engineering
Purdue University
West Lafayette, IN 47907
USA
Construction Materials: Fill and Soil

Jensen, M L
Department of Geology & Geophysics
University of Utah
Salt Lake City, UT 84112
USA
Igneous Ore Deposits: Formation Processes
Metamorphic Ore Deposits: Formation Processes
Sedimentary Ore Deposits: Formation Processes

Johnson, K S
Oklahoma Geological Survey
University of Oklahoma
Norman, OK 73019
USA
Fillers and Coatings: Sulfate Minerals

Jones, T S
US Bureau of Mines
2401 E Street NW
Washington, DC 20241
USA
Niobium and Tantalum Resources

Jorgensen, D B
US Gypsum Co
101 South Wacker Drive
Chicago, IL 60606
USA
Gypsum and Anhydrite

Kalyoncu, R S
Tuscaloosa Research Center
US Bureau of Mines
Box L
Tuscaloosa, AL 34586
USA
Construction Materials: Granules

Kamen-Kaye, M
1 Waterhouse Street (5)
Cambridge, MA 02138

USA
Hydrocarbons: Origin, Migration and Accumulation
Petroleum: Oil and Gas Fields
Petroleum: World Resources

Kaplan, R S
US Bureau of Mines
2401 East Street NW
Washington, DC 20241
USA
**Recycling of Metals: Technology*

Kirk, W S
US Bureau of Mines
2401 East Street NW
Washington, DC 20241
USA
**Cobalt Resources*

Koch Jr, G S
Department of Geology
University of Georgia
Athens, GA 30602
USA
Statistical and Computer Models in Mining and Exploration

Kopanda, J E
National Bureau of Standards
Center for Building Technology
Building 226, Room B348
Washington, DC 20234
USA
**Calcium Aluminate Cements*

Kotzin, E L
American Foundrymen's Society
Gold and Wolf Roads
Des Plaines, IL 60016
USA
Foundry Sands: Classification and Sources
Foundry Sands: Silica Sand and Nonsilica Minerals

Lefond, S J
Industrial Minerals Inc
29983 Canterbury Circle
Evergreen, CO 80439
USA
Abrasives: Precious and Semiprecious Minerals

Lehman, R L
Department of Ceramics
Rutgers University
Piscataway, NJ 08854
USA
**Whiteware: Component Minerals*

McCarl, H N
University of Alabama
School of Business
Birmingham, AL 35294
USA
Prices of Industrial Minerals: History

McCarthy Jr, J H
US Geological Survey
Federal Center
Denver, CO 80225
USA
Geochemical Exploration for Mineral Deposits

McKelvey, V E
[Deceased]
Mineral Resources: Definitions, Uses, Classification and Future Availability

McLaren, M G
Department of Ceramics
Rutgers University
New Brunswick, NJ 08903
USA
Traditional Ceramics: An Overview

Marchant, W N
US Department of the Interior
18th and C Streets NW
Washington, DC 20240
USA
**Electronic and Optical Minerals*

Marsden, R W
[Deceased; late of University of Minnesota, Duluth, Minnesota, USA]
Iron Resources

Mason, B H
The France Stone Company
PO Box 49
Waterville, OH 43566
USA
Construction Materials: Lightweight Aggregate

Miller, M H
US Geological Survey
Box 25046
Mail Stop 905
Denver Federal Center
Denver, CO 80225
USA
Antimony Resources
Bismuth Resources

Morgan, J D
US Bureau of Mines
2401 E Street NW
Washington, DC 20241
USA
Stockpiling in the USA

Murray, H H
Department of Geology
Indiana University
1005 E 10th Street
Bloomington, IN 47405
USA
Abrasives: Nonsiliceous Minerals
Binders: Clay Minerals
Binders: Industrial Minerals
Fillers and Coatings: Clay Minerals
Fillers and Coatings: Industrial Minerals
Filters, Sorbents ad Ion Exchangers
Filters, Sorbents and Ion Exchangers: Clay Minerals

Ness, H
National Association of Recycling Industries Inc
330 Madison Avenue
New York, NY 10017
USA
**Recycling of Metals: Technology*

Newcomb, R T
Department of Mining & Geological Engineering
College of Mines and Engineering
University of Arizona
Tucson, AZ 85721
USA
Economic Theory of Mineral Resources and Markets

Nielsen, J M
General Electric Corporate Research & Development
120 Erie Boulevard
Schenectady, NY 12305
USA
Silicon Dioxide Hazards

Owen, R M
Department of Geological Sciences
University of Michigan
Ann Arbor, MI 48109
USA
Ocean Resources and Mining

Patterson, S H
US Geological Survey
954 National Center
Reston, VA 22092
USA
Aluminum Resources

Patton, J B
[Deceased; late of Indiana Geological Survey, Bloomington, Indiana, USA]
Construction Materials: Dimension Stone

Peters, W C
Mining and Geological Consultant
5702E Seventh Street
Tucson, AZ 85711
USA
Geological Exploration for Mineral Deposits
Mineral Deposits: Examination and Evaluation
Open-Cut Mining
Underground Mining

Peterson, D T
Ames Laboratory
Iowa State University
222 Metals Development Building
Ames, IA 50011
USA
Thorium Resources

Proctor, P D
Department of Geology
Brigham Young University
Provo, UT 84601
USA
Silver Resources

Reddy, B
Charles River Associates
200 Clarendon Street
Boston, MA
USA
**Critical and Strategic Materials*

Reed, B L
US Geological Survey
Branch of Alaskan Geology
1209 Orca Street
Anchorage, AK 99501
USA
Tin Resources

Reid, J C
North Carolina Geological Survey
PO Box 27687
Raleigh, NC 27611
USA
Tungsten Resources

Sawkins, F J
Department of Geology
University of Minnesota
Minneapolis, MN 55455
USA
Gold Resources

Schwartz, M A
US Bureau of Mines
2401E Street NW
Washington, DC 20241
USA

Pigments: Industrial Minerals
Pigments: Titania, Iron Oxides and Other Minerals

Severinghaus Jr, N
Franklin Limestone Co
612 Tenth Avenue North
Nashville, TN 37203
USA
Fillers and Coatings: Carbonate Minerals

Shaw, D M
Department of Geology
McMaster University
Hamilton, Ontario L8S 4MI
Canada
Geochemical Distribution of the Elements

Sibley, S F
US Bureau of Mines
2401 E Street NW
Washington, DC 20241
USA
**Cobalt Resources*

Slade, M E
Department of Economics
University of British Columbia
997 1873 East Mall
Vancouver, British Columbia V6T 1Y2
Canada
Prices of Metals: History

Stanley, T W
International Economic Studies Institute
Suite 908
1625 Eye Street NW
Washington, DC 20006
USA
Stockpiling

Thayer, T P
3211 W Maritan Drive
St Petersburg Beach, FL 33706
USA
Chromite Resources

Tooker, E W
US Geological Survey
345 Middlefield Road MS 90A
Menlo Park, CA 94025
USA
Ores of the Future

Turner, F
Magcobar Group
Dresser Industries
PO Box 1407
Houston, TX 77001
USA
Well-Drilling Materials: Clay and Nonclay Minerals
Well-Drilling Materials: Industrial Minerals

Vine, J D
21736 Panorama Drive
Golden, CO 80401
USA
Lithium, Cesium and Rubidium Resources

Wedow Jr, H
[Deceased]
**Cadmium Resources*
**Zinc Resources*

Weeks, R A
7 Michegan Avenue
St Cloud, FL 32769
USA
Gallium, Germanium and Indium Resources

Welch, R A
Department of Geography
University of Georgia
Athens, GA 30602
USA
Remote Sensing in the Search for Mineral Deposits

Wiberg, G S
Canadian Bureau of Chemical Hazards
Environmental Health Directorate
Health Protection Branch
Ottawa, Ontario K1A 0L2
Canada
Asbestos Hazards

Wilson, D G
Room 3-447
Massachusetts Institute of Technology
Cambridge, MA 02139
USA
Recycling of Demolition Wastes

Zwartendyk, J
Energy, Mines and Resources Canada
Mineral Policy
580 Booth Street
Ottawa, Ontario K1A 0E4
Canada
Resource Appraisal

SUBJECT INDEX

The Subject Index has been compiled to assist the reader in locating all references to a particular topic in the Encyclopedia. Entries may have up to three levels of heading. Where there is a substantive discussion of the topic, the page numbers appear in **bold** type. As a further aid to the reader, cross-references have also been given to terms of related interest. These can be found at the bottom of the entry for the first-level term to which they apply. Every effort has been made to make the index as comprehensive as possible and to standardize the terms used.

C.P.C. ABERYSTWYTH
LLYFRGELL